Introduction to HUMAN FACTORS and ERGONOMICS for ENGINEERS

SECOND EDITION

Human Factors and Ergonomics

Series Editor
Gavriel Salvendy

Professor Emeritus
School of Industrial Engineering
Purdue University

Chair Professor & Head
Dept. of Industrial Engineering
Tsinghua Univ., P.R. China

PUBLISHED TITLES

Conceptual Foundations of Human Factors Measurement, *D. Meister*

Content Preparation Guidelines for the Web and Information Appliances: Cross-Cultural Comparisons, *H. Liao, Y. Guo, A. Savoy, and G. Salvendy*

Cross-Cultural Design for IT Products and Services, *P. Rau, T. Plocher and Y. Choong*

Designing for Accessibility: A Business Guide to Countering Design Exclusion, *S. Keates*

Handbook of Cognitive Task Design, *E. Hollnagel*

The Handbook of Data Mining, *N. Ye*

Handbook of Digital Human Modeling: Research for Applied Ergonomics and Human Factors Engineering, *V. G. Duffy*

Handbook of Human Factors and Ergonomics in Health Care and Patient Safety, Second Edition, *P. Carayon*

Handbook of Human Factors in Web Design, Second Edition, *R. Proctor and K. Vu*

Handbook of Occupational Safety and Health, *D. Koradecka*

Handbook of Standards and Guidelines in Ergonomics and Human Factors, *W. Karwowski*

Handbook of Virtual Environments: Design, Implementation, and Applications, *K. Stanney*

Handbook of Warnings, *M. Wogalter*

Human–Computer Interaction: Designing for Diverse Users and Domains, *A. Sears and J. A. Jacko*

Human–Computer Interaction: Design Issues, Solutions, and Applications, *A. Sears and J. A. Jacko*

Human–Computer Interaction: Development Process, *A. Sears and J. A. Jacko*

The Human–Computer Interaction Handbook: Fundamentals, Evolving Technologies, and Emerging Applications, Second Edition, *A. Sears and J. A. Jacko*

Human Factors in System Design, Development, and Testing, *D. Meister and T. Enderwick*

Introduction to Human Factors and Ergonomics for Engineers, Second Edition, *M. R. Lehto*

Macroergonomics: Theory, Methods and Applications, *H. Hendrick and B. Kleiner*

Practical Speech User Interface Design, *James R. Lewis*

Smart Clothing: Technology and Applications, *Gilsoo Cho*

Theories and Practice in Interaction Design, *S. Bagnara and G. Crampton-Smith*

The Universal Access Handbook, *C. Stephanidis*

Usability and Internationalization of Information Technology, *N. Aykin*

User Interfaces for All: Concepts, Methods, and Tools, *C. Stephanidis*

FORTHCOMING TITLES

Computer-Aided Anthropometry for Research and Design, *K. M. Robinette*

Foundations of Human–Computer and Human–Machine Systems, *G. Johannsen*

The Human–Computer Interaction Handbook: Fundamentals, Evolving Technologies, and Emerging Applications, Third Edition, *J. A. Jacko*

Handbook of Virtual Environments: Design, Implementation, and Applications, Second Edition, *K. S. Hale and K M. Stanney*

The Science of Footwear, *R. S. Goonetilleke*

Skill Training in Multimodal Virtual Environments, *M. Bergamsco, B. Bardy, and D. Gopher*

Introduction to HUMAN FACTORS and ERGONOMICS for ENGINEERS

SECOND EDITION

MARK LEHTO • STEVEN J. LANDRY

CRC Press is an imprint of the
Taylor & Francis Group, an **informa** business

CRC Press
Taylor & Francis Group
6000 Broken Sound Parkway NW, Suite 300
Boca Raton, FL 33487-2742

© 2012 by Taylor & Francis Group, LLC
CRC Press is an imprint of Taylor & Francis Group, an Informa business

No claim to original U.S. Government works

Printed on acid-free paper
Version Date: 20140320

International Standard Book Number-13: 978-1-4398-5394-8 (Hardback)

This book contains information obtained from authentic and highly regarded sources. Reasonable efforts have been made to publish reliable data and information, but the author and publisher cannot assume responsibility for the validity of all materials or the consequences of their use. The authors and publishers have attempted to trace the copyright holders of all material reproduced in this publication and apologize to copyright holders if permission to publish in this form has not been obtained. If any copyright material has not been acknowledged please write and let us know so we may rectify in any future reprint.

Except as permitted under U.S. Copyright Law, no part of this book may be reprinted, reproduced, transmitted, or utilized in any form by any electronic, mechanical, or other means, now known or hereafter invented, including photocopying, microfilming, and recording, or in any information storage or retrieval system, without written permission from the publishers.

For permission to photocopy or use material electronically from this work, please access www.copyright.com (http://www.copyright.com/) or contact the Copyright Clearance Center, Inc. (CCC), 222 Rosewood Drive, Danvers, MA 01923, 978-750-8400. CCC is a not-for-profit organization that provides licenses and registration for a variety of users. For organizations that have been granted a photocopy license by the CCC, a separate system of payment has been arranged.

Trademark Notice: Product or corporate names may be trademarks or registered trademarks, and are used only for identification and explanation without intent to infringe.

Visit the Taylor & Francis Web site at
http://www.taylorandfrancis.com

and the CRC Press Web site at
http://www.crcpress.com

Contents

Preface xvii
Acknowledgments xxi
Authors xxiii

Chapter 1 Guided tour of ergonomic design 1
About the chapter 1
Introduction 1
What is ergonomic design? 3
Human-centered design 9
Military equipment design 13
Ergonomic criteria 15
Models of human performance 16
    Helson's hypotheses 17
    Other models 18
    Model of software design 23
Macroergonomics 24
Carrots and sticks 26
Trends in industry that impact ergonomic design 28
Organizations and additional information on ergonomic design 32
Ergonomic methods 32
    Field studies 33
    Experimental simulations 34
    Laboratory experiments 34
    Computer simulation 35
    Differences in ergonomic methods 35
Final remarks 36
Discussion questions and exercises 36

Chapter 2 Human system 39
About the chapter 39
Introduction 39
Skeletal subsystem 40
    Extremities 40
    Joint-related disorders 42
    Spine 45
Muscles 47
    Muscle contractions and capabilities 47
    Role of oxygen in muscle actions 49

Muscle injuries and disorders 49
   Gender, age, and training effects on muscular strength 50
   Exercise as a means of CTD prevention 50
Anthropometry 50
   Predicting the stature of people 51
   Predicting the segment mass of the human body 53
   Other anthropometric relationships 54
Body movement 54
   Muscular–skeletal system as levers 57
Sensory subsystems 59
   Visual sensory subsystem 59
      Structure of the eye 60
      Eye movements 62
      Visual tasks 63
      Perception of color and brightness 64
      Visual acuity 65
      Other visual abilities 66
   Human perception of sound 67
      Position and motion sensing 71
   Other senses 74
Support subsystems 74
   Respiratory operations and mechanics 75
   Circulatory subsystem 76
   Metabolism 78
   Indirect calorimetry 81
Final remarks 82
Discussion questions and exercises 82

Chapter 3   Design of work areas, tools, and equipment 85
About the chapter 85
Introduction 85
Applied anthropometry 86
   Drafting templates 89
Design of work areas and stations 90
   Traffic areas 90
   Workplace dimensions and layout principles 92
   Design of seating 98
Office design 102
   Computer workstations 104
      Location and positioning of VDT terminals 105
      Keyboard positioning and layout 106
Design of tools and equipment 108
   Hands and handedness: Some initial design principles 108
   Other desired properties of grip design 113
   Other features of hand tool design 117
   Techniques for determining hand tool adequacy 118
   Power tools 119
   Power transmission hazards 121
   Point of operation hazards 122

Protective equipment for the operator 125
    Safety shoes 125
    Helmets 127
    Protective gloves 129
    Eye protection and spectacles 130
    Hearing protection 130
Accommodating people with disabilities 131
Final remarks 132
Discussion questions and exercises 132

Chapter 4 Assessment and design of the physical environment 135
About the chapter 135
Introduction 135
Cleanliness, clutter, and disorder 136
    General housekeeping and maintenance 136
    5S programs 137
Lighting and illumination 140
    Luminous environment and its measurement 140
    Lighting methods 143
    Some basic principles and lighting requirements 147
Noise 156
    Health effects of noise 156
        OSHA noise exposure limits 157
    Annoyance and other effects of noise 159
    Noise control strategies 160
    Hearing protection 161
Temperature and humidity 162
    Thermal regulation 162
    Heat transfer 163
        Metabolism 163
        Conduction and convection 165
        Radiative effects 166
        Evaporation 166
    Control strategies for hot and cold environments 167
        Engineering controls 167
        Administrative controls 169
        Protective clothing 169
    Assessing thermal conditions and comfort 170
        Temperature and humidity measurement 171
        Effective temperature 171
        Windchill index 172
        Heat stress index 172
        Wet bulb global temperature index 175
    Effects of thermal conditions on performance 177
Hazards and control measures 179
    Fall and impact hazards 179
    Hazards of mechanical injury 180
        Cutting/shearing 180
        Crushing/pinching 181

Vibration hazards and cumulative trauma disorders 182
    Vibration hazards 182
    Cumulative trauma disorders 183
Pressure hazards 183
Electrical hazards 184
Burn, radiation, and fire hazards 185
    Burn hazards 185
    Ultraviolet radiation hazards 186
    Fire hazards 186
Hazards of toxic materials 189
    Measures of exposure and exposure limits 190
    Protection from airborne contaminants 192
Final remarks 192
Discussion questions and exercises 193

Chapter 5   Work measurement and analysis 195
About the chapter 195
Introduction 195
    Therbligs 195
    Some probabilistic assumptions 197
Methods improvement 199
    Principles of motion economy 200
    Motion and micromotion study 202
Work-physiology-related principles 203
Some principles of manual assembly 208
Analysis of lifting 211
    Principles of lifting 214
Computer tools for analyzing lifting tasks 216
Learning 217
    Some applications of learning curves 218
    Modeling human learning 218
        Why use a model of learning? 219
        Performance criteria and experience units 219
    Some learning curve models 221
        Powerform learning curve 221
        Discrete exponential learning curve 225
    Fitting learning curves 228
        Powerform model 229
        Discrete exponential model 230
Comparing alternatives for learnability 230
Correct learning curve model 231
Forgetting curves 232
Time study 232
    Some purposes served by time study 233
    Selecting a procedure for timing 233
    Elements of a task 234
    Recording forms 235
    Selecting and training the observed operator 235

  Sample size and time statistics 236
Performance leveling 240
  Subjective ratings 241
  Multifactor ratings 242
  Synthetic rating 243
  Extracted basis of rating 244
  Alternatives to performance rating 244
  Determining allowances 245
    Machine interference allowances 246
Maintaining standards 248
Indirect performance measurement 248
Criteria other than time 249
Final remarks 250
Discussion questions and exercises 251

Chapter 6 Modeling physical human performance 255
About the chapter 255
Introduction 255
Synthetic data systems 257
  Why use a synthetic data system? 257
  Motion–time measurement 258
    Structural basis of MTM 258
    Time predictions of MTM-1 261
    Analytical procedure with MTM-1 261
    Assumptions associated with MTM-1 262
    Applications of MTM-1 263
    MTM-2 for simpler tasks and quicker analysis 263
  Work-factor system 265
    Structure of the work-factor system 266
    Assumptions of the work-factor system 268
  Mento-Factor system 269
  Iowa model 270
  Standard data systems 271
  Formula development 272
  Finding the relationship between predictive variables and normal performance 273
  Statistical computer packages for regression 275
  Computer modeling methods 277
Speed–accuracy trade-offs 279
  Fitts' law 280
Eye–hand coordination 283
Final remarks 284
Discussion questions and exercises 285

Chapter 7 Sampling methods in industrial ergonomics 291
About the chapter 291
Introduction 291
Activity sampling 292
  Sampling procedures 293

  Sampling theory and sample size 295
   Binomial sampling model 295
   Calculating sample size using the normal approximation of the binomial sampling model 296
    Use of nomographs to estimate sample size 297
  Developing the observation schedule 299
  Recording and analyzing the data 300
  Checking for stationarity 301
  Using work sampling to estimate performance standards 302
  Using work sampling in problem solving 304
Sampling strategies 305
  Stratified sampling 305
   Designing a stratified sampling strategy 306
   Analysis of stratified sampling data 308
Sequential Bayesian work sampling 308
  Beta distribution and its parameters 310
  Bayesian confidence intervals 311
  Finding tabled Bayesian confidence intervals 312
  Bayes' theorem and the Bayesian concept 312
  Managing Bayesian work sampling studies 313
   More extensive Bayesian confidence intervals 315
   Numerical example of Bayesian activity analysis 316
  Poisson activity sampling 319
Final comments 319
Discussion questions and exercises 319

Chapter 8 Macro-ergonomics: Task analysis and process mapping 323
About the chapter 323
Introduction 323
Ergonomic design principles 325
Traditional principles in ergonomic design 325
Visual graphs of operations 328
  Operations process charts 329
  Flow process charts 330
  Flow diagrams 332
  Multiple activity charts 333
  Precedence diagrams and matrices 334
  Link diagrams 335
  Task time-line analysis charts 337
  Fault trees 338
  Failure modes and effects analysis tables 339
  Cause-and-effect diagrams 341
  Decision flow diagrams 341
Analysis of tasks and jobs 343
  Describing the task 345
   Task analysis of making coffee 346
  Task flow diagram 347
  Hierarchical task analysis 348

Determining task requirements 349
Function allocation 352
Final remarks 353
Discussion questions and exercises 353

Chapter 9 Computer simulation of processes and tasks 357
About the chapter 357
Introduction 357
Essential elements of computer simulation 358
    Higher-level computer languages 359
    Verifying and validating a simulation model 362
Computer simulation in ergonomics 363
    Industrial applications of process simulation 364
    Case study of process simulation in a BOF shop 365
Operator-in-the-loop simulation (OLS) 366
    Training simulators 368
    Ground vehicle simulators 369
    Simulation fidelity 370
    Case study in operator-in-the-loop simulation 371
    Other operator-in-the-loop simulators 373
Design of simulation experiments 373
    Analysis of variance 374
    General experimental designs and their features 374
Final remarks 377
Discussion questions and exercises 378

Chapter 10 Modeling and evaluation of cognitive tasks 379
About the chapter 379
Introduction 379
Communication theory 380
    Information transmission 381
    Rate of information transmission 383
    Information partitioning and hypothesis testing 388
Signal detection theory 389
Human information processing 395
    Models of attention 395
    Role of working memory 397
    Mode of processing 398
    Declarative versus procedural knowledge representations 401
    Error types and forms 401
        Skill-based errors 401
        Rule-based errors 403
        Knowledge-based errors 403
        Judgment-based errors 404
Decision making 404
    Classical decision theory 404
        Non-compensatory decision rules 405
        Subjective expected utility model and prospect theory 406

  Human judgment and inference 407
    Human inference 409
    Heuristics and biases 411
    Brunswik lens model of human judgment 412
    Recognition-primed decision making and situation awareness 413
    Models of time pressure and stress 414
 Cognitive simulation 415
  GOMS: Cognitive work measurement of a skill-based task 415
    GOMS model of task performance 416
    Experimental evaluation of the GOMS model 417
    Natural GOMS language 418
 Final remarks 420
 Discussion question and exercises 420

Chapter 11 Control tasks and systems 423
 About the chapter 423
 Introduction 423
 Control systems 424
 Manual control 426
  Elementary model of manual control 428
  Tracking tasks 428
  Human controller 431
  Naive theory of adaptive control 434
  Positioning and movement time 437
  Control error statistics 439
    Partial means 441
  Numerical simulation of control activity 444
 Fuzzy control 446
  Fuzzy measurements 446
  Fuzzy logic 447
  Examples of fuzzy control 448
 Supervisory control 451
 Relation to manual control 453
 Final remarks 455
 Discussion questions and exercises 455

Chapter 12 Design of displays and controls 457
 About the chapter 457
 Introduction 457
 User-centered interface design 458
  Natural mappings 459
  Population stereotypes and stimulus–response compatibility 461
 Message meaning and comprehension 462
  Linguistic modeling 462
  Testing the comprehension of symbols and display elements 464
  Testing the comprehension of text 464
 Types of displays 466
 Principles of display design 468
  Sensory modality 469

  Display location and layout 470
  Legibility of display elements 472
  Information content and coding 477
 Design of controls 483
  Principles for better control–display relationships 488
 Final remarks 491
 Discussion questions and exercises 491

Chapter 13 Ergonomics of product quality and usability 493
 About the chapter 493
 Introduction 493
 Quality management and customer-driven design 495
  QFD: A framework for quality assessment and improvement 497
  Identifying customer requirements 500
  Specifying design requirements 505
  Rapid prototyping and testing 507
  Customer ratings and preference testing 508
 Usability analysis and testing 510
  Task analytic methods 511
  Expert evaluation 511
  Typical procedures in usability testing 516
  Interviews and postexperimental questionnaires 520
 Designed experiments 521
  Independent variables 521
  Dependent variables 523
  Basic experimental designs 524
 Final remarks 525
 Discussion questions and exercises 525

Chapter 14 Questionnaires and interviews 527
 About the chapter 527
 Introduction 527
 Questionnaire design 528
  Planning the questionnaire 528
  Sampling procedure 529
  Constructing the questionnaire 530
   Questionnaire response scales 531
  Pilot testing and data collection 535
  Data analysis 536
   Chi-square analysis 536
   Spearman rank correlation coefficient 537
 Wilcoxon sign test 540
   Reliability 541
 Interviews 542
  Interviewing methods 543
   Brainstorming and focus groups 544
  Interrater consistency 545
 Final remarks 547
 Discussion questions and exercises 547

Chapter 15 Quality control and inspection 549
About the chapter 549
Introduction 549
Some common types of inspection 551
    Quality assurance inspections 551
    Maintenance and safety inspections 551
    Screening of people and materials 552
    Detection of tampering 552
Human inspection 552
    Signal detection theory 553
        Nonparametric extensions of SDT 556
    Inspection strategies 558
    Viewing time and movement 559
    Task pacing and time pressure 560
    Individual differences 562
    Illumination levels 563
    Visual freedom and other issues 565
    Analysis and improvement of inspection tasks 566
Inspection economics 567
    Inspection costs 567
    Location of the inspection station 570
    Learning and quality improvement 572
Final remarks 573
Discussion questions and exercises 573

Chapter 16 System reliability and maintenance 577
About the chapter 577
Introduction 577
Reliability and availability 578
    Reliability analysis 578
    Fault avoidance and fault tolerance 579
Maintenance programs 585
    Assigning and coordinating maintenance activities 585
    Setting maintenance priorities and standards 588
    Maintenance schedules 588
    Storing and delivering technical information as needed 589
    Tools, component parts, and materials delivery 590
    Training for maintenance and repair 590
    Record keeping and measurement of maintenance program effectiveness 591
Reducing maintenance effort 593
    Fault and malfunction detection 593
    Diagnostic support 596
    Verifying the fault and the correction 602
Design for disassembly 602
    Fasteners and tools 607
    Work envelope 608

Final remarks 609
Discussion questions and exercises 610

Chapter 17 Occupational safety and health management 613
About the chapter... 613
Introduction 613
Some historical background 614
    Safety and health achievements and future promises 616
Fundamental elements of occupational safety and health 617
    Classifications of occupational injuries and illnesses.. 618
    Accident causation 619
        Multifactor theories 621
        Role of human error and unsafe behavior 622
    Risk management and systems safety 623
Occupational health and safety management programs 624
    Compliance with standards and codes 624
    Accident and illness monitoring... 626
        Accident reporting.. 626
        Calculation of incidence rates 627
    Hazard and task analysis 629
        Critical-incident analysis 629
        Work safety analysis 630
Control strategies 630
    Job and process design 630
    Hierarchy of hazard control 631
    Hazard communication 633
        Sources of safety information... 634
Final remarks 639
Discussion questions and exercises 639
Resources.. 639

Chapter 18 Personnel selection, placement, and training 641
About the chapter... 641
Introduction 641
Personnel selection and placement... 641
    Personnel systems 642
    Concepts of ability testing 646
    Individual differences on the job 650
    Economic considerations in personnel selection 652
Training 653
    Some training principles 654
    Transfer of training 655
    Other training strategies 657
Job aids.. 659
Final remarks 659
Discussion questions and exercises 660

Chapter 19 Work groups and teams 663
About the chapter..663
Introduction 663
Simple model of group effectiveness 664
Life cycle of teams and crews..665
Group performance and biases...668
    Measuring team–crew work performance 669
Industrial work teams 671
    Principles of team effectiveness 673
Research, development, and design teams 674
    Team roles... 676
    Communications within research, development, and design teams 679
    Some observations from NASA 680
Final remarks 681
Discussion questions and exercises 681

Chapter 20 Job evaluation and compensation 683
About the chapter..683
Introduction 683
Job evaluation systems 685
    Job analysis...686
    Job ranking methods 686
    Job classification methods 688
    Point systems 689
        NEMA point system..690
    Job factor comparison method 693
Wage administration 696
    Forms of compensation 697
    Establishing a wage structure 698
        Wage and salary surveys..699
        Wage calibration curves..699
        Wage curves..700
        Rate ranges..701
    Incentive plans...702
        Individual level plans 703
        Group level plans...704
        Organization level plans...705
Final remarks 705
Discussion questions and exercises 705

Appendix A: Some probability distributions 707
Appendix B: Tables of statistical distributions 711
References 723
Index 755

Preface

This book is designed to provide a coordinated and fairly comprehensive set of guidelines and principles for the design and analysis of human-integrated systems. Its focus is on application of these guidelines and principles to industry and service systems. In this second edition, several things have changed. The ordering of the chapters has been greatly altered to improve its organization and group together topics related to the physical and cognitive aspects of human-integrated systems. The material has also been updated substantially, but as in the first edition, emphasis is placed on designing the products with which people work (e.g., tools, machines, or systems), the tasks or jobs people perform, and environments in which people live. The common thread throughout the book is on how better human factors can lead to improved safety, comfort, enjoyment, acceptance, and effectiveness in all of these application arenas.

Human factors (also called ergonomics) is, broadly, about designing and analyzing systems with respect to the capabilities of the humans who interact with them. We generally view this as having two separate aspects—physical and cognitive—although they overlap quite a bit. On the physical side, which is covered primarily in Chapters 1 through 8, human factors engineers design and analyze systems to ensure that humans can operate them without injury or excessive stress, that the work processes used are efficient, and also learn how to model a system so that it can be analyzed or simulated. On the cognitive side, which is covered primarily in Chapters 9 through 20, human factors engineers design and analyze systems to ensure that the overall system performs well with the human in the loop. Human factors engineers and researchers study different concepts of how humans interact with systems, how they make decisions, how things are perceived, and generally how different designs influence behavior.

Human factors engineers generally work at the interface of the technology/system and the human operators. It is a fairly young field, so it is not yet driven by the same mathematical rigor as, for example, circuit design. So far, we are still driven mainly by principles rather than theory. Much of what we know, especially on the cognitive side, is discovered through field observation and human subject experiments.

We expect this book to be of use to both students of human factors, who are its primary audience, as well as practitioners. Indeed, we hope the students will retain the book and use it in their careers. The book is designed so that engineers can refer to it when faced with particular human factors problems.

The book is roughly divided into two parts, as noted earlier. Each chapter, however, addresses a particular topic, as follows:

- Chapter 1 is an introduction, providing an overview of ergonomic design.
- Chapter 2 provides necessary background material on the human system, including the muscular-skeletal system, anthropometry, and sensory/support subsystems of the body.
- Chapter 3 applies the principles from Chapter 2 to the design of work areas, tools, and equipment.
- Chapter 4 focuses on analysis and design of the environment of the workplace, including 5S principles, lighting, noise, and temperature/humidity.
- Chapter 5 introduces the analysis and measurement of work, using methods improvement, lifting analysis, learning curves, and time study—this chapter often forms the heart of what is commonly considered human factors in industrial engineering.
- Chapter 6 explores how physical human performance has been modeled, including the use of synthetic data systems and Fitts' law.
- Chapter 7 focuses on sampling methods that are used to determine the prevalence of certain events or activities within a work system.
- Chapter 8 expands the focus to examine "macroergonomics," or the study of the human factors of the overall system in which the work tasks are embedded—methods discussed include visual graphs of operations, task analysis, and process mapping.
- Chapter 9 builds upon the material introduced in Chapter 8 by introducing the use of computer simulation as a tool for modeling and describing systems that go well beyond traditional forms of task analysis and process mapping.
- Chapter 10 introduces several approaches used to model and evaluate cognitive tasks performed by people in a wide variety of industrial and consumer settings, as well as measures of human performance, such as the amount of transmitted information, receiver sensitivity, and human error, biases, or deviations from optimality.
- Chapter 11 focuses on how people use control devices to control or guide the activity of machinery, vehicles, or processes. Emphasis is placed on how to model and measure human performance of control tasks.
- Chapter 12 provides guidelines applicable in a wide variety of settings on how to design displays and controls. The chapter begins by discussing the importance of providing a good mapping of the relationships between displays, controls, and their effects. Attention then shifts to the important issue of display content and comprehension.
- Chapter 13 describes the emerging role of ergonomics and industrial engineering in product design, especially in regard to rapid prototyping and usability testing at all stages of the product design cycle.
- Chapter 14 then focuses on how to construct questionnaires and interviews as information-gathering tools commonly used by ergonomists during rapid prototyping, usability testing, and other forms of analysis.
- Chapter 15 moves on to the topic of quality control and inspection. Some common ergonomic issues are addressed, and suggestions are given for how to analyze and improve the design of inspection tasks.
- Chapter 16 focuses on the closely related topic of maintenance and repair and steps that industrial engineers and ergonomists can follow to increase the reliability of products and processes they design.
- Chapter 17 covers one of the most important aspects of macroergonomics—the management of occupational safety and health. The chapter includes a brief history of workplace safety and health, legal responsibilities of the employer, and methods of classifying causes of injury, illness, and accident prevention.

- Chapter 18 moves on to a second important topic in macroergonomics—personnel selection, placement, and training, and some of the important steps ergonomists can follow to help ensure an effective workforce.
- Chapter 19 extends the latter discussion of workforce issues by introducing readers to some of the problems and challenges that emerge when teams, crews, or other groups of people perform tasks or operations.
- Chapter 20, the final chapter, covers a third important area of macroergonomics—job evaluation and setting wages in a fair and consistent manner.

The breadth of coverage of the chapters is quite wide, going well beyond most traditional texts in the field. The book itself has sufficient material to be used in its entirety for a two semester sequence of classes, or in part for a single semester course, focusing on selected topics covered in the text. This breadth and depth of coverage will also make the book useful to engineers, designers, and other practitioners seeking a good reference source that is more approachable than a handbook on the topic of human factors and ergonomics and its application.

Acknowledgments

We would like to dedicate this edition of the book to the memory of Professor James Buck and Professor Jim Barany to acknowledge the many contributions and commitment they made at Purdue University over the past 40 years to educate thousands of engineers on the topics covered in this book. James Buck, of course, helped write the first edition and we are delighted to build on that work in this second edition. Jim Barany provided numerous suggestions on how to improve the book we found to be helpful, such as his broad division of the field into ergonomics from the head up versus head down. We also would like to acknowledge the many other contributors. Our particular thanks go out to Kerina Su who put in many hours of effort into the process of revising and developing figures for this edition of the book. Our particular thanks also go out to the many students and instructors who provided valuable suggestions over the past few years on how to improve the book.

Authors

Dr. Mark Lehto is a professor in the School of Industrial Engineering at Purdue University, where he is also the cochair of the Interdisciplinary Graduate Program in Human Factors and Ergonomics and has resided since completing his PhD at the University of Michigan in 1986. In addition to his teaching and research at Purdue University, Dr. Lehto has provided guidance to hundreds of companies and outside organizations on the topics covered in this book.

Steven J. Landry received his PhD in industrial and systems engineering from the Georgia Institute of Technology, Atlanta, Georgia, in 2004; his SM in aeronautics and astronautics from the Massachusetts Institute of Technology, Cambridge, Massachusetts, in 1999; and his BS in electrical engineering from Worcester Polytechnic Institute, Worcester, Massachusetts, in 1987. He is an associate professor and associate head of the School of Industrial Engineering at Purdue University, West Lafayette, Indiana, where he teaches courses in human factors and statistics. Previously, he was an aeronautical engineer at NASA Ames Research Center and a C-141B pilot, instructor, and flight examiner for the United States Air Force. He is the author of several book chapters on aviation human factors. His research involves air transportation systems engineering and human factors. Dr. Landry is a member of the Human Factors and Ergonomics Society, the American Institute of Aeronautics and Astronautics, the Institute of Electrical and Electronics Engineers, and the Institute of Industrial Engineers.

chapter one

Guided tour of ergonomic design

About the chapter*

This chapter introduces the reader to the field of ergonomic design of work and workspaces. An overview is given of activities that take place during ergonomic design, followed by a brief discussion of design criteria and some of the models used to predict how well designs will perform. Laws, principles, and techniques that are part of ergonomic design are then discussed. Attention is also given to management-related activities and the overall role of ergonomic design within the firm.

Introduction

People are one of the most important resources in an enterprise. In today's industrial and service operations, people perform physical tasks of assembling and handling materials, as well as sensory and cognitive tasks such as inspecting components, issuing tools, entering data, and managing people and operations. Simply put, it takes people to make products and provide services. The efficiency with which people can accomplish these tasks is a primary driver of productivity, and productivity is a primary driver of economic activity.

People are also prospective customers. If manufactured products and services meet the needs and desires of customers at reasonable prices, customers will buy them. In fact, the more attractive these products and services are, and the more competitive the price, the greater the sales. When sales improve, employment levels and wages increase. In a nutshell, that is what this book is about: improving both the effectiveness of people in industry and the products and services that industry sells.

Ergonomics is the study of people at work. This field got its name in the summer of 1949 when a group of interested individuals assembled in Oxford, England to discuss the topic of human performance. The group consisted of anatomists, physiologists, psychologists, industrial medical officers, industrial hygienists, design engineers, work study engineers, architects, illuminating engineers, and anyone whose work concerned some aspect of human performance.†

A proposal was put forth at that meeting to coin a new word for this emerging field, ergonomics, which couples *ergos*, the Greek word for work, with the word *nomos*, meaning natural laws. The group decided to adopt this term and called themselves the Ergonomics Research Society.‡ One of the advantages of this new term was that it did not imply that any of the contributing disciplines were more important than the others. Sometime later,

---

* Each chapter begins with a description of the chapter contents and the educational objectives for writing the chapter.
† The preface of Murrell's (1965) book states that the disciplines represented at the forming of this Society came from both scientific and application backgrounds and that the human sciences of psychology, anatomy, and physiology were all represented.
‡ In later years, the Ergonomics Research Society changed its name to the "Ergonomics Society." The term Ergonomics has become known internationally, although members of the Society had many spirited discussions during the 1970s about changing their name. The late E. Edwards (1993) covered some of this discussion.

the term "human factors" was coined in the United States for a society of similar purpose, but with less diversity of backgrounds. Recently, the U.S. society changed its name to the Human Factors and Ergonomics Society.

Some experts define the objective of ergonomics and human factors engineering as designing machines to fit human operators. However, it is also often necessary to fit operators to machines, such as in the form of personnel selection and training. It is probably more accurate, then, to describe this field as the study of human–machine systems, with an emphasis on the human aspect. This book takes such an approach, but limits its scope to

- Small systems of individual operators or small crews of people
- Industrial and consumer service settings, processes, and products
- Design and operating decisions associated with these systems

Confining its emphasis to these three concerns makes this book more specifically focused than most textbooks used in basic courses on human factors engineering or ergonomics. For the most part, those texts devote considerable attention to background sciences such as psychology and physiology and tend to stress military and nonindustrial applications because much of the research is driven by the military. In contrast, this book stresses applications of ergonomic analyses and the design of better person–machine systems.

A good introduction to the concept of ergonomics was given several years ago by Paul Fitts. Fitts described human performance as a function of five factors (labeled by the acronym, LIMET), which were

- Learning
- Individual differences
- Motivation
- Environment
- Task

These five factors identify many important aspects of ergonomics that also involve management and other groups within a firm. Learning is the objective of training, and individual differences are the underlying justification for developing personnel selection procedures intended to ensure companies hire people who can perform their jobs. Once a company attracts the desired personnel and trains them, it must concentrate on motivating them to perform their jobs well. This will permit the company to make the money necessary to meet the bills for operation, pay the wages of the personnel, and finance new projects for making products and services better, using safer processes. Worker motivation is promoted in part by a company environment that is safe, effective, and conducive to good work. It should also be emphasized that the tasks people perform sometimes need to be revised for more effective operations. Consequently, all of these factors contribute to better ergonomics.

But it is simplistic to think of them as independent variables because there are interactions between these factors that are not well understood. For example, elements that motivate people in the 95th percentile of performance may not be the same as those that motivate people in the 5th percentile.

Some other roots of this discipline go to the beginning of the twentieth century when Frederick Taylor led a movement he called *Scientific Management* (See Taylor, 1911). He started his movement while working in the steel industry in Philadelphia, where he conducted a series of experiments to design better types of shovels for moving sand into casting molds. His ideas caught on in many places, although most companies used them only partially.

Taylor was also a man who attracted others with his ideas and insight. As a consequence, many design ideas poured out of the Taylor school. These ideas and others generated by his contemporaries of that day started what is now known as *industrial engineering*. This group of pioneers placed little stress on product design; their primary interests were methods design, time study, and process design.*

This book goes well beyond that initial focus. Various principles of design are described and illustrated in the following chapters for typical products, including machine tools, power tools, hand tools, product units, and computer software. This book emphasizes these topics because industrial engineers and other ergonomic specialists are often involved in the design and evaluation of products and the environments they are used within. This involvement is likely to increase in the future, making it particularly important for prospective engineers to develop the skills, techniques, and knowledge needed to make a useful contribution in this area.

The word "design" appears in the book title and throughout the preceding paragraphs. Before going further, some clarification of the term is needed. Many students of civil engineering imagine themselves planning a beautiful suspension bridge over the Straits of Whatever. Mechanical engineering students think of laying out the plans for an automotive suspension system for a Monte Carlo or an Indianapolis race car. The dream of an electrical engineering student may be to design the world's most powerful computer. An architectural student envisions designing a prestigious building of great beauty and usefulness. Whatever the end product, the activity known as design involves planning, problem solving, formulating a method or plan, artistically and/or skillfully describing that end product on paper or on a computer, and devising a model of the product for illustration and testing. All of these activities are part of design, whether the end product is a structure, a process, a procedure, or something else. Since the design must be made into an operating object or process, designs must be quantitative in nature. But let us be more specific.

What is ergonomic design?

Ergonomists study people and how they operate equipment in the home, in commerce, in factories, and in governmental activities. It stands to reason, then, that ergonomic engineers design products. They also design jobs. More specifically, they design methods and tools for performing tasks, machine-tool interfaces, workplace layouts, inspection stations, process-control interfaces, machine-tool layouts, training programs, questionnaires, interviews, human performance predicting systems, material handling systems, machine adaptations for handicapped personnel, safety and health programs,† ways to resolve grievances, warning signs, abatement programs for undesirable environments, communication systems, error tracking systems, job aids, wage administration programs, teams of personnel, and programs for personnel selection and placement. While this is not a complete list, it gives some idea of the scope of ergonomic design.

---

* Books describing this set of topics include Barnes (1968), Mundell (1985), Nadler (1970), Neibel and Freivalds (2003), and Polk (1984).
† Safety and health programs in industry are a specialty within human factors and ergonomics. Some books in this area are Brauer (2006), Goetsch (2010), Heinrich, Peterson and Roos (1980), and Hammer and Price (2000).

Chapanis (1995) identified a number of activities that might occur during the ergonomic design of products and jobs. Certain activities have origins that go back nearly a century; others are much more contemporary.* Many of these activities are covered more extensively in later chapters of this book. The activities identified by Chapanis are

- Analysis of similar systems
- Activity analysis
- Critical incident study
- Functional flow
- Decision/action analysis
- Action/information requirements analysis
- Functional analysis
- Task analysis
- Failure mode and effects analysis
- Timeline analysis
- Simulation
- Link analysis
- Controlled experimentation
- Operations sequence analysis
- Workload assessment

Three activities that can be performed concurrently at the beginning of the design process are analysis of similar systems, activity analysis, and critical incident study. The *analysis of similar systems* establishes a background for making improvements where possible, and may involve obtaining

1. Descriptions of the operations and difficulties associated with them
2. Historical perspectives of the old product and its competitors from an ergonomic point of view
3. Procedures in maintaining products or industrial operations and specific problems in maintenance
4. Requirements for special skills
5. Numbers of personnel needed to operate the machine
6. Training requirements

The principal purpose of this analysis is to assemble a full background history from which the designers can work. An illustration of one such background history is shown in Box 1.1 for the case of beer cans.

Activity analysis or *sampling*† involves determining what tasks are performed and how much time is spent on each task. With this information, ergonomic engineers can

---

* Chapanis addressed this situation at the International Ergonomics Association meeting in Toronto, Canada in August 1994 and it was later published in *Ergonomics* (1995). No quotations marks are used, in order to simplify the discussion here. Alphonse Chapanis participated in founding the human factors and ergonomics society. He was an ergonomics researcher during World War II and afterward taught at Johns Hopkins University, Baltimore, MD.

† Activity analysis or sampling has been given a number of other names. Chapter 7 describes this technique as "work sampling," "occurrence sampling," and "ratio delay." Essentially, the technique requires observations on an operation where specific activities are identified as they are observed. Sometimes, the observing is continuous, and other times the operation is observed at randomly determined times.

BOX 1.1 FORM FOLLOWS FAILURE OR THE CASE OF BEER CANS

This unusually titled section provides insight into how design improvements might be found by looking for design failures. Actually, this section gives a historical account of canned meats, vegetables, and beverages. In this account, you will see that every major development in the evolution of canning occurred when someone capitalized on a failure in the existing situation. Consider this story.

The practice of preserving foods and beverages is only a couple of centuries old. The Parisian Nicolas Appert found that bottling meat could preserve it for long time periods if bottles were immersed in boiling water and then sealed airtight. The idea worked well, but the bottles would break if dropped or handled too roughly. That problem was remedied by Peter Durand of London who used tin canisters. This practice was continued by the British firm of Donkin and Hall. Unfortunately, they did not provide a way to remove the contents of the canisters. Many methods were tried. Soldiers in the United States during the War Between the States used knives, bayonets, swords, and even guns. Another shortcoming of the earlier canning efforts was the heavy weight of the container. Often the canister was heavier than the contents.

In 1898, Ezra Warner of Waterburg, Connecticut, patented a "can opener" that was part bayonet with a large curved blade. In use, the opener had to be jabbed into the can top near the edge, forcing the large blade under the top. After pulling up on the handle, the process was repeated until the entire lid was circumscribed by the opener. Opening a can consisted of a series of jerky motions rather than a smooth continuous action, and the process created a hazardous jagged edge. Twelve years later, William Lyman of West Meridan, Connecticut filed a patent for a new can opener that had a cutting wheel for making smooth continuous cuts. However, this opener required adjustment for the size of the can, and the user needed to pierce the can almost dead center for effective operation. Persons with lesser skills experienced difficulties with the Lyman can opener. Here is a case where the design depended on human skills, which were generally inadequate throughout the population of users. It was not until 1925 that the wheeled can opener that we use today was invented. Unfortunately, that design was only practical in opening cans that have a raised seam where the top meets the side metal. In addition, the sides had to be sufficiently thin to accept a serrated wheeled cutter. In spite of all this development, certain deficiencies continue to characterize the design of can openers today.

Early beverage cans consisted of a rectangular piece of sheet steel that was rolled into a cylinder and welded to two circular end pieces. Opening such cans frequently resulted in a large loss of contents, produced jagged edges around the opening, and also required auxiliary equipment. Here again, knives were used in the earlier times. Later, the famous "church key" opener came to the rescue.

However, in order for a church key to work, the can needed to have a strong edge to pivot the opener and make a pie-slice wedge cut near the can's outer edge. These keys were made with curved points so that excess gas could escape from carbonated beverages without losing the liquid contents, unless someone shook them too much. Over time, this manner of opening cans became easier through the use of stronger but less dense metal. The use of lighter metal took advantage of carbonation

(continued)

BOX 1.1 (continued)

(with CO_2). The pressure within the can puts a modest tensile force on the metal to give it greater strength.

Between that development and the earlier cans, there were numerous innovations. In the 1930s, the Continental Can Company looked at beer bottles that were easy to open with the church key by simply prying off the cap, ending up with a can with a funnel-shaped top and a regular bottle cap over the small funnel opening. This is a case of focusing on an acceptable solution and transferring it to the problem.

In the late 1950s, two major developments changed the design of beverage cans. Ermal Fraze of Dayton, OH, developed the first version of the pop-top can. His idea was to attach a tab that could be used to open cans without a church key. Actually, his invention was prompted when he forgot the church key one day and tried to use a car bumper as a make-shift can opener. The device he developed had a prescored tear strip and a connective ring. A user needed to flip up the ring, then grasp that ring and pull it to rip open the tear strip that became separated from the can. Because of the prescoring, the edges of those pull-strips were smooth. Drinking directly from the can was, therefore, not necessarily hazardous. However, those pull-off tab rings caused many problems in public areas. Most people dropped the ring tabs right where they opened the can. The litter was unsightly and in some cases caused injuries to people walking barefoot. By the mid-1970s, the public was highly concerned. This concern eventually led to a variety of openers that keep tabs on the cans. Evolutionary improvement now allows people to drink from such cans without cutting or scratching their face.

Other major developments in the design of beverage cans arose in the 1950s. Kaiser Aluminum Company developed an aluminum can. Soon afterward, Adolph Coors Company came up with the 7 oz can rather than the standard 8 oz can. At first those cans were quite thick to withstand handling and other stresses. One pound of aluminum made only 20 cans. Today, that same pound results in 30 cans. Thinner walls were made possible by creating the domed bottom to withstand pressures from inside the can. Tops of can sides were necked and a stepped top lid was created. Shapes were revised to minimize the metal content. Besides the savings of metal on each can, the public recovered more metal from the cans through recycling. While only about 25% of the cans were recycled in 1975, recycling in 1990 reached over 60% and in time it is expected to rise to at least 75%.

The message here is that "Form Follows Failures." It is clear that difficulties resulting from a particular product design inspire new inventions by prompting people to identify and solve those problems. Today's challenge is to think ahead to avoid as many of those problems as possible. Ergonomic design focuses on that objective for human users.

improve products and processes by making the more frequently performed tasks easier, in some cases at the expense of activities that occur infrequently. The results of activity sampling can help designers verify personnel requirements, in terms of how many people are required, at particular skill levels, to perform a job. Such data can also be used, in some cases, to assess operator workload and stress levels. Numerous other assessments can be made during activity analysis, as discussed in later chapters.

Critical incident study or *analysis* is another activity often performed at the beginning of the design process. In one of the first applications of this approach, which dates back to the late 1940s, aircraft pilots were asked to describe situations that almost resulted in an accident (near misses). The results of this investigation led to a number of modifications of aircraft cockpit design. From a more general perspective, critical incident analysis entails interviewing people who have worked with existing systems or products to discover maintenance problems, operating difficulties, and other concerns that might or might not need to be fixed in the design of the new product. Critical incident studies reveal the importance of *robust* products and processes, the performance and worth of which are not detrimentally affected by variations of use or by an environment outside of an expected range of values. Such products are impervious to reasonable variations.

Functional flow study and *decision/action analysis* are the next steps shown in Chapanis's list of activities. Functional flow study consists of breaking the operations down into component actions and identifying one or more sequences of actions that must be performed to properly operate the product. Identifying the required activities and their precedence relationships is part of this procedure. Decision/action analysis is a complementary procedure that documents the sequence of decisions and actions required in a complete application of the product. Chapanis presents the following steps for an automated teller machine (ATM):

1. Initiate transaction
2. Request customer identification
3. Provide customer verification
4. Verify customer's identity
5. Provide menu of transactions
6. Select transaction from menu
7. Specify amount of funds to be transacted
8. Check funds sufficiency

Note that activities are described at this stage without identifying how decisions are made, or who or what does what. Some of the missing details are provided in the next two analyses.

Action/information requirements analysis provides a more detailed view of what information is needed to make the necessary decisions and perform the required actions. Particular focus is placed on documenting what information is needed in order to know which action should be taken. Determining when and how the information will be provided is also important. In the earlier example of the ATM, the procedure needed to initiate the transaction must be obvious or a means of informing the customer will have to be provided. Requests for customer identification will have to be understandable. The same can be said for each of the remaining steps. If the customer feels intimidated by inquiries and gives an inappropriate response, it is particularly important that the ATM does not automatically lock up and call the local police.

Function allocation is the next stage in Chapanis's view of product design. A function, from this perspective, refers to an operation or activity that can be performed by either a person or a machine. The goal at this stage is to determine who or what performs each function. Some people also call these allocations "trade-off studies" because trade-offs are made when allocating functions. That is, a machine will probably outperform people on some criteria while people are better on others. However the allocation is made, the guiding principle is to trade better performance on less-important criteria for better performance on more-important criteria.

Each function that is not allocated to a machine has a corresponding set of tasks that will be performed by people. Ergonomic engineers need to remember that these tasks must be planned so that people have the strength, knowledge, information, or protection needed to perform them. Moreover, allocating functions to machines may not simply remove that function from the human; it may change the set of functions that need to be performed, such as adding the need for the human to monitor the automation.

Task analysis is an important component of the ergonomic design process* performed to ensure that the aforementioned objective is attained. Task analysis involves documenting tasks and task sequences, time requirements, information requirements, and the skills, knowledge, and abilities needed to perform the tasks. Training and other means of aiding performance will often be specified to help ensure that tasks match the capabilities of the people that will be performing them. Throughout the task analysis, various activities, such as the five activities described next, may also be necessary to help determine how well the tasks will be performed under expected environmental conditions.

Failure mode and effects analysis is the first of these five activities. In this approach, the effects of component failures (such as tire failures on a car) are systematically evaluated. Focus is placed on identifying and remedying failures that have severe consequences or a high likelihood of occurring. The second activity, *task-timeline analysis*, is useful for identifying conflicts between tasks that might interfere with performance or result in excessive workloads. This approach involves plotting the tasks over time, starting from triggering events and continuing until ending events for the particular activity. Task-timeline analysis arose out of aircraft design procedures that are more keyed to system events than most industrial processes, but are otherwise very similar.

Link analysis, the third activity that may be necessary during task analysis, was made famous by Chapanis in the early days of human factors. Link analysis focuses on identifying linkages (and their strength) between people and other system components. Linkage strengths often measure how frequently people use or otherwise contact items in their environment (such as the components of an automobile instrument panel), but can reflect importance in many other ways. Such data can guide system layout. Components with strong linkages should be placed closer together, if possible. Numerous forms of the fourth activity, *simulation*, have been devised. As discussed in later chapters,† both digital computer simulation and operator-in-the-loop simulation are appropriate to product design.

The fifth activity, *controlled experimentation*, is especially important in ergonomics. As the name implies, this activity involves interrogating nature under controlled conditions so that other influences do not interfere. The use of controlled experiments provides a way to directly invest in product quality by assuring product usability. To perform this activity during product design, ergonomists need to have the ability to rapidly prototype new product designs for assessing their qualities. The more rapid the prototyping, the faster the turnaround in design.

Operational sequence analysis and *workload assessment* are the final two elements of Chapanis's perspective. The focus here is on what happens after the product is sold and put into use. Numerous issues become relevant here, beyond those we can discuss in this section, but it is important to note that when the product is complex, training is clearly

---

* Chapter 8 discusses task analysis in more detail.
† Chapter 9 focuses on simulation techniques used in industrial product and process design.

part of the sale. With less complex products, such as computer software, training can be a built-in portion of the design. In those situations, it is often better to find a short demonstration phase after or during the sales process. One strategy is to allow users to play with the system and discover its many features. In such a case, the methods of use are described interactively. However, not all products can depend on owners operating a product safely without appropriate guidance. Some of these determinations can be made prior to putting a product on the market, from failure mode and effects analysis and elsewhere. However, obtaining feedback from customer surveys, field studies, and other sources is often critical.

Workload assessment is another important issue but is the least clearly developed portion of the Chapanis schema. Workloads vary among similar products (e.g., consider the case of hand or power tools), and this feature affects their long-term usability. While this aspect of product design has not received as much focus as it might deserve, it will likely develop and emerge in ergonomics practice in the future.

Human-centered design

As stated earlier, this book is about designing workplaces, tools, workstations, and much of the equipment people use in the manufacturing and service industries, as well as the products from these industries for use by people. Often, although not always, the focus is on design for people, or *human-centered* design. While the term human-centered implies a focus on people using the design, it also includes people who are responsible for the system being designed. Beyond a focus on the end products of design, this section seeks to show the reader that designers need guidance in terms of design principles. These principles invite the designer to think about design limitations, starting points, specific concerns, and the like. A number of such principles are outlined in the following.

Principle 1.1

Select people to fit their machines and jobs.

This principle really means that at times ergonomics specialists need to choose people for particular jobs because of individual differences in knowledge, abilities, and skills. Many people cannot tolerate assembly lines because of the monotony, while others like them because the mental demands are so small. Sometimes intelligent people like minimal demands because they want to concentrate on their own thoughts. Systems should be designed to account for these individual differences. Ergonomic specialists also need to modify employee knowledge, abilities, and skills through training. In general, the objective of ergonomic design is to enhance human talents in the roles people serve. That enhancement may be direct, through training, or indirect, through changes in the system in which people work.

The next two closely related principles address human limitations in opposite ways.

Principle 1.2

Take advantage of human attributes by expanding requirements for human abilities so that people can better perform their roles.

More simply stated, too much focus on human limitations may cause you to miss opportunities to improve the system. Besides this, many people prefer to work with systems in which their abilities shine. The next principle treats human limitations more as an opportunity.

Principle 1.3

Overcome human limitations so that those limitations do not become system limitations.

There are many clear and obvious ways to accomplish this goal. Power tools magnify human motor abilities in terms of speed, strength, and durability. Magnifying glasses enhance visual sensitivity, and special sensors convert electromagnetic waves, outside of human sensory abilities, into important information. Production control systems and associated software help identify feasible and near-optimum production schedules. Quality- and inventory-control systems help identify causes of lost productivity. These features need to be captured and harnessed to enhance human abilities, overcome human limitations, and to do these things in a way that promotes user–owner acceptance. Users want improved capabilities and management wants the benefits of those capabilities. It is the designer's role and responsibility to ensure that both parties are satisfied.

Principle 1.4

Be sure that the problem identified is the right problem and that it is formulated correctly.

It is too easy to ignore this fundamental principle. The story related in Box 1.2 describes the effect of not keeping one's mind on the real goal.

After formulating a design problem, the designer must create a feasible solution. Part of this activity depends upon the designer's creativity. But a large part of this activity comes from acquiring knowledge about the system under design, the *roles* of people, and the *goals* and objectives of those roles. Part of the knowledge collection can be conducted by asking people informally or through the use of questionnaires and interviews. Some notes on constructing questionnaires and conducting interviews are given later.* Additional techniques on data collection and representation to enhance design solutions are shown in other chapters. The following ideas and principles provide some of the front-end analysis to design that lead to an initial appropriate solution.

Principle 1.5

Consider the activities of interest as to whether or not people are required to exercise significant levels of skill, judgment, and/or creativity.

BOX 1.2   AUTHENTIC ROLES

They tell the story about an airplane pilot who called back to the navigator and asked him to determine where they were on the face of the earth. But he received no reply. A couple of minutes later he called again, and again there was no reply. On the third call the pilot was angry. So the navigator finally broke down and said that he could not figure out exactly where they were. The pilot then said, "Damn, we were making such good time too!"

The point here is that speed is unimportant when you do not know where you are or where you are going. So first assure that the objectives of the design and principal design requirement are clearly known by the whole design team.

---

* Chapter 14 addresses the design of questionnaires and interviews. It should be noted that the Analysis of Similar Systems, as shown by Chapanis's proposed product development, is often performed by questionnaire or by interviews.

Generally, the lower the requirements for human skill, judgment, or creativity are, the more the designers ought to consider automating the activity. When human operators contribute less in an activity, they usually care less about that activity, so it is likely to be forgotten in operation. A similar notion posed from the opposite end is as follows.

Principle 1.6

Find out the degree to which people enjoy being involved with these activities.

Professor Donald Norman* asserts that ergonomic specialists should design systems that people find enjoyable to operate. He makes the point that ergonomic designers need to rise above the standard of merely acceptable systems. This leads us to another principle of human-centered design.

Principle 1.7

Query human operators about their dissatisfaction to see if it is caused by (a) a need to "feel in control," (b) a desire for achieving self-satisfaction in task performance, or, (c) perceptions of inadequacies of technology for quality of performance, or ease of use.

Responses to these inquiries provide useful design information on how to correct the design. For example, if the need to be in control or to achieve self-satisfaction is *not* the central concern, the designer must determine if the perceived inadequacies of the technology are well founded. If they are, the designer should eliminate the functions in question from the candidate set; if they are not, the designer should provide demonstrations to familiarize personnel with the actual capabilities of the technology. When people are in crews or teams where each member depends on all the others, one or more members who do not perform their jobs properly create a danger to the entire crew. The same is true when some team members are machines, either with or without new technologies. In either case, the team analogy provides an important perspective for ergonomic designers.

When changes are made to an existing design in order to improve it, designers need to think clearly about the things that should and should not be changed. Some of the principles addressing this question are as follows.

Principle 1.8

To the extent possible, only change the system functions that personnel in the existing system feel should be changed.

Many designers tend to fix things that operators do *not* perceive as wrong. When the better features of the existing system are changed, the operators may perceive the designers as adversaries and be less cooperative with them. It is usually better to hold back further changes and let the operators identify other problems in later testing. However, that principle cannot always be followed. When designers cannot follow it, the following principle becomes important.

Principle 1.9

Consider increasing the level and number of activities for which personnel are responsible so that they will be willing to change the functions of concern.

When it looks like the job has to be changed in order to get the system working properly, one needs to maintain the perception that major changes are necessary. Making

---

* Professor Norman (1988), at the University of California, San Diego, CA, provides several examples, along these lines.

operators aware of the problem and encouraging them to get actively involved in fixing it may reduce resistance and the suspicion that changes are unnecessary or trivial.

Two more things to keep in mind when there are substantial changes for an operator are as follows.

Principle 1.10

Be sure that the level and number of activities (tasks) allocated to each person or team forms a coherent set of activities and responsibilities, with an overall level that is consistent with the abilities and inclinations of the personnel.

Principle 1.11

Avoid changing activities when the anticipated level of performance is likely to result in regular intervention on the part of the personnel involved.

It is often easy to collect a variety of tasks that do not naturally fall into a single job and hand them off to anyone who could conceivably perform the task. Those jobs that no one likes are often treated this way, and the result is to create high personnel turnover, which usually raises costs. Ergonomics may be the focus, but *economics* is the glue that makes ergonomics feasible.* Accordingly, designers must be careful of task allocations that might be analogous to glue solvents. Designers must also recognize that human intervention is one way to keep people in touch with the systems they operate. Familiarity with equipment and systems is particularly important when the human operators must intervene in an emergency. Suddenly dumping a failed system into the lap of an uninformed human operator is an invitation to disaster. It is far better to let operators see the potential for a failed system so that they can deal with the preventative problem rather than present them with corrective problems that are often vastly more difficult to solve.

The last principles of ergonomics design presented here concern the roles and goals of system operators and system limitations.

Principle 1.12

Assure that all personnel involved are aware of the goals of the design and know what their roles will be after the change.

Consistency between the goals of individual operators in the system and the overall system objectives is imperative. Otherwise, the problem of "not knowing where one is going" is prevalent, and that feature distinguishes teams or crews from similar collections of people. It is also often important to evaluate people's abilities and consider how consistent they are with attaining the overall goals of the system. This observation leads to the following principle.

Principle 1.13

Provide training that assists personnel in gaining any newly required abilities to exercise skill, judgment, and/or creativity and help them to internalize the personal value of these abilities.

Some of the training sessions might include simulations in which people play multiple roles so that they can better judge their own positions within the new system. While training is usually a good idea after the design is completed, an important principle prior to design completion is the following.

---

* It is assumed that most readers have some knowledge of engineering economics or the mathematics of finance.

Chapter one: Guided tour of ergonomic design

Principle 1.14

Involve personnel in planning and implementing the changes from both a system-wide and individual perspective, with particular emphasis on making the implementation process minimally disruptive.

Human cooperation and teamwork often fails to overcome human discord, but being part of the planning improves human acceptance; the secondary breakup of soil is often more difficult than the primary breakup. The final principle on human abilities and limitations is as follows.

Principle 1.15

Assure that personnel understand both the abilities and limitations of the new technology, know how to monitor and intervene appropriately, and retain clear feelings of responsibility for system operations.

These final principles on the ergonomic design of human-centered systems remind us that designers strongly affect the attitudes of operating personnel long after the design is completed.

After designing a solution to the formulated problem, designers need to refine and improve the initial solution toward an optimum. The activity involved at this stage is usually iterative. Additional examinations are needed to assure that the system does in fact meet the abilities and limitations of all its component roles. Designers need to assure that people in the system understand how it operates. In some simple systems, this procedure is elementary, but in many cases it is not, so various issues regarding the final design solution need to be examined.

Two obvious concerns about the solution are whether the system works and whether it works as planned. Some other obvious questions need to be addressed: Is the stated problem solved? Are all the requirements met? Demonstrations of the solution must be made to assure that the answers to these questions are correct and complete. Finally, designers need to determine if people use the system (acceptance) and if they use it as intended. Certainly, serendipity sometimes does occur, but it is *not* something a smart designer plans on. A number of specific techniques for finding answers to some of these questions are presented in succeeding chapters of this book.

Another issue on which designers need assurance is *viability*. That is, the benefits of the solution must be sufficiently greater than the costs, or management will not accept the design. Since finding a solution that meets all of these concerns or issues may be difficult, smart designers try to select problems that make a strong economic difference first. In that way they establish a strong success rate before embarking on more questionable challenges. As Paul Fitts has said, "The cream comes off the milk first." To be sure, that statement made much more sense in the days before homogenized milk, but although the statement is outdated, the idea is not. Ergonomic designers should first go after the projects where the potential payoff is greatest. In general, designers need to balance their efforts among potential projects such that expected marginal returns are about the same for each project.

Military equipment design

Although this book is not focused on the design of military equipment, system design in the U.S. military is more advanced and formalized than in the industrial sector, so a brief look at military design is in order. Military weapon systems are expensive and are used by many physically fit men and women between 18 and 45 years of age. These conditions justify a more formal and thorough design system than most industrial concerns can afford. Another principal difference between industrial and military design can be seen in the stringent requirements placed on military systems.

The seven phases in U.S. military procurement operations are

1. Basic research and development
2. Mission analysis and planning
3. Concept development and preliminary design
4. Competitive system demonstration
5. Full-scale engineering development (test and evaluation)
6. Production and deployment
7. Turnover to users

Between each phase is a reconsideration point where the next phase depends upon approval of the preceding phase. Needless to say, these phases are time and cost driven. Numerous analyses are made during each phase, similar to those shown for Chapanis's list shown previously.

Many of the questions that must be answered during these analyses are behaviorally oriented. Part of Meister's (1982) extensive list of questions is given in Table 1.1. Note that this list is based on the premise that a new military system is partially or totally replacing an older system. Although the relationships between these questions and particular forms of analysis are frequently fuzzy and rarely well articulated, they need to be faced and understood as well as possible.

Table 1.1 Behavioral Questions Arising during the Design of Military Systems

Phase	Behavioral Questions
Mission analysis	1. What changes are envisioned in the number and type of personnel? 2. What are the expected changes in personnel selection, training, and system operation?
Concept and competitive Demo	3. Which design alternative appears to be most effective in terms of human performance? 4. Will system personnel be able to perform all functions effectively? 5. Will an excessive workload develop? 6. What factors are expected to raise potential human error and can they be eliminated or reduced?
Competitive demo and full scale	7. Identify the better design alternative. 8. What personnel performance level can be achieved and do these levels satisfy system requirements? 9. Define personnel training requirements. 10. Are equipment and job procedures ergonomically proper?
Full scale and production	11. Can the system be produced as designed? 12. Have all the system variables been ergonomically redesigned? 13. Can personnel perform satisfactorily and does the system provide for personnel needs? 14. Do any behavioral problems still exist? 15. What are the problem causes? 16. What solutions can be recommended?

Ergonomic criteria

Ergonomic criteria are used to judge problem solutions and alternative designs. A variety of criteria are typically used in ergonomic design. Some relate to the job and others to human performance. One long-used criterion is *speed* or performance *time*. Speed is important to measurement of productivity, and it is almost everything in the design of sports products. For this reason, industrial operations are frequently time studied over several cycles of production. However, speed is unimportant unless other important criteria are satisfied. One such criterion is *accuracy*. When errors occur, accuracy declines. Just as speed and time are complementary, so are accuracy and error.

Speed and accuracy are often inversely related. You may have heard someone say, "If I had more time, I would have done it better." In ergonomics, this relationship is known as the *speed–accuracy trade-off*. That is, slower performance speeds improve accuracy and faster speeds cause errors. Many situations involve this trade-off, so ergonomic designers should be aware of it and be prepared to deal with it. This trade-off occurs often enough that many ergonomic and industrial engineers tend to expect it, although there are numerous situations where it does not occur. In fact, accuracy often drops when people perform a job at *either* a faster or a slower pace than normal. Walking at a slower or faster pace than one naturally walks increases energy consumption. The point here is, "Don't assume. Find out."

Another aspect of performance time is *time variability*. Since no one performs a task exactly the same way on every occasion, time variance occurs within each person. In addition to the time variability within persons, mean times vary between persons. The between-person variance of mean performance times is one commonly used measure of individual differences for a class of tasks.

Other important criteria in contemporary industry include *safety* and *health*. Workplaces must be safe places today. Years ago this was not an issue in many factories, where managers exhibited little concern for the health and welfare of their employees. With the creation of the Occupational Safety and Health Administration (OSHA), a direct governmental body now watches over these criteria. In addition, lawsuits frequently result from improper safety practices, and insurance costs increase for those who do not maintain safe practices. Even those who do not view safety and health as an important moral obligation are finding that economics alone make these criteria important. Economic leverage is particularly obvious in light of product liability costs, which have caused many companies to fail. Another ergonomic criterion is *ease*. When jobs are easily performed, *endurance* improves and fewer rest breaks are needed. Furthermore, when a job is made easier, it is usually safer. Accordingly, most of the preferred criteria are correlated most, but not all of the time.

The *amount* of learning required to use a product or perform a job and the *rate* at which this learning occurs are both important ergonomic criteria. In the best case, work methods are quickly and easily learned on-the-job, and products are easily used without requiring instructions or training. With faster learning, better performance occurs sooner and lasts longer over production cycles and the life of a product. When a job or product requires a large amount of learning, training may be needed. While training can be expensive, training costs are usually much less than the consequences of improper performance and can be reduced by developing good training methods. Errors in performance lower the *quality* of outgoing products or services, and also result in accidents. In today's highly competitive international marketplace, the profitability of most companies is determined by the quality of their products, making quality a critical criterion. Since many people define quality as the measure of successfully meeting customers' wants and needs, quality is often measured by greater *customer satisfaction*.* Other measures of product quality include fewer *product defects*, higher *product usability*, and reduced *product liability*.

Other criteria pertaining to the design of production or service systems describe how employees view system operations. Sometimes these criteria are combined under the general heading of *better morale*, but for the most part these criteria are measured through *employee turnover* and *absenteeism*. When employees do not like their jobs, their morale is down, they find excuses to avoid going to work, and they may actively engage in new job hunting. Unfortunately, these criteria are rather extreme. Accordingly, more sensitive measures of subjective evaluation are solicited through questionnaires and interviews.

The criteria identified earlier are some of the most important for industrial applications. They are not, however, the only ones. You will see other criteria introduced throughout this book. In particular, you will see *economic* criteria employed in the form of lower costs, higher revenues, greater net present worth, and greater returns on investments. Ultimately, the marketplace determines each company's fate. In order to understand the role of economics in ergonomics, consider the statement from Rose et al.[†]: "To succeed in introducing a new ergonomically better, working method, that method must also have economic advantage." The statement is clear. No matter how much moral right or might an ergonomist might feel, the money needed for investment will go to economically better alternatives, if such investments are available. Geoffry Simpson made the point a bit stronger, "What is the point in spending vast sums of money to create the healthiest and safest factory in the country, if the only people who work in it are liquidators?" The net result is simply that prospective ergonomic projects must be economically justified if they are to be funded. At the very least, prospective projects must be expected to return more economic value then the amount invested. Moreover, stockholders of a company invest in the stock to make money. If company A will not produce an ample return, they will sell their stock and invest in another company that will. In a capitalistic society, that is a fact of life. Nevertheless, economic evaluation is a difficult and challenging problem for ergonomic designers.

Models of human performance

Several years ago, Drury[‡] stated that, "In order to design, we must predict and to predict we must model." Various models that fit within Drury's perspective are presented later. While these models are useful, it is important to recognize their limitations as well. Professor George Box[§] stated that "all models are wrong, but some are useful." By this, Prof. Box meant that all models are simplifications, and, therefore, some aspect of the model is incorrect. It is, therefore, not the accuracy of the model that is of primary interest, but rather how useful the model is in a particular, well-defined context.

---

* Customer satisfaction is often measured with questionnaires and interviews, the topic of Chapter 14.
[†] Rose, Ericson, Glimskar, Nordgen, and Ortengren (1992) made the initial direct quote in the following. The second quote is from G.C. Simpson (1993) who was head of the Ergonomics Branch, British Coal Corporation, Burton on Trent, England at the time.
[‡] Professor Drury (1988), of the University of New York in Buffalo, NY, delivered these thoughts in a keynote address to the Ergonomics Society.
[§] Professor Box is a Professor Emeritus of Statistics at the University of Michigan. He is distinguished for his work on experimental design and is the son-in-law of Sir Ronald Fisher, who is considered one of the founding fathers of parametric statistics.

What Drury did not mention was that *in order to model well enough for the model to be useful, we must understand that which we are modeling*. This means we must first collect information about system goals and objectives, the roles people play in that system, and how the system changes over time. With this information, modeling, predicting, and designing can then begin.

While models are influential in design, science progresses from hypotheses to *theories*. As most readers may know, theories are explanations of phenomena to which there are *no known factual disagreements*. There may be multiple theories for the same phenomenon, and theories need not be useful nor serve any purpose other than explaining relationships. Models, on the other hand, are discarded if they are not practical, and are expected to predict phenomena quantitatively, although their precision varies greatly from model to model. Besides their roles in making predictions, models help people learn complex skills (e.g., troubleshooting) and they provide a framework for engineering design and for experimenting with alternative designs. For those reasons, models play an important role in ergonomic design. But models cannot be expected to be accurate under all circumstances. The reader should be forewarned that incorrect models can lead people to commit logical fallacies or misunderstand relationships. Models that are accurate for only a limited range of the variables in question can cause people to overgeneralize.*

Before going further with a discussion of relevant ergonomic models, it should be stated that *behavior* and *performance* are distinctly different notions. Behavior is what an individual does. On the other hand, performance is how well something is done relative to some standard. Different people do things differently, of course. Some people naturally use their right hands to write or to assemble a device; others use their left hand. Following their own inclination, people develop an average speed that is a measure of performance with regard to the time criterion. Both behavior and performance exhibit individual differences between people. Reversing behavior, for instance, by having people use their non-preferred hand, usually results in decreased performance. A principal point here is that performance cannot be adequately described without a prior description of behavior or an allowable class of behaviors.

Helson's hypotheses

A number of years ago, Harry Helson[†] developed what he called hypotheses, which are really generic models of human behavior and its performance implications. Helson worked many years during World War II on the human factors of antiaircraft guns aboard ships and on other military applications. Accordingly, he speaks with a great deal of empirical knowledge and wisdom. His *U-Hypothesis* deals principally with performance.

> U-Hypothesis: For most variables of concern in ergonomics, performance, as an inverted function of that variable, is U-shaped. The bottom of the U is nearly flat but the extremes rise almost vertically.

The clear implication of this hypothesis is that being precise in optimizing human performance is not terribly important because the benefits of doing so are small compared to

---

* See Chapanis (1961) for a fascinating paper extending this theme.
† Helson (1949) described a variety of what he called "hypotheses" based on a great deal of experience. While these hypotheses are actually models of human behavior and performance, they have important implications, which can be thought of as principles of ergonomic practice.

being very nearly optimum. But this hypothesis clearly shows that it is *crucial* to be in the immediate vicinity of the optimum. Thus, the quest to identify the optimum is exceedingly important in ergonomic design even though it is not important that the exact optimum is specified.*

The *par hypothesis* is another important idea developed by Helson.

> Par Hypothesis: Human operators set a performance level as their "par" or standard. When they see their performance drop below that par, they provide additional effort. If performance is above that par, they withdraw effort.

This hypothesis states that people monitor their own performance and modify it as a result of what they observe. The modifications are analogous to the operation of a mechanism known as a governor. Performance that follows the behavior described by this hypothesis would vacillate slightly but remain relatively consistent. This hypothesis also predicts that human performance on successive cycles will be negatively correlated. That is, a longer time on the nth repetition of the task would be associated with a shorter time on the $n+1$th repetition. Unskilled human operators often experience performance vacillations during the early part of learning, particularly when reproach follows poor performance and compliments follow good performance. While some ergonomists may disagree with parts of the par hypothesis, virtually none will disagree with the self-monitoring of performance by people.

Helson's third hypothesis concerned *anticipation* and *averaging*.

> Anticipation and Averaging Hypothesis: People learn anticipatory patterns of behavior. When the rate of operation exceeds the person's ability to exactly complete the pattern, an averaging form of behavior results.

People learn intricate movement patterns over time. At first, people tend to perform movements as a series of discrete steps. With practice, more complex sequences of movements, or movement patterns, are learned. Learning these patterns allows people to anticipate future movements within the pattern. People still follow the general outline of the pattern, when they are under time pressure, but start to leave out some of the details. An example of this situation is when a person is tracking a sine wave pattern. At low frequencies, the pattern is tracked with almost no error because it is so easy to anticipate. As the frequency of the sine wave increases, the tracking pattern may lag behind the input signal slightly, but tracking performance remains good on the average. With even higher frequencies, the peaks and valleys are still there but tracking accuracy decreases.

Other models

People are often modeled as having three general capacities. That is, people can obtain information from the environment using sensory organs, such as eyes or ears, process

---

* See Nadler (1970) for a small difference in philosophy. His principal thesis is that ergonomic designers should always be seeking the optimum or a solution as close to it as possible. In practice, this means that designers should not be satisfied with "acceptable" design solutions at the beginning of the design process, even though they may have to be later in this process.

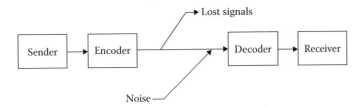

Figure 1.1 Simple communications system.

it with their brain, and then make physical responses using output devices, such as the hands and voice. Developing a basic understanding of these capabilities, as well as the interrelationships among them, is the primary focus of most introductory human factors or ergonomics courses (See Bailey 1989, Bridger 2003, Burgess 1986, Faulkner 1983, Gagne and Melton 1962, Huchingson 1981, DeGreene 1970, Kantowitz and Sorkin 1983, Konz and Johnson 2004, Megaw 1988, Murrell 1969, Sanders and McCormick 1987, or Wickens 1992).

This modeling perspective is particularly applicable to communication processes. Communication between people requires one person to use an output device, such as the voice, which is compatible with another person's input device, such as the ear. Commands may be communicated or questions may be asked by one person of another who responds by following the command or giving an answer. Every element in the communications sequence takes time and introduces the possibility of error.

Figure 1.1 describes a simple communications system linking the sender to a receiver. This communications model was developed from the *information theory* of Shannon and Weaver. The term "Uncertainty Analysis" is used here to describe communications fidelity. Within the figure, an encoder is a device that converts an intended message into symbols that are compatible with the receiver, which decodes the transmitted symbols. A message from the sender is verbally coded for the receiver who must hear those sounds and decode them into a meaningful message in order for communication to occur. Background noise can interfere with an accurate transmission, sometimes making the signal unintelligible. Although this model appears to be theoretically identical to one an electrical engineer would use to design a microwave communications system for telephone traffic, there are some subtle but important differences that the ergonomics specialist must keep in mind.

When humans deal with machines, problems of *communications* are even greater. People communicate to machines through input devices called *controls*. Switches, keys, and valves are examples of machine input communications devices. Communications from the machine to the person occur through *displays* of various kinds which are sensed by the human. Lights, mechanical devices, and computer monitors are examples of displays. This sequence of communications activities can be made faster and more accurate when the machine's input and output devices are designed to be compatible with the human operator. These machine controls and displays are referred to as the *operator/machine interface*, and one of the major objectives of engineering is to optimize this interface through design. This process involves the selection and location of the machine devices. Some people feel that ergonomic/human factors considerations stop at the immediate interface and that the nuts and bolts beyond the interface veneer are not part of this discipline. That idea is wrong! While it is true that a lot of nuts and bolts are irrelevant to ergonomic considerations, operators must have some *mental model* of the system they use. When that model is misleading, human errors will likely result, as explained earlier.

The implications for training are critical: *Be sure that the models used in training are accurate even if those models are not elaborate or highly precise.* Ergonomic designers are responsible for detailing the mental model.

Some simple but important probability models are also used in ergonomic design. The normal (Gaussian) probability density function is commonly applied, since human performance time is usually assumed to vary according to a Gaussian distribution. The *Poisson* and *binomial* sampling models are two other popular probabilistic models. As noted earlier, sampling is often more economical than direct observation and continuous monitoring. These sampling models and other associated distributions are discussed at length later.[*] The reader is invited to review the information on these topics contained in the appendices of this book.

Signal detection theory (SDT)[†] is another model used in ergonomics that is based on probability theory. SDT has been used to describe decision making by both humans and machines in a wide variety of applications. Visual inspection is one of the more common applications of SDT. SDT describes the activity that occurs when an inspector judges whether product units meet specifications, as a statistical decision process. It is assumed that the inspector will make decisions that minimize the cost of both (1) rejecting good items and (2) accepting bad items. The model also assumes that these decisions are subject to uncertainty. That is, the probability of a defect, after making an observation, will often be between 0 and 1.[‡]

A variety of other models describe individuals performing tasks either separately or concurrently[§] with others. Some combinations of two tasks cannot easily be performed at the same time, while other combinations can. The first case is illustrated by tasks requiring eye movements between widely separated locations. Other tasks, such as simultaneously rubbing your belly and patting your head, can be done concurrently with a little practice. A *limited resource* model can be used to describe behavior in these situations, and these models can often explain speed–accuracy trade-offs within a task or changes in performance between tasks.

Learning curves provide another, often mathematical, model of how performance improves with experience.[¶] One learning model assumes that people will improve at a constant ratio for each doubling of the number of cycles made. Another assumes that performance on each consecutive trial is a constant additive and multiplicative function of current performance. The different models have advantages and disadvantages and distinguishing the better-fitting model empirically is not always easy.

Other models describe human behavior as analogous to mechanical operations and may serve conceptual and computational needs in design. One such model is a *queuing system*, which represents a population of customers who arrive in some probabilistic fashion. After arrival the customers go to a service facility. Following the servicing, customers

---

[*] Chapter 7 deals with the subject of activity sampling.
[†] Signal detection theory originally was a tool of science and electrical engineering. A number of people from the University of Michigan, Ann Arbor, MI, including Wilson P. Tanner, Jr., David M. Green, and John A. Swets, used this form of statistical decision theory in several psychological applications.
[‡] A probability of either 0 or 1 means you are sure of something, and therefore there is no uncertainty.
[§] Concurrent tasking involves doing two or more things at once. Several psychological theories on how people perform concurrent tasks are taken directly from economics. Human attention is viewed as a limited resource. When attentional needs are greater than those available, performance drops. These theories are still in a formative state.
[¶] Learning curves are discussed in detail in Chapter 9.

Chapter one: Guided tour of ergonomic design 21

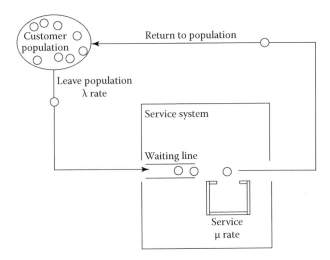

Figure 1.2 Abstract queuing system.

return to the population. Figure 1.2 provides an abstract description of a queuing system. The person operating a tool crib in a factory is analogous to a service facility. When a mechanic (customer) needs a jig, fixture, or special tool, the mechanic goes to the tool crib and asks for the necessary item. The tool crib attendant attempts to fill those needs, and the mechanic goes back to his or her regular job. There may be several tool crib attendants and a variety of ways that customers can queue up to the tool crib windows. Many other examples, particularly for the service industry, are available in basic textbooks on industrial engineering.

Another example of modeling of human behavior applies when a person uses a word such as a "young" person. Would you say that a person 4 years old is a young person? Your answer is probably yes, but you are probably more apt to think of a 10, 11, or 12 year old as being a young person. If queried further, you might say the term is appropriate, but less fitting, for people 16 or 17 years old. The point is that people use a term that has some fuzziness in its meaning and that some numerical values have a stronger "belongingness" to the notion than others. The theory surrounding this is called "fuzzy set" theory.* Concepts of this theory are covered in this book because they are useful in understanding approximate thinking and reasoning, especially in conjunction with the logic people use in controlling vehicles, processes, and tools. Fuzzy set theory provides a very precise way of describing human imprecision. When a lot of people perceive certain relationships in similarly fuzzy ways, this information can be used for design purposes, provided that it is sufficiently consistent. Some common commercial products on the marketplace today have control designs based on fuzzy logic. Fuzzy set theory also provides a way to investigate population stereotypes when they exist and a means of eliciting knowledge and meaning by individuals. For example,[†] this approach has been used to document how workers interpret phrases such as "Avoid prolonged contact."

---

* Professor L.A. Zadeh of the University of California at Berkeley conceived the idea of fuzzy set theory in the 1960s. Like many such ideas, it has taken time to find its way into applications.
† For more detail, see the article by Lehto, House, and Papastavrou (2000).

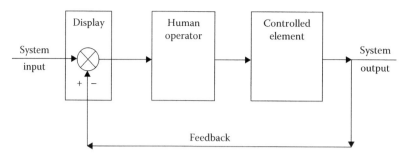

Figure 1.3 Abstract control system.

One other model that has been applied in a variety of situations is *analogue control logic*.* People driving vehicles follow paths such as roads or factory aisles. Obstacles along these paths are avoided by making control changes. Servo-mechanisms follow the same general control theory even though corresponding elements, operating rules, and parameters differ. Figure 1.3 describes one control model, often presented in elementary control theory books. Along these lines, consider a driver who looks ahead and sees a stop sign. That observation evokes a series of commands that guide the driver to start slowing down well in advance of the sign. If the rate of deceleration is inadequate, a greater braking force is applied. When the rate of deceleration is too much, the braking force is reduced. However, the notion of adequacy comes from personal goals, an accepted pattern, or from a command display. Those notions of adequacy are not always constant, as many engineers unfortunately assume in the design of servo-mechanisms. It is important to recognize that these models all have limited descriptive accuracy in describing human situations.

One of the more flexible forms of modeling uses the contemporary digital computer and simulation languages. While simulation programs can be written on computers without the use of contemporary simulation languages, those special languages *vastly* simplify the effort. As most readers probably know, any analogue model can be broken down into digital pieces and the pieces can be written in computer code. The special languages let one do this with a minor amount of code writing. Moreover, computer simulation models allow one to describe situations dynamically and the computer allows exact replications to an almost infinite degree, so that a whole distribution of outcomes can be seen. The use of driving simulators to identify safety concerns illustrates this approach. In applying this method, one must remember that accidents are very rare events, so evaluating the danger of having an accident requires a lot of replicating. For example, an automobile driver on a dangerous roadway might average one accident in 200,000 miles of driving. To observe even a single accident, assuming an average speed of 50 miles/h, one would expect 4000 h of direct observation on the road or an equal amount of that time on a driving simulator to be required. That is about 2 years of one-shift observation solely devoted to this single case. On the other hand, even a small computer today can replicate critical situations in a few hours. For this and several other reasons, simulation has become an

---

* As an engineering and a scientific subject, analogue control logic has some ambiguous beginnings. One of the first ergonomics applications was by Birmingham and Taylor in the 1950s. It has grown in mathematical sophistication in later years (see Wickens, 1992). Kelley (1968) provides one of the best philosophical books on control theory ergonomics.

Chapter one: Guided tour of ergonomic design

extremely important ergonomic design aid. Some illustrations of this methodology are shown later.*

Model of software design

Barry Boehm (1988) tried to characterize the design process he observed when studying how people design software packages for computers. His results appear in Figure 1.4. As the reader can see, software design entails many activities similar to those employed in designing other things; one goes from concept, to plan, to testing the plan. But the important feature Boehm shows in this figure is that there are several iterations of those steps, yielding refinements corresponding to each iteration. Many architectural and civil engineering design projects operate in a similar vein, beginning with concepts, making plans

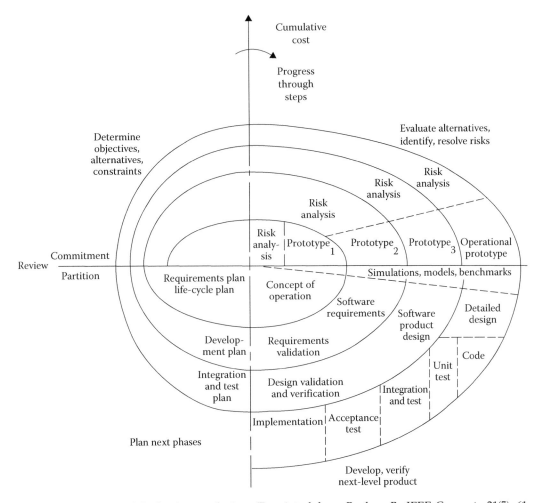

Figure 1.4 Spiral model of software design. (Reprinted from Boehm, B., IEEE Comput., 21(5), 61, 1988. With permission.)

---

* Chapter 12 discusses the use of simulation during ergonomic design. Experimental simulations and computer simulations are both illustrated as methods for applications such as large steel-making shops and mills, individuals driving vehicles, or people controlling industrial processes.

for implementing those concepts, checking those plans, and then restarting a new iteration with increasingly fine changes and refinements resulting from each iteration. Finally, the design deadline nears and the design is completed.

Some newer concepts in industrial products design are appearing in many industries today. Traditionally, a design engineer or industrial designer took a crack at the idea alone. He or she then presented the design concept to management, who would either give the go-ahead or tell them to go back to the drawing board. After management was told to go on with a design, the plans started their routing through a series of specialists. Eventually, the design came to the manufacturing engineers, who determined how it should be manufactured and made materials specifications. Nearing the end of the chain, the ergonomist received the plans and was asked to make the design as user-friendly as possible. By this time, the available budget greatly limited any but the most modest of changes. This system of design is referred to as *over-the-wall design* because the design plans move from one person to another, as in throwing the plans over the wall. Computer networks and graphics have greatly changed these design procedures, making it possible for plans to go to all persons simultaneously in what is now called *concurrent engineering design*. This means that people from different specialty areas can work on different features of the design simultaneously. This latter approach provides many advantages for obvious reasons.

Macroergonomics

A number of years ago, James Buck* encountered Figure 1.5 at a committee meeting on computer languages in industrial process control. This figure provides an excellent overview of ergonomics, and partly for that reason it has been included in this introductory section of the book. A more important reason for its inclusion is that this figure illustrates some of the many ways ergonomic designers interface with other people in the firm. Such interaction is critical to a successful ergonomics program.

The part of Figure 1.5 above the box marked "Ergonomics Engineering" is referred to here as *macroergonomics*. People who work in macroergonomics tend to work directly with others in the firm who are concerned about requirements and restrictions of the company. Those requirements and restrictions fall into meaningful categories such as those shown at the top of Figure 1.5. Each category represents a company concern but not necessarily a different job. These different concerns often result in different agendas and priorities.

The category *process* refers to the principal manufacturing or service-providing operations of the company. People representing this concern usually stress being the best in the industry, having the latest processing technology, being the most productive, and making the best products or providing the best service. Much ergonomic effort goes into that category, so there are usually very strong ties between macroergonomics and the process group.

People who represent the company's concerns and commitments to the environment address both external and internal potential hazards to well-being. Macroergonomics usually stresses internal plant hazards in an effort to improve industrial safety and health. In most firms specialists other than ergonomists address external plant hazards. While everyone recognizes the importance of safety and health in industry, most people do not realize that job-related injuries and illness cost the United States far more than AIDS or Alzheimer's disease and at least as much as cancer or heart disease. For example, in 1992,

---

* James Buck was professor and chair, Industrial Engineering at the University of Iowa.

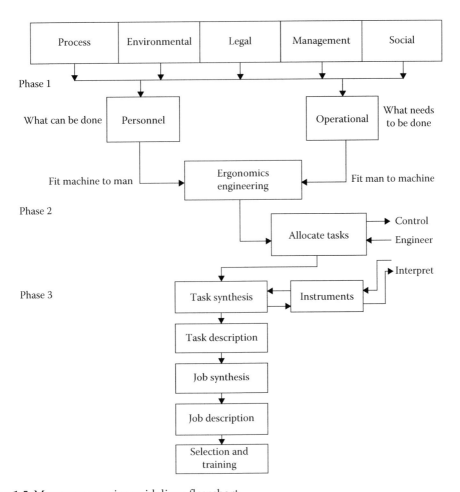

Figure 1.5 Macroergonomics guidelines flowchart.

6,500 Americans died and 13.2 million were hurt from work-related causes, averaging about 18 deaths and 36,000 injuries/day in industry. In terms of dollars, the estimated daily cost to U.S. industry is about $178 million in direct costs and approximately $290 million more in indirect costs, for a total of $468 million. Clearly, proper ergonomics could save a lot of money in safety and health alone. The reader should also recognize that U.S. industry is likely to be far safer and healthier than companies in less-developed countries.

Legal issues are a major concern of macroergonomics, and many of these issues are discussed in this textbook. One of the oldest and most important legal problems concerns relationships between management and labor. Other legal aspects of macroergonomics are discussed in more detail in the following section on "Carrots and Sticks."

Since macroergonomics includes many management functions, practitioners of macroergonomics will often interact with upper management. Part of that interaction includes showing managers that ergonomics has a scientific and an ethical basis. There are limits to what is known, and the solutions to some ergonomics problems must be found using empirical tests, field studies, and a myriad of other techniques. Economics is another critical issue when designers go to management with ergonomic proposals. For this reason, the role of economics in ergonomics is frequently addressed throughout this book.

Macroergonomics also addresses social considerations, a class of concerns that affects ergonomics design in a wide variety of ways. One clear application is in job evaluation. Other applications include shift design and setting levels of compensation (Box 1.3).

Each of these areas of macroergonomics discussed earlier has a major impact when setting *strategies* that best satisfy two principal roles of ergonomics: fitting jobs to people and fitting people to jobs. The latter role is not as obvious as the former, but it is important because it involves personnel selection and training.[*] These strategies rarely involve specific techniques, as far as macroergonomics is concerned, but they do reflect overall priorities. In some companies, ergonomics may have a very prominent role within the company, while other companies may treat this specialty as in-house consulting. Macroergonomics management is responsible for determining these priorities. It is also responsible for the record-keeping aspects of ergonomics, a role that is expanding in today's world.

Carrots and sticks

Over time, legislation has helped shape industry. Usually, laws are enacted by the legislature to right some wrong. For the affected people, laws usually function as a carrot, or encouragement to do something not previously done. For others who must change their way of performing their jobs, the law is more of a threat than encouragement.

The history of laws governing company behavior toward employees is strewn with the hulls of many well-meaning but misguided or inadequate laws. Some others had great impact in their day but are no longer an issue. Child labor laws are an example. One of the laws with largest impact in recent times was the one that brought occupational safety and health concerns into strong focus. While management may not enjoy the extra paper work, or the visits from government officials who enforce these laws, or the expense of design changes, this legislation has created areas where employees are healthier and safer. Not all areas of industrial safety and health have been improved, however. The legal emphasis on safety and well-being has helped many companies avoid the expense and time of unnecessary lawsuits.

In more recent years, legal threats have also been associated with products with regard to how they are designed, manufactured, and sold by the company. This carrot and stick category is usually called *product liability*. The principal notion of product liability is that any design defect that contributes to an accident or health problem and which could reasonably have been foreseen and corrected at the time of design is a company responsibility. Some product liability lawsuits have been settled for huge amounts. One noted example is the design of the gasoline tasks on the Ford Pinto model. The fact that other car models were designed in a similar fashion did not protect the Pinto designers. For every one of these well-known cases, thousands of other cases are not publicized.

Besides the laws on safety and health and those relevant to product liability, another legal requirement is for *no discrimination* on the basis of gender, race, religion, place of origin, and age. These antidiscrimination laws simply say that neither employment nor wage classification shall be determined by any of those categorizations. This law has been in place for many years, so damages can go back a long time. A company that is found to have discriminated, even if unintentionally, can face the cost of tracking down all previous employees. In some cases, the administrative cost can exceed the direct wage costs. The principal protection here is to be as objective as possible in job hiring and promotion and to establish a strong scientific basis for job differences.

---

[*] Chapter 18 addresses these subjects.

BOX 1.3 BRIEF EXAMPLE OF ERGONOMIC DESIGN

Ergonomics is often used in the design of buildings. For example, New York state building code section 1003.2.13.2 addresses enclosed exit stairways, stating that "an enclosed exit stairway, when it is part of an accessible means of egress, shall have a clear width of 48 in. (1219 mm) minimum between handrails."

Why are such standards necessary, and from where are the requirements derived? Most office buildings are designed to have their space leased, and stairways are not "leasable" space. Therefore, the building owners would prefer to minimize the amount of non-leasable space, and therefore minimize the size of stairways and other areas used for egress from the building.

Narrow stairways, however, are uncomfortable to use, particularly when there are people both coming up the stairs and going down the stairs. Moreover, in cases of emergency evacuation, you would like stairways and other egress areas to be wide enough to accommodate a large flow of people so that they can escape the building quickly. There is, therefore, a trade-off between usable (leasable) space and comfort and safety.

The required minimum stairway width is just about exactly twice the forearm–forearm breadth of the 95th percentile (largest) male. This means that with a stairway whose width is 48 in., two large males, and presumably anyone smaller than them, could fit side-by-side on the stairway without touching.

However, consider what occurred in New York in the World Trade Center on September 11, 2001. In that case, as people were trying to evacuate the building by going down the stairs, fully equipped firefighters, such as those shown in Figure 1.6, were going up the stairs to fight the fire. These fully equipped firefighters were much wider than a large male, and substantial difficulty was encountered in using the stairs, both to go down and egress the building, and to go up and fight the fire.

As a result, the National Institutes of Standards and Technology recommended that, in the future, "stairwell and exit capacity should be adequate to accommodate counterflow due to emergency access by responders." Human factors specialists fought to have the stairwells in the new building made wider than required by the building code in New York, and, in the end, the stairways were made 20% wider. The International Code Council now requires stairways in new sprinklered buildings to be 50% wider than according to the previous code.

The Americans with Disabilities Act of 1992* extends these previous laws by requiring companies to make reasonable accommodation in jobs for those with disabilities. That is, for people who can perform the essential functions of a job, but are prevented from holding that job without accommodation. Exceptions, of course, include people on illegal drugs or alcohol who are not undergoing treatment or those who would pose a direct threat to the health and safety of other individuals in the workplace. This law includes all firms with 15 or more employees. To meet their legal requirements, firms must make facilities readily accessible to and usable by disabled employees. Companies also need to schedule or structure jobs for those employees. Firms do have an escape clause, if providing such accommodation creates "undue hardship" for the firm, but it appears that the firm has the burden of proof. It is likely that it will take time and numerous law suits to refine

---

* Milas, G.H. (1992) provides some additional background.

Figure 1.6 Fully loaded firefighters evacuating a dummy during a fire simulation. (DoD photo by Dennis Rogers, U.S. Air Force, Colorado Springs, CO.)

the definitions of many of the terms of this act. To defend themselves, firms must create objective job-related bases for job performance. It also is necessary to document means of accommodation, as well as the ergonomic and economic basis for the actions taken. One real difficulty is envisioning all the potential disabilities that could be encountered and figuring out how those disabilities can be accommodated.

These different "carrots and sticks" should be viewed by ergonomics designers as opportunities to make important contributions to industry's humanity and competitiveness in world markets. Many of these opportunities can be found by identifying those constraints that are not absolutely firm and thinking about relaxing them. At the risk of seeming trite with short statements on how to design more effectively, please remember that "Simple is usually elegant!"

Trends in industry that impact ergonomic design

Many people have contributed inventions, operating methods, and new concepts that have changed the world. Many inventions have greatly changed industry and its products, often resulting in lower and more competitive prices and altered patterns of public consumption. Those changes in the public's purchasing behavior irrefutably affect how successful companies must operate. The principal point made by the brief clips of history that follow is that ergonomic designers should be at least on the proactive side of these changes, rather than reacting to them. It would be better still for ergonomists to be the inventors and entrepreneurs leading these changes. Most of the discussion that follows addresses current history and considers how ergonomic practices are changing in industry today.

One of the people who most affected industry is Samuel Colt. Many readers may be thinking about the development of the Western United States after the War Between the States and the role that the Colt pistol has played in many a Western story. However, Samuel Colt gave us an even greater invention: the first assembly line, in 1847. General Zachary Taylor ordered 1000 Colt 44s and Colt decided to assemble them by having individual workers put together a single part, then pass on the assembly to the next person. Colt's assembly line concept did not catch on in general industry at the time, partly because of the difficulty of securing easily interchangeable parts. As manufacturing specifications became better and machining processes became more repeatable, the notion of assembly lines became more feasible. Random Olds further developed the concept and installed it in his automobile factory in Lansing, MI, in 1899. Slaughterhouses worldwide had used the disassembly line for decades before the Olds assembly line. But it was not until Henry Ford developed his production of the Model T around the assembly line concept in 1924 that the economics of assembly lines became evident, and, consequently, their popularity rose dramatically. Along with these benefits came some constraints of assembly lines in the form of inflexibility. Henry Ford articulated one of those constraints by saying that the public could have a Model T in any color they wanted so long as it was black. Thus, the assembly line proved itself for high-quantity, low-variation production.

While Henry Ford did not invent the assembly line, despite popular belief, and his was not the first company to use assembly lines in automobile manufacturing,* he did make it famous, and it soon caught on throughout the entire automotive industry and elsewhere. What Henry Ford is most famous for is the concept of encouraging mass consumption through high employee pay and high productivity. The Ford Motor plants near Dearborn, MI, were paying $5 for an 8h day compared with other automobile manufacturers who paid half that rate for a 9h day back in the 1920s. Ford helped his employees buy new Ford cars too.[†]

Another traditional concept of industry today is standardization in the form of standard methods and standard times (also known as time standards). These concepts arose through the pioneering works of Frederick Taylor and Frank Gilbreth in the early days of the twentieth century. Taylor preached the use of the "scientific method" in industry to find the best method to perform all jobs and then to standardize those methods. The time it took a worker who was just acceptable to perform the job using the standard methods was the standard time. Among the companies where he consulted, Taylor was able to attract many followers to his ideas. Frank Gilbreth also had tremendous zeal for improved methods of performing jobs. As a bricklayer in his early days, Gilbreth virtually tripled the rate of laying bricks before he turned his energies to manufacturing industries. He too pushed for standard times based on the best methods. While Frank Gilbreth was not the charismatic individual that Taylor was, his reputation was enhanced by the help of his wife Lillian, who helped Frank in his work and carried on his work for decades. That work included the micromotion studies that later led to a system of predicting performance times synthetically, using well-defined motions and performance time values for those motions. Accordingly, standard times could be set for jobs using direct time studies of the prescribed best methods or synthetically predicted standard times.

The preceding historical accounts show why the notions of *standard methods* and *time standards* were firmly entrenched in U.S. industry in the late 1970s. Those procedures led to high productivity, especially when tried-and-true assembly-line methods were used. However, established methods often did not allow for much product flexibility, and when more product flexibility was introduced, the time standards were often inconsistent. In addition, industry's push for high production rates led to lower quality in many cases. In

---

* Random E. Olds was reported have used assembly lines in his Lansing, MI, plant making Oldsmobiles before Henry Ford made Fords on the assembly lines of Dearborn, MI.
† The story is that Henry Ford was not very happy when he saw an automobile in the employees' lot that was not a Ford. So he gave the workers a heavy discount for a Ford and a tough time if they did not buy one.

the 1970s, offshores competition provided products of greatly improved quality at reasonable prices. This competition challenged U.S. production methods, forcing U.S. industries to redirect their interests into achieving product quality. Some companies overreacted and threw out time standards altogether, only to find to their dismay that those standards served other important roles (e.g., in production planning and scheduling). Other companies found that some of the faster methods also produced some of the best quality. Thus, one of the contemporary concepts of industry is the continuous search for means of quality improvement. A secondary lesson that industry learned was the need for agility* in manufacturing capabilities and robustness in product designs. In many ways, U.S. industry was ahead of many overseas competitors in robust designs,† but there is need to achieve even more.

Another contemporary concept in industry is the notion of "just in time" (JIT). This concept was pioneered by Japanese companies under the name of "Kan-Ban." While there are many variants of the concept, the essential notion is that all materials, components, and subassemblies are to be delivered to industry just as the delivered material, component, or subassembly is needed. Clearly, the amount of inventory for these materials, components, and subassemblies can be substantially reduced by the JIT concept. This concept also cleans up the work area around the people assembling the product. JIT also causes production problems to surface because the concept calls for gradually working toward the JIT notion. As the delivery lead times get smaller and smaller, unexpected production problems occur. Often when those problems are solved and fixes are put in place, further efficiencies occur. In practice, JIT can be viewed as delivering to the production people tote boxes of parts in smaller and smaller quantities. Smaller boxes are easier to lift, and parts in the box can be reached more easily. However, because the deliveries of the tote boxes are more frequent, any negative ergonomic side effects create ergonomic problems where none existed before. Another issue is that JIT requires better time coordinations, so performance time estimations and lower time variations become more important, as do other production criteria besides performance time.

Contemporary industry also tries to reduce inventories of materials, components, and subassemblies by using *smaller production runs*. Yesterday's long production runs yielded the benefits of greater learning effects. That is, direct labor times become progressively less as people on the production operations gather greater skills, problems of coordination are ironed out, and production methods are sharpened. Shorter runs provide less opportunity for learning, but the inventories between production runs are reduced. The downside to shorter production runs is that they require more frequent setups of production. Consequently, more attention must be paid to developing efficient setup procedures.

One of the oldest concepts in industrial economics, called the make–buy decision, has also seen greatly renewed emphasis during the late 1990s. This decision involves determining whether it is less expensive to make the product yourself or to buy it from some vendor. The new slant on this decision is to define *core competencies* of the company, which should be kept in-house. Components that are not mainstream items are often outsourced, even if the company can make them for slightly less cost. The rationale behind

---

* Agility in manufacturing capabilities refers to the ability to change quickly, easily, and nimbly. See the concepts of JIT and concurrent design in the following.
† The term robust typically means something that is healthy, hardy, and strong. A robust design performs well under extreme or abnormal conditions. A robust analysis is insensitive to fundamental assumptions.

this decision is that these noncore activities reduce the company's ability to flexibly change manufacturing rates and processes.

One other trend in current practice is to use simpler *cellular manufacturing processes*. In this approach, parallel assembly areas, or cells, are provided where identical assembly of all products can occur. If the company is using a "pull system" (i.e., downstream operations dictate the rate of adding new starting assemblies), any stoppages downstream cause the product flow to shift to the parallel cell, along with needed personnel. This approach requires personnel to be able to handle a far wider collection of tasks and jobs than in a traditional assembly line. But it makes manufacturing operations insensitive to bottlenecks in the operations, even when the bottlenecks are only part-time.* It is also clear that parallel cells greatly reduce queuing line buildups due to variations in the service system or manufacturing operations. Many banks, airline ticketing centers, and other businesses follow this approach by having customers queue up in a single line and then go to one of several parallel service systems. Two other features that aid in cellular manufacturing include simplification of materials handling and the integration of all testing used in quality control. One of the principal benefits besides high flexibility is higher morale on the shop floor, because people performing the assembly can see their individual contributions much better than in highly automated assembly environments.

Another contemporary concept is *quality certification*. The traditional practice in receiving materials, components, and subassemblies from vendors was to perform acceptance inspection of the vendor products after the items were delivered to the plant. When using JIT procedures, deliveries are made more often, so acceptance inspection may become an ongoing process. To avoid conducting these inspections themselves, many companies employ certification procedures where the vendor performs the necessary inspections before delivery. Vendors and buyers have traditionally had adversarial roles, but this is changing because when vendors perform all the necessary tests, they can ship directly to the assembly stations where the components are needed. Hence, the inventory time of vendor items is negligible, and finished goods are free to be shipped as soon as ordered. Products with many different options are often received from the vendor as partially finished goods. Optional features are then added in a final assembly stage, just before shipping to the customer. This allows firms to maintain small total inventories and still meet rapid shipping requirements.

The movement toward *concurrent engineering design*† is another recent change in contemporary manufacturing practice. As noted earlier, traditional over-the-wall design procedures required the principal designer to complete much of the design before passing it on to other specialists. At this point, the manufacturing engineer would be asked to start designing the manufacturing processing operations. After this, a reliability engineer might be called on next to perform reliability calculations. The design might then come back to the principal designers when modifications were needed to meet minimum specifications. Other specialists, such as the ergonomics designer, would be called on later, but important design features would often be frozen at this stage. Contemporary computer technology permits designers to share the latest versions of designs with all members of the design team. Consequently, all members of the design team can make recommendations before most of the design is frozen. Concurrent engineering design also makes it easier for the

---

* Laughlin (1995) describes cellular manufacturing at Sun Microsystems Computer Co.
† Kusiak (1993) provides an interesting and far more complete version of concurrent engineering design. Also see Anderson (1990), Helander and Nagamachi (1992), and Turino (1992).

team to make logical trade-offs while generating alternative designs. Applications of these new design practices include the following objectives:

1. Design to have an economic *manufacturing process* plan
2. Design to meet long-term customer *reliability* and *testability* needs
3. Design to meet the customers' *quality* expectations
4. Design for *production-inventory* control and *materials* purchasing
5. Design for customer *performance* and *servicing* requirements
6. Design to be *assembled, shipped,* and *sold* legally
7. Design for overall *coordination, maintenance,* and long-term *renewal*

These contemporary concepts greatly enhance the role of ergonomic designers, by making them more responsible for product design both from the finished product viewpoint and from the standpoint of ergonomically sound manufacturing processes.* Meeting these requirements is a major challenge in contemporary and future ergonomic engineering.

Organizations and additional information on ergonomic design

Numerous organizations are associated with ergonomics and ergonomic design, and many of these organizations are excellent sources of information on this topic. A number of these organizations are listed in Table 1.2. Note that several of these sources are affiliated with governmental agencies such as NIOSH, OSHA, CSERIAC, and NASA. NTIS is an excellent source of technical reports submitted by researchers to governmental funding agencies, and other documents that can be found nowhere else. Professional societies such as SAE, IIE, HFES, and ASHRAE, and consensus-standard-making organizations such as ANSI and the ISO are also good sources of information on ergonomics. Some of the ANSI technical standards were developed by the aforementioned professional societies. Most of the aforementioned sources have web pages that allow easy access to the information that they are willing to provide.

Ergonomic methods[†]

Ergonomics designers often use the results of various studies as a basis for a design. Those results are obtained by using one or more of the general types of experimental methods described in the following. Typically, designers first try to glean the appropriate information from books, papers, reports, and other documents. On other occasions appropriate information may not be available, and the designer may have to do his or her own study in order to collect the information. Before this can be done, the designer must determine which method or methods to use. That is, designers need to be aware of how the methods used might impact their conclusions. The type and quantity of information obtained depends on the method selected, and the information obtained must address the problem posed. Selecting a method because it is familiar or expedient may actually change the problem being confronted, resulting in a correct solution, but to the

---

* Corlett (1988) describes some of the future of ergonomics he expects.
† This discussion was adapted from a chapter written by McGrath (1964). It is included here to help the student see the importance of various forms of empirical testing for design purposes rather than just for scientific purposes.

Table 1.2 Sources of Information on Ergonomics

	Organization Name	Address
ACGIH	American Conference of Government Industrial Hygienists	6500 Glennway Avenue, Cincinnati, OH 45221, United States
ANSI	American National Standards Institute	11 West 42nd Street, 13th Floor, New York, 10036, United States
ASHRAE	American Society of Heating, Refrigerating, and Air-Conditioning Engineers	1791 Tullie Circle, Atlanta, GA 30329, United States
CSERIAC	Crew System Ergonomics Information Analysis Center	AL/CFH/CSERIAC Wright-Patterson AFB, Dayton, OH 45433-6573, United States
Ergonomics Society	Ergonomics Society	Devonshire House, Devonshire Square, Loughborough, Leics, 11 3DW, United Kingdom
HFES	Human Factors and Ergonomics Society	P. O. Box 1359, Santa Monica, CA 90406, United States
NTIS	National Technical Information Service	5285 Port Royal Road, Springfield, VA 22161, United States
NASA	National Aeronautics and Space Administration	LBJ Space Center, Houston, TX 77058, United States
NSC	National Safety Council	444 North Michigan Avenue, Chicago, IL 60611, United States
NIOSH	National Institute of Occupational Safety and Health	4676 Columbia Parkway, Cincinnati, OH 45226, United States
OSHA	Occupational Safety and Health Agency	200 Constitution Avenue, NW, N3651, Washington, DC 20210, United States
SAE	Society of Automotive Engineers	400 Commonwealth Drive, Warrendale, PA 15096-0001, United States
ISO	International Organization for Standardization	1 Rue Varembe, Case Postale 56, CH 1211 Geneve 20, Switzerland
IIE	Institute of Industrial Engineers	25 Technology Park, Norcross, GA 30092, United States

wrong problem. The following discussion introduces ergonomic research methods and some of their features.

Field studies

One method used by ergonomic specialists is the field study. With this method, investigations take place within "real-life" or "natural" environments. In their most elementary form, field studies consist of making casual or anecdotal observations. That is, one simply goes where the desired action is, watches, and records observations. An example is studying expressway driving behavior by going to expressway overpasses and simply describing what is going on. Many surveys involve this form of field study. This approach is often used in the early stages of ergonomic design to get a better feel for the problem prior to applying other methods.

In some field studies, the experimenter systematically manipulates some factor or variable of the system under study. It is often quite difficult to do this, but when it is possible, this approach can be very effective. An example is where the researcher is observing an order-picking operation (such as a parts department where the spare parts are selected and collected to meet specific orders). While one cannot change the orders without creating a difficult production problem, one could alter the sequence of those orders so that the same person could be observed processing widely differing orders. Without the ability to change order sequences, the study would have to depend upon luck to achieve the desired conditions for a particular person.

Experimental simulations

Another study method is experimental simulation, more appropriately called *operator-in-the-loop simulation*. This method involves the creation of relatively faithful representations of the real world on a simulator, and then observing how people perform in simulated situations. Significant progress has been made in recent years to make simulators seem realistic. Sometimes computers are used to generate all of the external events so that they occur on an efficient schedule. In fact, experimental simulation can replicate a situation more exactly than using direct observations.

For illustration purposes, suppose that you want to study what drivers do when a traffic light turns yellow just as they enter an intersection. It would be hard to schedule this event in real life and would involve significant risks to the drivers and others. On the other hand, these events could be easily scheduled in a driving simulator with no significant risk. Another example along these lines is to use a simulated nuclear reactor in nuclear reactor control rooms for training purposes that emulates the reactor operations and returns data on the state of the reactor. No one would allow such testing on a real nuclear reactor because of the cost and inherent risks of a disaster. Other examples of simulation may differ greatly in their fidelity, ranging from the realistic simulation of a B747 aircraft cockpit, down to an abstract video game with a joy stick and a 12 in. monitor. The control provided by simulations allows an experimenter to study some very difficult situations.*

Laboratory experiments

Another study method is laboratory experimentation. The principal distinction between experimental simulations and laboratory experiments is that the latter method controls more variables to remove unwanted effects. Many variables that occur naturally in life are not of interest in design, and so they are disregarded in the laboratory experiment. Generally speaking, laboratory experiments are used to study fundamental human processes rather than realistic operations. As an illustration, suppose that the ergonomist was concerned with the effect of high heat and humidity on telegraphy. Now, many of the places on earth that are both hot and humid have a great number of bugs in the air. Adding those mosquitoes or other bugs to the study can create a substantial number of problems that will in turn add considerable noise to the data being collected. But the removal of variables tends to inflate effects being observed. While that inflation helps researchers observe effects more carefully, much like using a microscope, that inflation makes the effect appear

---

* Chapter 9 deals with simulation as a general technique in ergonomics studies.

much greater than it may actually be in practice. However, laboratory experiments serve some purposes splendidly.

Computer simulation

The final method described here is computer simulation and/or mathematical modeling of human behavior. The use of a computer program or mathematical equation to predict how long it will take someone to perform a task is one example of this approach. Along these lines, deterministic simulations develop their predictions from a task description and tabulations of mean performance times as a function of relevant variables. In Monte Carlo simulation, the time needed to complete subtasks within a task is assumed to be a random variable. Multiple runs are then made to analyze overall task performance. Mathematical models share many properties with computer simulation. One gets very precise results with a single run of either a mathematical model or a computer simulation; not so with the Monte Carlo variation. The use of mathematical models in ergonomics is, as one would expect, very infrequent. Some of the control situations of mechanisms have been modeled mathematically, but that is about it.

Differences in ergonomic methods

It is rather obvious that these study methods vary along a continuum of factors. Field studies focus on rather concrete, realistic, and uncontrolled situations. At the other end of this continuum is computer simulation, in which the situations are abstract, artificial, and highly controlled. While experimental simulation and laboratory experiments occupy separate points along the continuum, they are less distinguishable from each other in these respects. In addition to these features, there are other advantages and disadvantages associated with each method. Some of these advantages have already been identified earlier. One clear point is that no one method is superior in all respects. Rather than focusing on which method is best, it makes more sense to emphasize that the various methods are complementary to each other. In many cases, multiple methods should be used.

Some people advocate programmatic procedures using exploratory studies when little is known about the phenomena being studied. That first stage should usually employ a field study, to assure correct identification of the problem. A second stage often entails setting up and testing key hypotheses. Typically, experimental simulations or laboratory experiments are the appropriate vehicles for that purpose. Next, there is usually a need to elaborate and refine theoretical or operational models. At this stage, computer simulations can be extremely useful. However, computer simulations do not have high face validity because some assumptions associated with the model are buried in computer code. Thus, it is necessary to validate the model. Here again, laboratory experiments or experimental simulations are useful. Finally, cross-validation is needed from real-life cases to establish high face validity. Although few research investigations cover all five of these stages, this scenario provides perspective into how these analyses fit together into a larger whole. Problem solving in industry follows a similar set of stages, even though the stages are not formally identified.

Some other requirements of these study and problem-solving methods deserve mention. A method should be comprehensive, covering the entire region of interest in an accurate fashion. Clearly, field studies are comprehensive, covering the entire region of interest, but they are not *efficient*. Part of the reason for field study inefficiency is that a large amount of *noise* (i.e., outcome variability) obscures the results. With little or no control of the variables being studied, there is no way to prevent the intrusion of noise. On the reverse side, laboratory studies are very efficient, as noise is at a minimum, but they are not comprehensive. Experimental simulations usually fall somewhere between field studies and laboratory experiments with respect to both efficiency and comprehensiveness.

Final remarks

This initial chapter was written as an introduction to the field of ergonomic design, primarily as it is practiced in industry. This field is emerging out of the joint efforts of human factors and work design specialists, whose interests have included time-and-motion studies, methods design, and work simplification. Ergonomics is both a many-splendored thing and a many-edged sword. No profession works with a more complicated system—the human being. While that feature is part of the challenge, it is also part of the splendor. What is clear is that the role of this activity has vastly increased, both in breadth and depth. Ergonomic design needs a predictive capability to make good designs and justify them. This is particularly true considering the contemporary driving forces of international competition and the need to compete in these markets. Ergonomic design must satisfy both the needs within the company and the needs of customers who use the products produced. Ergonomic designers will need to be leaders in concurrent engineering design if they expect to steer product designs to be user-centered and to simultaneously be part of the manufacturing, product reliability, and the maintenance servicing of those products as well as their manufacturing operations. Since the customers of the firm's products determine perceived quality and sales, who is better equipped to help lead the concurrent design efforts? As in the past, these are changing and challenging times.

Discussion questions and exercises

1.1 Think of a specific product such as a camera. If you were to measure human performance in using this device, what criteria would be important? Identify and denote how you would measure each criterion.

1.2 Imagine a particular situation, such as driving a car to work. First determine how to measure the quality of performance in this situation. Second, suggest how you can identify variables that are likely to change performance measurements and why you feel performance would change.

1.3 Discuss why ergonomics designers are advised NOT to try to fix things that the operating personnel do not feel are broken, even if the designer is sure of it.

1.4 If an ergonomic intervention was estimated to increase sales by $80,000 within 2 years, what is the maximum investment if the cost of capital is 15%, the cost of sales is 50%, and the tax rate is 40%?

1.5 Why would an ergonomic designer want to know the standard deviation of the means of different people performing a specified task? If human tasks need to be performed between two successive machine actions, which is nearly a constant time interval, why is the average standard deviation of performance time *within-persons* an important piece of information in ergonomic design?

1.6 Think of the LIMET acronym and describe a situation that represents each interaction between LxE, IxT, and LxT.

1.7 What is meant by a "robust design" and what do critical incident studies have to do with it?

1.8 For each principle of human-centered design, think of an example where it is important. Explain why.

1.9 What are the implications of Helson's U-hypothesis with regard to developing designs that optimize human performance?

1.10 Suppose that you were asked to rate objects that were good examples of being "big" by simply rating the objects as 100% if they fully fit the concept, 0% if they don't fit it at all, and some intermediate value for the degree to which the objects exemplify the concept. Later you take the objects and rating, rearrange them to describe a function, from zero through intermediate values to 100% and back through intermediate values to zero. This description is a very precise way of describing imprecision. What topic in this chapter exemplifies this notion and how is it of any use in ergonomic design?

1.11 How can a study of design failures help one find a better design?

1.12 What are some of the U.S. laws that affect ergonomic design? Support your answer by showing the effect.

1.13 What is the biggest drawback of doing field studies in ergonomic design practice?

1.14 What is the principal difference between experimental simulations and laboratory experiments? What are some of implications of these differences?

chapter two

Human system

About the chapter

This chapter introduces the reader to human anatomy, physiology, and psychology from a system's perspective, and is intended to provide necessary background for the following chapters. The discussion includes a review of muscular–skeletal, sensory, and support systems. Some of the subtopics covered are movement and force exertion, work-related disorders, human variability in shape and size, strength, endurance, and work capacity. Two final objectives of this chapter are to familiarize readers with various measurement units employed in ergonomics, and to show how well the human body is engineered.

Introduction

Ergonomists must deal with the fact that people come in different sizes and shapes, and vary greatly in their strength, endurance, and work capacity. A basic understanding of human anatomy, physiology, and psychology can help ergonomists find solutions that deal with these issues and help prevent problems that can cause injury to workers. To provide some perspective regarding the scope of this problem, let us begin by considering some of the component systems of the human body. Figure 2.1 describes the human body in terms of four component systems:

1. The *sensory* systems (vision, hearing, position, touch, taste, and smell) are stimulated by energy sources (e.g., light, sound, or heat) or materials (e.g., airborne chemicals, acid on skin, and salt on tongue) in the outside environment.
2. The central *information processor* (brain and nervous system) processes information acquired from the sensory systems.
3. The *effector* systems (arms, hands, eyes, legs, etc.) are consciously controlled to modify the environment and acquire information.
4. The *support* systems (circulatory, digestive, metabolic, heat-regulatory, etc.) act in various ways to keep the other systems functioning.

One general observation is that the four major systems perform fundamentally different functions (i.e., sensing, processing, acting, and support). Each of these systems contains a sometimes overlapping set of subsystems. For example, people can use their fingers to read Braille (as sensors) and type (as effectors). The components of systems and subsystems are intricately interfaced throughout the body. For example, the skeletal and muscular subsystems, contain nerves and sensors, as well as muscles and bone, and are controlled by the central processor. The circulatory system, an important component of the body's support system, similarly connects with the effector system through veins and arteries that supply the muscles with nutriments. One of the more important points that should be clear from the previous discussion is that although each system performs a different function, overall performance involves a complex interaction between the different subsystems.

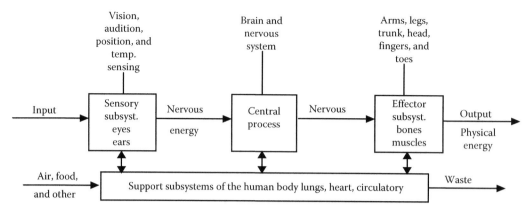

Figure 2.1 Subsystems of the human body.

The following discussion will first address the human skeleton, joints, and muscle systems. This sets the stage for introducing the biomechanical concept of levers, and their implications regarding human strength and speed of movement. Other systems of the human body will then be discussed. Focus will be placed on sensory operations, blood circulation, metabolism, and body heat regulation.

Skeletal subsystem

The human skeleton consists of two principal systems of levers, the arms and legs, which are flexibly connected to the central body structure consisting of the skull, spine, rib cage, and pelvis. Joints between these bones allow for body movements and the shape of these joints constrains those movements. The human skeleton contains over 200 bones and a corresponding set of joints. Most of these articulations are of little consequence to the ergonomic specialist who is designing work methods and products used by people. Analyses normally focus on the extremities (people's arms, wrists, hands, legs, and feet) and the lower back.

Extremities

Bones in the extremities are structured like pipes, with closed ends near the joints. Muscles are attached to the bones by tendons. A bone depression or protrusion is normally present at the spot where the tendon attaches. The surface layers of a bone are hard and dense, and tend to be smooth except for roughened areas where ligaments and tendons are attached. Several small holes allow arteries, veins, and nerves to pass into the soft and spongy interior of the bone.

Joints occur at the locations where bones come together, or articulate. Joints tend to be complex structures, made from many different materials beside bone. Within a joint, ligaments and muscles hold the bones together. Most ligaments and tendons are made from inelastic *collagen fibers*. Some joints (especially in the spine) are held together by stretchable ligaments made from elastic fibers. The contact surfaces of bones in a joint are normally covered with a thin, smooth, and very slippery layer of collagen fibers, referred to as *cartilage*. This cartilage layer acts as a shock absorber and also helps minimize friction forces.

Synovial and cartilaginous joints are two important types of joints found in the human body. The former make up most of the articulating joints of the upper and lower extremities,

Chapter two: Human system

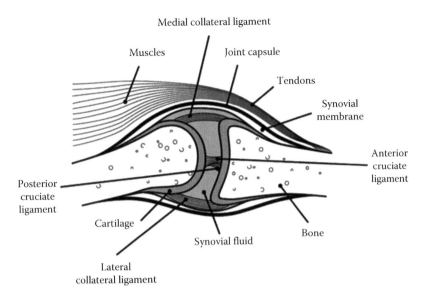

Figure 2.2 Synovial joint.

while the latter are found principally in the spine. In synovial joints, the mating ends of the two bones are covered with articulator cartilage as shown in Figure 2.2. A joint *capsule* made from collagen fibers surrounds the joint and helps sustain stresses. This capsule also has a synovial membrane lining that produces a lubricant and nourishing fluid.

Some joints, such as the elbows and fingers, have a synovial membrane around muscle attachments to lubricate that point of friction. Other joints, such as the knees, contain annular disks of fibro-cartilage, which are believed to further reduce friction.

The range of movement allowed by a joint is influenced by many factors including

- The shape of the articulation surfaces
- The distribution of the muscles and ligaments
- Muscle bulk

Joints at the ends of the fingers, knee joints, and elbow joints are known as *hinge* joints. Hinge joints have inelastic ligaments stretching down each side that prevent sideways movements. Other joints are less restrictive. *Gliding* joints allow two-dimensional movement at articulations in the wrists and ankles. The *saddle* joint found at the base of the thumb also allows two-dimensional movement. The hip and shoulder joints are examples of *spherical* joints (or ball-and-socket joints) similar to trailer hitches. Since the hip joint is a large ball-and-socket joint that is deep within the pelvis, it can carry heavy loads over a small range of movement. The shoulder joint is smaller and not nearly as deep within the shoulder bone, so it cannot take as great a load, although it has a greater range of movement than the hip joint.

Pivot joints are joints where a protrusion from one bone fits into a recess of another. For example, at the upper end of the lower arm near the elbow, a protrusion of the radius bone fits into a recess in the ulna bone. Pivot joints restrict movements in some directions. The range of movement allowed by particular joints can be measured using goniometers that are strapped to the outside of the human body at the location of the studied joint. Figure 2.3 illustrates bones and joints in the upper extremities, of particular importance in ergonomic

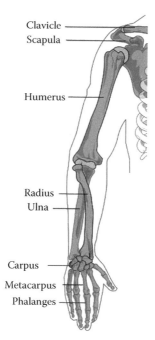

Figure 2.3 Bones of the upper extremity.

design. The single *humerus* bone of the upper arm connects to the *scapula* in the shoulder. The forearm has two bones, the *radius* and *ulna*, which connect the elbow to the *carpal* bones of the wrist. Note that the bone on the outside of the wrist is the ulna. The radius is interior to the ulna. The lower part of Figure 2.3 shows bones of the wrist and hand. *Gliding joints* are found where the carpal bones articulate with each other and with the radius. The joint at the base of the thumb where the carpal bones articulate with the metacarpal bones within the hand is called a *saddle joint*. The *phalange* bones form the fingers.

Because human joints are susceptible to a large variety of such disorders, designers of work methods and hand tools must take care that work operations do not cause joints to exceed their natural limitations in terms of force and direction of movement. While such disorders can be caused by single-event accidents, they are more commonly due to a cumulative building of trauma over time. Ergonomic designers must consequently be particularly concerned about device and job designs, which might lead to various types of cumulative trauma disorders.

Joint-related disorders

Synovial joints are very well engineered and work beautifully so long as they are not subjected to stresses beyond their design limits. One potential problem is that joints can be damaged when they are forced to move in directions that they were not designed for. Sudden forces may cause a *sprain* or *dislocation* of the joint, two types of injury that are common in sports medicine and that occur occasionally in industrial operations. Sprains occur when ligaments are torn or stretched beyond their limits. The more the ligaments are stretched, the less likely they are to withstand subsequent stresses. This increases the chance of incurring severe strains or dislocations in the future. *Tearing* of the protective

Table 2.1 Some Cumulative Trauma Disorders

Back Disorders
- Degenerative disk disease—Spinal discs narrow and harden during degeneration, usually with cracking of the surface
- Herniated disk—Spinal disk rupture causing the fluid center to spill or bulge out
- Tension neck syndrome—Neck soreness occurs, often because of prolonged tension in the neck muscles
- Posture strain—Continual stretching or overuse of neck muscles
- Mechanical back syndrome—Portions of the vertebrae known as the spinal facet joints start to degenerate

Elbows and Shoulder Disorders
- Epicondylitis—Tendonitis of the elbow, also called tennis elbow
- Bursitis—Inflammation of the bursa (small packets of slippery fluid that lubricates spots where tendons pass over bones)
- Thoracic outlet syndrome (TOS)—The nerves and blood vessels under the collarbone are compressed
- Radial tunnel syndrome—The radial nerve of the forearm is compressed
- Ligament sprain—Either a tearing or stretching of a ligament that connects bones
- Muscle strain—Overuse or overstretching of a muscle

Hand and Wrist Disorders
- Synovitis—Inflammation of a tendon sheath
- Tendonitis—Inflammation of a tendon
- Trigger finger—Tendonitis of the finger resulting in snapping or jerking rather than smooth movements
- DeQuervain's disease—Tendonitis of the thumb, usually at the base
- Digital neuritis—Inflammation of the nerves in the fingers, usually caused by continuous pressure or repeated contact
- Carpal tunnel syndrome (CTS)—Pressure on the median nerve

Leg Disorders
- Patellar synovitis—Inflammation of the synovial tissues within the knee joint
- Subpatellar bursitis—Inflammation of the patellar bursa
- Shin splints—Very small tears and inflammation of muscles stemming from the shin bone
- Phlebitis—Varicose veins and blood vessel disorder often caused by protracted standing
- Tronchanteric bursitis—Inflammation of the bursa at the hip
- Phlebitis fascitis—Inflammation of tissue in the arch of the foot

joint capsule is another common injury. This tearing allows the synovial fluid to leak out of the joint capsule, resulting in disorders such as water on the knee.

Continued misuse of joints over time can cause several types of *cumulative trauma disorders* (CTDs, also called overuse or repetitive-use disorders). Table 2.1 describes a variety of CTD associated with the back (neck to lower back), hand, lower arm, and legs, several of which will be discussed in more detail in the following. These disorders can involve damage to many different parts of the body, besides joints, including ligaments, tendons, tendon sheaths, bursa, blood vessels, and nerves.

Ulnar deviation is a case in point. This disorder occurs when ligaments are stretched to the point that they no longer hold the lower part of the ulna bone in place, resulting in permanent disability. People who work for long time periods with their hands in an unnatural position may be prone to this disorder. Cobblers (i.e., shoe makers of a past era) who manually put in the welts between the upper shoes and the soles often experienced ulnar deviation.

Work-related *tendonitis* is another CTD that can restrict movements in many different ways. For example, excessive gripping and twisting activity can cause *DeQuervain's disease*. This disorder involves inflammation of the tendon sheaths at the base of the thumb. This syndrome is painful and can severely restrict movement of the thumb.

Trigger finger is a related disorder due to overuse of the finger used in shooting guns, first observed among soldiers who fired rifles on the rifle range for many hours of the day. *Bursitis* is an inflammation of the bursa, which often contributes to tendonitis in the shoulder. Lifting, reaching, and various overhead tasks can cause this disorder. In advanced cases, patients lose their ability to raise the affected arm.

Thoracic outlet syndrome is a disorder resulting from compression of nerves and blood vessels beneath the collarbone, which can cause the arms to become numb and interfere with movement. This syndrome is associated with a wide variety of physical tasks, where people exert forces with extended arms.

Another CTD known as *carpal tunnel syndrome* involves both the hand and the wrist. Joints of the finger are operated by muscles in the forearm that attach to tendons encased in sheaths. These sheaths pass over the bones of the wrist in what is known as the *carpal tunnel*, which is formed by the U-shaped carpal bones at the base of the back of the hand and the transverse carpal ligaments forming the opposite side below the palm of the hand. A number of tendons in sheaths, nerves, and blood vessels run through that tunnel. A second tunnel, called the *Guyon canal*, is adjacent to the carpal tunnel away from the thumb side and provides a conduit for the ulnar nerve and artery. When the wrist and fingers are in their normal position, the sheaths, nerves, and blood vessels and arteries are nearly straight. However, when the wrist and fingers are bent, the sheaths are constricted. It consequently requires more force to pull the tendons through the sheaths, and damage may occur. The damage may cause irritation and swelling, which then causes secondary problems. Compression of blood vessels and arteries, due to swelling, is of particular concern because this reduces the blood supply. With oxygen and nutrient supply shut off, and the removal of waste products similarly stopped, muscles are less capable of sustained exercise and require longer periods of time for fatigue recovery. Bending of the wrist can also cause nerve compression, which can result in a variety of problems including carpal tunnel syndrome.

Carpal tunnel syndrome is one of the most common of the CTDs. People who have this disorder often feel numbness or tingling in their hands, usually at night. It is caused by swelling and irritation of the synovial membranes around the tendons in the carpal tunnel and that swelling puts pressure on the median nerve. Associated causes and conditions of carpal tunnel syndrome are repetitive grasping with the hands, particularly with heavy forces, and bending of the wrist. Many other conditions are associated with this disorder, including arthritis, thyroid gland imbalance, diabetes, hormonal changes accompanying menopause, and pregnancy. It is important to identify potential cases before they become serious. Early treatment consists of rest and the wearing of braces at night to keep the wrist from bending and some mild medications. In more severe cases, physicians often use cortisone injections or surgery. The best solution is to ergonomically design workstations to help prevent this disorder from occurring in the first place.

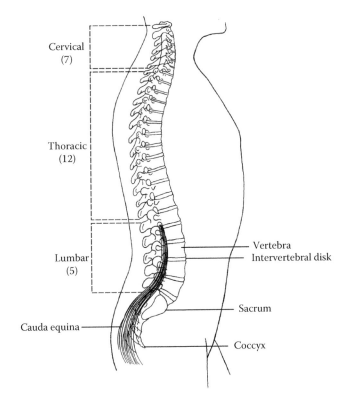

Figure 2.4 Side view of the spine.

Spine

The human spine is a complex, load-bearing structure. A side view of the spine, from the base of the skull downward, is shown in Figure 2.4. The bones of the spine progressively increase in size from the top to the bottom of the spine, corresponding to an increase in carrying capacity for the lower bones. The individual bones of the spine are called *vertebrae* and are divided into four different categories, depending on their location. That is, there are seven *cervical* vertebrae, counting from the top of the spinal column downward; 12 *thoracic* vertebrae in the middle region; five *lumbar* vertebrae below the thoracic vertebrae, in the next lower region; and a fused group of *sacral* vertebrae, just above the coccyx, or tailbone.

The specific vertebrae are numbered in terms of their location (i.e., C1–C7, T1–T12, and L1–L5). Most back problems related to the spine involve the vertebrae of the lumbar region, illustrated in Figure 2.5. The vertebrae are separated by disks, which have a fibro-cartilage exterior and a liquid center. These disks serve as shock absorbers between the bony vertebrae in the spine. Over the course of a day of standing, these disks are under compression, causing the spine to shrink about 1.25 cm. Those disks also begin to degenerate after middle age. Since the disks occupy about 25% of the spinal length, and disks compress as the person ages, older people tend to grow shorter.

Several muscles and ligaments run along the spine. One set of ligaments runs from the skull to the tailbone, along both the front and back of the spine. Individual vertebrae are joined by elastic ligaments and muscles that help the spine maintain its natural curvature and permit limited bending in multiple directions. Rotations of about 90° in the upper part of the spine and about 30° in the lowest part are also possible.

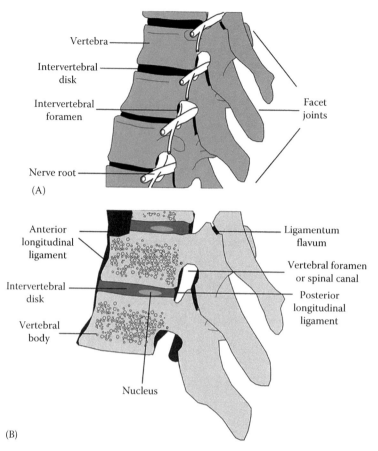

Figure 2.5 View of the lumbar region of the spine—(A) side view and (B) cross section.

The abdominal muscles in front of the spine create tension that tends to curve the body forward. Tensing the erector spinae muscles behind the back counteracts these muscle forces.* The latter muscles connect to bony, rearward projecting, protuberances on individual vertebra. Other muscles also connect to the spine, including the *trapezius* muscle of the upper shoulder and the *latissimus dorsi* of the lower back. The separate muscle attachments to particular vertebrae provide the spine a great deal of mobility. This happens because each vertebra is separately controllable. Needless to say, determining the sequence of muscle activations needed to reach a particular posture or make a particular movement is a complicated control problem.

Back disorders cost industry a great deal of money. *Low back pain* is a special concern because it often disables a person for an extended period. Some cases have resulted from moderately poor posture over long time periods. More often this condition is a result of lifting heavy loads or lifting in a poor posture. Unexpected load shifts and slips while exerting force are especially likely to cause back strains and other injuries. In some cases, a disk may be *ruptured*, resulting in severe pain due to pressure of the disk's liquid center on the spinal cord.

---

* The fact the muscles and ligaments in front and behind the spine are both exerting tension forces can result in large compression forces on the spinal column.

Degenerative disorders include gradual damage to the back muscles or to the back's elastic ligaments over time. *Osteoarthritis* is a degenerative spinal condition often suffered by elderly people. This disease occurs when the disks shrink enough that the vertebrae press against each other, resulting in stiffness, pain, and sometimes nerve damage. One particular nerve along the spine that is vulnerable to this condition is the *sciatic* nerve that extends down the back of the thigh. When a disk or other part of the spine is injured, putting pressure on this nerve, pain can move down the buttocks, thigh, and calf, and into the foot. Other causes of back pain, as well as pain in other parts of the human body, include muscle spasms and strains, as discussed in the following.

Muscles

Muscles are made up of body cells that have the ability to contract along a single direction. Microelements of a muscle include the tiny filaments called *actin* and *myosin*, which slide along each other during contractions and extensions. Collections of those elements make up *myofibrils*. Groups of myofibrils are wrapped inside protein filaments to make muscle *fibers*. Individual muscle fibers are wrapped inside a membrane called the endomysium to create *muscle bundles*.*

Muscles are classified into three different categories. Striped muscles connect to bones in the body† and are used to make voluntary movements and maintain posture. Smooth muscles are used within the organs to perform functions, such as constricting blood vessels. Cardiac muscles in the heart pump blood. Since striped muscles are particularly important from an ergonomic perspective, we will confine further discussion to them.

Some of the fibers within striped muscles are called *red* fibers (because of their myoglobin pigment). Such fibers have the ability to store oxygen received from the bloodstream. These special fibers serve the important role of supplying oxygen to the muscle bundle in times of blood supply insufficiency.

Muscle contractions and capabilities

It is important to realize that muscles are positioned in opposing pairs so that contraction of one muscle causes movement around a joint in one direction and contraction of the other muscle causes movement in the opposite direction. When fully contracted, the length of a muscle fiber is roughly half that of the relaxed length. Each fiber contraction creates a muscle force, with greater numbers of contracted fibers resulting in greater forces. Bigger muscles usually contain more bundles and have greater maximum force capabilities or force endurance capabilities than smaller muscles. Severe muscle bulk, however, limits contraction. Bundles composed of fibers that are mostly parallel to each other produce the greatest forces.

Individual muscle fibers are activated on an all-or-nothing basis by motor cells of the nervous system. When a very light muscle force is needed, only a small number of motor cells activate fibers in a muscle bundle. As greater force is needed from a particular muscle

---

* For additional details see Astrand and Rodahl (1977).
† Muscles are not directly attached to bones. Rather, the ends of muscles converge into collagen fibers that combine in bundles to form a tendon. Some of these tendons are barely visible while others are very long (e.g., those leading to the fingers). Tendons are attached either to roughened portions of the bone surface or to special bone projections. Muscle connections closest to the center of the body are called *origins*, while those connections more distant are called *insertions*. The ends of muscles are connected to two, three, or even four different bones.

bundle, the nervous system activates more motor cells in the bundle so that more fibers contract. If the force needs to be sustained more than momentarily, groups of fibers take turns contracting to delay fatigue. Thus, it is rare to have a muscle bundle contracting all the fibers simultaneously. The central nervous system monitors the muscle bundles and changes the positions of body parts in order to alter the forces applied by muscle bundles.*

Very Brief History of Physiology

Modern-day physiology, one of the scientific bases of ergonomics, originated in Europe before the middle of the nineteenth Century. One leading researcher was Claude Bernard of France who performed and published numerous experiments. In addition, three very famous Germans helped initiate this science. One was Carl Ludwig at the University of Marburg. Ludwig was probably more famous for developing devices to measure blood pressure changes, muscle motion, and other physiological phenomena. Johannes Muller acquired fame by writing the earliest textbooks on the subject. Justus von Liebig brought a strong interest and background in chemistry to the field of physiology. In England, Sir Michael Foster wrote a major textbook on the subject and founded the Physiology Society. Sir Charles Bell of England and Francois Magendie of France provided fundamental research that distinguished sensory and motor nerves and muscular sensations. Toward the latter part of that century, Henry Newell Martin brought Foster's methods to Johns Hopkins University and S. Weir Mitchell, a student of Claude Bernard, helped found the *American Journal of Physiology* in 1898.

As a general rule, a muscle contraction (ΔL) is a function of the maximum possible contraction (Max ΔL), the actual muscle force (F), and the maximum possible force from that muscle (Max F). This relation is described by the following equation:

$$\Delta L = \text{Max } \Delta L \left(1 - \frac{F}{\text{Max } F}\right) \qquad (2.1)$$

Contractions and relaxations of muscle bundles create small electrical signals that can be detected on the skin surface above the muscle bundles by electrodes and a signal amplifier. Signals of 50–100 mV can be detected over a time period of approximately 1 ms.[†] Cell membranes within a muscle bundle depolarize immediately prior to fiber contraction. Since 10–100 fibers connect to a single motor nerve cell that activates those fibers, a signal involves at least a single motor cell. More often, several motor cells activate many collections of muscle fibers at approximately the same time. Consequently, larger signals

---

* While there is little intent to expand discussion at this point to the nervous and sensory subsystems of the human body, it is important to briefly mention *proprioception*. Sensory receptors *within* body tissues, primarily in muscles near joints, provide sensory information on the location of body members. Proprioception can be used instead of vision to guide movements of the body. These receptors also sense the forces on those body members that are induced by motion dynamics. That is, proprioception cues, as well as visual and other sensory cues, augment those from the inner ear (i.e., vestibular sacs and the semicircular canals).
† EMG measurement has a long history (e.g., Adrian and Bronk, 1929). Jasper and Ballem (1949) are reported to give a very comprehensive description of EMG analysis. Sidowski (1966) provides a good overview of this and associated methods of electro-body-measurements.

correspond to greater muscle activation and larger muscle forces. This method of analyzing muscle activity is called *electromyography* (EMG). To be more correct, EMG signals measure motor nerve activity, which is highly correlated to muscle activity. Obviously, EMG detection is best used on larger muscles, which are nearer to the skin surface and separated from other muscles that one does not want to monitor.

Role of oxygen in muscle actions

Muscle fibers are activated by nerve impulses within the muscle bundles. These impulses trigger a complex series of enzyme and chemical reactions that result in muscle fiber contraction. Oxygen is needed in this process and is brought from the lungs to the muscle by its blood supply. Carbon dioxide resulting from chemical reactions during contractions of the muscle bundles is simultaneously carried away by blood to the lungs. When oxygen is not in sufficient supply, lactic acid builds up in the muscle bundle until more oxygen is available. Typically in sudden start-ups of an activity, the resident oxygen supply is used up and the current blood supply has not delivered enough oxygen, so some lactic acid builds up in the muscles. This is known as the *oxygen debt phase*.*

With increases in the blood flow, oxygen is delivered fast enough to reduce the level of lactic acid in the bundle. The increased blood supply typically continues after the physical effort has stopped in order to remove all of the lactic acid. This is known as the *repayment with interest phase*. Many capillaries located between the muscle fibers deliver blood to the activated fibers. Arterioles (small arteries) branch from the arteries to bring blood to the capillaries, and venues (small veins) pick up blood from these capillaries for transport back to the blood veins. Blood is directed to those arterioles and venues away from resting muscle bundles so that there is more blood available to working muscle bundles.

If working muscles are contracted too much or for too long, they cannot get an adequate supply of blood and oxygen. Lactic acid accumulates in the affected muscles, and they may become painful. *Work designs that call for the application of excessive forces over long time periods must, therefore, be avoided.* This is especially true for back muscles. Another issue is that rigid, unnatural postures can cause muscles in the back, neck, and elsewhere to become painful. For this reason, postural freedom is an important principle of workplace design. It also follows that when extensive physical exertion is necessary, rest periods are required to balance the muscular effort.

Muscle injuries and disorders

Repeated or heavy muscular exertion often causes muscles to become *hypersensitive*, or likely to contract more quickly than normal. It becomes difficult to relax a hypersensitive muscle because of residual activity during rest and possible tremors within the muscle. Locally tender areas due to intramuscular fluid on the fibers are often present. In some cases, whole muscle bundles may go into contraction simultaneously and cause a muscle *cramp* or *spasm*. Once a cramp occurs, the person usually tries to use unaffected muscles, which may overburden these other muscles and result in further pain or cramps. A common case of cramp or spasm is a *charley horse*. Spasms or cramps are a natural body defense against overexertion. Beyond discomfort and pain, muscle spasms can easily cause reduced range of motion for an extended period of time, possibly resulting in employee absences.

---

* This phase and the repayment with interest phase, as mentioned in the following, are discussed later in this chapter under the section "Metabolism."

Another issue is that sudden, unexpected, force exertions often cause muscle injuries. For example, a worker might make a quick violent movement to secure a slipping load. Muscle strains occur when muscles or tendons are stretched too far. In minor cases, this results in stiffness and soreness. In more severe cases, the muscle may tear or the tendon may tear loose from the bone. This results in intense pain and requires a long time for recovery.*

Poor posture often contributes to the aforementioned problems. Improvements can usually be made by redesigning the work method so that the loads are spread over more muscle groups. Often this is accomplished by allowing the operator to change posture between cycles of the task or by rotating personnel. After people are hurt, another challenging problem must be dealt with. That is, ergonomic specialists are often responsible for developing ways to accommodate people with temporary or permanent disability† due to a job-related injury.

Gender, age, and training effects on muscular strength

In general, women are not as strong as men of the same age, weight, and physical condition. The size of this difference depends on the task performed. Studies have shown‡ that women's ability to lift objects from the floor ranges from 50% to 65% of men's ability. Other fractions are reported for different tasks. In general, these studies verify the classic rule of thumb that women of the same age, weight, and physical condition are on average *about two-thirds as strong as men*.

Exercise as a means of CTD prevention

Exercise is beneficial because it builds up endurance and strength. Many people advocate warming up before starting a task and doing stretching exercises during regular work breaks. For athletes, these exercises provide a physical transition from the static to the dynamic state. Likewise, persons whose jobs tend to be very static are urged to do some stretching exercises during breaks. The use of exercise to condition the specific muscle bundles used heavily on the job is a more controversial topic.

Proponents of conditioning argue that building up these muscles will make them stronger and less susceptible to injury. Opponents argue that the conditioning exercises use the muscles and joints most at risk for CTD. Hence, performing the exercises would be expected to bring on the CTD sooner. The answer to these questions is still unclear.

Anthropometry

The study of body sizes and other associated characteristics is generally referred to as *anthropometry*. While the term typically refers to static space dimensions, such as length, width, and shape, other important anthropometric measurements include the weights and inertial properties of body parts. Anthropometric measurements are essential when designing devices and/or systems to fit the users or employees. For example, almost

---

* Associated cumulative trauma syndromes are discussed earlier in this chapter.
† Some aspects of designing to accommodate people with disabilities are included in chapters to follow. The recent ADA mandates industry and transportation providers to achieve accommodation if economically feasible.
‡ See Poulsen and Jorgensen (1971), Chaffin (1974), and Annis (1996) for more on gender differences in muscular strength and their effects for lifting or other tasks. Chaffin and Ayoub (1975) studied a more varied set of muscle force actions and gave fractions of about 50%–65%. Asmussen and Heeboll-Nielsen (1962) report fractional ranges in isometric muscle strengths of 58%–66%, whereas Davies (1964) reported a fraction of 65%.

Chapter two: Human system

everyone would expect doors in building to be well above 6 ft (1.83 m) tall, because we are well aware that many people exceed 6 ft in height. But how large should the diameter of a screwdriver handle be, if you want the human hand, including fingers and the thumb, to surround the circumference? Or suppose that you are designing eyeglasses and you want the hinges outboard of the glass frames to be slightly smaller than the width of the human head just above the ears. What size do you need? Clearly, people who design products for the human body need to know something about the wide variety of body sizes that are possible, and that is what this section of the chapter is about.

Much of the collected anthropometric data have been taken from selected subpopulations, rather than the population as a whole, partly because many studies were originally directed at some specific design question asked by clothing or footwear manufacturers. A great deal of the data were obtained for the military to help determine the sizes of uniforms and properly design equipment.

A historical note is that anthropometrists originally used tape measures and other devices to make direct measurements of people. Since that process was very time consuming, it was largely dropped in favor of working extensively with photographic records. The use of photographs for measurement also improved the repeatability of measurements because all measurements were made on a flat surface. More recently, the use of laser scanning devices, interfaced with computer data collection software, has greatly reduced the time needed to collect information on body shapes. This potentially revolutionary development is already leading to new applications of anthropometric data.

Predicting the stature of people

People from a specified population vary in height from person to person following a normal (or Gaussian) distribution. Figure 2.6 shows a bell-shaped curve obtained when the stature of British men (in millimeters) was plotted versus the relative frequency of each height for the population. Below the bell curve is a percentile scale, ranging from the 1st to the 99th percentile. These two points correspond to heights of approximately 148 and 202 cm. The median (50th percentile height) is about 175 cm.

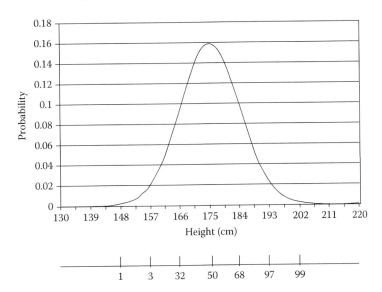

Figure 2.6 Normal distribution of stature in British men.

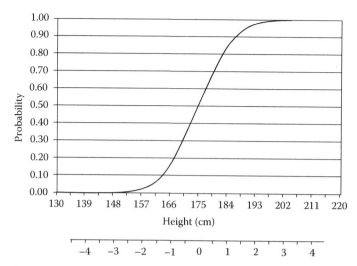

Figure 2.7 Cumulative normal distribution of stature in British men.

Figure 2.7 shows the cumulative normal function for the various heights. Here, the lower line in this figure represents the z-values of standardized normal variation or standard deviations.* The tail of the normal distribution to the right of a z score of 1.96 has an area of 2.5%. Accordingly, a range of z-values between −1.96 and +1.96 describes a 95% confidence interval around the mean. For the data shown, 95% of the population falls in the interval from 157 to 193 cm. The range from $z = -1.64$ to $z = +1.64$ corresponds to the middle 90% of the population and sizes ranging from about 160 to 190 cm.

Designers often use such data to fit their designs to selected populations. This approach is especially likely to be useful, when designers have some control over the subpopulations. For example, in military settings, pilot candidates who are unusually tall or short can be screened out.

One major complicating issue is that anthropometric measurements show consistent variation between members of different ethnic groups, between genders, by age, and over time.

The effect of age is shown in Table 2.2. Note that height peaks before age 40 and declines a few percent as the person continues to age. Children grow and older people shrink. But child growth differs between the genders. Table 2.2 shows approximate age effects on the stature of both genders. Note that at younger ages, females grow a bit faster than males, but males grow more overall.

NASA and other organizations have attempted to document ethnic and other differences in stature data. Much of the information they have assembled shows that human stature has been increasing worldwide since the mid-1800s. The average heights of military personnel worldwide have shown dramatic increases from about 172 cm during the early part of the twentieth century to 178 cm (or slightly more) in 1980. This increase represents a rate of growth in the human stature of about 1 mm a year. Although the accuracy of some of the older records may be debatable, nearly all countries with records going back very far have shown a continual upward trend in stature. Many people attribute this worldwide change to better diets.

---

* Also see Appendix B for standardized normal z scores.

Table 2.2 Changes in Mean Human Stature as a Function of Age

Age	Females		Males	
Years	Height (cm)	%Max	Height (cm)	%Max
1	74	46	75	43
5	110	68	111	64
10	138	86	138	79
15	160	99	168	97
20	161	100	174	100
35	161	100	174	100
40	160	99	174	100
50	159	99	173	99
60	158	98	172	99
70	157	98	171	98
80	156	97	170	98
90	155	96	169	97

Source: Kroemer, K.H.E. et al., *Engineering Physiology—Bases of Human Factors/Ergonomics*, 2nd edn., Van Nostrand Reinhold, New York, 1990.

These data also tend to show ethnic differences. For example, Norwegian and Danish statures have tended to be the greatest of many reported in the world. Among Europeans, the stature of the French tends to be the smallest historically. The stature of the Japanese tends to be very much smaller than those of European nationalities during the early part of the twentieth century, but this is changing quickly.

Predicting the segment mass of the human body

Human physical activities normally involve movement. The larger the mass of the body part that is moved, the greater the forces that are needed to move it. For this and other reasons* it is often useful to be able to predict the mass of different segments of the body. Some of the work at NASA provides an empirical base for predicting those mass magnitudes.† They made traditional body mass measurements and then ran statistical regression analyses in which they predicted the weight of the specific body segments as a function of the person's overall body weight (W).

Table 2.3 shows the regression equations developed by NASA, along with the standard error of the estimate, and the coefficient of determination. The masses and weight are recorded in kilograms. One caveat is that when the predicted weights for all the parts are added together, the sum is not exactly equal to the total weight of the human body. Fortunately, as can be easily verified by the reader, these differences are not very large.

---

* Knowledge of the mass and inertial properties of body parts is especially important when simulating the forces exerted on the human body in automobile crashes. In the latter situation, dynamic loading forces are of primary concern. Accordingly, over the years a number of anthropomorphic devices (dummies) have been developed for 5th, 50th, and 95th percentile males and females.
† Both Kroemer et al. (1990) and Roebuck et al. (1975) illustrate methods for determining the mass of body segments.

Table 2.3 Regression Equations Predicting Body Segment Weight in Kilograms from the Total Body Weight (W) in Kilograms

Segment	Regression Equation	Standard Error of Estimate	Coefficient of Determination
Head and Trunk	0.580 W + 0.009	1.36	0.968
Total Leg	0.161 W + 0.000	0.62	0.919
Total Arm	0.047 W + 0.132	0.23	0.883
Head	0.148 W − 3.716	0.20	0.814
Trunk	0.551 W − 2.837	1.33	0.966
Thigh	0.120 W − 1.123	0.54	0.893
Calf and Foot	0.165 W − 1.279	0.16	0.934
Calf	0.135 W − 1.318	0.14	0.933
Foot	0.009 W + 0.369	0.06	0.810
Upper Arm	0.030 W − 0.238	0.14	0.879
Forearm and Hand	0.168 W − 1.295	0.10	0.874
Forearm	0.119 W − 0.913	0.09	0.827
Hand	0.051 W − 0.418	0.03	0.863

Source: Clauser, C.E. et al., Weight, volume, and center of mass of segments of the human body, Report # AMRL-TR-69-70, Aerospace Medical Research Laboratory, Wright-Patterson Air Force Base, OH, 1999.

Other anthropometric relationships

Another role of anthropometry is to determine the relationships between the sizes of different body parts and other variables; this topic is discussed further in Chapter 4. As indicated earlier and shown in Figures 2.8 and 2.9, the size of different body parts is strongly related to a person's height. Taller people tend to have longer arms and vice versa although this is not always the case. Table 2.4 illustrates some other relationships for which correlation coefficients are reported. These coefficients tend to be consistent between males and females. Most of these relationships are also consistent with common sense. Examples include a general tendency for older people to weigh more, for taller people to be heavier, for heavier people to have larger waists, etc. More importantly, these numbers provide a starting point for thinking quantitatively about design problems. For example, a designer wondering how long to make a seat belt could make an initial estimate using the correlation in Table 2.4 and some knowledge of how heavy prospective customers might be. This would probably be easier than finding a representative sample of heavy people and measuring the circumference of their waists.

Body movement

An ergonomic designer must be familiar with how the human body moves, especially when designing workspaces. Classical physiology provides a good starting point. This approach focuses on rotational movement around different joints. Figure 2.8 depicts a selected set of movements often of interest in ergonomic design. As shown in the figure, movements around a joint can be measured in different planes.* (The names of the different planes of the body are shown in Figure 2.9.) Flexing and extending movements

---

* Other systems for body planes have been proposed. For example Roebuck's "global coordinate system" was proposed as a simpler alternative, but it is not yet fully accepted. See Roebuck, Kroemer, and Thomson (1975) for a complete description. Kroemer et al. (1990) provide a general description of this system.

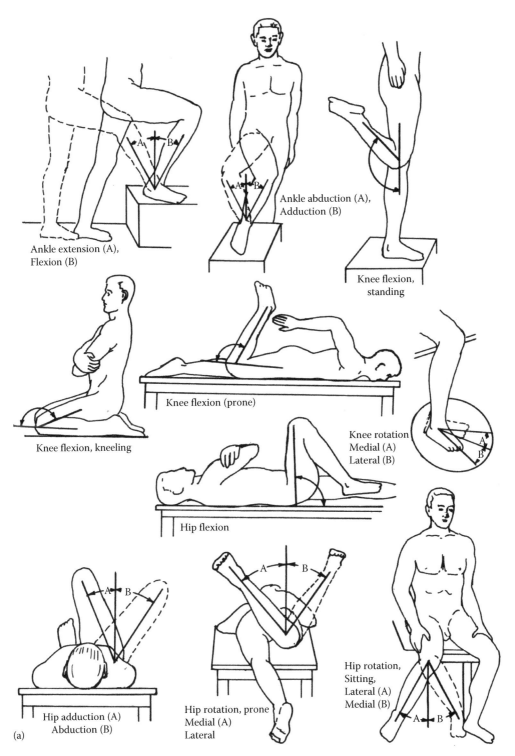

Figure 2.8 Movements of the human body. (From Van Cott, H.P. and Kincade, R.G., *Human Engineering Guide to Equipment Design*, U.S. Government Printing Office, Washington, DC, 1972.)

(continued)

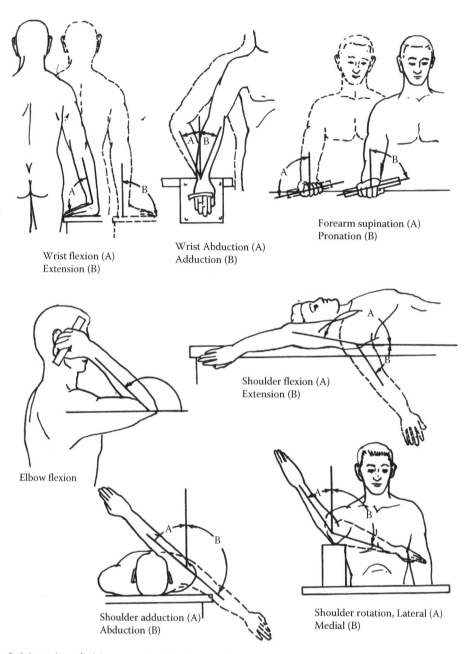

Figure 2.8 (continued) Movements of the human body.

are distinguished from each other, as are abductions (movements away from the body) and adductions (movements toward the body).

The classical approach is useful, because it provides a precise definition of mobility (i.e., in degrees) for a well-defined set of movements. This allows the movements of any given individual to be precisely measured. These measurements can be combined and statistically analyzed to describe a population. Variation within populations is a topic of particular interest to designers.

Chapter two: Human system 57

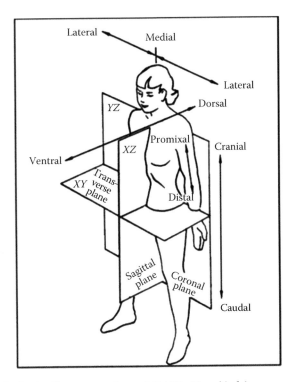

Figure 2.9 Planes of the body. (Image courtesy of NASA, New York.)

Table 2.4 Selected Correlations of Body Measurements of Adult Females and Males

Characteristic	A	B	C	D	E	F	G
A. Age	—	+.22	+.05	−.02	+.04	+.23	+.10
B. Weight	+.11	—	+.53	+.46	+.50	+.82	+.40
C. Stature	−0.03	+.52	—	+.93	+.91	+.28	+.33
D. Chest height	−.03	+.48	+.98	—	+.90	+.22	+.28
E. Waist height	−.03	+.42	+.92	+.93	—	+.24	+.31
F. Waist circumference	+.26	+.86	+.22	+.21	+.14	—	+.28
G. Head circumference	+.11	+.41	+.29	+.25	+.23	+.31	—

Source: Kroemer, K.H.E. et al., Engineering Physiology—Bases of Human Factors/Ergonomics, 2nd edn., Van Nostrand Reinhold, New York, 1990.
Data above the diagonal are for females and below the diagonal are for males.

Along these lines, Table 2.5 provides mobility data for joint movements shown in Figure 2.8 based on a sample of 200 people. The numbers in the table compare 5th, 50th, and 95th percentile males to females. Statistically significant ($\alpha = 5\%$) differences between 50th percentile females and 50th percentile males are also shown. Positive differences denote that the female mobility for that joint was greater.

Muscular–skeletal system as levers

Ergonomists often develop models of the forces exerted within the human body when people perform tasks. These models provide fundamental insight as to how different product

Table 2.5 Mobility of Adult Females and Males with Contrasts at the 50th Percentile

Joint	Movement	50th Percentile Female (Male)	Significant Difference	5th Percentile Female (Male)	95th Percentile Female (Male)
Ankle	Flexion	23.0 (29.0)	−6.0	13.0 (18.0)	33.0 (34.0)
	Extension	41.0 (35.5)	+5.5	30.5 (21.0)	51.5 (51.5)
	Adduction	23.5 (25.0)	Not signif.	13.0 (15.0)	34.0 (38.0)
	Abduction	24.0 (19.0)	+5.0	11.5 (11.0)	36.5 (30.0)
Knee	Standing flex.	113.5 (103.5)	+10.0	99.5 (87.0)	127.5 (122.0)
	Prone flexion	130.0 (117.0)	+13.0	116.0 (99.5)	144.0 (130.0)
	Medial rotation	31.5 (23.0)	+8.5	18.5 (14.5)	44.5 (35.0)
	Lateral rotation	43.5 (33.5)	+10.0	28.5 (21.0)	58.5 (48.0)
Hip	Medial rotation	32.0 (28.0)	+4.0	20.5 (18.0)	42.5 (43.0)
	Lateral rotation	33.0 (26.5)	+6.5	20.5 (18.0)	45.5 (37.0)
Wrist	Flexion	71.5 (67.5)	+4.0	53.5 (51.5)	89.5 (85.0)
	Extension	72.0 (62.0)	+10.0	52.5 (47.0)	87.5 (76.0)
	Abduction	28.0 (30.5)	−2.5	19.0 (22.0)	37.0 (40.0)
	Adduction	26.5 (22.0)	+4.5	16.5 (14.0)	36.5 (30.0)
Forearm	Supination	108.5 (107.5)	Not signif.	87.0 (86.0)	130.0 (135.0)
	Pronation	81.0 (65.0)	+16.0	63.0 (42.5)	99.0 (86.5)
Elbow	Flexion	148.0 (138.0)	+10.0	135.5 (122.5)	160.5 (150.5)
Shoulder	Adduction	52.5 (50.5)	Not signif.	37.5 (36.0)	67.5 (63.0)
	Abduction	122.5 (123.5)	Not signif.	106.0 (106.0)	139.0 (140.0)
	Medial rotation	110.5 (95.0)	+15.5	94.0 (68.5)	127.0 (114.0)
	Lateral rotation	37.0 (31.5)	+5.5	19.5 (16.0)	54.5 (46.0)

designs (i.e., hand tool designs), postures, and work methods might impact performance or cause injuries. In this modeling perspective, the skeleton is viewed as a system of linked levers. Bones become links in the system, which are normally loaded in compression. Joints become fulcrums. Muscles and ligaments exert opposing tension forces around each joint or fulcrum, resulting in moments or torques at each joint and compressive loads on the bones.

Three different classes of levers can be found in the human body. First-class levers consist of a fulcrum separating parallel opposing forces. Figure 2.10A illustrates one first-class lever found in the human body. As shown in the figure, an upward pulling force on the forearm is balanced by a counteracting force exerted by the triceps muscle where it attaches behind the elbow joint. The downward force of the weight of the forearm also acts against the upward pulling force. The effectiveness of a particular lever depends upon how much mechanical advantage or leverage is available. In Figure 2.10A, the muscle has a short lever arm compared to that for the load. Consequently, the muscle must exert a much higher force than the load. This results in a compressive force on the humerus (the bone within the upper arm). It also should be pointed out that the length of both lever arms changes as the forearm assumes different positions.

Second-class levers are distinguished by having the fulcrum at one end, an opposing force at the other, and the load in the middle. An example is given in Figure 2.10B, involving the lower leg and foot. Third-class levers are shown in Figure 2.10C using the arm. In contrast with Figure 2.10A, there is a downward load in the hand (at the end opposite to the fulcrum) and the opposing force in the center. Note that the mechanical advantage is greater in Figure 2.10C because of the use of a different muscle group (the biceps) to counterbalance the load in the hand.

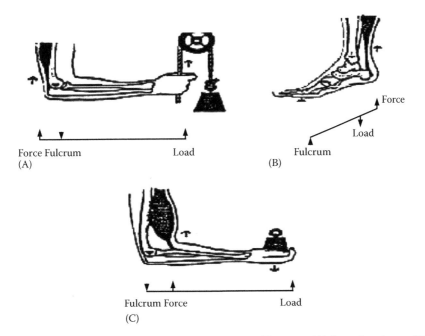

Figure 2.10 Examples of the human body as three types of levers—(A) first-class lever, (B) second-class lever, and (C) third-class lever.

Sensory subsystems

People use a variety of senses in their daily lives. Vision is probably the most-used sense, and audition is not far behind. Both of these senses are important to ergonomic designers. People also have senses of taste, touch, odor, heat, cold, body position, and body orientation, all of which are significant but often of less importance in typical ergonomic design situations. For that reason, less detail is provided on the latter topics.

Visual sensory subsystem

The visual senses are stimulated by electromagnetic energy at wavelengths (λ) between 360 and 760 nm (nm stands for nanometer or 10^{-9} m).* Accordingly, energy in this bandwidth is called light. Since the velocity (v) of light is constant in a vacuum at 3×10^8 m/s, the frequency of this energy (f) follows the physical relationship:

$$v = f\lambda \qquad (2.2)$$

The sun and other heavenly bodies are natural sources of light. Artificial lighting, including incandescent and fluorescent lamps, is another important source.

Light reflecting off objects in the environment contains large amounts of information. Vision is a complex activity in which this information is extracted. As the first step in this process, light must impinge upon our receptors of luminous energy within the eyes.

---

* Frequency measurements are also made in microns and Angstroms. In microns, the visual range is 0.36–0.76 µm. In Angstroms, the visual range is 3600–7600 A.

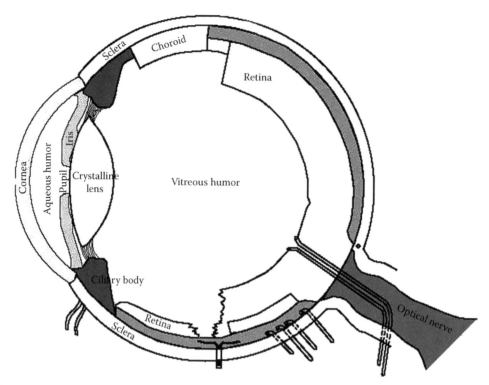

Figure 2.11 Schematic view of the human eye.

That is, light must enter the eyes through the cornea, and then go through the anterior chamber, the lens, and vitreous humor, before landing on the retina in the back of the eye. This information is then processed in several steps, beginning at the retina, and ultimately within the visual cortex of the brain. The following discussion provides a brief introduction to the structure and function of the human eye,* which will be expanded upon in later chapters in this book during the discussion of lighting in Chapter 4 and visual displays in Chapter 12.

Structure of the eye

As Figure 2.11 shows, the eye is roughly a sphere of about 2.5 cm in diameter. The outer layer, or sclera, sometimes known as the whites of the eye, covers around 85% of the outer eye surface.

The cornea is located at the very front of the eye. The cornea is about 1 mm thick and consists of five layers of transparent membranes. Since the cornea protrudes from the eyeball, it creates what is called the corneal bulge.

Directly behind the cornea lies the anterior chamber, which is filled with a saline solution called the aqueous humor. At the rear of this chamber are the iris and the lens. The iris is composed of nearly totally opaque membranes, which are connected to ciliary muscles that open and close the iris to change the size of the eye's pupil from about 3 mm to about 6.5 mm. These actions by the iris control the amount of light entering the eyes by opening the pupil to its largest diameter when light levels are as low, and closing the pupil to

---

* For more on the topic of vision, the reader is encouraged to consult a text such as *Foundations of Vision*, written by Wandell (1995).

its smallest diameter when light levels are high. It takes about 3–4 s for these changes to occur. Both pupils respond even if only one eye is exposed to a brighter light.

A flattened sphere of translucent tissue known as the lens is located immediately behind the iris. The ciliary muscles of the eye connect to the lens. Consequently, the shape and cross section of the lens can be controlled to change the focal length of the eye. This process allows near objects to be clearly focused on the retina, and is referred to as visual accommodation. The large interior or posterior chamber of the eye located behind the lens is filled with vitreous humor (a saline solution, also containing albumen). This liquid is kept at a positive pressure to maintain the spherical shape of the eyeball. The retina is located on the back inside wall of the eye. It contains an outer layer of pigmented cells that protect the photoreceptors beneath.

The visual axis of the eye extends through the centroid of the lens to a location in the retina where a small indentation known as the fovea can be found, as shown in Figure 2.11. The fovea is highly populated with sensory cells known as *cone receptors*, which are able to distinguish colors. The regions of the retina outside the fovea also contain *rod receptors*, which detect light but do not distinguish between colors. Cone receptors are distributed throughout the retina at a low density except for in the fovea, where they are concentrated in a very high density. The density of rod receptors increases from zero in the fovea to a maximum at locations around 20° on either side of the fovea. Rod receptor density then decreases in the periphery. The number of rods and cones at the eye's blind spot, where the optic nerve enters the retina and attaches to collections of photoreceptors, is of course zero.

If a person has adapted to high light conditions, vision is primarily provided by the cone receptors and is known as photopic vision. When adapted to low light conditions, vision is primarily provided by the rod receptors and is known as scotopic vision. Scotopic vision is nearly black and white. The relative sensitivity of each form of vision changes with the wavelength of the radiation as shown in Figure 2.12. Sensitivity peaks for light at a wavelength of about 550 nm for photopic vision and about 510 nm for scotopic vision. This change in sensitivity between photopic and scotopic vision is known as the "Purkinje shift," after its discoverer.

Within the retina, collections of photoreceptors are connected to nerves, which are eventually combined in optic nerve bundles leading from each eye to the visual cortex of the brain. The optic nerve bundles merge before entering the brain. The nerves connected to receptors in the nasal region of the retina crossover at this junction, so the left side of

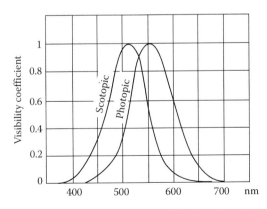

Figure 2.12 Relative sensitivity of the human eye in well-lighted conditions (photopic vision) and after dark adaptation (scotopic vision).

the brain is connected to nerves from the right eye and vice versa. Nerves from receptors in the temporal regions of the retina go to the same side of the brain, as the corresponding eye. Complex visual processing is done within the brain to align the visual axes and superimpose the views from the two eyes. Also, it is believed that rod vision and cone vision are both operative most of the time. Foveal vision, involving cone receptors, is used to focus on images in the primary visual task. Peripheral vision, involving rod receptors, provides a broader field of view, and is used to sense motion and orientation.*

Eye movements

Each eye is held into its eye socket by three pairs of opposed muscles. By tensing and relaxing opposed muscles, the eye can be rotated horizontally about 70° in each direction and vertically about 40° upward and 60° downward. Each eye can be covered by the eyelids. Blinks serve to lubricate the front of the eye. There is little or no evidence that blinks cause substantial visual interference. Eye movements may be coordinated with blinks, just as they are with head movements. Other minor eye movements (e.g., tremors, flicks, and drifts) are of little importance in ergonomics.

Extensive studies have been made over the years on how the eye moves from one fixation point (where the eye is stationary) to the next. These movements are called saccades; a movement of 5° can be made in about 30 ms, and 40° requires about 100 ms (from 167 to 400 ms). Movement rates range up to a maximum of over 500°/s. Eye movement velocity reaches its maximum shortly before the movement is half over. The velocity then decreases at a rate lower than the original acceleration rate during the remaining interval. Most saccadic movements from 15° to 50° are similar in their speed profile and maximum acceleration. Other studies have revealed that this acceleration pattern is similar to that followed by a mechanical servomechanism, which places equal emphasis on not undershooting or overshooting the target.

Yarbus[†] predicted the time to move one's eyes A_0 arc seconds as

$$T(\text{ms}) = 21 A_0^{0.4} \qquad (2.3)$$

Researchers have developed many ways of measuring eye movements.[‡] One approach is to record eye movements using videotape or motion pictures.[§] Another method is to shine a light (or an infrared beam) onto the side of the corneal bulge and pick up the reflection on a prism. Along these lines, MacWorth developed a camera-mounted helmet that used the corneal bulge reflection method. Other researchers have used reflective contact lenses. Another device floods the eye with infrared light, which is invisible to the eye, and then measures the infrared reflectance off the eye's sclera with photoreceptors on each side of the eye. As the eye moves right, more sclera is on the left side and less on the right side of the eye and so there is more reflection on the left than the right. The corneo-retinal standing potential method (or EOG) has also been used. This method involves placing electrodes on the vertical and horizontal axes of the eyes. The voltage between the electrodes changes when the eyes move toward and away from an electrode. These voltage changes provide another measure of eye movements but are not as accurate as the other aforementioned methods.

---

* Larish and Flach (1990) discuss velocity sensing in terms of "global optical flow." That is, the movement of images on the retina generates a perceived optical flow and a sense of velocity.
† See Yarbus (1967).
‡ Many researchers call eye movements a "window to the brain."
§ Fitts', Jones, and Melton (1950) recorded the instruments that pilots fixated upon during aircraft landings. These data are shown in Figure 3.7.

Eye-tracking systems are currently available from many vendors, as can be easily verified by a quick search of the web, using the keywords "eye tracking systems" in Google or other search engines. Economical PC-based head-mounted systems, such as the Polhemus VisionTrak eye-tracking system, currently used in the Purdue Work Analysis and Design Laboratory, enable a subject to have full head and eye range of motion while simultaneously collecting data on eye movements and fixations. Such systems include data collection software that organizes eye fixation locations, movement velocities, and accelerations in convenient tabular or graphical format. Many systems also include a playback mode in which the eye fixation location or point of gaze is superimposed on a video image of the object viewed by the subject while they perform a task. The latter feature is especially convenient for determining where a person is looking when performing a visual task, such as reading, as expanded upon later.

Visual tasks

Ergonomists have studied eye movements in a large number of visual tasks. One of the better-known examples is reading. When people read, their eyes move along a line of print from left to right in a sequence of fixations and saccades, until they reach the end of the line. At this point, their eyes normally move to the next line. Sometimes a missed or misunderstood word causes the eyes to go backward along the same line of print in what is known as a regression. Fixations occur on most words but rarely on articles (e.g., a or an), and longer words may require two or more fixations. Poor readers tend to make more regressions, and have longer, more frequent and overlapping fixations. A fixation normally lasts about 200–320 ms, depending upon the complexity of the reading task. Overall, fixations account for 85%–95% of the reading time.

Several related studies have focused on instrument reading tasks. Such tasks are similar to reading print, but not exactly the same. For example, aircraft pilots during landings spend about 97% of the time in visual fixations, varying from 400 to 900 ms per fixation and averaging 600 ms. As more time is spent on each instrument, less time is spent moving the eyes from instrument to instrument. These fixation times are much longer than those given earlier for reading from print. Part of the reason is that an aircraft instrument is quite complex.

Beginnings of Psychology

Probably most people would pinpoint the start of psychology with the work of Herman von Helmholtz who verified and expanded the theory of color perception. Helmholtz based his theory on the earlier work of Thomas Young who originally recognized the three-component theory of colored light and upon that of Sir Isaac Newton, Deme Mariotte, and Sir Charles Weatstone who identified visual depth perception, which we call psychophysics today. Ernst Heinrich Weber started work on the measurement of human sensations during the middle of the nineteenth century. Gustav Theodor Fechner developed the Weber–Fechner Law. Max von Frey and Erst Mach first identified the four skin senses. Sir Francis Galton and Karl Pearson also contributed to these theories. However, it was not until 1879 that Wilhelm Wundt of Leipzig University established the first research laboratory and founded the first scientific journal to start this new science.

Perception of color and brightness

People need to identify colors in many different tasks (see the Color Insert for figures related to the perception of color and brightness). For example, drivers need to identify the color of traffic control lights, and various forms of color coding are used in control rooms and in maintenance operations. Unfortunately, about 6% of the male population has some deficiencies in color identification, particularly with reds and greens. Less than 1% of the female population is similarly impaired.

One basic finding of visual perception research is that the perceived brightness of light changes depends on its wavelength or frequency (λ), when luminance or energy levels are held constant. In one of the first experiments addressing this issue, Gibson and Tyndal asked subjects to look through a tube that was split vertically into two different sections. The luminance of the light on the left side was held constant at a frequency different from the light on the right side. The subject was asked to adjust the luminance on the right side until its perceived brightness was the same as the left light. This experiment showed that the perceived brightness was lower for frequencies above or below 550 nm. This work eventually led to the photopic luminosity function (see Figure 2.15), which is used today to convert physical measurements of luminance into perceived brightness levels over the visual spectrum. The latter figure also shows a slight shift in peak sensitivity under scotopic and photopic conditions.

Another famous experiment was L. A. Jones's (1917) exploration of the magnitude of wavelength change that could be perceived correctly 50% of the time. Jones used psychophysical methods and found that people were most sensitive when the frequency was 490 or 585 nm. Changes of as little as 1 nm were detectable at these wavelengths. These wavelengths correspond to blue-green and yellow-orange colors. Sensitivity was poor at wavelengths below 420 nm or above 650 nm. Changes of at least 2 nm were required before they could be detected.

Today, color is usually measured using a chromaticity diagram developed by the Commission International de l'Eclairage (CIE), which is also known as the International Commission on Illumination (ICI). Colors are described using two coordinates:

$$x = \frac{X}{X+Y+Z} \tag{2.4}$$

$$y = \frac{Y}{X+Y+Z} \tag{2.5}$$

where the values of X, Y, and Z, respectively, correspond to spectrophotometer readings of red, green, and blue luminance.

Several versions of the chromaticity diagram have been developed over the years. The original CIE 1931 chromaticity diagram (see Color Insert 2) defines the color space by mapping combinations of x and y to particular colors. The updated CIE 1976 diagram, shown in Figure 2.13 (see Color Insert 1), specifies colors in terms of the vector [u', v'], where u' and v' are calculated* using values of x and y. A bright red light, such as a traffic stoplight, has approximate coordinates of [u', v'] = [0.5, 0.6] on the CIE 1976 diagram, whereas a green traffic light would be about [0.02, 0.5]. It follows that green is near the top of the CIE 1976 chromaticity diagram, violet-blue at the bottom, and white light or daylight is in the middle of the rounded triangular shape (about [0.25, 0.45]). Some light sources may contain

---

* To go from the [x, y] coordinates in the CIE 1931 diagram to the [u', v'] coordinates in the CIE 1976 diagram, use the equations: $x = 9u'/(6u' - 16v' + 12)$ and $y = 4v'/(6u' - 16v' + 12)$.

Chapter two: Human system 65

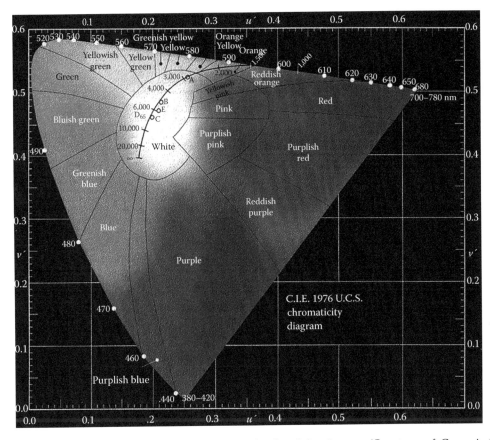

Figure 2.13 **(See color insert.)** C.I.E. diagram of colored luminants. (Courtesy of Commission Internationale de l'Eclairage, Vienna, Austria, 2004.) See related color inserts in the following pages.

one or more luminants of different colors or brightness. As indicated by Equations 2.4 and 2.5, the combined effect is additive. For example, a mixture of red and yellow luminants appears to be orange (around [0.4, 0.52]). Adding white light dilutes the purity of primary colors, making them less saturated.

Another characteristic of vision is that brightness intensities tend to follow the Weber–Fechner Law. That is, brightness intensities (I) and just noticeable differences in brightness (ΔI) are related as follows:

$$\frac{\Delta I}{I} = \text{const.} \tag{2.6}$$

One implication of this equation, which has been confirmed in numerous studies, is that people are more sensitive to light intensity changes when the background light intensity is low.

Visual acuity

The ability of people to visually resolve fine details has been extensively studied and several measures of visual acuity have been developed. The Snellan E or letter chart is one of the more commonly used tools for measuring visual acuity. Various configurations* of

---
* In its original form, the letter "E" was presented in different orientations.

this chart have been developed, which display rows of letters of the alphabet. The letters in each row are at different sizes, and are viewed from a specified distance. A person who can accurately read smaller letters has better visual acuity. Use of the Landolt ring chart is similar method for measuring visual acuity. A Landolt ring is a C-shaped figure. These rings are rotated to differing degrees within the charts. The smallest ring that the observer can consistently and correctly identify provides a measurement of visual acuity.

From a more general perspective, the smaller the visual angle (α), at which a stimulus can be first distinguished, the better the visual acuity. If the height of the smallest item on an eye chart that can be correctly identified is h, and the chart is at a distance d from the observer, then α can be found as

$$\frac{h}{d} = \tan \alpha \tag{2.7}$$

$$\text{Visual acuity} = \frac{1}{\alpha} \tag{2.8}$$

Visual acuity testing may be done with the visual test charts at various distances. People who have good acuity for close charts but poor acuity for distance charts are *myopic* or nearsighted. Conversely, people who have good acuity with the distance charts but not with near charts are *hyperopic* or farsighted.

Dynamic visual acuity (DVA) tests measure how well people can identify moving targets. Typical DVA tests move the target horizontally around the observer at an angular velocity of ω degrees per minute. As ω increases, the required visual angle α also increases, but there are large individual person differences in the changes in α that occur with increase in ω. People with poor DVA show large increases in α as a function of ω. People with good DVA experience little deterioration in α. In general, near-zero correlations are found between α values when $\omega=0$ compared with $\omega>0$. Accordingly, one cannot assume that people with good static visual acuity also have good DVA. Figure 2.14 describes some results Elkin (1962) found in a test that allowed people to track the target before it was exposed (called anticipatory tracking time) and then for another time period, called "exposure time," after the target appeared. Essentially, Elkin used the critical angle α as the lack of dynamic visual acuity (rather than the reciprocal) so that a lower DVA threshold indicates a better DVA. He showed that increasing both anticipatory tracking time and the exposure time improved DVA.

Other visual abilities

Visual *contrast sensitivity* is a visual ability measured using static targets of alternating black-and-white lines that are varied in their frequency. As the frequencies are increased or decreased, at some point the pattern blurs together. Myopic people tend to have blurred vision at low frequencies, showing that their vision usually is not good for either near or far targets. It is unclear whether visual acuity and contrast sensitivity measure the same traits.

Depth perception is another important visual ability that can vary greatly between individuals. This ability is extremely important in some jobs; for instance, when landing an aircraft. Tests of this visual ability often involve asking subjects to align rods (or other objects) located some distance away from the observer. Such tests provide only a partial measure of people's ability to perceive depth. It is well known that visual depth perception is done in most realistic settings using visual cues of interposition and texture, rather than by stereoscopically detecting the convergence angle.

Chapter two: Human system   67

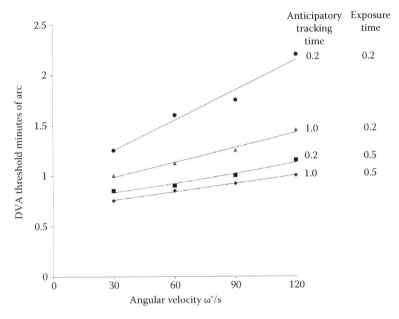

Figure 2.14 DVA as a function of angular velocity (ω) under alternative exposure times and anticipatory tracking times.

People's *dark adaptation* rate is the last visual ability that will be discussed in this section. People's ability to quickly adapt to changing light conditions is very important in certain jobs found in the military, such as watch duty or other forms of observation on ships and aircraft. Adaptation rates are also important for car or truck drivers operating at night. When the highway is almost deserted, drivers become partially dark adapted. For some dark-adapted people, the abrupt glare of headlights can create serious safety problems.

Human perception of sound

To a physicist, sound is nothing more than vibration within a bandwidth that people can perceive. When a body, such as a tuning fork, vibrates in the air, an element of the fork moves back and forth at a given frequency. The molecules of air compress as the fork moves one way and they rarify when it moves the other. An air pressure meter next to the fork increases above atmospheric pressure and then decreases below, over and over again as the fork continues to vibrate.

If these changes in pressure were plotted over time, they would appear as shown in Figure 2.15, where P_a is the maximum amplitude of air pressure variation relative to atmospheric pressure, f is the frequency of the vibration in cycles per second, $1/f$ is the time period of a cycle in seconds, λ is the wave length or distance a sound wave travels between cycles, and C is the speed of sound in the medium it travels within (1,130 ft/s in air, 4,700 ft/s in water, 13,000 ft/s in wood, and 16,500 ft/s in steel).

It follows that the relationship between the wavelength (λ) and frequency (f) is

$$\lambda = \frac{C}{f} \tag{2.9}$$

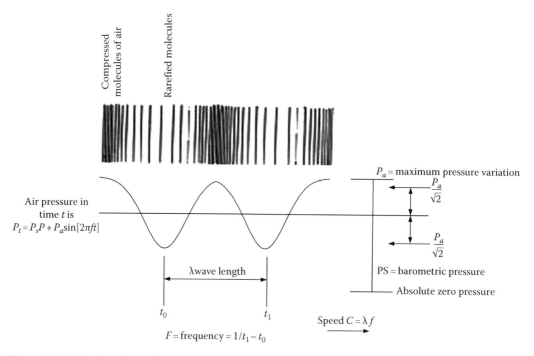

Figure 2.15 Physics of sound waves.

The air pressure variations over time are related to the amount of energy in the sound waves. The average effective air pressure variation (P) is the root-mean-square of the varying pressure or

$$P = \frac{P_a}{\sqrt{2}} \quad (2.10)$$

Typical units of measure for P are N/m² or dyn/cm² at the source of the sound.

Two sound sources have quite different effects if their frequencies are in or out of phase. Figure 2.16 describes several conditions where P_1 and P_2 are two sound sources that combine into P_r. Condition (A) occurs when both sources are equal in power, of the same frequency, and are in phase as zero pressures, peak positive and peak negative pressures all occur simultaneously. As a result, P_r is equal to the sum of $P_1 + P_2$. This corresponds to an increase in loudness of 6 dB.* Condition (B) shows P_1 and P_2 as equal, but 90° out of phase. Hence, the combination of the pair of sounds, P_r, increases in pressure by only 40%, and the increase in loudness is only 3 dB. Condition (C) shows P_1 and P_2 as equal in intensity but 120° out of phase. The summed sound wave has a maximum amplitude that is no different than for the individual sound sources. The last condition (D) has P_1 and P_2 equal, but 180° out of phase so that the high pressure of one corresponds to the low pressure of the other and the resultant pair, P_r, is zero. (Note that this is essentially the method used in "noise-cancelling" headsets.)

---

* Decibels (dB) are a commonly used measure of loudness, as described in the following in greater detail.

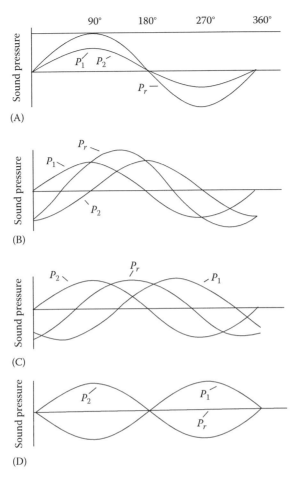

Figure 2.16 Result of P_R of combining two pure tones (P_1 and P_2) that are in phase (A), 90° out of phase (B), 120° out of phase (C), and 180° out of phase (D).

A sound source radiates sound energy, which will escape in all directions unless there is some constraint to radiation (e.g., a reflecting barrier). When there is no constraint, each sound wave moves away from the source at the speed of sound (C) in the medium. The wave front is part of a sphere, with a radius of r distance units from the sound source to the wave front. When the receiver of the sound energy is exactly r units of distance from the source, the receiver is impinged with sound energy. The amount of energy hitting the receiver depends upon the power at the source in watts (W), the distance between the source and the receiver r, the density of the medium within which the sound is traveling, and the area of the receiver. The sound intensity (I) is

$$I = \frac{W}{4\pi r^2} \quad (2.11)$$

in W/m² at a distance r meters away from a source with power of W watts. Note that the denominator of the preceding equation is the surface area of a sphere whose radius is

r meters. Thus, sound intensity decreases with the square of the distance. Also, power in W/m² is related to effective pressure as follows:

$$I\,W/cm^2 = \frac{P^2}{C\rho} \tag{2.12}$$

where
  C is the velocity of sound in air (343 m/s)
  P is the root-mean-square sound pressure in N/m²
  ρ is the density of the medium where the sound is traveling in kg/m³

The denominator of the right-hand side of Equation 2.12 is nearly a constant in air. It therefore follows that

$$\text{Log}\,(IK) = 2\,\text{Log}\,P \tag{2.13}$$

where $K = 415$ kg/m² s in air at normal temperature of 20°C.

The most common units of sound measurement are given in units relative to a reference level near the threshold of hearing, which is 10^{-12} W/m² = 10^{-16} W/m² for power units or 0.0002 dyn/cm² = 0.00002 N/m² for pressure units. Loudness is measured in decibels, and can be mathematically expressed as a function of pressure:

$$L_{sp} = 20\,\text{Log}\,\frac{P(\text{dyn/cm}^2)}{0.0002} = 20\,\text{Log}\,\frac{P(\text{N/m}^2)}{0.00002} \tag{2.14}$$

Loudness can equivalently be expressed as a function of power:

$$L_{sp} = 10\,\text{Log}\,\frac{I(\text{W/cm}^2)}{10^{-16}} = 10\,\text{Log}\,\frac{I(\text{W/m}^2)}{10^{-12}} \tag{2.15}$$

For a person to hear something, sound waves must enter the auditory canal of the ear and cause the eardrum to move (see Figure 2.17). Movements of the eardrum in turn cause vibrations in the bone structure that connects the eardrum to the oval window located in front of the vestibular canal. Because this canal is filled with fluid, waves result from these vibrating stimuli. These waves run up the vestibular canal and back down the tympanic canal where the round window absorbs the wave energy. Disturbances in the vestibular and tympanic canals cause the basilar membrane to move and the corti on that membrane to bend. The corti at each location give off signals proportional to their movement, which are then transmitted via the auditory nerve to the brain. The amount of movement at each location depends on the frequency, as well as the amplitude of the sound wave. Consequently, the location of the responding corti provides information about the sound's frequency. Intensity roughly corresponds to the number of different corti responding. More complex coding also occurs.

People are much more sensitive to sounds at certain frequencies than others. For example, older people and people who have been exposed to excessive noise, often have trouble hearing higher frequencies. The contour plots in Figure 2.18 show constant perceived loudness levels, for young subjects with normal hearing, over the frequency range

Chapter two: Human system

of 20–10,000 Hz. These contours show that people are most sensitive to sounds at frequencies around 3500 Hz. The effects of frequency are greatest in the low loudness range of sound pressure (dB).

Calculating Sound Intensity and Loudness

Sound received from 20 W source after traveling 5 m has an intensity (I), which is calculated using Equation 2.11, shown as follows:

$$I = \frac{20\,W}{4 \times 3.1416 \times (5\,m)^2} = 0.064\,W/m^2$$

If the receiver was only 1 m away, the intensity would be

$$I = \frac{20\,W}{4 \times 3.1416 \times (1\,m)^2 \times 100\,cm/m} = 1.59\,W/m^2$$

In the second instance, the power is 25 times greater, even though it is still quite small. But this example illustrates that distance attenuates sound power very rapidly and that sound power comes in small units.

A sound power of 2.65 dyn/cm² has a loudness (L), which is calculated using Equation 2.14, as shown in the following:

$$L_{sp} = 20\,Log\,\frac{2.65\,(dyn/cm^2)}{0.0002} = 82.44\,dB$$

If the acoustical pressure is cut in half, the loudness drops to 76.22 dB and it is again cut in half, the loudness is 70.40 dB. Accordingly, a reduction in sound pressure by about 6 dB occurs when the acoustical pressure is cut in half.

To account for this phenomenon, a sound measurement unit called the *phon* was developed. This measurement scale uses the L_{sp} value for sounds of 1000 Hz as a reference point, against which the perceived loudness at other frequencies is compared. That is, a 50 dB sound at 1000 Hz is 50 phon and so is another sound of a different frequency that sounds equally loud. For illustration, a 120 Hz sound of 60 dB is 40 phon and a 300 Hz sound at 30 dB is 20 phon. However, a sound of 40 phon is not perceived as being twice as loud as a 20 phon sound. Consequently, another measurement unit was developed, called the *sone*. One sone is defined as the perceived loudness of a sound of 40 phon. A sound of X sones is perceived to be X times as loud as a sound of 1 sone.

Table 2.6 shows sound measurement values for some typical situations. Note that the threshold of pain in hearing is about 125 dB and a whisper is between 20 and 30 dB. It is interesting to note that the threshold of hearing is at 10^{-16} W/cm² or 0.0002 dyn/cm²; these equivalent values are the numerators in Equations 2.14 and 2.15.

Position and motion sensing

People use visual and auditory cues for orientation but also use sensory organs within the inner ear to sense position and motion. Figure 2.19 depicts the anatomy of the human ear,

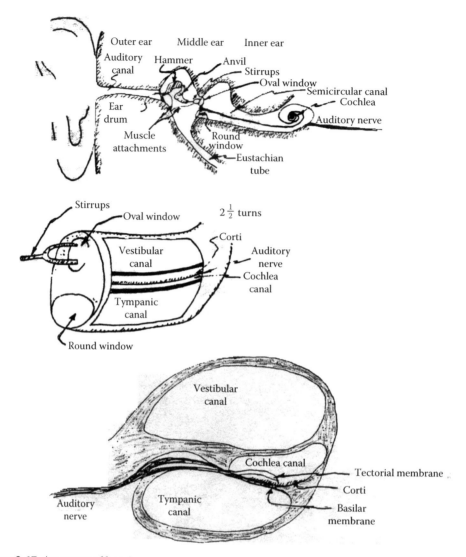

Figure 2.17 Anatomy of hearing.

including the vestibular apparatus and cochlea, the part of the inner ear containing the vestibular and tympanic canals. The vestibular mechanism contains the vestibule and semicircular canals filled with endolymph fluid (a saline solution). A gelatinous mass, called the ampulla, containing hair tufts called cristae, can be found at the end of each canal.

When the head is tilted, the pressure of the fluids changes at the ampulla. This pressure change is picked up by the cristae, which in turn excite a nerve. The common base of the three semicircular canals is the vestibule, containing the utricle and saccule, which are also filled with endolymph fluid and enclose otolithic sensing organs. Two forms of sensing occur. Within the utricle receptor are ciliated hair cells on the interior surface, which connect to nerve cells. In addition, a gelatinous pad contains otoliths, or small rocklike bodies, that are connected to nerve cells. Since the gelatinous pad is semirigid, it tends

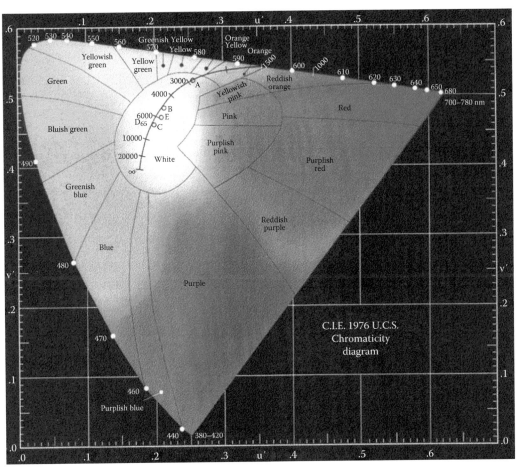

Figure 2.13 C.I.E. diagram of colored luminants. (Courtesy of Commission Internationale de l'Eclairage, Vienna, Austria, 2004.) See related color inserts in the following pages.

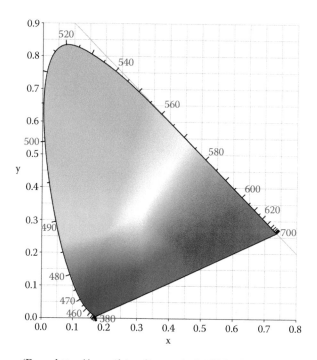

The CIE 1931 diagram. (From http://en.wikipedia.org/wiki/File:CIExy1931.png)

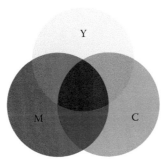

The Munsell color system. (From http://en.wikipedia.org/wiki/File:Munsell-system.svg)

The Munsell color wheel. (From http://en.wikipedia.org/wiki/File:MunsellColorWheel.png)

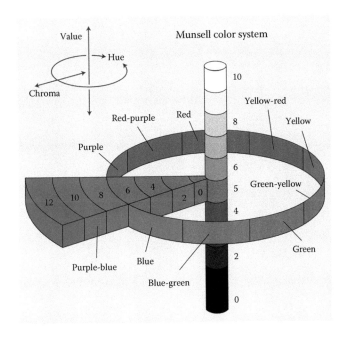

A RGB color cube.

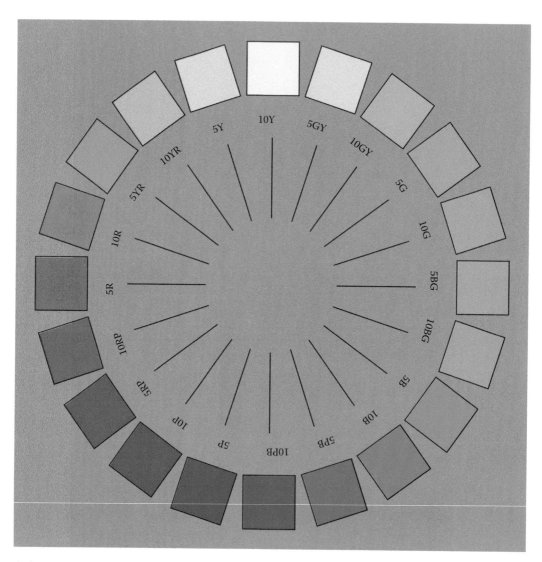

A diagram showing additive RGB color mixing commonly used in monitors

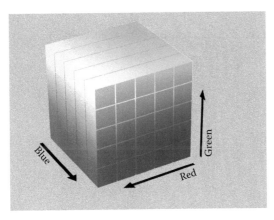

A diagram showing subtractive CYM color mixing commonly used in printers. (From http://en.wikipedia.org/wiki/File:SubtractiveColor.svg)

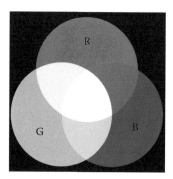

A diagram illustrating creation of color image in a CYM printing process. (From http://en.wikipedia.org/wiki/File:SubtractiveColor.svg)

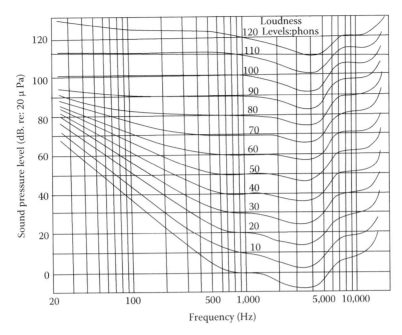

Figure 2.18 Equal loudness contours for pure tones at different frequencies. (Courtesy of OSHA, Washington, DC.)

Table 2.6 Typical Situations and Corresponding Pressure and Power Sound Measurements

Situations	Rustling Leaves, Whisper	Office Conversation	Store Vacuum Cleaner	Factory Noise, Heavy Traffic	Subway, Rock and Roll Band	Thunder, Pain Threshold
Lsp (dB)	20–30	40–50	60–70	80–90	100–110	120–130
Pressure (dyn/cm^2)	.003–.007	.040–.070	.4–.7	4.0–7.0	40.0–70.0	400–700
Power (W/cm^2)	10^{-14}–10^{-13}	10^{-12}–10^{-11}	10^{-10}–10^{-9}	10^{-8}–10^{-7}	10^{-6}–10^{-5}	10^{-4}–10^{-3}

to move as a whole. Accordingly, accelerations impose loads on the otoliths, which are interpreted by the brain as changes in velocity. As the semicircular canals lie in three-dimensional space, differences in accelerations are reflected in those dimensions. Hence, the device acts like a three-dimensional velocity and accelerometer, constantly sending the brain position, velocity, and acceleration information that the brain resolves with other cues it receives.

Fogel (1963) describes a number of effects that occur when information from the acceleration and position senses conflicts with visual information. Many of these effects and conflicts occur in aviation and driving simulations of situations, and are likely to cause motion sickness, particularly when visual and motion cues are out of phase.*

---

* In simulators, this type of motion sickness is often referred to as simulator sickness.

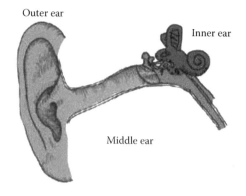

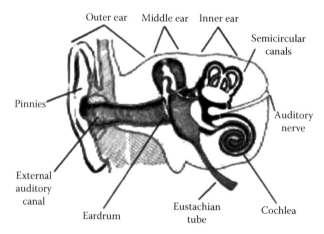

Figure 2.19 Anatomy of the ear.

Other senses

Other senses of the human sensory subsystem include the sensations of pain (free nerve endings), temperature and pressure (skin and underlying tissue), touch and vibration (mainly skin), smell (top of the nasal cavity), taste (tongue and mouth), position and movement (muscle and tendon endings), and kinesthesis (joints of the skeletal subsystem). Table 2.7 provides a summary of the many senses that people have. The last few senses are of particular importance in ergonomics as they allow people to perform numerous activities without constant visual monitoring or correction. However, visual and/or auditory feedback is often an essential element in such tasks, especially when people first start to learn tasks or with tasks that are infrequently performed and never highly learned.

Support subsystems

Support systems play a vital role in task performance, by maintaining the necessary equilibrium conditions, or in other words, a stable environment within which the earlier discussed subsystems can adequately perform their functions. Among the several subsystems that provide such support, two are particularly relevant to ergonomic design: the interrelated respiratory and circulatory subsystems. The associated topic of metabolism is also highly relevant and is discussed in the following.

Chapter two: Human system

Table 2.7 Summary of Human Sensory Processes

Sensation	Organ	Source of Stimulation
Sight	Eye	Electromagnetic energy
Hearing	Ear	Air pressure intensity and frequency change
Rotation	Semicircular canals of inner ear	Fluid pressure change in the inner ear
Taste	Tongue and mouth	Chemical substances in saliva
Smell	Top of the nasal cavity	Vaporized chemical substances
Touch	Skin (mostly)	Surface deformations up to 4000 pulses/s
Tactile (vibration)	Skin (mostly)	Vibration amplitude and frequency of mechanical pressure
Pressure	Skin and tissue below	Deformation
Temperature	Skin and elsewhere	Change in temperature—radiation, convection, and contact
Pain	Skin partially	Pressure
Position and movement	Muscle and tendon ends	Muscle contraction or stretching
Kinesthesis	Joints and ligaments	Unknown

Respiratory operations and mechanics

Breathing results from lung action. The diaphragm muscle pushes the air out of the lungs during the expiratory part of the breathing cycle. With the relaxation of that muscle, the partially collapsed lungs open up again during the inspiratory action. When the atmospheric pressure is greater outside than inside the lung, air moves in to fill up the partial vacuum. Air moves through nasal passages and/or the mouth down the trachea in the throat before entering the lungs. At the bottom of the trachea are the bronchi tubes that branch off to the two lungs. Once the air is inside the lungs, it passes through a series of smaller bronchiales tubes that diffuse the incoming airstreams to a large number of alveoli sacs. These sacs act as semipermeable membranes between different gases and liquid blood. At the alveoli, oxygen in the air is transferred to and exchanged for carbon dioxide in the blood. The rate of transfer depends upon the health of the person and the breathing volumes and rates.

Figure 2.20 shows various volume changes in the lungs and breathing passages. The so-called dead volume consists of the volume of the trachea and bronchi (neglecting nasal and oral cavities). These two semirigid tubes do not expand and contract during the breathing cycle. The minimum collapsible lung volume is called the residual volume. At a basal condition,* small volumetric changes are adequate to maintain the body at a very low level of metabolism. The volumetric change in that condition is known as the tidal volume. The difference between the residual and tidal volumes is the expiratory reserve volume. Also, the difference between the maximum lung volume and the tidal volume is known as the inspiratory reserve volume, which is approximately equal to the expiratory reserve volume. The sum of those two reserve volumes and the tidal volume constitutes the vital capacity, which is a measure of respiratory health. The bottom part of this figure describes inspiration and expiration phases in the breathing cycle, where the smaller ranges correspond to basal conditions and the larger ranges to doing very heavy work.

* A condition where the human body is awake but lying down and fully relaxed.

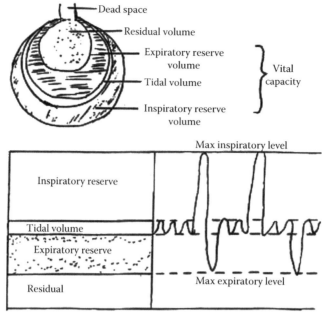

Figure 2.20 Respiratory air volumes and lung capacities.

Circulatory subsystem

Blood is pumped from the heart to the lungs and back for the purpose of exchanging carbon dioxide and oxygen at the alveoli of the lungs. Oxygen-rich blood from the lungs is then pumped from the heart to locations all over the body in order to distribute the oxygen and remove the carbon dioxide waste products. In effect, the heart consists of two parallel pumping subsystems in one organ. On the person's right side is the pulmonary system of circulation. Oxygen-starved blood is pumped by the right two chambers of the heart into the pulmonary artery, through the lungs, and then back to the other side of the heart through the pulmonary vein.

The other two chambers of the heart make up the systemic system that pumps oxygen-rich blood from the heart through arteries throughout the human body. Figure 2.21 illustrates the heart operation schematically and Figure 2.22 shows a semi-realistic description. In both systems, the heart acts as a double pump, each side performing in a similar manner but with substantially higher pressures in the systemic system. On each side of the heart, the blood flows first into a primary chamber called the atrium. The vein leading to the atrium and the atrium itself have elastic properties that smooth out pressure pulses. When the valve at the end of the atrium is stimulated by the nervous system, it opens and some of the blood in the atrium flows into the ventricle. The ventricle constricts, raising the pressure on the blood, at which point the valve at the end of the ventricle opens and blood enters an artery under pressure.

When a person is performing a task with low energy (metabolic) requirements, the heart pumps along at about 50–70 beats/min. This heart rate can be found easily by feeling for the artery in the wrist and counting the pulses for 15–30 s. Any other location where an artery is near the surface of the skin would work equally as well. When the task changes to one with higher energy requirements (i.e., greater metabolism), the heart beats faster in order to increase the blood flow. Other mechanisms in the heart also

Chapter two: Human system

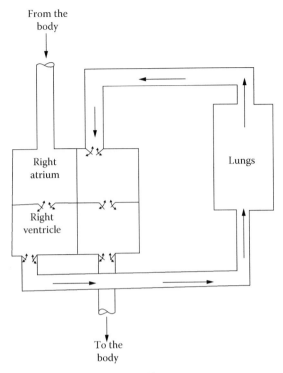

Figure 2.21 Schematic blood flow in the heart and lungs.

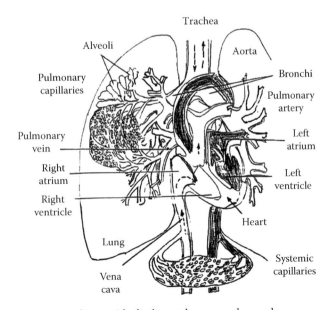

Figure 2.22 Blood flow relationships with the heart, lungs, and muscles.

increase blood flow in the body, so that more oxygen can be delivered to the muscles that are expending greater energy.

Physicians monitor heart operations with an electrical device called an electrocardiogram (ECG or EKG*). This device picks up small electromagnetic signals from the skin near the heart and records them on paper. When the heart is electrically stimulated by the nervous system to start the pumping cycle of the systemic system, the ECG shows what is called a P-wave. At this point, blood pressure begins to rise in the atrium and the valve to the ventricle then opens. Several events occur in the ventricle with the opening of the entry valve. These events are known as Q-, R-, and S-waves, occurring in close temporal proximity to start the heart constriction or systole phase. That phase ends with a T-wave, which opens the valve at the end of the ventricle and causes blood to flow out into the aorta to start its journey around the body. The heart goes into a relaxation or diastole phase before starting the next pumping cycle. These two phases account for the high and low blood pressures or systolic and diastolic pressures, the principal measures used to medically monitor heart functions. During the performance of low-energy tasks, blood spends nearly 0.5 s in the ventricle and about 0.3 s between successive ventricular inflows and outflows. Only about half of the blood in the ventricle is ejected with each stroke of the heart during low-energy tasks.

For high-energy tasks, the heart rate can more than double. Greater fractions of the blood held in the ventricle are also ejected with each cycle into the aorta. At rest, the cardiac output is about 5–7 L/min. Blood flow from the heart can be raised to levels four to five times higher when people perform strenuous activity. In addition to the adjustments of the heart mentioned earlier, the human body diverts blood flows away from bones, organs, and fatty tissues to the activated muscles. Consequently, the flow of blood and oxygen to the affected muscles can be greatly increased when needed.

Metabolism

Metabolism is the sum of those chemical processes that convert food into useful energy to support life and operate muscles. The three main components of food are carbohydrates, fats, and proteins. Carbohydrates and fats are the main sources of energy for physical activities. Proteins are used principally to maintain tissues. Ergonomics is primarily concerned with identifying the metabolic requirements of physical work. This chapter reflects this perspective, but for the sake of completeness, some discussion is included on the chemical processes that take place during metabolism, primarily with respect to the circulatory support system.

Basal metabolism is the rate of energy consumption of the body when it is at rest, expending energy only to maintain body functions. Basal metabolism increases with body weight and averages about 1.28 W/kg for males and 1.16 W/kg for females. At higher activity levels, work is performed, which consumes energy, resulting in greater breathing and heart rates to bring more oxygen to the affected muscles and carry away waste products and excess heat.

Since heat is a by-product of metabolism, metabolic rates can be directly measured by putting people in special chambers and measuring how much heat they generate when they perform a task. Because this method of direct measurement of metabolism is expensive, metabolic rates are often measured indirectly by determining the rate of oxygen consumption. The difference in the oxygen levels of the inspired and expired air, as well as the volume of air breathed per unit of time (at standard temperature), provides an

---

* The symbol EKG for electrocardiogram stems from the German spelling of heart. While ECG is more commonly used today, every so often the old EKG symbols appear.

approximation of the oxygen consumption rate. For greater accuracy, the volume of carbon dioxide production is also often measured.

Some Chemistry of Metabolism

During digestion, proteins are broken into amino acids and nitrogen is removed. When food passes into the digestive system, it is processed into a series of enzymes that are first digested by acids in the stomach and later by the alkali medium in the intestines. This process breaks the carbohydrates down into sugars, mostly glucose but also fructose and galactose, and these sugars are further decomposed into glycogen, the primary energy source in the blood for muscular activity. During muscle contractions an enzyme acts on adenosine triphosphate to break it down into a diphosphate. Glycogen restores the adenosine diphosphate to a triphosphate but creates lactic acid, which is poisonous in sufficient quantities. However, lactic acid turns to water and carbon dioxide in the presence of oxygen.

Many of the carbohydrates and fats are converted to glucose and then to glycogen that is stored in the muscles and liver. Some glucose stays in the bloodstream as a source of energy for the brain and some goes directly into fatty tissue for later conversion to glycogen. The liver traps tiny droplets of fat in the blood and converts those droplets into glycogen.

Figure 2.23 illustrates a metabolic rate curve over time, based on oxygen consumption during a period when a person is resting, then suddenly performs a task, and then rests again. Note that the metabolic rate during rest is near basal and that the supply curve increases but at a decreasing rate until reaching the energy requirement rate of the specific task. The lightly shaded area to the left of the rising supply curve is the energy debt, the amount of energy demanded by the task but not immediately supplied. After the activity is stopped, the supply

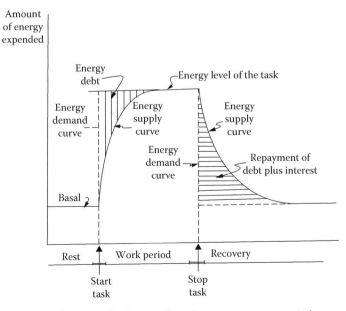

Figure 2.23 Consumption of oxygen during work and a rest recovery period.

curve drops exponentially during the debt repayment plus interest phase. Thus, the generated metabolism must totally cover the energy cost of the task plus any losses along the way.

The general nature of this supply curve over time is similar for most people over most physical tasks. However, individual differences occur that are related to age, gender, nutrition, physical condition, adaptation to atmospheric conditions, and other factors. Metabolic rate curves are also affected by environmental factors, including air temperature, humidity, concentration of oxygen in the air, and the presence or absence of vibrations. Metabolic rate is also influenced by the task energy demand rate, duration of the task, configuration of the body, and the particular muscular bundles used in the task. Tasks that involve a greater energy demand rate usually require longer times for the supply curve to reach the demand level and a longer time for the recovery before commencing a new task. The greater recovery time is why longer fatigue allowances should be given to jobs that entail more physical exertion.

Some typical energy requirements of tasks are shown in Table 2.8. Basal metabolism is added to these active metabolism rates. Both the active and basal metabolism calculations are given in terms of the body weight kilogram. The metabolic rates associated with walking or running are also often of interest. The metabolic rate associated with *walking* at a velocity of V km/h can be estimated using the following equation:

$$\frac{W}{kg} = 2.031 + 0.124V^2 \tag{2.16}$$

The metabolic activity rates associated with *running* can be estimated with a similar equation that also considers the runner's weight M in kilograms using the following equation:

$$\frac{W}{kg} = -\frac{142.1}{M} + 11.05 + 0.04V^2 \tag{2.17}$$

Table 2.8 Active Metabolic Rates Associated with Particular Activities in Watts per Kilograms of Weight

Watts/kg	Activity Corresponding to Metabolic Rate
0.6	Standing relaxed, typing on a computer, playing cards
0.8	Dressing and undressing
1.0	Driving a car or typing rapidly
1.2	Dishwashing or typing rapidly
1.4	Washing floors
1.6	Riding a walking horse, sweeping a floor with a broom
1.7	Playing golf or painting furniture
3.0	Cleaning windows
4.1	Skating
5.0	Riding a trotting horse
5.8	Playing tennis
6.6	Sawing wood with a hand saw
7.9	Playing football
8.5	Fencing

Metabolic rates are sometimes given in kilocalories (kcal) rather than watts. Watts can easily be converted to kilocalories using the relation of 0.86 W = kcal = 0.238 kJ.

Indirect calorimetry

Calorimetry typically involves placing a chemical or thermodynamics process in a chamber and then measuring how many heat units (kilocalories or British thermal units) go into or out of the process over time. However, as noted earlier, putting people inside of a chamber that is large enough to allow them to work, eat, and live in other ways is an expensive undertaking. Consequently, indirect calorimetry is preferred by ergonomists to measure metabolic rates as a means of determining physical difficulty in terms of physical energy used. Indirect calorimetry measurements are accurate and quite precise.

Measuring oxygen consumption rates to estimate metabolic rates is one indirect approach that can be used. A drawback of this approach is that wearing a mask and other equipment can interfere with performing some tasks. In cases where costs are too high or interference from the equipment is suspected, heart rate measurements are often used alone. Since the body must send additional blood to active muscles, the heart rate observed when performing a task provides an indirect measure of the associated metabolic rate. Heart rates are typically measured in one of three ways: counting palpations or beats per minute at arteries of the wrist or neck, listening for the heart beats with a stethoscope, or by using electrodes attached to the skin near the heart and an electronic recorder.

Some of the latter types of indirect measures of calorimetry are inexpensive and not at all intrusive. Modern telemetric devices now available are particularly useful for measuring heart rates over the entire workday in a wide variety of settings. While the use of heart rate measurements alone to infer the physical difficulty of human tasks is accurate and at least moderately precise, it is best used when heart rates are between 100 beats/min and about 150 beats/min. Corresponding energy requirements at the lower and upper heart rates are, respectively, 20 and 45 kJ/min. Below or above this range, there tends to be considerable unexplained variation in energy expenditure for the same pulse rate. But within the aforementioned interval of heart rates, the relationship between the heart beat rate and energy expenditure is very close to linear, with little unexplained variance. Accordingly, energy expenditures on tasks tend to escalate from a base of 20 kJ/min with a 0.52 kJ/min increase for every additional beat per minute. There are, of course, individual person differences, but such differences are of little consequence, if the heart rate measurements are taken in each task condition, and the comparison is within rather than between individuals.

This latter form of measurement is typical of many ergonomic studies that seek to determine which methods or tool designs require the least amount of energy expenditure. When the heart rate is within the interval of 100–150 beats/min, corresponding energy expenditures (i.e., metabolic rates) can be determined* with the following equation:

$$ER = 20 + 0.52[MHR - 100] \tag{2.18}$$

where
    ER is the energy required
    MHR is the measured heart rate above 100 beats/min
    Total result is in kJ/min (1 kJ = 4.2 kcal = 3.62 W)

---

* For the ranges stated, Kroemer et al. (1990) found a relationship between heart rate and energy expenditure (kJ/min) that was very close to linear with a very high coefficient of determination.

Subjective evaluation is another way of indirectly estimating energy expenditure. This procedure is usually less precise than heart rate measurements but quite accurate. Borg (1962, 1982) shows that heart rates corresponding to 100, 120, and 140 beats/min are subjectively rated by many people as medium, heavy, and very-heavy work, respectively.

Final remarks

This chapter provides an account of the general structure of the human body in order to help readers understand how the body works and how its dimensions and capacities affect task performance. This background will help in understanding the topics in the following chapters that address specific problems. Another objective is to show the need for knowledge about the body size and motion so that usable, safe, and efficient tools and products can be designed. Knowledge of human dimension is particularly important for the design of products that people wear, and layouts of assembly stations. Some of these applications are shown later in this book, with special attention paid to principles of motion economy, design of hand tools, and industrial safety. Ogden Nash's poem, "A London Taxi," provides a closing look at ergonomic design requirements from a user's point of view and Ogden's as well:

> The London taxi is a relic
> For which my zeal is evangelic.
> It's designed for people wearing hats,
> And not for racing on Bonneville Flats.
> A man can get out, or a lady in;
> When you sit, your knees don't bump your chin.
> The driver so deep in the past is sunk
> That he'll help you with your bags and trunk;
> Indeed, he is such a fuddy-duddy
> That he calls you Sir instead of Buddy.

Discussion questions and exercises

2.1 When a person has his or her lower arm in a horizontal position and the palm is facing upward, why can he or she exert a greater upward force than a downward force?

2.2 What is the two-thirds rule relative to arm and leg strength and what are its implications in ergonomic design?

2.3 Name three principal factors that affect muscular strength and state how muscular strength changes with each of these factors. Discuss how each of these factors affects designs of products that people use.

2.4 When metabolism is measured by the rate of oxygen consumed, the metabolic rate increases rapidly at the onset of a physical task in an exponential fashion to meet the task requirements. Describe briefly what happens to the metabolic rate immediately after the individual stops performing the task and rests. Explain how this change in metabolic rate affects the design of jobs and job requirements.

2.5 What are ligaments, their elastic properties, and the role they serve in body movements? How does this knowledge relate to how products or hand tools should be designed?

Chapter two: Human system 83

2.6 What is the technical term for total volume of the tidal, inspirational, and expirational reserves of the lungs at the maximum limit of air volume intake in a breath cycle. Why is measurement important medically? If this measure was used as part of personnel selection, what is the implication of high or low readings on the part of an applicant?

2.7 Physical human movements are constrained by the body joints required by it. Describe the axes of movement and rotation capability of a pivotal joint. How do joint constraints affect ergonomic design?

2.8 When a muscle is held in a contracted position continuously for long time periods, there is a reduction of blood flow to a bundle of muscles. What is the physiological explanation of why these muscles fatigue? What does this mean in evaluating work methods?

2.9 The head of the radius connects with the upper ulna. Describe movement about this joint. For each type of movement, is the degree of rotation unlimited, somewhat limited, or highly constrained?

2.10 Give an example of the following types of joints in the human body and associated movement characteristics: spheroid, pivot, hinge, and saddle joints.

2.11 If a partial load is placed on a muscle for a sustained time period, how does the body react in order to reduce fatigue? How does this body functioning affect the way industrial work methods are determined?

2.12 When a muscle contracts to pull up a bone structure from one end, using the other end as a fulcrum, and raising a load in the middle, then what class of lever is being used by that body part? How is it useful for designers to think about the human body as different types of levers?

2.13 How do the approximate dimensions of the human body differ between males and females? Discuss how these differences might impact design.

2.14 If the peak sound pressure is 0.0020 dyn/cm^2, what is the average sound pressure variation about the atmospheric pressure?

2.15 What is the difference between phons and sones in terms of what they measure?

2.16 If two sound sources are each 10 dB and they are of the same frequency, what is the difference in the resulting sound level between the situations where they are totally in phase and totally out of phase with each other?

2.17 How does one use Landolt rings to measure visual acuity?

2.18 What part of the eye regulates the amount of light entering and falling on the photoreceptors? How does it work?

2.19 What kinds of eye movements occur during reading and about what percentage of the total reading time is spent in each or those movements?

2.20 Two ways of measuring eye movements include a method where electrodes on the sides of the eyes disclose horizontal movements and another method where two infrared sensors at the sides of the eyes measure the amount of infrared light received to estimate the amount of horizontal eye rotation. How do these two mechanisms work?

2.21 What does the vector [x, y] denote on the C.I.E. diagram?

2.22 What additional sensory information does the brain receive about the orientation of the human body other than from vision and audition?

2.23 Discuss the advantages and disadvantages of each method of indirect calorimetry.

2.24 Discuss the advantages and disadvantages of designing for a fixed percentage of a population. When is this approach likely to be problematic?

chapter three

Design of work areas, tools, and equipment

About the chapter

This chapter focuses on the space requirements of people performing jobs in factories, commercial facilities, and offices. Several space-related issues related to workplace design are addressed to varying degrees, for example, issues related to the design of work and traffic areas, workstations, work surfaces, work layout, seating, equipment used in workstations, and hand tools. Some of the topics addressed include clearance and reach requirements, accessibility, visibility, accommodation of disabled people, and the need for adjustability. The chapter begins with a brief discussion of the important role of anthropometric data in workplace design, before introducing methods of using anthropometric data, such as drafting templates and computer modeling techniques. The discussion then shifts to selected application areas. As emphasized in the chapter, the topics covered are traditional areas of industrial engineering and ergonomics, but significant changes in the regulatory environment and other factors have resulted in new needs and design constraints.

Introduction

A workplace is the place where one works, be it a workbench, an assembly-line station, or a desk. In all cases, tools, parts, equipment, and other devices must be located in easily accessible locations, if people are to work productively in comfort and good health for protracted time periods. Tools and equipment must also be appropriately sized so that they fit the people using them. It does not take a lot of imagination to realize that seating, desks, hand tools, clothing, and personal protective equipment all pose potential problems if they do not fit the person using them. Workplace layout becomes especially important when workers are repetitively doing the same activity hour after hour. Part of the issue is that workers may need to maintain the same posture or a limited set of postures most of the time they spend working. The layout determines posture, and this determines a whole host of other factors, such as whether the job will be pleasant or unpleasant, fatiguing, or potentially harmful.

People must also have enough space to be able to easily move around work areas while they do their tasks. While the provision of adequate space does not guarantee proper performance, lack of it almost guarantees poor performance. A closely related issue is that traffic areas, such as aisles, passageways, doors, entrances, ramps, and stairs, must be properly designed to allow convenient, unimpeded ingress, egress, and movement around a physical facility. Inadequate or blocked passageways may impede ingress or egress from work areas or allow inadequate clearance for passing vehicles or people. At the very least, passageways must be adequate to allow quick egress under emergency egress conditions.

Unimpeded access of disabled workers to work areas is another concern that has been bumped up in priority by the passage of the Americans with Disabilities Act (ADA). Visibility is another important issue. It is important that people of all sizes be able to enjoy unblocked vision of the things they need to see while doing their day to day tasks or jobs, and especially so under demanding conditions where the failure to see things has serious consequences.

Reflecting such concerns, practitioners have traditionally focused on creating workplaces that fit the workers who use them. The designs that are specified must accommodate the majority of users, typically those within a range from the 5th percentile to the 95th percentiles of the population. When the population is normally distributed in size, the interval corresponding to 95% of the users is defined as $\mu \pm 1.96\sigma$ (where μ is the mean and σ is the standard deviation of a particular measure used to describe the fitted population). Special people in the lower and upper 2.5% of the population need to be addressed as well, but not necessarily with the same design. An old rule of thumb is to design large enough for a large man and small enough for a small woman. However, it clearly does not follow that one size fits all. When it does not, a number of different sizes are needed. With more sizes, users are fit better but the design is usually more costly. Accordingly, some trade-off is needed. Current practice also focuses heavily on providing means of adjustability to accommodate wide variations in human size.

All of these approaches rely heavily on anthropometric measures of various types.* Simply put, anthropometry provides a scientific basis for analyzing and designing elements of the workplace so that they fit people of different sizes, as expanded in the following section.

Applied anthropometry

The field of applied anthropometry relates basic measures of human size, strength, and bodily motion to very helpful design criteria used by designers interested in creating things that fit or otherwise better match the size or other aspect of the human body. Some of these basic measures and design criteria are shown in Table 3.1. The table also gives examples of the many different things that might be designed (referred to as designed elements in the table) to satisfy certain objectives or requirements associated with particular criteria and basic measures of the human body.

As quite apparent from the table, many different design criteria can be suggested. The first criteria, *fit*, is especially appropriate for describing more intimate types of items, which are worn by a person such as clothing (see Box 3.1), ear plugs, gloves, or helmets. As such, fit requires body shape to be considered. Body shape is one of the most difficult items to get good information on. However, data collection systems are becoming available that allow the shape of a person's foot, for example, to be scanned. This information can, in theory, then be sent to a manufacturer who will make a shoe or other clothing tailored to a specific person. This approach is rapidly becoming a reality.

Clearance requirements are often described in terms of the height and width of a rectangular opening. The necessary clearance describes how much of safety margin needs to be provided between the latter measures and particular measures of the body to make it easy to pass through the opening or prevent collisions from occurring. *Accessibility* is a somewhat overlapping criterion that refers to how easy it is to reach something. This

---

* Chapter 2 covers the topics of anthropometry, human body movement, and biomechanics in some detail.

Table 3.1 Examples of Anthropometric Measures and Criteria Mapped to Design Elements and Objectives

Design Criteria	Anthropometric Measure	Designed Element	Design Objective or Requirement
Fit	Body size and shape	Clothing	Degree of looseness, tightness, comfort, etc.
	Hand size and shape	Glove	
	Foot size or shape	Shoe	
	Head size and shape	Helmet	
	Nose size and shape	Glasses	
	Ear size and shape	Ear plug	Ability to reach fingers around the handle
	Finger length	Handle	
Clearance requirements	Body size and shape	Door of car or building	Egress or ingress
		Aisle or passageway	Clearance between users
	Standing height	Overhead objects	Collision avoidance (bumping head)
	Finger size or shape	Button or key on keyboard	
	Foot size or shape	Brake pedal	Inadvertent activation of keys Inadvertent activation of gas pedal
Accessibility	Height	Height of work surface	Object within reach
	Length of arm	Location on work surface	
	Length of finger	Access hole on engine	
	Length of foot	Location of brake pedal	Pedal within reach
Inaccessibility	Diameter of finger or hand	Guard	Prevent finger or hand from entering guard Openings into hazard zone
	Diameter of child's head	Distance between bars on baby crib	
	Length of arm or finger	Separation distance from hazard	Prevent head from entering though opening between bars Prevent hand or finger from reaching the hazard zone
Posture	Standing height	Height of work surface	Reduce bending
	Shoulder rotation	Height of shelf	Reduce extended reaches
	Wrist deviation	Relative height of desk to chair	Eliminate excessive deviation
Visibility	Standing eye height	Location of sign	Vision not blocked
	Sitting eye height	Height of seat in car	Person can see over hood of car
		Height of screen in theater	
		Relative height of auditorium seats	Person can see over head of person sitting in front of them
Mechanical advantage	Grip strength	Handle length of scissors or shears	Person able to exert enough force to cut object
	Finger length	Handle diameter	Person able to grasp handle tightly
Adjustability	Variability of eye height	Car seat	Adjust height of seat for shorter people
	Variability of length of lower leg	Office seat	Adjust height of seat for shorter person or higher work surface

BOX 3.1   BODY TYPE AND CLOTHING DESIGN

Clothing designers need to know something about the shape of the human body, before they can design clothing that fits (see Kemsley, 1957). Length dimensions, such as arm length and chest width, provide only an approximation of the human body's shape. One of the earliest methods of body-type classification was developed by Sheldon who referred to a lean, slender, and fragile person as ectomorphic; a strong, sturdy, and muscular person as mesomorphic; and finally, the stocky, stout, and round person as endomorphic. Sheldon hypothesized that ectomorphs, mesomorphs, and endomorphs were fundamentally different in character. This approach turned out to be a dead end, as a psychological theory, but his distinction is definitely relevant in anthropomorphic modeling. More recently, software developers have devised the categories of slim, athletic, regular, robust, corpulent, and extra corpulent. Most shoppers know that many clothing manufacturers produce different cuts of clothing for these different shapes of people.

---

principle is especially important in deciding where to locate tools, parts, or items in a work area. It also applies to access openings, guides decisions on where to locate controls and input devices. It is especially important, for example, to be sure that emergency stops can be reached at the time they are needed.

Inaccessibility requirements refer to situations where it is important to make sure something is far enough away to make sure it might not be accidentally contacted. This principle is applied extensively in the design of guards and barriers. *Postural* criteria refer to the way particular combinations of design elements and human bodily measures interactively affect objectives such as the need to reduce bending, twisting, or awkward sustained postures. *Visibility* criteria are for the most part concerned with blockages of vision as a function of expected eye positions and the location of obstacles. Both of the latter factors can be controlled to some extent if necessary steps are taken, such as providing lower seating in the front rows of an auditorium. *Mechanical advantage* criteria describe how characteristics of particular design affect the forces people can exert. Longer handles on cutting shears, for example, make it easier for people to exert high cutting forces. *Adjustability* criteria refer to how variability in particular bodily dimensions is accommodated by particular design elements. Examples include adjustable height work tables, chairs, or stretchable gloves.

As mentioned in Chapter 2, anthropometric data is available for a number of target populations. While such data is helpful, it often requires quite a bit of interpretation before it can be applied. For example, simply knowing that most people have a body width measured in centimeters does not tell us how wide an aisle needs to be. Much effort has been spent over the years to make this leap in judgment, resulting in recommended dimensions for a wide variety of settings and tasks. Such sources include handbooks (e.g., Van Cott and Kincaide, 1972; Woodson, 1981) and standards published by governmental groups including the U.S. military (i.e., MILT 1452), NASA (STD-3000A and others), and OSHA (see CFR 1910). The Society of Automotive Engineers (SAE) also publishes detailed recommendations for automobile interiors. It also should be mentioned that databases and scaled drafting templates are available from a variety of commercial institutions and consulting firms. Humanscale 1, 2, 3 is an example. Other commercially

Chapter three: Design of work areas, tools, and equipment

available databases include the Architectural Graphic Standards* and Time-Saver Standards. Architectural Graphic Standards, as implied by the name, are primarily used to specify the dimensions of architectural details such as stairs, ramps, lavatories, and sinks. Some of the factors considered in various sources include the number of people occupying a space, the fire safety requirements of a space, or the building codes of a location.† Required dimensions to provide adequate access to facilities by disabled people are also available in most such sources.

Drafting templates

It might be argued that the most traditional ergonomic method for designing spaces for people to work, play, or live, has been to use scaled drawings and drafting templates. Drafting templates are two-dimensional scaled models of the human body made of firm transparent plastic. The templates normally include pin joints that allow the body parts to be rotated into different positions, and come in different sizes. Figure 3.1 shows examples of articulated drafting templates providing scaled side views of 5th, 50th, and 95th percentile bodies. The first step of the process is to draw the side view of the design in the same scale used in the available templates. The templates then are overlaid onto the locations where people are expected to work. As part of this process, the different parts of the template can be moved in a way that duplicates how the people are expected to move.

Although such modeling helps determine, for instance, whether operators can reach all of the required objects and controls, use of drafting templates does not always provide an accurate answer. These devices show quite accurately whether operators can touch

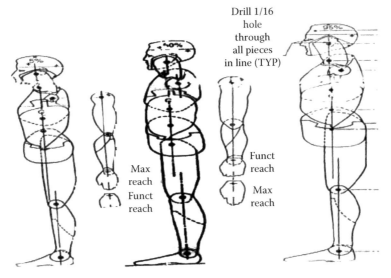

Figure 3.1 Drafting templates of human body shapes.

---

* The Architectural Graphic Standards, Ramsey and Sleeper (1932) are periodically updated. Also see Humanscale or other commercial sources for detailed information on this topic.

† The National Fire Protection Association (NFPA) publishes a collection of standards called the Life Safety Code, which addresses safety of occupants and safe egress. Other standards include model building codes and local codes.

particular objects, but whether operators can grasp objects or activate control devices is not always certain. As a result, these graphic methods of assessing many of the ergonomic features are only considered as a first phase approximation of the design. Later tests are frequently made of actual equipment or full-scale mockups in order to be surer of ergonomic adequacy in the design.*

Design of work areas and stations

A work area may be thought of as the area within a building or other facility within which people do their tasks. A work area is likely to contain one or more workstations along with equipment or tools. In some cases, several people may be using the same work area, but normally will be at different workstations. Some work areas, and buildings in general, will also have traffic areas, such as passageways, aisles, corridors, or stairs, through which people pass as they perform their activities. The following discussion will provide some guidelines regarding the design and layout of work areas and workstations, illustrating some of the many ways anthropometric data can be used. The discussion will begin with a brief overview of some typical requirements for traffic areas, before moving on to several more specific topics, including recommended workplace dimensions, work layout, seating, and computer workstations.

Traffic areas

A lot of information is available from governmental sources such as MILT STD 1452, OSHA (see 29 CFR 1910), or the FAA (HF-STD-001) containing a variety of general requirements for stairs, aisles, ramps, floors, and other traffic areas. Some of the more typical requirements found in such sources and elsewhere (also see Van Cott and Kincaide, 1972) are as follows:

1. Aisles and work areas should not occupy the same floor space to help prevent interference and collisions.
2. Providing necessary pull-outs or turning space in aisles for passage of wheelchairs or material handling devices or equipment.
3. Ensuring that adequate space is marked out and allocated for placing materials in storage or marshalling areas, to eliminate interference with work or passage.
4. Appropriate markings of aisles and passageways.
5. Adequate clearance dimensions for aisles and passageways (Figure 3.2A and B).
6. Appropriate dimensions of stairs (Figure 3.3).
7. Appropriate flooring materials free of protruding objects that might create tripping hazards.
8. Eliminating obstacles that might be collided with.
9. Avoiding blind corners.
10. Making sure there is connected, accessible path, by which disabled users can reach most of the facility.
11. Ensuring that doors do not open into corridors.
12. Avoiding one-way traffic flow in aisles.
13. Special requirements for emergency doors and corridors.

---

* See Pheasant (1991) in the *Evaluation of Human Work* (Wilson & Corlett, editors).

Chapter three: Design of work areas, tools, and equipment   91

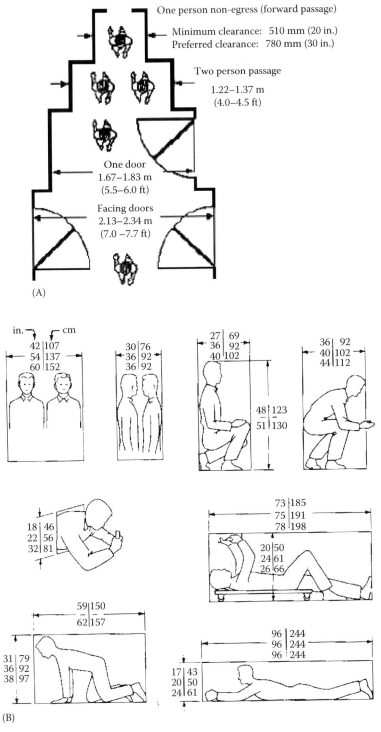

Figure 3.2 Clearance requirements in passageways. (From A: FAA HF-STD-001, 2003; B: Rigby, L.V. et al., Guide to integrated system design for maintainability, USAF, ASD, TR 61-424, 1961.)

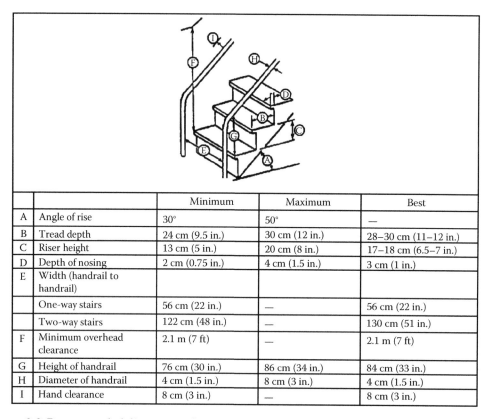

		Minimum	Maximum	Best
A	Angle of rise	30°	50°	—
B	Tread depth	24 cm (9.5 in.)	30 cm (12 in.)	28–30 cm (11–12 in.)
C	Riser height	13 cm (5 in.)	20 cm (8 in.)	17–18 cm (6.5–7 in.)
D	Depth of nosing	2 cm (0.75 in.)	4 cm (1.5 in.)	3 cm (1 in.)
E	Width (handrail to handrail)			
	One-way stairs	56 cm (22 in.)	—	56 cm (22 in.)
	Two-way stairs	122 cm (48 in.)	—	130 cm (51 in.)
F	Minimum overhead clearance	2.1 m (7 ft)	—	2.1 m (7 ft)
G	Height of handrail	76 cm (30 in.)	86 cm (34 in.)	84 cm (33 in.)
H	Diameter of handrail	4 cm (1.5 in.)	8 cm (3 in.)	4 cm (1.5 in.)
I	Hand clearance	8 cm (3 in.)	—	8 cm (3 in.)

Figure 3.3 Recommended dimensions for stairs. (From FAA HF-STD-001, 2003.)

Not all of these requirements are easily related to anthropometric qualities of the human body, but many of them are. For example, it is easy to see that the clearance requirements for passageways, as well as possible work locations (Figure 3.2A and B), are very much related to the width and orientation of the human body. The potential presence of wheel chairs, material handling carts, or other items is also a concern. Figure 3.4 shows some clearance dimensions for a traditional wheelchair. A quick comparison to the clearance dimensions in Figure 3.2A shows that wheelchairs require significantly more space to pass without risking a collision in a passageway than walking people do. Consequently, it may be necessary to provide turn around spaces or pull out areas for wheelchairs in corridors. Stair dimensions such as appropriate tread depth, riser height, handrail height, handrail diameter, and overhead height are also related to body size in important ways.

Workplace dimensions and layout principles

People do many of their tasks while standing or sitting in front of tables, work benches, desks, conveyer belts, or other flat work surfaces. In some cases, as when people write on a piece of paper, the activity is performed on the surface itself. In others, much of the activity is performed immediately above the surface, as when people pick up an object and manipulate it in some way, before setting it down again. If we wanted to describe the spatial element of the particular task setting, or workplace, accurately, several workplace

Chapter three: Design of work areas, tools, and equipment

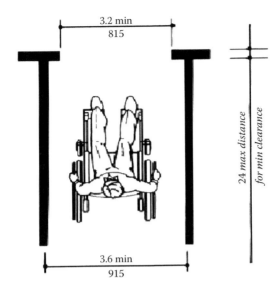

Figure 3.4 Wheelchair clearance dimensions. (From FAA HF-STD-001, 2003.)

dimensions might be recorded, such as (1) the height of the work surface, (2) the height of chairs, and (3) the location and distance of objects from the person doing the task. It turns out that each of these measures is important, as expanded upon later (see Figure 3.5 for related recommendations).

The height of a work surface above the floor has been shown to be an important factor that can influence how easily and efficiently people can do their tasks. Over the years, the consensus opinion was that table tops or other work surfaces should be slightly below the elbow of a person whose upper arms are hanging down naturally. Later, Grandjean showed that the table height should be lower when the physical work was heavier and required less precision. Grandjean's results lead us to a basic ergonomic design principle, stated as follows:

Principle 3.1
For high precision work, tables should be up to 10 cm (around 4 in.) higher than the normal elbow height and for heavy work as much as 20 cm below normal elbow height.

Grandjean also specified bench heights for standing work, as given in Table 3.2. Note that different recommendations are given for men and women. This follows, because men tend to be taller. Another observation is that the elbow height of seated people obviously changes, depending upon how high the chair is. This directly leads to a second design principle.

Principle 3.2
For seated people, the appropriate table height depends on chair height.

This principle also holds the other way around, but chairs that are too tall are obviously a problem. In general, chair heights above 26 in. are discouraged. Another comment is that since chair heights vary, and many can be adjusted, it is hard to guess where the work surface will end up relative to the elbow of seated workers in the real world. This, of

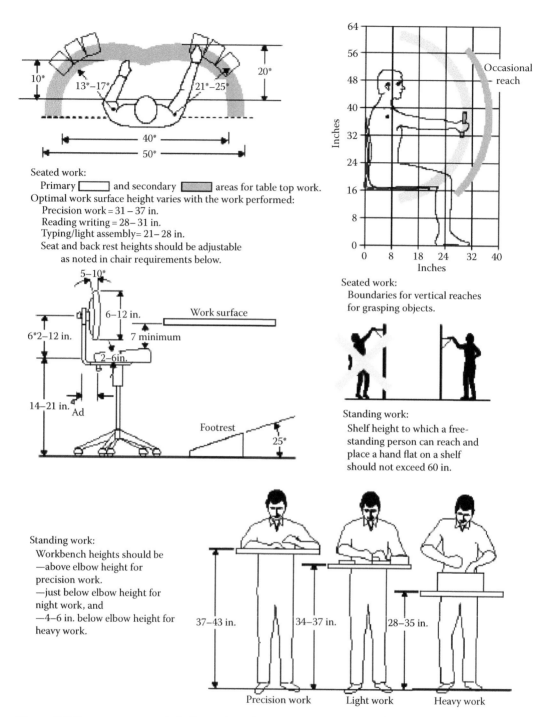

Figure 3.5 Some recommended dimensions for seated and standing tasks. (Courtesy of NIOSH, Washington, DC; adapted in part from Grandjean, E., *Fitting the Task to the Man: An Ergonomic Approach*, 1982.)

Table 3.2 Recommended Bench Height for Three Types of Work by Men and Women

Type of Work	Men Lower Limit	Men Upper Limit	Women Lower Limit	Women Upper Limit
High precision	100 cm–40 in.	110 cm–43 in.	95 cm–37 in.	105 cm–41 in.
Light work	90 cm–35 in.	105 cm–41 in.	85 cm–33 in.	90 cm–35 in.
Heavy work	75 cm–30 in.	90 cm–35 in.	70 cm–28 in.	85 cm–33 in.

course, implies that it will probably be necessary to observe people to see if there are any seriously mismatched chairs and work surfaces.

One other issue is that people will often be reaching out to different locations on the work surface and elsewhere while they do their task. The reach envelop defined by sweeping movements of either arm across the work surface can be described as two arcs in a horizontal plane centered at the shoulder joint of each arm. The length of those arcs depends upon the length of the arm, given a particular form of reach. Some of these same considerations occurred to the people who were first developing ergonomic design. In particular, Squires* proposed a reach envelope that extended from the elbows to the hand knuckles as the hands swung in an arc. Barnes' slightly different approach also included a maximum reach area defined by a completely extended arm.† Das and Sengupta (1995) updated this work, describing the maximum reach being determined by

$$R = \sqrt{K^2 - (E-L)^2} \qquad (3.1)$$

where
  R is the extended arm's radius in centimeters
  K is the arm length in centimeters
  E is the shoulder height in centimeters
  L is the elbow height in centimeters

Coordinates from the body center and the edge of the table, or any other coordinate system, can be computed by applying this formula and shifting the result within the desired coordinate system. The values required for the equation, and for shifting the result to the desired coordinate system, can be obtained for various populations from anthropometric tables such as those provided by NASA.

---

* Squires (1956) proposed what he called a normal work area, where most of the work would be performed. His approach assumes the elbows are close to the body. That is, the arms are not extended. However, the elbows are allowed to move naturally as the forearm pivots.
† Barnes (1963) assumed the elbow position was tighter to the body, which gives a slightly reduced normal area. Barnes also determined a maximum reach area. Since the shoulder joint is about 15 cm from the body's centerline and the length of the arm is another 60 cm, the maximum reach is approximately 75 cm about the body's centerline. This rotation occurs around the shoulder joint so the maximal value is obtained only when reaching straight off to the side, which is given by the length of the arm plus the distance from the person's centerline to the shoulder joint.

Farley (1955) identified the normal reach as given by the sweep of just the forearm, without the whole arm extended (i.e., the upper arm hangs down vertically; the hand is extended forward and sweeps in an arc). Squires (1956) indicates that in normal motion, the elbow does not stay fixed, resulting in a more complex shape for the normal reach. As generalized by Das and Behara (1995), the coordinates for the normal reach can be computed by the following equations:

$$X = S\cos\theta + J\cos\left[\alpha + \left(\frac{\beta}{90°}\right)\theta\right] \quad (3.2)$$

$$Y = S\sin\theta + J\sin\left[\alpha + \left(\frac{\beta}{90°}\right)\theta\right] \quad (3.3)$$

where
  S is the elbow to shoulder projection distance
  J is the distance from the elbow to the end of the thumb
  α is the angle made by the forearm with the horizontal
  $\beta = 180 - \alpha - \arccos(H/J)$, H is one-half the shoulder breadth
  θ is the angle in degrees for the arc swept out by the elbow

The application of these principles results in curves such as those shown in Figure 3.6, which shows these values for the right arm of 5th and 95th percentile U.S. females. Movements of

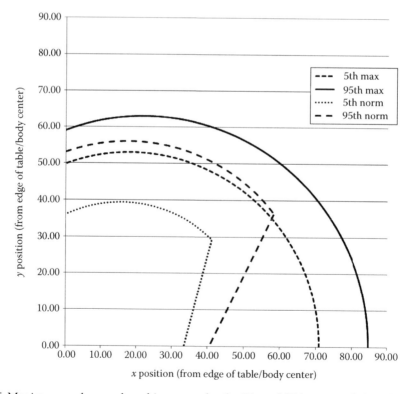

Figure 3.6 Maximum and normal working areas for the 5th and 95th percentile U.S. females.

the left arm are mirror images of these curves.* For a specific application, the ergonomics specialist should apply the anthropometric data for the population of interest.

The fact that it is easier for people to reach to particular locations provides a basis for several other principles on how the objects should be laid out on or about the work surface.† The first principle is as follows:

Principle 3.3

Tools, materials, and controls should be located close to the point of use. More frequently used items should be in the normal or primary area of the reach envelop. Less frequently used items should be placed in the maximum (or secondary) region if space is limited. Items that are very infrequently used may be placed outside the secondary region. However, extreme reaches should not be required.

This principle obviously follows, because longer reaches and motions by the hands take more time and effort. Bringing the items one must grasp closer to the usage point simply reduces the work that must be done by the operator. There is enough work to be done; do not make more. Another issue is that reaches outside the secondary region will require body movements. Locating all materials and tools in the normal and secondary reach areas keeps everything in front of the worker, which reduces the need for twisting motions. Extreme reaches to high shelves, for example, are potentially dangerous, as they may cause bodily strains or falls.

Another related layout principle is given in the following.

Principle 3.4

There should be a definite and fixed place for all tools and materials.

Providing definite and fixed locations for tools and materials reduces the need for visual search and eye–hand coordination. When this principle is satisfied, it usually does not take long before people can select tools and materials with little, if any, visual guidance.

Principle 3.5

Gravity feed bins and containers should be used to deliver materials close to the point of use.

This principle also follows directly from the idea that the parts and other components needed in assemblies should be located near the point of use. Shorter movement distances and time result in less effort expended per part assembled. Somewhat surprisingly, this idea was not so obvious when Mr. Charles Babbage first stated it in the 1830s.

Principle 3.6

Materials and tools should be arranged in the sequence of use.

Materials and tools should be located so that they can be easily and quickly accessed at each stage in the task sequence. Arranging items in the sequence of use reduces movement

---

* In using these curves, designer should remember that there are other sources of difference besides gender (e.g., nationalities). Nationality differences can be quite important when designing products that are used by different populations. In addition, the curves assume that workers obtain a particular position with respect to the work surface. Lastly, some things inside the normal reach area may not be easily accessible if they are too close to the worker—see Wang, Das, and Sengupta (1999) for a discussion of an inner boundary to the reach area.
† The layout principles given here are especially applicable to assembly tasks. Over the years, several standards have been developed for work place layout. One of the more complete standards pertains to visual display terminal workstations. See the most recent human factors standards for computer workstations (American National Standards Institute, 2007).

and transportation times, and also helps operators keep track of where they are in the operational sequence. This arrangement has at least two important side effects. One is that such an arrangement reduces learning time by eliminating the need to memorize the task sequence. The operator needs only to look at the materials and tools and the sequence is obvious. The second side effect is that after the sequence is well learned, people can go through the motions without thinking about each step. If they must consciously come back to overt motor control, for any reason, the sequence of materials and tools guides them to where the work left off.

Design of seating

Human anatomy plays an important role in the design of seating. People experimented with chair designs long before ergonomists focused on this issue. The fact is that people enjoy sitting down, as verified many years ago by Vernon (1924) when he studied a shop of women employees. Jobs in this shop could be performed equally well by a person standing or sitting.* When six different operations were observed, 89% of 76 women chose to sit when they did the job. There is no question that seating is needed; only the question of appropriate seat design remains.

A number of design principles for seating have been recognized over the years (see Figure 3.5 for NIOSH guidelines). Some principles appear to be simply common sense. However, over time, precise value limits and design recommendations have been developed that go well beyond ordinary perceptions. The very important first principle is the following:

Principle 3.7

The design should allow the sitter to change posture.

This principle appears in a number of different places and reflects the fact that posture changes are necessary for blood to flow properly to different parts of the body. Posture changes help muscles relax and prevent people from becoming stiff and sore.

Principle 3.8

The primary region of weight distribution should be in the buttocks, with the greatest distribution under the persons' bony protuberances of the pelvis.

Those bony protuberances are known as the ischial tuberosities. Seats that distribute the load to the thighs will inhibit blood flow in the legs. Most of us can recall instances where our legs have tingled or "gone to sleep" because of being cramped too long. The sensation results when the cramped position inhibits blood flow. Padding and shaped contours are ways of improving weight distribution. For example, consider the stool shown in Figure 3.7. The seat is hard and relatively flat; this may become uncomfortable over longer periods of sitting. Instead, one might mold the seat pan to fit the human buttocks and provide padding for comfort.

Principle 3.9

The height of the seat should be adaptable to specific users if possible.

Some anthropometric measurements relevant to seating are given in Table 3.3, separately for men and women. The values given include averages, and 5th percentile to 95th

---

* Vernon (1924) reported these voluntary sitting and standing statistics. Murrell (1969) shows additional information from Vernon's study.

Chapter three: Design of work areas, tools, and equipment

Figure 3.7 Stool for seating. (Photo courtesy of Tabouret Au Fil Du Bois Sebastian Rivory.)

Table 3.3 Some Statistics for Seated Males and Females

Dimension	Men 5th Percentile In. (cm)	Men Mean In. (cm)	Men 95th Percentile In. (cm)	Women 5th Percentile In. (cm)	Women Mean In. (cm)	Women 95th Percentile In. (cm)
Sitting height	33.5 (85.8)	36.0 (92.2)	38.0 (97.3)	29.5 (75.5)	33.5 (85.8)	34.5 (88.7)
Elbow height (above seat)	7.5 (19.2)	9.5 (24.3)	10.5 (26.9)	8.5 (21.8)	9.0 (23.0)	10.0 (25.6)
Buttocks (to the back of the knee)	17.0 (43.5)	19.0 (48.6)	20.5 (52.5)	16.0 (41.0)	18.0 (46.1)	19.5 (49.9)
Lower leg height	15.5 (39.7)	16.5 (42.2)	17.5 (44.8)	14.5 (37.1)	15.5 (39.7)	16.5 (42.2)
Buttocks width	13.0 (33.3)	14.0 (35.8)	15.5 (39.7)	12.5 (32.0)	14.5 (37.1)	16.0 (41.0)
Shoulder height (from the seat)	21.5 (55.0)	23.0 (58.9)	25.0 (64.0)	19.5 (49.9)	21.0 (53.8)	22.5 (57.6)

percentile ranges. All measures are in inches. Note that a chair with a seat height that adapts from 35 to 44 cm (i.e., 14–17.5 in.) fits about 90% of the women in the population and one that adapts from 38 to 47 cm (i.e., 15–18.5 in.) fits that percentage of men. However, a fixed height chair of around 41 cm (i.e., 16–16.5 in.) fits only about 50% of the women and 75% percent of the men.

Principle 3.10
More seat depth is needed for long-term seating than for short-term seating.

Shallower seats are easier to get into and out of than deeper seats. Deeper seats provide greater support. The recommended depth is about 43.5–51 cm (17–20 in.) for situations where the person will be sitting for a longer time. These estimates are about 5 cm (2 in.) shorter for short-term seating. In fact, some fast-food restaurants cut down on this seat dimension to encourage patrons to eat quickly and move on. The bottom of the seat normally is tilted slightly (3°–5°) from front to back. This helps keep the person's trunk from sliding forward. It also is usually recommended that the back of the seat be tilted backward. Chair backs are often tilted at an angle of 95°–110° from the floor. This backward tilt allows some of the weight to be distributed to the seat back. A related important principle about the seat back is as follows:

Principle 3.11

The chair should provide support in the lumbar region of the sitter's back.

Support of the lumbar region of the spine should be provided starting at a point about 20 cm (8 in.) above the bottom of the seat and extend upward at least 10 cm (4 in.) higher. Because of the curvature of the human body in the horizontal plane at the lumbar region, many people recommend that the portion of the seat back providing lumbar support be curved with an approximately 41 cm (16 in.) radius.

Principle 3.12

Space should be provided under the seat and in front of the person for their feet.

It is important to provide adequate space under the chair and under the bench or table in front of the operator. This allows a person to change the position of their feet. Foot room under the chair is most important for getting out of the chair. To demonstrate, observe yourself as you get out of a chair and then try to execute the same maneuver with both feet in front of the chair. Adequate leg room becomes a special concern when people must remain seated for long time periods. What seems to be more than adequate, when seated for a few minutes, is often far from adequate, after sitting for 8 h, as is often the case in airplanes, automobiles, and in certain jobs where motion is restricted.

Principle 3.13

Flat footrests should be provided for stools and high chairs.

Rounded footrests with a small diameter can cause leg cramps if the sole of the shoe is soft enough and the person applies sufficient force. Stools and high chairs are particularly useful in many industrial situations because they allow the operator to easily switch between sitting and standing. Rump rests, stools, and other support stands are useful when legroom is restricted. Those devices provide good temporary seating.

Principle 3.14

Chairs must be matched to the use related context. Typical multipurpose chairs are adequate for many applications, but do not meet the requirements of offices, comfort seating, or elderly users.

There are many different types of seating. Each has some special design elements that need to be considered. Multipurpose chairs are used in a variety of places including schools, waiting rooms, and sometimes in offices. Typically, these chairs are simple molded plastic units with metal or plastic legs and frames. Chairs in an office should satisfy the general design principles for seating discussed earlier. But office work often involves long hours typing, making data entries, or performing other kinds of sedentary activity. Some

> **BOX 3.2 SOME CURRENT GUIDELINES FOR OFFICE CHAIRS**
>
> 1. Allow forward and reclined backward sitting postures.
> 2. Provide a backrest with an adjustable inclination, which extends 480–520 mm (19–20 in.) vertically above the seat surface. The upper back part should be slightly concaved, with a breadth of 320–360 mm (13–14 in.) and be concaved in the horizontal plane with a radius of about 400–500 mm (16–20 in.).
> 3. Provide a well-formed lumbar pad that will fit between the third vertebra and sacrum, which is about 100–200 mm (4–8 in.) above seat height.
> 4. Provide a seat surface about 400–450 mm (16–18 in.) across and 380–420 mm back to front with a slight hollow in the seat and front edge turned upward 4°–6° to prevent buttocks from sliding forward. Light padding rubber about 20 mm thick and a permeable nonslip cover material.
> 5. Provide foot rests for shorter persons whose legs do not reach the floor.
> 6. Contain heights adjustable from 380 to 540 mm (15–21 in.), a swivel, a rounded front edge of seat surface, a chair base with five arms each containing castors or glides, and controls for adjustments, which are user-friendly.
> 7. Another desirable feature for an office chair is that the seat back to lowers slightly as the chair tilts backward.

of these considerations make office work more or less of a problem. Some current guidelines for office chairs are given in Box 3.2. For the most part, these guidelines are similar to the more general ones discussed earlier. The need for mobility, however, is unique to office settings. A chair with a swivel and castors allows mobility without requiring people to get up out of their chair. This added convenience, however, creates a potential stability problem that must be carefully addressed by the designer.

While office work often appears to be softer than factory work, there are other medical or safety aspects, which cannot be overlooked. Hunting, Laubli, and Grandjean (1981) measured the frequency of seat related complaints associated with various office jobs, as shown in Table 3.4. The number of complaints for the data entry and typing jobs are quite high compared to the traditional job. If adjustable seating helps reduce such complaints, their cost can certainly be justified. Given that these chairs have a life as long as 10 years, or 2500 working days, the added cost for adjustment features really amounts to pennies per day. Some of the design features of standard office chairs include providing a free shoulder region, lumbar support, a small tilt of the seat to prevent sliding forward, adjustable height, back-rest curvature, and clear space below the seat. Nevertheless,

Table 3.4 Seating-Related Complaints for Four Different Office Jobs

	Data Entry (%) ($n=53$)	Conversational (%) ($n=109$)	Typist Full Time (%) ($n=78$)	Traditional Job (%) ($n=54$)
Pressure in shoulders and neck	38	28	35	11
Pain or limited head mobility	30	26	37	10
Pain in forearms	32	15	23	6

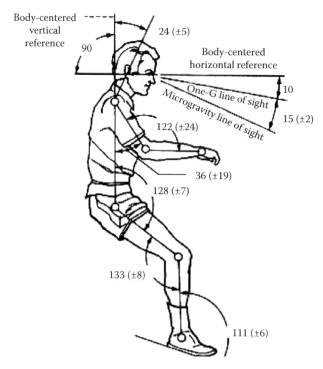

Figure 3.8 Neutral posture. (Image courtesy of NASA.)

despite all these features intended to improve comfort, people often complain about their seating.

Seats such as those used in automobiles must provide comfort for unusually demanding conditions. Automobile seats are generally lower, wider, softer and tilted back further than office chairs. The desirable position can be approximated by examining NASA's neutral body posture, shown in Figure 3.8; a good driving position is given by this position, rotated back approximately 20°–30°. Other forms of comfort seating, such as couches, are similar but provide less lumbar support. Cushioning in an automotive seat is usually designed to appropriately distribute the weight of occupants, and can help them to maintain good posture. Loads on the seat bottom are largest in the middle. Weights against the seat back are largest at the bottom, decreasing with greater height. Weight distributions are slightly different with comfort seating because of the greater seat tilts of about 112° with the floor.

Office design

Most offices contain desks, files, tables, and other types of furniture. In recent years, televisions, computers, monitors, connections to the Internet and intranets, scanners, copy machines, and fax machines have also entered the office. These devices permit people who are separated in time and space to communicate with text copy, drawings, pictures, sketches, voice, and even face to face. These office devices have already had impact on the design of offices and office furniture. A second force has been the changing needs of personnel to communicate within design or project teams. Team members often need to meet informally in small groups of two to five people, creating a need for more informal think for individuals and small groups. Some people feel that future offices will focus less on providing private spaces for individuals, where the size of the office has often reflected status, and more on alternative configurations expected to result in greater productivity, in which the formal organization hierarchy will give way to teamwork and status will become less important than mobility and adaptability. Numerous current examples illustrate these trends. At the corporate headquarters of ALCOA, walls were taken down and old offices were replaced with cubicles, kitchen-like gathering centers, and medium-size meeting rooms. A 1996 article in *Business Week* reports a concept for offices, which provides flexible use through individual-person spaces around the perimeter of a large room and an open central area where groups of various sizes can meet. Some of the individual workspaces are like small cylindrical booths with access openings that the individual can keep open or close for greater privacy. The space inside the booths accommodates a computer, communications equipment, files, and a whiteboard for drawing sketches. Other designs feature movable walls or lockers on wheels. By positioning these partitions around the perimeter of the individual's space, a new office may be formed or an old one reformed.

Chair Design Case Study*

Consider the problem of the elderly getting up out of a chair. Seats that are flat require that the person rise from a position in which the leg is bent 90°, which is difficult for those who do not have adequate leg strength or flexibility. A study was commissioned to examine how to design chairs for the elderly to help with this problem.

The study, which resulted in a chair whose seat pan lifted at the rear, involved the use of questionnaires, interviews, and television recordings of people using two models of these chairs and riser-mechanisms. The piston and associated mechanism was found to be one of the difficult features. Various design issues were identified by observing users while they learned how to use the chairs.

Some pistons were too forceful, particularly in certain phases of getting up or sitting down. The study also found that some people could not lean forward far enough to see the release button on the chair front. An alternative second release button was consequently recommended on the chair side. The researchers determined that users needed instructions on how to properly use this chair, so they recommended that a video tape be made showing entry and exit procedures with and without use of walking devices. They also prescribed improvements in the existing written instructions for how to make adjustments. Finally, they recommended a way marketing people could work effectively with ergonomics personnel, particularly in selling the chairs to people who to some extent are not independently mobile.

Another concept that is gaining popularity in many companies is hotelling, a plan to let office personnel work virtually anywhere they wish as individuals. Some people can work at home, or even from a vehicle. Others will work in different shared offices scattered across the company's operating boundary. Company meeting rooms of varying sizes and facilities are also provided at a variety of locations. Team leaders merely need to reserve these rooms. Hotelling has helped companies reduce office space in expensive locations and therefore reduce office budgets. The affected personnel find reduced travel

---

* See Mizrahi, Isaacs, Barnea, Bentur, and Simkin (1995).

times, improved communications, and reduced project cycle times in product designs and project developments.

Computer workstations

Most offices today include a computer keyboard and a visual display terminal (VDT), and will continue to house such equipment for the immediately foreseeable future. However, the needs of office users are quickly evolving over time. Recently purchased equipment becomes obsolete almost overnight. This makes it necessary for desks and associated furniture to have simple wiring paths for power and communications connectors so they can be flexibly moved, and so that replacement equipment can be easily installed.

Some typical guidelines and recommended dimensions of computer workstations are given in Box 3.3. However, there are differences of opinion concerning the layout of computer workstations. Briefly consider a study by Grandjean, Hunting, and Pidermann (1983) who observed the posture of 94 young people (both men and women averaging 28 years old) at VDT workstations in airlines, banks, and factories. These subjects could adjust their equipment within rather broad limits. Some of the observed results follow:

- Keyboards varied in height above the floor from 620 to 880 mm (24–35 in.).
- VDT screens varied in height above the floor to screen top from 900 to 1280 mm (35–50 in.).
- The VDT screens were back from the front edge of the desk from 400 to 1150 mm (16–45 in.) horizontally.
- Screens were inclined from an angle of 75° forward at top to 110° backward at top.

It is interesting to compare these results to the guidelines in Box 3.3. Many subjects chose settings outside the recommended ranges. This observation implies that what people

---

BOX 3.3 SOME RECOMMENDED DIMENSIONS OF COMPUTER WORKSTATIONS

Most ergonomists suggest guidelines for how computer workstations should be designed. Providing flexibility for the individual user is a particularly important concern. Some typical recommended dimensions are as follows:

- Keyboard height above floor to the home-row should be 700–850 mm (27–34 in.) and 100–260 mm (4–10 in.) behind the front desk edge.
- VDT screens should incline from 88° to 105° from the horizontal and the center of the screen should be 900–1150 mm (35–45 in.) above floor and 500–750 mm (20–30 in.) behind the front desk edge.
- VDTs without adjustable keyboard height and without adjustable height and distance for the screen are not suitable for continuous work.
- Controls for adjusting the equipment locations should be easy to handle, particularly if people work on rotating shifts (i.e., multiple people use the same equipment).
- Distances between the front desk edge and a back wall should not be less than 600 mm (24 in.) at knee level and not less than 800 mm (32 in.) at the foot level.

prefer is not what anthropometric and ergonomic experts suggest. On the other hand, it is possible that people do not know how to adjust their equipment properly. Some users may also find it too difficult or inconvenient. This, again, leads to the conclusion that adjustability features need to be user-friendly. Training may also be a good idea to make sure people know how to adjust their equipment properly.

Another issue is that recently designed desks for computers will often replace the old center desk drawer with a lower tray that provides a place to store and use the keyboard. These trays often do not provide a lot of adjustability in placement of the keyboard. Also, many desks designed for the VDT screen to sit in a fixed location do *not* meet the adjustability guidelines. Some controversy surrounds these guidelines.

Location and positioning of VDT terminals

Ankrum and Nemeth (1995) make the point that no one single fixed posture is likely to be right for a person, no matter how well intentioned the argument. Instead, viewers should position the VDT about 50–100 cm (about 20–39 in.) away from the viewer, at a height where the viewer will be looking downward at the display, 29° below the Frankfurt plane. Figure 3.9 shows the eye–ear and Frankfurt planes of the human head. The eye–ear plane goes through the center of the ear openings to the point where the upper eyelid meets the lower eyelid. It is more difficult to observe the Frankfurt plane without an x-ray because it goes through the center of the ear openings and through the lower edge of the eye socket. The Frankfurt plane is about 10° below the eye–ear plane.

The prescribed VDT location is downward in relationship to the eye–ear and Frankfurt planes. When the display is closer than 50 cm, the downward angle tends to be greater. Lower VDT positions have been shown to improve the viewer's posture and to reduce discomfort of the back. Kumar (1994) demonstrated that electromyography (EMG) activity in the trapezius and sternomastoid muscles was greater with higher VDT locations. Many head angles have been recommended. Hsiao and Keyserling (1991) recommended a forward head tilt of 13° from the vertical and many others have suggested values of 5°–20° below the Frankfurt plane. Lie and Fostervold (1995) recommend a downward angle of 15°–45° for proofreading. Lower VDT positions were observed to allow a wider range of head-neck postures without sacrificing visual comfort.

One other design issue is whether desktops should be flat or slanted downward toward the front edge. Eastman and Kamon (1976) conducted a study showing that people who worked at desks that slanted downward toward them maintained more erect postures

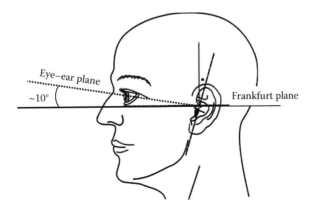

Figure 3.9 Eye–ear and Frankfurt planes.

than those at flat desks, and they exhibited lower (i.e., better) EMG recording in the trapezius muscles (back of the shoulder). People using the sloped-top desk rated it higher than the flat top when they were reading, but the reverse held when they were writing. Of course there are practical problems of pencils rolling off sloping desks too. Researchers such as Mandal (1984) advocate the sloped-top desk and higher chairs around it on the basis of better posture.

Keyboard positioning and layout

In 1926, a German named Klockenberg wrote a book on typewriters in which he described a keyboard, which allowed the two hands to assume more natural orientations. This keyboard separated the left-hand and right-hand keys. Note that this keyboard more closely corresponds to the neutral body position shown in Figure 3.8. Kroemer (1972) performed a study of this keyboard and showed several advantages. A recent "ergonomic" keyboard by Microsoft, shown in Figure 3.10, follows this principle—the keyboard separates the keys for each hand, which allows the hands to hold a more natural, inward slanting angle.

The shape of the keyboard is a different issue than the layout of the keys. The issue of keyboard layout is discussed in more detail in Box 3.4.

Recent years have brought about increased public awareness of some cumulative trauma disorders (CTDs) arising from computer office work. In some offices today, more than half of the computer users wear splints and protective-medical devices on their lower arms. Three factors have been identified as primary causes of CTD occurrence: repetition, posture, and force, with a heavy emphasis on the first factor. As noted earlier, data entry is one of the most highly repetitive tasks, sometimes requiring more than 10,000 keystrokes per hour. This task is consequently one of the highest in terms of CTD occurrence.

Sometimes jobs are such that it is difficult to change the repetition requirement, and in such cases, designers should try to offset the effects of repetition by improving postures and reducing the forces exerted. With regard to posture, most ergonomists recommend neutral (or straight) wrists, deviated neither toward the ulna or the radius bones nor much in extension or flexion (vertical plane deviations).

Figure 3.11 identifies some different hand and wrist postures associated with computer keyboard work. As one would expect, some movement of the wrist is acceptable, but

Figure 3.10 Microsoft ergonomic keyboard.

BOX 3.4 BRIEF HISTORY OF KEYBOARDS

Prior to about 1456, when the printing press was first developed and the Gutenberg Bible was first printed, monks used to hand-letter all books. The typewriter was not a practical solution until about 1873 when Christopher Sholes, Samuel Soule, and Carlos Glidden of Milwaukee, Wisconsin, perfected the first commercially manufactured machines. It is doubtful that anyone today would recognize the Sholes–Soule–Glidden machine as a typewriter. But around the turn of the twentieth century, mechanical typewriters were generally available for use in offices.

Keyboard layout is a classical topic in ergonomics. The traditional QWERTY layout (i.e., QWERTY corresponds to the top row of letters on the keyboard) was originally chosen because this design slowed down typing enough to keep the keys from jamming. The need to slow down typing is obviously no longer an issue for modern computer keyboards. However, the QWERTY layout became the standard that it is still followed today by almost everyone. The Dvorak arrangement has been shown to be faster and more accurate for the average person. In contrast to the QWERTY layout where the middle row is ASDFGHGJKL, the Dvorak layout has a middle row of AOEUIDJTMS. The Dvorak layout assigns the most frequently used key to the stronger fingers and the middle row of the keyboard. Note that all of the vowels are in the middle row of the left hand.

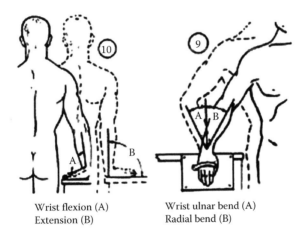

Wrist flexion (A)
Extension (B)

Wrist ulnar bend (A)
Radial bend (B)

Figure 3.11 Some hand–wrist postures.

beyond a certain degree of extension, flexion, or deviation in the radial or ulnar direction, pressures build up in the carpal tunnel so that the medial nerve is threatened. Extension should be limited to about 20° and flexion should be less than 15°.* Wrists deviations do not appear to be bad until they are near the extreme. Recognize, though, that standards and limits change over time. Since a neutral wrist position is difficult to maintain with conventional keyboards, attention has shifted to articulated keyboards or some of the other designs indicated earlier.

* See Gelberman, Sxabo, and Morternson (1984) or Rempel, Horie, and Tal (1994) for these limits.

A number of studies (e.g., see Hedge et al., 1996) have shown that more extension/flexion and ulnar deviation resulted from using a conventional keyboard and conventional tray as opposed to one where the height of the keyboard was adjustable (i.e., on a keyboard tray found today on most computer desks) and that used a wrist rest. In addition, keyboards that are flat require wrist flexion, which has been shown to contribute to carpal tunnel syndrome. As a result, the ergonomic keyboard shown in Figure 3.9 has a built-in wrist rest and is adjustable to have a slight downward tilt.

Design of tools and equipment

Tools and equipment help people perform tasks that they could not otherwise easily perform. Obviously, a person cannot take down a big tree without using an ax, saw, or some other tool. A person can scoop sand with their hands to move it from one place to another, but using a shovel is much more efficient. For some people with disabilities, a tool might perform the function of a damaged limb or other part of their body, allowing them to perform jobs, move about and otherwise behave like nondisabled people. Power tools supply forces of which any person can take advantage. Ergonomists will often focus on a variety of production-related issues related to tools and their use. Another issue is that the energy supplied and applied at the point of operation by power tools and many machines can be hazardous. To help ensure safe and effective operations, ergonomists may be involved in the design of a variety of safety devices and procedures, such as interlocks, guards, fail-safe devices, warnings, and instructions. Ergonomists will also be involved in the design and selection of various types of personal protective equipment for the worker.

Hand tools are used to extend the capabilities of the human hand by increasing grip strength, torque strength, impact strength, fine control, and the ability to handle larger volumes of contents. Hand tools can give the hand new capabilities as well. Design features and principles of design are given in the following.

Hands and handedness: Some initial design principles

Human hands differ in size between people. Women's hands tend to be smaller than men's hands. Designers often try to accommodate a range of hand sizes from the smallest woman to the largest man. Some hand dimensions are shown in Figure 3.12 and corresponding measurements are given in Table 3.5 for the 99th percentile male and the 1st percentile female. Note that the measurements for males are larger than for females. This observation leads us to the first design principle of hand tool design as stated in the following:

Principle 3.15

Fit the tool to the body sizes of the tool users.

Hand strength or grip is another important consideration in the ergonomic design of hand tools. This factor varies with the person's age, use of the dominant versus subdominant hand, how open the hand is when squeezing, the diameter of the handle being gripped, and the gender of the user. Tables 3.6 and 3.7 give grip strengths in kilograms for men and women of various ages for their dominant and subdominant hands.*

---

* These forces were found using two rods, which the person squeezed together. The distance between the rods corresponds to the opening size.

Chapter three: Design of work areas, tools, and equipment 109

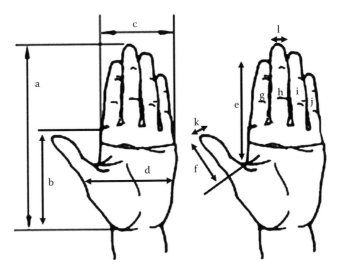

Figure 3.12 Dimensions of the human hand. Corresponding measurements are given in Table 3.5.

Table 3.5 Measurements of the Human Hand for the 1st Percentile Female and the 99th Percentile Male

Hand Measurement	Small Female	Large Male
Hand length (a)	165 cm	215 cm
Palm width (d)	97	119
Metacarpal breadth (c)	74	85
Tip to crotch (e)	43	69
Thumb length (f)	43	69
Second digit length (g)	56	86
Third digit length (h)	66	97
Fourth digit length (I)	61	91
Fifth digit length (j)	46	71
Thumb thickness (k)	16.5	23
Third digit diameter (l)	16.5	22.5

Figure 3.13 plots the data of Table 3.6. The figure shows the overall differences between males and females, and the trend in both groups for the dominant hands to be stronger and for older people to be weaker. Overall, the subdominant hand is about 20%–21% weaker than the dominant hand. This information leads to the following two design principles:

Principle 3.16

For greater hand strength, design the tool so that the dominant hand can use it. Better yet, design for use by either hand.

One example of a design strictly for use by the right hand is the old-style ice cream scoop with a thumb-actuated scoop-clearing mechanism. Left-handed people had difficulties with this utensil because the thumb-actuated mechanism was on the wrong side, forcing them to use their right hand, which of course was weaker. Consequently,

Table 3.6 Grip Strengths in Kilograms for Men and Women for a Variety of Handle Openings

Handle Opening in Millimeters	25	50	75	100
Male				
50th percentile	20	32	50	44
5th percentile	10	18	30	26
Female				
50th percentile	8	14	25	14
5th percentile	4	6	12	7

Source: Greenburg and Chaffin, 1977.

Table 3.7 Grip Strength in Kilograms for Males and Females Using Dominant and Subdominant Hands in Six Different Age Groups

Age (Years)	30	40	50	60	70	80
Male						
Dominant hand	44.6	42.5	42.0	39.0	36.3	32.5
Subdominant hand	35.3	35.3	33.7	32.6	28.0	25.0
Difference (%)	−21	−17	−20	−16	−23	−25
Female						
Dominant hand	29.0	28.0	25.0	23.0	19.0	16.3
Subdominant hand	24.0	22.7	21.8	17.3	14.7	14.0
Difference (%)	−17	−19	−13	−25	−23	−14

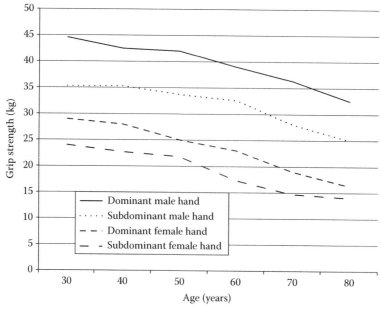

Figure 3.13 Dominant and subdominant hand grip strength of men and women as a function of their age.

Chapter three: Design of work areas, tools, and equipment 111

Figure 3.14 Ice cream scoop designed for use by either hand. (Photo courtesy of Evan-Amos.)

left-handed users had a hard time filling and clearing the scoop when the ice cream was very cold. An experiment revealed that alternative designs that enabled either hand to work the device equally well usually resulted in at least a 5% improvement for left-handed people, compared to their performance using the right-handed version. Figure 3.14 illustrates an ice cream scoop design that allows use by either hand.

Note that there is often a trade-off between a tool that is designed to be used by either hand and one that is most efficient for one or the other hand. So, many possibilities exist here, and a simple principle does not seem to handle all of the problems. Perhaps the best advice is to consider both approaches. It is also likely that the solution will depend upon the usage rate. The latter point leads to the next principle.

Principle 3.17
If hand strength is quite important, design the hand tool for the individual user.

Table 3.7 shows grip strengths for males and females as a function of handle opening sizes (in mm). This data shows that the size of the opening has a strong effect on how much gripping force a person can exert. Note that, in Table 3.7, the hand strengths differ greatly with gender. For example, the fifth percentile female exertion value is three times higher for an opening of 75 mm, than for an opening of 25 mm. Also, for each opening size the corresponding values for males are two to three times higher than those for females. This means that the 2/3 value discussed in Chapter 2 overestimates the grip strength of females. Because there is a lot of variation between populations of people, a designer must not assume a single tool serves all users. If multiple tools are needed for each particular population, the next problem is creating them.

The best type of grip for a tool depends upon its function. Power grips are used to hold hammers, mallets, saws, and other tools involving large force exertion. The hand is wrapped around the handle of the tool in all power grips, but the force may be applied in different directions relative to the forearm.[*] Regardless of the direction the force is applied in, a large portion of the hand surface is placed in contact with the tool handle. This allows forces and impacts to be distributed over the entire hand. Example use of a power grip[†] is illustrated in Figure 3.15.

---

[*] With saws, the force is parallel to the forearm. In hammering, the forces are at an angle with the forearm, and with wrenches, torque is delivered about the forearm.
[†] Other examples of power grips and their advantages are as follows: Pliers for greater grip strength, wrenches and screw drivers for increased torque strength, hammers and mallets for greater impact, and hand scoops for an increased ability to handle larger volumes of contents.

Figure 3.15 Power grip being used on a sledge hammer.

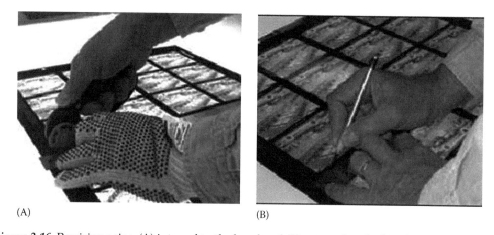

(A) (B)

Figure 3.16 Precision grips: (A) internal to the hand and (B) external to the hand.

Another form of grip is the precision grip. There are two recognized versions of this grip, which are shown in Figure 3.16. Internal precision grips involve a pinch grip with the tool handle held within the hand. A finger placed on the front of the tool provides guidance and tremor control. This form of grip is employed in the use of table knives, razor blades, and metal files. External precision grips also involve a pinch grip. The thumb and two fingers are placed near the front of the tool, which rests against the side of the hand near the thumb crotch. Both of these grips are natural. The best grip depends on the intended function of the tool.

Principle 3.18

When power is needed in a hand tool, design it for a power grip. If precision is needed, design it for a precision grip.

Chapter three: Design of work areas, tools, and equipment

Other desired properties of grip design

It is critical that the tool fit the individual user. As discussed earlier, people come in a wide variety of sizes and shapes and one size rarely fits all. For example, although the optimum diameter of a screwdriver handle for one person is often not much smaller or much larger than for another, it still differs, and this difference can easily outweigh the cost of the tool. A size that is optimal in one application may be bad elsewhere, because the orientation of the hands may change depending on how it is used.

When determining the proper size of a hand tool, it is important to test different sizes of people and tools for all of the foreseeable uses. Performance data and subjective appraisals should both be obtained. These appraisals must be obtained after the tool has been used for sometime in representative tasks. The styling of a hand tool can bias people and this can impact their rating before they actually find out how good it is.

When using certain hand tools, such as knives, screwdrivers, and chisels, people apply a forward force against an object. In those cases, the hand must be protected from slipping forward over the tool. This protection can be obtained by enlarging the grip cross section in front of the hand position. When users pull the tool toward them, similar features can be provided to keep the tool from slipping out of the hand if the gripping force is relaxed. That is, the grip behind the hand can be made larger. Figure 3.17 shows two screwdrivers. The top screwdriver contains a ratchet; so pushing the forward on the handle causes the screwdriver bit or a drill bit to rotate. This particular tool could benefit from enlarged sections on both ends of the handle. The second screwdriver is a common screwdriver. Most readers have probably experienced slips of this tool within the hand. A handle design that will prevent many such hand slips would be to increase the diameter of the handle and add a guard to the front of the handle.

Principle 3.19

Do not let the tool vibrations travel to the person.

A hand tool often carries vibrations to the hand. Grip materials are available that insulate the hand from vibrations. Either wood or medium hard rubber is recommended. Besides

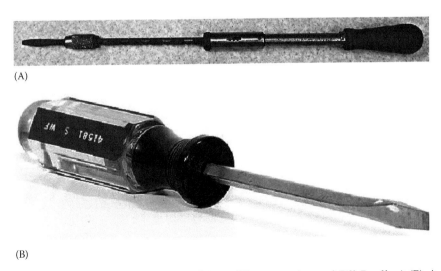

Figure 3.17 (A) A yankee ratcheting screwdriver. (Photo courtesy of Bill Bradley.) (B) A regular screwdriver.

insulating the hand from vibration, these materials are compressible, which helps distribute loads from the tool evenly over the hands. It is also important that the grip material be smooth and conform closely to the hand, because ridges tend to transmit vibrations to localized regions of the hand. Vibration damping gloves can be used effectively in some cases.*

Such gloves place vibration-damping materials over the palm of the hand and part or all of the fingers. Vibration damping materials normally do not cover the full finger region because doing so would reduce finger agility. If finger agility is needed for an associated task, the designer faces a trade-off between reduced human performance and greater protection.

Principle 3.20

In addition to vibration insulation, grips should also insulate from heat and electrical energy.

Here again, both wood and hard rubber are effective insulating materials. Since designers never know the exact situations under which hand tools are used, they should provide the extra protection as a precaution. Think of how many different ways you have used screwdrivers other than for turning screws.

Principle 3.21

Grip designs should create a sense of tool orientation from the feel of the grip.

This feature is particularly important when high productivity is needed. When users can sense tool orientation from the grip's feel, they do not need to use vision for this purpose. If vision is needed elsewhere and also needed to detect tool orientation, either the tool use must wait or the user must take a risk. The problems can be avoided by a proper grip design. For example, a claw hammer handle can be made in an asymmetrical oval cross section. When the narrower oval section is in the direction of the hammerhead, a correct orientation can be easily felt. Also, the broader oval section in the middle of the hand spreads the shock of the hammer blows more evenly across the hand.

Principle 3.22

Good hand tool design provides adequate mechanical advantage in applying force.

An important corollary is that forces should be exerted with the body part in its natural orientation. Numerous cases of CTD[†] have been reported when people are forced to work with their bodies unnaturally positioned. One of the more famous cases involved the hand tool used by shoemakers of yesteryear (i.e., cobblers). When they installed the shoe welt, one hand strongly deviated toward the ulna. After years of work, their wrists had been bent so much for so long that they would not straighten. Many other tools force the hand into unnatural positions with the wrist bent one way or the other. Pliers are a case in point. If the pinch point is directly in front of the operator, the wrist must be bent in order to use the pliers. However, if the handle of the pliers is bent, such as shown on the right-hand side of Figure 3.18, then the pliers can be used with the wrist in its natural orientation. Keeping the wrist in a natural (also called neutral) orientation reduces the possibility of CTD. Hence, the design principle is as follows:

Principle 3.23

Bend the tool's handle, not the wrist.

---

* Goel and Rim (1987) show that gloves can reduce the transmission of vibrations emitted by certain tools.
† See Chapter 2 for information about several specific types of CTD.

Chapter three: Design of work areas, tools, and equipment 115

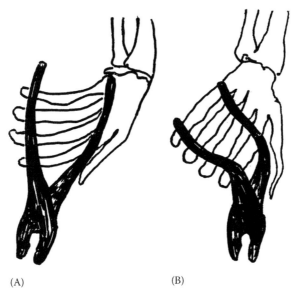

(A) (B)

Figure 3.18 An x-ray of (A) regular and (B) bent pliers in the hand. (Adapted from Damon, F.A., *Western Electr. Eng.*, 9(4), 11, 1965; Tichauer, E.R., *J. Occup. Med.*, 8(2), 63, 1966.)

Actually, the aforementioned principle is a little shortsighted, because this notion applies to all human joints, not just wrists. However, the principle is most often violated at the wrist.

Another advantage to bending the handles of hand tools is that doing so can improve eye–hand coordination. The Tichauer–Damon (see Box 3.5) pliers shown in Figure 3.19 illustrate this and other effects. Here a guard was added that allows the hand to apply more forward thrust, and protects against solder spatter. Heat and electrical insulating material were included to provide additional protection. An important advantage of this design is that bending the tool provides a clear line of sight to the user, which facilitates precise manipulation of the needle-nose pliers. Bending the handle also made it easier for a user to pull backward with greater force if needed. Another example showing how bending the tool can improve eye–hand coordination is Tichauer–Damon's design for a soldering iron shown in Figure 3.20.

BOX 3.5 TICHAUER AND DAMON

Edwin Tichauer left Europe as a young man and went to Australia where he worked for a number of years. He later immigrated to the United States where he joined the industrial engineering faculty at Texas Technical University, Lubbock, Texas. It was there that he and Fred Damon met and began their collaboration on the design of a variety of hand tools. Fred Damon worked at the Western Electric Company. This was the manufacturing unit for the Bell System in the United States where telephone instruments, cabling, and switching equipment were made. Most of the hand tool testing was done at the Kansas City, Mosby, plant in the 1960s. Later Tichauer worked at New York City University and Hospital.

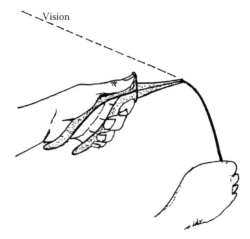

Figure 3.19 Damon–Tichauer pliers allowing a clearer line of sight.

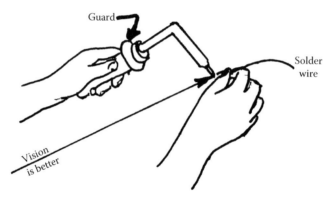

Figure 3.20 Soldering iron by Damon and Tichauer.

Bending the tool may also provide more power for certain orientations of the human body. This is particularly important in construction tasks. For example, consider the different orientations of the body when painting floors, ceilings, or walls. Hammering nails on roofs of various slopes requires similar variations in posture. In addressing the latter problem, Bennett experimented with different hammer handle designs. Subjects preferred handles, which brought the hammerhead, *forward* about 15°–17° in the hammering arc.

The bent or curved handle is only one development in the long and interesting history of toothbrushes (see Box 3.6). Prior to its development, people used to take a piece of cloth or a sponge, wet it, dip it in salt or baking soda, and brush their teeth using a finger to reach the surfaces of each tooth.

There is, however, a down side that ergonomic designers need to be aware of and investigate in their own circumstances. It has recently been shown* that bending some tools creates a loss of tool symmetry. Consequently, the bent-handle tool shows improvements in some orientations, but actually reduces effectiveness (e.g., range of motion) when used in other orientations. One specific finding is that bent handle tools

---

* Dempsey and Leamon (1995) describe cases where bent tool handles are ineffective.

BOX 3.6 INVENTION OF THE TOOTHBRUSH

William Addis was convicted of inciting a riot in London and was sent to London's Newgate Prison in 1790. While in prison he carved a handle out of a piece of bone and drilled holes in the opposite part of the handle where he stuck in hog-hair scrub bristles. This design was refined over time in many ways. When Addis got out of Newgate he sold these toothbrushes to wealthy Britons.

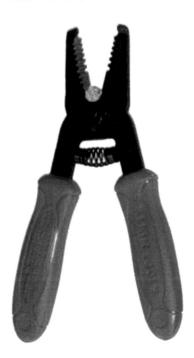

Figure 3.21 Pliers with a retracting spring.

are worse when force needs to be applied vertically. Accordingly, bent-handle tools tend to be task specific.

Other features of hand tool design

Many hand tools require the hand to open and close during their use. Pliers, scissors, and some ice cream scoops are all examples. As noted earlier, the maximum grip strength changes greatly with the size of the hand opening. Typically, the maximum strengths occur when the opening is between 65 and 85 mm (2.5–3.3 in.). These tests also show that the hand exerts more force in closing than in opening. Therefore, it usually helps to add a light opening spring. The latter feature saves people the time and effort of opening those tools. Figure 3.21 shows pliers with a retracting (opening) spring. The weight of the tool is another issue. Powered hand tools are especially likely to be heavy. Ergonomic guidelines often suggest a 5 lb limit on hand tool weight.*

---

* For example, Aghazadeh and Waly (1998) suggest this limit in their *Design and Selection Guide for Hand Held Tools*.

The way the controls are activated is another important aspect of hand tool design. For example, many soldering guns for home use are designed similarly to a pistol. The trigger activates the power to heat the tip of the solder gun. For the occasional home user, this is no problem. However, for people in industry who spend long hours soldering, repeatedly pulling the pistol trigger can cause a form of CTD called trigger finger. This ailment occurs when the sheaves carrying the tendons to the trigger-activating finger become inflamed from extensive use. The U.S. Army discovered this disorder back in World War I when soldier recruits spent long hours on the rifle range. The Army called it, "trigger finger." In industry, this problem can be avoided by activating the tool automatically when it is lifted from a holding cradle or by using a trigger strip, a long strip in the grip handle that activates the tool when it is held.

The assembly of miniature products, such as electronic boards that go inside small devices, presents one of the more difficult situations for hand tools. Because the fingers are poorly equipped to pick up tiny parts and insert them, tweezers of various kinds are needed to gain control of the parts and align them for insertion into the boards. Tweezer handles must be designed so that very precise control can be exercised. A slight finger relaxation should not release the parts being held and moved by the person. Otherwise, dropping of parts would be a problem. These applications often require the use of magnifying glasses, which further complicates eye–hand coordination during part handling. The design of the grasping part of the tweezers is another important issue. The shape of the object being grasped is an important consideration. Mechanical, pneumatic, and magnetic methods of grasping are used.

Some hand tools are inherently dangerous. Obvious examples include tools with hot parts or sharp edges. There is a natural tendency to try to reduce the danger when the tool is not being used. For example, a knife can be put into a sheath or the blade can be made retractable into the handle. Many other strategies can be followed that reduce the hazard to varying degree. Although following these strategies can be helpful, it must be realized that partial solutions sometimes make things worse. Things that appear dangerous are usually treated with respect, but things that appear safe can encourage unsafe acts. If the protection feature is off, or is not guaranteed to be fail-safe or fool-safe, injuries are likely to result, as is product liability. If it is not feasible to design out the danger, it is usually better to make it obvious and provide warnings than to create a semi-safe product. Sometimes the warnings invite attention to the danger.

Techniques for determining hand tool adequacy

Many techniques can be used during the design of hand tools. Cinegraphic, television, or related techniques such as cyclegraphs or chronocyclegraphs are all useful when studying how tools are used. These techniques show tool and body motions during use of the tool. This reveals whether the tool is used properly and whether or not the human body maintains natural positions and orientations during use of the tool. Body positions can be used to predict potential long-term problems. Remember that CTD develops slowly and it is related to unnatural body positions.

EMG is sometimes used to measure the amount of muscular activity of different muscle groups during tool use. Better hand tool designs reduce the amount of muscular activity and distribute them to the muscle bundles in proportion to their ability to render force. Infrared photography and scanning devices are used to show skin surface temperatures during tool use. Blood flows to portions of the hand in proportion to the activity level.

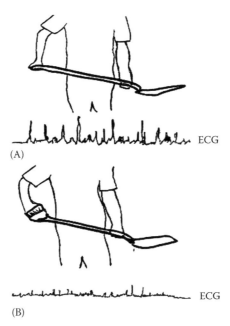

Figure 3.22 ECG heart records during use of (A) regular and (B) revised shovel designs.

Since blood is maintained at the core body temperature, the areas of increased flow in the periphery of the body become warmer. Often high temperature lines or areas appear where the hands come into contract with ridges or localized areas in the tools' handles or grips. Improved designs will minimize such effects by distributing the forces more uniformly over greater portions of the hands.

Tool designs can also be evaluated by measuring heart rate or metabolic rate.* These two measurements tend to be highly correlated, so both are not usually required. Either measure indicates the energy level exerted by the body during tool use. A better-designed hand tool will require less human effort. Some contemporary heart rate measuring devices are relatively inexpensive and accurate. Figure 3.22 plots the heart ECG recordings observed for workers shoveling with two different types of shovels. The bottom design is better as it involves less vigorous heart activity. However, energy expenditure measurement techniques are not very discriminating unless there are considerable differences between different hand tools.

Power tools

A power version is normally available for almost every type of hand tool, as can be verified by a quick look at the catalogue from any tool supplier. Power tools are normally powered by an electric motor or small gasoline engine. Pneumatic or hydraulic tools are also often used in industry. The primary advantage of using a power tool is the significant time

---

* Chapter 2 describes the direct and indirect measurement of metabolic rates. In many industrial cases, heart rate measurements are used as a rough measurement of metabolic rates. Heart rates are highly correlated with metabolic rates within an individual and heart rate measurements (ECG or EKG) are easily taken and inexpensive to acquire.

saving they provide. (An example is provided in Box 3.7.) Power tools also result in more consistent performance and can reduce fatigue and injuries, by eliminating or reducing the forces people must exert to perform a task. For example, power wrenches can be set to torque nuts at preset values and can quickly exert high torques to loosen rusty, corroded connections.

Success Story from Hendrick (1996)

Several years ago, Ian Chong from Ergonomics Inc. developed a poultry deboning knife with a pistol grip. This knife was designed to reduce tendonitis, tenosynovitis, and the incidence of carpel tunnel syndrome. When it was installed in a meat cutting factory, CTD incidences went down, resulting in an immediate savings of $100,000 per year on worker compensation premiums. Workers reported less pain and greater job satisfaction. Over the next 5 years, production line speeds increased between 2% and 6% and an additional $500,000 was saved on worker compensation premiums.

Monitoring Tool Use

Hand tools tend to be used in a wide variety of orientations. Their frequency of use also varies extensively from one-time, single use to very heavy usage over long time periods. For these reasons it is important for the ergonomics personnel to create a comprehensive program for monitoring tool use within their facilities. Tools, like many other things, are used in surprising ways that are not always apparent before evaluation.

Misuse of the tool is very common right after it is first introduced. Initial performance is usually worse even if there is good reason to believe that performance will ultimately be much improved by use of the new design. The introduction of a new hand tool should be treated similarly to introducing a new method. Prior to introducing new hand tools, ergonomists should examine the effect of different orientations and different tasks on the tool's effectiveness. Unfortunately, some hand tool designs are better for one type of task and worse for others. This forces a choice between one or more tools. If only one tool will be used, the tool best for more frequent or longer duration tasks is typically chosen. However, the choice probably should be an economic one.

One disadvantage of power tools is that they tend to be heavier and take up more space than hand tools. Electric cords and pneumatic or hydraulic lines are another problem. Part of the issue is that they can get tangled and present a tripping hazard. They also can be damaged by the tool itself, in potential electrocution or injury due to high pressure hydraulic fluids. This lends to the first principle of power tool design.

Principle 3.24

Support or counterbalance the weight of a power tool by including a fixture or mounting apparatus.

A quick visit to almost any manufacturing facility will reveal a variety of clever fixtures used to support the tool in a convenient location for the task to be performed. For example, power screwdrivers or wrenches might be suspended immediately above the

> **BOX 3.7   POTENTIAL TIME SAVINGS ASSOCIATED
> WITH USE OF POWER TOOLS**
>
> In manufacturing shops today it is almost rare to see a manually driven hand tool when power tools are available because labor cost savings will often pay off the cost of a power tool in an extremely short time. A typical worker will work an 8h shift 5 days per week for about 50 weeks per year. This adds up to about 2000 paid hours per year. When the allowances total 15%, work is being effectively performed 85% of the time, giving 1700 productive hours per year. If that operator is paid at a stated rate of \$K/h, the actual cost per hour over the year is \$K/h × 2000h/1700h = \$1.17647K Without including the cost of other benefits, the cost per actual hours worked is therefore about 18% higher than the hourly rate. A saving in performance time of a human operator is on the actual time worked, not the total time.
>
> If a worker assembles N units of a product per year uniformly over the year, and each unit takes T actual time units (in hours), the workers contribution to the cost of that item is simply $N \times T \times K'$ dollars where K' is the cost per worked hour. Fractional time savings through better methods or power tools are F times that contribution per year, where F is the fraction of actual time saved. The assumption made here is that this production is spread uniformly over the year.
>
> For example, consider a particular item that requires 2min per piece (0.0333h per piece). When 100,000 pieces are made per year, this product requires 3,333 actual assembly hours per year. At a rate of \$9/h, without including the cost of other benefits, the cost per year for this part is \$K35,294.12 or \$K0.353 per piece. If a \$200 power-tool could save 10% of the assembly time per part, the cost savings per year is
>
> $$0.1 \times 3333.33 \times (\$9/0.85) = \$3529.38/\text{year}$$
>
> The latter amount is likely to be much higher than the cost of a power tool.

work area on a retractable chain or cord. The appropriate design of such a fixture will minimize the need for the operator to reach and move the tool during the operation. Such fixtures will also minimize the effort needed to move the tool, by supporting its weight. The analysis and design of fixtures for power tools is one area that ergonomic designers will pin focus upon. Another issue is that power tools will often pose special safety concerns that must be addressed in various ways.

Power transmission hazards

The use of an external power supply makes power tools and machinery potentially more dangerous than hand tools. Hazards associated with the external power supply are of special concern to designers. Electric power carries the added danger of electric shock, and gasoline engines create fire hazards. Accordingly, most new designs of small electric-power tools have either grounded motors or double insulation, providing considerable protection so long as those tools are used in relatively dry environments. With gasoline powered tools, ergonomists must ensure that the ignition spark and exhaust system are well shielded from the human operator. For all power tools, interlocks should prevent the operator from contacting these moving parts. Many small engine repair shops wisely require repair-personnel to disconnect the spark wire before working on the device to prevent accidental starting.

Torque-controlled tools such as pneumatic wrenches often automatically shut off when a preset amount of torque is exerted. This assures precise tightening, but when the tool shuts off, inertial forces are applied to the tool operator's hand. These forces result in wrist extension or flexion, depending upon the handedness of the operators. Over time this repetitive loading can result in CTD. Including a torque arm that isolates the human operator's wrist from the inertial forces can reduce this loading. The torque arm keeps the tool handle from rotating, while allowing the operator to move the tool to and from the point of operation. In a study of torque producing tools, the researchers attached devices to the operator's wrist to directly measure the extension or flexion movement and force (Neelam, 1994). Use of a torque arm reduced wrist rotations very substantially. While use of a torque arm proved to be effective, there was considerable resistance on the part of the worker. A training program was implemented to overcome that resistance. Interestingly, use of an automatic tool retractor cut performance time approximately in half.

Point of operation hazards

At the point of operation, machinery and tools often pose a potential for abrading, cutting, shearing, or crushing injuries when people contact sharp or moving parts. Tools and machinery may also propel flying objects, resulting in eye or other injuries, and give off dusts and potentially hazardous materials. Other point-of-operation hazards include high voltages, nonionizing radiation, and hot or cold surfaces. Point-of-operation hazards account for a small, yet significant, percentage of injury claims in industry, reflecting the overall shift in the U.S. economy away from manual machining tasks. Much of the current focus of safety engineering and safety standards is placed on eliminating or reducing such hazards.

Many tool and equipment hazards can be eliminated by modifying their contact surfaces. For example, sharp edges, points, rough surface finishes, pinch points, and temperature extremes can be eliminated in some cases. Tools and equipment should also be designed to eliminate or minimize the effects of errors. Guards, interlocks, presence-sensing, and lockouts are common means of reducing the effects of operator errors or equipment malfunctions. Many types of protective equipment are also available. A complicating issue is that guards are often removed, especially during setup procedures, or when maintenance and repair activity is performed. While removal of guards may be necessary in these instances, operators may also remove guards and safety devices during normal operations. Some operators complain that the guards obstruct vision, or otherwise interfere with performance. Designers must be especially aware of this issue, and consider means of guarding that interfere as little as possible with use of the tool or machinery. Auditing or otherwise monitoring use of the equipment is often necessary to ensure operators are using the equipment properly. Some principles of tool and equipment suggested by the earlier discussion are given as follows:

Principle 3.25

Install a guard or barrier that makes it difficult for operators to contact the point of operation of the tool or machine.

OSHA and other regulatory agencies in the United States require machine guarding for power tools and other machinery. Many different forms of guarding are used in industry, especially for machining operations. In the latter situation, machine tool operators often put raw stock into machine tools and remove the finished pieces. During these loading and unloading times the operator is in danger if the machine is activated unintentionally. Guards are intended to keep the operators' hands and other body parts out of the machine tool during operations. One way to guard a machine is to install a metal or plastic barrier that prevents the operator from reaching into the machine tool. When the operator needs to put new stock into the machine, he or she must first remove the barrier and then replace it after the stock is properly positioned in the machine tool.

Interlock switches are often included that prevent the machine from being activated until the barrier is back in position. The latter arrangement usually works fine except in cases where part of the stock extends beyond the barrier while the machine is operating. Some companies devise barriers that have an opening precisely the shape of the stock. This allows the stock to be inserted through the opening to the point of operation, while keeping the hands away at a safe distance. In such instances, the operator may hold on to the piece while the machining operation is performed and then pull it out after the operation is finished.

Principle 3.26

Make it difficult for operators to activate the tool or machine when their hands or other parts of their body are in the danger zone.

A common strategy of this type is to use a pair of palm switches that the operator must press down at the same time to activate the machine (e.g., the power to the machine runs through both switches serially). This arrangement keeps both of the operator's hands well away from the point of operation at the time the machine is activated. A similar approach involves connecting two pull-switches to cuffs worn by the operator around both wrists. After the machine is loaded, the operator merely pulls both arms well back from the machine tool to activate the machine. Hold-out devices and sweep devices work in similar ways to keep the operator's hands away from the point of operation. The approaches discussed previously are likely to be effective for a machine, which activates quickly, such as a press-brake, die-press, a shear, or a similar acting machine where machine actuation times are quite short.

Principle 3.27

Use machine feeding mechanisms that keep hands or other parts of their body out of the danger zone.

The use of hand-feeding tools illustrates this approach. Tools of this kind are extensions of the operators' hands that enter the machine tool near points of danger. If the machine tool (e.g., die press) activates, the hand tool is endangered but expendable so long as the machine tool was not. Some designs of these hand tools are similar to large pliers. Others hold the part using a vacuum or magnet. In any case, the operator uses the tool to insert the raw stock, and then withdraw it after the operation is finished. Another example is the use of a push stick or block for easing lumber through a table or radial wood saw.* Automatic or semiautomatic feed and ejection devices play a similar role, by eliminating the need for operators to place their hands near the point of operation.

Principle 3.28

Deactivate the machine or provide a warning if people or body parts are in the danger zone.

---

* Radial arm wood saws are those with the saw blade and motor above the table, whereas bench wood saws have the motor below the table with the tip of the blade rising through the table. Both kinds of saws are safer when used with a push-stick.

A variety of presence-sensing devices are used to help keep operators and others out of the danger areas. These devices can be interfaced with other sensors to deactivate the equipment or trigger an alarm when people are potentially at risk. Presence-sensing mats detect when someone is standing in a particular location. Light curtains and other photo-electric devices detect when something interrupts one or more light beams. Some of these devices use an array of beams that spans a wider area, than would be possible with a single focused beam. Since the light is pulsed at a given frequency, stray light that is not at that frequency is rejected. Some presence-sensing devices detect disturbances in an electromagnetic field to locate the operator or determine whether a body part, such as the hand, is in the danger area.

Setup, maintenance and repair tasks are particularly problematic for many machines and tools, because many of the ordinary safety devices need to be deactivated during this phase, to allow access to internal parts. Jog controls, and low energy operation modes are especially useful because they allow the machine to be turned on and off quickly while testing it. Blocks and stops can be included to keep parts from moving while the machine is being worked on. Lock-out and tag-out procedures are particularly important to keep someone from powering-up the machine when it is being worked on.

Principle 3.29

Provide the operator a way to rapidly deactivate the machine under emergency conditions.

Emergency stop controls are important safety feature of many types of machinery found in the workplace. Emergency stops are often connected to a braking system, but in some cases just deactivate the power source. The location of emergency controls is a particularly critical design issue. The control must be both easy to see and easily activated at the critical time. The size, shape, location, and conspicuity of the control relative to the position of operator in the emergency condition must all be carefully considered to deal with this issue.

Principle 3.30

Warn the operator and others who may be in the danger zone of potential dangers.

It is often important to warn all operators and others that there is danger near machine tools and to indicate precisely where particular dangers exist. Places in a machine where the fingers, hands, or other body parts can get caught by the machine action are known as pinch points. These should be clearly pointed out so people will keep their hands away. Warning labels should be placed adjacent to the pinch points and in direct view of the operator. Awareness barriers and awareness signals are another way of warning operators or others that they have entered a potentially dangerous area.

As should be obvious from the earlier discussion, many ergonomic design issues arise with regard to all of the aforementioned approaches for making tools and machinery safer at the point of operation. An ergonomic designer must first determine what the potential hazards are. These hazards often are associated with particular forms of unsafe behavior, so the ergonomic designer must know what these behaviors are and why they occur. Safety measures that interfere with performance are especially likely to be disabled, so the focus needs to be placed on developing noninterfering solutions. Once potential solutions are developed, the ergonomist must evaluate and test them.

Some of the questions that might be addressed include: Are the opening in guards too large? Is it easy to use the equipment when the guard is installed? Does a pull-out or related device fatigue or otherwise stress the operator? Do the presence-sensing devices activate at the right time? Are the presence-sensing devices reliable? Do they provide too many false alarms? Are jog controls easy to use, and are the step sizes appropriate? Can emergency stops be reached quickly enough at the time they are needed? Does the emergency braking system stop the system quickly enough? Are lock-out and tag-out procedures being followed properly? Are people disabling the safety devices, and if so why?

Answering these questions can be difficult. Safety standards, such as those provided by the American National Standards Institute (ANSI) or OSHA, provide a good starting point. However, further analysis using some of the ergonomic design tools discussed here and in other chapters of this book will often be necessary. The design of protective equipment for the operator poses a closely related set of questions and will be discussed in the following.

Protective equipment for the operator

Today, a wide variety of protective equipment is provided to workers in industry. Protective equipment is needed to reduce the 8.4–8.5 injuries per 100 workers that occurred in the United States during the 1990s. If that equipment is to be effective, it must fit, it must be used, and it must protect. The converse is not true. Ergonomic issues must be considered during the design of protective equipment. Supervisors and other personnel need to remind all operators to use protective equipment and ergonomics personnel must make clear the importance of this equipment through signs, word of mouth, and maintaining consistency with safety concepts. Ergonomics specialists are also involved in selecting appropriate equipment for manufacturing operations.

Safety shoes

Safety shoes are one of the oldest forms of personal protection. Yet their invention does not go back many decades. Although there is far more to safety shoes than the steel-capped toes, that is one of primary features of safety shoes. Steel caps protect the foot from objects falling over on the toes. It is not uncommon during lifting or other materials handling operations to have the material slip and fall onto the feet of workmen and the most vulnerable location is the toes. Other desirable features of protective shoes include reasonable wear characteristics, and cushioning for the many hours of standing required in many jobs. Most companies require their personnel to buy their own safety shoes and either directly or indirectly compensate them. Some exceptions to that occur when special requirements are made such as needing special acid protected boots. In any case, safety shoes need to provide a good fit to the feet, which of course depends on the available of design information about the size and shape of feet.

Another important feature of safety shoes is how well the sole adheres to the floor or to climbing surfaces. In the United States for 1988, 17% of disabling worker injuries and 13% of worker fatalities were due to falls, according to the Bureau of Labor Statistics and the National Safety Council, respectively. Many falls occur while climbing or maneuvering on elevated surfaces. Stairs, escalators, ladders, scaffolds, and floor openings pose particular hazards. Most falls, however, involve slipping or tripping while walking, pushing, or pulling objects on level ground. The coefficient of friction (COF) between a shoe and floor surface is a primary determinant of slipping hazard. This coefficient varies between floor surfaces and shoe soles, and as a function of contaminants. Industrial flooring is frequently wet from water or lubricants and so the soles should offer slip resistance.

A realistic approach to slip and fall safety must give equal emphasis to both the *required* and *available* slip resistance. The required slip resistance is that amount of reactive horizontal force necessary to allow a person to complete a particular maneuver without slipping. The required slip resistance can be expressed as a fraction, which is the generated

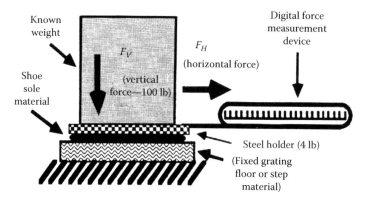

Figure 3.23 Apparatus for measuring slip resistance. (From Miller, J.M. et al., *Slip Resistance Predictions for Various Metal Step Materials, Shoe Soles and Contaminant Conditions*, Society of Automotive Engineers, Warrendale, PA, 1987.)

horizontal foot forces divided by the vertical foot forces at each instant during the performance of a task. These respective horizontal and vertical forces will be dictated by the particular task being performed, and can be measured using force platforms. The available slip resistance is the maximum ratio of horizontal to vertical force that a particular surface-sole-contaminant condition provides. This ratio has traditionally been called the COF.

Instrumentation is now becoming available to make the difficult determination of whether the slip resistance aspects of a climbing system are adequate. Figure 3.23 illustrates an example setup for measuring the available slip resistance. Such instrumentation and methodology permit a comparison of required versus available slip resistance. The term adequate slip resistance can be partially defined as being where available slip resistance is greater than required slip resistance for the normal activities. However, it must be recognized that too high of a COF increases the chance of tripping.

Force platforms are now available from a variety of manufacturers, and can be used to determine the required slip resistance in a variety of tasks, including the use of climbing systems to gain ingress and egress from high profile vehicles, such as trucks and heavy equipment (see Box 3.8). Many methods are also available for measuring the COF of flooring materials for common types of sole materials. Tabulations of such measurements are available from manufacturers of flooring materials and floor coatings. ANSI standards also are a useful source of information on testing methods and recommended COF values for various surfaces, including antislip surfaces used on stairs and other areas.

Wooden Shoes for Safety

A few years ago, a report from the European Community of Nations dealt with an attempt at creating standards on industrial protective equipment. As it turned out, in the Netherlands, some workers are still using wooden shoes for some specialized work and so those doing the testing in Holland, thought that they should include wooden shoes in the testing program. Since the Dutch history includes wooden shoes and they are a tourist item, most people took the message as a humorous one. But the laboratory reported good things about wooden shoes including the fact that they protected the whole foot, not just the toes. In this case, humor may have been the mother of invention.

BOX 3.8 EFFECT OF SHOE SOLE MATERIALS, TRUCK STEP MATERIALS, AND CONTAMINANTS ON SLIP RESISTANCE

Part of the task of operating large high profile commercial vehicles requires drivers to climb into the cab (ingress), out of the cab (egress) or around other parts of their vehicles. Vehicles having this high profile characteristic include agricultural machinery, construction equipment, industrial material handling trucks, over-the-road trucks and railroad cars. The climbing systems on these vehicles often include some type of fabricated metal step surfaces. Such surfaces can provide both durability and a degree of slip resistance, which can reduce slip and fall potential.

The relationship of slip resistance (or COF) to safe climbing system maneuvers on high profile vehicles has become an issue because of its possible connection to falls of drivers. To partially address this issue, coefficients of friction were measured for seven of the more popular fabricated metal step materials. Slip resistance on these step materials was measured for four types of shoe materials (crepe, leather, ribbed-rubber, and oil-resistant rubber) and three types of contaminant conditions (dry, wet-water, and diesel fuel). Results showed that COF varied primarily as a function of sole material and the presence of contaminants. Certain combinations of contaminants and sole materials resulted in COF values near zero (i.e., crepe rubber and diesel oil).

Unexpectedly, few effects were attributable to the metal step materials. Numerous statistical interactions suggested that adequate levels of COF are more likely to be attained by targeting control on shoe soles and contaminants rather than the choice of a particular step material. For more details on this study see Miller, Rhoades, and Lehto (1987).

Helmets

Helmets offer people protection from falling objects. Simple versions, such as caps and hats, are more of a fashion item than a practical form of protection, though some offer protection from the sun's rays. In days of old, helmets protected the wearer's head from the enemy's sword and arrows. Today, soldiers wear helmets to protect themselves from small caliber bullets and flying metal fragments. People working in construction and in many industrial settings wear plastic helmets for protection from falling objects. Impact hazards* are particularly significant in the construction industry, but may be a problem elsewhere. Anyway, the helmet is supposed to deflect a falling object or absorb the energy and distribute it over the entire head.

Most helmets include an adjustable band that circles the head at the crown. Webbing within the helmet goes over the head and connects to the band. The rigid outer shell of the helmet is designed to prevent penetration by sharp objects and resist impact. The outer shell is attached to the band and webbing in a way that provides space between the outer shell of the helmet and the wearer's head. Downward movement of the outer shell toward the wearer's head caused by an object striking against the helmet tightens the band around the head and transfers force to the webbing. Energy is absorbed during this process, and the forces are spread out over time and over a larger area. For smaller

---

* The National Safety Council refers to accidents of being hit by a falling object or the person falling due to poor footing and other causes as "fall and impact hazards."

impacts, this prevents the outer shell of the helmet from hitting the head. Most helmets are also designed to assure sufficient air space around the head to let air flow and cool the head. In colder settings, caps may be worn under the straps. Although it may seem otherwise, one size of helmet does not fit all people. The adjustability of the band reduces the number of different helmet sizes needed. Different helmet manufacturers change the amount of band adjustability and hence the number of different sizes needed for the population.

Accommodating Visually Disabled Web Users

Web designers have recognized the needs of visually disabled users and are trying to make web designs more accessible. Various forms of assistive technology have been developed. One approach is to make everything on the screen very large. Screen reader software is another particularly useful assistive technology that gives blind people access to computer systems by transforming the visual interface into a nonvisual representation. Screen reader software is often used in conjunction with output hardware to convert standard screen output into an accessible form, usually Braille or synthesized speech. Voice recognition software is increasingly common.

Screen reader software makes textual information more available to visually disabled users of the web, and undoubtedly will continue to play an important role in the future. A screen reader can also partially inform the visually disabled user about images and hyperlinks by reading their captions. However, the trend toward more graphically and complexly structured web pages is making it increasingly difficult for screen reader technology to effectively inform users of a web page's contents. This creates obvious problems for the user. One major concern is that this approach does not allow users to conveniently (i.e., nonsequentially) access those keywords, sentences, and paragraphs of interest while skipping over the parts of the text they do not care about.

Yang and Lehto (2001) discuss two experiments that explored the value of using image processing techniques to improve the understandability of tactile images by visually disabled users. In Experiment 1, blind subjects interpreted photographic images (typical of those found on the web) more accurately after image processing. Segmentation techniques were somewhat more effective than edge detection. The understandability of less complicated images found on the web, such as symbols, was not significantly increased by either method. In Experiment 2, blind users performed better on web browsing tasks after web pages were simplified using the segmentation method. These positive results are encouraging and support further research on this topic.

Helmets used in the construction and manufacturing sectors differ distinctly from helmets used in motor vehicles and those used in various sports. These differences reflect the direction and magnitudes of the expected striking forces. For example, motorcycle helmets offer protection for forces from all sides of the head, rather than just the top of the head.

Accommodation Can Save Lives

In recent years, some exciting advances have been made in prosthetic devices. Artificial limbs are now being developed that are actually intelligent. These limbs are similar to those used in robots. That is, they contain sensors and microprocessors interfaced to electric motors. In some cases, the motors move parts of the artificial limb. In other cases, they open and close hydraulic valves to allow movements around a joint at appropriate times.

The latter approach is used in some newer designs of artificial legs, to intelligently bend and straighten out the artificial leg as people walk. If you have ever observed someone trying to walk with a wooden leg, it quickly becomes obvious, that intelligent bending would help tremendously. This follows because the leg must bend at the knee to bring it forward, naturally. Users of a wooden leg have to stop after each step and extend the wooden leg out to the side before they can bring it forward. This process is obviously slow and fatiguing.

The New York Times, in their January 3, 2002 issue discusses the case of a Mr. Curtis Grimsley who was working on the 70th floor of the World Trade Center the day of the terrorist attack. Mr. Grimsley had lost his leg in an automobile accident a few years before, and had recently been fitted with an intelligent artificial leg. The new leg was a great improvement over the two previous models he had used before, which required him to stop after each step, replant his feet, swing the leg around, etc. To make a long story short, the elevators were not an option after the attack, so Mr. Grimsley needed to walk down 70 flights of stairs if he wanted to live. He made it, without stopping, at the same pace as everyone else. In his own words, "The C-leg made a world of difference between me being dead or alive because it allowed me to actually come down the stairs normally."

Protective gloves

Gloves are a piece of clothing particularly important in industry because people use their hands for so many purposes. Hands also need the same protection as other parts of the human body, in fact, probably more so because of their dexterity and probable exposure in maintenance and repair operations. In addition it is the hands that hold hand tools or handles of machines. Accordingly, hand anthropometry is particularly important to the design of gloves and those dimensions, in turn, are important to the design of hand tools, control devices, and machine handles, and gloves. Table 3.5 gives various measurements of the human hand, which are important in the design of gloves. These measurements were given in centimeters for the 1st percentile woman and the 99th percentile man, respectively.

Gloves keep peoples' hands warm in the cold environments and protect them from sharp, hot, or cold objects. Gloves also offer protection from the vibrations of tools and machines. Vibration arresting gloves have cushioning materials at specified locations within the gloves.*

---

* Raynaud's syndrome is a peripheral circulation disorder that is frequently associated with prolonged use of vibrating equipment. Prevention and treatment of this CTD is to keep the hands warm and away from air blasts. Hence, a glove coupled with vibration arresting materials provides protection against this disorder.

For some equipment (e.g., jack hammers), it is important to cover all or at least most of the inner hand, particularly the palm. Other tools may require cushioning at the palm of the hand. Although gloves improve safety, they can interfere with movements. They also increase the effective size of the hand, and in some cases can reduce the forces exerted.

Eye protection and spectacles

Eye shields, goggles, glasses, or spectacles are devices to aid and/or protect the eyes from flying particles, chemical sprays, ultraviolet light, and other hazards. Combining eye protection with visual correction provides an interesting challenge.* Spectacles place a transparent material in front of the eyes that refocuses images onto the human retina. Anything that holds that material in the correct orientation is a potential design solution for spectacle frames. Aside from fashion considerations, spectacles should

- Be comfortable
- Protect the eyes from the lens breaking
- Give freedom from fogging
- Minimally obstruct the visual field of view
- Allow easy, secure removal or installation, and safe handling
- Offer proper eye ventilation

Some people have additional needs. For example, athletes need highly secured spectacles because of the physical nature of many sports. Many older people need multifocal lenses. Different occupations have different visual needs.†

Anthropometric measurements are needed to design spectacles. The width of a person's face is obviously important. This value is approximately equal to the bizygomatic breadth.‡ Nose dimensions are important, as is the distance between the pupils (interpupillary breadth), and the distance between the outer extremes of the eye balls. Another important dimension is the distance along the sagittal plane from the back of the ears to the bridge of the nose. All of these dimensions are needed to design traditional forms of eye-spectacles.

Hearing protection

Ear plug, muffs, and other forms of hearing protection will be discussed in Chapter 4. As mentioned in that discussion, there are a number of ergonomic issues associated with the use of hearing protection. A particular concern is that human heads and ears vary in size and shape. This variability can make it difficult to fit certain people. A poor fit may allow much of the sound to leak in and also affects the comfort of hearing protection.

---

* This material was abstracted from Burgess (1986). While the reader may disagree with the Burgess design of spectacles, one has to agree that the design solution is intriguing.
† Carpenters often need to read plans, see objects at a distance, and also to perform close work immediately above the head. People working at computer terminals need only close and short-to-intermediate vision for the computer screen. Golfers need to write on score cards, read distances to the next hole, follow the flight of the ball (hopefully long), and strike the ball with the club. Musicians need to read the score and to follow the direction of the conductor who is probably a moderate distance away. The principal point is that these distance requirements are highly varied but so are the directions of the close-to-distance vision needs.
‡ This breadth is the maximum width of the head and it is measured from the tops of two bonal arches on each side of the head just in front of and slightly below where the top of the ears attach to the head.

Accommodating people with disabilities

Most people think of people with disabilities as being *physically disabled,* or, in other words, incapable of performing certain movements. In fact that is only one part of the story. Some people have *sensory disabilities*, or impaired senses. Anyone who uses spectacles or a hearing aid has some sensory impairment. People who are blind or deaf are more severe examples. Two additional categories that are equally valid, but often overlooked, are *intellectual disabilities* and *emotional disabilities*. While there are reasonable means for accommodating people with physical or sensory disabilities, unfortunately much less is known about how to deal effectively with these last two categories. Certainly, cognitively challenged persons can fill jobs, which require minimal intellectual skills.* In the case of emotionally challenged workers, some jobs can be sufficiently isolated to meet some needs of this group without creating problems with the other employees. Other emotionally challenged people are unable to handle isolation. Accordingly, the available design solutions for accommodating intellectually or emotionally challenged people are somewhat limited.

Two principal strategies characterize accommodation efforts. One is to diminish the sensitivity of the job to the disability, that is, fit the job to the person. Some examples of this approach include assuring wheelchair access to the job locations, altering job displays and controls to be within the reaching space, increasing visual brightness, or increasing loudness for the hearing impaired. An interesting example in agricultural settings involves the use of hoists that lift disabled operators into the seat of a tractor. Such solutions also may replace foot controls, such as a clutch with hand controls.

Most desks that are ergonomically designed are sufficiently adjustable that they can be quickly adapted for physically disabled people. A wheelchair can simply replace the chair, and the desk height can be adjusted to meet the individual person's needs.

A second accommodation strategy is to either enhance the remaining capabilities of a disabled person to meet job requirements or shift some modes of equipment operation so that nondisabled modes can be used. Some examples of this second strategy include prosthetic devices to replace missing limbs, hearing aids or eye glasses, screen and text readers that convert visual images or text into tactile form for blind or visually disabled people, or installing digital computer controls which a physically disabled person can activate. Regardless of the strategy used, there is a need for creativity in making accommodations. On the positive side, disabled employees who are accommodated are usually among the most faithful, appreciative, and consistent employees. Moreover, designing to accommodate disabilities for the extra ordinary person helps the designer learn how to better design for the ordinary person.†

There are other responsibilities in accommodating disabled people. One is the installation of bathrooms for disabled persons. People who use wheelchairs require different toilet heights, more spacious stalls around toilets, and handrails to aid mobility.‡ Some

---

* An older movie entitled "Bill" focused on a cognitively challenged person named Bill who was able to leave a mental institution to work in a small coffee shop at the University of Iowa in the 1960s and 1970s, helping both Bill and the university patrons.
† Newell and Cairns (1993) provide an interesting discussion of why designing to accommodate extraordinary persons improves designs for ordinary persons. They point out that many beneficial products on the general market today resulted from such accommodation. Also, with the aging U.S. population, more products are needed to fit the demands of this group of people. Almost all of us at one time or another are disabled.
‡ See Chapter 13 of Hutchingson (1981) for some additional background on the mobility of physically disabled persons.

suggest that every designer ought to spend a day in a wheelchair just to become aware of the constraints continuously imposed on wheelchair-bound persons. Doing so would quickly drive in the point that stairs and curbs are barriers, which impede the mobility of wheelchair users. In many cases the curbs are unnecessary and their removal becomes an option. Sometimes it is easy to build a ramp over a curb. Also, there are new wheelchair designs that offer greatly enhanced mobility for people.

Unfortunately, meeting building codes does not ensure that all forms of mobility aids can be used. Older buildings are particularly problematic, because they were designed before accommodation of wheelchair users was an issue. Entrances in older buildings normally include a flight of stairs, and changes in elevation often occur even on the same floor of the building. Solutions include elevators and ramps between floors of different buildings. Sometimes an elevator can be installed between adjoining buildings that has doors on both sides to accommodate different floor elevations between the two buildings.

Final remarks

Applied anthropometry in ergonomic design is the focus of this chapter. Applied anthropometry has been a traditional area of focus for ergonomists. However, as should be apparent from the preceding discussion, this area is rapidly evolving. New tools are now available that greatly extend the ergonomic designer's abilities. At the same time, the number of potential applications is rapidly increasing because of changing technology and the current regulatory environment.

Discussion questions and exercises

3.1 What are some basic anthropometric criteria and measures? Give some associated design objectives.
3.2 Why should bench height differ between work tasks? How does the normal area differ from the maximal area for males and females, respectively?
3.3 Why is it important in designing seating to allow operators to change their posture?
3.4 How do automotive seats differ from chairs at work and why should they differ?
3.5 What part of a person's back particularly needs support from the chair?
3.6 What is the Frankfurt plane and what is its significance?
3.7 What is the recommended height of the top of a computer VDT and how would you determine it?
3.8 At what height should the armrests of a chair be?
3.9 Why should chairs have open space beneath them and how much is needed?
3.10 If you wanted to place a foot-actuated pedal for a seated operator at a workstation, what range of positions in front of the operator would be the best to choose for toe actuation?
3.11 Differentiate between radial and ulna deviation of the hand and between extension and flexion of it.
3.12 What is the principal difference and advantage of the Klockenberg–Kroemer or Apple keyboards over the regular typewriter or computer keyboard? What is your opinion of the comfort keyboard relative to the Apple and regular keyboard?
3.13 What do you believe is the most important principle in the design of seating?
3.14 Why would you expect computer graphic models of people to be more useful to ergonomic designers than using templates?
3.15 Explain why there should be a definite and fixed place for all tools and materials. What advantages are provided by this principle?

Chapter three: Design of work areas, tools, and equipment

3.16 What does QWERTY mean and what is its significance?

3.17 Techniques that are used to determine a better hand tool design are chronocyclegraphs, infrared photography, EMG, subjective appraisals, television, timing devices, and heart rate or metabolic rate measurements. Which of the following statements correspond to those techniques? Discuss the most appropriate use of each technique to
   a. Compare the path of tool movements to a proper path and to track speed changes over the movement path to assure the movements are smooth.
   b. Determine advantages of a tool design for skilled, precise movements.
   c. Determine speed advantages of one particular design.
   d. Determine activity levels in individual muscle bundles and assure that the distribution is proportional to the bundle size.
   e. Show near-surface blood flows so that the distribution of force to the hand is balanced and no extreme localizations occur.
   f. Assess the mechanical advantages during use.
   g. Estimate the relative energy expenditures with different tools.
   h. Describe the perceptions of ease of use.
   i. Estimate the relative loads on different members of the human body (e.g., fingers, hands, arms, trunk, etc.).
   j. Define the population of users.

3.18 What are important considerations in the grip design of a ratchet-activated hand drill that converts vertical hand motion into a drill twisting motion?

3.19 Describe some of the principal differences between precision and power grips in hand tool design.

3.20 Describe the features that should be included in the design of soldering irons that are used for fine industrial assembly operations.

3.21 Explain why bending a tool's handle is not always a good ergonomic idea.

3.22 Why is it necessary to explain to users how a new tool should be operated during the introduction of a new tool design?

3.23 What are some methods for improving the safety of tools and machines at the point of operation? Discuss the role of ergonomics with respect to each method.

3.24 Discuss ergonomic issues that might arise for each form of protective equipment covered in this chapter.

3.25 A new manufacturing tool is expected to cost $150,000.00, but it would change the production rate from 15 assemblies per day (in two daily shifts) to 40 during a single shift. The power for the machine is expected to cost $0.25/h and maintenance is priced at $1000/year. Demand for this product is expected at the current rate for the next 10 years. Currently, three operators are needed for this process, but only one is expected to be required with the new machine. Assume that the company uses $25/h for an employee. The company uses a minimum attractive rate of interest at 10% annually or at an equivalent amount. Should you as the manager of the ergonomics department recommend this purchase and what is the net present worth?

chapter four

Assessment and design of the physical environment

About the chapter

The physical work environment can have a significant effect on productivity, safety and health, worker satisfaction, and employee turnover. Ergonomists and other specialists are often asked to determine whether the environmental conditions in particular settings are satisfactory. Dirty, cluttered, and poorly organized work, traffic, and storage areas are one common problem. Other potential concerns include exposure to hazardous materials, temperature extremes, inadequate lighting, or noise levels. Addressing these issues requires knowledge of how environmental conditions impact people, assessment methods, and a toolbox of solutions. Engineering solutions that involve altering the environment are the most fundamental approach but are often expensive. Less expensive solutions include administrative controls, such as job rotation, rest breaks, and employee selection, as well as implementing better methods of housekeeping. Providing protective equipment and clothing is another potential solution in some situations.

Introduction

The natural and man-made environments we live and work within vary greatly. In daily life we accommodate ourselves to these environments in a variety of ways. We employ heating systems and wear warm clothing when the atmosphere is cold and use cooling systems and wear light clothing when it is hot. We turn up the light when it is dark and shut our doors against loud and undesirable noise. Ergonomic designers must go beyond these commonplace procedures and either deal with environmental controls directly or through specialists in heating, ventilation, air-conditioning, illumination, and acoustics. In interacting with specialists, the ergonomic designer must coordinate the activities of various specialists and provide informed guidance to help develop cost-effective solutions.

At the most basic level, the goal of the ergonomic designer is to ensure that the work environment is safe, comfortable, and conducive to the tasks people need to perform. Man-made environments, such as those found within buildings or other structures, can often be modified to keep them safe and comfortable. However, many environments pose challenges that cannot be solved by implementing environmental controls, for technical or economic reasons. In the latter situation, ergonomic designers must focus on other means of protecting workers from dangerous or unpleasant environmental conditions. In some cases that means limiting exposure of people to environmental conditions, such as temperature extremes or noise. In other situations, special equipment or devices must be developed that allow workers to safely perform in unfriendly environments. Linemen and road crews, cleaning up after an ice storm, for instance, would naturally prefer to work in a more pleasant environment, but emergency operations must be maintained.

Other challenges to ergonomic engineers include developing and maintaining better means of housekeeping, specifying lighting that appropriately accommodates task requirements, and designing clothing and equipment to help workers more safely and comfortably endure the necessarily hostile work environment. Protecting people from excessive noise in the workplace is another important concern.

Cleanliness, clutter, and disorder

Dirty, cluttered, or poorly organized work environments can lead to health problems and accidents, reduce employee morale and productivity, and reduce the quality of the products and services produced by a manufacturer or service provider. Some of the many ways this can happen include the following:

- People can slip on spilled liquids or powders or trip over small objects or clutter on the floor, resulting in serious injuries. Scraps and small sharp objects may also cause cuts when people wipe them off surfaces or clothing.
- People's ability to move about their environment might be impeded by stacks of work in process (WIP) or other objects unnecessarily cluttering work or storage areas, aisles, and passageways. Such clutter also might block access to equipment and tools, obstruct visibility, or even create a fire hazard.
- Tools, parts, or other objects stacked on shelves or cluttering a work surface may use up much of the available space and significantly interfere with people's ability to do their tasks as intended, not to mention find something they happen to be looking for, leading to delays and problems such as using the wrong tool. Objects on crowded surfaces may also fall off onto people or be damaged when they hit the floor.
- Dirt and grime accumulated on light fixtures, windows, walls, ceilings, and elsewhere in the facility can greatly reduce the brightness of the work environment, interfering with people's ability to perform essential visual tasks, in general, creating an unpleasant effect.
- Toxic, irritating, allergenic, carcinogenic, tetragenic (potentially causing birth defects), or otherwise harmful substances may cause health problems when they contact the skin or are ingested when workers smoke or eat without first washing their hands. These substances may also be brought home by workers on their clothing, exposing family members, such as children or pregnant women, who may be particularly sensitive to their effects.
- Dusts, vapors, and gases may enter the air and be inhaled, resulting in serious health problems. They also might accumulate in significant quantities to create fire and explosion hazards or contaminate products that are being produced.
- Poor sanitation might lead to the spread of disease within facilities and is a special concern in health-care settings and food industries.

Other examples along these lines can be easily imagined. One traditional solution is better housekeeping and maintenance, as expanded upon in the following section.

General housekeeping and maintenance

As implied by the previous discussion, good housekeeping is important in almost any imaginable work facility, and is especially important when toxic or hazardous materials are present or used in the production process. Some general requirements and elements of an adequate housekeeping program are as follows:

1. Cleaning and maintenance should be scheduled on a frequent periodic basis to ensure that dirt and clutter do not build up over time to unacceptable levels; to minimize leaks from machinery, storage drums, and other sources; and to ensure that air filters and ventilation systems work properly.
2. Spilled liquids, dusts, and other objects should be immediately cleaned up using appropriate methods that do not add to the problem. In particular, toxic materials, acids, and otherwise reactive or hazardous materials should normally be neutralized or diluted before attempting to remove them. Also, centralized vacuum systems that exhaust their contents outside the facility are preferable to portable vacuum cleaners that recirculate air. Use of the latter systems, as well as sweeping, can cause dust to become airborne and increase the chance that potentially toxic materials will be inhaled.
3. Washrooms and showers should be provided to workers in dirty jobs. A common criteria is one water tap and shower for every three workers to ensure they do not have to wait too long to clean up.
4. Work and traffic areas should be clearly marked to separate them from temporary storage areas for work in progress (WIP). Otherwise, there is a tendency for WIP to accumulate and block work areas and aisles.
5. Convenient, easily accessible locations should be designated for storing tools, parts, and other essential items used in the workplace. These locations should be periodically monitored to ensure that items are placed in the designated locations and, if they are not, to determine necessary changes to ensure work areas will not be cluttered.
6. Waste containers or other disposal devices should similarly be provided in convenient locations. Waste containers should be emptied on a frequent periodic basis to ensure they are not filled beyond capacity.

Despite the obvious benefits of following good housekeeping practices, any ergonomist with significant industrial experience will agree that many, if not most, organizations have difficulty maintaining a clean, uncluttered work environment for their employees. This tendency is especially true for small manufacturing faculties, but even larger organizations devoting significant efforts to housekeeping often have significant room for improvement. Part of the issue is that housekeeping is often viewed as a janitorial task, separate from the day-to-day responsibilities of most employees. Another issue is that clutter has a tendency to build up over long periods. In our experience, it is not unusual to find tools, equipment, and parts that have been sitting around unused for years, sometimes even taking up valuable space on the shop floor!

5S programs

As part of the so-called lean revolution in manufacturing, many companies in recent years have been looking for ways to produce more with less. Implementation of a 5S-plus safety program is often the starting point for such efforts. 5S (Hirano, 1996) can be viewed as a systematic approach for continuously improving housekeeping that goes well beyond the traditional janitorial perspective of this topic, and in so doing addresses many of the root causes of a dirty, cluttered, disorderly work environment.

> **BOX 4.1 BENEFITS OF 5S**
>
> 5S is particularly interesting to many organizations because a successful 5S program can do much more than improve housekeeping. For example, consider the case of Wabash International, the largest manufacturer of truck trailers in the United States. At the end of its fiscal year in 2001, Wabash reported a loss of $232.1 million, and debt approaching half a billion dollars. A change in management took place in May, 2002, and a 5S program was implemented shortly afterward. The company turned a profit for the first time in several years of $25 million in the first half of 2004. In October, 2004, Wabash received the U.S. Senate Productivity Award. Shortly before receiving the award, Wabash invited a group of about 100 members of the Wabash Valley Lean Network to visit their main production facilities in Lafayette, IN, to see what they had accomplished. Some of the benefits of 5S reported to the group by Wabash International included (1) reduced space requirements, allowing them to roughly double production without investing in larger facilities; (2) an 80% decrease in accidents; (3) large improvements in productivity and product quality; and (4) greatly improved employee morale. During the plant tour, employees at the site enthusiastically discussed the before-and-after photographs prominently displayed throughout the site at individual work areas. More importantly, they all seemed to have plans for future improvements.

At the most basic level, 5S is a five-step process followed to continuously improve a selected work area, usually involving workers drawn from the particular work area selected for improvement and a 5S expert who facilitates each step of the process. An important part of 5S is to take photographs showing what the work area looked like before and after going through the process (Box 4.1).

These photographs are often posted at the work site to publicize what was done and maintain awareness of the need for continuous improvement.

The five steps in the 5S process are

S1—Sort
S2—Set in Order
S3—Shine
S4—Standardize
S5—Sustain

Note that the names given to the steps are rough translations of Japanese terms, reflecting the fact that 5S originated in Japan. Each of these steps will now be briefly reviewed.

Step 1 Sort. The first step of 5S is to take an inventory of all items currently in the work area. A red tag is placed on all items that do not need to be present in the analyzed area. The latter items are then moved to a temporary holding area. Unless someone claims that a red-tagged item needs to be returned to the work area within a preset time (often 2 weeks), it is discarded, sold, given away, or otherwise disposed.

It is easy to see that the first step of the 5S procedure can free up space and eliminate clutter. In practice, participants are sometimes reluctant to throw away familiar, but unused, items. A common comment is along the lines of, "you never know when (blank) might be handy." Placing the items in a holding facility for some time can often help modify such reactions. As time passes, most people become more accustomed to the fact

Chapter four: Assessment and design of the physical environment

that familiar, but useless, objects are no longer there, and begin to perceive the advantages of a less cluttered environment.

Step 2 Set in Order. The second step of 5S is to arrange the remaining items in an orderly fashion and clearly designate a correct location for each item. From the perspective of Hirano in his discussion of the visual workplace, it should be obvious when something is out of place or missing. This objective can be attained in many different ways. Typical examples include the following:

1. Hanging tools on boards on which the shape of each tool is prominently traced out in the appropriate location. This makes tools more visible and eliminates the tendency to pile up tools in jumbled heaps on work surfaces.
2. Placing parts in transparent legibly labeled trays.
3. Color coding and numbering schemes to designate logical groupings and ordering of manuals placed on shelves.
4. Marking the appropriate location where incoming parts and WIP should be placed.
5. Hanging signs that prominently identify particular areas and the location of equipment in the work facility.
6. Rearranging equipment to *open up* the work area, and otherwise improve the layout.

It is easy to see that performing the first two steps of 5S on a periodic basis can go a long way toward reducing clutter and preventing it from building up in the first place. This can lead to some obvious productivity improvements by making it easier for people to quickly find things when they need them and reducing the time needed for people and materials to move around the facility.

Step 3 Shine. This third step of 5S, often also called sanitize, is to carefully clean even traditionally ignored parts of work areas and equipment and then paint them white so that dirt or grime will stand out. An important element of this process is to assign cleaning responsibilities to specific workers to create accountability. Proper cleaning supplies must also be made readily available following 5S guidelines analogous to those for tools. Painting everything white reflects the original focus in Japan on creating the so-called visual workplace, where even minor problems stand out and become obvious. One of the advantages of this approach is that maintenance issues such as minor leaks in hoses and fittings become obvious long before serious problems occur. A second advantage is that the environment becomes much brighter. Many 5S participants find the often dramatic transformation from a dark, dirty environment to be a great improvement and are quite enthusiastic about what they have accomplished.

Step 4 Standardize. The fourth step of 5S often overlaps greatly with steps 1 and 2. This follows because the number of tools, dies, fixtures, parts, and types of equipment needed for a particular process can often be reduced by standardizing processes, tools, and the products they provide to customers. This, in turn, helps reduce clutter and can greatly increase productivity. Some examples of how the need for particular items might be eliminated through standardization are as follows:

- A company might switch from producing 10 colors of a product to the four most popular, thereby reducing both the type and amount of paint that must kept in inventory.
- A company might switch from using both screws and rivets to using rivets alone, thereby reducing the types of tools and parts needed in the work area.

Another issue is that completing step 2 of 5S can in some cases lead to new arrangements of production equipment, which might change the way particular tasks will be done. If so, Step 4 of 5S will also involve developing standard operating procedures for the new arrangements.

Step 5 Sustain. The last step of 5S is to develop ways of sustaining the improvements that have been made. Ideally, the first four steps of 5S are made part of each worker's job so that the improvement process is continued on a permanent basis. Companies also might publicize the 5S program with a newsletter and conduct periodic 5S inspections to demonstrate commitment. Inspections should use a standard evaluation method, so workers know exactly what is expected of them and how well they are doing. Some companies publicize the results of these inspections by presenting 5S awards to encourage friendly competition between different areas of the facility.

Lighting and illumination

Like other environmental conditions such as cleanliness and order, lighting can have a negative effect on human performance and safety. Part of the issue is that poor lighting can interfere directly with the tasks people perform because vision is one of the major senses guiding human activities. Hence, inadequate lighting can cause workers to misread signs or instruments, or make other mistakes, resulting in lower productivity and reduced safety.

The solution to inadequate lighting is not always *more* light but *better* light instead. As expanded upon in the following sections, there are many ways of improving lighting conditions that can be cheaper and more effective than simply investing in more light fixtures. Placement and maintenance of light sources is at least as important as the number of light fixtures. Well-chosen paints and finishes on ceilings, walls, floors, and work surfaces can also play an important role, as can modification of the task to reduce visual demands.

Luminous environment and its measurement

Much of the available light in a given environment might come directly from the sun or from a light fixture. Light also arrives after reflecting off of floors, walls, ceilings, and other surfaces. The amount of luminous energy falling on a surface is called the illumination. The latter quantity is measured in units of footcandles (fc) or lux (lx), where 1 fc is equal to 10.76 lx (Box 4.2). The illumination level of a particular surface depends upon both the intensity and location of the available light sources (Figure 4.1). The amount of light energy from a source that arrives at a particular surface decreases with distance and is also related to the orientation of the surface with respect to the light source.

Getting a bit more precise, the illumination (E) at the surface where the visual task takes place, assuming the light comes from a single point source, depends on the intensity (I) of the light source, the distance (d) between the light source and the visual task location, and the cosine of the angle (β) between a line perpendicular to the surface and the light source. This relationship is described by the equation

$$E = \frac{I}{d^2} \cos \beta \qquad (4.1)$$

Note that if the light source is directly overhead, the angle β is 0, resulting in the simplified equation

$$E = \frac{I}{d^2} \qquad (4.2)$$

BOX 4.2 TUTORIAL ON PHOTOMETRIC MEASURES

The measurement of light often appears a bit bewildering at first. To help clarify this topic it may be helpful to first think of an idealized situation where photons of light flow in all directions uniformly from a single point in space. The n photons radiated at any instant in time travel outward at the same speed, creating an expanding sphere. As the sphere expands away from the point source, the n photons on the surface of the sphere are spread out over an increasing area. At this point we are ready to consider some important photometric units, which are shown in Table 4.1.

To start with, the energy of the emitted radiation in the visible range is called luminous energy (see Box 4.3). The rate at which luminous energy is generated by the light source is called the *luminous flux* and is measured in units called *lumens*. Lumens measure the part of radiant power (in watts) that is in the visible range. That is, a lumen is the amount of luminous energy transmitted per second. The amount of luminous flux (in lm) radiated from the source in a particular direction is called *luminous intensity*. Luminous intensity is normally measured in candelas. A *candela* describes how much luminous energy is emitted per second in a given direction, in units of lm/sr.

A *steradian* (sr) is defined as the solid (two-dimensional) angle subtended by rays emanating from the center of a sphere that mark out a surface area on the surface of that sphere that is equal to the radius squared. (To get a better feel of what a steradian is, it may be helpful to take a quick look at Figure 4.1; the solid angle of 1 sr is described by the four lines subtended to the point source from each corner of a 1 ft^2 rectangle on the surface of a sphere with a 1 ft radius.) That is, 1 sr is the solid angle (in steradians) subtended by an area (A) on the surface of a sphere with radius r is equal to A/r^2. Doing some simple algebra using this relation shows that both areas shown in Figure 4.1 subtend a solid angle of 1 sr. The entire sphere subtends a solid angle of 4π sr, because the area of a sphere is equal to $4\pi r^2$ (i.e., $4\pi r^2/r^2 = 4\pi$ sr).

If we now assume that the point source in Figure 4.1 emits light with a luminous intensity of 1 cd in all directions, the luminous flux distributed over the surface of a surrounding sphere is 4π lumens. This follows because a 1 cd point source produces 1 lm/sr and there are 4π sr in a sphere. The 4π lumens are distributed over an area of $4\pi r^2$ for a sphere of radius r. The *illuminance*, or in other words, the amount of light falling on any area of the surrounding sphere is, therefore, equal to 1 lm/r^2. For a sphere with a radius of 1 ft, the illumination is 1 lm/ft^2, which by definition corresponds to 1 fc of illumination. If the radius of the sphere is 1 m, the illumination level becomes 1 lm/m^2, which in SI units corresponds to 1 lx of illumination. Note that 1 m^2 is about 10.76 ft^2. Accordingly, 1 fc equals 10.76 lx. *Luminance* is a measure of how much light is emitted by a surface and is normally measured in units of either ft-Lamberts or candela/m^2. Returning to the previous example, if the surface of the sphere was transparent (i.e., made of clear glass) and had a radius equal to 1 ft, it would have a luminance of 1 ft-Lambert. *Reflectance* and *transmittance* coefficients, respectively, describe the proportion of the luminous flux landing on a surface that is reflected or transmitted.

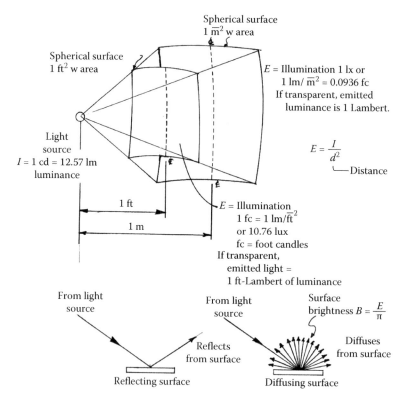

Figure 4.1 Luminance, illumination, and brightness measurements.

Table 4.1 Some Commonly Used Photometric Measures[a]

Photometric Quantity	Photometric Unit	Radiometric Quantity	Radiometric Unit (SI or SI-Derived Units)
Luminous energy	Talbot	Radiated energy	J
Luminous intensity	Candlepower cd = lm/sr	Radiant intensity	W/sr
Luminous flux	lm	Radiant flux	W
Illuminance	lm/m^2 = lx lm/ft^2 = ft-candles	Irradiance	W/m^2
Luminance	ft-Lambert = lm/ft^2 = 3.426 cd/m^2, cd/m^2 = nit	Radiance	(W/sr)/m^2
Reflectance	Dimensionless ratio	Reflectance	Dimensionless ratio
Transmittance	Dimensionless ratio	Transmittance	Dimensionless ratio

[a] Photometric measures can be viewed as radiometric measures adjusted for the capabilities and limitations of the human. Some of these capabilities and limitations are discussed in Chapter 2. See Box 4.3 for an example of how radiometric measures are adjusted to become photometric measures.

After the light from the source strikes an opaque surface, some of the light is absorbed, and the remainder is reflected. The amount of light reflected by a surface determines its brightness or *luminance*. The nature and amount of reflected light depends upon the surface color and its spectral properties. Light reflected off a shiny metallic surface such as a mirror leaves the surface at an angle of reflection similar to the angle of beam incidence, provided that the surface is *perfectly reflective*. Other materials, such as soft cotton battens, tend to reflect a light beam in almost every direction equally (i.e., *perfectly diffusing surfaces*). These two extremes of reflectance are shown on the bottom of Figure 4.1. Highly diffusing surfaces are often desirable. This reduces the possibility of unpleasant or even harmful reflections or glare.

The proportion of light reflected by a lighted surface is referred to as *reflectance*. That is,

$$\text{Reflectance} = \frac{\text{luminance}}{\text{illuminance}} \quad (4.3)$$

when luminance is measured in ft-lamberts, and illuminance in ft-candles. If the luminance is measured in cd/m^2, and illuminance in lx, the reflectance becomes

$$\text{Reflectance} = \frac{\text{luminance} * \pi}{\text{illuminance}} \quad (4.4)$$

For historical reasons, a flat, perfectly diffusing surface is called Lambertian. A unique feature of Lambertian surfaces is that they appear equally bright from any viewing angle. For example, if you move your head from side to side while looking at this page of the book, the brightness of particular locations on the page will not seem to change, assuming you are not casting a shadow on it. Things change completely when the surface is reflective. For example, a light reflected off a mirror or other reflective surface, such as a shiny table top, will change position when you move from side to side. As you move from side to side, you change the angle at which you are viewing any particular location on the surface. The reflection will appear at a location where your viewing angle happens to coincide with the angle at which most of the light is reflected from the light source (Figure 4.1). In other words, the luminance of a particular spot on the surface changes with viewing angle.

White, painted surfaces have a reflectance of around 90%. Newsprint and bare concrete are around 50%. Dark painted surfaces may be as low as 5%. The use of appropriate paints and finishes is an important design strategy for spreading reflected light evenly throughout the workplace in a way that can both reduce lighting bills and create a more pleasant environment. Standards such as *ANSI/IES RP-7-1979—Practices for Industrial Lighting* recommend reflectance values of 80%–90% for ceilings, 40%–60% for walls, 25%–45% for desks and bench tops, and 20% or higher for floors (Box 4.3).

Lighting methods

As mentioned earlier, the light in a given environment can be provided by a variety of sources. Natural lighting provided through windows and skylights is a good and economical option in many cases. Many different types of artificial lighting are also used in work environments and elsewhere. Methods of artificial lighting can be classified into at least four categories:

1. *Direct radiants.* In this form of lighting, 90% or more of the light is directed toward the viewed object, and is illustrated by the placement of an incandescent or fluorescent light fixture on the ceiling. Direct radiant lighting is the most commonly used method in offices and many work settings.

BOX 4.3 LUMINOUS ENERGY AND LUMINOSITY FUNCTIONS

Artificial sources of light produce luminous energy in a variety of ways. An incandescent light bulb, for example, produces luminous energy when electricity passing through a filament heats it to a high temperature, causing it to glow brightly. Fluorescent tubes, on the other hand, emit light when UV rays generated by passing electricity through a gas strike a mixture of powders coating the inside of the tube. Each type of powder fluoresces, or produces light, in a particular bandwidth when contacted by the UV light. Consequently, different mixtures of powders change the overall coloration and energy spectra of the emitted light.

If we plotted out the amount of energy produced over time by an incandescent light bulb at particular wavelengths in the visible spectrum, we would get a nearly continuous curve giving the power density or spectrum of the produced light. The curve for a fluorescent light, on the other hand, would show a separate bandwidth for each of the fluorescent powders. The luminous energy generated by a light source can be measured using a spectro-radiometric instrument as described by the following equation:

$$J = \int_{0.38\mu}^{0.76\mu} J_\lambda d\lambda \tag{4.5}$$

where J_λ is the *power density* function or, in other words, the power (J) of the light source at each frequency, over the visible spectrum of 0.38–0.76 µm. Note that the measurement defined by Equation 4.5 is purely physical in character. If the power density function is multiplied by the luminosity function that describes the human eye's sensitivity to luminous energy at various wavelengths, the result is a luminous power measurement which is tuned for human seeing. As discussed in Chapter 2, and as shown in Figure 2.15, the sensitivity of the human eye to light is described by two different bell-shaped functions over the visible spectrum, depending on lighting conditions. For well-lighted conditions in which the cone receptors of the eye are used, the sensitivity of the human eye is described by a curve that peaks at a wave length of 0.55 µm. Under low light conditions, when rod receptors are used, the curve shows maximum sensitivity at a wave length of 0.5 µm. After correcting photometric measurements for the sensitivity of the human eye, the luminous intensity (I) of a light source becomes

$$I = \int_{0.38\mu}^{0.76\mu} J_\lambda v_\lambda d\lambda \tag{4.6}$$

where v_λ describes the sensitivity of the human at each frequency (λ), over the visible spectrum of 0.38–0.76 µm, as described by the luminosity function.

2. *Indirect lighting.* Rather than directing light toward the viewed object, 90% or more of the light is directed upward toward the ceiling. The reflected diffuse light illuminates the environment more uniformly than direct lighting and eliminates direct glare. However, some light is lost when it is absorbed, instead of reflected, by the ceiling.

3. *Mixed direct radiant and indirect lighting.* In this approach, around 40% of the light is directed downward, while the remainder is directed upward toward the ceiling.
4. *Supplemental task lighting.* Lamps or other fixtures often located close to the viewed object are used to provide a high level of light in a localized area.

The light sources used for each of these methods are most commonly incandescent bulbs or fluorescent tubes. Other light sources used in some applications include mercury, metal halide, high-pressure sodium, and low-pressure sodium lamps. The advantages and disadvantages of particular light sources are for the most part related to economics and color rendering (Table 4.2), as briefly summarized here:

1. Incandescent bulbs are inexpensive, but by far the least energy-efficient option, providing up to 20 lm of light/W of energy used, and much lower levels for low wattage bulbs. They burn out typically in less than a year, and give off significant amounts of heat. Tungsten–halogen filament bulbs last longer, are more efficient, and provide a higher quality of light than regular tungsten filament bulbs. Incandescent bulbs produce light rich in the red and yellow spectrum, which most people find pleasant. However, this can make it difficult to accurately identify some colors, demonstrating that a high color-rendering index (CRI) is no guarantee that color rendering will be adequate.
2. Fluorescent tubes are 3–5 times more efficient than incandescent bulbs, and have a much longer life. Compact fluorescent lamps are now available that offer a cheap, highly efficient replacement for incandescent bulbs. Some modern fluorescent tubes produce light with an energy spectrum that closely matches that of natural light, and provide good color rendering. However, the quality of the produced light varies greatly between types of tubes. Standard spectrum fluorescent tubes are viewed as cold and unpleasant by some people, and produce only fair levels of color rendering. Fluorescent tubes also produce less light with age, or at cold temperatures, and have a tendency to flicker when they are old or defective.

Table 4.2 Efficiency, Life, and CRI of Some Commonly Used Lighting Sources[a]

Light Source	Typical Range of Rated Life (h)	Output Efficiency (lm/W)	CRI[b]
Incandescent bulbs	1,000–2,500	5–20	70–100
Fluorescent tubes	4,000–20,000	35–100	Warm white = 52
			Deluxe warm white = 75
			Deluxe cool white = 89
Mercury vapor	12,000–20,000	25–60	50–55
Metal halide	6,000–20,000	70–110	65–80
High-pressure sodium	10,000–20,000	80–130	22–75
Low-pressure sodium	15,000–18,000	100–190	0

[a] Based on information in *IESNA Lighting Handbook* (1993), supplier information, Answers.com, and other sources.
[b] Color rendering is a measure of how colors appear under an artificial light source compared to their color under natural light. With better color rendering, colors appear more vibrant or close to natural. The so-called CRI is a method developed by the CIE or International Lighting Standards Commission used to measure, on a scale of 0–100, the relative ability of a light source to render eight standard colors compared to a standard reference illuminant. A high-quality, incandescent light source is used as the reference when testing artificial light sources with color temperatures less than 5000 K. The CRI of incandescent light is the de facto standard, but is an imperfect measure that can be criticized since incandescent light is weak in the blue–green spectrum.

3. Mercury lamps have a long life and produce around 50 lm of light/W of energy used. On the negative side, the light produced is very weak in the yellow–red end of the light spectrum, making it very difficult to accurately perceive many colors under their light. Their efficiency also drops significantly with age and many people find their light to be unpleasant.
4. Metal halide lamps are more efficient than mercury lamps and have a fairly long rated life. However, they tend to produce only about 50% of their rated output toward the end of their rated life. Color rendering is fair to moderate.
5. High-pressure sodium lamps are very efficient, producing around 100 lm or more of light per watt of energy used. Lamp life can be around 5 years at average burning rates of 12 h/day. Color rendering is fair.
6. Low-pressure sodium lamps are most efficient, producing up to 190 lm of light/W of energy used. Lamp life is 4–5 years at average burning rates of 12 h/day. On the negative side, low-pressure sodium lamps produce a yellow light perceived to be unpleasant by many people, with very poor color-rendering properties.

It also should be mentioned that some two-lamp lighting fixtures use a high-pressure sodium lamp along with a metal halide lamp. This results in a very efficient hybrid light source. More importantly, the two light sources add together, resulting in a color more pleasing to most people than that given off by either source by itself.*

As implied by the previous discussion, the energy saved by switching to a more efficient light source can be very significant. Lighting accounts for about 20%–25% of the electricity consumed in the United States, and up to 50% of the electricity consumed in office environments. Many energy-saving strategies are available as pointed out by Cook (1998), and energy savings as high as 65% have been attained by motivated companies. Some of the strategies discussed by Cook and others include

- More efficient lamps, ballasts, and fixtures†
- Better placement and arrangement of light fixtures, shields, and shades
- Use of daylight
- Timers and occupancy sensors that turn lights off when they are not needed
- On-off and dimmer switches in convenient locations so that people can turn off or reduce unneeded lighting
- Photoelectric sensors that sense the amount of daylight in the room and either switch lamps on or off or adjust the lamp brightness accordingly
- Localized lighting in areas where more light is needed
- Reduced lighting levels in nonproduction areas
- Regular cleaning of lamps, fixtures, windows, skylights, walls, and ceilings
- Placing lights in better locations

It is easy to see that one or more of these strategies are likely to yield benefits in most work settings. Energy savings, however, is only one part of the issue. A more important

---

* The C.I.E. chromaticity diagram for colored luminants can be used to specify the result of combining two sources of light, as discussed in Chapter 2. The color of each source corresponds to a particular point $[x, y]$ on the C.I.E. chromaticity diagram. The combined light source corresponds to a point somewhere on a straight line connecting the $[x, y]$ coordinate values of the two sources. If both sources have the same luminous power, the combined light source lies halfway between the two sources.
† Fixtures with specular reflectors that direct most of the light downward can significantly increase light levels where it is needed. The reader interested in learning more about this topic is encouraged to consult the *IESNA Lighting Handbook*.

Chapter four: Assessment and design of the physical environment

consideration is that the amount and quality of the provided light must both be adequate. Otherwise, lighting will not be serving its intended purpose.

Some basic principles and lighting requirements

Countless hours have been spent over the years by engineers, ergonomists, architects, and others attempting to determine how much light is necessary for the visual tasks that people perform. They also have spent at least as much time on the issue of light quality. Both topics are complicated in many ways, but a lot of perspective can be obtained by thinking about some of the ordinary activities people do under natural lighting conditions (Table 4.3). It does not take a lot of imagination to map out which of these natural lighting conditions are most appropriate for particular activities. For example, direct sunlight is terrific for playing baseball, but most people find it far too bright for comfortably reading a book or sleeping. The light under a beach umbrella may still be a little bright for reading a book printed on glossy white paper but okay for reading a newspaper, and so on.

At some point, the lighting becomes too dim to easily do certain activities. For example, on a very dark day, people may decide it is dark enough to justify turning on a few lights before they start reading the paper or preparing food in the kitchen. Interestingly, people can perform certain activities quite adequately even at extremely low illumination levels. For example, some people can quickly find their way around familiar environments under starlight or even on an overcast night. The ability of people to function at this range of illumination is so normal that most people do not even think about it. However, it is truly impressive, given that direct sunlight results in a level of illumination about a billion times greater than that found on an overcast night (assuming we are not in an urban area where there is a lot of light pollution reflecting back from the clouds).

The previous discussion directly leads us to perhaps the most obvious principle of lighting design.

Principle 4.1

Provide enough lighting to ensure an adequate level of visibility for the tasks that will be performed.

To apply this principle we must know something about what types of tasks will be performed, and which visual details need to be perceived to ensure safe and efficient

Table 4.3 Approximate Surface Illumination Level (fc) for Various Natural Lighting Conditions

Lighting Condition	(fc)
Direct sunlight	10,000
Full daylight	1,000
Overcast day	100
Very dark day	10
Twilight	1
Deep twilight	0.1
Full moon	0.01
Quarter moon	0.001
Starlight	0.0001
Overcast night	0.00001

Source: See Van Cott and Kincade, 1972.

performance of the particular tasks. The critical details that must be perceived to ensure an adequate level of performance might be lines or characters traced out on a sheet of paper or presented on an LED display, a door knob, the edge of a table or sidewalk, or a small crack on the surface of a part.

We also will need to know something about how particular levels of illumination impact the *visibility* of critical details for the range of users who will be within the task environment. As discussed earlier in Chapter 2, visibility at a particular level of illumination depends on both the size of a visual detail (or element) and how well it stands out from its background. The first quantity is normally measured in terms of the *visual angle* (α) subtended by a visual element to the eye of the viewer. The second quantity corresponds to *contrast*, or how bright the viewed object is compared to the background.

By convention, visual angle is normally measured in arc minutes, where 1 arc min equals 1/60°. For small angles, the visual angle (in arc minutes) is given as follows, for a target of height (b) located at a distance (d) from an observer*:

$$\alpha = \frac{3438 * h}{d} \quad (4.7)$$

where h and d must be in the same measurement units. Contrast (C) is normally expressed as the following ratio:

$$C = \frac{B_L - B_D}{B_L} \quad (4.8)$$

where
B_L is the luminance of the brighter of two contrasting areas
B_D is the luminance of the darker area

In some cases, especially for light-emitting displays such as CRTs, contrast is calculated by setting B_L to the brightness of the background and the numerator is the absolute value of $B_L - B_D$. In the first approach, contrast varies between 0 and 1. In the second, it varies between 0 and much larger values (i.e., C approaches infinity as B_L gets closer and closer to 0).

Visual acuity is the ability of the eye to perceive fine details, as explained earlier in Chapter 2. Visual acuity can be measured in terms of the fineness of a line that can be seen, the diameter of a spot that can be seen, separations of lines or spots that can be detected, or the size of a gap or misalignment of line segments. For the most part, a person's visual acuity can be measured as the smallest visual angle a person can detect subtended from some visual element of an object, regardless of whether the visual element is a line, spot, space, circle, or the identity or orientation of an alphabetic character. If that angle is α, a reasonable measure of visual acuity is

$$VA = \frac{1}{\alpha} \quad (4.9)$$

Visual acuity is typically reported relative to "20/20," which is interpreted as meaning objects at 20 ft are seen as if at 20 ft. The decimal equivalent of 20/20 is 1.0. Based on this, a

---

* For example, the visual angle subtended by a side view of a mailbox (with a height of 6 in. and length of 18 in.) to an observer 100 ft away is .5 * 3438/100 = 17.2 arc min, which is well within the normal range of visual acuity. Note that we used height instead of length in the calculation, since the height of the image corresponds to a smaller detail than its width. Also note that we converted all units into feet before doing the calculation.

visual acuity of 2.0 is equivalent to seeing objects that are at 20 ft as if they are at 10 ft, and a visual acuity of 0.10 is equivalent to seeing objects that are at 20 ft as if they are at 200 ft.

Experiments performed over 60 years ago revealed that visual acuity improves when the level of illumination is increased. Visual acuity also increases when there is more contrast between viewed objects or symbols and their surrounding background. This leads us to a second, somewhat more specific lighting principle, expanded upon next.

Principle 4.2

Less illumination is needed when the visual acuity needed to perform the task is low, when the background luminance is high, or when the contrast between a visual detail and the background is high.

Illumination and contrast have an additive effect on visual acuity for dark objects viewed against a lighter background.* Experiments have indicated that, holding contrast constant, there is about a linear improvement in visual acuity as the background luminance is increased (Johnson and Casson, 1995). Holding luminance constant, increasing contrast increases visual acuity approximately linearly as well. The two effects are additive, and does not appear to be strongly affected by the original acuity of the person.

These results are largely consistent with earlier experiments, including those conducted by Blackwell (1959), in which 100 young observers viewed a 4 min (angle) luminous disc projected several times for 0.2 of a second against an illuminated screen. After each presentation of the disc, the observer adjusted the luminance of the disc upward or downward. This process continued until the observer felt the disc was just barely noticeable. The luminance values of the disc and background where the disc first became visible were then used to calculate a *threshold contrast* level. These steps were repeated several times over a wide range of illumination levels. The results were then plotted out, resulting in a curve that showed how the average threshold contrast changed with illumination level. Blackwell called this curve the *visibility reference function*. Similar curves were obtained in the original experiment and other studies by Blackwell for several different task conditions and target sizes.

An interesting finding was that the latter curves were shifted upward or downward from the visibility reference function but otherwise were quite similar. Curves for smaller targets and more demanding tasks were shifted upward, while larger targets resulted in a downward shift. All of the curves showed that more illumination was needed before the target became visible when tasks were more demanding. That is, the threshold contrast increased if (1) smaller targets were used, (2) the timing and location of the target on the screen was randomly varied, (3) the target was moving, or (4) presentation time was decreased.

The latter results and the fact that not all visual tasks performed in any given environment are identical, lead us to a third principle of lighting design.

Principle 4.3

Provide general illumination over a room for the most typical tasks and supplement with auxiliary light sources for more exacting visual tasks.

This principle goes a long way toward matching the amount of illumination to task-related needs. It also provides a way of reducing the overall cost of illumination, since there is normally no need to provide high levels of illumination everywhere people might go. This

---

* The latter situation is illustrated by the illumination falling on this page. That is, we have dark characters against a white background that receives and then reflects light.

basic idea is a key element of the IESNA general illumination recommendations (IESNA, 1993), shown in Table 4.4. The first step in applying the IESNA procedure is to determine which category a particular task falls within. A long list of example tasks for particular settings is provided in the *IESNA Handbook* that the analyst can consult to determine which category seems to be the most appropriate. Note that the examples shown in Table 4.4 give a general idea of each category that might be enough for some analysts to decide upon a category.

The next step is to determine what weighting factor to use. As indicated in the lower part of Table 4.4, this requires for all of the categories that the analyst know how old the occupants of the space will be. For categories A–C, the analyst must also estimate the average weighted room reflectance value (AWRF). This quantity is calculated as

$$\text{AWRF} = \frac{\sum_i A_i r}{\sum_i A_i} \tag{4.10}$$

where
   A_i is the area of the ith surface of the room (i.e., ceiling, floor, walls)
   r is the reflectance of the corresponding surface

The weighting factors for age and AWRF are then looked up from the matrix at the bottom of Table 4.4, and added together. The following rules are then applied to assign a high, medium, or low value within the range of illumination values suggested for the particular category. That is, if

```
sum < 0, assign the lowest value in the range
sum > 0, assign the high value
otherwise, assign a value half way between the low and high values
```

For example, assume that the relevant category is B and that the occupants are expected to be around 50 years old. Consequently, the age weighting factor is 0. If the AWRF happens to be 50%, the reflectance weighting factor is 0. The sum of the two weighting factors is 0. The range of illuminations for category B goes from 50 to 100 lx. Since the sum is 0, the recommended illumination is 75 lx, or simply the midpoint of the recommended range.

A similar process is followed for categories D–I, except that the analyst must now determine the reflectance value of the task background, instead of the AWRF. The analyst must also determine how important the speed and accuracy requirements associated with the task are. After determining each weight and summing them, a similar set of rules are used to assign a high, medium, or low value within the range of illumination values suggested for the particular category. That is, if

```
sum < -1, assign the lowest value in the range,
sum 2, assign the high value,
otherwise, assign a value half way between the low and high values.
```

The use of an age weighting factor in the IESNA guidelines suggests a fourth principle of lighting design.

Principle 4.4

More illumination and contrast is often helpful to people who are older or visually impaired.

   Not all people can see as well as the typically young subjects (with excellent vision) participating in most of the studies of vision. However, there is little doubt that older people

Table 4.4 Some General and Specific Illumination Recommendations

Lighting Type	IESNA Category	Illuminance (lx)	fc	Activity Areas
General illumination—simple visual tasks	A	20–50	2–5	Public areas with dark surroundings
	B	50–100	5–10	Short temporary visits—passageways
	C	100–200	10–20	Areas for occasional visual tasks—storage areas, entrances, lobbies, stairways
General illumination—moderate visual tasks	D	200–500	20–50	Viewing of high-contrast, large objects—simple assembly, inspection, or rough machining, reading high-contrast visuals, or handwriting
	E	500–1,000	50–100	Viewing of low-contrast, small objects—medium machining or assembly, reading low-quality reproduced copies, moderately difficult inspection
	F	1,000–2,000	100–200	Viewing of low-contrast, very small objects—hand engraving, color inspection, reading very poor copies, difficult inspection
Additional lighting—exacting visual tasks	G	2,000–5,000	200–500	Prolonged exacting visual tasks—fine buffering and polishing, very difficult assembly or inspection, extra-fine hand painting
	H	5,000–10,000	500–1,000	Very prolonged, exceptionally exacting tasks—extra-fine assembly machining or assembly tasks, exacting inspection
	I	10,000–20,000	1,000–2,000	Very special tasks of small contrast, very small objects—surgery

Task and worker characteristics	Weight		
	−1	0	+1
Age	Less than 40	40–55	Over 55
Speed or accuracy (only for categories D–I)	Not important	Important	Critical
Reflectance of room surfaces (for categories A–C)	Over 70%	30%–70%	Less than 30%
Reflectance of task background (for categories D–I)			

Source: IESNA, Lighting Handbook: Reference and Application, 8th edn., Illuminating Engineering Society of North America, New York, 1993.

need more illumination and contrast than younger people when they perform reading and other visually demanding tasks (Wright and Rea, 1984). Increased illumination and contrast can also compensate for poor visual acuity or contrast sensitivity. On the other hand, older people are more likely to be glare sensitive, so increased illumination levels might actually add to the problem for older people in some settings.

Principle 4.5

More illumination or contrast is needed when viewing times are short. This is also true if the critical visual details are on moving targets, or if their location or presence is unpredictable.

The Blackwell studies mentioned earlier provided strong evidence that people's visual acuity drops when viewing times are extremely short. The latter studies also showed that greater contrast and illumination improved performance at short viewing times. Some additional insight regarding this issue can be obtained from a fairly recent study that examined how visual acuity was influenced by exposure time and contrast (Adrian, 2003). A total of six young subjects viewed targets displayed in the center of a CRT display providing a background luminance level of 32 cd/m². The targets were viewed at eight different exposure times varying between 0.01 and 2.0 s. The target contrasts were 0.74, 0.46, and 0.17. It was found that the data were well described with the following equation:

$$VA = 0.57(\log C * t) + 1.705 \tag{4.11}$$

where
 t is the observation time in s
 C was expressed as $\Delta L/L$ with ΔL being the difference in luminance between the target luminance and background level (*L*) in cd/m²

The results are shown graphically in Figure 4.2. This equation shows why high levels of contrast are so desirable for reading and other visually demanding tasks where people

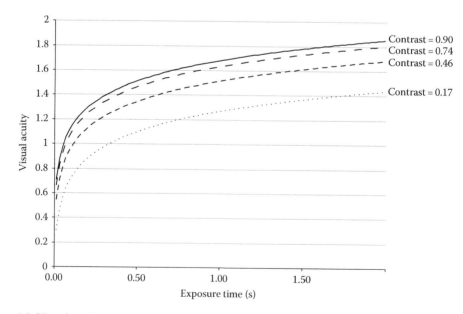

Figure 4.2 Visual acuity vs. exposure time for four different contrasts. (From Adrian, 2003.)

need to rapidly move their eyes from location to location. Simply put, at lower levels of contrast, the eye must remain fixated at a particular location for a longer time before the needed information is acquired.

Another principle of lighting design somewhat related to the latter observation is as follows.

Principle 4.6

Increasing either the contrast between an object and its background or the size of critical details can be a better strategy than increasing the amount of illumination.

This principle is well illustrated by considering a couple of examples. To illustrate the advantages of improving contrast, imagine that people in an office are having trouble reading the faded-out versions of a form produced by a poor copy machine. The advantages of providing better copies rather than more light are quite obvious. Another related strategy is to modify the color and brightness of the work surface so it provides good contrast with items placed upon it. Increasing the size of the viewed object is also a feasible strategy in many cases. For example, consider a situation where people are having trouble reading the small print on a label. The best solution here might be to use larger characters.

The advantages of improving contrast are especially significant for LED displays, computer monitors, and other types of luminous displays. With the latter devices, increasing the amount of background illumination will actually make it harder to see the displayed information. The reason this happens is that *ambient* light* reflecting off the surface of the luminous display reduces the contrast between the displayed characters and their background. This problem can be very severe. For example, those of us who have tried to use a cell phone outside on a bright day know that it is almost impossible to read the LED display under very bright conditions. The LED displays commonly used on automobile instrument panels also tend to wash out completely on bright days.

The effect of ambient light on the contrast of a luminous display is easily explained as follows. To start with, we know that the contrast C can be defined as $\Delta L/L$, where ΔL is the difference in luminance between the luminance of the displayed characters and the background luminance level (L) of the display. Increasing the amount of ambient light does not change ΔL because it increases the luminance of both the displayed characters and the background. However, since L is larger, the ratio $\Delta L/L$ becomes smaller, and, therefore, the contrast is lower. A large amount of ambient light causes the ratio $\Delta L/L$ to become nearly zero, meaning that there is essentially no difference in brightness between the characters on the display and their background. Consequently, the displayed values are severely washed out and nearly invisible.

The light reflected off the surface of the display is called *veiling glare*, because it can be thought of as a veil of light pulled over the viewed object. At this point, we are ready to declare another basic principle.

Principle 4.7

Ambient illumination of luminous displays such as CRTs and LEDs should be minimized to improve contrast. Light sources should be placed so they do not directly illuminate the surface of luminous displays.

---

* Ambient light is light from the environment.

A wide variety of strategies can be followed to attain this goal. Some of the steps that should be taken include

1. Providing shades and window curtains to keep direct sunlight off the surfaces of CRTs or other displays
2. Placing shades on lamps and light fixtures
3. Locating light sources well above the work surface so they are not directly casting their light on the screen
4. Providing dimmer switches so people can adjust the light levels downward
5. Changing the orientation of workstations to reduce reflections off of the display surface

These strategies, in slightly modified form, can also be applied to satisfy another important lighting principle.

Principle 4.8
Eliminate or try to minimize direct sources of glare or shadows by using indirect lighting or by appropriate positioning of light sources.

A little earlier in this chapter we talked about veiling glare, which is also sometimes called indirect glare. We all have a good intuitive feel for what direct glare is from experience. However, direct glare is a more complex topic than it might seem. For example, how is it possible that a light source that seems weak under sunlit conditions (i.e., a porch light or car headlights) seems so glaringly bright at night? There are two main parts to the answer. First, direct glare is related to how uniformly the light from the visual field is distributed. When the light is evenly distributed, it tends not to be perceived as glare even if it is very bright. On the other hand, if it is concentrated in a small part of the visual field, even a small amount of light can cause glare. The second part of the answer is related to what is called *dark adaptation*, that is, the human eye adapts to the level of ambient light. Once the eye is fully adjusted to a low level of light, it becomes much more sensitive to glare (Boxes 4.4 and 4.5).

Strong shadows are another common problem that can interfere with visual performance when they are cast over part of the visual field. Appropriate positioning of light sources can go a long way toward eliminating both glare and shadows. The use of indirect lighting and multiple light fixtures are also good ways of reducing glare and shadows. However, it should be mentioned that shadows do in fact provide useful

---

BOX 4.4 ORIGIN OF FROSTED LAMPS

The early version of the incandescent light enclosed the bulb filament with a clear glass bulb. Today almost all our bulbs are frosted. Although it was known for years that frosted bulbs produced less discomfort glare, no economical method of frosting the inside of the bulbs was known until Marvin Pipkin went to work for General Electric. Some of the young people Marvin first met at GE decided to play a joke on him by telling him it was his job to frost the inside of the bulbs. It was like sending someone out for a left-handed monkey wrench. However, Pipkin took the task seriously and experimented with acids until he found a combination that worked. In this case, humor was the mother of invention.

> **BOX 4.5 DISCOMFORT VERSUS DISABILITY GLARE**
>
> Ergonomic designers often distinguish between discomfort glare and disability glare. The former refers to high light levels causing discomfort, and the latter refers to observable effects on vision due to the presence of very bright sources of light in the field of view. A typical source of disability glare is sunlight near sunset or sunrise streaming into the windshield of an automobile, making it virtually impossible to see objects in the visual field near the sun. Disability glare can be reduced in many environments simply by removing highly reflective objects or giving them a matte finish to improve their light diffusion. Disability glare can occur in interior spaces when very bright light sources reflect upon the visual display tubes of computers, television units, or other visual devices.
>
> Discomfort glare was a major problem years ago when many homes had a single light bulb that hung on a wire from the ceiling. That single source was almost always in the visual field and was uncomfortable to look at. Discomfort glare has been largely eliminated in most modern facilities because of the advent of fluorescent lamps and contemporary luminaries that distribute light more evenly throughout the surrounding environment.

visual cues to people. This follows because the shadow cast by an object has the effect of increasing the contrast along the edges of the object.* Consequently, some people may find too much indirect lighting to be somewhat disorienting. This potential problem can be greatly reduced by providing a combination of indirect and direct lighting, rather than indirect lighting alone.

The next closely related principle follows.

Principle 4.9

Avoid or minimize extreme transitions in illumination levels between adjoining areas of the task or environment.

The apparent brightness of an environment increases greatly after people become dark adapted. For example, we all know that the outside environment on a summer day can be quite dazzling right after we step out of a cinema. After about 5–10 min, however, much of that dazzle disappears. Adaptation also works the other way around—that is, once people are adapted to bright conditions, such as full sunlight, moderately lighted environments seem much dimmer than normal. This can create serious visibility problems, in situations such as when a driver adapted to bright conditions enters a dimly lighted tunnel or underground parking garage.

Due to incredible range of brightness between sunlight and night conditions, it is very difficult to avoid large transitions when people move between indoor and outdoor environments. Architects place a special focus on designing transition zones to deal with this issue. For example, in many well-designed buildings, people enter through the main

---

* In some applications, shadows are intentionally enhanced to increase the contrast between small visual features and a low-contrast background. For example, light directed from the side that grazes the surface of an object makes it much easier to see small imperfections, such as cracks or scratches on a metal surface or loose threads lying on the surface of a cloth.

entrance into a sky-lit foyer (or transition zone) containing many windows. One advantage of this approach is that the foyer will be bright in the daytime and much dimmer at night.

Adaptation is also potentially an issue when people move their eyes between bright and dim areas of their task environment. For example, in night-flying conditions, pilots will occasionally switch their eyes from the dimly lighted visual horizon to more brightly lighted items on the instrument panel. Similarly, a person doing an inspection task may be moving their eyes back and forth from a brightly lit localized inspection area to other much dimmer areas in the surrounding area. As a general rule, it is commonly accepted that brightness ratios between a task area and the surrounding area are ideally about 1–1 and should not exceed 3–1. Ratios of 5–1 or higher are likely to be detrimental. Ratios of as high as 20–1 are considered to be acceptable between the task area and remote areas of industrial facilities.

Principle 4.10
Match the color and color-rendering properties of light sources to the task-related needs of the people using the environment.

Color and color-rendering properties vary greatly between light sources, as discussed earlier. Good color-rendering properties are a necessary requirement for some tasks, such as color inspection, visual and graphics design, and in a variety of environments such as makeup stations, fitting rooms, or showrooms. Personal preferences and vanity concerns are also important. A big part of the issue is that many people think they look better in warm light with good color-rendering properties. Such light gives skin a little more of a tanned outdoor tone, and is generally perceived to be more pleasant than cool (i.e., standard fluorescent) or harsh (i.e., low-pressure sodium) light.

Noise

Workplaces and other environments may contain equipment and machinery that produce significant levels of noise and vibration. Every operating machine tool, fan, or compressor in a building is a potential source of noise. Vibration is a less common problem, but it is a closely related issue that may be of concern when people drive vehicles, operate equipment, or use power tools.

Health effects of noise

Exposure to noise can cause a variety of health effects varying from insomnia and stress to hearing loss. Short-term exposure can cause a temporary loss of hearing, normally referred to as a temporary threshold shift. Prolonged exposure to noise over a period of years generally causes permanent loss of hearing. Figure 4.3 describes the incidence of hearing impairment for people of different ages exposed to varying levels of noise at work.* Work-related noise exposure level (dBA) is indicated at the bottom of the figure. Incidence of hearing impairment is indicated on the y-axis and corresponds to the proportion of people in a given population who suffer significant hearing loss. Each line describes the incidence rate for a particular age group as a function of noise exposure level (Box 4.6).

As shown in Figure 4.3, the proportion of people suffering hearing loss increased with greater noise exposure for each age group. Not surprisingly, older people had greater percentages of impairment at all levels of exposure. About 23% of the people in the older age group (50–59 years old) who did not work in loud environments were hearing impaired,

---

* This figure was developed by NIOSH several years ago using information from a variety of sources.

Chapter four: Assessment and design of the physical environment 157

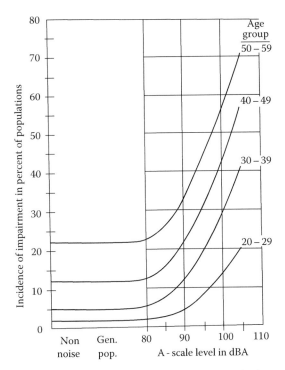

Figure 4.3 Prevalence of impaired hearing and sound levels at work. (From NIOSH.)

compared to only 2% for people of ages between 20 and 29 years. When people's jobs exposed them to noise levels of 90 dBA, 33% of the older group hearing was impaired, compared to 5% for the youngest group. At exposure levels of 100 dBA, the proportions were, respectively, 56% and 14%. At 105 dBA, they increased to 70% and 20%.

The sad conclusion is that the majority of older people working in noisy environments had significant loss of hearing. Younger people working in extremely noisy industries were about as likely to suffer hearing loss as the nonexposed people who were 30 years older (20% vs. 23%). These findings led to the establishment of OSHA noise exposure limits in 1981. OSHA noise limits are briefly summarized in the next section.

OSHA noise exposure limits

The OSHA permissible noise limits are shown in Table 4.8. The latter table specifies maximum allowable durations (in hours) over an 8 h workday to noise levels in dBA, for example, an 8 h exposure at noise levels at or below 90 dBA. This drops to 2 h at noise levels of 100 dBA. If exposure varies over the day, a noise dose (D) is calculated as follows:

$$D = 100 * \frac{\sum_i C_i}{\sum_i T_i} \tag{4.12}$$

where
C_i is the total time of exposure at a given level
T_i is the total time of exposure allowed

The maximum permissible noise dose is 100.

BOX 4.6 EXAMPLE CALCULATIONS OF NOISE DOSE

Consider the situation where an employee spent time in a number of different areas of the plant where he was exposed to a variety of noise levels as shown in the following. The TWAN exposure is computed as follows.

Exposure (h)	Noise Level (dBA)	Ci/Ti
0.5	100	0.25
2	95	0.50
4	90	0.50
		$\Sigma = 1.25$

It follows that the noise dose is 125, which exceeds the OSHA permissible limit of 100. Now assume that over the workday the employee exposures were as follows:

Exposure (h)	Noise Level (dBA)	Permissible Exposure (T)	Ci/Ti
0.5	100	2	0.25
1	95	4	0.25
2	90	8	0.25
4.5	80	32	0.14
			$\Sigma = 0.89$

The values of T were calculated using Equation 4.26. It follows that the noise dose is 89, which is within the permissible OSHA limits.

Note that the OSHA permissible noise levels in Table 4.5 require that the sound measurements be taken on the slow response dBA scale. OSHA also specifies that exposure to impulsive or impact noise should not exceed a 140 dB peak sound pressure level. One other element of the OSHA standard is that it mandates a hearing conservation program, whenever employee noise exposures equal or exceed that of an 8 h time-weighted average noise (TWAN) sound of 85 dBA. Required elements of a hearing conservation program include audiometric testing, exposure monitoring, hearing protection, training, and notification of the employee.

Table 4.5 OSHA Permissible Noise Levels

Duration (h/day)	Sound Level (dBA)
8	90
6	92
4	95
3	97
2	100
1	102
0.5	110
0.25	115

Chapter four: Assessment and design of the physical environment

The noise dose corresponding to an 8 h TWAN sound of 85 dB dBA can be calculated using Equation 4.12. The following second equation describes the allowable exposure time to noise levels outside the range of those given in Table 4.5:

$$T = \frac{8}{2^{(L-90)/5}} \quad (4.13)$$

where L is the noise level in dBA.

Annoyance and other effects of noise

Almost all of us will agree that loud noise can be very annoying. A related issue is that noise can greatly interfere with people's ability to communicate with each other. Some data addressing both topics are available from a study on noise annoyance and speech interference, in which subjects listened to words spoken at controlled levels in a variety of white noise conditions. The subjects wrote down the words if they could identify them and then rated how annoyed they were by the noise on a 5-point scale. The values and associated verbal anchors were

1—Not annoying
2—Lightly annoying
3—Moderately annoying
4—Quite annoying
5—Extremely annoying

The relationship between the annoyance ratings and noise level was determined by regression analysis to be as follows (Table 4.6):

$$\text{Annoyance rating} = -4.798 + 0.106 \text{ dBA} \quad (4.14)$$

The relationship between noise levels and subject reactions given by this equation is summarized in Table 4.9. Another equation was developed that describes the percentage of words missed as different noise levels:

$$\% \text{ Words missed} = -11.17 + 0.1989 \text{ dBA} \quad (4.15)$$

The reader should understand that Equations 4.14 and 4.15 are simply linear approximations of what happened in a particular setting. However, they are probably good enough for detecting potential problems in other settings. When the noise level was 85 dBA, the average annoyance rating was about 4.2, indicating that the subjects found it quite annoying,

Table 4.6 Relation between Noise Levels and Perceived Annoyance Levels

Level (dBA)	50	60	70	80	90	100
Annoyance	0.49 below tested range	1.55 barely annoying	2.61 almost moderately annoying	3.67 nearly quite annoying	4.72 almost extremely annoying	5.78 above extremely annoying

Source: Epp, S. and Konz, S., *Home Econ. Res. J.*, 3(3), 205, 1975.

which intuitively seems quite reasonable. Also, Equation 4.15 indicates that about 5.7% of the words would be missed due to noise interference if 95 dBA of white noise is present. Since the spoken language is very redundant, a 5% loss of words may not really interfere with communications very much, but it may be enough to require check procedures to assure understanding, such as repeating all critical orders, as required in naval ship handling. Another way to assure better communication of words in a noisy environment is to use the military word alphabet such as alpha, bravo, etc. for the letters of the alphabet. However, the need to take such measures would slow down the rate at which information is transmitted and undoubtedly add to the annoyance problem.

Noise control strategies

Ergonomics designers can and should do a number of things to reduce noise levels. Noise control strategies are given in the next section in the order of their probable effectiveness. Of course, the more of these that are implemented, the greater will be the effectiveness. Some suggested design principles, with comments, follow.

Principle 4.11

Reduce the noise level of the source itself.

This solution is without doubt the best choice to start with, and it is likely to be the most cost effective. If it is a question of machine noise level, could another, quieter-operating machine serve as well? If not, identify why the machine makes noise and then examine the possibility of making design changes to achieve those objectives. At times the effectiveness of the machine is tied to the noise it makes, and so, little can be done on this strategy.

Principle 4.12

Enclose the source.

This potential solution stops the sound emission at the source, but enclosures can make machine tending or maintenance operations far more difficult. Also, machine enclosures can cause the machines to overheat unless additional cooling is developed. Before enclosing machines, check with the manufacturers of the machines to see if there is any downside to enclosing. Sometimes enclosure walls are too thin and light and loud sounds come directly through them. Consider going to a heavier wall construction or a double wall construction so that structural members on the inside of the wall near the machines are separated from those in the wall away from the machines.

Principle 4.13

Increase the distance between the source and nearby people.

Moving the machine operation away from nearby personnel or moving the personnel away from the operation will help attenuate the sound before it reaches the people around it. Remember, noise levels decrease approximately with the square of the distance between the source and the receiver.

Principle 4.14

Place sound-absorbing and reflecting barriers in the noise path.

Absorbing materials will help reduce the sound level energy and reflecting barriers will direct some of that energy away from surrounding personnel so that the sound wave will need to go a large distance before encountering people and, hence, attenuate.

Hearing protection

Earplugs and earmuffs are commonly used in loud environments to protect people from excessive exposure to noise. Examples of both forms of personal protective equipment are shown in Figure 4.4. Earplugs are made out of soft materials, such as cotton, wool, plastic, or wax. When inserted into the ear, earplugs significantly reduce the amplitude of particular sounds. The amount of reduction in decibels (dB) is given by the noise reduction rating (NRR).* For example, if the NRR is 10 dB, then wearing the plugs in a 90 dBA noise environment would reduce the exposure to 80 dBA, assuming that the earplugs fit properly. The earplug should fit snugly in the outer ear entrance without leaving any openings around the plug through which sound might intrude. Earplugs also should have a retaining ring, protrusion, or other feature to both keep them from being inserted too far into the ear and make it easier to remove them. Along these lines, some designs attach a cord or bracket to the plug, which allows them to be easily removed.

Like earplugs, earmuffs often come in a wide variety of designs, as also shown in Figure 4.4. In some cases, earmuffs are combined with other forms of personal protections, such as helmets or face shields. Earmuffs also sometimes include speakers and microphones. Traditionally, earmuffs have provided a passive form of protection by acting as a barrier between the source of the noise and the exposed ear. More recently, the passive aspect of earmuffs has been supplemented with electronic features, such as noise cancellation and sound amplification (see Best's Safety & Security Directory—2003).

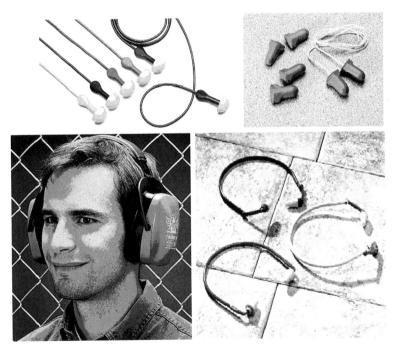

Figure 4.4 Some examples of hearing protection devices. (Reprinted with permission from Conney Safety Supply, Madison, WI.)

---

* The U.S. Environmental Protection Agency (EPA) requires that hearing protectors be labeled with the NRR, measured in decibels.

Both approaches process sounds from the environment before transmitting them over speakers within the earmuffs in a way that allows the wearer to hear normal conversation and other sounds they want to be able to hear, while simultaneously providing hearing protection. Noise cancellation becomes feasible when the sound is periodic or repetitive. This is done using sophisticated signal-processing circuits that cancel out periodic, predictable noise sources. Sound amplification is useful in environments where high-intensity, intermittent sound noise sources are present, such as stamping presses or gunshots. The latter systems use amplification circuitry within the earmuff to keep the sound levels within safe limits.

As implied by the previous discussion, there are a number of ergonomic issues associated with the use of hearing protection. One issue is that human heads and ears vary in size and shape. This variability can make it difficult to fit certain people, and may impact both the effectiveness and comfort of hearing protection. Another point is that wearing earplugs or earmuffs can interfere with verbal communication. As pointed out by Morata et al., interference with communication and the ability to do the job are two of the most common reasons given by workers for not using hearing protection (2001).*

Temperature and humidity

The atmospheric environments people work and live within vary greatly. In tropical settings, people face high temperatures and humidity. In polar settings, people deal with temperatures well below the freezing point of water. The human body performs remarkably well under various atmospheric conditions. The key issue is that the human body intelligently adapts to different thermal conditions to maintain the core body temperature within a narrow range of 36°C–37°C. This process is referred to as thermal regulation.

Thermal regulation

Thermal regulation involves several physiological responses taken to balance the heat generated within the body by metabolic processes against the flow of heat from the environment to the body. When the flow of heat out of the body exceeds that generated by metabolism, people start feeling cold, and the body makes several responses to *reduce* the net flow of heat out of the body. These responses include shivering, which increases the heat generated by metabolism, and reduced flow of blood to the extremities and skin, which reduces the rate at which heat is transferred out of the body. Conversely, when the flow of heat out of the body is less than that generated by metabolism, people start feeling hot, and the body makes several responses to *increase* the net flow of heat out of the body. These responses include lethargy, which causes people to be less active and, consequently, reduces the heat generated by metabolism, and increased flow of blood to the extremities and skin, which increases the rate at which heat is transferred out of the body.

If the physiological responses of the body are not adequate to maintain a balance between the heat coming in and the heat going out, the system will no longer be in equilibrium and there will be a net change in the amount of heat stored in the body. In the initial stages of this process, the physiological responses of the body still maintain core body temperature within the critical range by distributing the heat unevenly between the extremities and deep body tissues—depending on whether the heat storage is positive or negative, the outer extremities and skin will be, respectively, warmer or colder than the core body

---

* Workers in the Morata et al. study also mentioned itching, headaches, and other discomfort.

Chapter four: Assessment and design of the physical environment

temperature.* At some point, however, the core body temperature will eventually begin to change. A change of less than 1°C in core body temperature is generally viewed as the acceptable limit from a safety and health perspective. Larger changes can result in serious or fatal consequences.† Any situation where environmental conditions are severe enough to cause core body temperature to change significantly should be carefully monitored, and obviously will be uncomfortable or stressful to most people.

Heat transfer

As discussed herein, people produce heat and also receive heat from the air and other sources. The exchange of heat between the human and environment can be described with the following equation:

$$S = M \pm C_c \pm C_v \pm R \pm E \tag{4.16}$$

where
- S represents body heat storage
- M represents metabolic heat produced
- C_c represents conductive heat exchange
- C_v represents convective heat exchange
- R represents radiative heat exchange
- E represents evaporative heat exchange

When the body is in equilibrium with the environment, body heat storage is zero, and the air temperature is typically between 21°C and 32°C. Outside of this range, additional clothing, heating, or cooling may be required, depending on the individual and other factors.

Metabolism

The heat generated by metabolic processes within the human depends on how active people are.‡ As discussed in Chapter 2, people produce anywhere from 60 kcal/h when at rest to over 4000 kcal/h during heavy exertion. Metabolic rates for several different tasks or activities are given in Table 4.7. Note that the values in the table are given in met units. To calculate the metabolic rate of a particular person, doing a particular task, the values in the table must be multiplied by a conversion factor and the skin area of the evaluated person. (For an example, see Box 4.7.)

When performing these calculations, the DuBois approximation formula is frequently used to compute skin areas. This formula is as follows:

$$A_{skin} = 0.202 W^{0.425} H^{0.725} \tag{4.17}$$

where
- A_{skin} is in m²
- W is the person's weight in kg
- H is the person's height in m

---

* For example, in extremely cold environments, people's hands, feet, ears, and noses can freeze well before the core body temperature drops below the critical range. Similarly, when people are vigorously exercising or in hot environments their hands, feet, and skin can be much warmer than the core body temperature.
† Core body temperatures below 33°C or above 42°C are extremely dangerous and are likely to result in fatal consequences.
‡ The heat generated by the human body can actually be large enough to affect the thermal environment itself. The combined functioning of many metabolisms can generate sufficient heat to warm a cool auditorium!

Table 4.7 Metabolic Rates of Typical Human Activities Measured in *mets*

Activity	mets
Seated quietly	1.0
Walking on level surface @ 2 miles/h	2.0
Walking on level surface @ 3 miles/h	2.6
Using a table saw to cut wood	1.8–2.2
Handsawing wood	4.0–4.8
Using a pneumatic hammer	3.0–3.4
Driving a car	1.5
Driving a motorcycle	2.0
Driving a heavy vehicle	3.2
Typing in an office	1.2–1.4
Miscellaneous office work	1.1–1.3
Filing papers while sitting (standing)	1.2–1.4
Light assembly work	2.0–2.4
Heavy machine work	3.5–4.5
Light lifting and packing	2.1
Very light work	1.6
Light work	1.6–3.3
Moderate work	3.3–5.0
Heavy work	5.0–6.7
Very heavy work	6.7–8.3

BOX 4.7   CALCULATION OF METABOLIC RATE USING METS

Note that the metabolic values in Table 4.7 are given in met units. Met units reflect both the size of the person being evaluated and the level of activity. Size is an important consideration because larger people produce more heat. Size is closely related to skin area—that is, the larger the person, the larger the surface area. Consequently, skin area provides a way of measuring or estimating differences in size. A metabolic rate of 1 met corresponds to the basal or resting rate of energy production of a person, in units normalized by skin area. The maximum possible expenditure rates vary from about 12 mets for young men to 7 mets for 70-year-old men. The corresponding rates for women are about 30% smaller.

The metabolic rate (M) of a particular individual, performing a particular activity, is simply

$$M = (\text{number of mets for particular activity}) * CV * A$$

where
  the number of mets for a particular activity is obtained from a source such as Table 4.7
  CV is a conversion factor
  A is the surface area of the particular person

The conversion factor (CV) depends on the energy units used. The three most-commonly used cases are

$$CV_1 = 50 \text{ kcal}/\text{h m}^2, CV_2 = 18.4 \text{ BTU}/\text{h m}^2$$

$$CV_3 = 58 \text{ W}/\text{m}^2$$

The surface area of a particular person can be estimated using Equation 4.17, or by other means. Suppose that we wanted to estimate the metabolism of a typical young male who weighs 70 kg (154 lb) and has a height of 1.73 m (5 ft 8 in.). The first step is to estimate this person's skin area, using Equation 4.17, as shown here:

$$A_{skin} = 0.202 \ (70 \text{ kg})^{0.425} (1.73 \text{ m})^{0.725} = 1.8 \text{ m}^2$$

If this person is sitting quietly, his metabolic rate, M, is 1 met (see Table 4.7). Using conversion factor CV1, we get

$$M = 1.8 \text{ m}^2 (50 \text{ kcal}/\text{h m}^2) = 90 \text{ kcal}/\text{h}$$

Conduction and convection

The human body also gains or loses thermal energy from the environment through conduction and convection. Heat enters the body when hot air molecules are cooled by contact with the skin, and leaves the body when cool air molecules are warmed by the skin. This cooling and heating process causes convective air currents that greatly increase the rate of heat transfer. In this process, air molecules warmed by skin contact become lighter than the surrounding molecules and start to rise. As they rise, they bump into cool molecules. Some heat is conducted during each contact, and the formerly warm molecules begin to fall. This cycle of rising and falling molecules results in convective air currents that carry heat from or to the skin.

The rate of heat loss or gain through conduction–convection depends upon the exposed area of the person's skin, skin and air temperature difference, insulative values of the clothing covering the nonexposed skin, and the air velocity about the body. For a constant amount of skin exposure, the conduction–convection heat exchange rate is approximately given by the following equation:

$$T_C = kV^{0.6}(t_{air} - t_{skin}) \tag{4.18}$$

where
  T_C is the thermal exchange rate due to conduction–convection
  k is a constant*
  V is the air velocity
  t_{air} and t_{skin} are the temperatures of the air and skin, respectively

---

* Note that the constant k depends on the units used to measure air velocity and temperature, the skin area of the body, and the presence of clothing. When air velocity is measured in meters per minute and temperature in °C, the constant k is approximately equal to 1, if we ignore the effect of wearing clothes and assume a skin area of 1.8 m².

Equation 4.18 gives T_C in kcal/h when air velocity is measured in m/min and temperature in °C. Note that the exchange rate is a linear function of the temperature difference. Air velocity has a nonlinear effect—that is, increasing the air velocity by a factor of four roughly doubles the heat exchange rate. Another important consideration is that the thermal exchange rate increases when more skin is exposed to the air and with lower insulative clothing.*

Radiative effects

The human body emits and receives thermal energy in the form of infrared radiation. If the objects in the environment are colder than the surface of a person's skin, more heat is radiated out from the body than is received. The opposite situation occurs when hot objects in the environment radiate significant amounts of heat to the body. Almost everyone has felt the direct radiation of a bonfire on a cold night; the side of the person facing the bonfire is warmed and the side away from the bonfire is still quite cold.

Example sources of infrared radiation include blast furnaces used in the basic metals industries, infrared lamps, and the sun. The amount of radiation energy exchanged depends upon the cross-sectional area of contact. Consequently, the thermal exchange rate is considerably less when the source is overhead than if it is frontal. Radiative energy exchange is also greatly reduced when reflective clothing is worn. The rate of radiative energy exchange is approximately given by the following equation:

$$T_R = k(t_{source}^4 - t_{skin}^4) \tag{4.19}$$

where

T_R is the thermal exchange rate due to radiation in kcal/h
k is a constant
t_{source} and t_{skin} are the temperatures of the source and skin, respectively, in degrees Kelvin (k)†

A typical value of k is 4.92×10^{-8} when temperatures are in degrees Kelvin. It is obvious from Equation 4.19 that even small increases in the source temperature can cause large increases in the radiative thermal rate of exchange.

Evaporation

A third avenue of heat exchange between the human body and its environment is evaporation. Unlike the other two avenues of heat exchange, evaporation only results in a heat loss by the body. Evaporative heat losses depend primarily on the wetted skin area, air velocities, and the saturation deficit of the air. The saturation deficit is merely the difference between the existing humidity and saturated air. As each of these variables increases, greater heat losses occur through evaporation. The maximum rate of cooling that the human body can achieve by sweating is approximately given by the following equation:

$$E_{max} = 2.0\ V^{0.6}[42 - VP_a] \tag{4.20}$$

where

E_{max} is the maximum evaporative heat loss in kcal/h
V is the air velocity in m/min
VP_a is the ambient vapor pressure of water in mm of mercury‡

---

* The insulative role of clothing is discussed later in this chapter.
† The temperature in degrees Kelvin = 273 + °C.
‡ A vapor pressure of 1 lb/in.² is equivalent to 12.475 mm of mercury vapor pressure.

Chapter four: Assessment and design of the physical environment

Also note that the constant of 42 in this equation is the vapor pressure in millimeters of mercury of water on the skin assuming a skin temperature of 35°C.

Vapor pressure and other values of interest can be determined for particular combinations of temperature and relative humidity by reading a *psychrometric chart*, many variants of which are easily available. One such chart is shown in Figure 4.5. In Figure 4.5, vapor pressure values are plotted on the second-to-last vertical line on the right-hand side of the figure. The temperature values are plotted along the horizontal axis at the bottom of the chart, and the curved lines extending from left up and to the right are the relative humidity lines.* The upward slope of each relative humidity line describes how water vapor pressure increases with temperature when relative humidity is held constant. For example, the curved line toward the bottom of the chart shows that vapor pressure is about 3 mm of mercury at a temperature of 27°C and relative humidity of 10%.

Control strategies for hot and cold environments

Several different types of control strategies for hot or cold environments directly follow from the preceding discussion and can be roughly grouped into three categories: engineering controls, administrative controls, and protective clothing.

Engineering controls

The primary engineering controls used to modify hot or cold environments are adequate heating, ventilation, and air-conditioning. Heating and air-conditioning systems modify both air temperature and humidity and are obviously the first choice for improving many inside environments. However, it should be emphasized that heating and cooling costs can be a very important economic consideration when changes are contemplated for large facilities, such as warehouses. Determining the energy required to change environmental conditions from one state to another is an important step in deciding whether such changes are economically feasible.

The energy needed can be estimated for a perfectly efficient system by reading off enthalpy values for the compared conditions from a psychrometric chart, and then performing a few additional calculations. Enthalpy is the inherent amount of heat energy held by moist air, for different combinations of temperature and humidity, and is measured in units such as kilojoules per kilogram of dry air (as shown in Figure 4.5), the number of British thermal units (BTUs) per pound, or watt-h per gram of dry air. The difference in enthalpy between two environmental states multiplied by the mass of the dry air contained in the facility provides a measure of how much sensible heat must be added or removed to implement a proposed improvement. For pure *heating and cooling*, or, in other words, the addition or subtraction of heat without changing the humidity,† the change of state corresponds to a horizontal shift across the psychrometric chart from one temperature to another. Humidifying or dehumidifying without changing the dry bulb temperature analogously corresponds to a vertical shift across the psychrometric chart.

A second strategy is to increase the air flow around the people to increase evaporative cooling. The latter strategy is effective only at lower humidity levels. Effectiveness

---

* Relative humidity is defined as the ratio of the water vapor density (mass per unit volume) to the saturation water vapor density and is usually expressed on a measurement scale that goes from 0% to 100%. When the air contains no evaporated water, the relative humidity is 0. When air is saturated, the relative humidity is 100%.
† This type of atmospheric change is referred to in the heating and air-conditioning literature as sensible heating and cooling.

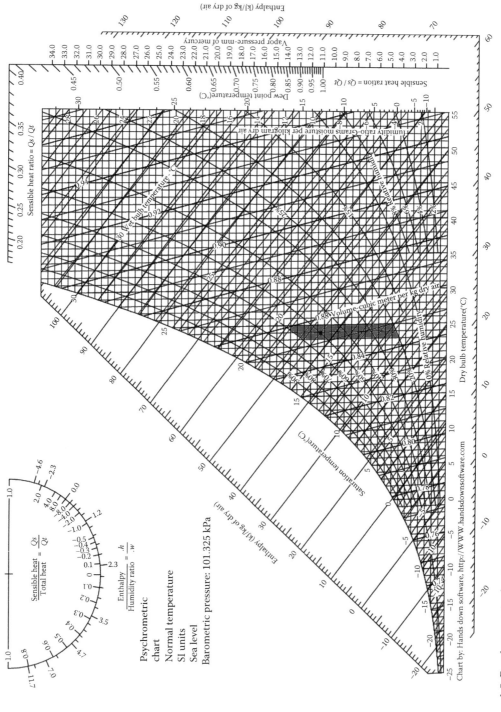

Figure 4.5 Psychrometric chart.

also depends upon how rapidly the air is moving when it contacts a person. The location of fans provided for cooling purposes is an important issue because air velocities drop off substantially with distance. A fan aimed directly at a person can lose half its air velocity with each doubling of the distance.* Another issue is that most fans do not generate any major change in air velocities for locations 15°–20° away from their axis of direction.

A third engineering control strategy is to place heat-producing equipment and processes in locations where most of the people do not go. Another common solution is to provide an air-conditioned space for assembly operations. Air-conditioned rest or break areas are also commonly provided. Other engineering strategies include metallic shielding to reduce exposure to a hot radiating source (e.g., furnaces) or providing tents, windbreaks, or temporary shelters for cold outside situations.

Administrative controls

A wide variety of administrative controls can be helpful when engineering controls are inadequate or infeasible. Developing ways of reducing the metabolic demands associated with the task is perhaps the most fundamental strategy. This can be done in many ways. Equipment can be provided that helps people perform some of the heavier tasks such as lifting. The number of personnel can also be increased to reduce the amount of demanding work each worker must do. A second strategy is to take steps to increase the tolerance of employees to heat. This includes taking steps to ensure employees in stressful conditions receive adequate water and salt, and are properly acclimated to the conditions. Other important administrative controls include scheduling rest breaks; job rotation, employee selection, and training are other common strategies.

Protective clothing

Protective clothing is an especially important solution in cold outdoor environments, but can also play a useful role in hot environments. Clothing protects people from the cold by providing one or more layers of insulation between a person's skin and the atmospheric environment. Although a person's core temperature is around 98.6°F (or 37°C), skin temperatures are lower, and sedentary metabolic rates are such that a nude sitting adult is comfortable at about 86°F (about 30°C). At lower temperatures, clothing must be provided to keep people comfortable. Over the years, methods have been developed for measuring the amount of insulation provided by clothing. The so-called *clo* unit corresponds to the amount of clothing necessary to make a typical adult comfortable in a normally ventilated room at 79°F (about 21°C) with a 50% relative humidity.

It can be shown that for a square meter of surface skin, 1 clo unit works out to about 0.155/W or 0.18/kcal/h. People in offices wear about 0.5–0.7 clo units. Men wearing cloth shirts and trousers over underwear are clothed in about 0.50–0.65 clo units. Add a sweater, and the clo units go up by around 0.30, but add a jacket instead of a sweater, and the clo units go up by about 0.40. A woman wearing a blouse and skirt over regular underwear has about 0.33–0.50 clo units. Add a sweater, and the clo units go up to 0.42–0.80. Changing from a blouse and skirt to a blouse and slacks brings the clo units up to 0.50–0.82. As heat discomfort becomes a problem, people can wear lighter clothing with less insulation, and that is exactly what happens in the summer months in the northern hemisphere.

---

* Although this relationship does not hold for long distances, it is a good rule of thumb for fans located up to 4–6 m (12–18 ft) away from a person.

In more extreme cold situations, special clothing is needed. Numerous situations require people to work outside in very cold to extremely cold situations (e.g., electrical linemen repairing storm damage). In other settings, people work in refrigerated areas. For example, people making ice cream and other frozen products work in hardening rooms with temperatures typically around −20°F or 28.9°C. In such situations, insulated clothing with high clo values is obviously necessary. Insulated gloves and boots are also important. In the design of hand protection from the cold, it is critical to maintain hand skin temperatures above 8°C (about 46°F) as tactile sensitivity drops badly below that level. In fact, manual dexterity is poor when hand skin temperatures drop below about 16°C (61°F) (see Fox, 1967).

A complicating issue is that gloves can interfere with finger movements and dexterity. A clever solution to this problem is illustrated by gloves commonly used by hunters that include a finger flap. The flap allows a person to uncover their trigger finger without taking the glove off of their hand. A very different solution is illustrated by gloves made of multiple thin layers of material, separated by air pockets. It has been shown that such gloves provide both good insulation and a high degree of dexterity.* The layering concept has a long and far-reaching history.† It also should be noted that mittens are warmer than gloves, but gloves allow better manual dexterity.

Several other types of protective clothing have been developed to protect people from heat stress. Workers in steel and other metal-making industries typically wear clothing with moderate insulative value and reflective properties to protect them from heat radiating from molten metals. Clothing with cooling capabilities is available. Examples include clothing that contains sealed liquids that absorb heat and are cooled by dry ice pockets (see Konz and Nentwich, 1969). Other types of clothing provide an evaporative cooling effect by including a layer of wetted fabric.

Many other types of specialized protective clothing have been developed for protecting workers. Clothing such as aprons provides protection from splashed liquids, often used when handling chemicals. The latter type of clothing is sometimes uncomfortable, because the impermeable fabric traps heat and moisture next to the body. Another example is flame-resistant clothing developed for welders and other workers exposed to sparks and flames. Reflective clothing providing protection against radiant heat is a related, but more specialized, type of clothing. The final example is *high visibility* vests for people who work outside and are exposed to oncoming traffic. The colors used and special reflective features of this clothing are selected to improve visibility. The loose-fitting design of most high visibility vests allows the vest to fit over coats, which is important on cold days. The loose, open fit also allows air movement, which makes it more comfortable when worn outside on a hot day.

Assessing thermal conditions and comfort

The previous discussion shows that at least seven factors impact heat exchange between the human body and the environment: (1) air temperature, (2) air humidity, (3) air movement velocity, (4) radiative heating, (5) exertion levels, (6) clothing, and (7) duration of exposure. Since so many factors impact heat transfer, it is not at all surprising that it can be

---

* See Riley and Cochran (1984). These authors examined different globes used in the foods industry, particularly meat processing.
† Not so long ago, people used multiple animal skins to keep warm. The Chinese used to refer to cold days by the number of suits one should wear to keep warm.

difficult to accurately predict how comfortable people will be in different thermal environments, not to mention effects on performance, and health effects.

Over the years, several different measures of thermal conditions have been developed. The simplest, and most commonly used measure, is dry bulb air temperature. The latter measure is often supplemented by measurements of relative humidity, obtained from the wet bulb temperature or use of other instrumentation. Radiant temperature (also called globe temperature) is another supplemental measurement. Several thermal indexes have been developed to predict thermal comfort and health effects, such as effective temperature (*ET*), windchill index, heat stress index (*HSI*), and wet bulb global temperature (*WBGT*). The latter measures take into account humidity, air movement, radiant heat, exertion levels, and other factors to a varying extent.

Temperature and humidity measurement

The starting point in the assessment of thermal conditions is to measure air temperature and relative humidity. More ambitious efforts also measure radiative heating effects and air movement velocity.

The *dry bulb* air temperature is measured using a thermometer shielded from sources of radiation (i.e., measuring outside temperature in the shade, rather than under the sun), and provides a useful starting point for evaluating thermal conditions. The *wet bulb* temperature takes into account the cooling effects of evaporation, and can be used to estimate the relative humidity when the dry bulb temperature is known. The wet bulb temperature has traditionally been measured for psychrometric purposes by placing a wetted wick over the bulb at the end of a mercury thermometer and then recording the temperature after the thermometer is rapidly swung through the air or air is blown over the bulb with a fan. Evaporation of air from the wetted wick causes the wet bulb temperature to drop compared to the dry bulb temperature, unless the air is fully saturated. Once both measures are known, the relative humidity can be determined using a psychrometric chart. For example, a quick glance at Figure 4.5 reveals that at a relative humidity of 25%, the dry and wet bulb temperatures are, respectively, 42°C and 25°C.

The *globe* temperature has traditionally been measured by a thermometer placed at the center of a thin-walled, 6 in. diameter, blackened copper sphere. The surface of the sphere absorbs radiant energy from the environment, causing the globe temperature to be elevated over the dry bulb temperature when significant sources of radiant energy are present. Convective heating and cooling of the surface of the globe also occurs. The globe temperature, consequently, measures a combined effect of radiant heating, ambient temperature, and air movement. An important point is that 20–30 min should be allowed before taking readings, to allow the temperature at the center of the globe to stabilize after being brought into a new environment.

Effective temperature

ET is a measurement scale intended to describe combinations of air temperature, humidity, and air movement that are perceived to be equivalent by people wearing light clothing, engaged in light activity.* In its original form, the ET scale assumed a standard condition of saturated, still air, with a mean radiant temperature (MRT) equal to the dry bulb temperature (Houghton and Yagloglu, 1923). In the early 1970s (Gagge, Stolwijk, and Nish, 1971),

---

* Other scales have been developed, such as operative temperature. The latter scale includes a radiative heating term and has been used extensively in France.

Table 4.8 Table of Windchill Index Equivalences Developed for the U.S. Army

	Air Velocities in m/s							
tdb C	2	4	6	8	10	12	14	16
+4	+3.3	+1.8	−5.0	−7.4	−9.1	−10.5	−11.5	−12.3
+2	+1.3	−4.2	−7.6	−10.2	−12.0	−13.5	−14.6	−15.4
0	+0.8	−6.5	−10.3	−12.9	−15.0	−16.5	−17.6	−18.5
−2	−2.8	−8.9	−12.9	−15.7	−17.9	−19.5	−270	−21.6
−4	−4.9	−11.3	−15.5	−18.5	−20.8	−22.5	−23.8	−24.8
−6	−6.9	−13.7	−18.1	−21.3	−23.7	−25.5	−26.9	−27.9
−8	−9.0	−16.1	−20.0	−24.1	−26.6	−28.5	−29.9	−31.0
−10	−11.0	−18.5	−23.4	−26.9	−29.9	−31.5	−33.0	−34.1

Source: Sipple and Passel, 1945.

a revised ET scale was developed that assumes a standard condition of 50% relative humidity rather than saturated air and is now used by the American Society of Heating, Refrigeration, and Air-Conditioning Engineers (ASHRAE).

The shaded region in the middle of Figure 4.5 gives a comfort region for lightly clothed, sedentary individuals in spaces with low air movement and where the MRT is equal to the air temperature. This comfort zone was established by ASHRAE who have supported considerable research over the years to establish scientifically based standards for thermal environments.

Windchill index

The windchill index was developed by Dr. Sipple for the U.S. Army. The intent of this index is to capture the effect of cold, windy conditions in outdoor environments. Anyone who has experienced cold winter weather has first-hand knowledge that increased air velocity makes it feel colder. Table 4.8 shows a moderately extensive windchill index. The numbers in the table give equivalent temperatures for different combinations of wind speed and dry bulb temperature. The equivalent temperatures correspond to dry bulb temperatures under low wind conditions of 1.8 m/s or less, which cools water at the same rate observed for the environmental conditions. For example, the table shows an equivalent temperature of −6.5°C, for conditions where the wind speed is 4 m/s and dry bulb temperature is 0°C. Note that this equivalent temperature is nearly the same as the value shown (−6.9°C) for a temperature of −6°C and wind speed of 2 m/s.

Heat stress index

Sometimes it is important for ergonomists to assess hot environments that expose people to conditions well outside the comfort zone. Some of the early studies addressing this issue were conducted by the U.S. military. One such study reported that people at minimal levels of exertion were able to tolerate dry heat at a dry bulb temperature of 200°F for around half an hour. Saturated air at temperatures between 118°F and 122°F, however, could only be tolerated for a few minutes, showing that humidity greatly reduces people's tolerance when dry bulb temperatures are high (Box 4.8).

Over the years, a large number of heat stress indexes (HSIs) have been developed that address the effect of temperature and humidity. One of the more well known is the HSI developed by Belding and Hatch (1955). This index relates the amount of heat that must

> **BOX 4.8 RELATIONSHIP BETWEEN THERMAL SENSATION AND ET**
>
> The relationship between ET and subjective thermal sensations (TS) was explored several years ago by Rohles, Hayter, and Milliken (1975). In this study, people were asked to rate thermal conditions on the following scale: 1—cold, 2—cool, 3—slightly cool, 4—comfortable, 5—slightly warm, 6—warm, and 7—hot. For conditions within the ASHRAE comfort region, the average rating was around 4.0, with a standard deviation of 0.7. As the observed standard deviation indicates, there were large differences in ratings between individuals. The authors of this study also developed a series of TS equations for ranges of ET. The equations and their respective ranges of prediction are as follows:
>
> $$TS = -1.047 + 0.158\, ET \quad \text{for} \quad ET < 20.7°C \quad (4.21)$$
>
> $$TS = -4.44 + 0.326\, ET \quad \text{for } 20.7°C < ET^* < 31.7°C \quad (4.22)$$
>
> $$TS = 2.547 + 0.100\, ET \quad \text{for} \quad ET > 31.7°C \quad (4.23)$$
>
> Figure 4.6 plots out the relationship given by these equations. It is easy to see from this figure that there is an S-shaped relationship between TS scores and ET. The subjects were most sensitive to changes in ET in the mid-region from slightly over 20°C ET to slightly over 30°C ET.

be dissipated to maintain thermal equilibrium to the maximum amount that can be dissipated by sweating using the following equation:

$$HSI = \frac{E_{req}}{E_{max}} \times 100 \quad (4.24)$$

where
   E_{max} can be calculated using Equation 4.20
   E_{req} can be calculated for a particular environment by solving for the value of E in Equation 4.16, which results in bodily heat storage of zero

The heat that must be dissipated by sweating to maintain thermal equilibrium is

$$E_{req} = M \pm C_c \pm C_v \pm R \quad (4.25)$$

Simplified nomograms for calculating the HSI are also available (McKarns and Brief, 1966). When the HSI is 100%, the required evaporative cooling capacity is equal to the maximum evaporative capacity. If this limit is exceeded, the net flow of heat is positive and body temperature begins to rise. The exposure time (in hours) resulting in a 1°C increase in body temperature can be roughly calculated as*

---

* As mentioned earlier, a change of less than 1°C in core body temperature is generally viewed as the acceptable limit from a safety and health perspective.

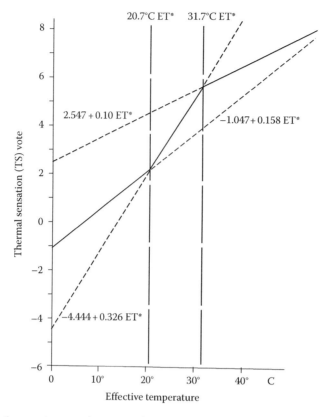

Figure 4.6 Thermal sensation as a function of ET*.

$$T = \frac{\text{body mass}}{E_{req} - E_{max}} \qquad (4.26)$$

where
  body mass is measured in kg
  E_{max} and E_{req} are measured in kcal/h

The latter equation assumes that the human body is entirely composed of water, and the equation follows since 1 kcal is the energy needed to heat 1 kg of water by 1°C.

Calculation of the HSI for a particular environment can help identify the need for potential corrective actions. Some physiological implications and corrective actions for particular values of the HSI given by Belding and Hatch are shown in Table 4.9. More importantly, calculation of E_{req} can provide helpful guidance regarding both the root causes of heat stress in a particular setting and the potential effectiveness of particular interventions. This follows because determining E_{req} involves separate determinations of metabolic heat, convection and conduction, and radiative transfers of heat. The relative influence of each component can be compared, and attention can then be focused on addressing the factors that have the greatest impact.

On the other hand, this index is a now a bit old and is not extensively used. One concern is that the assumption of a constant skin temperature of 35°C can cause the HSI to

Table 4.9 Physiological Implications of HSI Values

HSI	Physiological Implications
−20	Mild cold strain
−10	
0	No thermal strain
10	Mild to moderate heat strain. Decrements in mental functions, vigilance, and dexterity
20	No substantial effect on performance of heavy physical work
30	
40	Severe heat strain. A threat to health unless workers are physically fit
50	Break-in period required for unacclimatized workers. Medical selection
60	desirable to screen out people with cardiovascular or respiratory impairments, putting them at risk
70–90	Very severe heat strain. Only a small percentage of population may be expected to be qualified for such work. Adequate water and salt intake critical. Personnel selection should be subject to medical examination and trial period to establish adequate ability to acclimate to conditions
100	Maximum strain tolerated by fit, acclimated, young men

Source: Adapted from Belding and Hatch, 1955.

overestimate heat stress. This follows because skin temperatures can increase to around 38°C under high exertion conditions, which increases the rate of evaporative cooling. Another limitation stems from the fact that HSI was almost solely based on acclimatized healthy males between the ages of 30 and 40 years, a restriction that leaves out a large percentage of the population.

Wet bulb global temperature index

The WBGT is a more modern index that is both easier to calculate than the HSI and recommended by both the American Conference of Governmental and Industrial Hygienists (ACGIH) and the U.S. National Institute for Occupational Health and Safety (NIOSH). To calculate the WBGT, readings of dry bulb temperature (t_{db}), *natural* wet bulb temperature (t_{wb}), and MRT (t_r) must first be taken. Note that when the natural wet bulb temperature is measured, the wetted thermometer is held stationary. This procedure differs from that followed when the wet bulb temperature is measured for psychrometric purposes, because the natural wet bulb temperature is intended to measure the evaporative cooling due to natural air movement alone, rather than the maximum cooling effect possible due to evaporation (Box 4.9).

For sunny outdoor conditions, the WBGT is calculated as

$$\text{WBGT} = 0.7 t_{wb} + 0.2 t_r + 0.1 t_{db} \tag{4.27}$$

Indoors or outdoors without a solar load, the WBGT is calculated as

$$\text{WBGT} = 0.7 t_{wb} + 0.3 t_r \tag{4.28}$$

Once the WBGT is calculated, it can be used along with metabolic measures to specify work–rest requirements for the job (Figure 4.7). The latter figure contains ACGIH recommendations for different combinations of the WBGT index and metabolic demands. As shown in the figure, a higher WBGT and greater energy expenditures increase the amount of time that

BOX 4.9 EXAMPLE CALCULATION OF THE WGBT INDEX

Suppose that the following atmospheric information was available: $t_{wb}=87°F$, air velocity is 5 ft/min, t_r is 90°F, $t_{db}=89°F$. The job is done outdoors in a sunny environment and estimated to result in a metabolic rate of 350 kcal/h. The WBGT is as follows:

$$WBGT = 0.7t_{wb} + 0.2MRT + 0.1t_{db}$$
$$= 0.7(87) + 0.2(90) + 0.1(89)$$
$$= 87.8°F$$
$$= 30.9°C$$

A quick glance at Figure 4.7 reveals that the ACGIH-recommended work schedule is 15 min of rest per hour on the job.

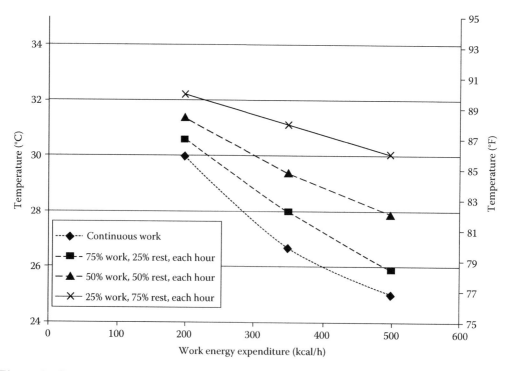

Figure 4.7 Permissible heat exposure threshold limit values. (Adapted from data contained in OSHA, 1999.)

should be allocated for rest. In the least-demanding case, the work can be done continuously over the 8 h workday. Three other cases are shown in the figure, respectively corresponding to 25%, 50%, and 75% of the time resting each hour. These recommended schedules assume that the worker will be resting in the same conditions they are working in. If the resting place is below 24°C (75°F), the guidelines reduce the necessary resting time by 25%.

Some other assumptions and details regarding the system are as follows:

- Continuous work includes 15 min work breaks each mid-half-shift (or every 4 h) and a mid-shift break of 30 min or more.
- Workers will receive 150 mL of water (1 cup) every 15–20 min during hot weather and adequate salt intake.
- A maximum allowable body temperature of 38°C (100.4°F), heart rate of 110 beats/min, and metabolic rate averaged over the workday of 330 kcal/h.
- Workers would be wearing light summer clothing.

A final comment is that the WBGT has been used by the military and other organizations for years, in ways other than discussed here, to both evaluate and set safety and health limits for thermal conditions. Additional information on this topic is available from several sources.* It should also be noted that a number of different researchers have established that other indexes, such as the Botsball index (BB), are highly correlated with the WBGT measurements.†

Effects of thermal conditions on performance

A number of studies have investigated the effects of the thermal environment on human performance. Typically these studies have not found strong effects until the atmospheric conditions are quite extreme. Part of the issue is that people have effective ways of coping with environmental conditions. Most studies of the effect of thermal conditions have also been conducted over short periods of time, perhaps insufficient to allow the effects to be accurately measured. The results of several older studies are briefly summarized in Box 4.10.

These and other studies (see Goldman, 1977) show that task performance is affected by atmospheric conditions, but performance decrements are task specific and subtle below 30°C ET (about 86°F ET). There is also some evidence that impairments in mental performance are related to how long people are exposed to warm thermal conditions. (See Box 4.10 for specific findings.)

Problems in human performance are also found when temperatures are low. For example, it was observed during World War II that sentries in the North Atlantic were not as vigilant after longer exposure to cold thermal conditions. Vigilance decrements occur after relatively small reductions in body temperature. When the body core temperature drops to about 35°C, people begin to show increasingly severe signs of mental confusion. The temperature of the extremities is another concern. People have difficulty manipulating hand tools or even striking a match when their hands are cold.

A final observation is that scarcely any of the studies reported here examined either seasonal or MRT effects on performance; so, little is known about the effect of these conditions. However, some population comfort effects have been identified in a series of studies sponsored by ASHRAE,‡ examining temperature preferences of people from several large cities in the United States. During the summer months, most people preferred an ET of 70°F–71°F. Most preferences fell within a range of 65°F–78°F. During the winter months, preferences of people from northern cities tended to fall within a range of 57°F and 68°F. These data show that people acclimatize, or change with the seasons, in locations where the average temperatures drop in the winter.

---

* A good source is Kroemer, Kroemer, and Kroemer-Elbert (1990, p. 143) in the second edition.
† Beshir, Ramsely, and Burford (1982) report that these two indexes correlate very highly, $r = 0.96$.
‡ ASHRAE is the American Society of Heating, Refrigeration, and Air-Conditioning Engineers.

BOX 4.10 EFFECTS OF THERMAL CONDITIONS ON TASK PERFORMANCE

Study Focus and Setting	General Finding	Source
Accidents in British coal mines	Higher accident rates when ambient temperatures were below 65°F or above 69°F.	Vernon and Bedford (1931)
Productivity of British factory workers	A reduction of 10%–15% in productivity in summer compared to winter. A 60% decrease when wet bulb temperatures exceeded 85°F.	Bedford (1953)
Errors made by British servicemen	Increased errors in recognizing Morse code telegraphy signals for atmospheric conditions at or above 87.5°F ET.	Mackworth (1946)
Impact of AC on test scores of U.S. Navy trainees	No significant differences in test scores between trainees who took tests in air-conditioned spaces (about 71.3°F ET) and non-air-conditioned spaces with wet bulb temperatures of 86°F–93°F.	Mayo (1955)
Impact of AC on U.S. government employees	Document search efficiency in the air-conditioned wing of building improved by 6.5%–13.5%, average productivity improved by 9.5%. Study conducted over the warm summer months in Washington, DC	Brandt (1959)
Student performance	Better academic performance in winter compared to summer months.	Nolan (1960)
Morale and health of Canadian students	Colds and upper-respiratory-tract infections increased sharply during the winter at relative humidity levels below 50%. Unchanged absenteeism.	Green (1974)
Accident rates and productivity of South African workers	Reduced productivity and higher accident rates for temperatures above 86°F (30°C).	Wyon (1977)
Performance of skilled European workers	Reduced performance for when ET was greater than 86°F (30°C) or when over 40% of the skin was wet.	Pepler (1958) and (1964)
Visual performance in laboratory setting	Visual tunneling with short-term exposure to 105°F dbt, 95°F wbt and 120 ft/min air flow. Greater effects for longer exposures.	Bursill (1958)
Pursuit tracking in laboratory setting	Poorer performance at temperatures above 94°F dbt and 84°F wbt.	Teicher and Wehrkamp (1954)
Perceptual performance	Tactile sensitivity best at about 72°F, visual–motor coordination best at 70°F.	Russell (1957)
	Vigilance begins to decline at oral temperatures around 35.6°C (96.0°F).	Poulton (1970)
Mental performance	Review of 14 different studies showed that impaired performance on arithmetic and memory tasks occurred more quickly at higher effective temperatures.	Wing (1965)

Study Focus and Setting	General Finding	Source
Arousal	Subjects in conditions 4°C above their preferred temperature were less aroused and more fatigued than at temperatures 4°C less than the preferred temperature. The average preferred temperature was 22.6°C. Mental task performance was not significantly affected by temperature.	Langkilde et al. (1973)

Hazards and control measures

In addition to lighting, noise, and temperature, engineers may need to develop a safety and health management program that actively identifies potential hazards and analyzes them for severity and frequency. Hazards are identified through a variety of methods. It is beyond the scope of this chapter to present some of the formal methods for analyzing hazards, such as failure modes and effects analysis. However, the first step in identifying potential hazards is to look at the injury and illness experience of your own company as well as those of others in your industry through accident statistics, as was discussed in the previous section.

In addition, hazards can generally be classified into categories. These categories can be used as a general framework for deciding which hazards are potentially applicable at a given work site. The remainder of this chapter presents a brief discussion of the common categories of hazards encountered in industrial workplaces. Typical hazards within each category are briefly described, and control measures are discussed.

Fall and impact hazards

Many accidents, such as falling or being struck by a moving or falling object, are referred to as fall and impact hazards. Many such accidents are easily preventable. As it turns out, falls and impacts account for about 34% of the accidents reported by the National Safety Council for which there was compensation for injury costs. These types of accidents are particularly common in the construction industry.

Falls often result from slipping or tripping. Wet floors are an invitation for slips and falls. Good housekeeping minimizes slip hazards, and marking wet floors alerts people to use caution. Keeping floor surfaces in good repair and clear of objects such as equipment and cords will reduce trip hazards.

Falls from elevated work surfaces, such as stairs, ladders, and scaffolds, are also a frequent type of accident in this category. Safety harnesses and guardrails are important safety equipment for minimizing falls from elevated work surfaces. In addition, it is important that ladders, stairs, and guardrails meet OSHA and architectural design standards. Figure 4.8 depicts some short portable stairs for short climbs. Note the side handrails that provide stability in climbing. Removable sections of handrails* should be marked and

---

* Many companies require removable handrails so that materials can be passed to an upper floor. If the removed rails require hand tools or fasteners for replacement, but these are unavailable, then the handrail could get replaced without being secured, creating a very dangerous situation that looks perfectly safe. A little design ingenuity is required.

Figure 4.8 Portable stairs with handrails.

designed so that replacements can be securely attached without using hand tools or fasteners. If falls from the front of the stairs are possible, a cross-handrail should be included.

Elevated work surfaces also present dangers from falling objects on those working below. While work situations in which workers are working physically above others are unavoidable in many construction and maintenance operations, extreme care should be taken to prevent hand tools and other objects from falling on those working below. Sometimes fine mesh fencing can be placed horizontally to catch falling objects. Regardless, hard hats should be worn for protection at all times.

Less obvious, but important to preventing falls, is the screening and proper selection of employees for working in elevated workspaces. There are psychological and physiological differences among people with regard to their ability to tolerate heights. Some people are afraid of being exposed to heights in which there are few barriers to falling or jumping. Such workers should be screened from jobs where work on elevated surfaces is required. New employees should always be accompanied by an experienced employee who can observe any adverse reactions to heights, or fear of falling, that might impair their safety. Also, people with colds or flu, or those who are recovering from those diseases, often have an impaired sense of balance and, therefore, should not be working in situations where momentary imbalance can affect their safety. Some medications for colds are suspected of impairing the sense of balance.

Hazards of mechanical injury

Machinery or tools pose a potential for cutting, shearing, crushing, and pinching injuries when people contact sharp or moving parts.

Cutting/shearing

Sharp cutting tools and machines are common hazards in many factories. Knives in the meat-packing industry is an example that easily comes to mind, but there are numerous sharp edges in companies that make metal, glass, lumber, or ceramic products. Almost all manufacturing firms have sharp machine parts.

A common cause of cutting and tearing accidents is powered cutting tools. Most of these tools have guards that prevent the human body from slipping into the rotating or reciprocating blades. While these guards are not infallible, removal of the guards should only be permitted in extreme situations where the job is impossible to perform with the guard in place. It is critical then that those guards are replaced immediately. The same is true for guards for shearing machines, which have been a source of many finger–hand amputations.

Crushing/pinching

Pinch points are locations other than the point of operations where it is possible for the human body to get caught between moving parts of machinery. These points are particular hazards to crushing of body parts. Other causes of crushing include hitting one's finger with a hammer or getting one's hands between two heavy moving objects such as objects suspended from crane cables. Besides crushing, many of these same accidents can lead to broken bones.

The use of machine *guards* and *safety devices* can provide protection from mechanical injury accidents. Guards are intended to prevent any part of the human body from entering a hazardous area of the machine, while safety devices deactivate the machine when body parts are in hazardous areas. Sometimes safety devices are hooked to machine guards to prevent the machine from operating without the guards in place.

Preferred guards and safety devices work automatically, impose little or no restrictions on the operator, and are productive. Good guards should also be fail-safe and prevent the operator from bypassing or deactivating them.* Some such guards totally enclose hazardous operations with limited, adjustable, or no access. Limited access guards allow small product parts to be inserted by the operator, but prevent entry of body parts. Adjustable access guards permit changes to the openings around the hazard area in order to adapt to different sizes of product parts. Unfortunately, there is a tendency for operators to leave the guard at the maximum size opening at all times and, hence, deactivate much of the guard protection.

A number of safety devices have been devised for worker protection. Optical screens, ultrasonic devices, and field-sensing devices are common safety devices that are used around presses and press brakes. Optical screens deactivate the machine whenever a light shield around the machine openings is broken. Ultrasonic devices and field-sensing devices detect any intrusion of the hands into a hazard area and terminate machine operation. Screens and sensing devices may need to be supplemented with special stopping mechanisms in certain types of machines that complete a cycle even after the power is cut off.†

Other safety devices are two-hand control devices and pullout devices. Two-hand control devices are really interlocks that require two control devices to be pressed‡ before the machine activates. Pullout devices often have wrist cuffs that are connected to small cables that pull the operator's hands clear of the danger area when the machine activates. However, pullout devices often restrict movement and productivity.

Another type of protective approach is to use mechanical means to feed product units into dangerous machines, thus, creating protective distance between the person and the

---

* There have been numerous instances of operators deactivating guards and other protective devices. These deactivations often occur when the operators are working on incentive systems and they see a way to make more money by keeping the safety device from slowing them down. Incentive systems should not be employed when there are dangers of these kinds in the shop. Disciplinary action is also needed when a safety device is knowingly deactivated and not reported.
† Presses and press brakes are examples of machine tools that frequently complete a cycle even after power is shut off. In such cases, special stopping mechanisms may be required.
‡ The two devices are presumably depressed with separate hands. However, when the two-hand switch devices are too close, an operator can elbow one switch and depress the second with the hand on the same arm.

hazard. In addition to guards and safety devices, there is need for supervision to thoroughly train operators and remind them not to wear rings or loose clothing that can catch in the machine and cause accidents.* All of these efforts have resulted in fewer industrial accidents with mechanical injuries, but the rate is still unacceptably high.

Woodworking industries present some particular challenges in mechanical injury. One of the reasons is that wood is a highly variable material, and another is in the nature of power woodworking machines. In sawing operations, for example, a knot in the wood can be caught in a saw blade in a way that forces the piece of wood back out toward the operator. To prevent some of these situations, newer models of woodworking power tools have antikickback devices. Basically, these devices force the wooden workpiece to bind if the piece suddenly starts to move in the reverse direction. Other antikickback devices consist of cam-like metal pieces next to the wooden workpiece that can move in one direction but will grip and dig into the workpiece when the direction of movement is reversed.

Mechanical injuries are a primary concern to ergonomic specialists involved in equipment design. A guiding principle is to design safety features that are simple and that minimally interfere with efficient operation of the equipment. Another principle is that danger points should be *labeled with an appropriate warning*.

Vibration hazards and cumulative trauma disorders

Vibration hazards and cumulative trauma disorders (CTDs) are common hazards in the workplace.

Vibration hazards

Most power tools are vibratory in nature and these vibrations are imparted to people handling them; chain saws, chipping hammers, jackhammers, and lawnmowers are a few that come to mind easily. These power tools tend to vibrate the operator's hands and arms. Also, some vehicles exhibit heavy vibration and people inside are subject to whole-body vibration.

It is known that whole-body vibration can increase heart rate, oxygen uptake, and respiratory rate. In addition, whole-body vibration can produce fatigue, insomnia, headache, and *shakiness* during or shortly after exposure. However, it is unclear what the long-term effects are from exposure to whole-body vibration.

In the case of hand and arm vibrations, there tends to be a vasospastic syndrome that is known as *Raynaud's syndrome* or *dead fingers*. This circulatory disorder is usually permanent. Typically, it takes several months of exposure for around 40–125 Hz vibration to occur, but there appear to be large individual differences among people relative to the onset. The two primary means of preventing or reducing the onset frequency of Raynaud's syndrome are (1) reducing the transfer of vibration from hand tools and (2) protecting hands from extreme temperatures and direct air blast. Three ways to reduce vibration transfer from hand tools are (1) to make tool–hand contacts large and nonlocalized, (2) to dampen vibration intensities at the handles with rubber or other vibratory-dampening materials, and (3) to require operators to wear gloves, particularly those with vibration-arresting pads.

---

* A number of companies appoint forklift operators or other roving operators to the additional duty of safety enforcement. As they go around the plant, it is their duty to remind anyone who is not following good safety practices to change behavior. Many people simply forget to wear safety glasses or hearing protectors, and simple reminders are good practice.

Table 4.10 Typical Cumulative Trauma Disorders, Associated Tasks, and Occupational Factors

Task Type	Occupational Risk Factors	Disorder
Assembling parts	Prolonged restricted posture	Tension neck
	Forceful ulnar deviations	Thoracic outlet syndrome
	Thumb pressure	Wrist tendinitis
	Repetitive wrist motion, forearm rotation	Epicondylitis
Manual materials handling	Heavy loads on shoulders	Thoracic outlet syndrome
		Shoulder tendinitis
Packing boxes	Prolonged load on shoulders	Tension neck
	Forceful ulnar deviation	Carpal tunnel syndrome
	Repetitive wrist motion	DeQuervain's syndrome
Typing, cashiering	Static or restricted posture	Tension neck
	Arms abducted or flexed	Thoracic outlet syndrome
	High-speed finger movement	Carpal tunnel syndrome
	Ulnar deviation	

Source: Adapted from Putz-Anderson, 1988.

Cumulative trauma disorders

CTDs have already been discussed in previous chapters, so only a very brief mention is made here. In a similar sense to vibration hazards, CTDs are frequently associated with certain occupational tasks and risk factors, of which ergonomic designers should be aware.

Table 4.10 summarizes some typical occupational tasks and risk factors along with frequently associated CTDs. While most CTDs do not lead to life-threatening situations, they do lead to missed workdays and considerable inconvenience.

Pressure hazards

Pressure hazards can be found in many industrial environments. There are a number of sources of highly pressurized gas around most industrial plants. Boilers are one common source. When the expansive force of an enclosed fluid exceeds the pressure vessel's strength, ruptures occur, often with explosive results and sometimes with great heat. In steam boilers, loss of water will create superheated steam with very high resulting pressures. Safety valves on most steam boilers are designed to trip at a safe upper pressure limit. While this practice reduces pressure hazards, the outflowing steam is also a safety hazard for burns.

Unfired pressure vessels, such as portable gas cylinders, air tanks, or aerosol cans, can also explode when excessive pressure builds. A major cause of such accidents occurs when these vessels are exposed to sunlight or other heating sources. Special care must be exercised to keep these pressure vessels in cool environments. Note also that portable gas cylinders often contain liquefied gases, which themselves become dangerous if released. These cylinders should be prominently marked as to their contents. Also, these cylinders should always be secured when in use, during transport, and in storage to prevent falling and resulting rupture. When a fall causes a valve break at the end of a cylinder, the cylinder can act like a flying missile.

Another pressure source that is common in industry is compressed air. Many industrial power hand tools, such as impact wrenches, use air pressure. Typically, a hose supplies compressed air to the hand tool. A typical source of accidents occurs when the hose accidentally uncouples from the tool and whips with great force. Long air hoses should be restrained in case of accidental uncoupling.

Electrical hazards

By far the most-common electrical injuries are shocks and burns. Burns cause destruction of tissue, nerves, and muscles. Electric shocks can vary considerably in severity. Severe shocks can cause temporary nerve center paralysis along with chest muscle contractions, resulting in breathing impairment and, if prolonged, death by asphyxiation. Severe shocks can also cause ventricular fibrillation, in which fibers of the heart muscles begin contracting in a random uncoordinated pattern. This is one of the most serious electrical injuries because the only known treatment is defibrillation, which requires special equipment and skills to administer.

Other electrical hazards include:

- Mechanical injuries from electrical motors
- Fires or explosions resulting from electrical discharges in the presence of dusts and vapors
- Falls that result from the electrical shock directly or as a result of human reaction afterward

Electrical current flows inversely with resistance. The outer layers of human skin have a high resistance to electricity when the skin is not ruptured or wet. However, wet or broken skin loses 95% or more of its natural resistance. Typically, dry skin has about a 400,000 Ω resistance, while wet skin resistance may only be 300–500 Ω. The following equation shows that current is equal to voltage divided by resistance or

$$I = \frac{E}{R} \qquad (4.29)$$

Using Equation 4.29, one can determine that if a person contacts a 120 V circuit, current is only 0.3 mA with dry skin, but 240 mA or more with wet skin.*

When a person comes into contact with an electrical source, current flows from source to ground. The path of this source-to-ground flow is critical. If as little as 10 μA reach the heart muscles, ventricular fibrillation can occur. One minute after the onset of ventricular fibrillation, the chance of survival is 99% if both a defibrillator and the people with the training to use it are available; otherwise, the chances are about 1/10,000 for survival.

One cause of electrical accidents is contact with bare power conductors. Ladders, cranes, or vehicles can strike a power line and act as a conducting agent to whomever is touching it. While insulated electric wires offer some protection as long as the insulation is intact, the insulation can break down due to heat or weather, or as a result of chemical or mechanical damage. As the insulation weakens, the hazard level approaches that of a bare conductor. In fact, there can be even more danger because the insulation makes the wire still appear to be safe. Insulation breakdown is often accelerated in high-voltage circuits. The corona around high-voltage wires often produces nitrous oxide, which becomes weak nitric acid in the presence of moisture; this compound further decomposes insulation.

---

* Amperage $I = 120\,V/400{,}000\,\Omega = 0.3\,mA$ with dry skin but $I = 120/500 = 240\,mA$ with wet skin.

Equipment failure is another cause of electrical accidents. Internal broken wires can sometimes ground out to the external casing of the equipment. To prevent this, manufacturers of electrical equipment have begun to provide double insulation or a direct-grounding circuit for internal wiring. In addition, ground-fault circuit interrupters are now widely used because these interrupters shut off the circuit almost instantly when the ground circuit exceeds the limit. Many building codes require ground-fault circuits in exterior electrical outlets and those in garages.

Electrical safety can be improved by designing electrically powered equipment with several features. Most importantly, all electric-powered devices should be designed such that they can be placed in a zero-energy state (i.e., power can be shut off). An associated second feature is the ability to lock out the power via a tag, cover, or hasp and padlock. It is one thing to turn off power; it is another to make sure it stays off, particularly during servicing or maintenance. A third safety feature is important in power equipment that utilizes energy-storing devices such as capacitors or accumulators. Such equipment should be designed so that an operator can safely discharge the energy from these storage devices without contact.

Sometimes fires and burns can result from electrical devices overheating. Electrical heating systems are particularly prone to overheating. Circuit breakers and thermal lockouts are common devices to prevent those occurrences. These devices may have thermally activated fuses, regular circuit breakers that are thermally activated, or circuit breakers that bend with heat and open the circuit at set upper temperature limits.

Another source of electric hazard is static electricity. *Static electricity* exists when there is an excess or deficiency of electrons on a surface. That surface becomes the source or sink of a very high voltage flow,* particularly when the humidity is low. Static electricity occurs often in papermaking, printing, and textile manufacture. A primary danger is that dust and vapor explosions can be ignited with the static electrical discharge.† Static electricity can be reduced by using materials that are not prone to static buildup. Another protection is to ground out equipment so that there is no buildup of static electricity. Other prevention measures consist of neutralizing the static electricity, and humidifying to reduce the buildup. Lightning is the natural discharge of static electricity. Large conductors and ground rods are used to safely ground out the discharges. Pulse suppressers are often used with computers and other equipment that are vulnerable to damage caused by power spikes, which lightning often produces.

Burn, radiation, and fire hazards

Burn hazards

While most industrial hazards in this category are heat related, it should be remembered that very low temperatures associated with freezing‡ or with cryogenic processes can also be a source of danger.

Burns are the principal result of accidents involving heat. The severity of a burn depends upon the temperature of the heat source and the duration and region of the contact. Milder burns, which usually only redden the skin, are *first-degree* burns. *Second-degree* burns are more serious and are usually associated with blisters on the burned areas of the skin. When the blisters break, there is a chance of infection. The most severe burns

---

* Typically, the static electrical flow has very low amperage but extremely high voltage.
† In the late 1970s, R.R. Donneley Company had a plant in Chicago that exploded due to static electricity. Paper dust in the overhead trusses was ignited by the static electricity caused by moving paper in the machines.
‡ Freezing is associated with food processing.

are *third-degree* burns, which penetrate all the layers of the skin and kill nerve endings. Third-degree burns can cause the affected tissue to appear white, red, or even a charred gray or black. Typically, part of the tissue, capillaries, and muscle are destroyed, and gangrene often follows. Freezing burns are similar to heat burns in many ways, including the degrees of severity.

Ultraviolet radiation hazards

Ultraviolet (UV) radiation can also cause burns. Eyes are particularly vulnerable to UV burns, as eyes are much more sensitive to UV radiation than the skin. UV radiation is emitted during welding, so special eye protection must be worn to protect the welder's eyes. UV rays can also bounce off light-colored walls and pose a hazard to others in the vicinity. For that reason, welding booths are recommended. Other types of equipment that pose similar hazards include drying ovens, lasers, and radar.

Fire hazards

A fire requires three principal elements:

1. *Fuel* (or reducing agent), which gives up an electron to an oxidizer
2. *An oxidizer*, which is the substance that acquires the electrons from the fuel
3. *A source of ignition*, which is anything capable of commencing the oxidation reduction reaction to a sufficient degree for heat or flame

Many substances are always fuels or always oxidizers, but a few will switch roles.* Fuels include regular heating fuels, solvents, and any other flammable liquids, gases, or solids.† Oxidizers include oxygen itself and compounds carrying oxygen.‡

The key to prevention and control is *keeping these three parts separated in time and/or space*. Airborne gases, vapors, fumes, or dust are particularly dangerous, since they can travel significant distances to reach ignition sources and often burn explosively.

A flammable material can be a gas, liquid, or solid. Flammable gases burn directly. In contrast to gases, liquids do not burn; they vaporize and their vapors burn. Solids often go through two phase changes before burning, first to liquid and then to gas, but that does not always hold. When the solid is in the form of a fine dust, it becomes much more flammable and even explosive.

Many gases are flammable in the presence of air when there is a sufficient concentration. The lower flammability limit (LFL) of a gas is the lowest concentration of the gas in air that will propagate a flame from an ignition source—below that it is too dilute. The upper flammability limit (UFL) is the highest concentration of a flammable gas in air that will propagate a flame from an ignition source—above that the concentration is too rich to burn. Both of these limits are measured as the percentage of gas by volume. Generally, the wider the difference between the UFL and LFL, the greater the range of burning conditions and the more dangerous the substance.

Flammable liquids burn when they vaporize and their vapors burn. Several variables affect the rate of vaporization, including the temperature of the liquid, latent heat

---

* Sometimes a substance can be either a fuel or an oxidizer, depending upon what the other substance is.
† Other examples of fuels are fuels for internal combustion engines or rocket engines; cleaning agents; lubricants; paints; lacquers; waxes; refrigerants; insecticides; plastics and polymers; hydraulic fluids; products of wood, cloth, paper, or rubber; and some metals, particularly as fine powders.
‡ Other examples of oxidizers are halogen, nitrates, nitrites, peroxides, strong acids, potassium permanganate, and fluorine gas, which is even stronger than pure oxygen.

of vaporization, size of exposed surface, and air velocity over the liquid. The *flash point* is the lowest temperature at which a liquid has sufficient vapor pressure to form an ignitable mixture with air near the surface of the liquid. The lower the flash point, the easier it is to ignite the material. Note that this term is distinctive from the *fire point*,* which is the lowest temperature that will sustain a continuous flame. The fire point is usually a few degrees above the flash point. Clearly, the lower these critical temperatures, the more sensitive that substance is to burning.

Thus, a liquid is classified as *flammable, combustible,* or *nonflammable* based on its flash point. A liquid is *flammable* if its flash point is below a specified temperature, and *combustible* if its flash point is above this level but below a higher specified level (and so somewhat harder to ignite). Liquids with a flash point above the higher limit (and so most difficult to ignite) are classified as *nonflammable*. Different organizations and agencies use different specified limits for these classifications.

Figure 4.9 illustrates the relationship between temperature, concentration, and flammability. Some better-known substances and their limits are listed in Table 4.11.

Another term of significance to fires is the *autoignition temperature*,† which is the lowest temperature at which a material (solid, liquid, or gas) will spontaneously ignite (catch fire) without an external spark or flame. Most organic materials contain the mechanisms for autoignition, which is also referred to as spontaneous combustion. Lower grades of coal are volatile and may self-ignite if temperatures build up sufficiently high. Hay, wood shavings, and straw undergo organic decomposition that creates heat internally and may cause self-ignition. Oily rags are particularly dangerous because of the large exposed area to air.

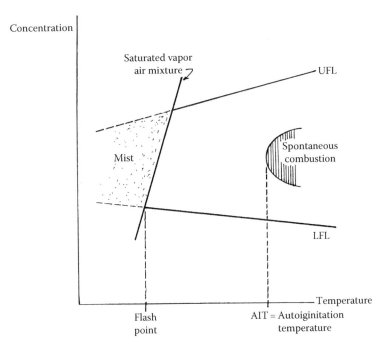

Figure 4.9 Flammability and other associated limits in combustion.

* A term that is equivalent to fire point is *kindling temperature.*
† Spontaneous ignition temperature or combustion temperature has the same meaning as autoignition temperature.

Table 4.11 Lower and Upper Flammable Limits for a Few Selected Compounds

Compound	LFL	UFL
Ethanol	3.3	19
Methanol	6.7	36
Methane	5.0	15
Unleaded gasoline	1.2–1.8	7.1–8.0
Jet fuel	1.3	8.0
Turpentine	0.7	—

When these rags are in a pile, the outer rags insulate those in the center and hold the buildup of heat there. Also, oily insulation around steam lines and heat ducts is another source of self-ignition. *Hypergolic reactions* are special cases of self-ignition where the fuel and oxidizer combust at room temperatures upon mixing. Some of the substances known for hypergolic reactions are white phosphorus, some hydrides, and many iron sulfides that are common wastes in mineral processing or oil fractionation.

Fires inside buildings often have poor ventilation so there is incomplete combustion, which in turn produces carbon monoxide. Carbon monoxide is both toxic and flammable. It is the toxicity of the carbon monoxide, more than the smothering effects of carbon dioxide and smoke, which is responsible for the highest percentage of fire-connected fatalities. Also carbon monoxide tends to reignite as more ventilation improves the combustion mixture. It is this reignition that results in explosions that frequently occur as fire fighters break windows or holes in the side of a burning building. Toxic conditions of carbon monoxide occur at as low a concentration as 1.28% for a 3 min duration.

Fire-extinguishing methods are specific to the type of fire. Class A fires involve solids that produce glowing embers. Water is the most common fire extinguishing recommended for this class of fires. However, water is not recommended for other classes of fires except as a spray alone or with special additives. Class B fires involve gases and liquids such as automotive fuels, greases, or paints. Recommended extinguishants of this class of fires are bromotrifluoromethane, carbon dioxide, dry chemical foam, and loaded steam. The administration of water to class B fires often spreads the fire. Class C fires are class A or B fires that involve electrical equipment. In this case, recommended extinguishants are bromotrifluoromethane, carbon dioxide, and dry chemicals. Care should be taken in class C fires not to use metallic applicators that could conduct electrical current. Finally, class D fires consist of combusting metals, usually magnesium, titanium, zirconium, sodium, and potassium. Temperatures in this class of fires tend to be much greater than in the other classes. There is no general extinguishant for this class of fire, but specific extinguishants are recommended for each type of metal.

One of the best strategies to protect against fires is, of course, to prevent their occurrence by keeping separate the elements that make up fires. A second strategy is early detection. Of course, different detection systems are based on different features that accompany fires. Some detectors have bimetallic strips that expand with a sufficient rise in temperatures to close an alarm circuit. Similar kinds of detectors enclose a fluid that expands under heat until a pressure-sensitive switch is activated. Another variety, called thermoconductive detectors, contains insulation between conductor wires. As the heat rises, the insulation value of this special material decreases to a sufficient extent that electrical power flows through that material, creating a new circuit that sets off the alarm. Accompanying the heat of a fire are infrared rays. Thus, infrared photoelectric cells are used in radiant energy detectors. Another form of detector is one based on light interference. These detectors have a glass or plastic tube, which the air flows through, and there is a light source on one side of the tube and a photoelectric cell on the other. When smoke comes from a fire into the tube, the light is refracted from the photoelectric cell and an alarm sounds. Finally, there are ionization detectors that measure changes in the ionization level. By-products of combustion cause the air to become ionized, which in turn activates these detectors.

Hazards of toxic materials

Toxic materials are poisons to most people, meaning that small quantities will cause injury in the average person. Those substances that only affect a small percentage of the population are *allergens*. Toxic substances are typically classified as

1. *Irritants* if they tend to inflame the skin or the respiratory tract
2. *Systemic poisons* when they damage internal organs
3. *Depressants* if they act on the central nervous system
4. *Asphyxiates* if they prevent oxygen from reaching the body cells
5. *Carcinogens* if they cause cancer
6. *Teratogens* if they affect the fetus
7. *Mutagens* if the chromosomes of either males or females are affected

A number of substances are well-known *irritants*. Ammonia gas combines with the moisture of mucous membranes in the respiratory tract to form ammonium hydroxide, which is a strong caustic agent that causes irritation. Acid baths in plating operations often use chromic acid that not only irritates but also eats holes in the nasal septum. Most of the halogens, including chlorine and phosgene, are strong irritants. Chlorinated hydrocarbon solvents and some refrigerants are converted to phosgene when exposed to open flames. Dusts of various origins* inflame the respiratory tract of most people and can cause pneumoconiosis. Long-term exposure to some dusts, even to low concentrations, can result in permanent damage to the health such as lung scarring.

Systemic poisons are far more dangerous than irritants. Lead poisoning from paint has been a well-publicized problem. Other metals, such as cadmium and mercury, and some chlorinated hydrocarbons and methyl alcohol, which are used as solvents and degreasing agents, are not well-known poisons.

There are a number of substances that are *depressants* or narcotics to the nervous system. While the effects of many are only temporary, depressants pose strong safety hazards because they interfere with human judgment. Best known as a depressant is ethanol, also known as ethyl alcohol, whose impairments to judgment and control are clear. Other depressants include acetylene (used in welding), as well as methanol and benzene, both of which are popular solvents. In addition to being depressants, methanol and benzene are also systemic poisons, while benzene is also a carcinogen (i.e., causes leukemia).

Simple asphyxiates are gases that displace inhaled oxygen, which in turn reduces the oxygen available in the human body. One simple asphyxiate is methane, which is a by-product of fermentation. Other simple asphyxiates include argon, helium, and nitrogen, which are used in welding, as well as carbon dioxide, which has numerous industrial uses and is a by-product of complete combustion. Problems arise with simple asphyxiates when people must

---

* Asbestos fibers, silica, or iron oxide are examples of irritants commonly found in industry.

work in closed spaces with large concentrations of these gases. More insidious are the *chemical asphyxiates* that interfere directly with oxygenation of the blood in the lungs and thereby create oxygen deficiencies. One of the best-known chemical asphyxiates is carbon monoxide, which is the by-product of incomplete combustion and is typically associated with exhausts of internal combustion engines. This gas has an affinity for hemoglobin that is over 200 times stronger than oxygen and is known as the cause of many deaths. The industrial insecticide hydrogen cyanide is another very dangerous chemical asphyxiate.

Carcinogenic substances are in the news frequently, but the degree of danger of these substances varies greatly. As mentioned previously, lead is a carcinogen. Another carcinogen is vinyl chloride. Vinyl chloride is extremely dangerous in other ways as it is flammable, giving off phosgene when burning, and it is explosive. Extreme care must also be exercised in handling these carcinogenic compounds as well as those that are *teratogens* or *mutagens*.

How is it possible to identify dangerous substances? Fortunately, both OSHA and the environmental protection agency (EPA) publish lists of products that should display warnings of danger. These lists also carry the chemical abstract identifier numbers (CAS no.) to help clarify chemicals with multiple names. Some firms use expert systems and other computer systems to identify these chemical hazards and the various limits of permissible exposure.

Measures of exposure and exposure limits

One measure of substance toxicity is the *threshold limit value* (TLV). A TLV is the maximum allowable concentration of a substance to which a person should be exposed; exposures to higher concentrations are hazardous. TLVs are used primarily with reference to airborne contaminants that reach the human body through the respiratory system.

Published TLVs indicate the average airborne concentration that can be tolerated by an average person during exposure for a 40 h week continually over a normal working lifetime. These values are published periodically. As new information becomes available, the American Conference of Governmental Industrial Hygienists (ACGIH) reviews this information and sometimes changes the TLVs. Therefore, TLVs are usually stated as of certain dates of publication. It should also be stated that the TLVs issued by ACGIH are *indicators* of relative toxicity; the actual danger depends upon other things as well, such as volatility, because more volatile substances generate more gas exposure. OSHA reviews the ACGIH publications of TLVs and may or may not accept those limits. Then OSHA publishes its *permissible exposure limits* (PELs). This term is used to distinguish OSHA's exposure limits from consensus standards by ACHIH or any other noted group. PELs and TLVs are highly correlated but not always identical. Many firms adopt the most restrictive limits. Note that some states publish more restrictive limits than those published by OSHA.

TLVs and PELs are generally expressed in milligrams per cubic meter. If a substance exists as a gas or vapor at normal room temperature and pressure, its TLV or PEL can also be expressed in parts per million (ppm). The relationship between concentrations expressed in units of mg/m^3 and those expressed in units of ppm (at 25°C and 1 atmosphere of pressure) is

$$\text{TLV in ppm} = \frac{24.45 \times [\text{TLV in mg/m}^3]}{[\text{gram molecular weight of substance}]} \quad (4.30)$$

For a few substances, such as asbestos, the TLV or PEL is stated in terms of units, or fibers, per cubic centimeter.

Emergency exposure limits (EELs) have been introduced more recently. These limits indicate the approximate duration of time a person can be exposed to specified concentrations without ill effect. EELs are important for personnel who work with toxic substance problems. There is also a *short-term exposure limit* (STEL) that denotes the maximum acceptable concentration for a short specified duration of exposure, usually 15 min. These STEL measures are intended for people who are only occasionally exposed to toxic substances.

Table 4.12 presents some of these limits for a few selected common substances. This table is presented for illustrative purposes only—PELs, TLVs, and STELs are updated regularly and can be found on the OSHA website.

Exposure can be measured at a single point in time or as a weighted average over a period of time. TLVs and PELs are generally stated as 8 h *time-weighted averages* (TWAs). An 8 h TWA exposure is computed as follows:

$$\text{TWA} = \frac{[C_1 T_1 + C_2 T_2 + \cdots + C_n T_n]}{8} \quad (4.31)$$

where
  TWA is the equivalent 8 h time-weighted average concentration
  C_i is the observed concentration in time period i
  T_i is the duration of time period i in hours
  n is the number of time periods studied

Table 4.12 Selected PELs, TLVs, and STELs from the OSHA Chemical Sampling Information File (2004)

Substance	Cas No.	Pel (mg/m^3)	Pel (ppm)	TLV (ppm)	STEL (ppm)[a]
Acetic Aid	64-19-7	25	10	10	15
Ammonia	7664-41-7	35	50	25	35
Asbestos	7784-42-1	0.1 fiber/cm^3	—	—	1 fiber/cm^3
Bromine	7726-95-6	0.7	0.1	0.1	0.2
Carbon dioxide	124-38-9	9,000	5,000	5,000	30,000
Carbon disulfide	75-15-0	—	20	—	10
Carbon monoxide	630-08-0	55	50	25	—
Ethanol	64-17-5	1,900	1,000	—	—
Fluorine	7782-41-4	0.2	0.1	1	2
Furfuryl alcohol	98-00-0	200	50	10	15
Hydrogen cyanide	74-90-8	11	10	—	4.7
Methyl alcohol	67-56-1	260	200	200	250
Nitric acid	7697-37-2	5	2	2	4
Pentane	109-66-0	2,950	1,000	600	—
Phenol	108-95-2	19	5	5	—
Propane	74-98-6	1,800	1,000	2,500	—
Stoddard's solvent	8052-41-3	2,900	500	100	—
Sulfur dioxide	7446-09-5	13	5	2	5
Uranium soluble	7440-61-1	0.05	—	—	—

[a] STELS are generally set by ACGIH, not OSHA, and should be compared to TLVs.

When a person is exposed to a single kind of substance hazard, one merely inputs the concentration and time data in this equation, and compares the answer with the PEL corresponding to that substance. If the computed TWA is less than the PEL, then conditions appear moderately safe and the legal requirements are met.

On the other hand, if there is exposure to multiple hazards, then a mixture exposure should be computed as

$$E_m = \left[\frac{C_1}{L_1}\right] + \left[\frac{C_2}{L_2}\right] + \cdots + \left[\frac{C_m}{L_m}\right] \qquad (4.32)$$

where
E_m is the equivalent ratio for the entire mixture
C_j is the concentration of substance j
L_j is the PEL of substance j
m is the number of different contaminants present in the atmosphere

Safe mixtures occur when E_m is less than *unity*. Even when the exposure to individual substances in a mixture is below each relevant PEL value, the mixture ratio can exceed unity and hence be dangerous; this is more likely as more substances are involved.

There are a number of different ways of finding the concentrations of various substances. Most typically, samples are taken at the site and these bottled samples or special filters are taken to a laboratory for analysis. For certain substances, there are direct-reading instruments that can be used. In addition, for some substances dosimeters are available, which can be worn by employees to alert them when they are in the presence of dangerous exposure levels. Miners frequently use dosimeters for the most common contaminants in mines, and people who work around radioactive materials constantly wear radiation dosimeters on the job.

Protection from airborne contaminants

The most common form of protection used for airborne contaminants is *ventilation*. The idea is to remove the substance from the air before it reaches people. Ventilating systems are typically designed by mechanical engineers who specialize in this application. An effective ventilating system consists of a contaminant collector, properly sized ductwork laid out to allow smooth airflow from the source to the collector, and exhaust fans that maintain a near-constant negative pressure throughout the system. A number of different type collectors are used depending upon the substance. Low-pressure cyclone collectors are often used for large particles and sawdust. Electrostatic precipitators, wet-type dust collectors, scrubbers, and fabric collectors are other forms of collectors used for different substances.

Personal protection devices are also used to protect people from airborne contaminants. Respirators with filters are commonly used in places where there is a lot of dust or vapor that can be neutralized by a chemical filter. Some types of respirators receive air from a compressor much like a diver in the sea. In addition to the respiratory tract, the skin must also be protected. Suits made of special fabrics are frequently used to protect the skin, along with gloves for the hands.

Final remarks

Ergonomic designers must address the design of the physical environment. Messy, uncomfortable thermal conditions, poor lighting, noisy environments, and various hazards cause

Chapter four: Assessment and design of the physical environment

safety problems, reduce morale, and affect task performance in many undesirable ways. To address these issues, a wide variety of factors must be taken into consideration to ensure that environments are properly designed. This chapter provides several strategies that can be used to meet this objective.

Discussion questions and exercises

4.1 Why is housekeeping such an important issue in most industrial facilities? How do 5S programs fit into this picture?

4.2 What is visual acuity and how is it measured? How is visual acuity related to task performance?

4.3 Suppose a point light source has 10 lm falling on a surface 4 m away from the source. What is the source luminance?

4.4 What is the luminosity function and why is it used in illumination design?

4.5 What is the effect of illumination on visual acuity? When might more illumination reduce visual acuity?

4.6 What is brightness contrast and how is it calculated?

4.7 How much illumination should be provided for a very difficult inspection task performed by a 30-year-old male, working on an incentive system that penalizes errors severely?

4.8 When and why might certain strategies improve visual acuity more than increased illumination? Explain and give examples.

4.9 How long does the OSHA standard allow people to be exposed to a sound level of 102 dBA? What is the noise dose of a person exposed for 7 h to 80 dBA noise and 1 h of 105 dBA noise?

4.10 What strategy of noise-level protection is usually the most cost effective when it is physically feasible?

4.11 Based on the results of Epp and Konz, how much noise would it take on the dBA scale to be quite annoying? At what dBA level would it be expected that 6.7% of the words would not be heard correctly?

4.12 If we compare the enthalpy at 70°F and 70% relative humidity with that at 75°F and 60% relative humidity, what is the enthalpy difference and what does it mean? Are any of these conditions in the comfort region for sedentary persons dressed in light clothing?

4.13 If the dbt is 80°F and the wbt is 70°F, what change in ET is caused by a fan blowing at 300 feet per minute (fpm) compared to one blowing at 20 fpm?

4.14 What are the primary sources of heat exchange by the human body, and which source is only a heat loss?

4.15 What does the *met* unit measurement mean? How many watts per square meter correspond to four mets? How many watts would be expended by a person who weighs 220 lb and is 6 ft 6 in. tall, exerting four mets? Is four mets very light, light, moderate, heavy, or very heavy work?

4.16 What is the relative humidity at 70°F dbt and 60°F wbt?

4.17 What is the difference in the moisture content of 40°F air and 70°F air, when both are at 50% relative humidity?

4.18 If the dry bulb air temperature is 65°F and the relative humidity is 80%, what is the highest temperature at which dew will form?

4.19 If the temperature is −6°C with air blowing at 12 m/s, what air velocity appears to be as cold when the temperature is +2°C, based on the windchill factor?

4.20 Explain how the HSI is calculated. What are the implications of an HSI of 100? What are the pros and cons of using the HSI?

4.21 If the current ET is 75°F and humidity is 80%, what would be the expected rise in ET with a large radiant heat source of 5°F MRT?

4.22 With the following conditions, 30°C wbt, 32°C dbt, and 33°C MRT, but with no solar load, what rest break schedule would be recommended by the Governmental Hygienists for work at 900 BTUs/h?

4.23 What does a *clo* unit measure?

4.24 Assume the following atmospheric environmental conditions: wet bulb temperature is 27°C, MRT is 34°C with solar load, dry bulb temperature is 29°C, workload is moderate at about 3 kg cal/min, rest is conducted in the same environment, normal lunch breaks are 30 min at noon, and there is a 15 min break at mid-morning and mid-afternoon. Assuming that there is plenty of water and salt intake and light summer clothing is worn, what is the appropriate work–rest cycle regimen?

4.25 Suppose that the wet bulb temperature is 84°F, the dry bulb temperature is 85°F, and the MRT is 94°F with direct solar heat present. How easy would a task have to be in terms of kilocalories per hour in order to work continuously over an 8 h shift under the permissible limits?

4.26 What is an expected effect on performance of a person looking out for icebergs when his oral temperature drops a few degrees below normal body temperature?

4.27 At what point does human performance begin to degrade when temperature and humidity are increased?

4.28 What is a TLV and what is the TLV for carbon dioxide?

4.29 What disease is associated with excessive exposure of the hands to vibration?

4.30 Denote the proper extinguisher for each class of fire.

4.31 What is Raynaud's syndrome? How is it caused and its effects reduced?

4.32 What voltage would be expected if a 5 μA current entered wet skin?

4.33 What happens as a result of excessive electrical energy to the heart?

chapter five

Work measurement and analysis

About the chapter

This chapter focuses on improving the methods followed when people perform tasks such as manual assembly, material handling, or even service tasks such as customer support. The purpose of the chapter is to familiarize the reader with a number of principles and techniques for designing and improving work. Specifically, the reader will be familiarized with motion analysis and learn principles for methods improvement, work physiology, and manual assembly. In addition, the reader will learn how to analyze lifting tasks using methods from the National Institute for Occupational Safety and Health (NIOSH). The chapter then describes learning and provides methods for modeling learning. Last, the chapter covers time studies, where the reader will learn to conduct time studies and interpret the output in a practical way.

Introduction

Assembly is a traditional topic in industrial ergonomics. Its roots go back to the first part of the twentieth century and earlier. In those early days, the primary focus was on improving productivity and operating speeds. This has changed in recent years, because the required precision of assembly has increased. Accuracy is now a more important issue.

This chapter starts with principles of motion economy and manual assembly. These principles focus on reducing unnecessary exertion to minimize fatigue and its effects. Understanding the physiology associated with such tasks is critical to making such tasks efficient and safe.

Because materials handling is a related activity that affects many people, ergonomic effort must similarly be focused on this activity. The variety of materials-handling situations is enormous; cases vary from simple lifting or manually carrying a load over a horizontal distance to using assistive equipment in manual handling or employing machines, such as a forklift truck. Ergonomic opportunities for improvement present themselves over the full spectrum of options.

Therbligs

The taxonomy of elemental motions conceived many years ago by Frank Gilbreth provides a good starting point for analyzing manual work. His idea was that tasks could be described using a smaller set of elemental motions and variables. Gilbreth's taxonomy of movements and sensory–motor actions was further developed in the next two decades after his death in the mid-1920s. Because of his many contributions to the field, the elements of this resulting taxonomy were named *therbligs*.[*]

---

[*] Therblig is Gilbreth spelled backward. This name was adopted in Frank's honor since he experimented so much with motions.

BOX 5.1   GILBRETH'S THERBLIGS

- Grasp object.*
- Reach with an empty hand.
- Move hand while holding an object.
- Release held object.
- Use a tool such as a screwdriver.
- Assemble, by placing one object into or onto another object.
- Select among two or more objects or by making other decisions. This therblig is not usually productive and when found should be questioned.
- Search, which is the act of hunting for an object using one's eyes or hands or both. This therblig can often be avoided by separately delivering objects to obvious locations.
- Find is an older therblig associated with the end of a search when the object is found and verified; today the find therblig is often combined with the search therblig, but the decision to do so is strictly a matter of convention.
- Hold is a therblig where the hand is used as a holding device. This therblig is not only nonproductive, it is often unsafe.†
- Inspect is the act of checking an object against specifications or checking the quantity of available objects for sufficiency.‡
- Disassemble is the antonym of assemble and is rarely productive because it can almost always be avoided.
- Preposition is the act of orienting an object prior to positioning it. A single positioning is often all that is needed. Almost all prepositioning is unnecessary.
- Delays are never productive, although they may be necessary in some situations. The idea is to create a job design that precludes the need for a delay.

* Kroemer (1986) identified other types of grasps, including the finger touch, palm touch, finger palmar grip (hook grip), and power grip.
† The admonition not to use the hand as a holder reappears later in this and other chapters.
‡ If the operator inspects the object while making it, then corrections can sometimes be made. The notion of coupling inspection into a regular manufacturing job corresponds to giving the worker real-time feedback. Removing the inspect therblig in such cases can lower product quality.

The first six therbligs shown in Box 5.1 are *productive* in nature. The remaining therbligs shown are often *nonproductive* elements of the task, especially when people do manual assembly in traditional manufacturing settings. The method of performing a task can be described with a two-handed or SIMO chart,* which lists the elemental acts and parameter values for each therblig. From the perspective of job design, nonproductive therbligs should be eliminated if possible. When nonproductive therbligs cannot be eliminated, they should be minimized or transformed to improve the method. Improvement can result in shorter average performance† times, smaller time variations (i.e., greater predictability), fewer errors, less effort, and safer actions. Eliminating unneeded therbligs provides a first pass at developing a better method. Other approaches are expanded upon in the following discussion.

* SIMO charts are a special case of multiple activity charts where the two hands perform the two activities. The word SIMO stands for simultaneously, although concurrently would be more accurate.
† Therbligs are basic elements in synthetic data systems, which are discussed in Chapter 6.

Some probabilistic assumptions

A principal objective of human performance measurement is to find the average value of the criteria of interest for a specified population. In some cases, the focus is on a specific kind of person. In either case, the measurement is carried out under specified conditions. In typical time studies, the focus is on a so-called standard operator, working at an average pace under typical environmental conditions, while using a specified method of performing the tasks. The particular time value arrived at in time studies is known as the normal time. Note that practically every specified statistic here is an average except for the method and the operator. A standard operator is often defined as a person of average ability on the tasks. Other definitions include people with minimum but acceptable training and capabilities, or somewhere between acceptable and average. Whatever an acceptable description of a standard person may be, the traditional approach compares all workers to this so-called standard operator. At times this practice leads to problems, particularly when large time variations are observed between different operators. Usability studies, where the designer may be interested in the average performance of the average person in the population, are really insufficient without accounting for the conditions of work and the goals of the individual at the time. In the following discussions, it is assumed that these specifications are met and that the criterion of interest is time. Measuring other criteria is similar, with some modifications.

It is often assumed that the distribution of observed time values for an individual performing a task in a consistent manner will follow the normal or Gaussian probability density function. Thus, the probability that the task will take a time (T) can be written as

$$p(T) = \frac{1}{\sigma\sqrt{2\pi}} e^{\frac{(T-\mu)^2}{2\sigma^2}} \tag{5.1}$$

where μ and σ are, respectively, the mean and standard deviation of the cycle time for the population. It is further assumed that different individuals from this population have average time values that also are normally distributed. Figure 5.1 describes this model of

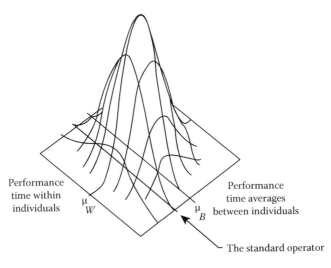

Figure 5.1 Performance time within and between individuals in the population of people under consideration.

human performance times on a specified task. In this figure, the axis on the left denotes the distribution of time values expected for an individual who repeatedly performs the task, cycle after cycle. The right axis denotes the relative frequency of the average values for the people in this population. The axes together represent performance time values expected of everyone in the population. Usability studies tend to focus on the mythical person at the location μB. Time study may focus on the same mythical person, but more often than not the standard operator is defined as a point below the population average, perhaps as illustrated in this figure. The normal-shaped functions in Figure 5.1 are really joint probability density functions that would need to be standardized (i.e., multiplied by a constant to make the integral equal to unity). The resulting probability density function depends upon which mythical person is the focus of the study.

The particular significance of these results is twofold. First, normal distributions have been found to describe human performance, and, second, the central limit theorem shows that sample means can be described by the normal distribution. This theorem states, in effect, that the sampling distribution of means is approximately a normal distribution with sample means $\hat{T}$ where

$$\mu_{\hat{T}} = \mu \tag{5.2}$$

$$\sigma_{\hat{T}} = \frac{\sigma}{\sqrt{N}} \tag{5.3}$$

as long as the mean and variance are finite and the population is at least twice the sample size N. In time study, an individual person is selected and timed over numerous cycles of the task. This practice corresponds to getting a large number of samples of the within-person normal distribution and adjusting for between-person differences using methods other than through sampling. As the within-person sample size N gets larger, the sample mean converges toward the population mean of the individual person being timed. The variance of time from that single subject gets smaller and smaller, thereby reducing the uncertainty in the estimate of that individual's population mean. However, the variance of time in Figure 5.1 has both within-person and between-person time components. The variance of the two components, X and Y, is

$$V(X,Y) = V(X) + V(Y) + 2\rho_{X:Y}\sqrt{V(X)V(Y)} \tag{5.4}$$

where the symbol $\rho_{X:Y}$ is the correlation between the within-person and between-person times. In time study, it is assumed that the performance times within and between persons are statistically independent. When this is true, $\rho_{X:Y}$ is zero. Traditional time study recognizes the need to consider the variation of within-person performance times. The mean time estimate also must be adjusted for between-people differences to define the standard human operator. This correction process is known as performance rating or leveling. This involves the use of a correction factor that makes the performance speed of the observed individual equal to that of the standard operator.

Performance leveling is not used in usability studies. Instead, a number of different people are observed in order to obtain the population variability. It, consequently, becomes necessary to find a sufficient number of people to represent that population. This can be very difficult in a small-to-medium-size company. Moreover, most industrial applications cannot economically justify repeated training and performance measurements over

Table 5.1 Presgrave's Observations of Differences in Performance between Individuals

$\mu-3\sigma$	$\mu-2\sigma$	$\mu-\sigma$	μ	$\mu+\sigma$	$\mu+2\sigma$	$\mu+3\sigma$
0.61	0.74	0.87	1.00	1.13	1.26	1.39

Source: Presgrave, R., *The Dynamics of Time Study*, McGraw-Hill Book Company, Inc., New York, 1945.

several people. However, the use of performance leveling entails the assumption that the performance time of a standard operator is proportional to the performance time of the observed operator.

Many years ago, Presgrave looked into the question of between-person differences in performance (Presgrave, 1945). He performed many tests on human performance time over very large numbers of people. His studies verified that the normal distribution accurately described individual differences in performance. He specified an average operator's performance in a population as 1.00. Faster people could do the same task in a fraction of the average operator's time. Slower operators took an increased fraction of time to do the same tasks. His studies show that these fractions tended to be as shown in Table 5.1.

In Table 5.1, σ is a standard deviation of between-person performance, so about 68% of the people fall into the interval $\mu-\sigma$ and $\mu+\sigma$ and about 95% fall into the approximate interval $\mu-2\sigma$ and $\mu+2\sigma$. Notice that the fractions less than unity are symmetrical to those values greater than unity and that the fraction improvement per standard deviation (σ) is about 0.13.

Methods improvement

A method is a prescribed procedure for performing a task or a job. It is a behavioral description of precisely what one does. Once the work requirements and goals of a job are known, ergonomic specialists typically identify what is done and record it. Photographs and/or videotapes frequently accompany method descriptions. Once the job is studied and a particular method is established, that method is known as the standard method for a company. Ideally that method is the easiest to perform and teach to new personnel, requires the least time, results in the fewest errors, and provides the greatest safety. But methods that are optimal for all desirable criteria are rare, if they ever exist at all.

Usually, the initial standard is the way people did it in the past. The job may be restudied, comparing the initial standard to alternative methods and either the initial standard is retained or a new standard is created. Restudies occur when new equipment is purchased that alters one or more tasks. The personnel who perform the jobs normally suggest many potential methods, and many companies encourage such suggestions with financial rewards commensurate with the cost savings. Other potential methods are generated by the ergonomics personnel. Since the same staff often evaluates all of the alternative methods, it is important that the staff exercises professional ethics and evaluates all proposed methods with objective fairness. These evaluations typically require empirical tests to be run on the various methods and measurements to be made on each criterion of concern.

Management should select the method that does the best on the most criteria as the standard method. In practice, there is often little difference between the alternative methods. This information, too, is helpful, because the performing person can elect one that he or she thinks is better. There are often motivational advantages in these cases. It is vital to keep the personnel who are going to perform the work involved in the method selection process. If one uses the operating personnel during the empirical testing, they can give their opinion at the end of the testing.

Origins of Human Factors

After World War II broke out in Europe in 1939 and the United States became involved a few years later, many people studied ways of developing military hardware that better fit human operators. Harry Helson studied features that made naval and antiaircraft gunnery more effective, and Wes Woodson studied many aspects of naval ergonomics. Alphonse Chapanis, Professor at Johns Hopkins University, studied the visual detection of objects such as trip wire lengths and pilots in the ocean. Ian Smith of England, Ross McFarland of Harvard University, and Paul Fitts studied aircraft safety. Fitts became a professor at Ohio State University and the University of Michigan after the war. At the war's end in the mid-1940s, many of these people were responsible for founding the Ergonomics Research Society in England and the Human Factors Society in the United States of America. Chapanis and Woodson are past presidents of the Human Factors Society, and one of the most prestigious awards of this society is named after Fitts.

Not all jobs have standards. Jobs that are infrequently performed and that typically involve very small order quantities do not need standards in the normal sense of the word, but their labor content or time requirement needs to be estimated so that scheduling and costing can be done. Other jobs of slightly higher frequencies and larger order quantities may have time standards based only on previous production data, and there may be little or no reason to establish a standard method. Jobs without standard times create problems in production control operations because there is little basis for planning future operations or predicting when orders will be ready except through historical records. Unfortunately, introducing new machine tools and materials-handling equipment make historical records invalid for planning purposes. Jobs without standard methods create difficulties for setup personnel because the new job has to be planned on the spot or remembered.

Method design and selection affects and is affected by the economics of the company and the welfare of company personnel. Care should be used to gauge the prospective benefits of redesigning methods. The bulk of ergonomic effort should be invested on those that pose safety and health concerns or have the greatest economic potential.

Principles of motion economy

Principles of motion economy have emerged over the years that can be helpful to analyze and improve task methods. As reflected by its name, motion economy is focused on saving time by making motions more efficient. These principles of motion economy are not algorithms that guarantee optimum performance. They are simply guidelines that usually result in better performance except in unusual situations. There also is no guarantee that one principle will not contradict another. To resolve contradictions, the designer must understand the underlying basis of the conflicting principles. Support for each principle is consequently provided in the discussion that follows, if warranted. Some of the principles are primarily related to the human body, others pertain to the workplace, tools, or equipment.*

---

* Most of the principles of motion economy and work design presented here were formulated long ago. Some more recent principles are included that were formulated and stated by Corlett (1988) and others.

The following two principles focus on the workers' hands.

Principle 5.1
The two hands should begin and complete their motions at the same time.

When the hand movements begin and end at the same time, the motions tend to counterbalance each other. Uncoordinated movement may require compensating body movements that will make the tasks more fatiguing and reduce efficiency. While some tasks are by nature fatiguing, ergonomists must do their best to minimize fatigue. Another reason for the motion of the two hands to begin and end at the same time is that if they do not, one hand is doing more than the other. When one hand is doing little, a new task sometimes can be assigned to it. Doing so can help minimize the total time required.

Principle 5.2
The two hands should not be idle at the same time, except during rest.

The basic idea here is that the activities should be distributed evenly to both hands, so that one hand does not have to wait for the other. The exception is that the preferred hand should be used to do the more complex control actions. For obvious reasons, the method should allow either hand to be the preferred hand.* Reducing idle time for either hand is analogous to balancing an assembly line. By allocating the tasks to be done between the two hands or between the several assembly-line stations, one minimizes the cycle time of either the hands or the assembly line.

The next two principles are closely related to each other and deal with the issue of posture.[†]

Principle 5.3
A person should be able to maintain an upright and forward facing posture during work.

Principle 5.4
All work activities shall permit the operator to adopt several different natural, healthy, and safe postures without reducing capability to perform the work.

These two principles both focus on postural freedom. The first principle stresses the importance of eye–hand coordination in many jobs, both for safety and efficiency. When people cannot directly face the work at hand, coordination is impaired. A person who is not directly facing their work must twist their body to do the job. The body will have to counterbalance twisting movements or twisted postures. This will result in extra stress and strain on the body. This situation not only forces a person to do more physical work, but also prevents the person from balancing the work over the body.
A fifth principle related to the human body is as follows.

Principle 5.5
Motions of the arms should be made in opposite and symmetrical directions simultaneously.

One reason behind this principle is, again, body balance. Movements that are NOT opposite and symmetrical create an imbalance that the body must correct. When the body

---

* Most people have a preferred hand that is stronger, faster, more flexible, and more controllable. Ergonomists must keep that in mind during the task allocation.
† These principles are part of those expressed by Corlett (1988). The ones stated here are paraphrased from the author's original statements.

is balanced, motions require less effort and the body is less vulnerable to fatigue. Another reason for this principle is movement compatibility. It is hard to perform distinctly different motions using different arms and hands. Combining distinctly different motions requires more coordination* and, therefore, more learning and more time.

The next three principles are straight out of a basic physics or engineering mechanics class.

Principle 5.6

Hand and body motions should be assigned to the lowest classification that can perform the task satisfactorily.

A lower classification refers to a body member with a lesser mass. The lower the mass moved, the lesser the required physical work. As mentioned earlier, Newton showed us that $f = ma$ where f means force, m is mass, and a is acceleration. If accelerations are the same, a lower mass means a lower force, and this means less effort is required. While this reasoning comes directly from elementary physics, it should be mentioned that body members with lower classifications (i.e., lower mass) usually have less strength. Accordingly, the limit to lowering the classifications is set by the strength demands of the task.

A second principle, straight from Newton, is as follows.

Principle 5.7

Momentum should be used to assist whenever possible, and should be reduced to a minimum, if it must be overcome by muscular effort.

If Mr. Charles Babbage were alive today, he would give a standing ovation to this principle. The natural force of momentum obviously has both positive and negative effects. A very simple practical example of this principle is to use bins with gravity chutes to deliver component parts (e.g., screws and bolts) to the people doing assembly tasks.

Following the same theme from physics, the third principle states the following.

Principle 5.8

Smooth continuous curved motions of the hand are preferred to straight-line motions involving sudden and sharp changes in direction.

It is obvious from physics that sharp changes in direction require added physical forces to overcome momentum and sustain high rates of acceleration. Remember, "A body in motion tends to remain in motion unless acted upon by an outside force." The muscles of the arm must provide the forces needed to make sharp movement changes by the hands.†
The relationship $f = ma$ shows that changes in direction take more force, effort, and time.

Motion and micromotion study

Motion study is useful in situations that require very highly learned manual skills or when many different people need to be trained in a particular skill. Motion study is especially likely to be cost effective for long production runs involving a lot of manual labor. Motion study requires a fairly large commitment in time and expense, and so it should

---

* Added coordination requires higher skills and higher skills necessitate longer learning.
† The reader may recall Newton's law, which reads: If an unbalanced force acts on a body, the body will accelerate; the acceleration magnitude is proportional to the unbalanced force and the direction of the acceleration is in the direction of the unbalanced force. Thus, $f = ma$.

not be undertaken lightly. Motion study usually involves video recording the operation of concern. By studying the video one can see where good principles of motion economy are not being followed. Many of the principles cited earlier are applicable for this purpose. Much of the focus is on identifying motions that are unnecessary or which can be changed to make them more efficient. Micromotion studies are conducted at a more detailed level by a frame-by-frame picture analysis.

These methods have found additional use in sports medicine. People trying to perfect their golf or baseball swing might find such analyses to be helpful. As you might guess, increased speeds correspond to greater distances between successive frames of the same points on the body, which means that those parts are accelerating. This information identifies when movement forces are greater, demonstrating again that the previous principles are related.

Perhaps one of the more important forms of motion and micromotion study involves measuring the differences between experts and novices. The study of experts helps identify the motions that really contribute to skilled performance. Studies of novices tell ergonomic specialists what beginners are likely to do, and the differences between novices and experts show where learning and training are needed. Data from persons with intermediate skills indicate what skills are learned more easily and which skills are more difficult to acquire.

Motion study and micromotion study are very useful when people must make very precise movements. This approach has been used to study and improve the performance of dentists, masons, carpenters, and athletes. In one of the earliest applications of its use, the study of bricklaying caused Frank Gilbreth to get interested in ergonomics.*

Work-physiology-related principles

Four related principles are based on work physiology[†] and take into consideration the particular joints involved, the muscle bundles used to perform the task, the location where the work is performed, and the adequacy of rest.

Principle 5.9

Work activities should be performed with the joints at about the midpoint of their range of movement.[‡] This principle is particularly important for the head, trunk, and upper limbs.

Joints in the human body are almost always strongest at the midpoint of their range of movement in the specified direction. Work activity should be designed accordingly. However, this principle does not tell the methods designer the preferred direction of movement, a question that can be important when the ergonomics designer is acting in concurrent engineering design.

---

* Frank Gilbreth would have called his interest "methods study" or "design of work methods" rather than ergonomics or human factors, as these terms had not yet come into use.
† These principles are part of the reason why the Chapter 2 background was provided before this chapter. Although the technique of time study is not discussed until Chapter 7, time studies are often used in conjunction with method design as a way of identifying better methods in terms of performance time. These principles are also part of those provided by Corlett (1988).
‡ Chapter 2 provides background on important joints of the body. The maximum amount of movement by a body part in a particular direction is referred to as the "range of movement." One of the problems in industry is that arthritis and other diseases of the joints tend to occur in older persons. These diseases can and do limit job capabilities unless they are accommodated for, as is often required by the American Disabilities Act.

Principle 5.10

Where a muscular force has to be exerted, it should be by the largest appropriate muscle groups available and in a direction colinear with the limbs being moved.

One of the ideas behind this principle is that the larger muscle group can produce the greater forces. With less exertion, there usually is more endurance. The mechanics of this principle follow common sense. If a muscle group exerts a force that is not completely colinear, or aligned, with the direction of the desired movement, the exerted forces are only partially effective. Depending on how the angle used to describe the degree of non-colinearity is defined, only the sine or cosine component is the effective force. In addition, the complementary muscle group must be engaged to offset the opposite component of the primary muscle bundle activated. Hence, losses in linearity reduce the effectiveness of the muscle groups.

Principle 5.11

Work should not be performed consistently at or above the level of the heart; even occasional performance of tasks where forces are exerted above heart level should be avoided. When light handwork must be performed above heart level, rests for the upper arms are required.

Working with one's hands held high is physically very tiring, even when the physical loads are light. Barbers in training often complain about holding their hands up above their waist, before they become accustomed to the strain. Rather than forcing people to work in this position, it is preferable to either lower the workpiece or elevate the workman. Even in cases that seem to require workers to use their hands above the heart, such as a mechanic in a service station lubricating an automobile, adjustments can often be made.

Principle 5.12

Rest pauses should reflect all of the demands experienced in work, including environmental and informational loading,* and the length of the work period between successive rest periods.

Scientifically establishing adequate rest pauses is a complex and difficult issue. Murrell provides an approximate basis for setting rest allowances based on the average energy expenditure of the person's job. In Figure 5.2, the energy expenditure per minute is plotted against recovery time requirements on the horizontal axis.† Murrell (1969) argued that people could work steadily without rest when their energy expenditure on the job results in a heart rate of 30–35 beats/min above their basal rates. In this approach, if a person's job requires an energy expenditure rate of b kcal/min (where $b > 5$), rest should be provided for recovery in proportion to how much b exceeds the standard rate s. Murrell's approximation equation is

$$a = \frac{w(b-s)}{b-1.5} \tag{5.5}$$

where
  a is the minutes of recovery time required per shift
  w is the shift work time
  s is the requirement of a standard task (Murrell estimated this to be 5 kcal/min)

---

* Chapter 10 deals with some models for measuring cognitive-information-processing loads. Unfortunately, the notion of environmental and informational loads is fraught with controversy because the science is not well developed and practitioners must rely on expert judgments.
† This figure is very similar to Figure 2.23, which describes aspects of metabolism.

Chapter five: Work measurement and analysis

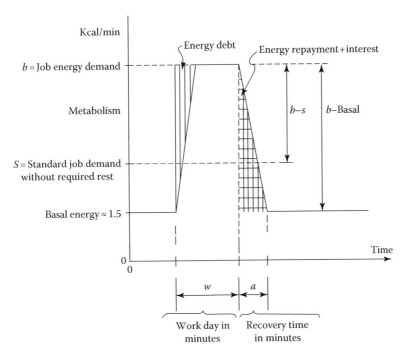

Figure 5.2 Murrell's approximation model for required resting time.

The value of 1.5 kcal/min used in the equation is the basal metabolic rate. Note that if both sides of Equation 5.5 are divided by w, the left side is the ratio of rest-to-work time. The right side is the ratio of how much b exceeds the standard task compared to how much b exceeds the basal demand.

Of all the principles associated with the human body, the one that is probably most controversial is the following.

Principle 5.13

Ballistic movements are faster, easier, and usually (but not always) more accurate than restricted or controlled movements.

Example of Murrell's Approximation

Suppose that the task in question requires 3800 kcal/shift of 8 h. Over the 480 min/shift less 15% for allowance, 3800/408 min yields about 9.3 kcal/min. The total required resting time per shift is

$$a = \frac{w(b-s)}{b-1.5} = \frac{408(9.3-5.0)}{9.3-1.5} = 225 \text{ min}$$

If one can assume that the allowances for personal needs, fatigue, and delays are suitable for resting and that those allowances are 24 min each (or a total of 72 min), the total extra rest required is 225 − 72 min = 153 min.

Kinematics is the study of motion without considering mass or the causes of motion. Kinetics is the study of the forces and actions that change the motion of masses. The study of dynamics considers both of these specialties and focuses on forces that are not in equilibrium. The whole notion starts from Newton's law that states the following.

Principle 5.14

If an unbalanced force acts on a body, the body will accelerate; the magnitude of acceleration is proportional to the unbalanced force and the direction of acceleration is in the direction of the unbalanced force.

The famous equation in engineering $f=ma$, stated several times previously, comes directly from this law. Motion starts with a force being applied. If this force exceeds other friction or other restraining forces, acceleration changes from zero to a positive value and the body starts moving in the direction of the force at some velocity. If no additional force is applied, motion continues at a fixed velocity. When the velocity is constant, the distance traveled is equal to the velocity multiplied by the movement time. This case, and the more general case where velocities change over time, is both described by the equation that follows:

$$s = \int_{t_0}^{t_1} v_t \, dt \tag{5.6}$$

In this equation, v_t is the velocity at time t and s is the distance moved. A second elementary relationship between velocity (v_t) and acceleration as a function of time (a_t) is

$$v_t = v_0 + \int_{t_0}^{t_1} a_t \, dt = v_0 + a(t_1 - t_1) \tag{5.7}$$

The far right-hand side only holds with a constant acceleration. As can be verified by the preceding equation, accelerations cause velocity to change. The reverse logic shows that

$$a_t = \frac{dv_t}{dt} = \frac{ds_t^2}{dt^2} \tag{5.8}$$

When acceleration is constant over the time interval, the distance moved is

$$s = v_0 t + \frac{a}{2} t^2 \tag{5.9}$$

and the time required with a constant acceleration a is

$$t = \frac{-v_0 + \sqrt{v_0^2 + 2as}}{a} \tag{5.10}$$

Figure 5.3 describes the relationships between speed, time, and distance when the acceleration is constant. Accordingly, knowledge of the velocities and accelerations over a path

Chapter five: Work measurement and analysis

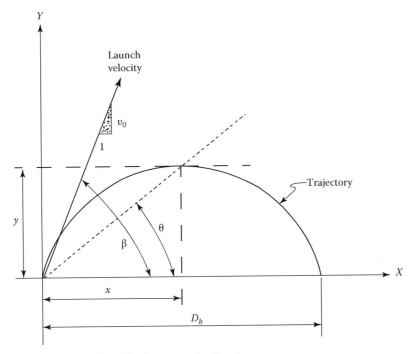

Figure 5.3 Path of a projectile with a launch velocity of v_0.

of motion can be used to determine forces exerted on the human body and the directions of these forces. This follows from the relationships between force, mass, and acceleration shown in the preceding equations.

The path of the projectile as a function of time t is described in the following. It also follows that the maximum horizontal and vertical distance achieved is

$$d_h = \frac{v_0^2 \sin^2 \beta}{a} \tag{5.11}$$

$$d_v = \frac{v_0^2 \sin^2 \beta}{2a} \tag{5.12}$$

Also the time required for the projectile to reach the highest point of travel is

$$t_v = \frac{v_0 \sin \beta}{a} \tag{5.13}$$

While ballistic motions by the hands will not duplicate these equations of movement because numerous constraints have been ignored (e.g., the hand–arm connection), those equations give a first-order approximation. In reality, movements will have an acceleration phase in which velocity increases, followed by a deceleration phase in which velocity decreases. Figure 5.4 provides a simple example of such movement.

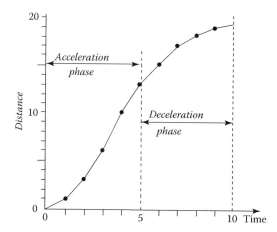

Figure 5.4 Distance as a function of time when acceleration is not constant.

Another issue is that ballistic movements often involve fewer muscle groups than controlled movements. In controlled movements, some muscle bundles work against others so part of the effort is wasted. The use of opposing muscle forces allows for more precise positioning in tasks such as turning an automobile steering wheel. However, ballistic movements are sometimes faster and more accurate than controlled movements.

Principle 5.15

Work should be arranged to permit a uniform and natural rhythm where possible.

The speed at which an individual performs if left alone often corresponds to a natural rhythm of the body. Most people have a natural gait.* When people walk at their natural gait, the effort required is minimal and their speed is very consistent. Tests have shown that the metabolic rate of a person increases when they are forced to walk faster or slower than their natural gait. Depending on the person, this gait varies from a speed slightly below 2.7 miles per hour (mph) to above 3.0 mph. Not all rhythms are necessarily natural. The physics of this issue are quite complex and involve resonance and other effects. Even unnatural rhythms can be learned. Once established, a rhythm reduces the planning and overt mental control required to perform a task.

Principle 5.16

Required eye fixations should be as few and as close together as possible.

This principle is related to eye–hand coordination; where the eyes go, so do the hands. Smoother visual paths result in smoother and less fatiguing hand motions. In our opinion, it is not as important to have eye fixations close together as it is to have them follow a smooth, obvious path.

Some principles of manual assembly

Assembly remains primarily a manual task in most of industry today. Some noteworthy exceptions include the mechanical assembly of electronic boards, and the assembly of large ships or locomotives. However, when assembly is within human capabilities, it is

---

* This is also true for animals. For example, horses can trot, walk, or gallop.

very difficult to find a more adaptable assembler than a person, whether on an assembly line or at a simple workstation.

In the more-developed countries of the world, many companies have been concerned about the growing cost of manual labor, and have sought alternatives. One alternative is to ship parts to locations in the world where manual labor has a lower cost, assemble the product there, and ship the finished or semifinished product back to the countries where the sales are greatest. The latter practice requires a low cost of shipping to be cost effective.

Another alternative which a number of companies have considered is to use robots or similar devices to assemble component parts into products. The latter approach was popularized in the 1970s, but at the time, robots were not well developed. Difficulties implementing this approach inspired research on how to design product parts for more economical assembly by robots.

One of the more noted studies on making product parts more compatible for robotics assembly was done by Boothroyd and Dewhurst (1983). These and other investigators identified conditions that made robotics assembly more difficult as well as those that made robotics assembly more effective.* Other studies reported that those features in the design of component parts that facilitated robot assembly usually facilitated manual assembly and that when those features were implemented, manual assembly was often less expensive.[†] Accordingly, the robotic assembly principles that they developed are also manual assembly principles. Those most appropriate for manual assembly are described subsequently.

Many of the following principles for facilitating manual assembly (and often robotics assembly as well) seem like simple common sense. While they are indeed commonsensical, they are not simple because it took a lot of people a long time to discover them. Other principles seem counterintuitive at first glance. A primary principle for most assembly tasks is the following.

Principle 5.17

Use a fixture, which is both simple and reliable.

A good fixture greatly aids the assembler in starting out the assembly process. Most fixtures have some clamping mechanism that holds the component after it is inserted. A foot- or leg-activated clamp is often recommended. It is important that the clamp not become a hazard to the assembler (e.g., catching fingers or a hand), especially during the insertion procedure. It is also important that the clamping device hold the component firmly without allowing movement, so that the next component can be easily attached without using the other hand as a holding device.

Principle 5.18

Eliminate or minimize the types and sizes of fasteners.

Since most fasteners require tools for assembly (e.g., bolts require wrenches and screws require screw drivers), reducing the types of fasteners reduces the number of tools needed by the assembler. Using fewer sizes of fasteners can also reduce the number of tools needed for inserting or tightening fasteners. With fewer tools, less pickup and release actions are required of the person.

---

* Genaidy, Duggai, and Mital (1990) compare the performance of people and robots on assembly tasks.
† See Helander and Furtado (1992) for more on this topic.

Principle 5.19

Design component parts that are easy to grip.

Avoid small parts that are not easy to grasp, such as washers and small pins. Small parts are not easy to hang onto and are frequently dropped during assembly. Also, avoid parts that tend to tangle and snag each other when they are placed in bins. Some springs, for example, have hook wires on the ends that are intended to connect with opposing units. These hooks create entanglements that slow up assembly.

Principle 5.20

Avoid parts that are weak, dimensionally unstable, or easily bent.

Weak parts break or otherwise deform. When a component part breaks during assembly, extra actions are needed by the assembler to get rid of the broken part, select and pick up a new part, and restart the assembly operation. Dimensional instability can result for many reasons. When a critical dimension is changed, the part no longer fits and the same result occurs; the old part must be removed and discarded and a new one is needed. Bent parts have the same fate. Avoiding these unfavorable conditions permits more repeatable assemblies.

Take advantage of the shape of component parts. Shape can make a considerable difference, as the next five principles illustrate.

Principle 5.21

Make mating parts self-locating.

When the male parts are beveled and the female parts are chamfered, the mating starts with very low tolerance restrictions and progressively allows for a tighter fit when the parts are pushed together. This leads directly to the next principle.

Principle 5.22

Lower tolerances induce faster and easier mating.

Higher tolerances require more precise alignment of parts during mating. More precise alignments require more eye–hand coordination, and as predicted by Fitts' Law, more time. Blockage of vision and other factors can add to the sensory requirements. Consider the case of an assembly worker who wears bifocal glasses. This person would have clear vision of close objects only when looking downward. If the assembly was overhead, the worker would have to tilt their head way back to see, and undoubtedly would find the situation to be uncomfortable and frustrating.

Principle 5.23

Try to make the parts to be assembled symmetrical relative to the grasp.

Parts such as rods or cylinders are normally easy to grasp around their longest axis. When the part is symmetrical, it will not need to be rotated before insertion, making the task easier and saving time. When the part is not symmetrical, then the next principle applies.

Principle 5.24

Provide orienting visual aids for asymmetric features.

Think of the so-called farmer matches (i.e., wooden-stem matches with the striking part at one end). The two ends differ, so matches are asymmetrical (more appropriately called semi-symmetrical). It is not easy for a robot, with no vision, or a person, who is not using vision, to know which end the match head is on. A worst case is when both ends appear very similar. More errors will necessarily occur as the similarity becomes greater.* Thus, coding with colors or symbols can assist the assembler greatly. Tactile elements can sometimes help the person holding tell which end is which. This idea leads to the next principle.

Principle 5.25

Provide tactile and visual feedback and maintain visibility throughout the assembly.

The visual sense is usually very important when assembling objects because of the eye–hand coordination needed. If assemblers cannot see the object during a critical phase, it is likely that they will either act without visual guidance or be delayed. In either case, performance speed is reduced and there is greater possibility of making an error.
A closely associated principle of assembly is given in the following.

Principle 5.26

The component parts should have a "snap fit" when inserted.

This type of component-to-component connection makes an almost instant connection, thus reducing assembly time. Snap connections also provide audible feedback so that the assembler knows the connection has been made and exactly when it was made and can therefore release control and begin the next action in the sequence. Audible feedback reduces the need for eye–hand coordination.
The last of these principles is as follows.

Principle 5.27

Feed parts to the assembling operator every time at the same location and in the same orientation.

Component parts should be delivered to the assembler repeatedly at the same place and in the same manner. When this is done, the assembler has less need to use vision to adjust the object before grasping it. These adjustments require time and increase the chance of making errors. When the component parts do not feed directly into the place needed, vibrators can be used along with gravity. The idea is not to take thinking out of jobs, but instead take it out when it does no good.

The next two sections address materials handling. Lifting is a particularly important part of materials handling and this is where the discussion starts. Following that, other modes of materials handling important to ergonomic design are discussed.

Analysis of lifting

In its most elementary form, lifting involves elevating a mass. Figure 5.5 shows two principal styles of lifting a box from the floor. The style to the left uses bent legs but a near-straight back, and the style on the right requires a bent back but straight legs. Most people assume that the bent-knees and straight-back style is better, as this lifting style has been widely publicized by safety advocates. It is true that the bent-knees and straight-back style generally puts a lower amount of torque on the back and thus lowers the possibility of back problems. But this style of lifting can only be used if the object is sufficiently small to fit between the legs.†

---

* See Chhabra and Ahluwalia (1990) for more on this notion.
† See Chapter 7 in Eastman Kodak (1986) for more information on lifting styles. In the case of very light objects and a lifter with no history of back problems, bending over to grab the item is likely to be faster and easier provided that the body does not twist.

212   *Introduction to human factors and ergonomics for engineers*

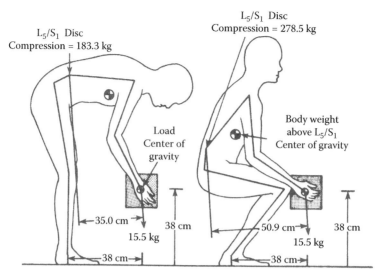

Figure 5.5 Two distinctive styles of lifting. (Reprinted from Park, K. and Chaffin, D., *AIIE Trans.*, 6, 105, 1974. With permission.)

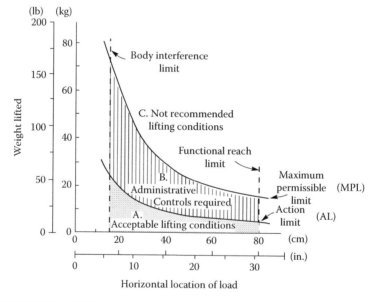

Figure 5.6 NIOSH (1989) guidelines on occasional human lifting.

Guidelines for lifting are available from various sources. Figure 5.6 shows the 1981 NIOSH guidelines for occasional lifts in the sagittal plane. That is, lifting that is performed vertically directly in front of the operator without any twisting. It is assumed in this guideline that lifts are 76 cm (30 in.) and not more frequent than 5 min apart. Figure 5.6 defines weight limits as a function of load center distance in front of the person lifting. Region A in this figure shows those conditions, which are acceptable for most people. Note that picking up an object places a torque on the human body. The greater the distance between the load center and the person's body, the greater the moment arm. Greater moment arms increase the torque produced by the load. Accordingly, action limits (AL) along the top boundary of region A, decrease with an increasing load center distance. Region B in Figure 5.6 defines lifting conditions that are acceptable for some trained workers who are in good health. That region is acceptable under administrative controls. Region C of the figure defines unacceptable lifting conditions.

Updated NIOSH guidelines were developed in 1991. The revised lifting equation considers asymmetrical lifting tasks and the effect of less than optimal couplings between the object and the worker's hands. These guidelines specify a recommended weight limit (RWL) for a variety of task conditions. The RWL is intended to approximate the weight that nearly all healthy workers could lift over a substantial period of time (e.g., up to 8 h) for the specified task conditions without an increased risk of developing lower back pain (Waters, Putz-Anderson, and Garg, 1994).

The RWL (in pounds) is calculated using the following equation:

$$RWL = LC * HM * VM * DM * AM * FM * CM \qquad (5.14)$$

where
- Load constant (LC) = 51 lb
- Horizontal multiplier (HM) = $(10/H)$
- Vertical multiplier (VM) = $1 - (0.0075 [V - 30])$
- Distance multiplier (DM) = $0.82 + (1.8/D)$
- Asymmetric multiplier (AM) = $1 - (0.0032A)$
- Frequency multiplier (FM) is obtained from Table 6.2
- Coupling multiplier (CM) is obtained from Table 6.3
- H is the horizontal location of the hands away from the midpoint between the ankles, in inches. If the measured horizontal distance is less than 10 in., then HM is set to 1. The maximum value of H is 25 in.

V is the vertical location of the hands above the floor, in inches or centimeters (measured at the origin and destination of lift). The maximum value of V is 70 in.

D is the vertical travel distance between the destination and origin of the lift, in inches. D is at least 10 in. and always less than 70 in.

A is the asymmetry angle measured in degrees and is defined by the location of the load relative to the worker's midsagittal plane at the beginning of the lift. If the worker has to position, regrasp, or hold the load at the end of the lift, the angle should be measured there as well. A varies between 0° and 135°. If $A > 135°$, then AM = 0, which results in RWL = 0.

F is the lifting frequency measured as the average number of lifts made per minute, over a 15 min period.

The value for the frequency multiplier is obtained from Table 5.2 and depends upon the average number of lifts/min (F), the vertical location (V) of the hands at the origin, and the duration of continuous lifting. The coupling multiplier (CM) depends on how good the hand-to-object coupling is. Table 5.3 shows how different handle and object configurations influence the hand-to-container coupling classification. The latter value (i.e., good, fair, or poor) is used in Table 5.4 to assign the coupling multiplier.

Table 5.2 Frequency Multiplier Table for the Updated NIOSH Lifting Equation

Frequency (F) (Lifts/min)	$D \leq 1$		$1 < D \leq 2$		$2 < D \leq 8$	
	$V<30$	$V \geq 30$	$V<30$	$V \geq 30$	$V<30$	$V \geq 30$
$F \leq 0.2$	1.00	1.00	0.95	0.95	0.85	0.85
0.5	0.97	0.97	0.92	0.92	0.81	0.81
1	0.94	0.94	0.88	0.88	0.75	0.75
2	0.91	0.91	0.84	0.84	0.65	0.65
3	0.88	0.88	0.79	0.79	0.55	0.55
4	0.84	0.84	0.82	0.82	0.45	0.45
5	0.80	0.80	0.60	0.60	0.35	0.35
6	0.75	0.75	0.50	0.50	0.27	0.27
7	0.70	0.70	0.42	0.42	0.22	0.22
8	0.60	0.60	0.35	0.35	0.18	0.18
9	0.52	0.52	0.30	0.30	0.00	0.15
10	0.45	0.45	0.26	0.26	0.00	0.13
11	0.41	0.41	0.00	0.23	0.00	0.00
12	0.37	0.37	0.00	0.21	0.00	0.00
13	0.00	0.34	0.00	0.00	0.00	0.00
14	0.00	0.31	0.00	0.00	0.00	0.00
15	0.00	0.28	0.00	0.00	0.00	0.00
$F \geq 15$	0.00	0.00	0.00	0.00	0.00	0.00

Principles of lifting

A number of principles can be employed to minimize injuries and lower back pain due to lifting. Some of these are derived from elementary mechanics, human anatomy, and others come from common sense. The NIOSH equation discussed earlier also implies a number of strategies for almost any setting where lifting is done. The first two principles discussed in the following are basically common sense.

Principle 5.28

Try first, to design lifting out of the task. Failing that, minimize its frequency, weight, and torque on the body.

Eliminating the potential problem is the first option that should be considered. The next strategy is to reduce the effect of lifting. With less energy, there is less loading and less chance of injury. In many cases one can use mechanical means to perform the lifting. Another principle, related to the first, is the following.

Principle 5.29

Lift smart, not hard.

Some simple recommendations can make lifting safer. That is, lift smart, not hard. Clear away all items along the path before you lift. Leave nothing to slip on or trip over. Open all doors that you will need to pass through. Releasing a hand to turn a door knob can cause the load to become unstable and result in a sudden jerk in your back. Be sure

Chapter five: Work measurement and analysis

Table 5.3 Hand-to-Contour Coupling Classification

Good	Fair	Poor
1. For containers of optimal design such as some boxes, crates, etc., a "good" hand-to object coupling would be defined as handles or handhold cutouts of optimal design (see notes 1–3)	1. For containers of optimal design, a "fair" hand-to-object coupling would be defined as handles or handhold cutouts of less than optimal design (see notes 1–4)	1. Containers of less than optimal design or loose parts or irregular objects that are bulky, are hard to handle, or have sharp edges (see notes)
2. For loose parts or irregular objects, which are not usually containerized, such as castings, stock, and supply materials, a "good" hand-to-object coupling would be defined as a comfortable grip in which the hand can be easily wrapped around the object (see note 6)	2. For containers of optimal design with no handles or handhold cutouts or for loose parts or irregular objects, a "fair" hand-to-object coupling is defined as a grip in which the hand can be flexed about 90° (see note 4)	2. Lifting nonrigid bags (i.e., bags that sag in the middle)

Notes:
1. An optimal handle design has 0.75–1.5 in. (1.9–3.8 cm) diameter, ≥4.5 in. (11.5 cm) length, 2 in. (5 cm) clearance, cylindrical shape, and a smooth nonslip surface.
2. An optimal handhold cutout has the following approximate characteristics: ≥1.5 in. (3.8 cm) height, 4.5 (11.5 cm) length, semi-oval shape, ≥2 in. (5 cm) clearance, smooth nonslip surface, and ≥0.25 in. (0.60 cm) container thickness (e.g., double-thickness cardboard).
3. An optimal container design has ≤16 in. (40 cm) frontal length, ≤12 in. (30 cm) height, and a smooth nonslip surface.
4. A worker should be capable of clamping the fingers nearly 90° under the container, such as required when lifting a cardboard box from the floor.
5. A container is considered less than optimal if it has a frontal length >16 in. (40 cm), height >12 in. (30 cm), rough or slippery surfaces, sharp edges, asymmetric center of mass, unstable contents, or requires the use of gloves. A loose object is considered bulky if the load cannot easily be balanced between the hand grasps.
6. A worker should be able to comfortably wrap the hand around.

Table 5.4 Values of the Coupling Multiplier

	Coupling Multiplier	
Coupling Type	$V < 30$ in. (75 cm)	$V \geq 30$ in. (75 cm)
Good	1	0
Fair	0.95	1
Poor	0.9	0.9

you have proper shoes for lifting; that is, flat nonslip soles and definitely no high heels. Use special equipment for large or heavy objects or for lifts above the shoulders.

Principle 5.30

Keep the load to be lifted close to the body. Assure that the feet are stable, close to the load, and pointed perpendicularly to the plane of lifting. Lift the load in that plane using primarily the legs.

Part of this principle follows directly from the mechanics. The closer the load is to the human body, the shorter the moment arm, and the lower the torque on the spine. The other part comes from the fact that the legs are strong and that twisting while lifting is almost an invitation to back problems. Twisting or bending must be avoided while lifting.

The next principle deals with the individual's physical fitness, because people who are not in good physical condition are more prone to back pain from lifting.

Principle 5.31

Avoid attempting difficult lifts or lowerings, unless you are in excellent physical shape.

One should always test the load before lifting. If it is too heavy, get a mechanical aid or someone else to either perform the lifting or assist in it. When the lifting-person perceives that he or she can lift the load alone, the following principle applies.

Principle 5.32

Assure that ample space is available for lifting and that you have a strong grip on the load before starting the lift.

Loads should be handled with smooth, flowing motions. Newtonian mechanics shows that smooth movements imply small accelerations. Instructions and training should be given about the proper way to perform the lifting.

Lifting is often done shortly after the person arrives somewhere. This is especially true with delivery or maintenance personnel. In such cases, one should not go quickly from a sedate or resting situation into lifting; so the next principle applies.

Principle 5.33

Walk around a bit before lifting.

The basic idea is to warm up before doing the lifting. Another lifting principle to keep in mind is the following.

Principle 5.34

Avoid uncertain timing changes in lifting or carrying.

When several people are lifting and some let loose without warning, those who remain to hold the weight can experience severe dynamic loads unexpectedly, and this can result in injury. The same thing can happen when an individual is carrying an object and has to remove an arm to open a door. Always be prepared for such unexpected occurrences.

Computer tools for analyzing lifting tasks

Several computer tools are available for analyzing lifting and other aspects of materials handling. Some of these tools compute the compressive and shear forces on the spine that occur during lifting and for other tasks. These forces are often calculated at the locations of discs between certain lower lumbar vertebrae. Compressive loading values between the L3 and L4 vertebrae are often focused upon in such analyses, because injuries often occur at this location. The computations are normally based on a static biomechanics model of the human body. Actual measurements tend to show higher compressive forces due to dynamic loading but nearly equal shear forces in the lower back. These studies also verify greater forces with the bent back style.*

---

* See Ayoub and El-Bassoussi (1978) for more discussion on this topic.

One program used to perform such an analysis is the three-dimensional static strength prediction program (3DSSPP).* This program has a graphic-user interface (GUI) used to show how people interact with other objects in the work environment.

A static biomechanical model is used in 3DSSPP to compute forces on critical joints of the worker during job segments.† The computed forces are reported as percentiles of the male and female American working populations expected to have sufficient strength to handle the load. NIOSH recommends strength limits that can be met by 99% of the male population and 75% of the female population. NIOSH also recommends a maximum back compressive force of 765 lb (3400 N) for safety-related reasons.

Another, more sophisticated, program is Jack, which was developed at the University of Pennsylvania. It has since been commercialized, and is currently owned by Siemens. Jack is a sophisticated digital human modeling software package, capable of simulating a human involved in any number of tasks. It has been used by the military to help design and assess vehicle designs, and was used in virtual prototyping for the U.S. space shuttle. Its outputs include joint angles and forces among many others, and its visualization is far superior to that of 3DSSPP.

Learning

In the next section, the speed of accomplishing a task will be determined. However, before determining this time, it may be necessary to determine if the workers are still learning the task. If so, the time it takes to accomplish a task may be longer than after the workers learn the task.

Learning is a phenomenon where performance improves with experience. As tasks are repeated, elements of the task are better remembered, cues are more clearly detected, skills are sharpened, transitions between successive tasks are smoothed, eye–hand coordination is more tightly coupled, and relationships between task elements are discovered. The aggregation of these effects results in faster performance times, fewer errors, less effort, and more satisfied workers and customers.‡

Learning effects must be distinguished from other changes in performance. Many other things besides learning can cause performance to improve. For example, performance might improve because of method improvements, new tools, better equipment or more automation, new incentive systems, better instructions and job performance aids, or even better supervision. Thus, after detecting improvements in performance, the analyst must always consider other explanations before assuming that learning has occurred. Like a theory, learning can never be proved; it can only be disproved. After learning is detected, mathematical models, called learning curves, can then be fitted to the performance data. After the analyst determines which model best fits the data, the learning curve becomes the basis for performance prediction.

Crews, teams, departments, companies, and even industries learn with experience. As a result, similar mathematical models are often used to describe organizational learning.

---

* Professor Don Chaffin and his colleagues developed the 3DSSPP at the University of Michigan (see Beck et al., 1993). As of this writing, version 6.0.5 of the software was available at http://www.engin.umich.edu/dept/ioe/3DSSPP/.
† The strength model in 3DSSPP is based on a static biomechanical model. The model does not account for accelerations as a dynamic model might. Hence, a safety factor has to be factored into the calculations.
‡ Several historical studies on individual person learning includes Barnes and Amrine (1942), Knowles and Bekk (1950), Glover (1966), Hancock and Foulke (1966), and Hancock (1967).

The term *progress curve** is often used in cases involving assembly lines, crews, teams, departments, and other smaller groups of people. The term *experience curve*† is sometimes used when larger organizational groups such as companies and industries are involved. The latter improvement curves include the effect of changes in how people interact and coordinate their activities. Progress or experience curves both include improvement due to engineering changes and other factors. Regardless of the term used, the mathematical models remain the same. In our discussion that follows, learning and progress curves for individuals and groups will both be considered.

Some applications of learning curves

Learning and progress curves serve numerous important purposes. One of the more obvious applications is in determining the cost of production operations. Since part of the cost of production is direct labor, learning over the production run can decrease the direct labor costs. A number of U.S. governmental organizations, consequently, require learning curves to be used when pricing services and products. Learning and progress curves are also used to evaluate production methods and jobs. When more than one method is available, one method may have an advantage for shorter production runs, while another method may be better for longer production runs. In such cases, one needs to find the break-even run size, and establish a decision rule for method selection based on run sizes.

A third application of learning curves is in production control. In production control, engineers need to identify operations where bottlenecks occur, and then solve the problems causing the bottlenecks. To do this, they need to know how performance changes for each operation over the production run. A learning curve is used to estimate this break-even point. Quality control is another area where learning and progress curves are used. Some defective product units are created because of human errors during production. The magnitude and number of these errors change with experience. Consequently, learning and progress curves are needed by quality control engineers to predict the defective product units and to help reduce these and other causes of defectives.

Modeling human learning

Modeling human behavior is complex but interesting and modeling human learning behavior is more so. As a consequence you might ask, "Why then go to the effort?" The answer is simple: to obtain quantitative estimates of performance with more or less experience.

Typically, both speed and accuracy are poor the first time a task is performed. When the task is performed repeatedly, people think of ways to improve and their behavior follows their thinking. Accordingly, learning models show decreasing performance times as a function of more task repetitions. This change is precisely what one would expect for performance time averages. However, performance continues to fluctuate from trial to trial, and people do not exhibit a constant performance time decrease as most models predict.

---

* Konz (1990) champions this terminology without necessarily distinguishing between learning by individuals and by small groups of people.
† See Conley (1970) or Hax and Majluf (1982) for studies of experience curves for different industries.

All of the commonly used models show decreases in performance time with experience. Each model has parameters to describe how quickly this happens. Some models show constantly decreasing performance times, which eventually approach zero. This situation is ridiculous. Other models show performance time decreasing to a preset level. Such models determine the amount of learning in addition to the rate of learning. These more realistic models still show constantly decreasing performance times down to the limit. Observations over time often reveal periods of constant performance, or learning plateaus, but no known models provide this realism. As with other models, learning models are simplified descriptions of reality.

Why use a model of learning?

It might be easily argued that data containing learning effects can be used instead of a learning model. So, why use a learning model? There are a number of reasons, four of which are discussed in the following:

1. Most operational data are influenced by factors other than learning. If such data are to be relevant in a new situation, it is necessary to show that these other factors will be present in the new situation. The past is rarely duplicated, so it may be difficult to justify this assumption. A learning curve model can be stripped of all extraneous effects except learning, which reduces this potential problem.
2. It is difficult to make predictions from operational data, without developing some type of predictive model. A learning model can often be fit to various products, within the same product family, by making minor adjustments to model parameters.*
3. Learning rates can be adjusted in a way that reflects expected differences between situations. More learning is to be expected for products with more complex manufacturing requirements. Models that vary learning rates in expected ways have high face validity that in turn helps gain acceptance of the model. It is difficult to directly match past data to new situations, which is another reason to develop and use a model instead.
4. Learning models simplify the issue. Even if a model oversimplifies the situation, it can be used as a first-order approximation. Further adjustments can then be made after observing the process. It is much more difficult to adjust a complex description to a simpler situation.

For these and other reasons, some learning curve models are described in this chapter, along with methods of fitting the models to data.

Performance criteria and experience units

Performance time is the most common criterion used in industry today. Production cycles are the most commonly used variable for denoting experience. If t_i is the performance time on the *i*th cycle, then a learning curve should predict t_i as a function of *n* cycles. Since

---

* As will be discussed in Chapter 6, the performance time required to produce the members of a product family can be often described by a single model with parameter values that distinguish the different products in the family. The predictions of such models can then be adjusted for learning. In order for this approach to work well, the production of products within a family must involve similar activities that vary in ways easily distinguished by different combinations of model parameters. A model without enough parameters may be inadequate. It is more difficult to develop a good model for a product family with many very different products.

learning implies improvement with experience, one would expect $t_i \; t_i \; i+1$ and $t_i - t_i + 1.0$ over the $i = 1, 2, \ldots, n$ cycles.

The cumulative average performance time on the ith cycle, or A_i, is also often used to describe performance. Cumulative average times consist of the sum of all performance times up to and including the nth cycle divided by n, or

$$A_n = \frac{\sum_{i=1}^{n} t_i}{n} \tag{5.15}$$

Note that for $n = 1$, $A_1 = t_1$. With learning, t_i tends to decrease with i and so does A_n. However, A_n decreases at a slower rate than t_i as i increases. This effect can be shown by the first-forward difference of A_n, which is

$$\Delta A_n = A_{n+1} - A_n \tag{5.16}$$

$$\Delta A_n = \frac{\sum_{i=1}^{n+1} t_i}{n+1} - \frac{\sum_{i=1}^{n} t_i}{n} = \frac{\sum_{i=1}^{n} t_i + t_{n+1}}{n+1} - \frac{\sum_{i='}^{n} t_i}{n}$$

$$= \frac{n \sum_{i=1}^{n} t_i}{n(n+1)} - \frac{\sum_{i=1}^{n} t_i}{n} + \frac{n t_{n+1}}{n(n+1)}$$

$$= \frac{n \sum_{i=1}^{n+1} t_i}{n(n+1)} - \frac{(n+1) \sum_{i}^{n} t_i}{n(n+1)} + \frac{n t_{n+1}}{n(n+1)}$$

$$= \left[\frac{n}{n+1} - \frac{n+1}{n+1} \right] A_n + \frac{t_{n+1}}{n+1}$$

$$\Delta A_n = \frac{t_{n+1}}{n+1} - \frac{A_n}{n+1} = \frac{t_{n+1} - A_n}{n+1} \tag{5.17}$$

So long as t_{n+1} is less than A_n, ΔA_n is negative and the cumulative average time continues to decrease.

Another criterion of interest is accuracy. The complementary measure of accuracy is the error rate. It is more common to measure errors in production than accuracy. The sequence of production errors is $e_1, e_2, \ldots, e_i, \ldots, e_n$ over n serial cycles. If the person is doing a single operation on a product unit, either an error is observed or it is not. Observations over a production sequence become a series of zeros and ones, respectively corresponding to no error and an error. When more than a single operation is performed, say M operations, up to M errors may be observed for each cycle. A common practice is to define e_i as the observed number of errors divided by the M possible errors in a cycle. When this approach is followed, e_i falls in the range of 0–1, and corresponds to an error rate. It also follows that a learning curve could be fit to the series of e_i values or to the cumulative average error rate. If learning is present, then one would expect to see a general decrease in e_i with increases in i. The cumulative average error rate would also decrease, but more slowly than e_i.

Performance time and errors per piece produced are complementary measurements of speed and accuracy. Speed is the inverse of the cumulative average time, A_n. For example, if $A_n = 0.1\,h$, speed $= 1/0.1 = 10$ units/h. Note that the aforementioned estimate is the average speed over the entire n cycles. If learning is present, the individual cycle times vary over the interval. Speed, accordingly, varies over the individual cycles. When errors are measured as a fraction of the correctly performed operations, the complementary fraction is the accuracy. For example, an error rate of 0.1 corresponds to an accuracy of 0.9. So there is also some advantage to using an error rate, instead of counting the number of errors. The product of accuracy and speed gives the correctly produced pieces per unit of time.

Some learning curve models

Learning curves originated as elementary mathematical functions that fit trends in serial data. This restriction to elementary functions tends to make the model fitting easier. Also, elementary functions tend to be more robust, but simplicity does not guarantee robustness. However, simpler models have fewer parameters and are more limited in their functional form. This may cause a poorer fit to the data. Such is the traditional dilemma in model selection.

Powerform learning curve

Powerform models have been used to fit learning curves in industrial settings, in numerous instances.* A few relevant examples are summarized in Table 5.5, which shows typical values of powerform parameters. This commonly used model predicts the time of the nth cycle to be

$$t_n = t_1 n^{-C} \tag{5.18}$$

where C is the learning rate parameter. In the earliest reference to learning curve models known to the authors, Snoddy reported that learning was approximated by this function.

Table 5.5 Some Reported Examples of Learning in Industrial Settings

Improvement (%)	Number of Cycles	Task and Source
68	1,000	Test and adjust—Conway and Schultz (1959)
73	11,000	Electronic assembly—same as above
79	1,000	Service time—IBM electronics
80		Man-hours per airframe—Wright in WWII
85	2,000	Labor hours per gun barrel on a boring machine—Andress (1954)
86	12,300	Cost/model-T—Ford
90	—	Bench inspection— Nadler and Smith (1963)

* A few examples are illustrated in Table 5.5; Conway and Shultz (1959), Glover (1966), and Hancock (1967). Konz (1983) provides more extensive lists of past studies.

Table 5.6 Effect of Parameter C in the Powerform Model

C	0.074	0.152	0.234	0.322	0.415	0.515	0.621	0.737
t_{2n}/t_n	0.95	0.90	0.85	0.80	0.75	0.70	0.65	0.60

Later, Wright* used the powerform model as a progress function in the aircraft industry. There is a particular property of this powerform model that led Wright to it. That property is that the fraction t_{2n}/t_n is constant for any n. In other words, when n is doubled, the time needed decreases by a constant proportion.

Consider the following time series generated using the powerform model $f(n) = 5n^{-0.322}$:

$n \rightarrow 1, 2, \ldots, 9, 10$

$t_n \rightarrow 5.0, 4.0, 3.5, 3.2, 3.0, 2.8, 2.7, 2.6, 2.5, 2.4$

$$\frac{t_{2n}}{t_n} \Rightarrow \frac{4}{5}, \frac{3.2}{4}, \frac{2.8}{3.5}, \frac{2.6}{3.2}, \frac{2.4}{3.0} = 0.8$$

This powerform model keeps generating smaller t_n values, which eventually approach zero, if n is large enough. Accordingly, zero is the mathematical asymptote of this model. The values of t_{2n}/t_n in the preceding example should all equal 0.8. Any variation is due to round-off error. These ratios show how performance changes with each doubling of the number of cycles performed. That is, the time per cycle after each doubling of the number of cycles is 80% of the previous time. Accordingly, this curve is known as the 80% powerform curve. Other t_{2n}/t_n ratios result with other values of the learning rate parameter C as shown in Table 5.6.

Table 5.6 shows that larger values of C result in a smaller change with each doubling of the cycles. When the functional powerform is substituted for t_{2n} and t_n, we have $\left(5(2n)^{-C}\right)/5n^{-C}$, which equals 2^{-C}, and that in turn equals t_{2n}/t_n. Parameter C then follows by conversion to logarithmic form, as

$$C = \frac{-\log\left[t_{2n}/t_n\right]}{\log 2} \quad (5.19)$$

where Equation 5.18 relates parameter C to any ratio of t_{2n}/t_n rather than just those shown in the preceding table.

For completeness, it should be stated that this model can be used to describe the cumulative average time, A_n (see Equation 5.15). Note that when there is a good fit of t_n to the powerform curve, A_n does not follow the powerform series. It follows in reverse that when cumulative average time data are fit well by the powerform models, serial times t_n do not fit the powerform model well. As a side observation, note that it is not easy to generate the sum $t_1 + t_2 + \cdots + t_n$ from the basic model shown in Equation 5.17 because the summation

$$\sum_{i=1}^{n} t_i = t_1(1^{-c} + 2^{-c} + 3^{-c} + \cdots + n^{-c}) \quad (5.20)$$

* Wright was working for an aircraft manufacturing company in Wichita, Kansas, when he noticed that the direct labor in building successive aircraft reduced by a constant fraction with each doubling of the number of aircraft produced. This characteristic is a property of the powerform model, which Wright identified.

Chapter five: Work measurement and analysis

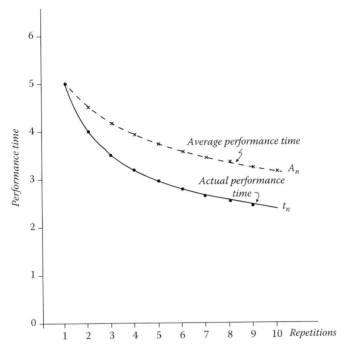

Figure 5.7 Time series and cumulative average time series for the powerform learning function.

contains a complicated power series. This makes it difficult, but not impossible, to compute A_n. Figure 5.7 contains a plot of serial time and cumulative average time values for the powerform model. Note that the cumulative average lags the serial cycle time, but both have the same asymptotic value, and gets large. Other curves are discussed later.

A variant of the powerform model was devised at Stanford University and is now commonly called the Stanford model.* This model is

$$t_n = k(d+n)^{-C} \qquad (5.21)$$

where
    k is a parameter similar to t_1 in the regular powerform model
    d is an additional parameter that causes the beginning portion of the learning curve to change more slowly

Note that Equation 5.20 simplifies to the traditional powerform model (Equation 5.17) when $d=0$ and $k=t_1$.

If a Stanford model were devised with $d=0.1$, $C=0.322$, and a fit was made to $t_1=5.0$ by setting $k(d+n)^{-C} = 5$, then the predicted time values and ratios of t_{2n}/t_n would be as shown in Table 5.7.

---

* Garg and Milliman (1961) report on this variation of the powerform model.

Table 5.7 Values of t_n and t_{2n}/t_n for the Stanford Model

n	1	2	3	4	5
t_n	5.00	4.06	3.58	3.27	3.05
t_{2n}/t_n	0.812	0.806	0.805	0.803	0.803

Table 5.8 Stanford Model when $d = 1.0$

n	1	2	3	4	5
t_n	5.00	4.39	4.00	3.72	3.51
t_{2n}/t_n	0.878	0.848	0.835	0.828	0.823

The Stanford model makes predictions that are close to those made by the traditional powerform model (Equation 5.17), since t_{2n}/t_n is about the same for each value of n. Using the first two values of n from Table 5.7, it follows for the Stanford model that

$$\text{Log } 5 = \text{Log } K - C \text{ Log } 1.1$$

and

$$\text{Log } 4.06 = \text{Log } K - C \text{ Log } 2.1$$

A simultaneous solution of these two equations that result in parameters k and C are 5.16 and 0.3221, respectively. It is interesting to observe that the Stanford model gives a series that is quite distinct from the powerform model when d is equal to 1.0, as shown in Table 5.8.

The results for this second Stanford model, in contrast to those in the first example (Table 5.7) and the traditional powerform model (the Stanford model with $d = 0$), show that t_n decreases at a slower rate for larger values of d. Also, the data in Table 5.8 show that the doubling sequence ratio changes over time. These tendencies are particularly obvious early in the sequence. Figure 5.8 plots the two Stanford model learning curves relative to the original powerform curve. The conclusion is that the Stanford model trades off some of the simplicity of the powerform model, to provide a potentially better fit to observed rates of learning, by including an additional parameter that allows the ratio t_{2n}/t_n to change over time.

Another variation of the powerform model was devised by De Jong.* He recognized that powerform models have an asymptote of 0. That is, as n approaches ∞, t_i approaches 0, in the traditional powerform model. This tendency obviously is unrealistic, because people will always require some time to perform a task. To get around this difficulty, he added a parameter m, which he called the incompressibility fraction. His idea is that the asymptote of this function should be the product of m and t_1. Learning over time occurs for the complementary part of performance or the product of $(1 - m)$ and t_1. Thus, De Jong's version of the powerform model is

$$t_n = t_1\left(m + (1-m)n^{-C}\right) \tag{5.22}$$

---

* De Jong (1957) studied learning in a number of Dutch industries. In one study, he observed people making windows for homes and other buildings over a period of 5 years. Learning took place over the entire period.

Chapter five: Work measurement and analysis 225

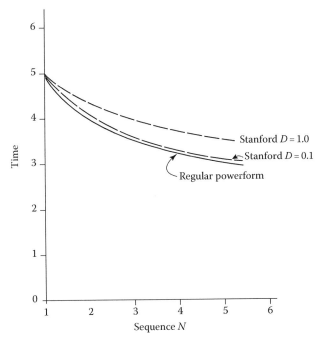

Figure 5.8 Regular powerform time series and two Stanford variations of the powerform models when $d=0.1$ and $d=1.0$.

Discrete exponential learning curve

The discrete exponential learning model is an alternative to the powerform model that can also be fit to serial time data, which shows a progressive decrease in values of t_n from t_1 down to a specified asymptote value t^*. The functional form of the discrete exponential model is

$$t_n = \alpha^{n-1}(t_1 - t^*) + t^* \tag{5.23}$$

where α is the *learning rate parameter*, which is restricted to $0 < \alpha < 1$. Pegels* first identified this model of learning and he noted its natural asymptote was t^* rather than the zero asymptote of the powerform model. Later† it was shown that this model could be derived from a constant-parameter forward-difference model. The latter model describes changes in performance proportional to current performance, shown as follows:

$$t_2 = \alpha t_1 + \beta$$

$$t_3 = \alpha t_2 + \beta = \alpha(\alpha t_1 + \beta) + \beta = \alpha^2 t_1 + \alpha\beta + \beta$$

---

* Pegels (1969) initially recognized this functional form and the fact that it had a natural asymptote; an element that was missing on the regular powerform model.
† Buck, Tanchoco, and Sweet (1976) reported on the structure of this model. A more detailed exposition of difference models is shown by Goldberg (1961).

More generally,

$$t_{n+1} = \alpha^n t_1 + \beta \sum_{i=0}^{n-1} \alpha^i \quad (5.24)$$

Since the portion within the summation is a geometric progression where α is less than unity, but greater than zero, this summation is equal to $(1-\alpha^n)/(1-\alpha)$. Accordingly, the preceding equation becomes

$$t_{n+1} = \alpha^n t_1 + \beta \frac{1-\alpha^n}{1-\alpha} \quad (5.25)$$

When $\beta/(1-\alpha)$ is defined as t^*, then Equation 5.24 becomes the same as Equation 5.22. It is interesting to see that the difference between t_n and t_{n+1} in this exponential series is

$$\Delta t_n = t_{n+1} - t_n = \alpha t_n + \beta - t_n = (\alpha - 1)t_n + \beta \quad (5.26)$$

The exponential series, becomes a linear equation when $\alpha = 1$. In this case, the rate of change, Δt_n, is equal to β.* It should also be stated that when the exponential series is converted to logarithmic form, the resulting curve is S-shaped.†

One advantage of the exponential learning curve is that the summed time of $T_n = t_1 + t_2 + \cdots + t_n$ can be easily computed to find the cumulative average times A_n. The basic formula for this is

$$A_n = \frac{1}{n}\sum_{i=1}^{n} t_i = t^* + \left[\frac{t_1 - t^*}{n}\right]\left[\frac{1-\alpha^n}{1-\alpha}\right] \quad (5.27)$$

Note that in Equation 5.27 the first term on the right-hand side (t^*) is the steady-state time value after all learning is complete. The remaining portion of the expression accounts for changes in performance time as people learn. In Box 5.2, the steady-state contribution is 18 units or 82% of the total time; leaving a transitory contribution of 3.9375 units, or 18% of the total. As n increases to 25, the learning curve more closely approaches the asymptote and the steady-state portion of time increases to 75 time units out of a total of 79 time units and so only 5% of the time is transitory. Figure 5.9 shows a slower exponential learning curve with α equal to 0.9, rather than 0.5, but with the same starting and asymptotic time values. The sum after 6 product units is 2.2. The 1.8 time units required at steady-state performance constitutes 66% of the total. After 25 trials, the total time is 9.4 units, and the 7.5 steady-state time units constitute 80% of the total.

---

* Goldberg (1961) provides much more detail about this constant-coefficient difference-equation. In particular, he shows that the series converges from t_1 to t^* so long as $0<\alpha<1$ because α^n diminishes with n. With $t_1<t^*$, convergence is upward. When $\alpha<0$, the series oscillates, because α^n is negative with an odd n and positive with an even n. The case of $a>1$ causes α^n to increase exponentially with a larger n and so the series diverges away from t^*. Accordingly, only the case of $0<a<1$ is useful for learning curves in the sense of traditional applications.
† Both Jordan (1965) and Muth (1986) observed S-shaped logarithmic functions in practice.

Chapter five: Work measurement and analysis

BOX 5.2 SHORT EXAMPLE OF LEARNING ECONOMICS

Some data from an assembly task were fit to a discrete exponential learning curve model using the time in hours to measure successive hours for each assembly. The fitted model for the standard operator was

$$t_{n+1} = \alpha t_n + \beta = 0.9794 t_n + 0.0423$$

The initial and asymptote values were $t_1 = 4.00$ and $t^* = 2.0534$ h/assembly, respectively. Orders come in batches of 100. The question faced was whether to allocate a single inexperienced person to the assembly or use multiple inexperienced people. If n people are used, the learning curve will apply to $100/n$ cycles. It is easy to compute the total working time (exclusive of allowances) for this assembly, for various values of n, by multiplying the average time per cycle, as given in Equation 5.26, by the number of cycles $100/n$. This gives the following results:

For one person,

$$T_{100} = 100(2.0534) + (4 - 2.0534)\frac{1 - 0.9794^{100}}{1 - 0.9794} = 288.05 \text{ h}$$

For two people,

$$T_{50} = 50(2.0534) + (4 - 2.0534)\frac{1 - 0.9794^{50}}{1 - 0.9794} = 163.79 \text{ h}$$

For three people,

$$T_{33} = 33(2.0534) + (4 - 2.0534)\frac{1 - 0.9794^{33}}{1 - 0.9794} = 114.71 \text{ h}$$

For four people,

$$T_{25} = 25(2.0534) + (4 - 2.0534)\frac{1 - 0.9794^{25}}{1 - 0.9794} = 89.67 \text{ h}$$

The total time needed to produce the 100 assemblies in each case is given as follows:
For one person, total time needed = T_{100} = 288.05 h
For two people, total time needed = $2T_{50}$ = 2 * 163.79 = 337.58 h
For three people, total time needed = $3T_{33} + t_{34}$ = (3 * 90.63) + 3.03 = 347.16 h
For four people, total time needed = $4T_{25}$ = 4 * 89.67 = 358.68 h

The single-operator case is much more economical than the others because the single operator does more on-the-job learning. The analysis assumes that each operator is inexperienced at the start of the process. Learning effects, of course, will be smaller for the next batch, if the same people do the assembly. Also note that in the three operator case, t_{34} was included as part of the total time, because one operator has to do the task 34 times. The other two do the task 33 times. Equation 5.24 was used to calculate t_{34}, shown as follows:

$$t_{34} = \alpha^{33} t_1 + \beta \frac{1 - \alpha^{33}}{1 - \alpha} = 0.9794^{33} * 4 + 0.0423 \frac{1 - 0.9794^{33}}{1 - 0.9794} = 3.03$$

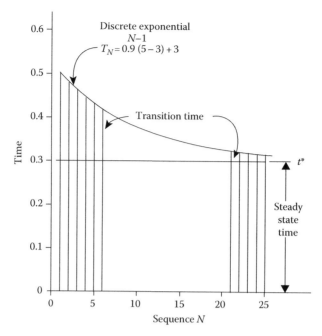

Figure 5.9 Steady-state and transitory times of the discrete exponential learning model.

Continuous exponential learning curve models have also been developed.* The latter model is given by the following equation:

$$t_n = t_1 - (t_1 - t^*)e^{-bt} \qquad (5.28)$$

where
 t is continuous running time while the task is being performed
 b is the rate parameter

In one version of the model, † is the time duration until learning is half-completed or $(t_1 - t^*)/2$, and b is $1/†$, which is similar to a half-life value. One other learning curve model that has been used over the years is the quadratic polynomial model.†

Fitting learning curves

The earlier discussion showed the mathematical basis of learning curves and some of the ways they differ. As might be expected, the different functional forms are also fit differently to data. Part of the reason is that different mathematical operations and conveniences become appropriate.

---

* Bevis, Finnicat, and Towill (1970) developed this model. Hutching and Towill (1975) extended this model and applied it to describe changes in performance accuracy.
† Carlson (1973) suggested this model. It is also shown by Nanda and Adler (1977).

Powerform model

The powerform model is linear when converted to the logarithmic form. This makes it easy to fit the powerform model to data. That is, converting Equation 5.17 to logarithmic form results in the expression

$$\log t_n = \log t_1 - C \log n \tag{5.29}$$

As shown by Equation 5.29, when the powerform model is valid, the logarithm of t_n is a linear function of the logarithm of n. Consider the previous numerical example of the powerform model shown in Table 5.9.

The data from this table are plotted in Figure 5.10. Note that the points form a straight line, confirming that a linear relationship is present. A simple linear regression can be made on these logarithmically transformed data. The negative slope of the regression line provides a direct estimate of parameter C. Since this plot was made using data that were generated by a powerform equation, the fit is perfect.

For De Jong's variation of the powerform model, an identical data-fitting procedure can be used if the incompressibility parameter M can be estimated first. Once M is estimated, the product of M and t_1 is subtracted from every value of t_i and the procedure continues

Table 5.9 Example Performance Data with Learning

n	1	2	3	4	5	6
t_n	5	4	3.5	3.25	3.125	3.0625
Δt_n	−1	−0.5	−0.25	−0.125	−0.0625	—

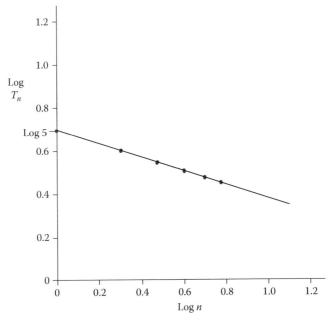

Figure 5.10 Log–log graph of the powerform example with log T_n as a function of log N, which can be used for fitting this learning curve.

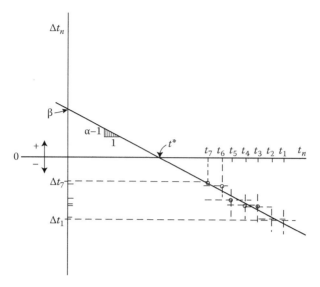

Figure 5.11 Fitting a discrete exponential model to the data.

as before for the powerform curve. Following this procedure will give an estimate of the compressible portion of De Jong's model. One only needs to add back the constant product of M and t_1. However, it is not altogether clear how one should estimate the fraction M. One approach is to extrapolate from the ultimate performance of similar tasks; perhaps using laboratory studies or past tasks that are most similar.

Discrete exponential model

In the discrete exponential model, and as shown in Equation 5.26, the first-forward difference Δt_n is a *linear* function of the performance magnitude t_n. Accordingly, the best linear fit provides estimates of parameters α and β. A linear regression of Δt_n as a function of t_n, has the slope of $(\alpha - 1)$. The ordinate intercept of that regression is β and the abscissa intercept is t^*, the asymptote. For example, consider the case where performance is as given in Table 5.9.

A regression of the data in Table 5.9, where Δt_n is determined as a function of t_n, yields a slope of −0.5, an ordinate intercept of 1.5, and an abscissa intercept of 3. When the slope value is equated to $(\alpha - 1)$ and solved, it is seen that $\alpha = 0.5$. Figure 5.11 shows a plot of these regression data.

The aforementioned example of fitting the discrete exponential model to data resulted in finding parameters α and β. Parameter t^* is a direct result of these parameter values. With the values of these parameters substituted in Equation 5.23, only the value of t_1 is needed and every value of the series can be predicted. One could use the first data point for t_1. But if there is any error in this initial value, then the entire series of predictions is affected, potentially leading to poor predictions.

Comparing alternatives for learnability

Ergonomists are often interested in developing products that can be used proficiently at the earliest possible time. To help ensure this objective, the usability of new designs can be tested for users unfamiliar with them. The learning curve of naive users provides one basis for assessing usability. In terms of the powerform model, the learning rate is specified by the parameter C. The discrete exponential model describes learning in two ways; parameter α shows the learning rate and the difference between t_1 and t^* describes the amount of learning expected. Unfortunately, the expected amount of learning is not available but a similar estimate can be obtained by computing t_{1000} and subtracting it from t_1. A design that results in quick learning is obviously desirable in any situation.

Some recent studies from Arzi and Shatub (1996) and Dar-El and Vollichman (1996) have shown two interesting results. The first is that tasks with higher cognitive content tend to have lower learning rates. An immediate implication is that quicker learning (or lower performance times) occurs when cognitive content is stripped from the task. A second important finding is that forgetting occurs between practice sessions, but people quickly return to the point on the learning curve they were at when they stopped practicing. This often corresponds to a brief warmup period, with slightly lower performance.

Correct learning curve model

One often hears the question, "How do you tell which is the real learning curve model?" The answer to that question is, "No true learning curve model is known to exist. No one model always describes learning better than another model." For some people and tasks the powerform model describes performance time and error changes very accurately. In other situations, the discrete exponential model fits the data better. Buck and Cheng* report a discovery task where most of the subjects were better described by a discrete exponential model. However, the powerform model was more robust in fitting the empirical time data. The critical point is, human learning is not fully understood and none of the models of learning are really derived from that process.

It would appear proper to test the data for certain properties noted earlier when deciding between models for learning. Thus, a powerform would predict a constant doubling ratio of t_{2n}/t_n or A_{2n}/A_n. That model would predict a straight-line plot when t_n values were converted to logarithms. Consequently, it would appear that when these phenomena were present, the powerform model would provide the better description. While that observation appears to be true, it should be clearly pointed out that the ratio $(t_{2n} - t^*)/(t_n - t^*)$ equals α^n when the data are truly exponential. Unless α is very close to unity, α^n can be very close to zero, even for a moderate value of n. When t^* is additionally very small, then there is not a lot to distinguish these models. But then, there is not much need to. Consider the differences between the two aforementioned numerical models where the first three serial values are identical. In these examples, the regular powerform is heading toward the zero asymptote and the exponential model is leveling off to approach the asymptote of 3.00. The results are as shown in Table 5.10.

Table 5.10 Comparison of Powerform and Discrete Exponential Models

	$n=$	1	2	3	4	5	6
Regular powerform	$t_n =$	5	4	3.5	3.2	3.0	2.8
Discrete exponential	$t_n =$	5	4	3.5	3.25	3.13	3.06

* Buck and Cheng (1993) report this effect for a discovery learning task.

Frequently, the series generated using a powerform model and a discrete exponential model are very similar. In that case it may not be important to distinguish the better-fitting model.

Forgetting curves

Some people have advocated the use of forgetting curves in addition to learning curves. Learning curves are described over a period of activity such as a production run. Forgetting curves, on the other hand, are described over periods of inactivity between sequential activities. If significant time passes by between successive production runs, then forgetting may occur during these intervals. In effect, the use of forgetting curves is similar to placing an inverted learning curve between successive learning curves. The forgetting curve describes losses in production efficiency for a particular product that increase for longer time periods between successive production runs.

In theory, the composite of forgetting and learning curves provided the modeling needed to optimize production planning by showing the costs one accepts for each extra day of delay in restarting the production of a product in contrast with the benefit for extending a production run one more day. Economic analysis would provide the answer to the optimum production run by coupling the aforementioned marginal costs with the marginal cost of added inventory. However, forgetting curves present several complications. One complication occurs when considering the ending of production for product A and the start of production of product B. Since the startup product is not A, the forgetting curve between identical products does not apply. When product B is very different from product A, one would expect the initial time values for product B to be at a maximum. However, when product B is very similar to product A, that initial time would be expected to be considerably less than the maximum. Hence, the complication is that one would need a forgetting curve between all pairs of products that are produced. A second complication is the potential change in a forgetting curve with changes in personnel. As with other performance time estimating techniques, forgetting curves would be focused on the standard operator. It would remain then to find out how forgetting curves shift with persons over the population of potential employees. A third complication is knowledge of how forgetting curves differ with the accuracy criterion and how variations of the speed–accuracy trade-off may affect forgetting curves. Because of these and other complications, forgetting curves are seldom used in industry today and it is usually assumed that performance times on a new production run are independent of the product made before so long as a different product was produced on the previous run. Also, it is assumed that the initial performance times during restarts of the same product are either new starts at the maximum (unlearned) rate or continuations of the current learning curve.

Time study

Much of the remaining material in this chapter focuses on time study. More on usability is covered later in Chapter 13.

The typical procedure followed in time study consists of 10 steps. These 10 steps are as follows:

1. Describe the method to be used in detail.
2. Decide the procedure for timing a cycle of the job.
3. Break the job cycle into elements that are to be separately timed.

Chapter five: Work measurement and analysis

4. Select forms to record time data on the elements over observed cycles.
5. Select an operator who is typical of the standard operator.
6. Train the selected operator to perform the job using the specified method.
7. Check for learning and continue training until a suitable level of performance is obtained by the operator.
8. Study that operator for n cycles of the task.
9. Rate the observed operator's performance relative to the standard operator.
10. Document the study.

Note that we are assuming that the method has already been specified, using the principles discussed in the previous chapters. When conducted in this way, time study gives you a "should take" time—the amount of time that a worker should take when performing the task in the recommended way. If one were to skip step 1 and just conduct the time study without specifying the method, that is, just timing the method used traditionally by workers, then the time study records a "did take" time.

The following discussion focuses on the steps assuming that the desire is to identify a "should take" time. However, determining and describing the method was discussed earlier and so it is assumed here that the method is known. Consequently, our discussion starts with step two. Also, please note that except for steps 5 and 9, the procedures for measurement of performance time are applicable to many situations besides time study.

Some purposes served by time study

Some of the purposes of time study or areas where the results of a time study can be used in industry include

- Planning production (e.g., finding the number of operators needed in various jobs)
- Selecting an appropriate method for performing human work
- Determining if a disabled person can perform a task safely and economically
- Evaluating a device or tool used by people
- Checking training progress
- Estimating direct labor costs
- Monitoring production progress over time
- Operating some of the wage-payment plans that depend upon performance

While this list is far from complete, there are several important differences between items on this list that can affect the procedures and analyses followed in a time study. At the same time, economic considerations provide for a push for consistency. It is of utmost importance that the ergonomic specialists know which procedures to follow.

Selecting a procedure for timing

Time study involves the use of a stopwatch or other timing device to measure how long it takes to do a task. In one method of time study, the timer is activated at the start of the session and continues running for the entire recording period. Time values on the stopwatch are recorded at the end points of each task element. This procedure is known as the continuous method of timing. Do not be surprised if you experience a little trouble in correctly reading these time points while the watch is running. This is a practiced skill.

Once Upon a Time

What a strange thing "time" is! Unlike space, time has no known beginning or ending, except in religious senses. It has neither symmetry nor asymmetry, except in déjà vu. Yet one is apt to describe it graphically as a line, similar to space. Time is one of the last simple concepts a child learns. Yet, some students and professors have a lot of trouble with this concept.

People slowly learned to measure time over the centuries. Ancient Babylonians looked to the sky to measure time. The position of the sun told time during the day and the moon denoted the months. Later, the Egyptians developed the sundial. But cloudy days posed problems in time measurement. The Greeks and Romans developed the water clock in which water would drip from one vessel to another and the rate of dripping formed a measurement of time. In China, a damp rope was burned to measure the time it took to burn from one knot to another. Later they noticed that candles and oil lamps burned at near constant rates, so marked candles and lamps were used to measure the passage of time. Hour glasses were invented in the 1300s. One still sees these devices used today in cooking eggs and timing the cooking of other foods. Toward the latter part of the 1300s, Europeans invented the mechanical clock where a weight on a chain was enacted by gears and ratchets to a pointer on a clock face.

Galileo's discovery of the pendulum in 1637 led to its later application in clocks by the Dutchman, Christian Huygens. In 1735, an Englishman, John Harrison invented the first chronometer, that could indicate time accurately aboard a ship in the rolling seas. Spring-driven clocks were employed until recent times. Most watches today are electronic.

In the snapback method, one or more watches are simultaneously reset with each time mark. Typically, three watches are held on a clipboard with a mechanism that sets each of the watches to one of three states (i.e., run, stop, and reset to zero). Each time the lever is depressed, one watch is stopped with an elapsed time recording on it. Each successive time the lever is pressed, the watch displaying the elapsed time shifts over one watch. Accordingly, the snapback method of timing automatically provides the elapsed time of an element of activity. The continuous method produces only cumulative time readings that must later be reduced to elapsed times.

Many timing devices are in use today. Electronic timing instruments are available that include buttons that make it easier to record time data for each element and cycle. Some devices even provide a means to load these data on a computer for analysis. Photoelectric, infrared, and pressure-sensitive devices can be interfaced with computers to record the times automatically. These automatic methods provide more consistent estimates than human analysts but are not necessarily more accurate.

Elements of a task

The objective of time study is to estimate the performance time of an entire task. However, practitioners almost invariably break up the task into elements that they then time individually or in small collections. One of the most important reasons for this practice is that identifying the elements allows the analyst to check that the proper method is being used.

When the method is not followed consistently, error is introduced into the timing procedure and the resulting estimate. Therefore, those elements that are not part of the desired method can be taken out of the prediction to insure a more accurate estimate.

Foreign elements can creep into the method for a variety of reasons. An example is when the operator who is being time studied drops and retrieves a tool. Adjustments are made later to these data to create time standards, but fumbles and other foreign elements should not be incorporated into the time-study data. If only the entire cycle was timed, then a fumble wipes out the entire observation. If the elements are timed separately, a fumble on a later element does not affect readings on the earlier elements. Hence, the use of elements allows one to use a portion of the interrupted cycle.

Another reason for timing task elements rather than simply the whole cycle is that some elements often are machine paced. Machine-paced elements tend to have substantially reduced time variability compared to human-paced elements. Also, shorter human-paced elements tend to have lower time variability than longer human-paced elements. It follows then that greater statistical precision can be obtained from a fixed number of observations when the elements are timed separately.

These reasons for breaking up the whole cycle of activities into a series of nonoverlapping elements also give clues to what those elements should be. First of all, the elements need to be long enough to be accurately timed. They should be 2.4 s or longer when using a stopwatch. When using other timing equipment, one should find the lower practical limit of precision in that case. Next, the proper method must be distinguished from alternative methods that the operator might follow, to help safeguard the data from foreign methods. Human- and machine-paced elements need to be separated from each other, and steps must be taken to ensure that the end points of each element can be consistently detected. Distinctive sounds may occur at end points, as when the workpiece hits the fixture. The latter situation is particularly desirable, as sounds can help the analyst detect the end point of a task element, even if they are looking elsewhere at the time.

Recording forms

The forms used to record time-study data are normally matrices with the rows representing successive cycles and columns denoting successive elements. Table 5.11 contains a simple example of a time-study form. The row sums give total cycle times and the standard deviations of the row sums provide estimates of the within-person time variability. Usually, special symbols are used to denote fumbles and out-of-sequence elements. Anything out of the ordinary should be noted at the time of the study on the form. These notes are important during later analysis because they can help explain unexpected results.

Selecting and training the observed operator

Time study requires the collection of extensive time data on a single subject. Therefore, it is important to select a subject that is as near the defined standard operator as possible. This feature reduces the need to rate (or level) that person's speed. As rating is the weakest link and the most criticized point of this approach, the less rating required the better. Some point along the axis denoting the between-person performance times in Figure 5.1 denotes the standard operator. The selected operator should approach that point as closely as possible.

In order to make the collected data as accurate as possible, the selected subject needs to feel as natural as possible. This natural feeling is difficult to achieve right away. It is

Table 5.11 Simple Time-Study Form

TIME-STUDY of:_____

Date _____

Analyst _____ Task Name: _____

Method Description: _____
_____
_____

Time for Task Components

Cycle	Component 1	Component 2	Component 3	Summed Time
1				
2				
3				

helpful if the selected person truly volunteered for the study and feels that the time data will be taken in as fair and objective a way as possible. People tend to respond in a positive way when they are treated in a respectful and truly professional way. It is usually best to explain everything to the person so that they fully understand, and can cooperate.

Next, time-study personnel must thoroughly train the selected operator to follow the prescribed method. Usually, it is best to visually record the first few attempts in performing the task. In this way, errors in performing the prescribed method can be immediately shown to the operator as corrective feedback. It is well known that immediate feedback assists in more rapid learning. Another suggestion is to fit a learning curve to time data in order to estimate when an acceptable learning level is achieved. Learning curves are described later.

Sample size and time statistics

After an adequate level of performance is achieved, the operator is observed over N cycles of the task. One initial issue is that of determining what N should be. The sample size needed reflects the acceptable degree of statistical precision for the mean time of the operator.

Let T_{ij} be the elapsed time on the jth element of the ith cycle of this task. The entire set of observed cycle times is expected to form a normal distribution with a mean value $E(T_j)$ and a time variance of $V(T_j)$. The mean and variance of performance time are calculated using the following two equations:

$$E(T_j) = \bar{T}_j = \frac{\sum_1^N T_{ij}}{N} \tag{5.30}$$

$$V(T_j) = S_j^2 = \frac{\sum_1^N (T_{ij} - \bar{T}_j)^2}{N-1} \tag{5.31}$$

Now, if performance times of successive task elements are related, then the correlation between those time values is calculated with the following equation:

$$r_{1,2} = \frac{\sum_{j=1}^{N}[(T_{1j} - \bar{T}_1)(T_{2j} - \bar{T}_2)]}{(N-1)\sqrt{V(T_1)V(T_2)}} \quad (5.32)$$

According to Helson's par hypothesis, this correlation should be negative.
It also follows that the time variance over two successive elements is

$$V(T_1 + T_2) = V(T_1) + V(T_2) + \rho_{1,2}\sqrt{V(T_1)V(T_2)} \quad (5.33)$$

which is a special case of Equation 5.4, where X corresponds to T_1, Y corresponds to T_2, and $\rho_{1,2}$ is the population correlation coefficient. This equation shows that when the times of successive cycles are negatively correlated, the time variance of both elements is less than the sum of the two variances.

For example, suppose that $E(t_1) = 7$, $E(t_2) = 10$, $V(t_1) = 4$, $V(t_2) = 9$, and $r_{12} = -0.7$ are the statistics for two elements of a cycle. The cycle mean is the sum of the two expectations or 17 time units. The variance of the cycle time is

$$V(t_1 + t_2) = 4 + 9 + 2(-0.7)(2)(3) = 4.6 \text{ time units}$$

If the time on the two successive elements is independent, the cycle variance is simply the sum of the component variances, or 13.

What is desired here is the value of the true mean time μ_j. $\bar{T}_j$ is an estimate of that value. In fact $\bar{T}_j$ is a random variable that may differ from μ_j. The distribution of $\bar{T}_j$ is related to the sample variance $V(T_j)$, where

$$V(\bar{T}_j) = \frac{V(T_j)}{N} \quad (5.34)$$

The latter quantity is the variance of the sample mean of the jth element over the N observations (i.e., cycles of the task element). The whole cycle has a sample mean of

$$E(\bar{T}) = \sum_{j=1}^{m} \bar{T}_j \quad (5.35)$$

over the m elements within the cycle. Let us specify an acceptable statistical error in the sample mean $\bar{T}$. The difference between the sample mean and the true mean μ is the error e. For N cycles, the needed probability statement is

$$p(\mu - e \leq \bar{T} \leq \mu + e) = 1 - \alpha \quad (5.36)$$

where α is an acceptable probability of being wrong. The probability density function of this event is a normal distribution.

Note that any deviation from μ with a normal distribution can be divided by the standard deviation of time in order to standardize the deviation as

$$p\left(\frac{\bar{T}-\mu}{\sqrt{V(T)}}\right) = p(z) = p\left(\frac{e}{\sqrt{V(T)/N}}\right) \qquad (5.37)$$

where
  z is the number of standardized normal units
  $\sqrt{V(T)}$ is the sample standard deviation of cycle times over the N cycles observed

If we rearrange Equation 5.37 slightly, the result is

$$z = \frac{e}{\sqrt{V(T)/N}} \quad \text{or} \quad N = \left(\frac{z\sqrt{V(T)}}{e}\right)^2 \qquad (5.38)$$

Therefore, given an acceptable error magnitude e and an acceptable probability of an error α (which is specified by the standardized normal deviate z), the sample size N can be found through Equation 5.38. Note that once an acceptable probability is determined, z can be found from a normal standardized deviate table (Appendix B) where the corresponding z value has a cumulative normal integral equal to $1 - \alpha/2$. The remaining $\alpha/2$ error is at the other end of this distribution. This procedure for calculating sample size is illustrated in Box 5.3.

BOX 5.3 EXAMPLE CALCULATION OF SAMPLE SIZE

To help clarify these ideas, suppose that you are willing to accept an error in the sample mean of 0.1 min and a probability of error equal to 10%. This yields a confidence of 0.9 and a corresponding z value of 1.645. If the standard deviation of cycle time is $S = 0.5$ min, the number of cycles can be computed as

$$N = \left[\frac{1.645(0.5)}{0.10}\right]^2 = [8.224]^2$$

$$= 67.65 \approx 68$$

Now consider the necessary sample size N changes when some of the other values provided earlier are changed.

If $\sqrt{V(T)} = 0.65$, or the time variance is larger, then $N = 98$, meaning a larger sample is needed.

If $e = 0.2$, or a larger error is acceptable, then $N = 17$, meaning a smaller sample is needed.

If $\alpha = 0.8$, or a lower confidence is acceptable, then $N = 42$, meaning a smaller sample is needed.

The relationship between these various features and the sample size follows from the preceding illustration.

The difficulty in estimating the sample size using the previous approach indicated is that there often is no information available about the time variability before starting the time study. This situation is easily corrected by taking a few observations, and then using them to provide an estimate of the variance. When more data are collected, a more precise estimate can be made.

If N' is a preliminary sample size and it is some integer multiple of 4, the results of these N' observations can be ordered from the smallest to the largest values. The first quartile value is the $N'/4$th ordered value, denoted as Q_1, and the third quartile value is the $3N'/4$th ordered value or Q_3. An empirical method for estimating the sample standard deviation from these quartile measurements is

$$\sqrt{V(T)} = \frac{3(Q_3 - Q_1)}{2} \tag{5.39}$$

This value can then be substituted into Equation 5.38, to estimate N, or the total number of observations required. Since N' observations have already been taken, $N - N'$ more observations are needed.

A conservative estimate of the error in the value of the sample standard deviation calculated with Equation 5.39 is given by Equation 5.40:

$$\sigma_{\sqrt{V(T)}} = \frac{\sqrt{V(T)}}{\sqrt{2N}} \tag{5.40}$$

Note that this equation gives the expected variation in the sample standard deviation, as a function of sample size. This variation is used to estimate upper and lower bounds of the required sample size as illustrated in Box 5.4.

Some other procedures for estimating the standard deviation of time from a few samples are discussed next. The range method involves first taking N' preliminary samples and finding the range. The range (R) is equal to the maximum time–minimum time in the sample. A range table is then used to estimate how many standard deviations the range spans. Table 5.12 shows the expected number of standard deviations with the observed range, given that N' observations have been taken.

For example, suppose that 10 observations were taken. The longest measurement was 1 min and the shortest was 45 s. The range is then 15 s. Given that 10 observations were taken, it is expected that

$$3.078 * \sqrt{V(T)} = 15 \text{ s so } \sqrt{V(T)} = \frac{15}{3.08}$$

Note that the calculated value would have been larger if the same range was observed for only five observations (i.e., 15/2.33).

A second, purely empirical, method was developed by General Electric years ago based on the duration of the task. Table 5.13 shows values of N for different task durations. Tasks with short cycle times require more observations, as shown in the table. Note that the products of T and N in Table 5.13 are all about 0.5 except for the last one. A similar procedure developed by Westinghouse Corporation includes a stipulation about yearly activity. A few elements of the Westinghouse system are shown in Table 5.14. The recommended sample size increases with yearly activity as shown there.

BOX 5.4 CONSERVATIVE CALCULATION OF SAMPLE SIZE

In the aforementioned numerical example, where the sample standard deviation

$$\sqrt{V(T)} = 0.5$$

Assuming a sample of size 40, Equation 5.40 gives the sampling error in the value of the sample standard deviation as

$$\sigma_{\sqrt{V(T)}} = \frac{\sqrt{0.5}}{\sqrt{2(40)}} = 0.0559$$

The actual standard deviation is, therefore, likely to be in the range 0.5 ± 0.0559 or 0.4441–0.5559. When these two values are plugged into Equation 5.38, a lower and an upper limit on the sample size is calculated as

$$N = \left[\frac{1.645(0.444)}{0.10}\right]^2 = 53.37$$

$$N = \left[\frac{1.645(0.5559)}{0.10}\right]^2 = 83.62$$

The value of 68 calculated earlier in Box 5.3 falls within this range.

Table 5.12 Expected Number of Standard Deviations (d) in the Observed Range for a Sample of Size N

	Sample Size N						
	5	7	9	10	15	20	25
d	2.33	2.70	2.97	3.08	3.47	3.74	3.93

Table 5.13 General Electric's Empirical Estimates of Required Cycles in Time Study

	Duration of Task (T) in Hours			
	$T < 0.002$	$T \approx 0.008$	$T \approx 0.02$	$T \approx 0.12$
N (cycles)	200	60	30	10

Performance leveling

Up to this point in the procedure, time-study personnel have taken a lot of observations on a single human operator, knowing full well that the person under observation is not the standard operator. Accordingly, the mean time would be expected to be above or below that of the standard operator if the person observed was respectively slower or

Table 5.14 Westinghouse's Basis for Replicating Cycles in Time Studies

Activity/year	Approximate Duration of the Task in Hours				
	0.002 h or less	0.008 h	0.020 h	0.12 h	0.50 h or more
Less than 1,000	60	25	15	6	3
1,000–10,000	80	30	20	8	4
Over 10,000	140	60	40	50	8

faster. Performance leveling is a procedure for accounting for that fact based on rating the observed operator relative to the standard operator. This rating is simply a correction factor that converts the sample mean of the time of a particular operator who was observed in the time study to the prescribed standard operator. When the average time of the observed operator is multiplied by the correction factor, the product is the normal time of the standard operator. Before discussing various procedures of performance leveling, it should be pointed out that performance leveling has been almost solely applied to speed (or performance time) leveling of sensory motor tasks. However, most, if not all, of the following techniques can be used with other criteria (such as performance accuracy) and with other kinds of tasks.

Subjective ratings

By far, the most popular method of leveling is to subjectively rate the operator's performance relative to the standard operator. The time-study personnel usually do this. In this case, time-study personnel rate the speed of the operator under observation relative to the standard operator. As the person who does the rating views the operator being studied, he or she describes the speed of the observed operator as being 100% if that operator has exactly the same speed as the standard operator. If the observed operator were moving at a rate 10% faster than the standard operator, then the rating would properly be 110%. If that person was observed to take 2.2 min/cycle and have a speed rating of 110%, then the normal time is

$$2.20 \text{ min} * 1.1 = 2.42 \text{ min normal time}$$

Ratings of less than 100% produce the opposite effect of making the normal time less than that of the observed operator. When the observed operator is slower than the standard operator, the standard operator should require less time.

The analyst must provide accurate and reliable ratings. Accuracy is a lack of bias, yielding an average rating that is correct. Reliability means the analyst will consistently give the same situation the same rating. Some people are just not very good at making such judgments. Others simply need training and occasional feedback. The Society for the Advancement of Management (SAM) produces a variety of training films that show people performing a task at different rates. Time-study personnel can view the film, make their ratings, and then receive feedback on how correct and consistent their ratings were. Another approach involves asking two or more time-study personnel to make ratings of jobs, which are then compared. These practices will help assure ratings that are both accurate and reliable.

Multifactor ratings

A number of different multifactor rating systems have been used over the years. Perhaps the most famous is the old Westinghouse system, which uses four different rating factors: skill, effort, conditions, and consistency. Each factor is rated on a six-point scale, with additional solutions. Table 5.15 shows the associated correction amount for the ratings. After each factor is rated, the corrections are found from the chart and algebraically summed. When this sum is added to unity, the result is the rating factor.

If the time-study person rated the operator's skill as *E2* (the lower end of a fair value), effort as *B1* (the higher end of excellent), conditions as excellent, and consistency as average, then the rating factor is

$$1.00 + \left[(-0.10) + (+0.10) + (+0.04) + (0.00)\right] = 1.04$$

The overall rating is about 2% faster than for the average operator. Note that this table provides adjustments both above and below a standard operator. For the skill and effort factors, there are six levels above the average rating. For the other factors there are three levels above and only two below.

It is interesting to note that the factors in Table 5.15 are not equal in their effect. Skill and effort ratings have more impact than conditions and consistency. Another observation is that the correction values are not symmetrical. Except for consistency, the negative maximum is always greater than the positive maximum. Since the people at Westinghouse who developed this system did so by collecting a lot of data in which these four factors varied from case to case, the results provide some basis for indicating the pace effect due to these factors. These data show that the working conditions alone only affect pace by about 6% or 7%. Another noteworthy point is that the sum of the positive maxima is +38% and the sum of the negative maxima is −50%. These two values are the upper and lower bounds on possible ratings using this system.

This old Westinghouse multifactor system is useful in other ways. For example, the skill factor provides a basis for estimating the potential benefits of training if one can estimate the change in skill resulting from the training program. Similarly, the effort factor might be useful to evaluate a prospective incentive program, and the conditions factor allows for an estimate in performance due to changes in working conditions.

Table 5.15 Old Westinghouse Multifactor Pace Rating System

Rating	Skill	Effort	Conditions	Consistency
Super	A1 +0.15	A1 +0.13	A +0.06	A +0.04
	A2 +0.13	A2 +0.12		
Excellent	B1 +0.11	B1 +0.10	B +0.04	B +0.03
	B2 +0.08	B2 +0.08		
Good	C1 +0.06	C1 +0.05	C +0.02	C +0.01
	C2 +0.03	C2 +0.02		
Average	D 0.00	D 0.00	D 0.00	D 0.00
Fair	E1 −0.05	E1 −0.04	E −0.03	F −0.02
	E2 −0.10	E2 −0.08		
Poor	F1 −0.16	F1 −0.12	F −0.07	F −0.04
	F2 −0.22	F2 −0.17		

This multifactor rating system has been criticized by people who say that these factors are correlated. Usually people with lots of skill exhibit that fact by working very consistently and not often indicating much effort on the part of the operator. As a consequence, many companies use only the first two factors of this system. Another multifactor system of rating is the objective rating system that Marvin E. Mundell developed years ago based on both a pace rating and a job difficulty rating (see Mundell, 1985).

Synthetic rating

Synthetic time prediction systems, as expanded upon in Chapter 8, predict the performance time of standard operators doing a number of specific tasks as a function of particular variables of the task. Therefore, at least some elements of the task being observed can be predicted for the standard operator through synthetic prediction. In some cases, the prediction of the synthetic rating system is used as a standard task equivalent against which task performance can be compared, to develop what is called a synthetic rating factor. If T_s and T_a are the predicted synthetic time and the actual timed means for the same element, respectively, the synthetic rating factor is

$$R_S = 1 + \frac{(T_S - T_A)}{T_S} \tag{5.41}$$

If several elements are used in the determination of the pace rating factor, a weighted average is used, with the various T_s values constituting the weighting values.

Conducting an Experiment to Measure Performance of Alternative Designs

When performance is measured to assess the best of two or more alternative designs, the typical procedure is to select a number of different people and have each person operate or try each design alternative in a randomized sequence. The idea behind randomizing the sequence is to assure that the sequence itself does not affect performance. For example, with three alternative designs called A, B, and C, six different sequences are possible: ABC, ACB, BAC, BCA, CAB, and CBA. Another consideration often given to performance measurement is subject selection based on factors that may affect performance differently. Typical factors are age and gender. Differences in strength, sensory abilities, and preferences are associated with both of these factors. If one selects three age groups, such as 15–39, 40–64, and 65–89 years, and both genders, there are 3×2 or 6 groupings. With the additional sequence factor, that is 6×6 or 36 different groups. Even without a replication within each group, such an experiment would require 36 different human subjects to perform it so that all sequences, ages, and genders are represented in the experiment.

If the ergonomist can assure that there is no learning embedded in the performance measurements and was only interested in the population average, only a single measurement would need to be taken from each of the 36 people who were randomly chosen, subject to the age and gender restrictions. This experimental arrangement of 36 performance measurements would give 36 measurements on each design alternative, 18 measurements for each gender, 12 measurements for each age group, and 6 measurements for each sequence.

Thus, direct tests on those additional hypotheses can easily be made. However, there might also be a situation in which older males exhibit an effect on some sequences but not others. These so-called interactions can be most troublesome, but if one design alternative possessed the best performance regardless of gender, age, or sequence, those interactions are of little consequence. Otherwise, one design may be best for one gender or some age group or groups, and another design for other cases.

For these reasons, data from experiments of this kind are usually evaluated using analysis of variance (ANOVA) to test if any of the main effects (designs, gender, age, or sequences) or the interactions are statistically significant. When only some of the main effects and lower-order interactions are significant, regression models can be fitted to the generated data for prediction purposes. The point here is that measuring human performance for comparing alternative designs is considerably more complicated and more expensive than traditional time study.

When one also wants to estimate standard deviations of performance within and between persons, additional measurements need to be made. Multiple observations of each person are also needed if learning is to be evaluated. Experimenters who want to control for learning effects often train each subject until the person performs at an acceptable level prior to taking performance measurements.

Extracted basis of rating

A number of physiological measurements are available that reflect the physical workload of the task. Typical measurements include heart rates or metabolic rates. To use this basis of measurement one needs a standardized task such as walking at about 2.75 miles/h (4.42 km/h or 73.76 m/min) and a basal condition as a resting level of the physiological measurement. If these two measurements on the operator being time studied are, respectively, h_s and h_0, and the corresponding measurement during the task being time studied is h_w, then the rating factor is

$$R_e = 1 + \frac{(h_w - h_0)}{(h_s - h_0)} \tag{5.42}$$

Sometimes, an additional adjustment factor is used. In either case, this form of rating tends to be rather expensive in terms of the time required of the time-study personnel. Consequently, this form of rating is rarely used. It does provide a way to check other forms of rating that may be contested, and it does provide an objective measurement.

Alternatives to performance rating

One obvious alternative to rating performance relative to a standard operator is to specify a particular person as the standard operator. In that way, Jane or John (not both) could always be called upon when a time standard needs to be set. This approach requires that Jane or John be available whenever standards need to be set. Another alternative, which is more objective, is to use a collection of different people sampled from the

Chapter five: Work measurement and analysis 245

population. This requires one to decide how many to use and the number of observations (i.e., cycles) to observe for each person. If d different people are chosen randomly from the population and each is observed over N cycles, then there are $d * N$ observations. Conditions occur where the choice of d and N are important even when the product $d * N$ is the same. When people differ significantly from each other, every added person increases the between-person variability. Hence, the total variability increases. This means that greater sample sizes are needed to meet precision requirements, and data collection becomes more expensive.

Determining allowances

A time standard is often defined as "the time required by the average fully qualified and trained operator who is working at a standard pace and performing a typical operation cycle under normal working conditions." Under normal conditions, the operator is engaged in nonwork activities some of the time. Coffee breaks, occasional trips to the bathroom, resting from fatigue, getting special tools or fixtures for a machine tool, and awaiting parts needed in an assembly are all typical cases where work is not going on. Allowances account for reasonable nonwork-related activity, and are used to adjust the normal time found in the time study to standard time. Thus, Equation 7.18 follows:

$$\text{Standard time} = [\text{Observed time}] * [\text{Performance rating}] * [\text{Allowance correction}] \quad (5.43)$$

The first two factors on the right-hand side of Equation 5.43 give normal times, which are then adjusted with an allowance. Three kinds of allowances that are typically used are personal allowance, fatigue allowance, and delay allowance. Some firms combine two of these categories and others break them into smaller categories. Since the allowances correspond to time away from productive work, allowance percentages are subtracted from the time available. Allowance corrections may be computed directly by percentage fractions or as an acceptable number of minutes for an allowance class. For a typical 8h shift of 480 min, the allowance correction can be made as Equation 5.44 shows:

$$\frac{480 \text{ min}}{(480 - \text{allowance minutes})} = \frac{100\%}{(100\% - \text{allowance}\%)} \quad (5.44)$$

Once the allowance percentages are determined, they may be used directly in Equation 5.44 on the right or converted to minutes for computing the correction allowance factor of Equation 5.44 on the left.

Table 5.16 shows averages and ranges of allowances typically used in industry. Each category is about 5% in the United States. These values tend to be traditional rather than arrived at through scientific means or by negotiation. Percentages do not differ significantly between unionized and nonunionized companies, but there tend to be differences between different industries.

Some have argued that it is improper to determine the total allowance percentage by simply allocating a given percentage for each class of allowance and merely summing up these three allowance percentages. Their argument is that people should

Table 5.16 Typical Ranges of Allowances

Allowance	Average (%)	Range (%)
Personal	5.0	4.6–10.0
Fatigue	5.1	3.8–10.0
Delay (unavoidable)	5.3	4.0–10.0

not be given a fatigue allowance when on personal allowance time, or while delayed unavoidably, and that the current practice of summing the allowances allows this to happen. These people would further state that a personal allowance should be deducted from the shift time before computing a fatigue allowances. After removing the fatigue allowance, delay time should then be removed. Note that this argument is not against the particular percentages used. The problem is with the practice of summing those percentages. Following the alternative prescribed philosophy, the total allowance time could be computed as

$$AL = p + f + d - p(f + d) - fd \tag{5.45}$$

where
  AL is the total allowance time
  $p, f,$ and d are the personal, fatigue, and delay allowances

The difference between the computation of an 8 h day using Equation 5.45 and the simple sum of the three categories, assuming 5% each, is about 2.4 min/shift. This difference is small enough that most analysts use the simpler method.

Except in rare cases, it is typical industrial practice to give all jobs the same allowances, although it is easy to argue that different jobs are likely to differ in both physical difficulty and unavoidable delays, and that these differences should be reflected in different fatigue and delay allowances. A number of variable allowance systems have been proposed, but the practice remains controversial. Theoretically, variable allowances should be preferred over constant allowances, but there is concern that they may be misapplied. Part of the criticism stems from the fact that there appears to be considerable unexplained or difficult-to-explain variability. An example of a variable allowance system is given in Table 5.17 that focuses on physiological bases for increasing variable fatigue allowances for more demanding tasks.

Machine interference allowances

One other form of allowance that applies to some kinds of operators is given for machine interference. In some jobs, an operator loads and starts a machine and then walks over to another machine, and so forth. When the first machine completes its running-time cycle, the operator goes back and unloads it before reloading for the second cycle. Often the operator's activities are not synchronized with machine operations, so the operator must wait some time before the running-time cycle is finished. Assigning more machines to the operator, so the machine would complete its running-time cycle and sit idle before being attended, might reduce this waiting time. However, it is likely that machine idle time is more expensive than operator idle time.

Economics of Performance Measurement

Ergonomics, like every discipline, requires economic justification in order to compete for added investments by companies. For example, suppose an assembly operation is using a manually operated screwdriver to insert six screws in each assembly of a particular job. The total assembly takes 6 min using the regular manually operated screwdriver. An ergonomist proposes substituting a power screwdriver for the manual one, a change that affects only the installation of the screws. The time is measured for the operator to put in the six screws manually, including the time it takes to pick up the tool and lay it back down. This requires about 5 s/screw for a total of 30 s plus about 10 s for pickup and drop-off of the screwdriver. Then the ergonomist tries a power screwdriver and finds that it takes only 4 s/screw with the power screwdriver and the pickup and drop-off time was about the same. The apparent savings derived from the power tool is only 6 s/assembly, but the measurement fails to consider the effect of greater fatigue when using manual screwdrivers. After taking fatigue into consideration, the power tool yields an average saving in direct time assembly of 3 s/screw or 18 s/assembly.

Since the screw insertions are only part of the total assembly time of 6 min/assembly, the total time saved per assembly is 0.3 min. At $12.00/h, the savings appear to be only $0.06/assembly. This calculation is not quite accurate, however, because it fails to consider the assembly in the context of the typical work shift. If the expected working time per shift is 408 min (480 min/shift excluding 15% for allowances), the production rate of a manual process is 408 min/6.0 min/assembly or 68 assemblies/shift. With the new power screwdriver, the rate would be 408 min/(6.0 − 0.3 min) or 71.58 assemblies/shift. This calculation shows that 3.58 more assemblies would occur during the shift if power tools were used. Now, for a production run of 10,000 assemblies, the manual method takes 147.06 shifts @ 68 assemblies/shift, but with the power screwdriver, it would only take 139.70 shifts. The savings over the production run are as follows:

$$147.06 \text{ shifts} \times 8 \text{ h/shift} \times \$12.00/\text{h} = \$14,117.76$$

$$139.70 \text{ shifts} \times 8 \text{ h/shift} \times \$12.00/\text{h} = \$13,411.20$$

According to these calculations, the total cost savings during the production run is $706.56, not counting any costs. There also is a required investment of about $150.00 for a heavy-duty industrial-grade power screwdriver that occurs at the beginning of the project, and the cost of the electricity over the life of the project must also be calculated. Using a $0.05/min cost for power, the total cost over the production run is

$$[24 \text{ s/assembly}][71.58 \text{ assemblies/shift}][139.7 \text{ shifts/run}]$$
$$[1 \text{ min}/60 \text{ s}][\$0.05/\text{min}] = [\$199.99/\text{run}]$$

The $706.56 in cost saving for a single production run is about twice as much as the cost of the tool and power combined. Most companies would be convinced at this point that the proposed project was economically sound, providing the estimates are accurate.

Table 5.17 Variable Posture and Motion Allowance Percentages

Category	Allowance%
Sedentary work, no muscular strain (light assembly, packing, or inspection)	0
Sitting, slight muscular strain (operating low-force foot pedal)	1
Standing or walking; body erect and supported by both feet on the ground	2
Standing on one foot (operating press foot control)	3
Unnatural postures (kneeling, bending, stretching, or lying down) Light shoveling or holding unbalanced loads	4
Crouch. Working with manual restraint (restricted tool movement in confined areas)	5
Awkward posture in conjunction with heavy work (bending or stooping in lifting heavy weights, carrying heavy loads on level ground or down slopes; one or both hands as convenient)	7
Carrying or moving awkward and heavy loads over rising ground. Climbing stairs or ladders with heavy loads. Working with hands above shoulder height (e.g., painting ceilings)	10

Source: Williams, 1973.

Maintaining standards

After standards are set, they need to be maintained by industrial engineering or time-study departments. Over time, many of these standards become outdated as a result of equipment changes, the creation of new methods, and changes in materials that affect task pace. Companies should install an automatic system that alerts the people in work-time measurement when there is a reason to recheck the time standard and indicate the reason. In addition to these follow-ups, periodic audits should be made to determine if there should be any changes in time standards. Some of the potential pitfalls of not maintaining standards are illustrated in Box 5.5.

Indirect performance measurement

The term direct human performance measurement is used here to mean that some criterion of performance (normally time) is measured under the immediate supervision of an ergonomics specialist. This term usually implies real-time observations and measurement by the specialist. Time study is a particular case in point. Other approaches to measuring performance time are indirect, such as occurrence sampling. In the latter approach, the fraction of time that a particular activity occurs is estimated by observing what the operator is doing at particular times. The proportion of observations is multiplied by the number of hours of the sampling study divided by the number of job units produced and provides an estimate of the time a particular job requires to produce a unit of a specific product. This form of indirect time study (really partially indirect) is also known as activity analysis or work sampling.

Another form of indirect study uses available time data to develop formulas that can be used to predict performance times. Such approaches are referred to as standard data systems and synthetic time systems, respectively. Some ergonomists are likely to refer to real-time recordings as indirect studies when the recordings are made with no one present. Others would not agree unless one or more automatic measuring devices are present and the collected data require a variety of inferences to be interpreted. Hence, the terms

> **BOX 5.5   ANECDOTE FROM AFT (1997)**
>
> A major camera repair facility was faced with imminent closure. Labor costs were apparently out of control and corporate management was making plans to relocate the facility out of the country to reduce these costs. The facility manager, in an attempt to save the operation, enlisted the support of experienced industrial engineers. Their findings were as follows.
>
> Although work standards were being used to allocate labor hours, they were not related to the current products being repaired or the current repair methods being used. The standards had been set prior to the dissolution of the corporate industrial engineering group several years earlier.
>
> Professional study of the existing methods resulted in standards that accurately reflected the work being performed. The actual labor costs were less than half those being charged under the outdated standards. Accurate reporting of the standards and the labor hours required to repair the cameras resulted in an annual saving of several hundred thousand dollars.
>
> Although the work content did not change, the accurate measurement system showed the real cost of repairing cameras. As a result, numerous jobs were saved.

direct and indirect are rather fuzzy when it comes to the use of recordings and automatic data measurement. A great deal of artistry is involved in the latter approach, because it is difficult to obtain all of the relevant data without including many false signals. Accuracy may become a problem because it is difficult to screen out false signals economically. On the other hand, automatic methods can precisely measure even very short duration events.

Criteria other than time

Other performance criteria can be measured either directly or indirectly. Some of these criteria are discussed extensively in later chapters, including rate of learning, endurance, safety and health, ease and/or comfort, and flexibility and/or forgive-ability. In some cases, direct measurements are difficult or likely to be inaccurate for a number of reasons. One reason is that some criteria are themselves inferred. For example, speed is inferred. For illustration, suppose that a shaft that drives a wheel of your car has four magnets at each quadrant of the shaft diameter and an electronic device measures the number of magnetic clicks/second. If that wheel has a tire about 64 cm diameter, each rotation drives the wheel about 201 cm or approximately 2 m, so a magnetic click occurs nearly every half meter of automobile movement. The speedometer merely needs to report to the car driver the number of clicks/second multiplied by 1.8 to compute the speed in kilometers/hour. This scheme updates the measurement of speed every second, and shows up in an official manner on the speedometer. However, it is not necessarily true that the current rate of wheel rotation determines the speed of the car!

Production speed is more difficult to measure directly. One can make periodic observations to count both elapsed times and completed numbers of production units, and then estimate the speed of production by dividing the cumulative number of produced items by the cumulative elapsed time. However, partially completed product units present a problem here. Another approach is to measure how long it takes to make a single product unit. The reciprocal of that time is the production speed or rate.

Accuracy of performance is perhaps the second-most common performance criterion (speed, of course, is the most commonly measured). Performance accuracy is often described as a proportion of error-free trials, as discussed further in Chapter 10. In today's industry where a lot of process controls are connected to computers, it is often possible to use computers to track human errors over time and over numerous human operators. The speed–accuracy trade-off is a well-recognized effect that must be considered when measuring performance times. Higher speeds induce more errors (i.e., lower accuracy) and lower speeds allow people to perform the tasks with fewer errors (i.e., higher accuracy). While ergonomic practitioners must always be alert for this trade-off, it does not always occur. When it does, it may only be significant if there is a strong negative correlation between performance times and errors. If that correlation holds and both criteria are important in a design, both criteria must be measured together. A speed–accuracy operating characteristic (SAOC) describes these trade-offs, as will be discussed in Chapter 6. Changes in instructions or biases induced by past tasks or experiences will cause differences in these two criteria, and these SAOC diagrams can describe that effect when it occurs.

Final remarks

Two important applications addressed in this chapter were manual assembly and materials handling. Both topics deal with people in motion. A taxonomy of human motions was introduced. Principles of motion economy were discussed along with some background information. Motion and micromotion analyses of manual assembly were additionally addressed in some detail. As was shown, accuracy in movements is important as is movement precision and trade-offs with speed.

A variety of models have been suggested for use in measuring learning effects. It should be noted that it is extremely hard to find best-fitting models, in many cases, because most of the models make similar predictions. Quantifying the learning effect and using it appropriately is normally more important than the model used. This suggestion follows much stronger remarks by Lord Kelvin who stated: "When you can measure of what you are speaking and express it in numbers you know that on which you are discoursing. But if you cannot measure it and express it in numbers, your knowledge is of a very meager and unsatisfactory kind."

In addition, some principles of lifting are shown and means of analyzing lifting tasks are illustrated. In particular, the NIOSH method of analyzing lifting tasks to determine their safety and suitability were presented.

Time study is used to estimate performance times of prescribed methods. The resulting time estimates serve many purposes in industrial engineering, not the least of which are planning production operations and helping to determine an appropriate method of performing a job. The principal purpose of time study is in setting time standards. That task is traditionally performed by using direct timing devices and a single human operator. Elements of the task are timed over many cycles to assure that the means fall within an acceptable statistical confidence interval.

Either after or during the time study, the observed data are leveled by rating the observed operator. A number of alternative methods of leveling have been proposed and used over time. The product of the observed operator's average time and the leveling factor gives the normal time. Allowances are established for personal time, fatigue, and unavoidable delays. The normal time and an allowance correction factor are multiplied to give the standard time for that job.

Chapter five: Work measurement and analysis

This procedure of directly measuring performance times serves many purposes in contemporary industry. However, it might not be sufficient when time variances are important. It is interesting to observe that ergonomists in aviation design use variants of time study in conjunction with the technique of task–timeline analysis. While it would seem that variability in human performance time should be especially important in flying aircraft or in tending multiple machines, it does not appear that many ergonomists are very concerned about time variability. This oversight could and should change, particularly as just-in-time (JIT) procedures become more pronounced in industry.

Other criteria of human performance are also measured in ergonomics design. In some cases, these measurements are made directly, and in others it is easier to make indirect performance measurements. Errors are one criterion that can be directly measured, as expanded upon later in this book.

Discussion questions and exercises

5.1   What useful purposes are served by identifying the sequence of therbligs a person performs in a task?

5.2   What major criticisms can be leveled at the collection of principles of motion economy relative to an optimum method?

5.3   What is the difference in human effort when simultaneous movements are made by both hands symmetrically about the human body in contrast to motions made asymmetrically about the body center?

5.4   Computer software is available to assist in conducting time studies. In the past, however, methods such as cyclegraphs, chronocyclegraphs, and micromotion studies were used. Research these methods and Match the given example descriptions with the following equipment and technique concepts:

   Micromotion _____ Cyclegraph _____ Chronocyclegraph _____

   (a) A photograph of a person taking a golf swing with small lights attached to various locations of the body members
   (b) A motion picture of a person lifting boxes, with strobe-interrupted lights attached to body members
   (c) A motion picture camera at about 50 frames/min
   (d) A motion picture camera at 1 frame/h
   (e) A motion picture camera at 16–20 frames/s
   (f) A still photograph with a strobe-interrupted light circuit for lights attached to body members
   (g) A motion picture camera at 1000 frames/min showing an assembly task under a strobed-ultraviolet lamp

5.5   Why does part symmetry aid assembly operations?

5.6   Explain each term used in the revised NIOSH lifting guide and how it impacts the RWL. In what major ways does the revised guide differ from the original?

5.7   Why are the maximum acceptable load limits for lifting greater when the horizontal load center of the load being lifted is closer to the body?

5.8   What is the maximum recommended weight to be lifted with low-frequency lifts over a full 8 h shift? If lift frequencies are about 5/min over a full 8 h shift, what is the maximum recommended weight to be lifted? At a high frequency of 12 lifts/min and a work period of 20 min or less, what is the maximum recommended weight?

5.9 A box of tools weighing 20 lb is to be lifted (occasionally—maybe twice per day) from the floor to a cart that is 46 in. high. (No control is required at the destination.) The handle of the box is 10 in. high (i.e., the handle is located 10 in. from the floor). Due to the width of the box, the worker must reach 24 in. in front of his or her ankles to grasp the handle, but does not need to twist.
  (a) How would you rate the coupling type (good/fair/poor)? Justify your answer.
  (b) Fill in a NIOSH lifting table and compute an RWL.
  (c) What is the lifting index?
  (d) How would the preceding analysis change if the lifting was done with one hand?

5.10 What is the ultimate cycle time if the performance-time learning curve is well described by an exponential function with parameters $\alpha = 0.9$, $\beta = 0.30$, and $t1 = 5$ min?

5.11 What is the total expected time over 25 cycles for the aforementioned learning curve?

5.12 A powerform learning curve was fit to the cycle time which has $t_{2n}/t_n = 0.90$ for all n.
  (a) What is the cumulative average time on the fourth cycle if the time $t_1 = 10$ min?
  (b) Indicate which model, for which set of parameter values, results in the following behavior of the ratio A_{2n}/A_n and explain why.
  (i) Increasing at an increasing rate
  (ii) Increasing at a decreasing rate
  (iii) Remaining constant over the series
  (iv) Decreasing at an increasing rate
  (v) Decreasing at a decreasing rate
  (vi) Oscillating with increasing magnitudes
  (vii) Oscillating without changes in the magnitudes
  (viii) Oscillating with decreasing magnitudes

5.13 Consider the following data and fit a powerform learning curve to these data. Denote the learning rate, the learning rate parameter, and the performance time on the 25 trial.

$n =$	1	2	3	4
$t_n =$	18.00	13.50	11.41	10.13 h

5.14 Fit a discrete exponential learning curve to the following data. Denote the ultimate performance time, the performance time in the 10th trial, and the cumulative performance time over the first 20 trials.

$n =$	1	2	3	4
$t_n =$	15	14.5	14.05	13.645

5.15 Suppose that workmen are assembling a particular product for 40 working hours. Then they work on other products for 60 working hours before returning to the original assembly task for another 40 h and so forth. What is implied by recent research on the repeated production learning curves?

5.16 A person requires 10 min to complete the first assembly and 7.2 min for the second. Based only on these data and knowledge that the powerform learning curve is the appropriate model, what is the learning rate parameter?

5.17 Suppose that a powerform learning curve model was fit to the average person for a specified task. It was found that the average subject made 20 errors in the first cycle of the task, 15 errors in the second cycle, and 11.25 in the third. Based on these data, what is the expected number of errors on the 10th cycle?

Chapter five: Work measurement and analysis

5.18 How many cycles of a job need to be directly measured to assure that the measured performance mean ± 0.05 min includes the true mean of the observed operator's cycle times with a probability of 0.95? The standard deviation of observed cycle times is 0.30 min?

5.19 Consider the following data and the situation described. What is the normal cycle time for this task based upon the old Westinghouse method of pace rating?
- Average cycle time of 1.50 min
- Excellent consistency as verified by a low variance of cycle times
- The operator displayed good (almost but not quite excellent) effort
- Conditions under which the timing took place were ideal
- The operator was classified as an average skilled person

5.20 You wish to conduct a time study for a computer repair task. An initial set of random data is taken and is shown in the following.

	Task Element				
Cycle	1	2	3	4	Total
1	182.0	200.0	194.0	184.0	760.0
2	225.0	205.0	190.0	180.0	800.0
3	190.0	205.0	230.0	205.0	830.0
4	195.0	180.0	200.0	235.0	810.0

(a) About how many cycles of data should you take on the actual time study, given that you want the estimate to be accurate to within 10 s of total time?

(b) From your time study, you get an estimate of 800 s/unit for the repair task. You wish to adjust the time using the Westinghouse multifactor pace rating system. You rate the skill of the participant as "Good (C1)," the effort as "poor (F1)," the conditions as "Good," and the consistency as "Average." Compute the estimated time to complete the task as adjusted for these ratings.

(c) Using 4% for personal allowance, 3% for fatigue allowance, and 3% for delay allowance, compute the estimated time to complete the task as adjusted for these allowances.

(d) How many units could a person repair in an 8 h shift?

5.21 A time study was made on an assembly task, and the average cycle time was found to be 1.87 min. The time-study rating of the observed operator's performance was 90%. This particular task has a 23 min fatigue allowance, a 10 min/shift delay allowance, and a 24 min/shift personal allowance. What is the standard time for this task?

chapter six

Modeling physical human performance

About the chapter

Direct measurement of human performance, such as used in Chapter 5, is often the most accurate way of obtaining quantitative information, but may be infeasible or prohibitively expensive in some situations. This chapter describes several other approaches that can be used to predict human performance at various tasks. Synthetic data systems are databases of predetermined times for particular motions, which can be aggregated into an overall prediction of how long a particular task can take an operator to perform. Various methods have been employed for this purpose, and this chapter covers several of them that differ in significant ways. Regression models can be used in conjunction with, or instead of, these methods in certain circumstances. Advances in computing technology have led to the development of very powerful computer models of humans that can be used to visualize and predict human performance. All these methods, however, are influenced by a trade-off between speed and accuracy. This is particularly noticeable in a subset of tasks, namely, moving and reaching tasks, which have been found to be well-predicted by the information content of the task.

Introduction

Ergonomists must be able to compare alternative tools, methods of assembly or other operations, and in other ways predict various important criteria of concern associated with proposed ergonomic projects. This chapter is about alternative ways to make these predictions. Both direct empirical measurements and the use of past records of performance provide methods for such predictions. A number of approaches are required because the procedures using past data are often more economical to use in terms of the designer's time or money. But past data may not be available or may be suspected to be wrong for a variety of reasons including changes in equipment or procedures. In addition to design purposes, performance prediction supports numerous other functions in industry, particularly for those companies that attempt to operate using just-in-time practices and other contemporary manufacturing procedures.

As mentioned earlier in this book, there are many kinds of human performance of interest to ergonomic designers. The principal criteria discussed in this chapter are speeds, accuracies, consistencies, and individual differences. Speed is usually measured by the time used to perform some unit of a task or collection of tasks (e.g., minutes per cycle), but speed is also the number of cycles per time unit. Accuracy can be measured by observing the number of errors per unit of the task or time on the task (e.g., errors/1000 entries or per hour on the task). Fewer errors indicate greater accuracy. The consistency of either speed or accuracy is frequently measured by the standard deviation of time or errors. Individual differences can also be measured by the standard deviations. Learnability is frequently measured by the time required for the average person to achieve a specified proficiency in a task or with a particular piece of equipment.

Answers to the questions "Why measure human performance?" and "Whose performance is measured?" are related to each other. The relationship is embedded in the design and selection of a procedure or method. In order to select an acceptable method for performing a job in industry, a designer first needs to identify some alternative methods. The recommended procedure for selecting the best industrial method requires designers to predict performance on each criterion for each method. Each method has a profile of predicted performance values such as speed, accuracy, consistency, and individual differences on these criteria. If one method has the fastest speed, greatest accuracy, most consistency, and the least individual differences, that method would be a perfect choice, but a single method does not usually dominate in all important criteria. In the latter case, designers must choose among alternative methods based on their advantages and disadvantages. Some trade-offs are made because certain criteria are considered far more important than others and so there is a priority weighting of the criteria. Another issue arises when individual differences appear between methods. If the cause of those individual differences is handedness, it may be easy to modify the method so that a left-handed person performs the job with some differences, but just as well as a right-handed person. In other words, it may be good to use more than one method. That is one of the ways the why and who questions may be related.

A second reason for measuring human performance is to evaluate the devices that people use. Those devices that do not give an acceptable performance profile are unacceptable designs. In some cases where the device is intended for anyone to use (e.g., hand tools), all of the previous criteria are important, including individual differences, because those criteria affect the potential sales of a product. When products are targeted for specific subpopulations (e.g., devices for disabled persons or sports equipment for the professional), the relative importance of different criteria can shift drastically. But then the who question is specified.

A third application of measuring human performance is in planning production. Production operations need to be properly planned for effective operations, and they, in turn, depend on proper assignment of people to the component activities to avoid production bottlenecks. Also, different production jobs need to be scheduled in various shops in order to avoid delays in later production operations. Knowledge of performance times and how they change with learning is extremely important in determining how many people are needed to get the job done.

Human performance prediction is very important when estimating the cost of producing a product. Many companies must bid on jobs, and to prepare the bids, estimates must be made of direct labor costs. When the cost of labor is a large part of a bid, winning or losing a contract or winning a contract and making money on it can be a function of the prediction accuracy. Even when not bidding on a project, many companies estimate standard labor times for accounting purposes. Accounting departments often report differences between the amount of time used and that predicted as an accounting variance. These variances are a form of feedback to management and supervision, telling them how things are going compared to their predictions.

The previous paragraphs tell us much about why we need to predict human performance. Human performance is typically measured using a procedure known as time study, but a much wider variety of procedures are available to the analyst. Typically, time study does not address criteria other than time, but it is not uncommon in time studies to measure accuracy and learning rates. Procedures for measuring performance using newly designed devices are often called usability studies. While time study is typically used for method selection purposes as well, that practice has questionable validity and there is some controversy over that practice. Time study typically uses a single subject and is usually much less costly than conducting a usability study. Usability studies maintain a much greater concern for individual differences, thereby requiring more subjects and causing costs to be higher.

Synthetic data systems

Synthetic* prediction systems are models that can be used to predict human performance of a prescribed task. These systems contain a taxonomy of human activities, which the modeler specifies in a task analysis. Parameters associated with each kind of activity correspond to variables that affect performance. For example, one activity is reaching out with one's hand, and a variable that affects performance time is the distance reached. Another variable is the care that the person must exercise in the reach. These variables or parameters are used in the system to predict performance times. Such predictions are for a standard operator defined by each particular system.

There are a number of such systems in use today. Two of the better known synthetic time systems are the motion-time-measurement system (MTM) and the work-factor (WF) system. Other synthetic time systems include motion time analysis, basic motion time study, dimensional motion times, Maynard operation sequence technique, modular arrangement of predetermined time standards, and master standard data. The two principal systems discussed in this chapter yield predictions of performance times that are very highly correlated with each other but are structured in different ways. The two systems are further distinguished by their different definitions of the standard operator. MTM system's standard operator is slightly less skilled so that system yields a greater average time prediction than WF predictions. Numerous counterexamples can be found, but the differences tend to be very small. These and other distinctions for these different types of systems will be discussed. The principal concentration, however, is on the MTM system. References to the WF system are mainly made to describe some of the structural system differences.

Why use a synthetic data system?

The reader may well wonder, "What can synthetic prediction systems do for me?" The foremost answer to this question is that they permit one to predict performance without making direct measurements on an operator performing the task. Direct measurements can be expensive because they entail training an operator in the prescribed method and observing that operator through the performance of numerous cycles of the task. While direct performance measurement involves rating or leveling, synthetic prediction avoids this potential bone of contention with organized labor. So, one answer is that synthetic systems are often less expensive and less disruptive than direct measurement. When highly precise measurement is not an issue, synthetic systems with lower degrees of prediction precision are even more economical.

Additionally, the use of a synthetic data system allows performance estimates to be obtained directly from a task analysis. This feature is particularly useful in the early stages of design when alternative methods of performing a task may be under consideration and the designer wants to find out which method is faster or whether a particular feature of design is causing a performance problem. In contrast to standard data systems, synthetic

---

* Therbligs are basic elements in MTM and other synthetic prediction systems.

prediction systems are more expensive in terms of analysts' time. However, standard data systems cannot be used unless there is a reasonably large collection of past jobs in the same job family. When such data are not available, synthetic systems can be used.

Motion–time measurement

The origin of motion–time measurement (MTM) is frequently traced to a book of the same title, written by Maynard and his coauthors in 1948. In fact, research on this subject began much earlier with the work of the Gilbreths and numerous others who helped identify the therbligs of motion analysis, many of which are now used in MTM. The original MTM system has been revised over time and predictions are made differently than in the past. However, the essential structure of that original MTM model is quite similar to the present detailed version known as MTM-1. The designator "1" is used to distinguish it from approximate versions that carry other designators.

The original MTM version was devised from motion pictures taken in numerous factories in the eastern United States. This form of direct factory data gave strong face validity to the study. The data were all leveled to a standard operator using the old Westinghouse rating method. Later on, the MTM Association was formed. Since then, the data have evolved over time.

Structural basis of MTM

A variety of physical actions or activities performed by people constitute the principal elements of the MTM system. These actions for hand-arm activities include such gestures as Reach, Move, Turn, Apply pressure, Grasp, Position, Release, and Disengage. Many of these actions speak for themselves. However, reaching is distinguished from moving by the fact that reaching is done without an object in the hand. A move, of course, involves a hand-carried object of some sort. In grasping, the hand gains hold of an object, whereas in releasing, the hand relinquishes the object. Positioning refers to orienting a production part, such as a pin or key, with respect to a machine tool, fixture, or another part.

Several cases are distinguished for each of the basic actions. For example, case A involves a reach to an object in a fixed location. It also occurs for a reach to an object in the other hand or to an object on which the other hand rests. Case B involves a reach to a single object in locations that may vary slightly from cycle to cycle. Case C involves a reach to an object jumbled with other objects in a group (also see cases D and E in Table 6.1). These three cases have very widely different visual requirements. When the object in question is in one's hand, vision is often not required. Similarly, the positions and orientations of objects that are always in the same location can be easily learned so that little or no eye–hand coordination is needed. Visual requirements needed to guide the hand increase with greater variability in the location-to-location positioning of the objects. The most extreme condition is when objects are jumbled together.

Specific parameters are associated with many actions. One parameter is the distance to be reached. More time is required for longer reaches. The time needed increases some more when the hand must start from a resting condition. Simple principles of physics tell us that bodies in motion tend to remain in motion unless acted upon by an outside force. For that reason, different effects are observed when the hands are already moving in the direction of the reach compared to when they are moving in a different direction or at rest. The Move action includes both the weight of the object and the distance moved as parameters that affect human performance times.

Table 6.1 MTM Application Data Card

Table I – Reach – R

Distance moved inches	Time TMU				Hand in motion		Case and description
	A	B	C or D	E	A	B	
3/4 or less	2.0	2.0	2.0	2.0	1.6	1.6	A Reach to object in fixed location, or to object in other hand or on which other hand rests
1	2.5	2.5	3.6	2.4	2.3	2.3	
2	4.0	4.0	5.9	3.8	3.5	2.7	
3	5.3	5.3	7.3	5.3	4.5	3.6	B Reach to single object in location which may vary slightly from cycle to cycle
4	6.1	6.4	8.4	6.8	4.9	4.3	
5	6.5	7.8	9.4	7.4	5.3	5.0	
6	7.0	8.6	10.1	8.0	5.7	5.7	
7	7.4	9.3	10.8	8.7	6.1	6.5	C Reach to object jumbled with other objects in a group so that search and select occur
8	7.9	10.1	11.5	9.3	6.5	7.2	
9	8.3	10.8	12.2	9.9	6.9	7.9	
10	8.7	11.5	12.9	10.5	7.3	8.6	
12	9.6	12.9	14.2	11.8	8.1	10.1	D Reach to a very small object or where accurate grasp is required
14	10.5	14.4	15.6	13.0	8.9	11.5	
16	11.4	15.8	17.0	14.2	9.7	12.9	
18	12.3	17.2	18.4	15.5	10.5	14.4	
20	13.1	18.6	19.8	16.7	11.3	15.8	E Reach to indefinite location to get hand in position for body balance or next motion or out of way
22	14.0	20.1	21.2	17.9	12.1	17.3	
24	14.9	21.5	22.5	19.2	12.9	18.8	
26	15.8	22.9	23.9	20.4	13.7	20.2	
28	16.7	24.4	25.3	21.7	14.5	21.7	
30	17.5	25.8	26.7	22.9	15.3	23.2	
Additional	0.4	0.7	0.7	0.6			TMU / in. over 30 in.

Table II – Move – M

Distance moved inches	Time TMU			Hand in motion	Wt. Allowance			Case and description
	A	B	C	B	Wt. (lb.) Up to	Dynamic factor	Static constant TMU	
3/4 or less	2.0	2.0	2.0	1.7				A Move object to other hand or against stop
1	2.5	2.9	3.4	2.3	2.5	1.00	0	
2	3.6	4.6	5.2	2.9				
3	4.9	5.7	6.7	3.6	7.5	1.06	2.2	
4	6.1	6.9	8.0	4.3				
5	7.3	8.0	9.2	5.0	12.5	1.11	3.9	
6	8.1	8.9	10.3	5.7				B Move object to approximate or indefinite location
7	8.9	9.7	11.1	6.5	17.5	1.17	5.6	
8	9.7	10.6	11.8	7.2				
9	10.5	11.5	12.7	7.9	22.5	1.22	7.4	
10	11.3	12.2	13.5	8.6				
12	12.9	13.4	15.2	10.0	27.5	1.28	9.1	C Move object to exact location
14	14.4	14.6	16.9	11.4				
16	16.0	15.8	18.7	12.8	32.5	1.33	10.8	
18	17.6	17.0	20.4	14.2				
20	19.2	18.2	22.1	15.6	37.5	1.39	12.5	
22	20.8	19.4	23.8	17.0				
24	22.4	20.6	25.5	18.4	42.5	1.44	14.3	
26	24.0	21.8	27.3	19.8				
28	25.5	23.1	29.0	21.2	47.5	1.50	16.0	
30	27.1	24.3	30.7	22.7				
Additional	0.8	0.6	0.85					TMU / in. over 30 in.

Table III A – Turn – T

Weight	Time TMU for degrees turned										
	30°	45°	60°	75°	90°	105°	120°	135°	150°	165°	180°
Small – 0 – 2 lb	2.8	3.5	4.1	4.8	5.4	6.1	6.8	7.4	8.1	8.7	9.4
Medium – 2.1 – 10 lb	4.4	5.5	6.5	7.5	8.5	9.6	10.6	11.6	12.7	13.7	14.8
Large – 10.1 – 35 lb	8.4	10.5	12.3	14.4	16.2	18.3	20.4	22.2	24.3	26.1	28.2

Table III B – Apply pressure – AP

Fully cycle			Components		
Symbol	TMU	Description	Symbol	TMU	Description
APA	10.6	AF+DM+RLF	AF	3.4	Apply force
APB	16.2	APA+G2	DM	4.2	Dwell, minimum
			RLF	3.0	Release force

Table IV – Grasp – G

Type of grasp	Case	Time TMU	Description
Pick-up	1A	2.0	Any size object by itself, easily grasped
	1B	3.5	Object very small or lying close against a flat surface
	1C1	7.3	Interference with grasp on bottom and one side of nearly cylindrical object Diameter larger than 1/2"
	1C2	8.7	Diameter 1/4"–1/2"
	1C3	10.8	Diameter less than 1/4"
Regrasp	2	5.6	Change grasp without relinquishing control
Transfer	3	5.6	Control transferred from one hand to the other
Select	4A	7.3	Object jumbled with other object so that search and select occur Larger than 1" × 1" × 1"
	4B	9.1	1/4" × 1/4" × 1/8" – 1" × 1" × 1"
	4C	12.9	Smaller than 1/4" × 1/4" × 1/8"
Contact	5	0	Contact, sliding, or hook grasp

Effective net weight

Effective net weight (ENW)	No. of Hands	Spatial	Sliding
	1	W	$W \times F_C$
	2	$W/2$	$W/2 \times F_C$

W – Weight in pounds
F_C – Coefficient of friction

Table V – Position* – P

Class of fit		Symmetry	Easy to handle	Difficult to handle
1-Loose	No pressure required.	S	5.6	11.2
		SS	9.1	14.7
		NS	10.4	16.0
2-Close	Light pressure required.	S	16.2	21.8
		SS	19.7	25.3
		NS	21.0	26.6
3-Exact	Heavy pressure required.	S	43.0	48.6
		SS	46.5	52.1
		NS	47.8	53.4

Supplementary rule for surface alignment
P1SE per alignment > 1/16 ≤ 1/4"   P1SE per alignment ≤ 1/16"
*Distance moved to engage–1" or less.

Table VI – Release – RL

Case	Time TMU	Description
1	2.0	Normal release performed by opening fingers as independent motion.
2	0	Contact release

Table VII – Disengage – D

Class of fit	Height of recoil	Easy to handle	Difficult to handle
1-LOOSE–Very slight effort, blends with subsequent move	Up to 1"	4.0	5.7
2-CLOSE–Normal effort, slight recoil	Over 1" – 5"	7.5	11.8
3-TIGHT–Considerable effort, hand recoils markedly	Over 5" – 12"	22.9	34.7

Supplementary

Class of fit		
1 – LOOSE	Care in handling	Binding
2 – CLOSE	Allow Class 2	One G2 per Bind
3 – TIGHT	Allow Class 3 Change Method	One APB per Bind

Table VIII – Eye travel and Eye focus – ET and EF

Eye travel time = $15.2 \times \dfrac{T}{D}$ TMU, with a maximum value of 20 TMU
Where T = the distance between points from and to which the eye travels
D = the perpendicular distance from the eye to the line of travel T

Eye focus time = 7.3 TMU

Supplementary information
– Area of normal vision = Circle 4" in diameter 16" from eyes
– Reading formula = 5.05 N where N = The number of words

(continued)

Table 6.1 **(continued)** MTM Application Data Card

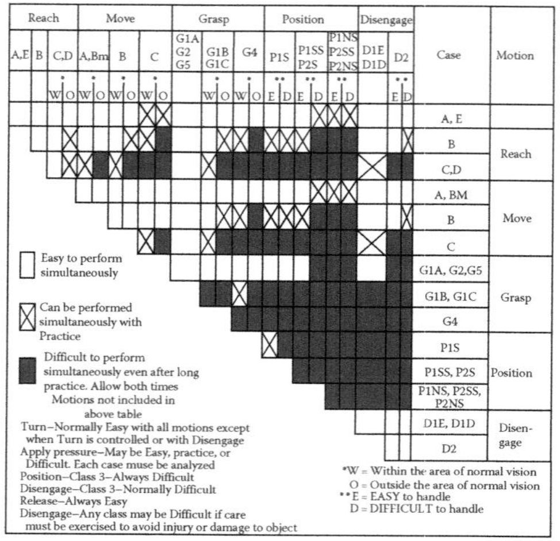

Supplementary MTM Data
Table 1–Position–P

Class of fit and clearance	Case of symmetry	Align only	Depth of insertion (per ¼")			
			0 >0≤1/8"	2 >1/8≤¾	4 >¾≤1¼	6 >1¼≤1¾
21 .150"–.350"	S	3.0	3.4	6.6	7.7	8.8
	SS	3.0	10.3	13.5	14.6	15.7
	NS	4.8	15.5	18.7	19.8	20.9
22 .025"–.149"	S	7.2	7.2	11.9	13.0	14.2
	SS	8.0	14.9	19.6	20.7	21.9
	NS	9.5	20.2	24.9	26.0	27.2
23* .005"–.024"	S	9.5	9.5	16.3	18.7	21.0
	SS	10.4	17.3	24.1	26.5	28.8
	NS	12.2	22.9	29.7	32.1	34.4

*Binding–Add observed number of apply pressures.
Difficult handling–Add observed number of G2's.
*Determine symmetry by geometric properties, except use S case when object is oriented prior to preceding move.

Table 1A – Secondary engage – E2

Class of fit	Depth of insertion (PER ¼")			
	2	4	6	
21	3.2	4.3	5.4	
22	4.7	5.8	7.0	
23	6.8	9.2	11.5	

Table 2 – Crank (light resistance) – C

Diameter of cranking (in.)	TMU (T) per revolution	Diameter of cranking (in.)	TMU (t) per revolution
1	8.5	9	14.0
2	9.7	10	14.4
3	10.6	11	14.7
4	11.4	12	15.0
5	12.1	14	15.5
6	13.7	16	16.0
7	13.2	18	16.4
8	13.6	20	16.7

Formulas:
A. Continuous cranking (Start at beginning and stop at end of cycle only)
$$TMU = [(N \times T) - 5.2] F + C$$
B. Intermittent cranking (Start at beginning and stop at end of each revolution)
$$TMU = [(T + 5.2) F + C] N$$

C = Static component TMU weight allowance constant from move table
F = Dynamic component weight allowance factor from move table
N = Number of revolutions
T = TMU per revolution (Type III motion)
5.2 = TMU for start and stop

Methods–time measurement MTM-I application data

1 TMU	= 0.00001 h	1 h	= 100,000.0 TMU
	= 0.0006 min	1 min	= 1,666.7 TMU
	= 0.036 s	1 s	= 27.8 TMU

Do not attempt to use this chart or apply Methods-Time Measurement in any way unless you understand the proper application of the data. This statement is included as a word of caution to prevent difficulties resulting from misapplication of the data.

MTM ASSOCIATION FOR STANDARDS AND RESEARCH

1111 E. Touhy Avenue
Des Plaines, IL 60018
Phone: 847/299-1111
FAX: 847/299-3509

MTM *association*

©Copyright 1997 MTM Association for Standards and Research

PRINTED IN U.S.A

Table IX – Body, leg, and foot motions

	Type	Symbol	TMU	Distance	Description
Horizontal Motion	Leg-foot	FM	8.5	To 4"	Hinged at ankle
		FMP	19.1	To 4"	With heavy pressure
	Motion	LM_	7.1	To 6"	Hinged at knee or hip in any direction
			1.2	Ea. add'l in.	
	Side	SS_C1	17.0	<12"	Use reach or move time when less than 12"- complete when leading leg contacts floor
			0.6	Ea. add'l in.	
	Step	SS_C2	34.1	12"	Lagging leg must contact floor before next motion can be made
			1.1	Ea. add'l in.	
	Turn	TBC1	18.6		Complete when leading leg contacts floor
		TBC2	37.2		Lagging leg must contact floor before next motion can be made
	Walk	W_FT	5.3	Per Foot	Unobstructed
		W_P	15.0	Per Pace	Unobstructed
		W_PO	17.0	Per Pace	When obstructed or with weight
Vertical Motion	Body	SIT	34.7		From standing position
		STD	43.4		From sitting position
		B.S,KOK	29.0		Bend, stoop, kneel on one knee
		AB.AS,AOK	31.9		Arise from bend, stoop, kneel on one knee
		KBK	69.4		Kneel on both knees
		AKBK	76.7		Arise from kneel on both knees

Table X – Simultaneous motions

Reach			Move				Grasp			Position				Disengage		Motion
A,E	B	C,D	A,Bm	B	C	G1A,G2,G5	G1B,G1C	G4	P1S	P1SS,P2S	P1NS,P2SS,P2NS	D1E,D2	D1D			
W O W O						W O W O			W O E D E D E D			E D		A,E		Reach
	⊠	⊠												B		
		⊠												C,D		
														A,Bm		Move
														B		
														C		
														G1A,G2,G5		Grasp
														G1B,G1C		
														G4		
														P1S		Position
														P1SS,P2S,P2NS		
														P1NS,P2SS,P2NS		
														D1E,D1D		Disengage
														D2		

☐ Easy to perform simultaneously
⊠ Can be performed simultaneously even after long practice. Allow both times
■ Difficult to perform simultaneously even with practice. Motions not included in above table

Turn - Normally Easy with all motions except when Turn is controlled or with Disengage
Apply pressure - May be Easy, practice, or Difficult. Each case must be analyzed
Position - Class 3 - Normally Easy
Release - Always Easy
Disengage - Any class may be Difficult if care must be exercised to avoid injury or damage to object

*W = Within the area of normal vision
**O = Outside the area of normal vision
E = EASY to handle
D = DIFFICULT to handle

Source: MTM Association for Standards and Research, 2006. Reprinted with permission.

Time predictions of MTM-1

Table 6.1 shows application data of the MTM-1 system of synthetic time prediction. Symbols at the top of each action column denote the abbreviations of those actions; R for reach, M for move, T for turn, G for grasp, and so forth. Added symbols are used to show the case and parameter values. For example, R20A means a 20-in. reach of class A. P2SSD means a position action of class 2 (close, requiring light pressure), where the positioning is semi-symmetrical and the handing is difficult. Three cases of symmetry are distinguished: (1) a symmetrical object can be fit into position without rotation after it is grasped, (2) a semi-symmetrical object may have to be flipped over, and (3) an asymmetrical object fits in only one orientation, and may require precise rotation to fit into place. It also follows that T75M means a turn action of 75° where medium force (2.1–10 lb) is needed.

In Table 6.1, time is measured in time measurement units (TMUs). One TMU equals 0.0006 min or 0.00001 h. TMUs are more easily converted into hours than minutes. For example, action R20A requires 13.1 TMUs, which equals 0.472 s, 0.0079 min, or 0.000131 h; Action P2SSD requires about 0.911 s, 0.0152 min, or 0.000251 h. In addition to hand motions, MTM specifies times for movements of other body parts, including foot motions, leg movements, eye travel, and eye focus.

Some movements are easier than others to make at the same time. Subtable X of Table 6.1 addresses this issue. This subtable shows possible combinations of each type of motion in a matrix. The symbols E, P, or D found in each cell of the matrix indicate combinations that can be easily performed, performed with substantive practice, or are very difficult to perform at the same time. Initially, and for short production runs, only easy combinations can be performed at the same time. Later on, both easily performed and performed-with-practice tasks can be performed concurrently. The transition times from one state to another are not clear from this table, but most people regard easy as right away and difficult as never.

Analytical procedure with MTM-1

A task analysis is first performed in order to identify all of the actions required along the sequence of actions. This analysis is done for each hand separately following the prescribed method of task performance. Typically, these actions are recorded in two vertical columns in a two-handed simo (simultaneous) chart. Table 6.2 illustrates a two-handed simo chart with left- and right-hand actions.

After the tasks have all been identified in the correct order for the method denoted, symbols corresponding to each action are typically denoted in separate columns toward the middle of the table. Predicted time values associated with each of those actions are found from Table 6.1. Those values are recorded in columns at the outer edges of Table 6.2.

BOX 6.1 MTM-1 EXAMPLE OF WALKING

Subtable IX in Table 6.1, for instance, shows that walking takes about 15 TMUs/pace by the standard operator. Notice that this time value leads to a prediction of the travel speed by that standard (MTM) operator with a 30-in. (76.2 cm) pace as follows:

$$\frac{30 \text{ in./pace}}{5280 \text{ ft/mile}\left[12 \text{ in./ft}\right][15 \text{ TMUs/pace}]0.00001 \text{ h/TMU}} = 3.16 \text{ mile/h or } 5.07 \text{ km/h}$$

Table 6.2 A Two-Handed SIMO Chart for MTM-1

Left-Hand Actions			Right-Hand Actions		
TMUs	Action Description	Symbol	Symbol	Action Description	TMUs
12.9	Reach hand to bin	R10C	RL1	Release assembly	1.0
8.7	Grasp washer	G1C2	R4C	Reach to bin	8.4
11.8	Move to fixture	M8C	G1C2	Grasp pin	8.4
5.6	Position washer	P1S	M4C	Move to fixture	7.3

At this point in the MTM analysis, one should refer to the simultaneous motions subtable to determine when actions of the left hand and of the right hand can be performed simultaneously. The latter term is frequently abbreviated as simo in MTM. When the left- and right-hand tasks can be performed at the same time, then only the longer of the two times is encircled; otherwise, both times are encircled. In Table 6.2, the corresponding TMU values are indicated with one asterisk when the two tasks are not easily compatible according to the simultaneous motions chart. Note that the release action is performed easily with the other actions, and so the first combination is the only one not marked in Table 6.2. The normal time is obtained by summing up the times in a way that accounts for simultaneously performed elements. In the example of Table 6.2, the normal time is 63.1 TMUs. This value is obtained by summing the times marked with an asterisk along with 12.9 TMUs for the reach task R10C that can be done simultaneously with the release task RL1.

Assumptions associated with MTM-1

A principal assumption of MTM is that the time estimates describe performance of an average human operator with minimal skill performing the job with the prescribed method at an acceptable productivity level. Various examples illustrate what constitutes an acceptable productivity level: one such example is the ability to deal out four hands of bridge in one-half minute.

It is further assumed that there is no mental activity associated with these actions outside of that, which is specified in the MTM case descriptions. Those descriptions only account for visual-perceptual activities. Of course, some mental activities can be performed simultaneously, or nearly so, with physical actions. But when these mental activities become substantial, performance decrements result in one or both of the tasks. Concurrent Tasking theory explains why this happens. Unfortunately, theories such as this can explain, but they do not predict.

Another associated assumption is that the times specified by the MTM system are independent of the sequences in which the tasks are performed, that is, the time to perform both tasks is assumed to be no different if task A is performed before or after task B. Years ago, Schmidtke and Stier reported on physical tasks where this assumption was violated. One of the principal points made in this report is that many industrial tasks do not permit a lot of sequence changes in fundamental physical actions because actions normally follow a sequence of reaching, grasping, moving, positioning, and then releasing. A psychological phenomenon, called the repetition effect, has been observed in which repeated tasks take less time. A number of studies reveal this finding to be quite controversial. Some studies with cognitive tasks show that the effect occurs while others do not. MTM currently adjusts for some sequence effects. For example, the time is lower if the hand is in motion before the next reach. However, repetition effects may occur in certain circumstances, so the designer must be aware of this possibility.

A further assumption of MTM is that the performance time of an action remains constant regardless of other actions that accompany it. More specifically, this assumption implies that reach, grasp, and move times do not change with the number of positions, applied pressures, or other actions that occur. While this assumption appears to be more plausible for physical motor tasks, there is no reason to believe that it holds for perceptual-cognitive tasks.

Another assumption is that performance times do not change with various forms of pacing, that is, whether the person is self-paced or speed-paced, performance is assumed to remain unchanged. There is evidence, however, to support the claims that performance does change in different pacing conditions. Machine pacing in industry usually involves rather constant machine times. In other situations, the time between tasks varies randomly. The effect of random pacing on performance times is not well understood. It is known that faster paces involving unknown durations between successive human tasks are more stressful and that stress, when experienced over time, results in reduced performance. It is not at all clear, however, that different forms of pacing would have any effect on performance when the pace approximates the self-paced ratio.

Synthetic time systems in general, and MTM in particular, also make the so-called modularity assumption, that is, performance times are assumed to be proportional to the summed performance times of the actions. The users of MTM assume that a collection of actions, called an action or task module, will have a total performance time equal to the sum of the times on those actions that cannot be performed concurrently. This assumption does not hold for some simple cognitive tasks. The differences between sensory-motor tasks and cognitive tasks result from the ways separate actions link to the next successive action. Most sensory-motor tasks link in a rather consistent and fixed manner: reach, grasp, move, position, and release. Cognitive tasks are more complex.

Applications of MTM-1

MTM and other synthetic time-predicting systems help identify which methods are faster (i.e., produce the shortest performance times). This test can be made as the alternative methods are considered and revised. Method improvements under consideration can be immediately tested for any speed improvement. Synthetic time systems can also be used in time study to develop a speed rating, as discussed in Chapter 5.

Perhaps the most obvious application of synthetic time predicting systems is in setting a time standard. This approach can be used instead of direct operator timing. Many companies use synthetic time system to estimate the long-run performance time values. Once that time is established, they will estimate a learning curve for a particular type of work. In these calculations, it is sometimes assumed that 1000 repetitions of the task will yield the learned performance time. The use of 1000 cycles appears to be an ad hoc or an agreed-upon value rather than empirically determined.

MTM-2 for simpler tasks and quicker analysis

A few years after MTM-1 was developed, a less detailed system called MTM-2 was developed. Other less detailed systems are also available. MTM-2 combines many of the therbligs used in MTM-1, which simplifies and speeds up the analysis at the cost of some precision. MTM-2 is not recommended for tasks lasting less than a minute or for highly repetitive task performance over long periods. Niebel (1988) shows that the error in MTM-2 predictions, relative to those of MTM-1, changes with the cycle time of the task. Ranges of

Table 6.3 Error in MTM-2 Predictions Relative to MTM-1 for Different Cycle Times

	Cycle Duration				
TMUs	1000	2000	3000	4000	5000
Hours	0.01	0.02	0.03	0.04	0.05
Minutes	0.6	1.2	1.8	2.4	3.0
Low percentage	5.0	2.3	2.0	2.0	2.0
High percentage	11.8	7.3	5.7	5.5	5.0

expected predictive error variations are shown in Table 6.3. If the aforementioned loss of precision is acceptable, the simplified MTM-2 system can be used effectively.

There are slightly fewer elemental actions in MTM-2 than in MTM-1. The two primary actions used in MTM-2 are the GET action, which combines the therbligs of reach, grasp, and release used in MTM-1, and the PUT action, which combines the therbligs of move and release. Most simple assembly tasks involve a series of GETs and PUTs. For both of these actions, performance time increases with the distance the hand is moved. The distances for GETs and PUTs are measured in inches from the end of the fingertips for finger motions, and from the knuckle at the base of the index finger for hand movements. For further simplicity, rather than using the exact measurements, GET and PUT distances are coded into one of the five categories shown in Table 6.4.

MTM-2 distinguishes three different ways of grasping objects in the GET action. Case A corresponds to a contact grasp where the fingers simply push an object across a flat surface. Case B is a more complex grasp where the fingers surround the object and press against it. Any other type of grasp is classified as case C. Time requirements for the three cases at each of the five distance codes are shown in Table 6.5. The time needed to perform PUT actions is determined in a similar manner as shown in Table 6.6. Corrections are almost always required to assure correct positioning when objects are being aligned. PUTs normally require corrections toward the end of the movement, just prior to inserting an object into an opening. A special case exists when parts are being engaged. If the distance exceeds an inch, an additional PUT action is included to account for the engagement of parts.

The weight of the manipulated object is considered for both actions. MTM-2 assigns 1 additional TMU/kg of the object per hand used in a GET. For example, if two hands are

Table 6.4 The Five Distance Codes Used in MTM-2

Distance (in.)	0–2	2–6	6–12	12–18	Over 18
Code	2	6	12	18	32

Table 6.5 MTM-2 GET TMUs for Grasp Case and Distance Code

	Distance Code =	2	6	12	18	32
GETA	No grasp	3	6	9	13	17
GETB	Finger grasp	7	10	14	18	23
GETC	Other grasp	14	19	23	27	32

Chapter six: Modeling physical human performance 265

Table 6.6 MTM-2 PUT Prediction Time (in TMUs) for Correction and Distance Codes

	Distance Code =	2	6	12	18	32
PUTA	Smooth motion	3	6	11	15	20
PUTB	One correction	10	15	19	24	30
PUTC	More corrections	21	26	30	36	41

used, and the object weighs 8 kg, then 4 TMUs are added to the predicted time. For PUTs, MTM-2 assigns 1 TMU for every 5 kg weight for objects up to 20 kg. Thus, 4 TMUs can be added due to object weight.

Other actions used in MTM-2 include: REGRASP (R), APPLY PRESSURE (A), CRANK (C), EYE ACTION (E), FOOT ACTION (F), and BEND & RISE (B). The action REGRASP is similar to that defined in MTM-1, but with the assumption that the hand retains control throughout the change in grasp. An action involving a REGRASP is assigned an additional 6 TMUs. Another action from MTM-1 is APPLY PRESSURE. An additional 14 TMUs are added when a hand or any other body member is used to apply force on an object, as might happen for a tight fit. Very little body movement is assumed to occur (less than 0.25 in.). The action CRANK is not part of MTM-1. In MTM-2, it occurs when the hands move an object in a circular path of more than 0.5 revolutions. A revolution of less than 0.5 is a PUT, rather than a CRANK. When the resistance to making a revolution is low, CRANKs are allowed 15 TMUs/revolution. As resistance to cranking increases, 1 additional TMU is allowed for every 5 kg of force required, just as for the weight requirement in PUTs.

An EYE action occurs when the eyes must be moved to an object and dwell there long enough to recognize a distinguishing characteristic. Another instance is when the eyes move across a work area in order to determine what to work on next. Each eye action is assigned a value of 7 TMUs, so long as they are independent of hand, leg, or other body movements. When they are part of the movement, the additional value is not added.

A FOOT action is assigned a value of 9 TMUs. A value of 18 TMUs is added for each step. A foot action occurs when the foot is moved less than 12 in., without moving the trunk of the body. A step involves movement of the truck of the body and is often more than 12 in. in length. The BEND & ARISE action corresponds to changes in body position such as standing up, sitting down, and kneeling. A value of 61 TMUs is given when this movement occurs.

MTM-2 also addresses situations where different cases of GETs and PUTs might be performed simultaneously. A special SIMO chart is provided that denotes which combinations can be performed easily (E), with practice (P), and with considerable difficulty (D; Table 6.7). This chart works similarly to the one used in MTM-1, but considers fewer combinations. In summary, the MTM-2 system is easier to use and takes less time than analysis with MTM-1, but there clearly are trade-offs in precision. Despite these trade-offs, the trend in industry is definitely tilted toward the use of MTM-2 and other even less detailed systems.

Work-factor system

The Work-Factor Company, Inc. introduced the WF system in 1938. While the origins of the WF system predate MTM by about 10 years, one must remember that World War II was going on for over half of these 10 years. The Work-Factor Company conducted applied

Table 6.7 SIMO Chart Showing Combinations Performed Easily (E), with Practice (P), and with Difficulty (D)

CASE	GETA	GETB	GETC	PUTA	PUTB (Outside Vision)	PUTB (Within Vision)	PUTC
GETA	E	E	E	E	P	P	P
GETB	E	E	E	E	P	P	D
GETC	E	E	D	P	D	D	D
PUTA	E	E	E	E	P	P	P
PUTB	P	P	D	P	D	P	D
PUTC	P	D	D	P	D	D	D

research for 4 years before making their system available to the public. They used a wide variety of means to collect their data, including stopwatches, micromotion analysis, and special photoelectric timing devices. Later on, this company was renamed the Science Management Corporation, and WF became a registered trademark. Part of the WF system is described in the following section for a limited set of physical movements. As was the case for MTM, a wide variety of approximate versions are also available.

Structure of the work-factor system

The WF system of synthetic time prediction is similar in many ways to the MTM system, but differs in a few important aspects. One difference is that WF assumes an average experienced operator with good skill and effort. As noted earlier, MTM assumes an acceptable operator; so the time predictions in MTM are about 20% slower than in WF. WF uses a very similar set of activities to MTM, but instead of distinguishing between cases, WFs that affect the difficulty of the movement are identified.

WF measures time in time units (TUs) where one TU equals 0.0001 min. Table 6.8 shows the time units associated with transport acts. Note that transport corresponds to a combination of MTM elements (i.e., reach, grasp, and move). The distance moved is a parameter of this system, just as in the MTM system. As shown in Table 6.1, the time needed for a transport depends on the task WF. Another parameter of the WF system is the weight transported. Separate WFs are assigned for men and women. Note that MTM also has a weight parameter for the Move action, but does not adjust for the gender of the operator. While there are physiological bases for gender differences, as discussed earlier in this book, there is also controversy about gender discrimination. Designers who wish to avoid this controversy must be careful to assure that any adjustment for gender differences is supported by research.

The predicted times in WF are adjusted for the specific body member used in a movement. WF also adjusts for the amount of manual control needed during movements. This feature is a principal distinction between WF and MTM systems. This is evaluated in terms of the following four control elements:

1. Care or precaution should be exercised with a dangerous or fragile object.
2. Directional control or steering to a target location or to maneuver the object to a specific location or to maneuver it through an area with limited clearance.
3. Changing direction significantly to avoid an obstacle, or to bring the hand to the opposite side of a panel or fixture.
4. Stopping at a relatively exact location.

Table 6.8 Example Predictions of Transport Time in WF

Distance Moved (in in.)	Basic	Difficulty Factor 1	Difficulty Factor 2	Difficulty Factor 3	Difficulty Factor 4
1	18	26	34	40	46
2	20	29	37	44	50
3	22	32	41	50	57
4	26	38	48	58	66
5	29	43	55	65	75
6	32	47	60	72	83
7	35	51	65	78	90
8	38	54	70	84	96
9	40	58	74	89	102
10	42	61	78	93	107
11	44	63	81	98	112
12	46	65	85	102	117
13	47	67	88	105	121
14	49	69	90	109	125
15	51	71	92	113	129
16	52	73	94	115	133
17	54	75	96	118	137
18	55	76	98	120	140
19	56	78	100	122	142
20	58	80	102	124	144
—					
22	61	83	106	128	148
24	63	86	109	131	152
26	66	90	113	135	156
28	68	93	116	139	159
30	70	96	119	142	163
—					
35	76	103	128	151	171
40	81	109	135	159	179

Distance moved is in inches; time in TUs.

Note that these control elements are not exclusive. If none of them are present for the task, the basic motion time is appropriate. If one or more elements are present, a factor of difficulty is assigned, equal to the number of controlled elements. As shown in Table 6.8, the assigned transport time values increase with increases in the difficulty factor.

Other elements in the WF system and corresponding tables include the following:

- Grasp
- Assemble
- Disassemble
- Release
- Pre-position
- Use
- Mental process (i.e., read, identify, inspect, or others)

These other elements are assigned types in a manner similar but not identical to the way that MTM assigns movement cases. For example, the grasp element encompasses the following types:

- Simple grasp, involving a single motion with easy-to-grasp objects.
- Manipulative grasp, requiring multiple finger motions to gain control of a single object or orderly stacked objects.
- Complex grasp, using multiple motions and sometimes forearm changes in order to get hold of an object that is jumbled with other similar objects.
- Special grasp, which entails transferring an object from hand to hand.

Different types of releases are also considered. These include contact release, gravity release, and an unwrap release.

Not all of the elements noted here are self-explanatory. The pre-position element involves turning or orienting an object in getting ready for the next element. For instance, consider a carpenter who is nailing a mopboard with finishing nails. If the nails are grabbed from a can, about 50% of the time the finishing nail must be reversed so that the pointed end is in the right direction. These nail reversals are an example of pre-positioning. Assembling elements involves joining two or more objects together. One object is referred to as a receiving object, the other an injecting object. When the tolerances get close, the time and difficulty increase dramatically. Also, the shape of the inserting object relative to the receiving object contributes to that difficulty and time. This effect is largely, but not entirely, related to asymmetry. As in MTM, more time is needed when rotation is necessary.

Assumptions of the work-factor system

The WF system assumes that the standard operator is fully skilled and experienced. Some people would suggest that the term fully skilled should be replaced with the term highly skilled, because it will take people a lot of practice time to reach the performance levels predicted by WF.

Many of the same assumptions mentioned earlier for the MTM system are also made in the WF system. Clearly, the time required to perform the sequence of actions A then B is assumed to be the same as the sequence B then A. The other assumptions discussed previously are likewise relevant.

Work-Factor Example of Walking

WF assumes that walking takes 80 TUs for each pace. If a pace covers 30 in., the rate in miles/h for general walking is calculated as follows:

$$\frac{[60 \text{ min/h}][30 \text{ in./pace}]}{[80 \text{ TUs/pace}][0.0001 \text{ min/TU}][5280 \text{ ft/mile}][12 \text{ in./ft}]} = 3.6 \text{ mile/h}$$

This prediction contrasts with about 3.2 mile/h using the MTM system (as shown in Box 6.1). This difference is quite consistent compared to other differences between the two systems.

Mento-Factor system

The Mento-Factor system of synthetic time measurement is part of the WF system. After the initial system was devised for physical tasks, Joseph Quick saw the need for another system, which pertained to mental tasks. He started development in 1949. Quick is a cofounder of Wofac Company, a Division of the Science Management Corporation, which holds proprietary rights to those synthetic time predicting systems. The Mento-Factor system contains descriptions of about 500 individual mental process time values and is comparable to the detailed WF System in use.

The first step when using this system is to do a task analysis. This involves identifying the mental activities involved for the task along with factors that influence performance. The next is look up the predicted times for the particular activity. The Mento-Factor system has extensive tables containing time prediction for a standard operator. As in the WF system, times are measured in TUs where one $TU = 0.0001$ min.

The Mento-Factor system considers 12 different mental tasks, as well as eye movements and fixations. Although eye motions are essentially physical rather than mental activity tasks, they were included in the Mento-Factor system because so many mental tasks occur right after an eye movement. Simply put, people see something and then respond to it. The activities considered in Mento-Factor are as follows:

- Focus and shift eye movements.
- See or the act of looking at an object.
- Conduct or the sending of nerve pulses to the brain.
- Discriminate or make useful distinctions between objects.
- Span or group objects together as a whole rather than treating each component part sequentially.
- Identify or distinguish an object as a particular kind that differs from other classes of objects.
- Convert, or change from one code to another.
- Memorize, or put into memory.
- Recall, or recollect from memory.
- Compute, or perform an arithmetic operation.
- Sustain, or increase mental effort with added computational requirements.
- Decide, or determine what to do next.
- Transfer, or shift attention from one thing to another.

After the task analysis of the mental actions has been completed, the physical actions are analyzed using a system such as the WF system. The two analyses are then combined to give a complete set of predictions.

Mento-Factor has been used to evaluate jobs with significant mental activity. For example, the R. L. Polk & Company used Mento-Factor to set time standards for many of their jobs several years ago. Polk produced directories for over 7000 U.S. and Canadian cities. Many of the human operator activities there involved reading and typing into a computer various data from source documents. Then the operators printed out the computer data and compared it to the source documents. Later they updated data, made corrections, and checked complications of these corrections. In addition to these proofreading, compiling, and editing activities, Mento-Factor has been advocated for visual inspection, counting operations such as those performed with cycle counters, computer programming, reading and listening to instructions, and a variety of other jobs.

Elementary Example of the Mento-Factor System

Consider the case of an operator who must SHIFT his or her eyes 17.4° and FOCUS on an object (a ratio of 1.20 for far-to-near focus is needed). For this set of parameters, the system predicts an eye-shift time of 8 work-time units and a focus time of about 40 work-time units. Next there would be a SEE task which is about 8 work-time units with these large visual arcs of 18 min but with low visual contrast ratios of 60%. Then a visual CONDUCT IN activity from the eyes to the brain, which is predicted as 1 work-time unit.

At that point, the human operator would need to IDENTIFY which of four different action signals the object represents. It is assumed here that different objects are given a different processing sequence on the machine tool and the operator has been given a rule to the effect that, "If the object is of type A, press the button for process 1." In this case the IDENTIFY task is predicted to require 24 work-time units. As it turns out, the DECIDE task also takes the standard operator 24 work-time units.

The operator must then CONDUCT OUT a nerve impulse to the hand, requiring 2 work-time units. Lastly there is the physical act of pushing the appropriate machine tool button, which is estimated from a physical synthetic time system to be 174 work-time units. So a cycle of this total task is the sum of the individual component mental and eye movement times of 107 work-time units plus the physical 174 work-time units for pushing the buttons or a total of 281 work-time units, or 0.0281 min/cycle (or 169 ms).

Iowa model

The Iowa Model was developed to predict time and accuracy means and standard deviations for the following mental tasks:

- COMPARE: "Is this part the same as that one?"
- LOCATE: "Find the location of the requested item."
- VERIFY: "Determine if minimum specified features exist."
- IDENTIFY: "Name the object in location X."
- CLASSIFY: "Find the frequency of some feature."
- ARITHMETIC: "Perform specified arithmetic operations."

A series of studies were performed during this effort. One question that was closely addressed in the laboratory studies was whether performance times and errors in any of the different types of tasks performed by a person changed because of any of the following:

- The effect of combining tasks
- The effect of performing particular task sequences
- The effect of pacing

The results have interesting implications to practitioners interested in synthetic time prediction using systems such as Mento Factor. The overall results are quite promising for synthetic time estimation and prediction. Less promising results were obtained for predicting performance accuracy. The following points briefly summarize some of these findings.

Performance time values were found to be nearly normally distributed or logarithmic-normally distributed. More ambiguous results were obtained for the accuracy criterion.

Performance time statistics changed in a nearly linear fashion as a function of several variables. The results were more ambiguous with regard to task accuracy. Speed and accuracy both improved on arithmetic tasks involving fewer digits. It took longer to do multiplication than addition, and there were more errors. Speed and accuracy both dropped when the tasks required search of a larger set of symbols.

People who were faster on one type of information-seeking task tended to be faster and more consistent on the other information-seeking tasks. No trend of this type was found for accuracy measures.

People were able to perform these information-seeking tasks while simultaneously performing other tasks, including simple vehicle control tasks or manual tasks. Performance times of the information-seeking tasks did not change, but control errors increased significantly and in a distinctly different manner depending on the type of information-seeking tasks performed.

Performance times and errors were not influenced by increased uncertainty as to which task would come next in the task sequence.

When people had to make responses to different types of tasks on different response devices, performance times were increased by an almost constant amount of time.

Tasks that were initiated in response to auditory signals were faster by a nearly constant value compared to tasks triggered by visual signals.

Time and error statistics did not change substantially between different forms of pacing.

The task sequence normally had a minor impact on performance time or accuracy. However, if a second task used information generated by the first task, task times dropped substantially, but the number of errors also increased.

The latter finding is especially relevant. Most human reliability models assume that errors on successive subtasks are statistically independent (Swain and Guttman, 1983). It follows that if errors are a major concern, it is better not to modularize tasks when there is a single response to component tasks.

Standard data systems

Standard data systems can be used to predict human performance directly or to determine general performance standards within families of similar products. Some examples of product families are the following:

- Grinding rotor blades for jet engines where different blades are distinguished by types and dimensions.
- Bending steel angle iron where the jobs differ by the size of the angle iron and the bending angle.
- Bearing races that differ in diameter, width, and other dimensions.

Similar products involve similar manufacturing tasks, and so their performance times and accuracies will be influenced by some of the same variables. Standard data systems offer several advantages over setting separate performance standards for each product.

One advantage is that the developed formula eliminates some of the variation due to the subjectivity of time study performance leveling. Another advantage is that these data systems provide insight into the job characteristics. Some tasks are particularly well suited to a standard data system, such as custodial work, where floor sweeping performance times are a function of the size of the area swept and the type of floor surface. Similarly, in data entry tasks, the number of errors might be a function of the number of data fields entered into the computer.

Development of a standard data system requires an abundance of prior data. In practice, most companies have available data on previously manufactured products that can be used. An equation fitted to this data yields a formula for estimating performance times or errors made on jobs previously performed. This formula can also be used to predict performance times on new, but similar, tasks.

Linear regression analysis is often used to develop the predictions. The following discussion assumes some familiarity with some or most of the concepts of linear regression.

Formula development

Developing a formula that accurately predicts performance times or errors is both a science and an art. One artistic aspect is deciding on which variables to use or consider including in the model, that is, one must define the function

$$\text{Performance} = f(X_1, X_2, \ldots) \tag{6.1}$$

where $X_1, X_2, \ldots$ are selected variables that differ for each product. Typical variables include dimensions, surface finish, numbers of corners, or particular features. The process of identifying equivalent variables requires some knowledge of the principal manufacturing parameters likely to be good predictors. In order to develop a formula that fits the needs of model users, some general principles should be followed.

Principle 6.1

Minimize the number of variables in the equation.

Usually three or fewer variables are desired for the equation under development. As more variables are added, descriptive power increases, but there is greater risk of redundancy due to correlation. Adding highly correlated variables will not increase the descriptive power of the formula, but it will increase user effort and may make the predictions worse, that is, the regression coefficients may become unstable.

Principle 6.2

Each variable should have an obvious effect on performance time or errors.

People find that formulas are easier to accept when they describe relationships that can be visualized. Obvious relationships help to establish face validity; one must convince users that a formula is reasonable. It is useful to ask people who perform manufacturing operations what factors they think affect performance time or errors. Usually, the first variables to be mentioned are the most obvious. For example, in cutting threads on steel rods, cutting times increase with a longer thread. Other variables, such as harder steel, will affect thread-cutting speed, but not as much as the thread length. On the other hand, harder steel may result in a higher reject rate than the length of the thread. The relevant variables and their importance change with different performance criteria.

Principle 6.3
The variables selected should be distinct from each other.

If the variables are similar, they are usually correlated. This redundancy results in less explanatory power. Each distinct variable should describe a different feature of the relationship. The more completely the relationship is covered, the better the descriptive relationship.

Principle 6.4
Selected variables should have values that can be easily found from drawings and specifications.

Variables that are difficult to measure might be great in theory but cause problems in practice. Because of the pressure of day-to-day operations, values, which are hard to find, will probably be poorly estimated or guessed. Either case will result in performance prediction errors.

Principle 6.5
The resulting formula should be simple but not too simplistic.

Precision in time prediction is situation specific. Using a more complicated formula to gain this precision is acceptable when the situation merits it. Adding unnecessary complexity is pointless. There is no reason to use a second- or third-order polynomial regression model when a linear model fits well. Such higher-order equations should only be used when the added sophistication is worth the complexity. There is nothing wrong with a simple formula, unless it does not meet job needs. It is often useful to develop multiple models, and let the users choose which one to use, depending upon their differing precision needs.

Principle 6.6
Format the variables, so they all have the same directional effects, as the effect occurs on performance.

Most variables can be restated either as a negation, an inverse, or otherwise. For example, if women do the job more quickly than men, the variable gender might be assigned a 1 when a male does the task and –1 for a female. The basic idea behind this principle is to keep the system consistent. Consistent variables make the equation more understandable, and can reduce errors during use due to people misunderstanding the variables.

Principle 6.7
Start simple.

Keep the most important variables in the formula. Add variables that improve the predictive ability of the formula. When the variables are added in order of decreasing importance, less important variables can be dropped from the prediction process, when the added precision is not needed. It is useful to list the average values of the variables, so that unused variables can be converted to an added constant.

Finding the relationship between predictive variables and normal performance

A useful technique when starting out is to plot the data and examine it. Plots typically show performance time as a function of the variable in question. Once the variables are plotted, it is useful to draw a vertical line through the average of the variable plotted on the *x*-axis and a horizontal line through the average of the variable plotted on the *y*-axis. Those two added lines separate the data into four quadrants. Then you only need observe

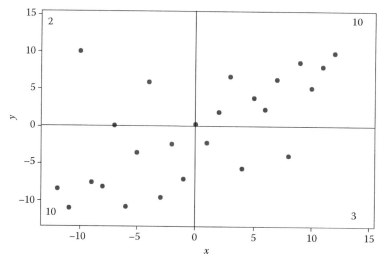

Figure 6.1 Correlated data ($r = 0.6$) showing higher frequencies of observations in the lower left and upper right quadrant.

the number of data points in each separate quadrant of this graph to determine if the variables are correlated. If the number of points in each of the quadrants is quite different, the variables are correlated. Figure 6.1 illustrates such a graph with data plotted in the four quadrants. Note that there are more points in the northeast and southwest quadrants than in the southeast and northwest quadrants. This indicates that the two variables are correlated. A similar graph is shown in Figure 6.2. The two variables in Figure 6.2 are also partitioned at their means into four quadrants. In the case where variables x and y are uncorrelated, one should expect approximately an equal number of plotted data points in each quadrant of the graph. Notice that the mean approximately splits each variable's values in half. Unless there is a relationship between these variables, there should be no correlation between them, and the data should be distributed similarly to that shown in Figure 6.2.

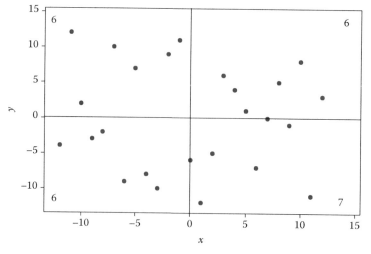

Figure 6.2 Uncorrelated data are nearly uniformly distributed in all four quadrants.

Data plots should also be examined for linearity or curvature. Linear models are simpler, but some cases may be inaccurate. Computer-generated plots of data can facilitate tests for linearity. Alternatively, a rough analysis of linearity is to estimate the standard deviations of each variable. Then draw two partition lines in addition to the previous quadrant divisions, about 2/3 standard deviations above and below (or left and right of) the means shown. These four lines (two vertical and two horizontal) give 16 cells over the plotted data. If the data are linear and positively correlated with the variable, most of the data should be in the diagonal from the lower left to the upper right. In addition, the amount of data in each of those four diagonal cells should be nearly equal. If the data are not linear, some simple transformations may take care of the problem.

Plotted data should describe most of the data variation. This feature is described in regression analysis by the coefficient of determination r. Note that

$$r^2 = \frac{\text{explained variance}}{\text{total variance}} \tag{6.2}$$

Returning to Figure 6.1, since $r = 0.6$, r^2 is about 0.36. This information implies that about 36% of the variance of performance time is reduced by knowledge of this formula. Keep this key point in mind when interpreting these coefficients of determination.

Statistical computer packages for regression

A number of statistical packages are available for computing multivariate regression equations, such as Minitab, SAS, SPSS, and S. These programs compute many statistics beyond those that are shown in this chapter. It is expected that you will be using these packages in some of your professional work, and it is critical that you understand the meanings of these statistics.

Let us illustrate a few points about these statistical packages through some SPSS analyses of an experiment. The subjects of this experiment were performing a tracking task and simultaneously activating switches of different types and locations. Tracking is analogous to an aircraft pilot holding the plane on a straight and level course against air turbulence. In most cases, the pilot is holding the joystick with the right hand and the throttle with the left. To actuate a switch to the left, the subject had to release the throttle, reach to the switch, activate the switch setting as instructed, and then bring the hand back in contact with the throttle or joystick. Actuating switches on the right requires the subject to exchange hands at the joystick while continuing to track. Then the pilot must reach with the right hand to the switch, actuate it, return the right hand to the joystick, exchange hands again at the joystick, and then return the left hand to the throttle.

A computer was employed during the experiment to separately measure the time required to reach, activate the switch, and return the person's hand to the throttle or tracking task. After the data were collected, they were categorized by the hand used and the type of switch. The tracking task was run at two levels of complexity. Variable x_1 corresponded to the tracking level complexity, and x_2 to the duration of the switch activation task. Other variables described the location and nature of the switches that were activated. Variable x_3 was the reach distance to the switch and x_4 was the visual angle between the center of the tracking display and the center of the switch activated. Variable x_5 was the logarithm to the base two (i.e., information content) of the number of switch activation points, and x_6 was the amount of change made in the switch. One of the first results obtained from SPSS for

Table 6.9 An SPSS Printout of the Variables in the Experiment and Their Statistics

Variable	Mean	Standard Deviation	Number of Cases
X_1. Tracking level	1.4914	0.5003	757
X_2. Event duration	26.3752	19.7145	757
X_3. Switch distance	16.3612	5.7568	757
X_4. Visual angle	54.7371	25.5522	757
X_5. Information content	2.5293	1.1705	757
X_6. Change information	1.4015	1.3307	757
X_3SQ. Switch distance squared	301.1136	192.9359	757
X_4SQ. Visual angle squared	3648.2061	2617.5716	757
$X_3 X_4$ Interaction of X_3 and X_4	757.7834	223.3366	757
Y = Log of event duration	1.4212	1.2948	757

this part of the experiment was a tabulation of the variables, their means, standard deviations, and number of cases. A few of those statistics are shown in Table 6.9.

Some additional results that this package provides include simple correlation coefficients between different variables. A few selected values from this table are shown in Table 6.10. Table 6.11 shows statistics for the independent variables as they enter the model during the stepwise regression. Appendix E provides more detail on some of these tests and confidence intervals.

Table 6.12 shows several coefficients of the regression equation with their 95% confidence intervals. Note that when confidence intervals of the regression slopes include the

Table 6.10 Correlations between Pairs of Variables in the Experiment

Variables	X_1	X_3	X_4	X_5	X_6	XSQ_1	XSQ_3	XSQ_4
X_4SQ	+0.016	−0.933	+0.982	+0.22	+0.241	+0.016	−0.941	—
$X_5 X_6$	+0.008	−0.180	+0.126	+0.92	+0.982	+0.008	−0.15	+0.235
Y	−0.039	+0.202	−0.201	−0.029	−0.070	−0.031	+0.20	−0.209

Table 6.11 Selected Variables Entered in the Stepwise Regression

Step	Variable (Entered/Removed)	F to Enter/Remove	Significance	Multiple R	Simple R
1	X_4SQ	34.2	0	0.208	−0.208
3	$X_4 X_6$	7.67	0.006	0.233	−0.146
4	X_3SQ	3.77	10152	0.243	+0.207

Table 6.12 Confidence Interval Check on the Regression Equation

Variable	Coefficient B	Std. Error	Student t	Lower 95%CI	Upper 95%CI
X_4SQ	−0.00194	0.000442	−0.440	−0.0003	−0.0011
$X_4 X_5$	−0.00116	0.00097	4.907	+0.00565	+0.076
X_3SQ	+0.00941	0.001918	4.907	+0.00565	+0.013
X_3	−0.26443	0.0660	−4.406	−0.38226	−0.14659
X_4	+0.02991	0.006388	+4.683	+0.01737	+0.04245

zero value, the effect of the independent variable on the dependent variable cannot be determined with statistical confidence. When both confidence limits are of the same sign, however, the corresponding positive or negative slope can be statistically assured. In such cases, the corresponding F test values are large, confirming that the regression trend is significant.

After developing a regression equation, the residuals should always be plotted to determine whether there are systematic trends in the error values. A residual plot shows each data point and the corresponding estimate of the regression equation. The difference between these two values is the error or residual. If the residuals are not normally distributed, a basic assumption of regression analysis has been violated. In some cases, introducing additional terms into the model can solve this problem.

Computer modeling methods

Several computer graphic models of people are available that can be used in computer-aided design (CAD). These systems were developed to assure that people can fit into available space comfortably, reach all the needed features, and perform all tasks effectively. Boeing engineers made one famous and early development in the United States, called BOEMAN, in 1969. This model represents a person as a series of connected solid geometric shapes (Figure 6.3). Other early U.S. computer graphic models include Rockwell Corporation's BUFORD and the U.S. Air Force's COMBINMAN. Perhaps the earliest computer graphic model of people is SAMMIE, which is an acronym for System for Aiding Man-Machine Interaction Evaluation. Two other early European versions are Germany's FRANKY7 and Hungary's OSCAR.

A wide variety of such models are now used prior to any mockup testing because they are inexpensive, relative to using actual human testing. Modern computer software

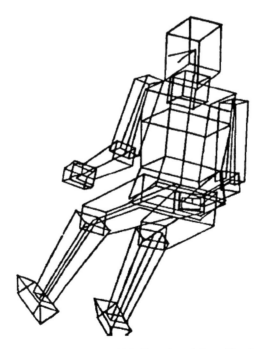

Figure 6.3 Human shape modeling in BOEMAN. (Reprinted from Dooley, M., *IEEE Comput. Graph Appl.*, 2(9), 17, 1982. With permission.)

packages, such as Santos or Jack, allow users to sketch or input a CAD model of the design layout. A solid anatomical model of the human body can then be stepped through some example tasks for the design layout to test its adequacy. These models can be scaled to duplicate the lower 5th percentile woman, the 95th percentile man, or even children. Most models include the capability for the modeler to view the scene from the operator's perspective so that the modeler can see what that person would see. This latter capability is particularly important when modeling tasks that require eye–hand coordination, such as the use of hand tools or vessels that pour liquids.

Jack (and Jill), mentioned in Chapter 5, is an excellent computer graphic modeling system. Jack contains an extensive anthropometric database describing men, women, and children, and provides a wide variety of analyses and outputs. Outputs include three-dimensional images and video simulations, lower back spinal forces, metabolic energy expenditures, strength predictions, reach envelopes, comfort assessments, and visibility reports.

Recently, a team of researchers at the University of Iowa have developed the virtual human model SANTOS™ (Abdel-Malek et al., 2004; Yang et al., 2005). The goal of the team was to develop digital humans that see, move, touch, and grasp like real humans, much like Jack. SANTOS is an anatomically correct model of the human, quite realistic in appearance. The SANTOS model makes movements in its virtual environment while performing tasks, such as lifting, carrying, climbing, walking, entering or leaving vehicles, and operating equipment. Movement patterns of SANTOS are determined using a variety of optimization algorithms and techniques that minimize performance measures such as discomfort and satisfy constraints such as the need for SANTOS to focus his eyes on a viewed object. The SANTOS team has made a number of interesting innovations, including the development of user-friendly methods for both specifying anthropometric percentiles for a given analysis and stepping SANTOS through a sequence of tasks and subtasks. Some other important features of SANTOS include dynamic modeling of balance, instability, and collision avoidance; realistic modeling of shape and posture; a sophisticated model of the human hand; and the modeling of the effect of clothing on shape, posture, and movement.

The potential advantages of SANTOS over some of the earlier modeling methods have attracted a great degree of interest from a wide variety of groups, including manufacturers such as Honda and Caterpillar, not to mention the U.S. Army. One of the main advantages of models such as Jack or SANTOS over mannequins or drafting templates is that they can provide quite realistic descriptions of human movement and posture.

Part of the issue is that many joints in the human body are quite complex. Consequently, the simple articulation concepts used in the drafting mannequins are at best only partially approximate. Computer models such as SANTOS or Jack include many more body segments than drafting mannequins. For example, the Jack digital people have 69 segments, including a 17-segment spine and 16-segment hands, which obey joint and strength limits derived from NASA studies. This arrangement results in nearly realistic postures for the modeled tasks.

Another feature of digital human models, such as SANTOS, is that they can describe different body shapes or somatypes. This latter feature is a very significant advance over the use of drafting templates or tables.

SANTOS also contains algorithms that control the various joints and body segments during reach movements. Simple algorithms of this type result in movement similar to a one-arm robot reaching toward an object. While the hand moves toward an object, trajectories are computed for both the wrist and elbow. These models are useful in workplace applications because it is important that person–machine interfaces do not constrain reaches. Some systems also contain clash routines that detect when two solid objects try to occupy the same space at the same time.

One of the more useful features of this technology is its ability to easily model workplaces and other features of the operating environment using a common computer language and database. As continued progress is made in digital human modeling technology, one should expect even greater integration within simulations of workspaces. This approach offers an opportunity to eliminate some of the more time-consuming steps of ergonomic design—that is, there will be less need to develop mockups and run experiments with human subjects, if digital modeling in virtual environments leads to good results.

Speed–accuracy trade-offs

In most settings, and particularly in industrial assembly situations, both speed and accuracy are important. Greater speeds reduce required performance times and lower the cost per item produced. On the other hand, when accuracy is not maintained, tasks must be repeated (taking more time), and poor product quality may be the result.

Pew invented the speed–accuracy operating characteristic (SAOC) curve to describe how people trade off speed and accuracy. The SAOC shows performance time on one axis and performance accuracy on the other, as illustrated in Figure 6.4.

Ideally, people will maintain adequate performance on both criteria, but in some situations they might choose to increase their speed, at the cost of some accuracy. In other situations, accuracy might become more important than speed. In some cases, this trade-off is linear, that is, a straight line as shown in Figure 6.4 describes the trade-off between speed and accuracy. The slope of the trade off line determines the bias of the person toward speed versus accuracy. A large slope in this example means that the person is willing to

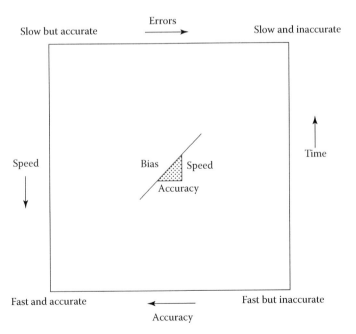

Figure 6.4 The speed–accuracy operating characteristic (SAOC). (Adapted from Pew, R.W., *Acta Psychol.*, 30, 16, 1969.)

slow down a lot to be a little more accurate. More complex, nonlinear, trade-off functions obviously are possible in realistic settings.

Fitts' law

About a century ago, Woodworth (1899) researched positioning accuracy for tasks such as aiming a rifle. He showed that people often make an initial moderately rapid movement followed by several smaller readjustments. Figure 6.5 shows some data for hand movements toward a target that illustrates this effect. Woodworth showed that movement time was related to the ratio of the movement distance and positioning tolerance.

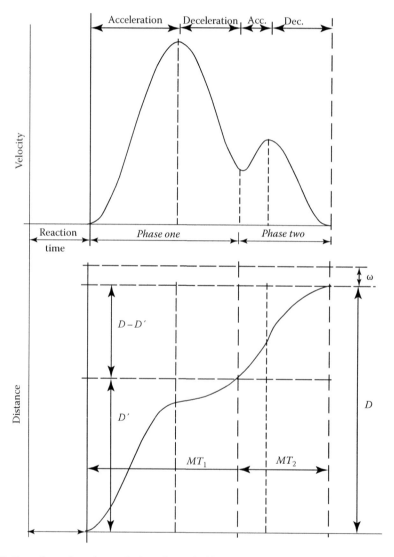

Figure 6.5 Path and accelerations of a hand in a ballistic start and a controlled completion as suggested by Woodworth (1899). (From Woodworth, R., *Psychol. Rev.*, 3(13), 1, 1899.)

Brown and Slater-Hammel (1949) later verified this relationship for discrete hand movements and defined it in terms of a speed–accuracy trade-off. Faster speeds require larger tolerances or shorter movement distances.

A few years later, Fitts (1954) investigated the relationship between the distance D to the target, target width W, and movement times MT. Fitts developed an index of difficulty I_d, based on information theory shown as follows:

$$I_d = \log_2 \left[\frac{2D}{W} \right] \qquad (6.3)$$

Figure 6.6 shows his initial experimental arrangement in which subjects moved a pointer from one target to the opposite target. The target center-to-center distance was recorded as D units, and each target was W units wide.

When Fitts manipulated D and W, and measured the resulting MT, he discovered the following relationship:

$$MT = a + b \log_2 [I_d] \qquad (6.4)$$

where
  a and b are empirically determined constants
  I_d is as defined in Equation 6.3

This relationship is now known as Fitts' Law. Although this relationship is not exactly a law in the scientific sense of the word, it has been verified in a vast array of circumstances. So it is at least Fitts' theory. Parameter a in Fitts' Law corresponds to an initial reaction time, whereas the primary movement and secondary adjustment times are represented by parameter b and the factor that Fitts referred to as the index of difficulty. His use of the logarithm to the base 2 was due to his analogy with information theory, where he viewed the movement distance D as analogous to the signal plus noise $(S+N)$ in Shannon's theorem and the tolerance $W/2$ as analogous to noise (N) in this theorem. This result showed that Fitts' index of difficulty was akin to a signal-to-noise ratio. Crossman and Goodeve (1983) verified Fitts' tapping test results and derived the same model from iterative corrections of movement rather than information theory.

Fitts and Peterson (1964) extended these investigations using the setup shown in Figure 6.7. This work verified Fitts' Law for single movements in randomly selected

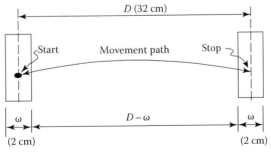

Figure 6.6 Layouts of Fitts' (1954) tapping experiment.

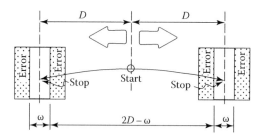

Figure 6.7 Fitts' and Peterson's (1964) experimental setup.

directions. They also found that the elapsed time before movement commenced and the movement time itself were relatively independent of each other. Langolf et al. (1976) further verified that the law held for manipulating parts under a microscope, Card et al. (1978) verified it for key-reaching in keyboard tasks, and Drury (1975) used Fitts' Law in the design of foot pedals. These studies all speak to the robustness of Fitts' Law.

However, a number of researchers have challenged the model used to explain the relationship in Fitts' Law. Even Woodworth implied that the first phase of movement is ballistic in character, while the second phase is controlled movement. If that is true, distance would primarily affect the ballistic part, and the residual correction would likely be affected by the tolerance width (W). Hence, movement times and velocities would resemble those in Figure 6.5, and D and W would be expected to have separate effects on movement time. Along these lines, a more appropriate model might be as follows:

$$MT = a + bf_1(D) + cf_2\left[\frac{1}{W}\right] \tag{6.5}$$

where
    a, b, and c are constants
    f_1 and f_2 are functions

Note that when $b = c$ and the functions are logarithmic, this equation is equivalent to Fitts' Law. The Zellers' model (Zellers and Buck, 1997) is based on an initial ballistic movement followed by a series of controlled movements for the remaining distance to the target. Consequently, f_1 is linear and f_2 is logarithmic or

$$MT = a + b(D) + c\log_2\left[\frac{1}{W}\right] \tag{6.6}$$

where a, b, and c are constants. Another model was developed by Kvalseth (1981) to describe a series of submovements in a tapping task, similar to that used in the earlier Fitts' study. Kvalseth's model is

$$MT = aD^b W^c \tag{6.7}$$

in which a, b, and c are constants. If this model is put into logarithmic form, it becomes similar to Equation 6.6. Note that Kvalseth's model is similar to Fitts' when parameter b is positive and parameter c is negative. Another model based on a series of ballistic movements was developed by Meyer et al. (1988). This model is

$$MT = a + b\sqrt{\frac{D}{W}} \tag{6.8}$$

Two other models describe the time needed to capture a target moving at the velocity V. The first was developed by Jagacinski et al. (1980). This model is

$$MT = a + bD + c\left[\frac{V+1}{W-1}\right] \tag{6.9}$$

Another model was developed by Gann and Hoffman (1988), which, like the previous, was based on a first-order continuous control model. This model is

$$MT = \frac{1}{c}\ln\left[\frac{A+V/c}{W/2+V+c}\right] \tag{6.10}$$

where
A is the initial movement distance
c is a constant

When A equals D, V is zero, and the target is static, the latter model becomes proportional to Fitts' model. As might be expected, these models often make very similar predictions.

Eye–hand coordination

People often reach out to a location outside of their normal vision or to an unfamiliar or infrequently used location. When they do this, people need to use vision to help guide the hand to the appropriate location. Also, the action at the end of the reach often cannot be performed without visual guidance. An important example is when people activate switches or change their setting. To activate or change a switch, one must bring the hand to the switch, change the state of the switch, and then move the hand back. The three components of switch actuation are frequently referred to as follows:

- Reach
- Activation
- Return

Aircraft pilots must perform such tasks many times during flight while they are simultaneously holding the aircraft on a straight course using the joystick. Some data on left and right switch activation during a simulated flying task are given in Table 6.13.

In this study, the pilot's right hand was normally on the joystick, so any switching activity using the right hand required the pilot to grasp the joystick with the left hand while the right hand did the switching. The reach started when the left hand moved, and the return ended when the left hand was back at the throttle on the left side. Switch activation statistics were measured separately for the two hands for the reasons shown earlier.

Table 6.13 Left- and Right-Hand Switch Actuations during Simulated Level Flying with Turbulence

Hand	Statistic	Reach	Activate	Return	Sum	Cycle	Correlation with Activate
Right	Mean	0.733	1.130	0.613	2.476	2.476	−0.258 with reach
	Std Dev	0.548	0.942	0.423	1.169*	1.220	−0.268 with return
	Variance	0.300	0.888	0.179	1.367	1.488	
Left	Mean	1.445	0.848	1.438	3.731	3.731	−0.349 with reach
	Std Dev	0.498	0.600	0.569	0.965*	1.143	−0.340 with return
	Variance	0.248	0.360	0.324	0.932	1.306	

Right-hand switch activations required changing hands, yielding longer performance times as the statistics in Table 6.13 verify. The table also shows correlations within the same switch-activation cycle. A negative correlation indicates that shorter reach times in a cycle were followed by longer activation times, and longer activation times were followed by shorter return times. In other words, there were compensating changes in performance between these three switch-actuation components. If the component times were statistically independent, the cycle standard deviation of time would equal the square root of the variance sum (the values denoted with an asterisk [*] in Table 6.13). Actual cycle standard deviations were 4% and 18% greater than the latter time statistics.

Performance times depended on the location of the switch, as one would expect. The fact that they increased with reach distances was no surprise to anyone. Performance time also increased for larger visual angles with respect to the center of the windshield. Television recordings revealed that large visual angles required a lot of eye–hand coordination, and that increased the time to reach out to the display. The number of settings on the switch affected the time of activation, as expected, but it also affected the hand returns. Switch activation times were distributed log-normally. These three variables discussed previously affected the time variance as well as the means. Vision was often used while the hand was in motion to direct the hand to the correct switch. It was used again when setting the switches and checking their settings. Activating switches took vision away from the flying task, causing performance to drop during the activation cycles. This effect increased for switches with more settings.

Final remarks

Synthetic time systems provide a basis for predicting human performance times without making direct observations. Moreover, those predictions do not require subjective rating. However, this form of prediction is a source of controversy in the minds of some people due to many embedded assumptions that are discussed previously. As mentioned, often taking actual data, as described in Chapter 5, can provide more accurate results when taking such data is feasible.

Much of the focus of the beginning of this chapter was on MTM and WF. MTM and WF tend to provide similar predictions. However, the standard operator defined by each of these systems differs by a small extent so that slightly longer time predictions tend to occur with MTM systems.

Both of these fundamental systems have multiple versions that entail less predictive precision but allow easier and faster computation. A major difference between MTM and

WF systems is that the former system has some capability to determine when concurrent tasking can be performed easily, with practice, or with great difficulty.

Among the numerous assumptions that underlie these synthetic prediction systems, some are questionable and unsupported by research. In some cases, performance differences with and without these assumptions are small. It should therefore be recognized that synthetic time prediction is only approximate, although those systems are sufficiently accurate and precise for many applications. Additionally, these assumptions may be reasonably adequate for some kinds of tasks but not for others.

So far, these synthetic time systems are restricted to predicting times for a mythical standard operator. Time variations over the population are not predicted by the methods used by industry today. Despite this limitation, the principal point to make with regards to physical tasks is that these synthetic systems have proved to be effective in many situations.

Some newer attempts to expand synthetic prediction systems to cover simple cognitive tasks and to predict other criteria than time are also covered in this chapter. Although a number of difficulties face these models, they have yielded some successes. In the end, as mentioned in Chapter 1, a model is a good model if it is useful.

One of the difficulties in these models is that a significant development effort may be required. Another difficulty may be the precision of prediction. However, as a number of people have stated, as a tool for use in the early design stages, the designer needs a technique to identify better design alternatives from worse ones. Accuracy and consistency are important in that context, but precision is not.

Standard data systems offer tremendous benefits when they are appropriate to the situation, and many of those situations have been identified here. Moreover, the procedures for examining data associated with possible uses of these standard data systems are illustrated. Standard data systems can also be used effectively with indirect work. In spite of the fact that all of the examples here related to performance times, these techniques are just as appropriate for using error data to predict accuracy, or time/error variance to describe consistency, and many of the other criteria used in ergonomics.

Discussion questions and exercises

6.1 Which of the two synthetic time measurement systems do you find to be (1) easier to apply, (2) more reliable in prediction, or (3) providing greater face validity? Why?

6.2 Describe the concept of synthetic time measurement in a manner suitable for a worker in one of the shops or a write-up in the Sunday newspaper.

6.3 Criticize the bases of MTM and WF systems of time prediction and give some everyday examples of why the predictions of these systems may be wrong. Defend against those criticisms.

6.4 Using MTM-1, what is
   (a) The time required for the hand to move 10 in. to an exact location carrying a 10 lb weight?
   (b) The time required in minutes if the left hand is reaching 12 in. to an object jumbled with other objects and the right hand is positioning an easy-to-handle object that is symmetrical to a close positioning with light pressure?

6.5 Determine the number of hours required for 1000 cycles of the task shown as follows using the MTM-1 time system without regard to learning or allowances.

Left Hand	Right Hand
R16A	R10C
G1A	G4B
M20C	M6A
PSS1 easy	PS3 easy
RL1	

6.6 During the time study of a task, a portion of the task was timed separately and found to average 0.070 min. This separately timed activity consisted of a 22 in. reach (by a previously stationary hand) to an object in a fixed location, a grasp of a 3/4 in. diameter rod (weight 1/2 lb) with some interference. These actions were followed by a 20 in. move to an approximate location, an exact positioning with heavy pressure required into a hole which could take the rod in either of two orientations, and a normal release. All of these activities could be performed by the preferred hand. The entire assembly task, of which the separately timed elements were only part, was timed to be 1.25 min. What is the normal time for this assembly task based on MTM synthetic rating system?

6.7 A task is composed of the following elements. Estimate the time required to accomplish this task. (Express your answer in TMUs, seconds, and minutes).
  1. Reach for fastener 10 in. away (considered case D)
  2. Grasp (pick up) the fastener (considered case 1B)
  3. Move the object 20 in. (considered case C)
  4. Position the fastener onto a fitting (considered case 2SS, difficult to handle)
  5. Turn the fastener 180° (weight is less than 1 lb)
  6. Repeat step 5 ten times
  7. Release the fastener (considered case 1)

6.8 Identify which of the following assumptions about human behavior is involved in the two synthetic time systems shown in this chapter and explain why it is or is not an underlying assumption.
  (a) The average performance time to perform elemental task A is unaffected by whether task A precedes or follows elemental task B.
  (b) The variance of performance time by a person is constant for each type of elemental task.
  (c) Between-person performance time variances are constant for each type of elemental task.
  (d) Performance time averages for elemental tasks are not affected by the number or relative frequency of different kinds of elemental tasks within a job.
  (e) Variables that affect performance time averages are the same for all people but the magnitude of the effect varies between people.
  (f) Sensory-perceptual requirements of elemental tasks do not affect average performance times.
  (g) Differences in the average performance times between people performing a collection of elemental tasks are inversely proportional to the speed over each task.
  (h) Elemental tasks that are performed concurrently do not affect the average performance time regardless of the person's bias toward those concurrent tasks.

6.9 Consider how you might construct a synthetic performance accuracy predicting system and describe how you might go about designing it. Denote some experiments that might need to be made in order to finalize the design.

Chapter six: Modeling physical human performance

6.10 Suppose that you were doing a direct time study of a person and one of the elements of this direct time study consisted of five repetitions of the following MTM elements: R20A, G1B, M10A (the objects were less than 2 lb), and RL1. What is the rating (leveling) factor for this subject based on MTM-1?

6.11 Describe how you would incorporate time variations within-persons and time variations between-persons into a synthetic time system.

6.12 In selecting variables for a standard data system of a particular family of jobs, what are some of the desired properties of the selected variables?

6.13 What is (are) the principal advantage(s) of a standard data system with respect to time studies, jobs not previously studied, and indirect labor?

6.14 Which of the following statements regarding the use of independent variables in a standard data system are true?
   (a) Each independent variable should be as highly (positively) correlated as possible with every other independent variable.
   (b) Each independent variable should be as negatively correlated as possible with each other independent variable.
   (c) Each independent variable should be as uncorrelated as possible (nearly zero) with each other independent variable.
   (d) Each independent variable should be as highly (positively) correlated as possible with the dependent variable.
   (e) Each independent variable should be as negatively correlated as possible with the dependent variable.
   (f) Each independent variable should be as uncorrelated as possible (nearly zero correlation) with the dependent variable.
   (g) The independent variable should make a clear and obvious relationship with the dependent variable.
   (h) The independent variable relationship with the dependent variable is inconsequential.
   (i) The regression coefficient of determination should be as close to unity as possible.
   (j) The regression coefficient of determination should be as close to –1 as possible.
   (k) The regression coefficient of determination should be as close to zero as possible.
   (l) The functional relationship between the collection of independent variables and the dependent variables should be as simple as possible to meet the needs of prediction precision.

6.15 A multiple linear regression analysis of y on x_1 and x_2 was found to be $y = 0.2668 + 0.7333\, x_1 + 1.0333\, x_2$. The variability in y is reduced by knowledge of the values of x_1 and x_2. Calculate this reduction, given that the following were calculated:

$$\sum x_1 y = 29,\ \sum x_2 y = 92,\ \sum x_1^2 = 14,\ \sum x_2^2 = 116$$

$$\sum y^2 = 200,\ \text{and}\ \sum x_1 x_2 = 38$$

6.16 The task of activating a wide variety of switches of various types and locations relative to the human operator was examined. Numerous factors of the switch type and location were found to affect the activation time required by a typical operator and a regression equation was obtained to predict the required time. Also the actual time values were found to have a moment-product correlation of 0.8 with the regression predictions. If the standard deviation of time by all subjects on all switch types and locations is 300 time-measurement-units (TMUs), then what is the expected standard deviation of performance time given that the factors of the regression equation are known?

6.17 A multiple linear regression analysis is used to determine the normal time of a family of jobs as a function of several independent variables. Answer the following:
(a) What does the r^2 coefficient of determination mean?
(b) How does the confidence interval for the predicted normal times change when an independent variable is varied from its minimum to maximum value? Is the confidence interval at its minimum, maximum, or average width when all the independent variables in the equation are set to their average values.

6.18 What did Schmidtke and Stier do that caused such a controversy in the field of predetermined time systems? Why didn't the whole notion of predetermined time predicting systems collapse as a result? Discuss the implications of the Iowa studies to this issue.

6.19 The following data show the results of a mouse-clicking task, where the person had to move the mouse a lateral distance D and click within a box of width W.
(a) Given these data, plot the information content of the task (in bits, according to the Fitts' law formulation) against the response time (RT) of the participant.
(b) Conduct a simple linear regression on the data to identify the regression equation and R^2 for the regression.
(c) Is any linear relationship suggested by the plot?

RT	D	W
0.66	−85.5	78.75
0.84	163.5	39
0.91	245.25	38.25
0.87	−168.75	23.25
0.66	−152.25	63
0.69	−218.25	79.5
0.56	−206.25	51.75
0.78	56.25	74.25
0.77	−145.5	36
0.71	264.75	66
1.03	−261.75	20.25
0.68	−372.75	70.5
0.80	−308.25	75.75
0.86	217.5	72
1.21	56.25	12
0.83	−204.75	41.25
0.68	−65.25	69.75
0.69	−325.5	74.25
0.80	−70.5	30
0.45	−74.25	78
0.50	−98.25	69
0.76	−358.5	69
0.80	−331.5	57.75
1.21	253.5	17.25

(*continued*)

RT	D	W
0.75	−176.25	55.5
0.83	−320.25	55.5
0.89	139.5	23.25
0.70	14.25	78
0.57	−57.75	59.25
0.88	232.5	63.75
1.18	−292.5	17.25
0.67	−24.75	39
0.87	−134.25	13.5
0.80	−370.5	68.25
0.87	148.5	33
0.86	230.25	28.5
0.58	−47.25	36.75
0.73	−195.75	36
0.86	153.75	78.75
0.69	66.75	37.5
0.78	256.5	36
0.83	4.5	24
1.11	−79.5	53.25
0.91	40.5	27.75

chapter seven

Sampling methods in industrial ergonomics

About the chapter

This chapter describes sampling methods used by ergonomists and engineers to estimate how often certain activities or system states occur. One objective of this chapter is to provide enough detail for the student to know when to use sampling methods and when not to. A second objective is to give the student the ability to use these methods and to suggest some typical applications. The chapter includes a brief review of the statistical basis of sampling, before discussing more advanced topics such as stratified sampling strategies and sequential Bayesian methods. An understanding of the statistical basis of sampling is required to determine the necessary sample size and to avoid situations that violate the underlying assumptions of the procedures followed. In some cases, it becomes necessary to increase the sample size, or to collect additional data to confirm the preliminary results. Stratified sampling strategies and sequential Bayesian sampling methods can greatly reduce the number of observations, making them very efficient, compared to traditional statistical methods.

Introduction

A number of sampling methods are used in industrial ergonomics to acquire information. The basic approach is very simple. It involves recording what is going on in a system at different times over some longer observation interval, and then using the observations to draw inferences about the system. The analyst classifies each observation as corresponding to a particular activity (such as setup, assembly, or maintenance tasks) or a system state (such as idle, down, operational, safe, unsafe, wearing or not wearing eye protection, etc.). The proportion of observations in each category is then used in many different ways. For example, the amount of time spent by a worker on a particular activity can be calculated by multiplying the observed proportion of time spent on the activity by the total time worked.

Sampling has both advantages and disadvantages over continuous observation. Sampling methods typically require more time to get data with statistical significance. But sampling is usually a much less expensive form of data collection than continuous observation. Sampling methods also distribute the observations over a longer interval, which balances out the effect of many uncontrolled factors. For example, observations might be taken at random times over several work days to estimate how much time the workers spend on a particular task. A time study, on the other hand, might involve an hour of data collection for a single operator in the morning on one particular day. If performance in the morning was different than at other times, the latter approach would contain this effect as a hidden factor. The ability to control for such effects gives sampling methods greater face validity and helps ergonomic designers sell their proposals to management and labor.

The key question is whether the activity or system state can be easily and accurately identified from each observation.* If so, sampling methods become more feasible.

Activity sampling

Activity sampling is a very commonly used technique in ergonomics. Other names for essentially the same procedure include *occurrence sampling*,[†] *activity analysis*,[‡] and *work sampling*.[§] Regardless of the name used, the technique consists of making observations of the system to determine whether a specific activity occurred. Usually these observations are made at randomly selected times. When enough observations have been taken, statistical conclusions can be made about the fraction of time the specified activity occurs. The relative frequency of specific activities is critical design information. Activities that occur more often require the greatest refinement. Often a small improvement on an activity that occurs frequently will net greater gains than a great deal of improvement on one that seldom occurs. One of the more common applications of activity sampling is to estimate the fraction of time a process is *unavoidably delayed*.[¶] Some other typical measurements include machine interference delays, idle time, down time, unproductive activity, and unsafe conditions (such as failure to use safety equipment, unsafe lifting methods, unsafe postures,** etc.).

These and other estimates obtained in activity sampling are often used when setting performance standards for current operations. Sampling can be employed in many cases at considerable savings in cost compared to time study and other methods of assessing performance that were discussed earlier in Chapter 5. Another important application is in crew and maintenance operations. Sampling can help determine potential problems in scheduling crews such as excessive waiting times and crew interference. Many workers prefer the use of sampling because it avoids performance rating and other sometimes controversial elements of traditional time and study.

Besides these industrial applications, activity sampling has been used in hospitals and in numerous other institutions to measure very similar quantities. Activity sampling serves other ergonomic design needs as well. Designers cannot simply shove their designs onto people, merely hoping they will work properly. The use of new or modified designs can be sampled for a period of time to determine if the designs are working as intended. In some cases, the data will reveal that the design needs some added tweaking or more radical correction. Thus, sampling can be a guide to design management.

---

* For example, it is obviously easy to see whether a worker is wearing their helmet or what tool they are using. However, it might be hard to tell whether the worker is waiting for the machine to finish a processing step (an unavoidable delay time) or simply relaxing (idle time). Along the same lines, a traffic engineer sampling drivers stopped at traffic lights might find it easy to observe the use of shoulder belts, but much harder to tell if the driver is composing a text message on their cellphone.

† Konz (1990) in his book *Work Design: Industrial Ergonomics* uses the phrase *occurrence sampling*, which is more accurate than work sampling but less conventional.

‡ Chapanis, A. (1959) refers to some of the traditional work sampling methods as *activity analysis*. Many ergonomis specialists with psychological backgrounds continue to use the latter phrase.

§ Tippett (1935) is credited as being the inventor of Binomial and Poisson work sampling methods. Tippett worked in the textile industry in the midlands of England. He stated that the idea came to him when he was sitting and listening to the looms operate and felt that he could tell in an instant by the sound when something was out of kilter.

¶ An unavoidable delay is the amount of time the activity is delayed through no fault of the operator, and is one of the traditional allowances used when setting time standards.

** See Mahone (1993) for an example involving cumulative trauma disorders in jobs.

Running a sampling study involves several steps. Assuming that the analyst has identified the operation to be analyzed, and is ready to begin the study, the following steps need to be followed:

1. Deciding upon the sampling procedure
2. Determining the necessary number of observations
3. Deciding the time interval the study will be conducted over
4. Determining when each observation will be taken
5. Recording and ultimately analyzing the data

These steps are briefly discussed in the following sections.

Sampling procedures

A variety of sampling procedures are commonly used in industrial applications. One distinction between procedures is in how the interval between observations is determined. Most of the time, samples are taken at random intervals. Another approach is to periodically sample at fixed (or constant) intervals (i.e., every 15 min). In stratified sampling methods, the sampling rate is changed over time or when certain conditions are met to make the study more efficient. Sampling methods also differ in how the observations are made. Manual observation involves a human observer who records the data of interest at the appropriate time. In some cases, the operators themselves do the recording. Cameras, computer software, and other automated observation techniques are also used. Some of the advantages and disadvantages of these methods are discussed in Table 7.1.

The principal reason for taking samples at random times is that doing so can improve the accuracy of estimates in certain cases. Operators are more likely to be following their normal behavior at the time of the observation, simply because the observation times are less predictable. Also, random observation times will not be correlated with performance cycles, and systematic differences in performance over time will be averaged out. On the other hand, random data collection times may create problems for the analyst. The locations being observed are often located at sizable distances from each other and the observer's office or base location. It takes time to get from one place to another and make an observation. In some cases, the randomly assigned intervals are too short for the analyst to get from one location to another. Another practical issue is that it may be difficult to remember to take the observations. Quite often, the analyst will have other responsibilities, besides conducting the work sampling study. Random reminder devices are available that can be carried in a pocket. These devices buzz or vibrate at the time a random observation should be taken. Some devices are also equipped to record the ongoing activity.

When events take place very quickly or seldom occur, random sampling may be very inefficient. Also, in some cases, more precise estimates are needed for certain more critical activities. Stratified sampling procedures are useful in such situations. Samples can be stratified in many different ways, as discussed later in this chapter.

A related issue is that in some situations the analyst must remain on location when making the observations. The analyst may have nothing to do during the time between observations. In such cases, continuous observation or periodic observations at high frequencies make better use of the analyst's time. Periodic sampling procedures also distribute the observations more evenly, which can reduce the number of observations needed to reach an acceptable sampling error for all the activities. However, if the system has cycles or events that occur at some natural frequency, periodic sampling may lead to very biased

Table 7.1 Advantages and Disadvantages of Sampling Procedures

	Advantages	Disadvantages
Random interval sampling	Less predictable to open operator when the observation will be made The effect of cycles and time variant behavior is balanced out	Very disruptive to the analyst's schedule Observations unevenly distributed More observations may be necessary
Periodic sampling	More convenient and less disruptive to the analyst Observations more evenly distributed over time	Predictable by operator (may result in changed behavior) Potential harmonic distortions or bias due to cycles and time variant behavior
Stratified sampling	Focused observations much more efficient, especially to measure short-duration events Provides additional information	More complex stratification may be biased
Manual observation	People are better at identifying what is happening	Presence of a human observer may change the subject's behavior Potentially expensive, time consuming
Self observation	Operator knows what they are doing Eliminates need for the analyst to make the observations Eliminates the need to track operators between locations	Potential for manipulation
Automated observation	Quick, cheap Larger data collection effort becomes feasible	Difficulty identifying event or activity May be disruptive

results. This is especially of concern for machine-paced tasks. In the most extreme case, periodic sampling in phase with the natural frequency will lead to only one activity ever being observed, making the results completely meaningless.

In many situations it is preferable to let the operators doing the work be their own observers. Usually this approach to activity analysis is used in professional settings such as hospitals, universities, or in technical industries. This procedure eliminates the need for the analyst to travel to the activity site, and the operator may also feel less intruded upon. On the other hand, there is definitely some potential for manipulation. For example, some people might understate the proportion of time they spend on break. The primary advantage of this approach is that the person performing the activity can identify activities that cannot be easily observed. This is especially important when people are performing cognitive tasks. Also, these studies do not have to be restricted to recording activities, and one can request that the location of the activity and that of the activity be identified (e.g., consulting with a colleague can be performed over the phone, in a conference, or face to face in an office). The pocket-carried reminder/recording devices mentioned earlier are especially useful in such settings. Some of these devices allow items from a list of expected activities to be selected by pressing a button. As an added feature, people in the study can rate their own productivity on the activity. They can also be asked to identify the ways of boosting productivity and surmounting obstacles to better productivity.*

---

* Lovetinsky (1998) describes a study of this type with engineers and technicians working in a company that makes aviation electronics.

Cameras, computer software, and other automated observation techniques are also used to collect sampling data. As for manual forms of the observations, the observations must be accurate and obtained without bothering the people operating the system. If these objectives can be achieved economically, continuous observation is a potential option. Huge amounts of data can be obtained very quickly with automated observation techniques. The major problem is that interpreting the data can be very difficult for the analyst. In some cases, it takes more time to interpret the data than would have been required to measure it manually. It is surprisingly difficult to tell what the operator was doing from photographs and even video. Trial runs must be done to ensure that the data can be interpreted properly in a reasonable time.

Sampling theory and sample size

Statistical sampling theory is built on the assumption that the characteristics of a larger population can be captured to varying degrees of precision by taking samples of varying size. The Binomial and Poisson sampling models are both used in activity sampling studies. The binomial model describes the number of events observed after taking a certain number of observations. The Poisson model also describes the number of events observed, but over a given time period instead of for a given number of observations. The two techniques are used in very similar ways (see Buck et al. 1996 for details). The binomial model is more commonly used in activity sampling studies, and will be focused upon here.

Binomial sampling model

The binomial sampling model assumes a situation analogous to selecting marbles of different colors one at a time from a bowl. Each marble in the bowl is assumed to have an equal chance of being selected. An observation involves selecting a marble and noting its color. The selected marble is placed back into the bowl after each observation. The bowl is then thoroughly shaken before the next sample is taken. Consequently, future samples are in no way affected by the previous samples. If red marbles make up the fraction q in the bowl, then the probability of selecting a red marble on each sample is q in every case. The probability of selecting a non-red marble will always be $1-q$.

When the sampling situation corresponds to sampling marbles from the urn, and a particular color constitutes an event, the probability of obtaining X events from a sample of n observations is given by the binomial probability function or using the following equation:

$$p(X) = \frac{n!}{X!(n-X)!} q^{X} (1-q)^{n-X} \tag{7.1}$$

where
  The values of $X = 1, 2, 3, \ldots$ up to a maximum of n
  q is the fraction $(0 \leq q \leq 1)$ of marbles corresponding to that event

In this case, selecting a marble of any other colors is considered a non-event. The mean and the variance in binomial sampling is as follows:

$$E(X) = nq \tag{7.2}$$

$$V(X) = nq(1-q) \tag{7.3}$$

It should also be noted that X in Equation 7.1 is an integer, but the mean and variance of X (i.e., first and second central moments of X) need not be integers.

Industrial applications are often analogous to the previous marble sampling situation, but there are some important differences. People who are designing and operating the sampling study should be aware of those differences in order to make the study conform as closely to the theory as practical. One obvious point is that the person making the observations must be able to correctly identify the events. This requirement is no different than assuring that the person who observes the color of the marble is not color blind. Clear, concise, and accurate observations of each and every object being observed are essential, as the theory is based on unequivocal and unique observations.

The binomial model also assumes that the fraction q is constant over the entire period of observation. This assumption is violated if taking an observation affects the behavior of the things being observed. Marbles clearly cannot change their behavior when an observation is about to be made; but, people obviously can, and do, even when there is no possibility of the study affecting their welfare. When people know that they are under observation, they are sometimes pleased and motivated, and at other times they are suspicious of the observers or the motives of management. People must follow their normal patterns of activity for the results of the study to be accurate. In some cases, it may simply take a little time to get the people accustomed to the observers' presence. Other cases pose more severe problems.

When some features of the situation violate these assumptions, there should be concern over the accuracy and precision of the study results. Obviously, an effort should be made to satisfy these theoretical assumptions. When one cannot achieve perfect correspondence, tests can be done to check how serious the deviations are.

Calculating sample size using the normal approximation of the binomial sampling model

During the 1940s, a renowned group of statisticians at Columbia University were working on applied statistical problems under the leadership of Professor Abraham Wald. One of their goals was to find conditions under which the binomial distribution was closely approximated by the Gaussian or normal distribution. They found that a close approximation occurred when

$$nq(1-q) \geq 9 \tag{7.4}$$

At minimal close-approximation conditions, the equality in Equation 7.4 holds, and the smallest n occurs when $q=0.5$. That smallest n is therefore 36.

Suppose that we start our consideration of the required sample size with the standardized normal random deviate z_α and the assumption that the binomial is closely approximated by the normal distribution. The normal random deviate is

$$z_\alpha = \frac{(X-\mu)}{\sigma} \tag{7.5}$$

where
  α is the probability level corresponding to the normal deviate z
  μ is the mean of the normal distribution
  σ is the distribution's standard deviation

Chapter seven: Sampling methods in industrial ergonomics

Using a Normal Distribution to Fit the Binomial

Equation 7.4 shows when the normal distribution is a reasonable approximation of the binomial distribution. As q deviates more from 0.5, the value of n meeting the constraint of Equation 7.4 gets larger, shown as follows:

$q =$	0.1	0.2	0.3	0.4	0.5
$n' =$	100	57	43	38	36

where n' is the smallest integer meeting the conditions of Equation 7.9. When q is greater than 0.5, the same exact trend occurs.

Since the mean and variance of the binomial are nq and $nq(1-q)$, respectively, the expected value (i.e., mean) can be substituted in Equation 7.5 for μ and the square root of the variance can be substituted for σ. These substitutions give the following result:

$$v_\alpha = \frac{(X - nq)}{\sqrt{nq(1-q)}} \tag{7.6}$$

Since the sample size n will certainly be much greater than zero, we can comfortably divide both the numerator and denominator of Equation 7.6 by n. The result of that substitution and some algebraic simplification is

$$z_\alpha = \frac{((X/n) - q)}{\sqrt{q(1-q)/n}} \tag{7.7}$$

Notice that X is the number of observed events in the total sample of n observations. The ratio X/n is the sample estimate of q. The numerator on the right-hand side of Equation 7.7 describes the difference between the sample estimate of q and the true value of q. That difference is the sampling error e. If we replace the numerator with e, Equation 7.7 can be rewritten, giving an equation for the required sample size n.

$$n = \left(\frac{z_\alpha^2}{e}\right) q(1-q) \tag{7.8}$$

When the sample size is equal to the value calculated with Equation 7.8, an error larger than e occurs with a probability of α. To estimate the required sample size, a normal random deviate $z_{2\alpha}$ corresponding to an acceptable level of risk α (see Box 7.1) is entered into Equation 7.8, along with the acceptable sampling error e and the proportion q. As easily determined by inspection of the equation, the required sample size increases when the acceptable error is small.

Use of nomographs to estimate sample size

Nomographs have been developed for determining the required sample size, which eliminate the need to go through the previous calculations. The latter approach is particularly useful for people who are less familiar with statistical concepts. An example

BOX 7.1 EXAMPLE OF SETTING STATISTICAL RISKS

With $\alpha = 10\%$, the z_α value is 1.645 for a single-sided test. For a two-sided test, $z_\alpha = 1.96$. When specifying the largest acceptable error, e, α refers to the probability of exceeding that error magnitude. When the error can occur in both positive and negative directions, it becomes necessary to calculate the probability of $\pm e$, giving a total error probability of 2α.

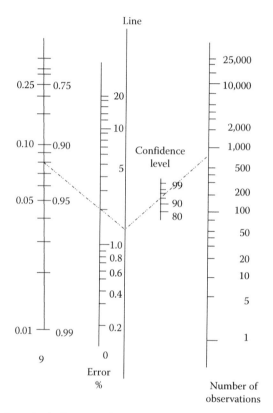

Figure 7.1 Nomograph for calculating sample size.

nomograph is illustrated in Figure 7.1 for determining the required sample size as a function of q, e, and $1 - \alpha$.

To use the nomograph, the analyst must first mark the point on the left-most vertical line of the nomograph corresponding to an appropriate value of q or $1-q$. Note that the numbers on the left- and right-hand sides of this line, respectively, correspond to values of q and $1-q$. The error tolerance e, measured as a percentage, is then marked on the next vertical line, and a straight line, running through the first two points, is extended over to the third line. This line is called the turning line. After drawing the turning line, the analyst must locate the point on the fourth vertical line corresponding to the desired confidence level $(1-\alpha)$. From the point where the turning line intersects the third vertical line, a straight line is extended through the point on the fourth vertical line corresponding to the desired confidence level all the way to the fifth vertical line, to obtain the required number of observations.

The turning line shown in Figure 7.1 corresponds to $q = 8\%$ and $e = \pm 2\%$. The line leading up from the turning point passes through the point corresponding to a 95% confidence level and denotes a sample size of about 726 observations. Direct calculation of the sample size using Equation 7.6 results in

$$n = \left[\frac{1.96}{0.02}\right]^2 0.08(1-0.08) = 707 \text{ observations}$$

As indicated by the previous comparison, the estimates obtained with nomographs tend to be a little conservative, that is, they inflate the sample size slightly. Some people recommend calculating the sample size using Equation 7.6 and then using the nomograph to double check the calculations.

Developing the observation schedule

Once the sampling procedure and the necessary number of observations have been decided upon, a detailed schedule must be developed that gives the exact times for making each observation. The first part of this process is to decide how long the study should last. The answer is almost entirely situation specific. The decision partially depends upon when the study information is needed. The availability of people to do the study is another relevant issue. In some cases, the analyst may have a particular observation rate in mind. If so, the length of the study is determined by dividing the required number of observations by the observation rate. This number can be compared to other situation-specific constraints. The length of the study, the sampling rate, or even the number of observations* can then be modified accordingly.

Once the number of observations and the study period are established, the exact time for making observations readily follows. The problem is obviously very straightforward, when samples are taken continuously or on a periodic schedule. When samples are taken at random intervals, the most common case, several procedures are available for scheduling the observations.

One such procedure is illustrated in Figure 7.2. Observation times are shown on the x-axis, and an ordered set of random numbers is shown on the y-axis. Note that the earliest time in the study period coincides with the smallest possible random number. Suppose there are N observations to be made. Then N random numbers should be looked up, each with sufficient digits so that everyone is unique. This set of random numbers ordered in numerical sequence becomes a set of points along the vertical axis (i.e., ordinates). The next step is to draw a line connecting the latest time in the study period to the largest random number (the dotted line in the figure). Lines parallel to the dotted line are then drawn from each point on the vertical axis down to the horizontal axis. The intersection points along the horizontal axis specify the times during the study period when observations should be made.

A mathematically equivalent procedure is to first assign one of the random numbers to each observation. The observation times are then obtained by multiplying the duration of the study period by the assigned random number and then dividing by the maximum random number.

---

* Reducing the number of observations is always an option, but can be risky, as doing so increases sampling error.

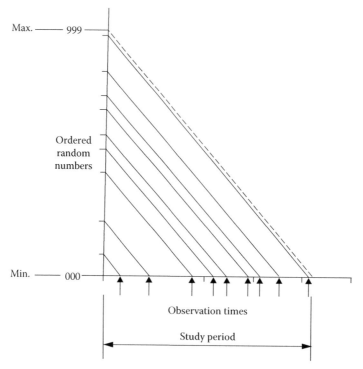

Figure 7.2 Example mapping of random numbers to observation times.

Both procedures comply perfectly with the theory of uniform sampling, that is, every possible time in the study period is equally likely to be selected as an observation time. If an observation time is too close to another for practical purposes, the offending time can be removed. Additional random numbers can then be created to fill those gaps and meet the sample size requirements. Note that this procedure does not guarantee that there will be an equal number of observations during each day, hour of the day, or other interval of interest in the study period. Assuring any or all of these conditions is a form of sample stratification, a procedure that is discussed later in this chapter.

For a variety of reasons, analysts often want to take the same number of samples each day of the study. One way of doing this is to first divide the total number of observations by the number of days in the study period. This gives the number of observations for each day. The earlier described procedures are then used to assign the observation times separately for each day. Although this procedure is not quite consistent with uniform sampling theory, the effects are normally not a concern. The primary objective after all is to create enough randomness, so no one can predict the time of the next observation. It is very doubtful that this minor constraint on randomization will make the observations too predictable.

Recording and analyzing the data

Before the data can be collected, data forms or other recording instruments must be decided upon. Data forms used in activity analysis are usually simple matrices with rows that show the observation times and separate columns for each different activity of interest (Table 7.2). Additional columns may be included on the form for recording variables

Table 7.2 Example of a Simple Form for Activity Analysis

Task_____ Observer_____ Date_____

Period	Idle	Working	Comment

used to stratify the data, such as gender, location, machine or tool used, etc. Computer spreadsheets have the same format, so such forms can be quickly generated using computer programs such as Microsoft Excel.

The data can be manually written on the form at the time of each observation, and then re-entered into the spreadsheet program. In some cases, the data might be entered directly into a spreadsheet running on a small handheld computer. Several commercial products are also available for recording observations and then downloading the data to a data analysis package.

Analysis of the data can be easily done with a spreadsheet program. Typically it involves little more than tabulating the frequencies of each activity. In some cases, cross tabulations are made to check for effects of stratification variables. For example, the frequency of idle time might be compared for different work categories. In consumer settings, seat belt use of males might be compared to that of females. Chi-square tests can be made to determine whether the observed differences are statistically significant.

Checking for stationarity

A primary assumption in sampling theory is that the fraction q in the binomial Equation 7.1 is constant. In some applications of sampling, q might be the fraction of time that a shop is idle or nonproductive. Since q is constant during the study period, the analogous assumption is that the probability of this shop being idle is the same any time during the study period. In reality, the factors that cause an idle shop may be different over different hours of the working day, or different days within a typical week, or within various weeks during the month. Accordingly, a test of the constancy of q would be to examine estimates of q (that is, X/n) at various time segments such as early mornings or late afternoons during different days of the week. If successive observations of about four or more were combined as periodic estimates of q and these estimates were plotted sequentially over the study, a visual observation would typically reveal any trends or show repeated patterns. A more formal procedure is to plot the estimates of q on a control chart. Control charts show confidence bounds around the mean value of q given by the following equation:

$$\text{Bounds} = \frac{X}{n} \pm z_\alpha \sqrt{\frac{q(1-q)}{n}} \qquad (7.9)$$

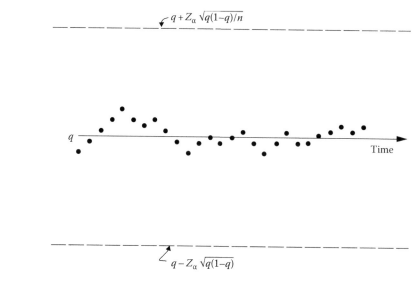

Figure 7.3 Using a control chart in a work-sampling study to identify changes.

where z_α is the standardized normal random deviate defined earlier. When values of q for small samples of size n are plotted over time on a control chart, as illustrated in Figure 7.3, a further investigation of the constancy of q will be suggested if q goes outside of the confidence bounds.

Using work sampling to estimate performance standards

One of the advantages of sampling over long time periods is that factors that temporarily impact performance are balanced or averaged out. With direct continuous observation, as in time study, it becomes necessary to compare the factors unique to the observation. Methods of performance rating are used to remove the variations due to differences in skill, effort, conditions, and other factors. When time standards are developed from sampling data, there is far less need for rating, although ratings can be effectively used when only a few different human operators are involved. To develop a standard or normal time, the sampling data needs to be combined with other data normally available in most companies. That is,

$$\{\text{normal time}\} = \frac{(\text{total time worked})[\text{fraction time active}](\text{performance rating})}{\text{total number of pieces produced}} \quad (7.10)$$

where

- The time worked is estimated from payroll time records or other sources
- The fraction of active time is estimated from the sampling study
- Performance ratings are made by the analyst if there are only a few different operators
- The total number of product pieces produced may be obtained from the production records

To convert normal time to a standard time, as explained in Chapter 6, the normal time must be multiplied by an allowance correction for personal needs, unavoidable delays and for other sources of down time, which are not due to the operator.

Numerical Example of Setting Time Standard Using Work Sampling

Suppose that 5 operators worked 9500 min over 4 days. A work-sampling study shows them to be actively working 95% of the time, when not on work break. The observer gives them an average performance rating of 105%. If production control showed that 6000 pieces of that product were produced during that time and an allowance correction of 15%, then the standard time for the observed operation is

$$\text{std time} = \frac{9500 \min \times 0.95 \times 1.05}{6000 \text{ pieces}} \times \frac{1}{1-0.15} = 1.86 \min$$

When performance ratings are made, the procedure is called rated work sampling. That is, when the observer comes onto the premises, they first identify the ongoing activity and then rate the speed subjectively within 5% intervals above or below 100%. A standard operator works at 100%, so ratings above 100% denote percentages faster than the standard operator's speed, and ratings below 100% indicate percentages slower. Generally, ratings are assumed to vary with the Gaussian (or normal) distribution as long as there are enough ratings and people to make that assumption.*

Observers will often need added training before they can accurately rate performance. However, when only a few operators are performing a job and the pace of the job may differ from standard operations, this added information is very useful. Some of the advantages of using sampling to produce standard times, compared to using direct time study, are that sampling

- Is less costly, typically by 50% or more
- Is more flexible
- Includes daily and weekly variations
- Is not affected by interruptions
- Has lower training requirements
- Does not need stop watches or other special equipment
- Is preferred by most operators
- Usually does not require performance rating
- Often highlights the causes of problems

However, sampling studies are rarely performed quickly, so it takes a long time to develop a standard if sampling methods alone are used.

---

* A paper by Buck, Zellers, and Opar (1998) describes how to use the Bayesian approach to develop speed ratings.

Using work sampling in problem solving

Work sampling can be very helpful for identifying problems in existing systems. Once some of the causes of ineffectiveness have been identified, focus can then shift to analyzing some of the following issues:

- The relative magnitudes of the various causes of ineffectiveness
- Any changes in the relative magnitudes over time
- The effectiveness of policies used to correct the causes

Clearly, those causes with the greatest effects on productivity need immediate attention. Problems with lesser effects can be handled later. Hence, sampling results help in the detection of problems and show ergonomic designers how to handle the problems to be faced.

In maintenance operations, for example, people must read the complaint, go to the site of the problem, diagnose complaints, get technical information on the particular equipment needing repair, obtain parts and materials, test out the corrective action, and complete any needed paperwork associated with the complaint. Usually a number of specialty personnel such as electricians, plumbers, and machinists are involved, as are support personnel who maintain the stored parts and materials, people who plan the maintenance activities of various crews, coordinators between the maintenance shop, managers, and other support persons (e.g., person doing drafting).

Figure 7.4 shows the results of an activity sampling study of maintenance personnel. High or low fractions of certain activities indicate problems. Changes in those fractions over time can suggest the existence of particular problems that require further investigation. For example, a large fraction of time spent waiting for assignments is an indication of lack of adequate planning. Such changes can also show the effects of corrective actions.

Sometimes small experiments can be made, such as temporarily assigning an extra person to parts storage or trying out a new technical information system. Work sampling

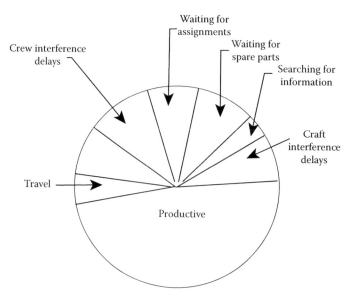

Figure 7.4 Results of work-sampling maintenance crews.

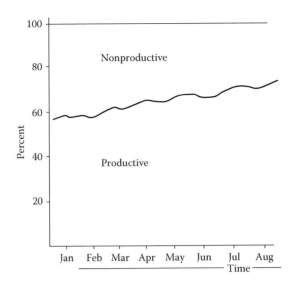

Figure 7.5 Productivity charting over time.

then can be used to test the effects of the changes. Experiments that clearly improve some but not all features of an operation will allow management to determine whether changes are useful.

While it is useful to know the causes of productivity losses so that corrections can be made, it is also important to know if those causes are changing over time, and if productivity is suffering or improving. Time plots, such as the one shown in Figure 7.5, provide additional information for ergonomics management.

Sampling strategies

Sampling is usually a more economical way of collecting data than continuous observation or sampling the entire population of interest. The cost of conducting a sampling study depends on how many observations are taken. Ideally, no more observations are taken than necessary to obtain the desired conclusions with a required level of precision. Taking more samples than needed results in wasted effort. Taking too few observations, on the other hand, results in an excessively high sampling error. Stratified and sequential sampling strategies provide two different ways of dealing with this issue, as expanded upon next.

Stratified sampling

In many cases, a sampling strategy is followed in which the observations taken are stratified or partitioned into subsets, frequently called strata. For example, a week might be divided into days in which particular activities occur. Observations are then taken within each subset or stratum, often in a way that keeps the number of observations taken within each strata the same (see Box 7.2). Stratified sampling may be done for at least three different reasons. One reason is to increase the representativeness of the observation. A representative sample has characteristics similar to the population being sampled. For example, if the population of interest is 50% male, observations containing an equal number of male and female subjects would be more representative than one where most of the subjects are females. A more representative observation increases face validity and may

BOX 7.2 EXAMPLE APPLICATION OF STRATIFIED SAMPLING

Stratified sampling may be appropriate when the suspected causes of a problem are associated with the days of the week. It may be very logical to separate Monday, Wednesday, and Friday into one stratum of 24 shift hours (i.e., 3 days and 8 h/day) and Tuesday, Thursday, and Saturday mornings into the other stratum of 20 shift hours (i.e., 2 days at 8 h/day and 1 day at 4 h). It would make sense to sample the first stratum at about a 20% greater rate than the second stratum in order to keep the number of observations per shift hour equal.

In another day-of-the-week illustration, suppose that a university has 8 h of 50 min classes on Monday, Wednesday, and Friday and during the 4 h of Tuesday and Thursday (T & Th) afternoons. However, on T & Th mornings, classes are 75 min long. If you want to obtain the same expected number of observations for each class regardless of the day or time of day, sampling rates need to be lower on T & Th mornings. Let Z be the rate (observations/h) on T & Th mornings and Y be the rate other times. During T & Th mornings there are three classes and during the other mornings there are four classes.

Assuming that the total morning times are the same, the sampling rate on M, W, & F, at Y observations/h must be 1/3 greater than the sampling rate on T & Th, if all classes are to have the same number of observations. If observations are made on some aspect of those classes and six observations are made per hour on T & Th mornings, then make eight observations per hour during the other time periods.

make the results more accurate. A second reason for using stratified sampling is to obtain information. For example, if there are real differences in seat belt use between males and females, stratifying observations of belt use by gender will show this effect assuming a large enough sample is taken. A third reason for stratifying is to increase the efficiency of sampling—that is, a stratified sampling strategy can reduce the total sample size needed to obtain results of the same statistical precision for each strata. A stratified sampling strategy can also increase the precision of the obtained results for particular strata of interest.

Designing a stratified sampling strategy

The design of a stratified sampling strategy involves determining (1) the number of strata, (2) the boundaries of each stratum, and (3) the number of observations taken within each stratum. Each of these quantities depends on why a stratified sampling strategy is being considered. Simply put, the strategy might reflect a desire to provide a representative observation, obtain some specific information about a subpopulation, or a need to keep the number of samples within a reasonable limit. The variables used to specify how the samples will be stratified also play an important role in designing a sampling strategy. In the simplest case, a single discrete variable (such as gender) is used to specify the strata. In other cases, a continuous variable (such as age or years of work experience) might be used. Combinations of discrete and continuous variables (such as experienced males) are another possibility. The key point is that the number of possible strata defined by a considered set of stratification variables is potentially very large. In fact, the number of possible strata becomes infinite when a continuous variable is involved. It consequently becomes necessary to define each stratum in terms of a range of values on some stratification variable. For example, we might group the sampled workers in terms of work experience into three categories (low, moderate, and high). Note that each category or stratum, in this case, can be defined as a range of years worked (i.e., low might be defined as less than 1 year of work experience).

Determining an appropriate number of strata and boundaries for each stratum is often done on the basis of subjective experience or after collecting an initial sample of data that reveals potentially important variables impacting the observations. For example, a observation of observations taken at random times over a period of several weeks might indicate that a particular machine is more likely to be inoperative late in the workday. To evaluate this issue, the analyst might design a sampling plan in which an equal number of observations are taken in each of the three intervals: (1) 8:00 a.m.–12:00 p.m., (2) 1:00 p.m.–4:00 p.m., and (3) 4:01 p.m.–5:00 p.m. An interesting aspect of this plan is that an equal number of observations (n) are taken in each interval, but each interval is a different duration. The total number of observations taken in this plan is $3 \times n$, which is a considerable reduction ($8 \times n - 3 \times n$) in the number of observations when compared to a plan in which n observations are taken in each 1 h period over the work day. Besides saving observations, the stratified plan measures the quantity of interest (i.e., down time) in the critical 4:00 p.m.–5:00 p.m. period using the same number of observations n, as the plan that uses uniform 1 h interval over the work day.

Note that in the previous example, we assumed that there was an advantage to having an equal number of observations in each stratum. It turns out that this is a good rule of thumb, but there are cases where it may be desirable to take more observations in certain strata. To get a better feel for this issue, it may be helpful to return to our earlier discussion on calculating the required sample size (n) for binomial sampling of a variable X with known variance $V(X) = nq(1-q)$, and an acceptable sampling error e. As pointed out there, if we use a normal approximation of the binomial distribution, the required sample size n was as follows:

$$n = \left(\frac{z_\alpha^2}{e}\right) q(1-q) \tag{7.11}$$

Now if we have m strata, and we assume binomial sampling of a variable X, and that for each stratum m we take n_m observations, we get the following result for each stratum:

$$V(X) = n_m q(1-q) \tag{7.12}$$

Using Equation 7.11 we then get a required sample size n_m for each stratum of

$$n_m = \left(\frac{z_\alpha^2}{e}\right) q(1-q) \tag{7.13}$$

If we then assume prior to designing the stratification strategy that q and the acceptable sampling error are the same for each stratum, it then follows that

$$n_{m-k} = n_m \quad \text{for} \quad 0 < k < m \tag{7.14}$$

also,

$$n_m = \frac{n}{m} \tag{7.15}$$

An equal number of observations should be provided for each stratum, if we are assuming ahead of time that q and the acceptable sampling error are the same for each stratum.

This assumption is a reasonable one in many situations and corresponds to a null hypothesis that the stratification conditions do not have an effect.

Analysis of stratified sampling data

As mentioned earlier, activity and work sampling studies generally collect binomially distributed data, such as the number of times an observed machine is idle. In some cases, this data is supplemented with various measurements collected at the same time the activity observation is made. In many situations, the strata are used as independent variables in various types of analysis of the data. When the supplemental data are measured on a continuous scale (e.g., lengths, amounts, etc.), it can be used as a dependent variable in various forms of ANOVA analysis. Contrasts can also be done with student t-tests.

Binomial data (such as the number of times a machine is idle or not) should not be analyzed with ANOVA or t-tests. A number of other approaches have been developed for analyzing binomially distributed data that can be used to analyze whether there are significant differences between stratification conditions.* For example, the analyst may be interested in determining whether machine idle time varies significantly between different days of the week. A general format for organizing such data is illustrated in Table 7.3. As shown in the table, the data are recorded for each stratum. The observed outcomes are in the two columns to the right. Each row of the matrix in Table 7.3 records results for a particular stratification of the observations.

If an equal number of observations are made for each stratum, the procedure illustrated in Box 7.3 can be followed to check for significant differences.† The first step is to organize the data as shown in Table 7.3, and then search the last two columns for the smallest value over all rows of the data. The value does not have to be unique. For our discussion here, let us assume that x_h^* is smallest. The next step is to first find the row in Table 7.4 that best matches the sample size n_h, and then look for a value in that row that matches x_h^*. The corresponding number at the top of the column in which the value was found gives the minimum statistically significant difference, which is referred to here as contrast c. Any value in the column that contains x_h^* of the data matrix (Table 7.3) that differs by c or more units from x_h^* is significantly different ($\alpha = 0.05$) from x_h^*.

Sequential Bayesian work sampling

In the earlier discussion of activity sampling, it was assumed that a sample size was determined ahead of time, before the data were collected. At the end, the fraction q is estimated and

Table 7.3 Stratified Sampling Data

Strata Number	Sample Size	Occurrences Observed	Non-Occurrences Observed
1	n_1	x_1	$n_1 - x_1$
...	...	...	...
H	n_h	x_h^*	$n_h - x_h$
...	...	...	...
L	n_L	x_L	$n_L - x_L$

---
* See Natrella (1963) and Allen and Corn (1978) for further details.
† Other methods, such as chi-square analysis can be used if sample sizes are not equal.

BOX 7.3 NUMERICAL EXAMPLE OF STRATIFIED SAMPLING AND POOLING

Consider the situation in which observations are made at randomly selected times of a room in a building. The observer records whether someone is present. If no one is there, the room is idle. An equal number of 16 observations are made each day over the normal working hours. The results are stratified by different days of the week as shown in the following:

Day	Idle	Not Idle	Sum
Monday	9	6	16
Tuesday	1*	15	16
Wednesday	8	8	16
Thursday	5	11	16
Friday	7	9	16

The following procedure can be followed to determine if the use of the room is significantly different on certain days. The smallest number of observed or unobserved events is the one observed idle event on Tuesday. Entering Table 7.4 on the row marked 15–16 for the sample size, move rightward along that row until you find the second column where the value 1 occurs, corresponding to the smallest value found in the data. Then move upward to find the minimum significant contrast of $c = 6$. Hence, $1^* + 6 = 7$, and any day with 7 or more observations of idleness is (statistically) significantly different from Tuesday in the utilization of that room. Therefore, room use on Monday, Wednesday, and Friday differs significantly from Tuesday's utilization. Note that when the data of Tuesday and Thursday are removed from the dataset, the smallest remaining number is 7 and there is no significant contrast with 7; therefore usage on Monday, Wednesday, and Friday is not significantly different. Also note that use on Tuesday does not differ from use on Thursday. Suppose that we pooled data for Wednesday and Friday into a single category, and did the same for Tuesday and Thursday. The result would be

Strata	Idle	Not Idle	Sum
Tue. & Thurs.	6*	26	32
Wed. & Fri.	15	17	32

In this pooled-data case, one must enter Table 7.4 with two sample sizes, 30 and 40, go to the smallest number 6* in each row, find the corresponding two minimum contrasts, and interpolate those contrasts. It turns out in this example that both sample sizes show a minimum contrast of 9, and so $6^* + 9$ or 15 is the minimum value significantly different from 6. Since the Wednesday–Friday stratum has 15 observed events, the use of the room on Wednesday and Friday is significantly less than on Tuesday and Thursday.

Table 7.4 Relationships between Sample Sizes, Smallest Observations, and Minimum Significant Contrasts ($\alpha = 0.05$ One Side) for Binomial Stratified Sampling

Sample Size	Minimum Statistically Significant Difference										
	5	6	7	8	9	10	11	12	13	14	15
7–9	1										
10–11	0	1–2									
12–13	0	1–3									
14	0	1–2	3								
15–16	0	1	2–4								
17–19	0	1	2–5								
20	0	1	2–5	6							
30		0	1–2	3–5	6–10						
40		0	1–2	3–4	5–7	8–15					
50		0	1	2–3	4–6	7–10	11–19				
60		0	1	2–3	4–5	6–8	9–13	14–24			
70		0	1	2–3	4–5	6–8	9–12	13–18	19–28		
80		0	1	2–3	4–5	6–7	8–11	12–15	16–23	24–33	
90		0	1	2–3	4–5	6–7	8–10	11–14	15–20	21–31	32–37
100		0	1	2–3	4	5–7	8–10	11–13	14–18	19–25	26–42

Source: Natrella, M., *Experimental Statistics, Handbook 91*, Superintendent of Public Documents, Washington, DC, 1963..

tested for significance. The following discussion of Bayesian work sampling uses some of the same reasoning. The major difference is that the sample size is not determined prior to data collection. Instead, a stopping condition is specified for when to stop sampling. Bayes' theorem is used to describe how the precision of the estimated value of q changes over the sampling period. Once the estimate of q is sufficiently precise, the sampling process is stopped. The practical advantage of this form of sampling is that it provides a much greater efficiency in sampling. As the reader will see in the following discussion, a user of Bayesian statistics can factor in prior knowledge about the system under investigation. The greater the prior knowledge, the fewer the future samples needed. The use of this knowledge is one reason this approach is efficient compared with classical statistics where no prior knowledge is assumed. A second reason for greater efficiency in Bayesian sampling is due to the iterative process followed to combine the effect of each new observation with the previous conclusions. This Bayesian approach also avoids using a normal approximation of the Binomial data.

Beta distribution and its parameters

The beta distribution, or more specifically the beta-1 distribution, describes the probability of q as follows:

$$p(q) = \frac{(a+b+1)!}{(a-1)!(b-1)!} q^{a-1}(1-q)^{b-1} \tag{7.16}$$

where the symbols a and b are parameters of the beta distribution.

The expected value of q and its variance for the beta distribution are as follows:

$$E(q) = \frac{a}{a+b} \qquad (7.17)$$

$$V(q) = \frac{E(q)[1-E(q)]}{a+b+1} \qquad (7.18)$$

The mode of q (i.e., the value of q with the greatest probability or the peak in the probability density function) is given as

$$M(q) = \frac{a-1}{a+b-2} \qquad (7.19)$$

If the mean or variance is known, Equations 7.17 and 7.18 can be used to estimate values for the beta parameters a and b. For example, suppose that the expected q and its variance were estimated to be 40% (0.4) and 1% (0.01), respectively. It follows that when

$$\frac{a}{a+b} = 0.4 \quad \text{and} \quad \frac{0.4[1-0.4]}{a+b+1} = 0.01$$

the calculated values of these parameters are $a = 9.2$ and $b = 13.8$.

Bayesian confidence intervals

Confidence intervals for the beta distribution are not usually symmetric about the mean. The beta distribution is symmetric about the mean only when the parameters a and b are equal. The Bayesian confidence interval for the beta distribution is defined as the interval under the distribution from the lower point q_1 to the upper point q_2 such that

1. The integral from q_1 to the upper point q_2 is equal to $(1-\alpha)$ where α is the acceptable sampling error.
2. The probability densities at the end points are equal. That is, $p(q_1) = p(q_2)$.

In other words, the true value of q falls within the range $q_1 < q < q_2$ with a probability equal to $(1-\alpha)$. Table B.2 in Appendix B provides the endpoints q_1 and q_2 for Bayesian confidence intervals at confidence levels of 0.90, 0.95, and 0.99, for various values of the parameters a and b. Note that these confidence intervals, respectively, correspond to α values of 0.10, 0.05, and 0.01. More complete tables of Bayesian confidence intervals can be found elsewhere.* Figure 7.6 illustrates an example Bayesian confidence interval for an asymmetrical beta distribution. Note that the two tail areas (the cross-hatched regions) are of equal area and sum to α. The area between the points q_1 and q_2 is equal to $(1-\alpha)$ and corresponds to the confidence level. Note that the distances between the mean estimate of q and the lower and upper bounds (q_1 and q_2) are not necessarily equal to each other. This is why the confidence interval is called asymmetric.

---

* See Bracken (1966), Schmitt (1969), or Buck (1989).

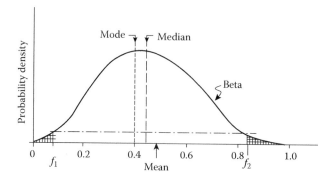

Figure 7.6 Bayesian confidence intervals (high density interval) of the asymmetric beta distribution (i.e., showing positive skewness).

Finding tabled Bayesian confidence intervals

In the case of the beta with parameters $a=9$ and $b=14$, a 95% Bayesian confidence interval can be found using Appendix B. This appendix does not contain all possible combinations of a and b (since this would result in an extremely large table). The two closest combinations are shown as follows:

a	b	q_1	q_2
9	11	0.241	0.661
9	21	0.145	0.462

By using linear interpolation, the confidence interval is found to be $q_1=0.212$ and $q_2=0.601$. Accordingly, there is only a 5% probability that q is smaller than 21% or larger than 60%. Note that since $E(q)=39\%$, the difference between $E(q)$ and q_1 is 18%. The difference between $E(q)$ and q_2 is 21%. This reflects the asymmetry of Bayesian confidence intervals. The mode of q can be found using Equation 7.19. In this example the mode is 0.38 or 38%, which is slightly smaller than $E(q)$.

Bayes' theorem and the Bayesian concept

Rev. Thomas Bayes developed this statistical theorem that bears his name. This theorem describes how the probability of q changes when evidence regarding its value is available. The theorem is

$$p(q|E) = \frac{p(q)p(E|q)}{\int_0^1 p(q)p(E|q)dq} \quad (7.20)$$

Bayes' theorem states that the probability density function of q after adjusting for new evidence is related to the probability density function before the new evidence (called the prior or a priori distribution) and the sampling distribution are available. This constant assures that $p(q|E)$ has an integral over the full range of q which equals unity, as all probability functions must. When $p(q)$ is a beta distribution and the distribution $p(E|q)$ is binomial, then the probability of q after adjusting for the new evidence is also beta, but with parameters A and B, which are

$$A = a + x \qquad (7.21)$$

$$B = b + n - x \qquad (7.22)$$

where
    a and b are parameters of the prior beta distribution
    n and x are the parameters corresponding to sample size and the number of observed occurrences used in the binomial sampling distribution

These parameters were discussed earlier in the *Binomial sampling model* section. Full algebraic developments of these concepts as well as extensions are shown elsewhere.* Note that A and B are parameters of the posterior beta distribution, which is obtained after a sample is collected.

In practice, people can often provide an initial estimate of q or a 95% confidence interval. Parameters a and b of the Beta can then be calculated from these estimates. For example, the analyst might say they are 95% confident that q is greater than 0.25 and less than 0.66. A quick glance at Appendix B shows that this belief is almost exactly fit by a beta distribution with $a = 9$ and $b = 11$.

It is natural for a reader to question whether or not people can do a reasonable job of estimating the prior probabilities needed in Bayes' equation, particularly because these sampling techniques often call for the manager of the sampling study to subjectively evaluate some statistics or statistical properties. Since the whole technique of the Bayesian work sampling scheme depends on the answer to this question, let us consider a little background on this subject.† The essence of those discussions is that people are quite accurate at estimating probabilities; their precision is within 5%–10% for mid-range probabilities. However, they do have strong biases at times. People are also quite good at estimating statistics subjectively, and estimating the first two statistical moments does not appear to be a problem, although they are more accurate at means than variances. Accordingly, human limitations in statistical estimation do not appear to be a major issue.

Managing Bayesian work sampling studies

Management activity involves determining a prior distribution, collecting binomial sampling evidence, and changing the prior distribution to a posterior distribution. This procedure is then repeated with the last posterior becoming the prior distribution for the next sample. Data collection should continue until the posterior Bayesian confidence is acceptable. Note that this criterion for stopping the study is analogous to the required sample size computations done in traditional work sampling, except that now one uses the Bayesian confidence interval. In addition, because this interval is asymmetric about $E(q)$, it allows two tests of an acceptable error e, corresponding to $[E(q) - q_1]$ and $[q_2 - E(q)]$. Once an acceptable error is reached, the sampling can be discontinued.

---

* See Hadley (1967), Raiffa (1968), or Buck (1989) for development of these ideas. These sources show how assuming a beta prior distribution and a binomial sampling distribution leads to beta posterior distribution with the aforementioned change in parameter values.
† Chapter 13 describes the results of several studies that describe the quality of people's subjective probability estimates.

Determining the prior distribution to begin the study involves the selection of an accurate but conservative prior distribution of the fraction q, which describes current knowledge. When people have worked near a situation for a long time, they can usually describe a conservative interval of q values that is reasonable. One merely needs to collect what knowledge exists about q and then construct a beta distribution. In this distribution, the average q is represented by the beta mean, which the individuals describe. Past data or subjective estimates of the mean can be used. If there is a great deal of disagreement among people, the analysis can start, with parameters $a = b = 1$. Actually, a prior beta with both parameters equal to unity is a uniform distribution over the entire interval $0 < q < 1.0$; this indicates that any reasonable value of q is just as meaningful as any other value. However, if a conservative but accurate prior beta can be constructed with larger parameter values, fewer samples are needed in the future study. If the start of the sampling yields results that are quite discordant with the prior beta, a revised and more neutral prior beta* should be considered.

If the initial prior beta has parameters a_1 and b_1, then the initial expected value of q, or $E(q)$, is $a_1/(a_1 + b_1)$. The next n binomial observations will give a first posterior beta with parameters of A_1 and B_1 and that beta becomes the prior beta with parameter $a_2 = A_1$ and $b_2 = B_1$, and so forth as the sampling goes on. Figure 7.7 illustrates this sequence. This progression of betas generates a sequence of points on the parameter map that heads up and to the right, yielding more rightward movement with a larger $E(q)$ value. It follows that the final posterior beta will have parameters A_r and B_r where the Bayesian confidence interval for a sufficiently high level is as short as or shorter than acceptable errors. If the approximate acceptable error is

$$e' = Z_\alpha \sqrt{\frac{V(q)}{N}} \quad (7.23)$$

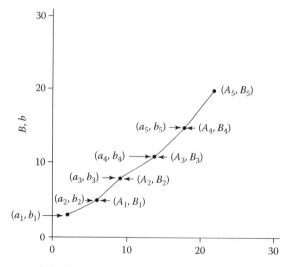

Figure 7.7 Plotting sequential study results over time.

---

* A more neutral beta is one where the parameters a and b are smaller and more nearly equal.

Chapter seven: Sampling methods in industrial ergonomics   315

where $V(q)$ (Equation 7.23) and $E(q)$ (Equation 7.24) are based on the initial prior beta, then the expected number of observations needed for completing the study is initially

$$N = \frac{E(q)[1-E(q)]}{e'^2/Z_\alpha^2} - a_0 - b_0 - 1 \qquad (7.24)$$

It then remains to plan the start of the sampling and continue until the exact Bayesian confidence interval meets all acceptable error tests.

More extensive Bayesian confidence intervals

One problem in determining Bayesian confidence intervals is that the tables do not show all combinations of the beta parameters, a and b. Some approximate ways of estimating Bayesian confidence intervals are given in Table 7.5 that can be used in such situations. These equations were obtained by regression analysis of many beta Bayesian confidence intervals.

To apply this approach, each of the regression responses in Table 7.5 is solved using the following expressions:

$$W = \frac{S(q)}{[q_2 - q_1]} \qquad (7.25)$$

$$= I + C_1(\gamma) + C_2(\gamma)^2 + B'/\min(A,B) \qquad (7.26)$$

$$Z = \frac{S(q)}{[E(q) - q_1]} \qquad (7.27)$$

$$= I + C_1(\gamma) + C_2(\gamma)^2 + B'/\min(A,B) \qquad (7.28)$$

where
  $\min(A, B)$ is the smaller of the two beta parameters (A or B) for the sample
  $E(q)$ is as defined before
  $S(q) = V(q)^{0.5}$
  γ is the skewness of q

the remaining parameters are as shown in Table 7.5

Table 7.5 Regression Equation for the Extended Procedure Used to Compute Bayesian Confidence Intervals

Confidence Level (%)	Regression Response	Intercept Constant (I)	Skewness Coefficient (C_1)	Skewness Squared Coefficient (C_2)	1/Smaller Beta Parameter (B')	Regression R^2
90	W	0.301	—	0.0321	0.0155	0.99
	Z	0.605	−0.0437	0.0997	−0.0156	0.98
95	W	0.253	—	0.0139	0.0398	0.99
	Z	0.508	—	0.0902	0.1010	0.99
99	W	0.196	−0.0082	−0.0076	0.0810	0.97
	Z	0.382	+0.0649	+0.0642	0.238	1.00

To calculate the confidence interval, the first step is to compute the skewness using the final posterior parameter values of the beta A and B. The skewness α_3 of the beta (the third central moment of the beta distribution) is

$$\gamma = \frac{2(b-a)\sqrt{a+b+1}}{(a+b+2)\sqrt{ab}} \tag{7.29}$$

The next step is to select the significance level desired (90%, 95%, or 99%). The remaining coefficients are obtained from Table 7.5, for the desired significance level. These coefficients are then used in Equations 7.26 and 7.28 to estimate the quantities W and Z. The final step is to solve for the values of q_1 and q_2 using Equations 7.25 and 7.27. This gives the end points of the Bayesian confidence interval. The following example goes through this procedure.

Numerical example of Bayesian activity analysis

Consider a prior beta of $a=9$ and $b=14$ based on an expected fraction q at 0.4 and $V(q)=0.01$ (i.e., an expected variation in q or standard deviation $S(q)=0.10$ or 10%). Suppose that it was determined that an acceptable tolerance or error is 1% below the mean and 2% above it at the 95% level where $\alpha=5\%$ for the combined tails. This tolerance constraint specifies $q_1 = E(q) - 0.01$ and $q_2 = E(q) + 0.02$. Since the prior beta has a 95% Bayesian confidence interval from 0.21 to 0.60, a considerable amount of sampling is clearly needed to reduce the final posterior beta to an acceptable size. An approximate sample size can be estimated from Equation 7.24 with the smaller acceptable error as

$$n = \frac{0.4[1-0.4]}{[0.01/1.96]^2} - 9 - 14 - 1 = 9196$$

where $z_{\alpha/2}=1.96$ for $\alpha=0.05$. With this sample size, it would take over 90 days at 100 observations per day to complete the study. However, it was discovered that there were enough observer resources available during the next 3 working days to take 20 observations each, and it was decided to do this in order to re-examine the adequacy of the prior beta distribution. The prior beta was plotted on the beta parameter map shown in Figure 7.8, corresponding to the point $[a=9, b=14] = [a_0, b_0]$. During the next three working days, utilization of the machine tool was observed 20 times each day. An event was said to occur if the tool was in use during an observation. The tool was observed to be in use seven, nine, and eight times over the 3 days observations were taken. These results created the three successive posterior beta as shown in the following:

Prior	Posterior
$[a_0=9, b_0=14]$	$[A_0=a_1=9+7=16, B_0=b_1=14+20-7=27]$
$[a_1=16, b_1=27]$	$[A_1=a_2=16+9=25, B_1=b_2=27+20-9=38]$
$[a_2=25, b_2=38]$	$[A_2=a_3=25+8=33, B_2=b_4=38+20-8=50]$

After the 3 days of sampling, the expected value of q is computed as

$$E(q) = \frac{33}{33+50} = 0.398$$

Chapter seven: Sampling methods in industrial ergonomics

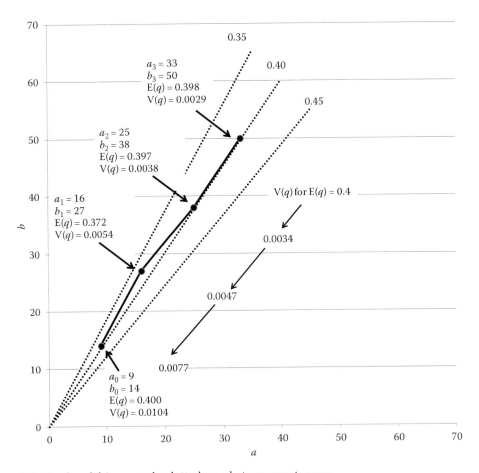

Figure 7.8 Results of this example plotted on a beta parameter map.

which is very close to the initial estimate of 0.40, so the initial beta appears to be reasonable. The variance at this stage of sampling is

$$V(q) = \frac{0.398(1-0.398)}{33+50+1} = 0.00285$$

and the standard deviation is 0.0534. Figure 7.8 plots these sequential samples and statistics on a beta parameter map. Clearly there was still too much uncertainty, and so it was decided to take 200 more observations.

After making those 200 observations, it was found that the last posterior plot on the parameter map was at the point [90, 133], where $E(q) = 0.40$ and $V(q) = 0.0018$. If 9000 more observations were taken, the expected final posterior beta on the parameter map would be at

$$A_r = 90 + 0.4(9000) = 3690 \quad \text{and} \quad B_r = 133 + 0.6(9000) = 5533$$

The mean, variance, and skewness of q at this expected final beta would be

$$E(q) = \frac{3690}{3690+5533} = 0.40009$$

$$V(q) = \frac{0.40009[1-0.4009]}{9224} = 0.000026, \quad \text{or} \quad S(q) = 0.0051$$

$$\gamma = \frac{2(5533-3690)[3690+5533+1]^{0.5}}{(3690+5533+2)[(3630)(5533)]^{0.5}} = 0.00849$$

Figure 7.9 describes the expected future sampling situation as viewed from an intermediate, or current, sampling perspective. It follows from Equations 7.25 and 7.26 for a 95% level of confidence that

$$\frac{0.0051}{q_2 - q_1} = 0.253 + \frac{0.0398}{3690} + 0.0139(0.00849)^2$$

Note that the previous two terms are very small. By solving the expression, we get the result

$$q_2 - q_1 = \frac{0.0051}{0.25301} = 0.02016$$

It also follows from Equations 7.27 and 7.28 for a 95% level of confidence that

$$\frac{0.0051}{0.40009 - q_1} = 0.508 + \frac{0.101}{3690} + 0.0902(0.00849)^2$$

A little more algebra reveals that

$$q_1 = 0.39005 \quad \text{and} \quad q_2 = 0.41021$$

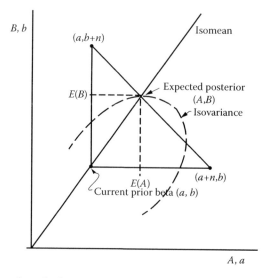

Figure 7.9 Projecting ahead on the beta parameter map from the current sampling point.

Note that $q_2 - E(q) = 0.41021 - 0.40009 = 0.01012$ and $E(q) - q_1 = 0.40009 - 0.39005 = 0.01004$ so both subintervals of the Bayesian confidence interval for $\alpha = 0.05$ are approximately 1%, which satisfies the desired level of precision.

Poisson activity sampling

The Poisson sampling approach is an alternative to binomial sampling that works in a similar way (see Buck et al. 1996 in the following for details). A principal difference is that Poisson sampling distribution* is used instead of the binomial sampling distribution. Also, a gamma prior probability density function is used instead of the beta. When these two functions are used in Bayes' equation, the posterior probability density function turns out to be another gamma distribution with parameters, which changes with the sample size and results. Accordingly, Poisson Activity Sampling allows posterior statistics to be incrementally calculated, just as with binomial sampling. Poisson sampling is not as efficient as binomial sampling. Nevertheless, it can be an excellent information collecting procedure if the data under observation fit the Poisson sampling paradigm.

Final comments

Sampling, in general, is an effective way of collecting information for many problems. Traditional work sampling is a binomial form of information collection. However, Poisson sampling methods are a viable alternative to binomial sampling. Both approaches have their advantages, so Poisson sampling methods should not be overlooked. In either case, Bayesian methods tend to be much more efficient than the traditional approach and are more adaptable to a variety of situations. In some cases, stratified sampling techniques are used to assure representative samples. The latter approach can also provide detailed information not easily obtained with other methods. When one needs proficiency measurements in sampling, rated or self-observing work sampling should be considered. A number of references given at the end of this book provide more detail and useful background information on these and other methods.

Discussion questions and exercises

7.1 If one considers an observation as an experiment, which is successful when the activity of concern is observed and unsuccessful otherwise, then the number of different sequences of observing four successes out of six observations is _____.

7.2 If an activity occurred 30% of the time, then:
   (a) What is the probability of observing this activity on four out of six randomly selected observations?
   (b) What is the probability of observing the activity more than four times out of six observations?

---

* The Poisson distribution was named after Simeon Denis Poisson who lived in France from 1781 to 1840, and spent most of his professional life at the Ecole Polytechnique in Paris. His research covered numerous fields, including probability theory. He developed this distribution in his *Recherches sur la probabilite des judgements*. Mood (1950) is one of the few sources that describe the Poisson paradigm. The Poisson process became more popular in Operations Research as the basis of waiting line or queuing theory. Also, readers who are familiar with the Pearson family of distributions will recognize the related gamma distribution as the Pearson type III with a zero origin.

7.3 If an activity occurs 30% of the time and six observations are made at randomly selected time points, then what is the expected number of observations during the activity and the variance in the activity being observed?

7.4 What is the smallest number of observations needed for a close approximation of a normal to a binomial if the activity occurs 30% of the time?

7.5 For a normal distribution of $\mu = 0$ and $\sigma^2 = 1$, what is the probability of finding a value from a variable following this distribution that is greater than 1.7?

7.6 How many observations are required to be 90% sure of being within ±3% (i.e., an error) of the population mean for an activity, which occurs 30% of the time? How many more observations need to be taken to increase one's confidence to 95% certainty?

7.7 If it required 3 min to make an observation from your normal location and all of the observations had to be completed in 4 working days' time, what is the smallest number of random digits in a random number that would permit any point in time to be observed?

7.8 Suppose that an operator at work stations completed 840 assemblies over a four-shift time period. Actual working times averaged 7h per shift. Random observations of this operation indicated that this operator was working on component A 40% of the time, component B 25% of the time, and component C the remainder of the time. What is the normal time required for each component's assembly based upon these observations?

7.9 The 8h working day shift was partitioned into three time periods: (a) the first half hour of a shift start or a lunch break and the first quarter hour after each of two coffee breaks per shift, (b) the last half hour of the shift, and (c) the remaining time. If a time proportional stratified sampling plan of 160 observations were to be made, how many samples should be made during each time period?

7.10 What are the advantages and disadvantages of using a fixed time interval between observations, compared to the same number of randomly selected observation times?

7.11 Your belief in the percentage of time that an activity occurred needs to be described by a fitted beta distribution. Suppose that you expect the mean percentage to be 20% and the most-likely value of that percentage to be 18%. What are the beta parameters?

7.12 Suppose that your prior beta that described the fraction of machine utilization has parameters 8 and 20, respectively, for parameters a and b. Then you perform a binomial experiment where 10 observations are made, and on 4 of them the machine tool is observed to be working. Compare the mean and variance of the prior distribution to those of the posterior distribution.

7.13 The average buffer inventory in standard lots in front of each machine tool was estimated to be 1.75, and the standard deviation was estimated at 0.5. It was decided to fit a probability density function to those estimated statistics and then to use sampling and Bayesian methods to update those statistics periodically over the sampling procedure. Three different samples were taken, each consisting of five observations, and the respective results were 8–10 observed events.
(a) What is the prior mean and variance?
(b) What are the three posterior parameter map points?
(c) How much variance reduction occurred after all of the sampling relative to the prior variance?

7.14 In a sequential Bayesian work sampling study, the mean and variance of the utilization were estimated to be 0.8 and 0.01, respectively. Afterward a sample of 10 observations was made and the machine was busy 7 of the 10 times observed.

What is the posterior mean and variance? What are the approximate 95% Bayesian confidence bounds around the mean?

7.15 How could parameter maps aid a person managing a sampling study? How could parameter maps be used in parameter estimation when one or more people are expressing their subjective opinions of the mean, mode, median, standard deviation, and or variance of the probability (density) function?

7.16 Suppose that you are going to perform a Bayesian work sampling study and your prior beta description of the occurrence of a machine utilization has parameters $a=8$ and $b=20$. You then perform a binomial sampling experiment in which 10 observations are made and on 4 of these the machine tool is observed to be in operation. What are the prior and revised means?

7.17 A rectangular probability density function has parameters a and b. The mean is $(a+b)/2$ and the variance is $(b-a)2/12$. Construct a parameter map for this probability density function. What special case of the beta function is equivalent?

chapter eight

Macro-ergonomics: Task analysis and process mapping

About the chapter

Much of the previous chapters have dealt with individual tasks or actions. However, ergonomic design is also concerned with the design of systems—of flows of material and information within socio-technical systems. In designing or analyzing such systems, ergonomics specialists will typically conduct some form of task analysis.

"Task analysis" is a general term, referring to the analysis of any task—from individual tasks as discussed in the previous chapters to an entire production or service task, which may involve many individuals and machines. This chapter introduces how such analysis is done and how it is documented.

Documentation of task analysis typically involves some kind of chart, diagram, or graph that conveniently describes the task or system. This chapter describes a variety of these graphic devices and discusses each in terms of its particular strengths and weaknesses. Most of these graphic techniques reflect some fundamental ergonomic design principle, so the chapter also provides a general discussion of these principles.

Introducing the reader to these principles is a primary objective of this chapter, along with developing knowledge of how the diagrams and charts support these principles. It is also important that designers understand how the charts and diagrams differ so that the proper form can be selected.

Introduction

Ergonomic design typically begins with the recognition of some unmet needs. Those needs may be for process improvement or the creation of a new product. While the first case is more common, the second is less constrained. In fact the former is often thought of as problem solving rather than design, and the latter is referred to as designing with a fresh slate. The fact of the matter is that both cases involve problem solving and design, as most experienced engineers will confirm. When you ask such people how they perform this problem-solving-design work, they generally tell you that they begin with an initial situation or concept. The description of that initial situation or concept may be a task analysis, a task time-line graph, a flow diagram, or some other description of the current operation or initially proposed plan. These graphic descriptions show specific features of the concept so that other members of the design team can understand what is there now (if anything) or what the essential features of the initial concept are. Some of the graphs, charts, and diagrams used initially to describe the concept also help identify problems and potential improvements.

The next step ergonomic designers consider is how to improve the initial design. Improvements are made by applying design principles. Some of the principles are applicable at the start of the process, and some are best used once the design is really finalized.* (Some typical design principles are described later in this chapter.) Experience helps designers learn which principles are most effective for given situations, and how to refine or extend them.

Designing and problem solving both require access to a great deal of relevant data. However, if that data are to be useful, it must be provided during the design process in such a manner that the designers can access the precise data when they need it and in the form they want. Since the problem solver or designer is often using heuristic principles to improve an initial or intermediate conceptual plan, the available data need to be in a form that is compatible with those principles. A large amount of data are often required to do this; so the data need to be provided in a compact format. Several charts, graphs, and diagrams have been developed that help satisfy these needs of designers.

Some companies and industries use certain charts, graphs, or diagrams because they happen to be the ones they have used in the past. The latter situation obviously is not optimal. The designer should be familiar with the entire set of available tools and their respective advantages. They then can select the tool that best fits the problem, or modify a tool to fit their needs. A variety of suggested charts, graphs, and diagrams are shown in this chapter along with comments about Box what each one emphasizes and ignores. Many of the design principles discussed in later chapters are conveniently applied using particular graphs, charts, and diagrams introduced here.

Stages of problem solving[†] in design are shown in Box 8.1. As indicated there, a satisfactory solution depends upon the goals one has set, the criteria used to judge whether the goals are met, the expectations developed, and any time constraints imposed on the designer. It also follows that goals may be a bit fuzzy, particularly when designing new products. As expanded upon later in Chapter 13, one of the first steps in product design is to interview prospective customers or send them questionnaires to identify what potential customers want and their attitudes toward the company's current or proposed new products.[‡]

BOX 8.1 STAGES OF PROBLEM SOLVING

- Define the class of concepts to be manipulated, limits of manipulations, and goals sought.
- Interpret the physical world, its measurement, and the organization of those interpretations into some structure.
- Determine various courses of action open to the problem solver.
- Judge the effectiveness of alternatives of action and select the most promising.
- Verify the results for the selected action.
- Stop when a sufficiently satisfactory solution has been found.

---

* Newell, Shaw, and Simon (1958) studied strategies used by problem-solvers. They note that problem solving often involves the use of heuristics. It is interesting to note that these strategies are similar to those used in chess. Playing chess involves an opening gambit, a middle game, and an end game. Similarly, different design principles are applicable at particular stages of the design process.
† See R. Gagné (1959) for greater elaboration of this definition.
‡ Chapter 11 describes how to develop and use questionnaires and interviews to obtain such information. Chapter 15 includes a related discussion on how to systematically assess and analyze customer needs and wants to help ensure that products satisfy the customer. The latter chapter also discusses several other important aspects of product design including rapid prototyping and usability testing at all stages of the product design cycle. Some the specific topics addressed include quality function deployment, requirements analysis, expert assessment, usability tests, preference testing, and the use of designed experiments.

The same is true when the focus is on the design of tasks and processes—the topic of the current chapter. Knowing how to construct these information-seeking tools allows designers to be more productive and proficient.

Resistance to Change

Many textbooks contain sections that berate human "resistance to change." MacLeod (1995) tells us about a person he photographed carrying cans between a workbench and a chute. That person dumped the contents with difficulty. When MacLeod showed the film to ergonomists, their proposed solutions including the following:

- Getting a mechanical dumper
- Lowering the chute opening height
- Using a smaller and lighter can
- Moving the bench nearer to the chute

MacLeod said that he took the film to show a potential cause of back problems and found these potential solutions were quite good. However, he later discovered that the product was also made in another department, which had safer and more efficient equipment for making it. The department, where the film was taken, was supposed to stop that operation, but resisted that change. The supervisors did not discover that fact and both departments continued to make the part. Resistance to change led to a problem in this example. However, in many cases, the resistors may be right. Some past practices were well studied by smart people before they were adopted, so new methods are not always better.

Ergonomic design principles

Ergonomic design principles can help designers create better designs and solve particular design problems. There is no guarantee that a particular principle will work, and there is no formula for determining which principles are more important. In fact, one can encounter situations where following one principle will cause another to be violated. In such cases, one should do the best one can and develop a few conceptual designs. Those conceptual designs can be evaluated later to see which is best. Ideally, prospective customers or end-users will be involved at this stage.

Another issue that should be pointed out at the start of this discussion is that some of the principles mentioned later seem so obvious that one wonders how they can be overlooked. In practice, however, it is easy to develop tunnel vision and focus on one aspect of the design or on one application, while ignoring the system as a whole. Therefore, even obvious principles are included in the following discussion.

Traditional principles in ergonomic design

Some classical design principles for improving layouts and methods are as follows:

- Eliminate unnecessary elements.
- Simplify elements.
- Keep things natural.

- Combine compatible elements where possible.
- Group elements in their sequence of use.
- Locate those elements with the most frequent use in the most favorable locations.

These principles and a few more are discussed individually in the next section.

Advice from Some Famous Architects

The eminent architect Frank Lloyd Wright recognized the advantages, as well as the beauty, of nature when he said, "Imitate nature." He said he used this principle in designing the Imperial Hotel in Tokyo, Japan, to be earthquake resistant long before earthquake resistance was really understood in building design. Frank Lloyd Wright worked in Chicago as a young apprentice for the eminent architect Louis Sullivan. Sullivan invented the principle "form follows function," which means the designer should recognize the function that each part plays in a design and make it obvious. Wright advocated the Sullivan axiom, and added the imitation of nature principle.

Principle 8.1

Eliminate unnecessary elements. (Identify and eliminate waste in all its forms.)

The key word here is unnecessary. Unnecessary design features add to the cost of the design without benefiting the user. Since the user is often the buyer, products without apparent advantages are not likely to sell. But it is not always easy to recognize unnecessary features. Part of the difficulty is that some design features are perceived by users as giving fewer and fewer added benefits. For example, consider the many so-called improvements that are now included with computer word processing software. This type of thinking led to the concept of product simplification, an approach that examines each function of every product component and questions whether or not something simpler could be used instead.

The application of this simplification principle to the industrial process design requires designers to ask themselves if each activity independently contributes to the overall design effectiveness. If the answer is no, then a particular process activity may not be necessary. For example, a disassembly activity might be performed on a part that is purchased preassembled. The original assemble activity could possibly have been avoided, unless shipping the component would have endangered a unit that is not fully assembled. However, even though some activities do not contribute directly to system performance, they may be required by other activities that do. In that case, the noncontributing activity cannot be removed.

Principle 8.2

Simplify where possible.

Simple solutions to a design problem are elegant because they satisfy the need at lower cost and frequently produce longer life cycles. Simplifying helps to remove *functional fixedness** and lets the designer proceed. In the case of a problem with constraints,

---

* Functional fixedness is an old term from the psychological literature. It means that people view objects in terms of their traditional functions, and they are likely to reject their use for a new function (see Woodworth and Schlosberg, 1954).

one simplification is to remove the constraints. Once a simple solution is found without constraints, a similar solution can sometimes be found for the constrained problem. The Latin motto for this is *Simplex Munditis*, which means that there is elegance in simplicity.*

Frank and Lillian Gilbreth

At about the turn of the twentieth century, Frank Gilbreth was a bricklayer and contractor in Boston, MA. He studied bricklaying methods and improved productivity of the job greatly and became rather famous for it. Later he used the same concepts of motion and micromotion study to improve industrial tasks in his consulting. His wife, Lillian, helped him develop some of those methods. Frank was invited to London to speak to a prestigious society. However, he died of a heart attack before leaving for England and Lillian went in his place and delivered the address. In the 1930s, Lillian joined Purdue University where she taught in the departments of Industrial Engineering and Home Economics. Lillian was honored posthumously by a stamp issued in her honor. The book and motion picture entitled *Cheaper by the Dozen* is the story of their life with their 12 children. In the movie, Clifton Web played Frank and Myrna Loy played Lillian.

Principle 8.3

Use, do not oppose, nature. Compatible activities feel natural and natural activities are rarely incompatible.

Natural activities are compatible because people intuitively avoid actions that feel awkward. The meaning of the word *natural* in design stems from the advice of Charles Babbage† to use nature's help in industry. Babbage often referred to the use of gravity to deliver materials from one location to another. Natural methods are usually much less expensive than unnatural methods, and they are often more reliable over time. For example, consider the case of a simple gravity feed compared to a motorized conveyor.

Principle 8.4

Combining activities can often improve operations, provided that the activities being combined are compatible.

There are obvious limits to combining activities and care must be taken to avoid exceeding those limits. In one sense, combining activities means doing two things at once. If the performance time of the combined activity is less than the sum of the times required of each activity separately, a speed advantage is achieved. Another clearly related approach is to use the same hand tool or machine tool for two separate activities. In this case, one saves the time and effort of changing hand tools or moving from one machine to another.

Principle 8.5

Revise operational sequences to reduce incompatibilities between successive activities.

---

* The Honda Automobile Company stresses the simplicity concept throughout the corporation and they have met considerable success as a result.
† Charles Babbage was a leader in England during the early nineteenth century. He was also the father of the digital computer even though he never got his mechanical computer to work. Oddly and Babbage also advocated the idea of interchangeable parts, and some say that the reason his mechanical computer failed was because it did not have such parts.

Simultaneously rubbing one's abdomen and patting one's head is a surprisingly difficult activity for most people. Concurrent activities interfere when they use the same sensory, cognitive, or physical resources. Hence, performance on one task results in deterioration of performance on the other. Sequential interference often occurs when the ending activities of one task make the beginning activities on the next task more difficult. For example, if task A requires a person to start at location U and end at location V, and task B starts at location W and ends at location U, then the sequence B-to-A is more compatible than the sequence A-to-B. In this way, changes in sequences can reduce interference. In certain cases, the nature of the job may dictate the sequence of tasks. For such situations it is advisable to organize components, parts, and tooling in a natural sequence.

Principle 8.6

Keep the flow of work moving along smooth curves over space, minimizing backtracking, extra movement, delays, and needless inventory.

Most readers will recognize this as a basis for the *Just in Time* concept of work philosophy that stresses proper work flows over time and space. Other readers will identify this principle as a variation of the recommendation to cut out unnecessary activities. Both interpretations are right.

The following discussion introduces some tools that are very helpful when applying the aforementioned principles. These tools are normally used during various forms of analysis, such as task analysis, activity analysis, process flow analysis, critical path analysis, link analysis, methods analysis, reliability analysis, hazard analysis, failure modes and effects analysis, and others. Some of these techniques will be introduced in this chapter. Others will be discussed in detail in the following chapters.

Visual graphs of operations

Some of the more common charts, graphs, and diagrams used in ergonomic design are introduced in the following discussion. Many have traditional roots in industrial engineering and others are much more contemporary. There is no attempt here to be exhaustive. It should also be noted that software is now available from a variety of vendors for developing charts and diagrams that can be very useful when analyzing tasks or processes. For example, Microsoft Visio is a diagramming program for developing flowcharts and other task diagrams often used to document processes in manufacturing, health care, and other industries. Simulation packages are also now available that can be used to develop task or process simulations from Visio flowcharts.*

Task Analysis

Task analysis is one of the most well-known techniques in ergonomic design. Some early uses of this technique are attributed to Frank Gilbreth and Frederick Taylor. At the time of Gilbreth and Taylor, investigators would observe and identify each specific action undertaken by the person being studied. The precision needed to identify the actions varied from time to time, but most investigators insisted that the descriptions be precise enough for the exact

---

* Simulation provides a powerful tool for analyzing tasks and processes, as expanded upon later in Chapter 12. Not all simulation packages are compatible with Visio, but most include a graphical interface for entering a flow diagram or chart describing the simulated process or task.

procedure to be repeated by someone else at another location. The units of measurement that Frank Gilbreth devised were later refined, expanded, and renamed "THERBLIGS" in his honor, as discussed in Chapter 5. Both Gilbreth and Taylor often studied human jobs away from the factory floor in order to experiment with improvements without interfering with production operations. Accordingly, task analysis originated with the creation of task descriptions obtained by observing the job as it was normally performed. The procedure is usually the same today. First find out what people do and then investigate how an ergonomic designer can help improve the task.

Operations process charts

Frank and Lillian Gilbreth first published the basic idea of the operations process chart in 1921.* This chart conveniently describes the sequence of tasks performed in an operation. It also shows how materials and product components are introduced into the production sequence. This chart developed by the Gilbreths had two kinds of nodes: *squares* represented people performing inspections and *circles* represented other activities. Some versions now use an expanded set of node symbols, as discussed further in the following section on flow process charts. A vertical line to immediately preceding and successive activities connects nodes. *Horizontal lines* show where a material or a product component enters the process. A material or component is normally required at the start of the process, so a horizontal line is almost always attached to the first node of the chart.

The Gilbreths advocated that the elapsed time of all activities should be recorded next to the corresponding nodes. They also suggested labeling each node with phrases summarizing the work performed during the activity. Time information is particularly important for finding the critical path through a network and for identifying the nodes that take the longest times. A well-known engineering principle is that the elements with the greatest potential for improvement should be focused on first. In most cases, these will be activities that take the longest time or create the greatest number of errors. Activity nodes corresponding to operations can be labeled O-1, O-2, O-3, …, in an increasing sequence, to provide a key or indicator on the chart that is cross-indexed to a detailed description provided elsewhere. Inspection nodes similarly might be labeled I-1, I-2, I-3, …. The materials or components that are introduced into the operation also are normally indicated, following a similar scheme.

Figure 8.1 illustrates an operations process chart for the assembly of a home furnace thermostat controller. Note that each of the three major paths of this chart shows a material or component being introduced prior to performing the first activities. Machining and other operations are performed on each path before they come together. At this point, an assembly operation typically follows. This figure illustrates the emphasis of operations process chart on the sequence of necessary activities.

In using these charts, designers ought to ask themselves about the necessity of each activity and about possible benefits of alternative sequences of activities. In the case of Figure 8.1, a designer might question the need for so much inspection, and ask if inspection I-2 might be more effective, if made after operation O-11. When expected time values for each operation are known, one can find the critical (i.e., longest cumulative

---

* This particular kind of chart has been around for many years and is still very popular in industry. Frank and Lillian Gibreth were mentioned earlier as pioneers in the field of industrial engineering and ergonomics. See Chapter 8 for more about these fascinating people.

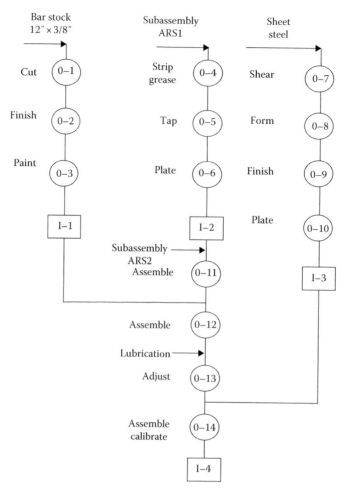

Figure 8.1 Example operations process chart.

time) path leading to operation O-14. Quicker operations along the critical path shorten the thermostat-making operations. Thus, this knowledge is most helpful.

Charts similar to operations process charts, called *fabrication charts*, are sometimes used to describe the sequence of activities in assembling a product. The latter charts often do not distinguish between vertical and horizontal lines.

Flow process charts

The flow process chart is a related diagram that also describes a sequence of activities to be performed. Flow process charts are often used to help identify inefficient operations, by focusing attention on activities that do not contribute to the end product or are not necessary, and go beyond operations process charts by including activities such as storage, transportation, and waiting. In order to describe the additional activities, mentioned previously, flow process charts use an expanded list of symbols. Any delays in the process are indicated by a fat-D symbol. Storage activities are shown with triangles. Some delays involve storage for short time periods. For example, parts may accumulate in a buffer as they wait for a machine to become free. Movements of materials, components, or assemblies are shown with arrow symbols. A circle within a square represents combined activities, typically, an inspection combined with other activities. A committee established by the American Society of Mechanical Engineers (ASME) developed this expanded set of symbols many years ago. Some companies use this expanded list for the operations process chart and list all activities, not just the essential ones. Neither vertical nor horizontal lines are used in flow process charts, but the sequential listing of the activities indicates the sequence of activities.

Figure 8.2 illustrates a typical flow process chart. This particular example describes the process of assembling a timing gear cover plate. As illustrated by this example, standard forms are typically printed ahead of time containing columns of symbols. The person

Flow process chart				MAN/Material equipment Type				
Chart no.2	Sheet no.1		Of 1	S u m m a r y				
Subject charted: *Used bus engines*				Activity		Present	Proposed	Saving
				Operation ○		4	3	1
				Transport ⇨		21	75	6
Activity: *Stripping, degreasing and cleaning prior to inspection*				Delay D		3	2	1
				Inspection ☐		1	–	1
				Storage ▽		1	1	1
Method: Present/proposed				Distance [m]		237.5	1500	87.5
Location: *Degreasing shop*				Time (max-min)		–	–	–
Operatives:	Clock nos.	1234 571		Cost Labor Material				
Charted by: Approved by:		Date:		Total		–	–	–
Description			Qty.	Distance (M)	Time (Min)	Symbol ○ ⇨ D ☐ ▽		Remarks
Stored in old-engine store				–	–	●		
Engine picked up						●		⎡ Electric
Transported to stripping day				55		●		⎨ hoist on mono-
Unloaded on to engine stand						●		⎣ rail
Engine stripped						●		
Transported to degreaser basket				1		●		By hand
Loaded into basket						●		Hoist
Transported to degreaser				15		●		"
Unloaded into degreaser						●		"
Degreased						●		
Unloaded from degreaser						●		"
Transported from degreaser				45		●		"
Unloaded to ground						●		
Allowed to cool						●		
Transported to cleaning benches				6		●		"
All parts cleaned						●		
All parts collected in special rays				6		●		
Awaiting transport						●		
Trays and cylinder block loaded on trolley						●		
Transported to engine inspection section				75		●		Trolley
Trays slid on to inspection benches and blocks						●		
On to platform						●		
Total				150		3 15 2 – 1		

Figure 8.2 Example flow process chart showing flow of materials. (From ILO, 1979.)

constructing the chart circles or draws lines between the appropriate symbols to show the sequence of activities that occurs in a particular process. Often, though not always, the movement distances are marked on the form when a transport activity occurs. Figure 8.2 shows distances as a separate column. This particular version of the flow process chart includes a summary table that denotes how often each type of activity is performed, as well as the sum of the transportation distances. It also includes a description of how parts are moved (in this case by hand) and how many parts are moved at a time.

Flow diagrams

Flow diagrams typically use an expanded set of ASME symbols similar to those used in flow process charts. However, flow diagrams do not list these symbols sequentially, as in flow process charts, or as a rectangular network of nodes and lines, as in operations process charts. Instead, flow diagrams display symbols on a pictorial layout of the plant where the activities occur. Plan views of the plants are shown with building partitions and machines are shown in scale. Dashed lines typically show movements of materials and components.

An example of a flow diagram is given in Figure 8.3. This example illustrates the manufacturing process of sheet metal stove pipe, except for the step of crimping one end of each pipe length. Note that sheet metal is first removed from storage in the warehouse and moved to a location next to the shear machine. The delay there (see the fat-D with the 1 inside it) indicates a relatively short wait, before the sheets of flat metal are cut into rectangles. After shearing, these cut pieces of flat metal are moved to a spot adjacent to the next operation where they wait again. The next machine function is to roll these flat pieces of metal into tubes. After this rolling process is done, these tubes are moved next to one of the three punches and

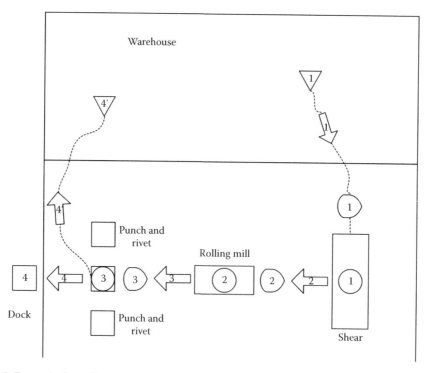

Figure 8.3 Example flow diagram.

Chapter eight: Macro-ergonomics: Task analysis and process mapping

rivet machines to await that operation. When this last operation is complete, the pipe sections are either moved to the loading dock for delivery to a customer or placed in storage.

Unlike flow process charts, flow diagrams do not usually display transport distances, because distances can be approximated from the scale drawing of the facility layout. Also, operation times are not usually shown. Flow diagrams are normally used to determine how the layout affects the process. Any backtracking or traffic congestion effects are also emphasized. Note that placing flow diagrams on top of each other shows potential interference between concurrently performed processes. Those interferences might be due to competition for machine tools or congested routes of transportation flow involving conveyors, forklifts, or other materials handling equipment.

As the name implies, flow direction and smoothness is the focus of the flow diagram. While it is helpful to smooth out the flow of all processes, it is especially imperative for those processes that involve the greatest manufacturing quantities.

Multiple activity charts

Several versions of multiple activity charts have been developed for various uses. These charts are given various names depending upon their use and the activities being described. Multiple activity charts usually contain a number of parallel vertical columns. Each individual column denotes different activity centers. Some examples include left-hand and right-hand charts, gang charts showing different members of a work crew, and Gantt charts showing stations of an assembly line. The vertical dimension is time. The horizontal lines are typically drawn within each column to separate different activities within an activity center. Those separate activities are either coded or labeled for identification purposes.

Figure 8.4 illustrates a five-station bicycle assembly line. Each station assembles a different subassembly of the bicycle that then goes to the next station. Within each station,

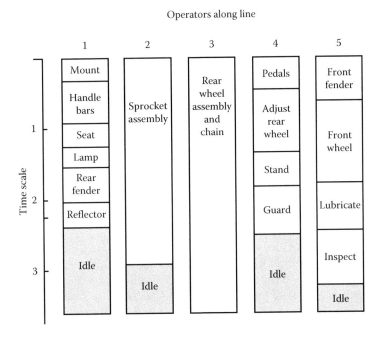

Figure 8.4 A multiple activity chart.

the component activities and their time requirement are shown. Idle time occurs at a particular station, when all the assigned work is completed but the next bicycle assembly has not yet arrived. In assembly lines, the cycle time for moving the assembly between successive stations is at least the sum of the greatest times of all stations. In the figure, Station 3 has the longest assembly time, so all of the other stations are temporarily idle during each cycle.

Assembly line balancing is often done with the aid of multiple activity charts. The person balancing the line shifts activities among stations in order to minimize the cycle time, subject to precedence requirements. Since Station 3 is a single activity which is system limiting, no improvement in cycle time can be made in this example as long as there are five stations. With fewer stations, however, the tasks could be combined differently to reduce the amount of idleness. The most radical solution is to replace the assembly line with five work cells. This reduces idle time to zero, but the worker in each cell will need to know how to assemble the entire bicycle. More support fixtures would also be needed. The additional training and support costs might outweigh the benefit of reducing idle time. On the other hand, some workers might prefer the expanded set of tasks. Analyzing this latter set of issues would be the role of the ergonomic specialist.

As indicated by the previous comments, multiactivity charts are very useful for identifying system-limiting activities among the specified activity centers. These charts also help identify interference, idleness, and imbalance among activities. For these reasons, multiactivity charts are used to describe members of work groups, people working with one or more machines, and the simultaneous left- and right-hand activities of an individual. In each case, time is a critical resource for a person, machine, or hand. It is also of note that the kind of chart that Henry Gantt advocated for scheduling is really a multiactivity chart turned on its side, where each row is a separate project being scheduled.

Precedence diagrams and matrices

Precedence diagrams are networks of circular nodes and branches. The nodes represent specific activities and one-way branches (i.e., arrows) show activity sequences. Following this logic, no activity is performed before itself, so there are no arrows emanating from a node and immediately returning to it. In fact, no path returns to a node. Thus, for every pair of nodes, say nodes A and B, either A precedes B, or A follows B, or A is unordered with B.

Figure 8.5 illustrates an elementary precedence diagram for five activities and a corresponding matrix representation of that diagram. In this matrix, the following symbols are used:

1	Indicates that the row activity must precede the column activity
2	Indicates that the row activity must follow the column activity
0	Indicates that these activities are unordered with respect to each other

Zeros appear on the major diagonal of the matrix, because it is assumed an activity cannot both follow and precede itself. Note further that if one folded the matrix of Figure 8.5 along the major diagonal, the sum of all of the overlaid symbols would be either three or zero. The reason for that is quite simple. If A precedes B, then B follows A. The first relation is assigned a 1, the other a 2, so they sum to 3. Similarly, if A is unordered with respect to B, B is unordered with respect to A. Both cases correspond to 0s, so the sum is also 0.

Chapter eight: Macro-ergonomics: Task analysis and process mapping

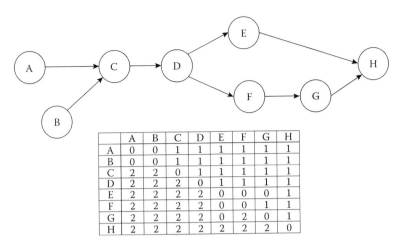

Figure 8.5 An elementary precedence diagram and matrix.

The principal purpose of precedence diagrams is to emphasize *sequencing* requirements. Since it is very difficult to remember all of the sequencing relationships in many production processes, these diagrams and matrices serve as useful memory aids, and provide useful checks for consistency. Note that performance times can be included for each node in a precedence diagram. The longest time path through that diagram is the *critical path*, and the times along that path denote the *shortest time* of total operation performance.* Some designers who need precedence information prefer looking at the graphs to the matrix. The precedence matrix is most useful to those who primarily want to do precedence testing on a computer and only need to be informed when violations occur.

Link diagrams

Link diagrams show the relationship between machines and people.[†] Links between elements show how strongly they are related. Some of the typical relationships described with links include (1) the frequency of travel between locations, (2) the amount or rate of flow of materials, (3) the frequency and sequence of monitoring particular displays or other items, or simply (4) importance, with perceived relative importance as the strength.

Figure 8.6 illustrates a simple link diagram of a home kitchen layout where the links describe the relative frequencies of travel between locations. This illustration came from the work of Lillian Gilbreth, who studied the layout of homes and commercial kitchens[‡] in the 1930s and 1940s.

Another illustration of a link diagram appears in Figure 8.7, which came from a study performed by Fitts, Jones, and Milton (1950).[§] The links in Figure 8.7 are the relative

---

* Readers who are aware of the program evaluation and review technique (PERT) will recognize the notion of the critical path and its length. PERT is used for project management purposes.
† One of the earliest references to this form of diagram was by Chapanis (1959).
‡ It is believed that Lillian Gilbreth came up with this idea while she was Professor of Home Economics and of Industrial Engineering at Purdue University, in West Lafayette, IN. Besides working on family home food services, Professor Gilbreth did a lot of work on institutional food services.
§ The experimenters performed this study by putting a mirror above the windshield inside the cockpit of larger aircraft, which allowed them to photograph aircraft pilots' eyes during landing operations. They calibrated the photographs by having the pilots look at each visual display. Those calibrated photographs aided in identifying the sequence of displays observed over time.

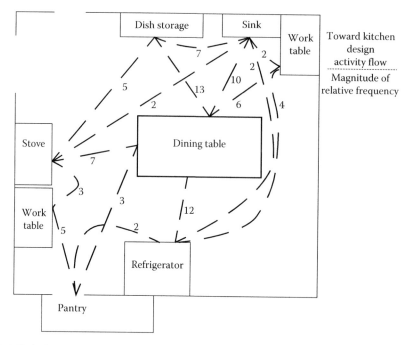

Figure 8.6 A link diagram of a home kitchen layout developed by Lilian Gilbreth.

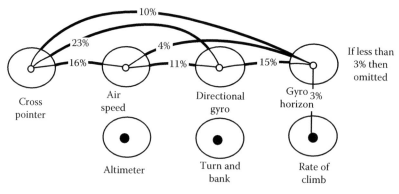

Figure 8.7 A link diagram of pilots' eye movements during landing. (From: Fitts, P.M. et al., *Aeronaut. Eng. Rev.*, 9 February 24, 1950.)

frequencies of pilot eye movements between pairs of specific visual displays during landings, a phase of flight where accident rates are especially elevated. It follows that if strongly linked displays were grouped together, less time would be spent on eye movements between displays. Chapanis gives other examples in his book on human factors techniques.* As illustrated there, and by our previous examples, link diagrams are a good supporting display both for direct use by designers and for illustrating to management the results of a redesign.

---

* See Alphonse Chapanis (1959) for link diagrams of Combat Intelligence Center (CIC) layouts. The CIC is part of warships where all the firepower is directed and it is the center of communications outside of the ship. Similar layouts have been tested at the special training and research center on Point Loma, San Diego, CA.

Chapter eight: Macro-ergonomics: Task analysis and process mapping   337

Task time-line analysis charts

The analytical technique known as *task time-line analysis* was originally developed for use by aircraft designers interested in issues such as pilot workload and how to allocate the tasks during flight phases. Aircraft pilots are required to perform a series of specified tasks within separately designated aircraft system phases, such as takeoff, mid-flight, approach to the airfield, and landing. Pilot workload varies greatly between flight phases, and the tasks must be allocated accordingly. Ergonomic specialists in the aircraft design and manufacturing industry must respond to the question, "Can the pilot and crew perform the tasks assigned within each phase of flight?" One of the traditional rules of thumb is that aircraft crewmembers shall NOT be more than 80% busy during any phase.*

The critical issue is that the tasks must be done or dire consequences will be incurred. This means a person must be assigned to perform each task, at an appropriate phase of flight. A diagram referred to as the task time-line graph (TLG) is useful during this process.

Figure 8.8 illustrates some basic elements of a TLG. Notice that the lower part of the graph shows the percentages of time that each operator is engaged in a task. By displaying both task requirements and operator capacity, this visual display also tells designers which tasks may theoretically[†] be reassigned to another operator.

Although task time-line analysis is used almost exclusively in aircraft design, it is relevant and very useful in many other situations. For example, human operators working

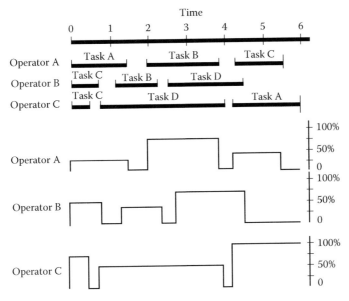

Figure 8.8 A task time-line analysis diagram.

---

* Parks (1979) identifies this rule of thumb used in the aircraft industry. While use of such rules simplifies the design process, they are controversial. Some critics charge that the 80% figure is an arbitrary choice. One argument is that the rule is based on agreement by experienced people, rather than scientific evidence. Others argue that some tasks are naturally more variable in performance time and it would seem that a smaller acceptable percentage should be used when the sequence of tasks contains more tasks with higher time variances.
† There are clearly other relevant considerations that go beyond determining how much time is needed to do a task, but this approach provides a starting basis, especially if there are numerous operators.

with machine tools or performing assembly operations are often constrained by the time requirements of a machine tool or process during various phases of the operations. This type of visual display may be particularly appropriate for those situations.

Fault trees

Fault tree analysis is a technique developed in the early 1960s to evaluate causes of accidents. This approach involves the development of fault trees to describe how or why accidents happen. When enough data are available, the fault trees are used to predict the probability that an accident will occur. A fault tree* resembles a decision tree. That resemblance accounts in part for the name. The accident or end result is shown at the top of a fault tree. This results in a hierarchy, where the conditions that must be present for an event to happen are shown below each event. The logical relationships between conditions and events are shown with two types of nodes (or logic gates):

1. AND-gate nodes require that *all* of the conditions immediately below the event must occur before the event can happen.
2. OR-gate nodes indicate that the event will occur when *any* of the conditions are present.

Figure 8.9 illustrates a simple fault tree containing both an AND-gate and an OR-gate. A quick look at this figure reveals that the end result can be prevented or its likelihood considerably lessened by removing any of the conditions beneath the AND-gate. Facilitating such diagnosis is the principal intent of fault trees. The specific example of Figure 8.9

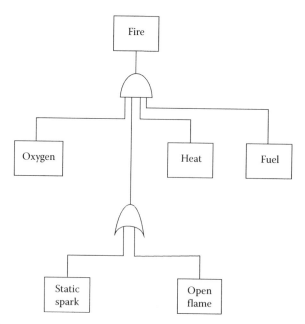

Figure 8.9 A very elementary fault tree.

---

* C.R. Asfahl (1990, 2nd edn.) tells us that engineers at Bell Labs invented fault tree analysis in the early 1960s to help avoid potential disasters involving missile systems.

> **BOX 8.2 NUMERICAL EXAMPLE OF FTA**
>
> As an elementary example, suppose the probability of a spark is 0.4 and that of an open flame is 0.3. If those two events (i.e., a spark or an open flame occurring) are statistically independent, the probability of both events occurring is given by basic probability theory as
>
> $$p(\text{spark or open flame}) = 0.4 + 0.3 - 0.3 * 0.4 = 0.58$$
>
> The probability of a fire is given as
>
> $$p(\text{fire}) = p(\text{oxygen}) * p(\text{fuel}) * p(\text{heat}) * p(\text{spark or open flame})$$
>
> If oxygen, fuel, and heat are always present, the probability of a fire becomes
>
> $$p(\text{fire}) = 1 * 1 * 1 * 0.58 = 0.58$$
>
> Note that if the probability of fuel being present is cut in half, the probability of the end result is also cut in half to 0.29.

shows that a fire can start only when four conditions (noted by the *and* node) are present: oxygen, heat, fuel, and either a spark or an open flame (which is exemplified by the *or* node). Note that eliminating an open flame reduces the chance of a fire, but does not eliminate it, because a spark is another possible ignition source, as indicated by the OR-gate.

Fault trees can be used to compute the probability of specific end results given that the relationship between events and their causes is known (Box 8.2). These calculations directly follow reliability computations using the *and–or* logic that is well known in the literature.* Figure 8.10 is a fault tree analysis diagram for an actual alerting system.

A related form of diagramming, known as integrated definition (IDEF), was devised several years ago by the U.S. Air Force for a myriad of applications.† Several versions of IDEF have been devised. The specific version is specified with a number (e.g., IDEF-1). IDEF diagrams are input into a computer. The diagrams allow AND- and OR- (inclusive or exclusive) gates to be used. IDEP diagrams can include links to show precedence relationships and object flows over time. Links can also show relative frequencies and importance. Consequently, IDEP can be used to generate fault tree diagrams, flow diagrams, precedence diagrams, and link diagrams. While the use of IDEP goes beyond the scope of this book, this approach is mentioned here to make the reader aware that there is software available for creating almost all of the previously discussed forms of graphs and for performing some of the discussed forms of analysis.

Failure modes and effects analysis tables

Fault tree analysis is often closely associated with a technique called failure modes and effects analysis (FMEA). FMEA is a systematic procedure for documenting the effects of

---

* See Barlow and Lambert (1975) for more details on fault tree reliability computations.
† Ang and Gray (1993) describe an early form of IDEP. Kusiak, Larson and Wang (1994) show its use in manufacturing, and Kusiak and Zakarian (1996) describe IDEP in reliability and risk analysis.

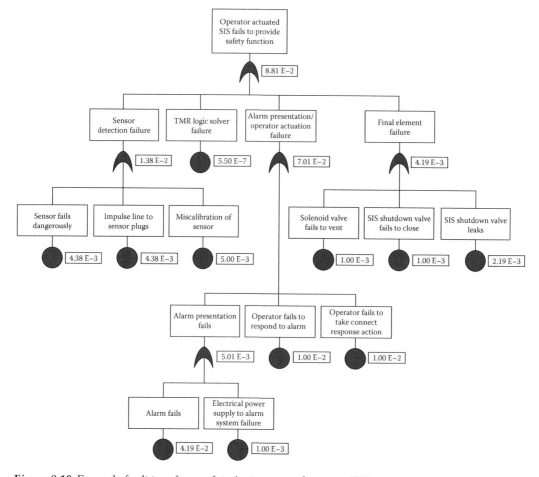

Figure 8.10 Example fault tree for a safety instrumented system (SIS).

system malfunctions on reliability and safety (see Hammer, 1972, 1993). Variants of this approach include preliminary hazard analysis, and failure modes effects and criticality analysis. In all of these approaches, worksheets are prepared that list the components of a system, their potential failure modes, the likelihood and effects of each failure, and both the implemented and the potential countermeasures that might be taken to prevent the failure or its effects. Each failure may have multiple effects. More than one countermeasure may also be relevant for each failure.

The idea behind FMEA is to trace elemental failures that combine in various ways and lead to a catastrophic accident or failure. Companies planning a new product can use this approach to identify potential causes of accidents. By keeping that knowledge in mind, designers can examine those results and find ways to make their designs safer by separating critical elements. Clearly, the information FMEA provides about the elements most sensitive to failures and how they might cause accidents is extremely valuable in preventive maintenance operations. Approaches very similar to fault tree analysis and FMEA can be used to estimate the reliability of human operators as well.*

---

* This example is from Gitlow et al. (1989).

Cause-and-effect diagrams

Cause-and-effect (C&E) diagrams arose in the field of quality control, rather than ergonomics, but they are equally useful in either application. Professor Kaoru Ishikawa, then President of the Musashi Institute of Technology in Tokyo, Japan, invented the concept because he found that many plant personnel were overwhelmed by the number of possible factors that influenced a process. At the time, the notion of *brainstorming* was quite popular, and this diagram was devised to support this process. But its applications go well beyond brain storming alone. A C&E diagram is also called a fish bone diagram because of its appearance and because the inventor's name sounds like fish.

Let us begin by showing how a C&E diagram might be used during a brainstorming process. In brainstorming sessions, a group of people is supposed to first generate causes of, or solutions to, a problem without evaluating them. C&E diagrams serve the former use by organizing possible causes for a problem. The diagram starts with the problem identified in a box on the far right. On the left are causes shown as subcauses. A horizontal line, referred to as the spine, is drawn that points directly at the effect. Major causes are placed in boxes that connect to the spine. The lines that connect causes to the spine are called bones. Note that the major causes are really general categories. If the brainstorming group is presented with a skeletal diagram showing the problem and some major causes, someone can write the suggested subcauses under the appropriate cause as one of the bones in the C&E diagram. As a subcause is identified, others will think of further subcauses and sub-subcauses. When everyone runs out of steam and the C&E diagram is rather full, evaluation can start. During the evaluation, the more likely causes can be encircled to highlight them.

Figure 8.11 shows an example of a C&E situation dealing with the problem of bad emergency medical service run reports. Numerous potential causes are identified in the chart. The analysts would next try to order the potential causes by likelihood and determine if they are contributing to the problem or not.

Decision flow diagrams

Decision flow diagrams are often used in the military, and some industrial settings, to describe how decisions are made within tasks. A primary application in the military is to show the proper way to operate specific kinds of equipment. Most of these diagrams are complicated because the equipment is complex. A simple decision flow diagram is shown in Figure 8.12, which shows the process followed in starting an automobile. Information needed by the designer is indicated with triangles. Decisions are indicated with circles, and actions with rectangles. The diamonds in Figure 8.12 indicate branching points in the task. A more complex example, using a slightly different format, is given in Figure 8.13.

When new equipment is designed, it is especially important that the designers ensure that the person operating it will have enough information to correctly make the necessary operating decisions. Furthermore, this information must be provided at the time it is needed. A decision flow diagram helps address these issues by showing what information is needed for each action the human operator performs in the operating sequence. The diagram shows how the actions cause the system to change its status, and also shows the corresponding flows of information back to the operator. When the system does not adequately inform the operator of its status, corrections can be made to address that deficiency. For example, status indicators (such as lights) might be included in the design. Decision flow diagrams can also guide the design of instruction and repair manuals,

342 Introduction to human factors and ergonomics for engineers

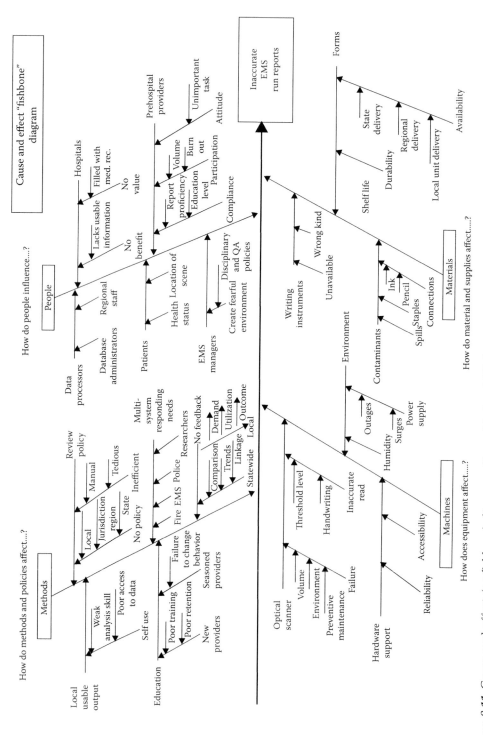

Figure 8.11 Cause-and-effect (or fishbone) diagram. (From National Highway Transportation Services Administration (NHTSA), *A Leadership Guide to Quality Improvement for Emergency Medical Services Systems*, 1995, Available online: http://www.nhtsa.gov/people/Injury/ems/leaderguide/ (accessed 11 July, 2012).)

Chapter eight: Macro-ergonomics: Task analysis and process mapping 343

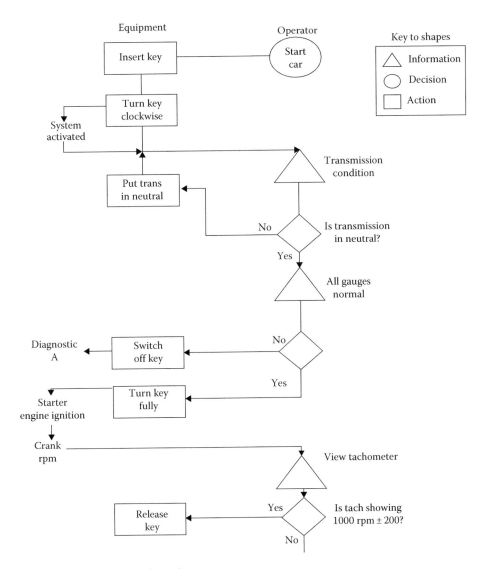

Figure 8.12 A simple decision flow diagram.

trouble-shooting procedures and diagrams, training materials, and other means of providing information to operators.

Analysis of tasks and jobs

Tasks involve activity performed by a person in order to attain some goal. These activities may be sensory, mental, or physical in nature. The bottom line is that performing the activities results in something productive. These activities follow some sequence and may be contingent on the results of other activities. Smaller units of task are often referred to as subtasks and components of subtasks may be called sub-subtasks. Theoretically, one can define more and more refined components down to the level of elements.

A specific job is defined by the collection of tasks that could be assigned to a human operator. Generally, a class of jobs requires a specifiable minimum level of knowledge,

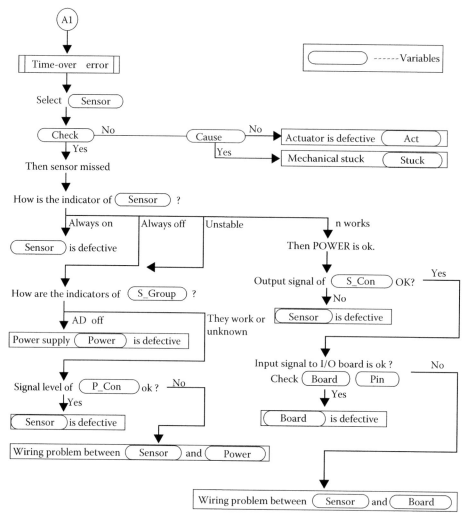

Figure 8.13 Decision flow of designer's trouble-shooting strategy for a chip mounting machine. (Reprinted from Naruo, Lehto, and Salvendy, 1990. With permission from Taylor & Francis.)

experience, and set of skills. Some jobs require greater levels of knowledge, experience, and skills. Jobs containing tasks with knowledge, experience, and skill requirements within a predefined range are combined into a common job class.

A *position* is a grouping of jobs that are treated as approximately equal. Positions consist of one, two, or more job categories for which the requirements in terms of knowledge, experience, and skill are very similar. In hiring, the determination is frequently made to hire a certain number of people in one position and a certain number in another. Since the jobs in a position are nearly equal, so are the wages. Defining positions helps in budgeting. Collections of tasks focused on specific objectives are referred to as *methods*. When a method is accepted by a company as the procedure to be followed, it is known as a *standard method*, and unless it is otherwise stated, it is assumed that method is followed.

The latter form of analysis began when the Gilbreths first proposed that tasks be described in terms of basic body motions. As time went on, several deficiencies were observed in the Gilbreths' original form of task analysis and some scientific improvements were proposed.* This history demonstrates clearly that the notion of a task is rather arbitrary. From a practical viewpoint, a task is a clearly observable, goal directed, unit of activity. It is also an activity given to an individual operator and carried out at that person's initiative. Consequently, activities requiring more than a single person are divided into two or more tasks, which may or may not be performed at the same time. Another feature of a task is that it is associated with a goal or subgoal—that is, ergonomic designers focus on tasks performed for some nontrivial purpose.†

While the goals of a task are not always important in understanding the task itself, one should never lose sight of the intentional and purposeful nature of a task. Some of the simplest goals of task performance are to achieve a weekly or biweekly paycheck. Ergonomists normally focus on operator goals that are specifically related to task performance. Doing so is particularly important when analyzing how people perform cognitive tasks, which involve problem solving and decision making. As applications become more and more cognitive, it is critical that designers document what the operator needs to know to perform the task correctly. These requirements can then be compared to the operator's knowledge, goals, and rules of conduct to assess performance for different users of particular designs.‡

A general overview of this process follows. More detail on particular ways of doing task analysis will be provided in almost all of the chapters. The first step in the process is to describe the behavior that will take place in the task. The next step involves determining task requirements. These requirements can then be analyzed to see if they are reasonable. When the requirements exceed the capability of the targeted population, the product or process may need to be redesigned. Task analysis also may involve allocating functions between people and machines.

Describing the task

The first step in describing a task is to identify the component activities. Once the component activities are known, focus can be shifted to analyzing sequences of activity. Task descriptions vary greatly in their level of detail. At one extreme, very detailed descriptions are developed (e.g., reach 19 in. in a direction 25° right of the body center to a location on the work bench for objects 3/8 in. diameter and 1 in. long). Such descriptions are very useful in some settings (such as for highly repetitive assembly tasks), but it takes a lot of effort to develop task descriptions at this level of detail. More importantly, this approach does not consider the cognitive aspects of job performance.

---

* Crossman (1956) added perceptual-motor skills and advocated the need for mental activities. Miller (1962) started his quest for a scientifically based taxonomy. A few years later, Seymour (1966) described the use of skills in these various tasks. About the same time, Annett and Duncan (1967) started the concept of hierarchical tasks. Those notions appear to have lain dormant for several years until there was a series of developments by Piso (1981), Shepherd (1989), and Stammers, Carey, and Astley (1990).
† Some approaches for taking such observations are discussed in Chapter 7. Making a video recording of the task is often the easy part. Anyone who has tried to analyze video tapes of task activity knows well that this can be time-consuming, tedious, and difficult. Fortunately, computer tools, such as the Interact software available from Mangold International (http://www.mangold-international.com), are now available to make it easier to analyze video recordings of task activity. The latter software allows the analyst to view digital video recordings of the task on a PC, and provides a convenient interface for marking the start and end of particular activities, which are then automatically recorded in a convenient spread sheet format. The interface also provides the analyst the ability to rapidly move back and forth at different speeds through the sequence of activities to specific frames of interest, which is a great improvement over trying to figure out exactly where a task starts or ends using a standard video player or movie projector!
‡ Chapter 10 goes into more detail on the use of GOMS to predict performance time for word-processing tasks.

The broad-brush approach takes much less time, and often gives more insight into cognitive and other issues at the early design stage of a product or process. On the other hand, the broad-brush approach still requires someone to go back later and supply the details, after the designers are more familiar with the design. Many designers prefer working with the broad-brush approach during the early stages of design. More refined descriptions of the task are then developed, once the design is nearly completed.

Task analysis of making coffee

To illustrate this initial stage of task analysis, consider the rather simple task of making coffee with a contemporary coffeemaker, given that filters, a water faucet, and a can of ground coffee are close at hand, and assuming that all elements of the coffeemaker are clean and sitting in their typical location. As a first pass at describing how to make coffee with this coffeemaker, the following three steps were identified:

1. Fill the coffeemaker with a pot of water.
2. Put a filter and a sufficient amount of ground coffee in the filter holder.
3. Turn on the coffeemaker switch and wait for coffee.

A block diagram showing these three steps is shown in Figure 8.14. The readers' likely feeling at this point is that the description is too coarse to be very beneficial. An expanded list for the first set of activities might be as shown in Figure 8.15. Similarly expanded lists could be developed for the filter holder activities, and the switch activities.

Note that each level of detail fits within the level of detail discussed earlier, creating a hierarchy of finer and finer details. If one used a hyperlink computer program, each level would fit within the one mentioned earlier so that a designer can move down to the desired level of detail. Virtually any reader who has made coffee with a similar type of coffeemaker will easily recognize how this particular procedure works, although that reader

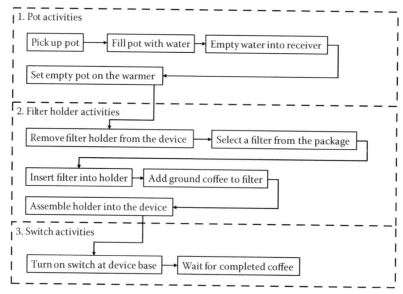

Figure 8.14 Task flow diagram for making coffee.

Coffee pot activities
1. Pick up the pot
2. Grasp the handle of the pot with right hand
3. Move the pot off the warmer
4. Transport the pot to the kitchen sink
5. Place the pot under the cold water faucet
6. Fill the pot with water
7. Turn on cold water faucet and let the water fill to the mark
8. Turn off the cold water faucet
9. Transport the filled pot to the coffee maker
10. Empty the pot into the receiver of the coffee maker
11. Tip the pot and pour the total contents into the water receiver
12. Untip the pot
13. Carry the pot to the lower level part of the coffee maker
14. Insert the pot back onto the warmer

Figure 8.15 A more refined view of making coffee.

may argue about the particular subtasks, and he or she could easily point out that steps 1 and 2 could be done in either order.

Task flow diagram

The activities performed in a task are often described with flow diagrams very similar to those used to describe computer programs. When there is only one accepted way to perform the task, the task flow diagram (TFD) is a simple sequence of boxes denoting the sequence of activities (see Figure 8.14). More complex TFDs include decisions (see Figures 8.12 and 8.13). When the design team starts with the currently existing procedure as the initial concept for finding improvements, the typical procedure is to first find the old TFD. The degree this description fits the current situation is then evaluated. If there is enough time available, a person from the design team may go to the site and observe people doing the task. If the observed activities correspond to the description in the TFD, it can be assumed to be current. Further verification may involve asking the operator performing the task about the accuracy of the description. The latter verification helps if the task includes sensory or mental activities that are not always obvious from direct observation.

When the observed activities do not correspond to the TFD, direct observational procedures can be used to develop the task description. These forms of direct observation may be continuous or periodic, and they may be performed by a human observer making on-site observations or by using some recording device such as digital video or photography. Here again, it is a good idea to take the resulting TFD from the observational study and ask human operators performing the tasks about its accuracy.

Another technique that is sometimes used is called a "talk aloud" procedure, which asks the human operator to talk about what he or she is doing while performing the task. The use of the talk aloud procedure is controversial because the acts of talking and performing the task can interfere with each other, and the result may be misleading, especially if performance time measures are taken at the same time. No matter which combination of procedures is used to determine an initial TFD, the primary purpose is to answer the questions: who, what, where, when, why, and how about the tasks and job.

Hierarchical task analysis

As should be apparent from the earlier example for making coffee, tasks and subtasks normally fit into a natural hierarchy. In some instances there are so many subtasks that it makes sense to structure the analysis hierarchically. Hierarchical Task Analysis (HTA) is not a single methodology, but rather a composite of techniques with common features. An HTA requires the tasks to be broken down into subtasks. That is, the main task may be broken down into three or four subtasks at the lower end, and up to eight or nine subtasks at the upper end. If the subtasks are well defined, each subtask corresponds to a goal, so the hierarchy of tasks and subtasks is equivalent to a hierarchy of goals and subgoals. The GOMS method discussed in Chapter 8 illustrates the latter approach.

When performing HTA, it is important that the subtasks be defined at the necessary level of detail. Coarser subtasks usually simplify or reduce the amount of analysis, but a level of detail that is too coarse can miss some critical actions. As the process continues, the analyst must decide whether to perform a depth-first approach by taking an initial subtask to the lowest level before moving on to the next subtask or doing a breadth-first analysis. Many analysts favor the breadth-first approach, because it allows one to balance the needed depth across the entire task. Other analysts prefer to alternate between these two strategies.

The ending stage of HTA involves checking for consistency. Various numbering systems are used to help in this process. Sub-tasks resulting from the first level of decomposition are often coded numerically with the main task designated as 0. The next levels are coded hierarchically. For example, subtasks 1, 2, and 3 of task 3 might be coded as 8.1, 8.2, and 8.3. Some people prefer to use alphabetical characters in the second hierarchy level, which results in codes such as 8.A, 8.B, and 8.C.

An example of HTA is shown in Figure 8.16 that describes tasks involved in an aircraft missed approach. In Figure 8.16, the tasks are arranged hierarchically, with task 1 consisting of two subtasks (1.1 and 1.2) and with subtask 1.2 consisting of three sub-subtasks (1.2.1, 1.2.2, and 1.2.3).

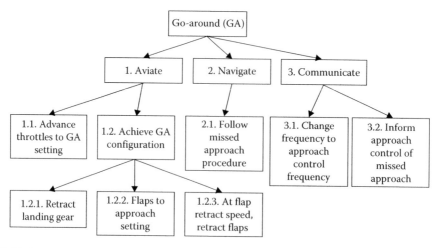

Figure 8.16 An example hierarchical task analysis for an aircraft missed approach.

A final comment is that software packages for performing HTA are available that can be very helpful to the analyst. For example, the TaskArchitect task analysis software, now used in the Purdue Work Analysis and Design Laboratory, provides a convenient interface for quickly entering, organizing, editing, and renumbering tasks and subtasks. It also includes the ability to automatically generate clearly laid out task diagrams.

Determining task requirements

Simply describing the task is sufficient in some cases. A good task description serves many useful functions, such as guiding the development of instruction manuals and other forms of product information. It also provides a starting point for further analysis of product features that may impact product usability and quality. From a traditional process-oriented view, a good description of the tasks performed is essential for management purposes. Once the tasks in a job are described, a job description, time standards, and improved methods directly follow. A good description of what is done can also guide the selection of personnel and help determine what training might be necessary for a particular job. It also can be used to decide upon appropriate compensation levels.

To attain these latter goals, something must be known about the task requirements. The first approach is to identify the time required to do the tasks, and in some cases compare these requirements against how much time is available. This approach was discussed in Chapters 5 and 6. A second approach focuses on the physical and physiological requirements of the task, which were addressed in Chapters 3 through 5. A third approach focuses on the information processing requirements of the tasks. Numerous taxonomies have been proposed that describe information-processing activities that occur in particular tasks. These activities can be classified into sensory, cognitive, and motor stages. One of the more extensive taxonomies (Miller, 1962; also see Fleishman and Quaintance, 1984) of this type is referred to as "25 Task Functions Involved in a Generalized Information-Processing System." Cognitive activities defined in this system are as follows:

- Input Select—Selecting what to pay attention to next.
- Identify—What is it and what is its name?
- Code—Translating the same thing from one form to another.
- Categorize—Defining and naming a group of things.
- Count—Keeping track of how many.
- Compute—Figuring out a logical/mathematical answer to a defined problem.
- Plan—Matching resources in time to expectations.
- Filter—Straining out what does not matter.
- Test—Is it what it should be?
- Decide/Select—Choosing a response to fit the situation.

The activities shown here are in some cases identical with those in the taxonomies of others (Berliner et al., 1964; Christensen and Mills, 1967). To apply this approach, the information-processing activity within each subtask is identified, and then analyzed further in terms of its difficulty. Table 8.1 illustrates this approach, without any analysis, for a small portion of the coffeemaker example discussed earlier in this chapter. Note that this description is more detailed than the earlier descriptions shown in Figures 8.14 and 8.15. Approaches such as MENTO-Factor and the Model Human Processor start from such descriptions to predict time requirements for the activity.

Table 8.1 Sensory, Cognitive Processing, and Motor Activities during the Use of a Coffeepot

Sensory Activities	Cognitive Processing	Motor Activities
Visually find the pot handle	Identify the pot handle	Reach the right hand for the pot handle
Visually find a removal path (not needed after learning is completed)	Verify that the path is correct and clear	Grasp the pot handle with the right hand
Visually monitor pot position		Move the pot to the sink
		Position under the cold water faucet

The ability requirements approach distinguishes between tasks on the basis of the skills and abilities required to perform the task. Extensive past work has been directed toward isolating the physical and mental requirements of tasks.* Table 8.2 lists cognitive abilities identified by Fleishman and Quaintance. Note that they also identified sensory and physical abilities, giving a total of 54 items.

Cognitive engineering describes such requirements in terms of knowledge and information-processing resources. Since knowledge can be acquired, the amount of knowledge a person has describes a skill. Information-processing resources, on the other hand, correspond to abilities because they cannot be modified by experience.

For the purposes of cognitive work measurement, and to evaluate the use of consumer products, it is often more helpful to analyze knowledge requirements. Knowledge requirements can be divided into the categories of procedural (operations) and declarative (facts) knowledge, which together can be organized into a knowledge/goal structure. A knowledge/goal structure is similar to a traditional task method in that it determines the sequences elemental tasks follow within a task. Such structures can be hierarchically depicted as networks of connected goals and subgoals that can be implemented as a set of rules within a production system. Hierarchical descriptions pose advantages of simplicity and are especially convenient for representing tasks.†

For detailed forms of analysis, it may be necessary to consider the demands of particular subtasks for specific information processing resources and the influence of fundamental cognitive abilities on task performance. Several researchers have used resource allocation models to describe information-processing resources and their influences on behavior. Although resource allocation models are likely to be useful during methods analysis, their detailed implementation is difficult. It may be easier to describe the general influence of resource conflicts as constraints in models that simulate cognitive task performance. Example constraints include the limited capacity of short-term memory, single channel operation of effectors, and the focus of attention. AI models, as exemplified by production systems, provide convenient ways of modeling such constraints in cognitive simulations (see Navon and Gopher, 1979; Sternberg, 1979; Wickens, 1984).

Many other task analytic approaches exist.‡ For example, the task characteristics approach emphasizes properties of the task, which influence performance. Typical of past efforts in the area, Farina and Wheaton (1973) described tasks in terms of 19 characteristics,

---

* Fleishman and Quaintance (1984) provide an excellent survey of such work.
† See the discussion of GOMS in Chapter 10.
‡ Simulation will be discussed in Chapter 12.

Table 8.2 Cognitive Abilities

Factor	Definition
Flexibility of closure	The ability to hold a given visual percept or configuration in mind so as to dissemble it from other well-defined perceptual material
Speed of closure	The ability to unify an apparently disparate perceptual field into a single concept
Verbal closure	The ability to solve problems requiring the identification of visually presented words when some of the letters are missing, scrambled, or embedded among letters
Associational fluency	The ability to produce words from a restricted area of meaning
Figural fluency	The ability to draw quickly a number of examples, elaborations, or restructurings based on a given visual or descriptive stimulus
Expressional fluency	The ability to think rapidly of appropriate wording for ideas
Ideational fluency	The facility to write a number of ideas about a given topic or exemplars of a given class of objects
Word fluency	The facility to produce words that fit one or more structural, phonetic or orthographic restrictions that are not relevant to the meaning of the words
Induction	This factor identifies the kinds of reasoning abilities involved in forming and trying out hypotheses that will fit a set of data
Integrative processes	The ability to keep in mind simultaneously or to combine several conditions, premises, or rules in order to produce a correct response
Associative memory	The ability to recall one part of a previously learned but otherwise unrelated pair of items when the other part of the pair is presented
Number facility	The ability to perform basic arithmetic operations with speed and accuracy; this is not a major component in mathematical reasoning or higher mathematical skills
Visual memory	The ability to remember the configuration, location, and orientation of figural material
Memory span	The ability to recall a number of distinct elements for immediate reproduction
Perceptual speed	Speed in comparing figures or symbols, scanning to find figures or symbols, or carrying out other very simple tasks involving visual perception
General reasoning	The ability to select and organize relevant information for the solution of a problem
Logical reasoning	The ability to reason from premise to conclusion, or to evaluate the correctness of a conclusion
Spatial orientation	The ability to perceive spatial patterns or to maintain orientation with respect to objects in space
Spatial scanning	Speed in exploring visually a wide or complicated spatial field
Verbal comprehension	The ability to understand the English language
Visualization	The ability to manipulate or transform the image of spatial patterns into other arrangements
Figural flexibility	The ability to change sets in order to generate new and different solutions to figural problems
Flexibility of use	The mental set necessary to think of different uses for objects

Source: Fleishman, E.A. and Quaintance, M.K. *Taxonomies of Human Performance*, Academic Press, Orlando, FL, 1984.

which were respectively associated with task goals, responses, procedures, stimuli, and stimulus responses. Several questionnaire-based methods are also available, such as the position analysis questionnaire developed by McCormick.

Function allocation

The idea of *function allocation* is that a designer should first designate the functions that are required to operate the system and then decide if a person or a machine will perform each function. In this usage, the word *function* is a task or operation. Part of the origin of function allocation stems from Fitts (1951) who observed that people were superior to machines of the day in some ways and that machines were superior in others. He compared how well people and machine performed a wide variety of sensing, motor, and cognitive tasks. Later this paper became known as the Fitts List.

The idea at the time was that designers could use this list to determine if a person or a machine could better perform the function required by the system. Since then, many people have attempted to modify the list including Gagné (1962), Meister and Rabideau (1965), and Jordan (1963). Table 8.3 describes the Fitts List with Gagné's modifications. An abbreviated* version states (1) humans are best at detection, pattern recognition, flexibility, long-term memory, inductive reasoning, and judgment; and (2) machines are best at speed, response precision, consistency, short-term memory, deductive reasoning, and multichannel performance.

Some people argue that Paul Fitts may have intended the original list as a Trojan horse, as a means of smuggling ergonomics into the design process. Whatever his reason was for inventing this list, throwing a hornet's nest into a society meeting would have caused less commotion. It should be emphasized that this topic is introduced here primarily for historical reasons, as few ergonomic designers can ever remember allocating functions using the Fitts List. Most designers view the concept of function allocation as an academic invention that may be useful when evaluating a design after it has emerged, but not during the design process. A great many things influence the design process to a much greater degree, such as regulations, new or improved technologies, market demands, and company or organizational philosophies.

Despite the serious questioning of this concept, as a first pass it is not a bad approach, especially during the evaluation of designs (see Jordan, 1963), and it has to be done at some level if a design is to be developed. One often stated concern is that function allocation is not a distinct or coherent process—that is, tasks are allocated, but no one specifically addresses the roles of various people as an overt act in the design process (see Fuld, 1993). Another issue is that people are extremely flexible, but they are also the least understood part of the process. Consequently, designers do not always know whether or not a human operator is a better choice for certain tasks. It used to be the case that human operators were much better when flexibility was imperative, but in our current age, computerized systems can provide a very high degree of flexibility. Also, one criterion may favor the choice of a human operator and another may favor a particular machine. In retrospect, a fundamental issue is that design is by its very nature iterative and "multi-criterial." These iterations are often evolutionary, and do not necessarily mesh well with the revolutionary approach of allocating functions on the first pass only.†

---

* A comprehensive overview of Task Analysis methods can be found in Luczak (1997).
† An exception to this statement is the use of function allocation methods to improve where task swaps improve system performance.

Table 8.3 Fitts' List as Modified by Gagné (1962)

Functions	Human Limitations	Machine Limitations
Sensing display	Limited to certain ranges of energy Change affecting human senses Sensitivity is very good	Range extends far beyond human senses (x-rays, infrared, etc.). Sensitivity is excellent
Sensing filtering	Easy to reprogram	Difficult to reprogram
Identifying display		Can be varied over relatively wide range of physical dimensions. Channel capacity is small varied only in very narrow range of physical dimensions Channel capacity is large
Identifying filtering	Easy to reprogram	Difficult to reprogram
Identifying memory	Limits to complexity of models probably fairly high, but not precisely known	Potential limits of capacity are very high
	Limits to length of sequential routines fairly high, but time consuming to train	Potential limits of routines are very high
Interpreting display	Same as identifying	Same as identifying
Interpreting filtering	Easy to reprogram. Highly flexible, that is, adaptable. May be reprogrammed by self-instruction following input changes contingent on previous response (dynamic decision making)	Difficult to reprogram. Relatively inflexible
Interpreting shunting	Can be readily reprogrammed to lower levels of functioning	Difficult to reprogram
Interpreting memory	Limitations to rule storage not known. Speed of reinstatement of rule sequences relatively low (as in computing). The use of novel rules possible (inventing)	Limits of rule storage are quite high. Speed of using rules fairly high (computing). Limited use of novel rules

Final remarks

This chapter has described a few of the more well-known principles and procedures designers use when problem solving during the design of products and processes. Some fundamental design principles were presented, followed by a description of various graphs, diagrams, and charts commonly employed in ergonomic design. Each of these visual techniques emphasizes certain important design principles. This chapter also introduced the topic of task analysis. This discussion included a brief overview of the steps involved in developing a good description of the task, determining task requirements, and allocating functions.

Discussion questions and exercises

8.1 Identify all of the appropriate chart(s) for each of the following conditions listed. (Note that some charts are appropriate more than once.)

8.2 Showing the locations of congestion and traffic interferences.
   (a) Denoting the constraints in revising the sequence of operations.
   (b) Showing the introduction of materials and sequences of activities.
   (c) Evaluating the allocation of tasks to a crew of people with respect to idle time.
   (d) Showing the relationship between personnel and machines and among personnel.
   (e) Uses numbered circles to correspond to activities and lines between the circles to signify required activity sequences.
   (f) Contains circles, squares, triangles, and arrows respectively for productive activities, inspections, storage, and transports, with elements arranged in vertical columns to describe the current procedure.
   (g) Uses connecting lines between combinations of circles and machine-shapes to signify relative importance, flow frequency, or other types of relationships.
   (h) Shows machines by their plan shape and such other shapes as circles, arrows, or triangles to represent particular activities, transports, or storage, and uses lines to denote the path within the facility layout.
   (i) Denotes required activities as circles and inspections as squares with enclosed numbers for their sequence and horizontal lines of material input.
   (j) Emphasizes layout effects, backtracking, and congestion in flows of materials or work-in-process.
   (k) Highlights nonproductive activities occurring in a particular method.
   (l) Shows those activities that can be combined in workstations along an assembly line in contrast to those, which cannot be combined due to constraints.
   (m) Shows imbalances of simultaneous activities.
   (n) Shows the effects of errors or failures (result at top of tree and can calculate the probability of the end result).
   (o) Shows logical relationships between conditions and events.
   (p) Determine critical elements to separate in order to reduce the probability of catastrophic failures.
   (q) Determine most likely causes of failure.
   (r) Specifically shows where information is needed to make decisions.

8.3 Suppose that someone made a precedence matrix of all the tasks that are part of fabricating a complex assembly, and then decided to create an assembly line with three stations. The respective stations along the assembly line were treated in an equal manner so that the precedence matrix for these assembly line stations was as follows:

	I	II	III
I	0	2	2
II	1	0	2
III	1	1	0

8.4 If the tasks were to be assigned to Station I, II, or III, how could the computer, using the data mentioned earlier, assess the assignments to find if precedence constraints were violated or not? Is there a way to assign them?

8.5 If the tasks that a person has to perform only require 70% of the time available, why is this either adequate, inadequate, or uncertain? Are similar assumptions included in some of techniques described in this chapter? Discuss situations whereby 70% is inadequate.

Chapter eight: Macro-ergonomics: Task analysis and process mapping

8.6 Suppose that you had some steel rings that represented specific equipment or people and you have springs that you connect between those rings to represent the link. If the spring selected for each linkage had a spring constant that was proportional to the importance of that linkage, what would the organization of the rings resemble when some machines or people were put into fixed locations?

8.7 There are general functions that need to be performed in most systems. It was suggested at one time that to most effectively design the system, designers ought to consider the function and compare performance of people with that of machines. The better performer would then be assigned to that task. What is this proposed practice called, and what was the original basis for this comparison?

8.8 What are some of the methods for determining task requirements? How is this approach different from describing the task?

8.9 What does *Simplex Munditis* mean? List some examples.

8.10 Suppose that you have borrowed a friend's iPod, which you want to listen to during a walk. Imagining that you have never used an iPod before, describe how you would take the main task of selecting and playing your favorite song, and break that main task down into subtasks that achieve the announced goal. Denote the subgoals of the subtasks that you identify and show the task diagram. (In writing these directions for a first time user, this should include a description of how to use the menu and button functions.)

8.11 You have just completed a letter to a friend on your laptop computer and you think of another friend to whom you want to write, and the news you want to send will be quite similar to that of the first letter. What are some of the subtasks you need to consider in order to get started on the second letter? Organize these tasks and subtasks within a hierarchy.

chapter nine

Computer simulation of processes and tasks

About the chapter

Using today's modern programming tools, all it takes to develop a computer simulation of a process or task is a good process chart or flow diagram. Computer simulation is, consequently, becoming an increasingly important approach relied upon by practitioners seeking to improve and better understand tasks and processes followed in manufacturing, health care, and other settings. This chapter begins by introducing forms of computer simulation in which the task or process is simulated entirely on the computer. For example, a computer simulation might describe the process in which inventory moves into and out of a warehouse or the arrival and completion of flying tasks performed by a pilot. This includes a discussion of how computer simulations are developed, modeled, verified, and validated. Examples are also given of how simulation is used by ergonomics practitioners. The chapter then moves to the topic of operator-in-the-loop simulation (OLS). Here, people interact with a computer simulation of the modeled system often referred to as a system simulator.

Introduction

The word *simulation* has a number of meanings including to pretend, to make a pretense, to feign, to deceive, to make believe, to mimic, or to make the appearance of something without really doing it. As a pretense, simulation has various degrees of fidelity (realism). These degrees of realism are analogous to the fidelity of musical recordings. Some recordings sound much more realistic than others, and in fact it was in musical records that we first heard the words *high fidelity*. Every emulation, except doing the real thing, is a simulation, and some simulations differ from the real thing more than others in realism, intent, and the methods of creation.

There are many different forms of computer simulation that are of importance to human factors or ergonomics practitioners. *Process simulation* (PS) is the most commonly applied method of computer simulation and can be viewed as a supplementary step easily performed as either part of or as a follow-on to traditional forms of task analysis both to (1) gain better understanding of the analyzed tasks or processes and (2) evaluate the effectiveness of potential improvements. For example, a PS of the tasks and activities performed by the operators in different workstations might be used to study the effect of different layouts and other factors on measures such as system idle times and production rates (see Figure 9.1). In *discrete event simulation*, events take place at particular times, such as when a customer arrives at a service counter, or a piece of equipment fails. In *continuous simulation*, a parameter of interest changes continuously over time. For example, the position of a forklift changes continuously over time while it is being driven. In practice, many computer simulations include both discrete events and continuous processes.

Figure 9.1 Example of work cell simulated using FlexSim. Simulated elements include the tasks performed by each operator at each station, movements of operators and product between stations, and work in progress (WIP).

In operator-in-the-loop simulation (OLS), people are included in the simulation who play an active role by interacting with the simulated system. For example, a person might perform a flying task in an aircraft simulator. In some cases, this interaction may include multiple people. For example, a pilot might also interact with a human flight controller while using the aircraft simulator. The simulator itself will contain a discrete or continuous simulation of the task or process, but will also allow human inputs to alter the state of the underlying model while it is running. For example, a process simulator might attempt to predict what will happen to the system or plant after the operator makes certain responses.

These different forms of simulation serve important practical needs. PS tends to relate operator activities to system performance as more of a top-down form of analysis, whereas OLS is typically more of a bottom-up procedure that examines how individual operators' methods, equipment, and procedures affect those of other operators and resulting system behavior. Compared to other methods used to study human performance, such as field studies or laboratory experiments, both forms of computer simulation are much safer, often less costly, and typically more controllable. However, the development of a good computer simulation will often require information about the particular operators who may be performing the jobs that is not immediately available. Consequently, a good task analysis or process mapping often will be the first step in developing a computer simulation model. The resulting computer simulation can then serve a variety of purposes, especially when real-world observation might be dangerous. Simulation, and OLS in particular, can also play an important role in training operators to better use the actual system.

Essential elements of computer simulation

Any meaningful computer simulation will contain an underlying model of a real system or process. Many different types of simulation models exist (Law and Kelton, 2000). For the purposes of this chapter, we will focus for the most part on what are called discrete-event simulation models. The latter models are commonly used to model a wide variety of systems and processes. In this modeling perspective, the system is normally defined as a group of entities that interact over time (Schmidt and Taylor, 1970). For example, if we were to model use of an ATM machine, customers and the ATM machine would both be entities. To model the behavior of such a system, the computer simulation must contain some form of bookkeeping system to keep track of the entities (in this example, different customers and ATM machines) and their activities over time. It is important that the bookkeeping system show how much time is spent by each entity in each activity over the operation of the system or over typical cycles of that operation. The bookkeeping system should also keep track of the time spent in each entity state (e.g., busy or idle).

Events of a computer simulation (CS) correspond to changes in an entity's activity or state. Event-sequenced programs look ahead to find the time of the next occurring event. The program then advances to that time. Differences between the current time and the subsequent event time are then recorded and put into the bookkeeping system. This procedure keeps going until either a closing time or a closing set of conditions is found. On the other hand, time-sequenced simulations move the program forward a fixed, specified time interval and then make changes on the events. All events in the latter type of simulation are assumed to occur at the end of a time interval. The simulation runs faster when the fixed-time interval is lengthened, but the analysis is less detailed.

As implied by the preceding discussion, a computer simulation will also include variables that influence how the simulated system behaves. Returning to the previous example, one variable might be the arrival time of customers to the ATM. A second might be the time required to complete a transaction. The variables used in a simulation are sometimes set to constant values (e.g., in a particular simulation, an ATM might always display an account balance 5s after the customer enters an account number). Random variables are also commonly used. Random variables are normally specified in terms of probability density functions, and associated parameters, such as means and variances. Returning to the preceding example, the simulation might assume that the time between customer arrivals is exponentially distributed, with a mean of 5 min. During the simulation, the actual customer arrival times would be generated by sampling random numbers and then converting them into arrival times. This sampling and conversion process is often referred to as Monte Carlo simulation.*

Higher-level computer languages

In recent years, many higher-level languages have been developed that greatly simplify the process of developing a computer simulation. A shortlist of some of the better known modern fourth-generation languages for PS includes FlexSim, Anylogic, AutoMod, MicroSAINT, and ARENA.[†]

All of the aforementioned modern, higher-level languages for discrete event simulation have sophisticated graphic user interfaces (GUIs) that make it easy to quickly develop simulations. Tasks and processes are described using networks of connected nodes, where nodes correspond to either resources that perform activities or networks

---

* John von Neumann and his colleague Ulam are often credited as the inventors of Monte Carlo simulation. von Neumann's work on the use of random numbers to generate output corresponding to particular probability density functions (von Neumann, 1951) was particularly influential.
† Information and student versions of each of these languages can be easily found using Google.

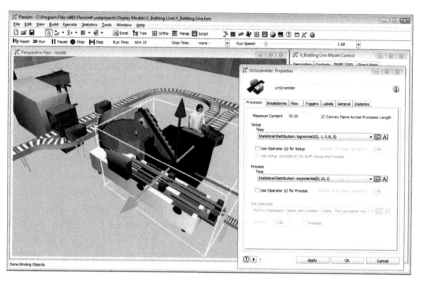

Figure 9.2 Entry of process parameters for an element of a FlexSim simulation model.

of activities performed. Some languages, such as FlexSim and AutoMod, focus more on defining networks of resources, while others, such as MicroSAINT, focus more on specifying networks of activities. Illustrating the first approach, FlexSim contains several predefined generic resources commonly used in modeling production facilities, such as conveyors, operators, cranes, forklifts, elevators, ASRS vehicles, robots, and others, which can be assigned times and linked into networks. For example, Figure 9.2 shows a network of resources including an operator, conveyors, and an unscrambler. The right side of the figure shows a menu for entering some basic statistical parameters for the unscrambler. In particular, values are shown specifying the type of distribution, mean, and standard deviation for setup and processing times. The figure also shows packages (referred to as a type of load in FlexSim) moving on the conveyor. Loads correspond to what we have called an entity in our earlier discussion and are moved by people, vehicles, and other resources from location to location. This approach is especially convenient for modeling flow, as the resources can be placed on a CAD diagram of the physical facility. The resulting diagrams provide a concrete easily understood description of the process that can be animated (Figures 9.3 and 9.4). That is, as the simulation runs, mobile resources such as people, carts, or forklifts will move from location to location as the process and tasks are performed.

The development of the simulation model using the MicroSAINT approach is more focused on specifying a network of tasks. This approach is quite convenient for describing the different tasks a person might do when sitting in a stationary location, such as when operating a piece of equipment, or the sequence of operations in an assembly line. One of the nice things about MicroSAINT is that the nodes in the task network can be hierarchically specified; consequently, for a simple sequence such as

1. Task 1
2. Task 2

Chapter nine: Computer simulation of processes and tasks 361

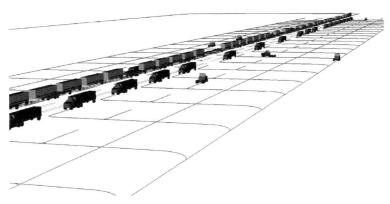

Figure 9.3 Portion of CAD model in AutoMod of trucks unloading and loading containers on a train. (From Anoorineimi, A. and Lehto, M.R., Simulation of an auto transport facility, Unpublished technical report, Purdue University Technical Assistance Program, 2010.)

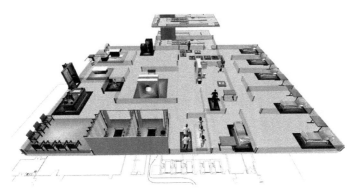

Figure 9.4 Resources in FlexSim placed on a 3D CAD drawing of a portion of a health-care facility.

Each of the tasks can be further specified in terms of subtasks, such as for Task 1

a. Subtask 1.a
b. Subtask 1.b

This feature allows the model to be described and displayed at different levels of abstraction, which is very helpful if there are many different tasks in the model, as it allows some of the extra detail to be hidden and displayed only when needed.

A general conclusion is that organizing the model as a network of resources is particularly good for modeling systems where mobile resources move around in the facility, such as for order-picking tasks, movement of materials from a warehouse to machines, etc. On the other hand, a network of tasks is especially good for modeling situations such as operating equipment where people perform multiple tasks. In each of these approaches, the time taken to perform an activity is an important aspect of the developed models. The time required to complete an activity at each node is in some cases assumed to be constant but often is described statistically. The exponential and normal distributions are often used in the latter case. Branches from one node to another indicate the sequencing of activities, and their logical relationships, within larger operations. Branching is often probabilistic. In the latter case, random number generators contained in the program determine which branches are followed. The network diagrams are useful during model development and make it easier to evaluate model correctness. If the model is wrong, the modeler needs to know what corrections to make, and the network diagrams are helpful here as well.

Many other developments in higher-level simulation languages make them even more effective. One is the more accurate and efficient generation of random numbers from particular probability density functions (Kachitvichyanukul, 1983). This feature avoids the use of approximation methods and makes modeling easier. Another advance is the development of data bases for collecting and analyzing statistics (Centeno and Standridge, 1993). These developments reduce the amount of work needed when creating the model. Including methods of output analysis within the simulation language is another associated improvement (Standridge et al., 1986). In recent years, researchers have also greatly improved the ability of CSs to communicate system dynamics to users and experts.

Perhaps the most important development is that simulation languages are much more user-friendly than they used to be. As mentioned earlier, modern systems now have interactive GUIs that allow the users to construct the model quickly. They also have animation features that allow users and other interested people to see how the system operates both in real time and at an accelerated pace (Box 9.1). Features for helping the analyst conduct designed experiments to test the significance of simulation results is another highly useful feature found in many modern simulation languages.

Verifying and validating a simulation model

While modeling is fun, it is insufficient by itself, and an elegant but inaccurate model is worthless. Therefore, it is necessary to assure the accuracy and precision of the model. To do this, models need to be verified and validated as much as possible. *Verification* refers to steps taken to ensure that the simulation model operates as intended or, in other words, to debugging the computer simulation program (Law and Kelton, 2004). In many cases, this process involves tracing the movement of entities through the simulated system step by step, and checking for errors. Animation features that show the flow of entities in the simulated system can help make errors in the code obvious. Simulation programs also include many other features to help debug the code, including run controllers, and methods of displaying variable values at selected points in the execution of the program.

Errors in the simulation code can also be identified by comparing the simulation results to the actual system. Time studies and work sampling provide data that can be used for this purpose. The simulation can also be compared to video recordings of the actual process. The latter approaches help validate as well as verify the simulation model.

Validation refers to the process of measuring the degree of similarity between the simulated model and the real case. One of the most common approaches for validating a simulation is to compare measures of human behavior and performance obtained in the real and the simulated environment. The simulated system should show performance times and other measures that are consistent with known measures of performance for the real system. Another approach is to have people familiar with the real system or process note any perceived differences between the real and simulated systems. In the best case, the experts will find it difficult to distinguish between the two cases, and the perceived difference will be small. A third approach to validation is to have people rate the

BOX 9.1 HUMAN FACTORS OF DEVELOPING COMPUTER SIMULATIONS

Any textbook on simulation will include a not always brief discussion of the benefits of simulation, which will pretty much always mention that simulation is a helpful tool for developing understanding of the simulated process. This focus is reflected by the heavy selling of the benefits of visual animation on the web sites for modern tools for PS such as AutoMod, AnyLogic, Arena, MicroSAINT, FlexSim, and many others.

All it takes now to get a quite realistic animation of a process is to input the equivalent of a traditional process chart or flow diagram into the modeling language. Once this information is entered using a series of menus, languages such as AutoMod or FlexSim can display high-fidelity images of forklifts and people moving around the plant, packages moving on conveyors, or cranes lifting objects. One of the interesting questions that arises is *"Does animation help people better understand the simulated process?"*

The answer is that for someone unfamiliar with the process the answer would seem to be an unqualified yes. This follows partially from our personal experience in working with clients, who clearly seem to catch on much more quickly when they view an animation of the model. Abundant evidence exists showing that novices will find it easier to learn and visualize a system from a concrete problem representation, and there is no doubt that a good animation of a process will normally be more concrete than the equivalent flowchart or diagram. There is experimental evidence documenting the value of animation in teaching "How-it-works." In a recent study we found that including animation significantly helped undergraduate engineering students understand how systems operated compared to diagrams or text alone (Lee 2008; Lee and Lehto, 2011). From an applied perspective, there is no doubt that this is likely to also be the case when the audience is a manager or someone else unfamiliar with the simulated process.

degree of fidelity. The key to success for the latter approaches is that the evaluators must have adequate experience with both the real and the simulated systems. Designed simulation experiments can also eliminate sources of bias in interpreting simulation results. An important point is that complete validation is impossible in most situations because this requires all possible system behavior to be tested, and usually that is impossible. On the other hand, an unverified and invalidated simulation is probably not worth the cost of using it.

Computer simulation in ergonomics

Much of the pioneering work in using simulation for human factors purposes was done for the military (i.e., Knowles, 1967; Siegel and Wolfe, 1969). A number of the Siegel and Wolf studies dealt with the use of nuclear systems. Computer simulation provided the needed safety with a reasonable degree of realism for that era.

A number of people subsequently used computer simulation to describe single persons performing military operations. Wherry (1971, 1976) attempted to model aircraft pilots in flight. He called this model *HOS* (human operator simulator). It was a monumental, many man-year effort. Others have created simpler aircraft pilot models based on

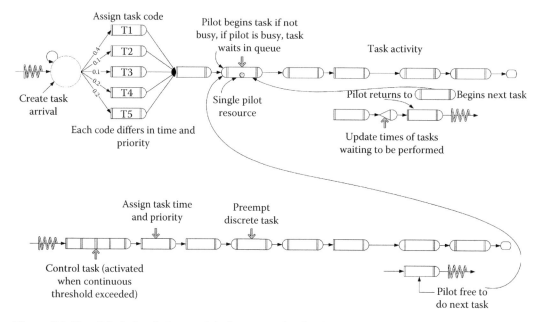

Figure 9.5 Simplified simulation model of an aircraft pilot during the mid-flight phase.

a closed-loop flow of tasks to a pilot (Grose, 1979). An example is shown in Figure 9.5, for pilot tasks occurring in the mid-flight phase. The top part of the figure corresponds to the discrete part of the simulation model. The discrete tasks arrive at random time intervals. A task is randomly assigned to one of five different categories. In this particular example, 40% of the arriving tasks are in category T1. The lower network in Figure 9.5 is the continuous aircraft control model. Note that the continuous control activity is triggered when a control-error-related threshold is needed (i.e., when the plane goes off course, the pilot has do to something). Once the error measure is within tolerance, the simulated pilot switches back to the next discrete task.

Quite sophisticated models have been developed along these lines. Much of this work focused on developing simulations of tasks performed by military personnel using a simulation language called SAINT (see Wortman, Duket, and Seifert, 1977). For example, Laughery, Drews, and Archer (1986) used this approach to evaluate the effects of different interfaces in military helicopters on pilot workload. Many other examples of using computer simulation to measure human workloads exist (Chubb, Stodolsky, Fleming, and Hasound, 1987). For example, Schmidt (1978) investigated the workload of air-traffic controllers, using a queuing model where tasks flowed to the controller, who in turn acted as a service facility. Others have employed simulation to test and evaluate various strategies of person–machine automation (Greenfield and Rouse, 1980). Wang (1986) used digital computer simulation to identify the precise procedure followed when performing elementary cognitive tasks. NGOMSL and other methods of cognitive simulation have also been applied to model tasks in other settings, as discussed in Chapter 10.

Industrial applications of process simulation

PS is commonly conducted in industrial and commercial settings to evaluate issues related to system performance, such as idle time of operators and equipment, the effects of adding new equipment or additional operators, line balancing issues resulting in bottlenecks in production, system downtime, and overall production rates. In one of the earliest examples of PS, Munro, Martin, and Roberts (1968) conducted a computer simulation to determine the appropriate number of operators in a production process. The problem here is that having too few operators can create a bottleneck in operations because they are not able to keep up with job demands, but including too many operators will create an unnecessary expense. Another well-known earlier study was conducted by Proctor, Khalid, and Steve (1971) to examine bottlenecks in production, related to the need for a machine to wait for the human operator to load it, and the need for the operator to wait for the machine to cycle before safely loading the machine. This problem was difficult to describe mathematically, because the inherent variability of both human operators and machine operations caused shifting bottlenecks to occur. Simulation provides a way to examine such systems, particularly when a single human operator must attend a number of machines. In another study, Kochar and Wills (1971) simulated systems controlled by two people interacting with each other. As you can imagine, the behavior of such systems is very complex because of the variability among people in their abilities to coordinate actions. Computer simulation has also been used to evaluate system reliability and availability in the health-care industry (Klein et al., 1993).

To illustrate a typical use of PS to improve an industrial process, a case study is given below in which computer simulation was used to improve the performance of a crew making steel ingots from molten metal and scrap iron in a basic oxygen furnace (BOF) shop (Kachitvichyanukul, Buck, and Ong, 1987).

Case study of process simulation in a BOF shop

A large steel manufacturer in Northern Indiana was interested in the effect of introducing new technology on system performance and utilization of personnel in a process where steel ingots were cast using a BOF. The principal element of the BOF shop is a set of very large vessels lined with ceramic bricks. Large cranes load scrap iron into an empty vessel and then pour flux and molten metal on top of this scrap. Molten metal arrives in specially built railroad cars that are shaped like torpedoes. After the vessel is charged with those contents, a long pipe, called an *oxygen lance*, is placed in the vessel. When oxygen is forced down that lance, the heat of the molten metal ignites the oxygen, causing an intense heat that burns carbon out of the molten metal and melted scrap.

Computer simulation was used to compare the current (as is) case to several proposed modifications. The current *as is* case was examined by making seven computer replications of a week's operation of the BOF shop and slabbing mill. Several modified configurations were also studied in a series of four simulation experiments. The first simulation experiment used two oxygen lances and placed no restriction on the rate of oxygen use. This differed from the existing method that limited the rate of blowing oxygen and used a single oxygen lance. The second simulation experiment investigated the effect of making a crew of operators and a crane always available when it was time to charge a vessel. In the existing method, a crane or crew might be occupied elsewhere on other tasks. The third simulation experiment studied the effect of simultaneously implementing the modifications used in the first two experiments. It was suspected that the latter simulation would reveal an interaction between the first two sets of modifications. This issue was evaluated by checking whether the sum of the differences between the *as is* case and performance in experiments one and two equaled the difference between the *as is* case and performance in experiment three. A fourth experiment tested the effect of introducing a new device, called

Table 9.1 Heat-to-Heat Time Statistics for Each Experiment and Comparison to the *As Is* Case

Experiment	Average Time (min)	Average Standard Deviation	Difference from *as Is* Case (min)	Ninety-Five Percent Confidence Interval of the Difference
As is case	44.96	10.71	—	—
No oxygen constraint	42.73	10.78	−2.23	−1.55 to −2.90
No crew or equipment constraint	43.67	17.68	−1.29	−0.45 to −2.12
Neither constraint	40.63	13.56	−4.33	−3.64 to −5.01
Sensor lance available	37.85	4.63	−7.11	−6.50 to −7.71

Source: Kachitvichyanukul, V. et al., *Simulation,* 49, 199, 1987.

the sensor lance. This lance continuously monitored the temperature of the molten metal and the amount of carbon. Use of this lance eliminated the need for the BOF crew to tip the vessel to collect a sample of the molten metal. Table 9.1 shows the heat time average, standard deviation, and 95% confidence interval for these four experiments and the *as is* case.

It is easy to see from the results in Table 9.1 that all of the proposed improvements resulted in better performance than the *as is* case. In the case of experiments one and two, the sum of their cycle time differences compared to the *as is* case is −2.23 and −1.29 min, which yields −3.52 min. This gives the expected difference for experiment three if there were no interactions. Instead, the difference between the *as is* case and experiment three is −4.33 min, which shows an interaction to be present. In other words, using an extra oxygen lance put more demands on both the charging equipment and the pulpit crew, so relaxing the latter constraint yielded an additional nonadditive benefit. Finally, the experiment on the sensor lance yielded a fantastic improvement to the BOF operations. Introducing the sensor lance technology improved the BOF shop capacity by about 18%. Beyond that, the latter change greatly reduced the standard deviations of performance time.

Data was also collected on the active time percentages of the human operators of a pulpit crew. Table 9.2 reports these percentages for the *as is* case and for experiments one and four. Note that changes in the percentage of active time in experiment one are positive for all operators, generally reflecting the higher productivity of the BOF and the reduced heat-to-heat times. On the other hand, introducing the sensor lance reduced the percentage of active time for some jobs and increased it for others.

Operator-in-the-loop simulation (OLS)

OLS involves the creation of relatively faithful representations of the real world on a simulator, and then observing how people perform in simulated situations. Significant progress has been made in recent years to make simulators seem realistic. For example, driving simulators have been developed that allow data to be collected without incurring significant risks to the drivers and others. Another example along these lines is to use a simulated nuclear reactor in nuclear reactor control rooms for training purposes that emulates the reactor operations and returns data on the state of the reactor. No one would allow such testing on a real nuclear reactor because of the cost and inherent risks of a disaster. Other examples of simulation may differ greatly in their fidelity, ranging from the realistic simulation of a B747 aircraft cockpit, down to an abstract video game with a

Chapter nine: Computer simulation of processes and tasks   367

Table 9.2 Active Time Percentages of Some BOF Operators for the *As Is* Case and in Two Experiments

Operator	*As Is* Case Time	Experiment 1 Actual% Change%	Experiment 4 Actual% Change%
Furnace operator	35.2	37.6 + 2.4	20.6 – 14.6
Furnace operator #2	49.8	53.3 + 3.5	36.5 – 13.3
Furnace operator #3	39.6	42.4 + 2.8	25.9 – 13.7
Crane operator	29.1	30.4 + 1.3	32.7 + 3.6
Ladle operator	66.5	69.7 + 3.2	73.6 + 7.1
Hot metal operator	22.3	23.5 + 1.2	25.2 + 2.9
Lab analyst	37.4	39.8 + 2.4	42.4 + 5.0
Skimmer	13.5	14.2 + 0.7	15.1 + 5.0
Scrapman	17.1	18.2 + 1.1	19.2 + 2.1
Scrap stocker	13.7	14.6 + 0.9	15.4 + 1.7

Source: Kachitvichyanukul, V. et al., *Simulation*, 49, 199, 1987.

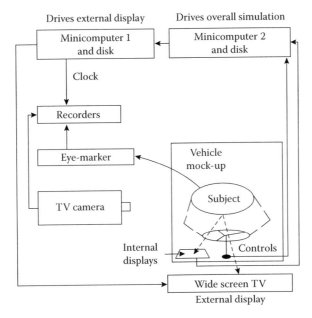

Figure 9.6 Schematic of an OLS.

joystick and a 12 in. monitor. The control provided by simulations allows an experimenter to study some very difficult situations.

The laboratory setup for an example OLS is shown in Figure 9.6. The figure depicts a vehicle mock-up along with auxiliary equipment needed to collect data and for other purposes. Some of the important elements of the OLS shown in the figure include

1. Generating the events that will occur and their times of occurrence
2. Generating and updating the external scene viewed by the user
3. Accepting and recording user inputs from a variety of control devices (such as steering wheel, brakes, accelerator)
4. Recording user eye movements and fixations

OLS differs from a pure computer simulation in that it involves one or more people as active participants in the simulation. In a pure computer simulation, the computer does everything once it is started. People are simulated with models of human behavior and performance. Replacing those human models with people gives OLS more face validity* than a pure computer simulation, provided that the people used are representative of the target population. This improved validity is a principal reason for using a simulator. However, a good experimental design is needed to develop statistically supportable statements from the OLS. If the simulation itself is inadequate, the testing capability in the laboratory is immaterial. If the simulation is used for training purposes, however, the story is different because the truth of the simulation lies in its transferability to the actual situation—that is, if people can demonstrate skill acquisition more quickly when the simulation is used, the story becomes one of comparing the cost of the simulation and alternatives to it.†

Many purposes are served by OLSs. Perhaps the most common purpose is *training*: crew proficiency training, part-task trainers, recognition training, and training of emergency procedures are all forms of training that use OLSs. Other principal purposes served are *design, development*, and *research*. These three purposes really describe a continuum that involves problem solving, determining design principles, and finding scientific facts. Military missions and NASA have used numerous OLSs for these design, development, and research purposes. Generally, OLSs are tools for designing, developing, or conducting research when the actual application is dangerous or expensive. Through simulations, people are removed from potentially dangerous or costly situations.

Many different types of simulators have been developed, and a number of companies specialize in making simulators. Link-Singer Co., Silver Spring, MD‡ designs and constructs all kinds of military and commercial simulators. Other companies include Systems Technology, Inc., Hawthorne, CA; Microsim, Washington, DC; and Doron, Binghamton, NY. Most of the manufacturers of nuclear reactors, including Westinghouse, General Electric, and Combustion Engineering, have simulators for training control room personnel. Some examples of simulators and their use are discussed next.

Training simulators

The famous Link Trainer, used during and prior to World War II, was one of the first motion-based simulators. It consisted of a very small airplane, with short wings and a tail, mounted on a base. The cockpit occupied about the center half of the trainer. The entire unit rotated around on its base and the plane revolved about its longitudinal axis to assume different attitudes of an aircraft in flight. The principal purpose of the Link Trainer was to teach pilots how to fly using instruments. A metal unit covered the cockpit so that the trainee had to trust the instruments. Since that time, aircraft simulators have been developed for many types of aircraft (see Figure 9.7). The military

---

* Where face validity is the direct believability of the results.
† A variety of studies over the years have considered how the fidelity of a simulation impact learning and behavior. See Gagné (1954); Muckler, Nygaard, O'Kelly, and Williams (1959); Smode, Hall, and Meyer (1966); Huff and Nagel (1975); and Williges, Roscoe, and Williges (1973).
‡ Link-Singer is located just outside of Washington, DC. This company specializes in making operator-in-the-loop simulators for training purposes. Of course, many military, nuclear power, airlines, and other organizations in the United States and abroad have simulators made for training new personnel and retraining existing personnel on new equipment and procedures.

Chapter nine: Computer simulation of processes and tasks 369

Figure 9.7 Student pilot operating a Bombardier CRJ-700 flight simulation. (Photo courtesy of School of Aviation Technology, Purdue University, West Lafayette, IN.)

led the way, but the commercial aircraft industry was not far behind. American Airlines maintains simulators of Boeing commercial aircraft used by the company in the Dallas area where their training headquarters are located. Fighter-pilot simulators in the military establishment are used for training in evasive maneuvers and tactics, and helicopter simulators are used extensively for similar training objectives, as well as for low-level search-and-recovery procedures (Laughery et al., 1986).

Both the U.S. Merchant Marine and the Navy have ship simulators for various kinds of ships. Some ships are exceedingly slow-acting systems, which require operators to anticipate the effect of their actions far in advance. Training people to maneuver those crafts provides considerable challenge, particularly when tight maneuvers are required, as in ports or during refueling at sea situations. The U.S. Navy also maintains submarine simulators and simulators for operating combat intelligence centers on ships of various types.

Ground vehicle simulators

Ground vehicles are more difficult to simulate than aircraft or ships, mostly because the visual requirements of a ground vehicle simulator are more complex. Ships and planes can move in almost any direction and that poses some modeling issues that do not have to be considered for automobiles, which generally travel along roads and highways. However, automobiles travel very close to many objects, so a highly detailed visual scene needs to be generated in the simulation if it is to be realistic. The visual scene around ships is kilometers and kilometers of water. The scene from an aircraft consists of objects at a considerable distance, rarely under 1000 ft or 305 m, so that detail is minimal.

The U.S. Army has tank simulators in the Detroit area that are remarkable engineering achievements because of their very high mass requirements and because tanks involve difficult tactical maneuvers in highly varying environments. One of the better-known tank simulators is the low-frequency simulator called SIMNET. Numerous other ground vehicle simulators exist as well. Both Volkswagen (in Germany) and Ford (in the United States) have moderate-level automobile simulators. Daimler Benz, the maker of Mercedes Benz cars and trucks, has one of the world's best automobile simulators. Mazda (in Japan) has built a simulator and so has Australia, at Monash University, Clayton, Victoria. Sweden has a moderate fidelity automobile simulator in Uppsala, home of the Volvo, and Krupp of Germany has also built a truck training simulator. However, most of the ground vehicle simulators were built for research and development needs, including the truck simulator MICROSIM and the Turner Banks automobile simulator in Washington, DC. The simulator of System Technology, Inc., Ann Arbor, MI, however, serves both research and training purposes. In addition to these ground vehicle simulators, there are a number at universities and in government use. Virginia Polytechnic Institute, Blacksburg, VA, has a moderate-fidelity, fixed-base automotive simulator. The University of Michigan's Transportation Research Institute, Ann Arbor, MI, has a low-fidelity, fixed-based simulator, and the University of Central Florida, Orlando, FL, has a moderate-fidelity simulator in their Institute for Simulation and Training. Some of the best university motion-based automobile simulators are at the University of Toronto and the University of Iowa (see Buck et al., 1993 for further details). More recently, the U.S. government built the national advanced driving simulator (NADS) operated by the University of Iowa, which is currently the most advanced ground vehicle simulator in the world (see Buck and Yenamendra, 1996).

Simulation fidelity

Most OLSs are very specific to a situation. Simulation fidelity becomes particularly important because OLSs must be adequately realistic to the people using the simulator. This is not to say that every OLS must be of the highest fidelity. In fact, one advantage people bring to a simulation is their imagination that enables them to fill in many missing details. If the simulator is serving tentative design needs or initial training needs, a low-fidelity OLS can serve the goal very well. When a final design solution is needed, perhaps because of a governmental policy, or for training operators to a high degree of proficiency in emergency situations, the issue of fidelity becomes extremely important. In the end, it is economics that drives decisions. The object or system being simulated also affects fidelity requirements. Some systems (e.g., large ships or nuclear reactors) are very slow acting and can, therefore, run much longer before updating. In computer jargon, slow-acting systems require a smaller control bandwidth. Mechanical engineers speak about this as response frequency.

The fundamental question is how faithful does one need to be? Consider the case of automobile simulators. Figure 9.8 shows a minimally detailed visual scene that might be presented to the driver. This scene would be similar to that viewed when driving in a rural region in the western part of the United States. The question is, "Would this lack of visual detail cause people to behave differently than in normal driving situations?" A related question is, "How important is it to provide acceleration and motion related cues?" The answers to these questions are still under investigation, using much more ambitious simulators, such as the NADS. The NADS simulator provides a very high degree of fidelity by mounting a vehicle shell or an entire car within a dome placed on a moving track (Figure 9.9) that allows movements over an area larger than a basketball court. The dome also allows nearly 360° of rotation in either direction. The net effect is that the driver will

Chapter nine: Computer simulation of processes and tasks 371

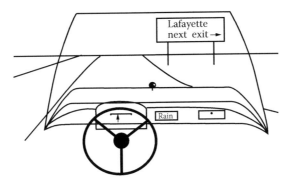

Figure 9.8 Visual elements in a low-end OLS driving simulator.

Figure 9.9 National advanced driving simulator (NADS).

feel acceleration, braking, and steering cues much as if he or she were actually driving a real vehicle. The NADS simulator also uses modern visual and audio technology to provide highly realistic visual scenes and sounds. One of the objectives of the simulator was to provide convincingly realistic scenarios of impending accidents without endangering subjects that could be used to conduct research on a variety of topics, such as how people react to a variety of potential roadway hazards, the benefits of active safety devices (i.e., antilock brakes, stability or rollover prevention devices, or collision avoidance systems), or the effects of drugs and alcohol on driving performance.

Case study in operator-in-the-loop simulation

A few years ago, a study was done using a high-quality driving simulator that examined several issues related to the use of an automated highway system (AHS) in the United States (Buck, 1996). The focus of the study was on the effects of adding a third lane to be

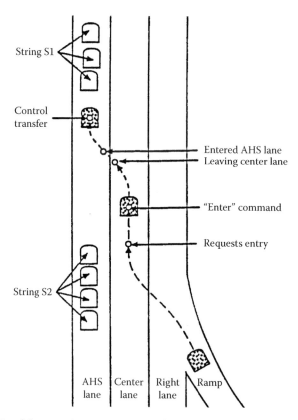

Figure 9.10 Schematic of the entering maneuver in the experiment.

used exclusively by AHS-outfitted vehicles to the two-lanes (each way) normally used on expressways. Figure 9.10 illustrates how this third lane was supposed to work. As shown there, strings of AHS vehicles travel in the left-most lane, with gaps between successive strings. The right two lanes contain non-AHS vehicles and AHS vehicles either waiting to join the other AHS vehicles, departing from the AHS lane for an exit ramp, or simply choosing to remain in manual mode for one reason or another.

The schematic also illustrates an entering AHS vehicle coming from the on-ramp to the right lane of the expressway, and crossing over to the center lane before attempting entry into the AHS lane. It was assumed that the entry procedure would require the driver to request entry, whereupon the central computer would check the vehicle for adequacy to drive in AHS mode. If the checkout was adequate, the computer would direct the vehicle to enter immediately behind a passing string of AHS vehicles. The driver of the entering vehicle would be instructed to accelerate as rapidly as possible after entry until the computer could take over. Two modes of transferring control from the driver to the central computer were studied. In the first, the transfer occured automatically as soon as the first wheel crossed into the AHS lane. In the other mode, the driver manually transferred control by pushing a button on the steering wheel. Both moves were studied because it was felt that there might be legal reasons for a manual mode of control transfer even though that mode might be slower.

Experiments in an automobile simulator verified that both older and younger drivers could perform the maneuvers of entering and exiting the AHS. However, almost all subjects experienced difficulties during exiting, particularly at higher AHS design speeds. The problems encountered during exiting were almost always due to difficulties in seeing slower manually controlled vehicles up ahead in the middle lane, because the vehicles in the AHS lane immediately ahead of the exiting car obscured the exiting driver's forward vision. This difficulty was compounded when there were greater speed differences between the AHS design speeds and those allowed for manually controlled traffic because this caused the exiting driver to catch up more quickly with other vehicles after they exited the AHS lane. It, consequently, was concluded that some form of vision aiding would be needed to help drivers exit more safely.

Other operator-in-the-loop simulators

Many industrial engineering situations can benefit from an OLS for training personnel who operate control processes. A number of years ago, a group of us put together a general-purpose process-control simulator (Buck et al., 1978). Despite the now-antiquated equipment, many of the concepts and applications appear just as relevant today. The aim was to create a very flexible simulator suitable for different processes, but to limit the visual feature to computer-driven displays. This simulator had four major subsystems:

1. An *executive* subsystem that controlled the other subsystems and advanced the PS in time
2. A *support* subsystem that collected statistics during the simulation and provided simulation and evaluation assistance
3. A *process* subsystem that described the process behavior under control
4. An *operation interaction* subsystem that contained files to hold various display dialogues needed during the simulation and handled operator inputs and outputs

This general-purpose simulator was called the unit task simulator (UTS). After building this system and debugging it, the UTS was used in the steel industry.

Design of simulation experiments

One reason for creating a computer simulation model is to run experiments on it to see how the simulated system operates under different conditions and determines which factors influence performance. For example, a simulation of an industrial process might allow production rate for different layouts of the facility to be compared in terms of both average performance and the normal variation that reflects the day-to-day operations of the facility.

When the simulated system is complex, it is often helpful to design a simulation experiment to determine the importance of various factors relating to the object, device, or process being designed. In order to do this, the factors need to be set to specific values (typically called levels) in the simulation model. The outputs of the simulation model when the factors are at particular levels then needs to be analyzed. The particular experimental design selected by the ergonomist reflects his or her interest in the factors and economic constraints being faced. After the simulation runs are completed according to the experimental design, simulation outputs are usually analyzed by an analysis of variance (ANOVA) as expanded upon below.

Analysis of variance

Simulations include variability in order to appear realistic. This is done through the use of random number streams that are employed in the simulation model. The resulting outputs, consequently, will vary between runs, so they need to be analyzed statistically. The usual analytical procedure is to perform an ANOVA. Different forms of ANOVA are used, depending on the experimental conditions. A fairly typical situation in ergonomics is where specific treatments (T_i) are made on a process, such as using a new technology or power tools compared to manual tools. Also in typical situations, alternative methods (M_j) are used such as having workers perform two needed operations or having one worker perform one operation and another perform the second. An important aspect of these different treatments and methods is the measured productivity (Y_{ij}). The model of an ANOVA for this experimental situation is

$$Y_{ij} = \mu + T_i + M_j + TM_{ij} + \varepsilon_{ij} \tag{9.1}$$

where
  μ is the average value overall experimental conditions
  T_i is the $i = 1, 2, \ldots$, treatments
  M_j is the $j = 1, 2, \ldots$, methods of operations
  TM_{ij} is the interaction between treatments and methods
  ε_{ijk} is the normally distributed error for the k replication of observations

The terms T_i and M_j correspond to the main effects of treatment and methods, respectively.
 The interaction of these two variables becomes meaningful when a particular method coupled with a specific treatment produces changes in the criterion that are quite different than for the same method coupled with the other treatment. For example, if a particular grinding compound (e.g., method A) produces especially good results with Grinder X, and especially poor results with Grinder Y, an interaction is present.
 Typically, replications are included in the simulation to account for a variety of things including the particular random number stream used. In some cases, a separate term is included in the model to account for the effect of replications. More often, the replicate term and its interaction with treatments and methods are pooled to estimate the error term ε_i. When the ANOVA model parameters T_i or M_j or T_iM_j are large relative to ε_{ijk}, those factors are said to be significant. The change in Y_{ij} needed to reach significance depends upon the analysis. Experimenters are always urged to plot the data and examine the results. If there are effects caused by the treatments or methods, plots such as in Figure 9.11 should indicate a change in Y_{ij} when there is a change in T_i or M_j as Figure 9.11B shows but not Figure 9.11A. If there is a separate effect on Y_{ij} due to both T_i and M_j, the effect should appear as shown in Figure 9.11C, although the changes in the criterion may be going downward rather than upward. When the effect of treatments T_i and methods M_j cannot be described as simply the sum of the separate treatment and method effects, an interaction occurs and the TM_{ij} parameter takes on value and the plots may appear as those in Figure 9.12.

General experimental designs and their features

One of the most important features of a designed experiment is the number of factors involved. In a full factorial design, all combinations of factors are contained in every replication of the design. In ergonomic studies, many designs completely replicate the

Chapter nine: Computer simulation of processes and tasks

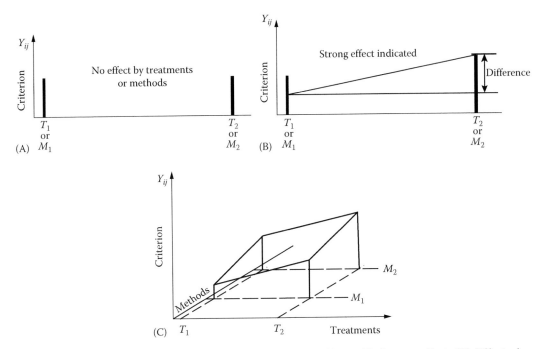

Figure 9.11 Different effects tested in ANOVA (A) No effects, (B) Strong effect, (C) Effect plus interaction.

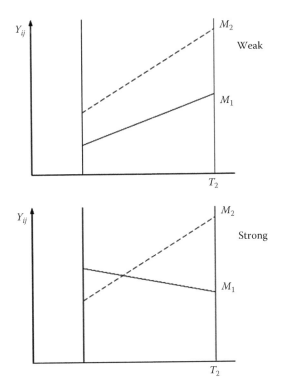

Figure 9.12 Strong and weak interactions tested in ANOVA.

conditions for each subject. Doing so allows between-subject differences to be separated from the other factors being tested (Box 9.2). In simulation, a separate run using a different random number stream is frequently used as a replication. Note that all of the parameters of Equation 9.1 can be estimated from a full factorial experimental design, and that feature is one of the reasons for their popularity.

Experimental designs are said to be balanced when an equal number of observations is taken for each possible combination of the independent variables. For illustration, suppose you are working on an urban automotive safety study. You are interested in comparing how often male and female drivers wear seat belts. The type of vehicle driven is also of interest. The levels of the first factor are obviously male versus female. Suppose the second variable is pickup truck versus sedan. An example of a balanced design would be to take 1000 observations of belt use for each of the four possible combinations of vehicle and gender. However, suppose that most of the drivers of pickup trucks were males, so only 300 observations of belt use were obtained for females driving pickup trucks, compared to 1000 for each of the other three combinations.

BOX 9.2 RANDOMIZATION PRINCIPLE

An important principle in experimental design is the following.

Principle 9.1

Variables of interest should be part of the experimental design. Those not of interest should be randomized.

The idea of randomizing a variable means that typical values of the variable are used, but the specific value is randomized so that there is no systematic effect from it. That condition can be achieved by allowing a variable to be included in the simulation with randomized levels that include the variable but at typically varying conditions. Randomizing precludes systematic manipulated conditions regardless of how the other experimental factors are varied.

The unequal number of observations in the latter case complicates analysis. An obviously inefficient solution would be to throw out most of the data and do ANOVA with 300 observations for each cell, or a combination of the independent variables. This situation is very common, for a variety of reasons, so most statistical packages include less drastic methods for dealing with unbalanced designs, which take advantage of the entire data set. In some cases, experimenters intentionally design the experiment to be unbalanced.

Confounding is another issue, which is often related to whether the design is balanced. When confounding is present, the effects of the independent variables cannot be independently estimated. For example, suppose that in the previous case, 1000 observations were taken of females driving sedans and another 1000 observations were taken of males driving pickup trucks. This, of course, is an unbalanced design since no observations are taken of females driving trucks or of males driving sedans. In this, quite obviously flawed design, the effects of gender and vehicle are confounded, that is, it is impossible to determine whether the differences between the two groups are due to gender, vehicle, or both. Partial confounding is a more complex issue, often intentionally included in complex experimental designs.

Returning to the previous example, suppose that 1000 observations were also taken of males driving sedans. This allows the effect of vehicle to be estimated by comparing the two groups of male drivers. It also allows the effect of gender to be estimated by comparing males driving sedans to females driving sedans. However, both estimates are confounded with the interaction between gender and vehicle. If it is safe to assume that this interaction is not significant, then this design provides valid conclusions and also reduces the amount of the data that must be collected. Both effects can be calculated and there is no need to collect data for the fourth combination (females driving pickups). The critical issue, of course, is whether the assumption of no interaction is valid. Simply put, *analysis is more difficult and less precise* when there is confounding in the design.

Many different experimental designs that involve various degrees of confounding are available (see Montgomery, 2004). Full factorial designs avoid confounding and are very useful and popular because any factor or combination of factors can be estimated. But full factorial designs are expensive in terms of experimental resources. For example, consider the case where each factor has two levels. Assuming a single replicate, the number of experimental conditions to be examined in this case becomes 2^d, where d denotes the number of factors. As the number of factors and levels increase, the number of observations needed quickly becomes quite large. More generally, if n is the number of levels, the number of observations for a single replicate of a full factorial design is n^d.

Hence, full factorial designs require a lot of simulation runs and analysis becomes expensive. Alternative experimental designs reduce the number of experimental conditions in each replication. So there is a lower cost in terms of experimental resources, but that cost reduction is achieved at a price of either not being able to make estimates of all of the experimental design parameters or reducing the precision of some of those estimates. Examples of such designs can be found in most textbooks on experimental design (i.e., Montgomery, 2004).

Final remarks

In recent years, general practitioners of human factors and ergonomics have discovered simulation to be an effective method for discovering and rectifying problems in a wide variety of applications. Simulation has traditionally been used in aviation ergonomics, where it has historical roots dating back to the use of Link Trainers in the World War II era. Today, ergonomists have expanded the use of simulation greatly in industrial, military, and automotive settings. Simulation provides a very useful method of evaluating system performance and human behavior at the same time, and is especially useful in training pilots and others who are performing difficult and potentially dangerous tasks. Some of the different forms of simulation can be effectively combined for ergonomic design purposes. Simulation also allows ergonomists to study systems during emergency conditions. Although simulation is certainly not a panacea for all ergonomic problems, many problems can be handled with these tools. The many varieties of simulation now available offer opportunities that are perfectly feasible today but were not a decade ago.

Discussion questions and exercises

9.1 Two principal forms of simulation are discussed in this chapter for use in ergonomics design.
  a. What are the principal differences between these forms of simulation?
  b. What are some of the more important ergonomic design purposes served by computer simulation?
  c. What are some of the more important ergonomic design purposes served by OLS?
9.2 What are some of the important weaknesses of computer simulation relative to its use in ergonomic design?
9.3 Give examples of the use of computer simulation in ergonomic design. In your opinion, which applications have the greatest face validity?
9.4 What are some advantages and disadvantages of OLS in contrast with computer simulation?
9.5 What are some of the problems of doing a field study to investigate the effect of increased automation in contrast to using computer simulation or OLS?
9.6 Identify the appropriate technique for the following descriptions:
  a. A small computer is used to mimic process situations while a person controls the process off-line. The person's performance times and accuracies are measured during these tests for the purpose of improving the interface design.
  b. The tasks and processes involved to make a product are simulated using a computer. The purposes of these tests are to estimate total operation effectiveness with different designs and to examine the effects of those designs on individual operators.
9.7 In the early days of computer simulation, this kind of simulation was performed on a computer by calling up values of a uniform random variable, converting that number into a value from a specified probability density function, and then repeating this operation over and over again. Because of the similarity to some typical gambling situations, what was this form of simulation called?
9.8 One of the earliest ergonomic applications of computer simulation was used by the military to describe crews of military personnel. Later, Robert Wherry led an effort to model individual aircraft pilots during flight. What was this individual human operator simulation called, about when was it done, and what are some of the difficulties faced in this modeling?
9.9 What is the difference between verifying and validating a simulation? What are some of the techniques used? What are some of the check procedures that can be used to assure that the simulation model is reasonable?
9.10 Where and why is good experimental design important in simulation?
9.11 How might CS and OLS become part of an integrated strategy for product and process design?

chapter ten

Modeling and evaluation of cognitive tasks

About the chapter

This chapter introduces several approaches used to model and evaluate cognitive tasks performed by people in a wide variety of industrial and consumer settings. These approaches provide several new measures of human performance, not considered in the earlier chapters of this book, such as the amount of transmitted information, receiver sensitivity, and biases or deviations from optimality. The approaches also provide helpful guidance into how knowledge and goals influence behavior and some of the underlying reasons for human error. The chapter begins by introducing the topic of communication theory. An important contribution of this theory is that it provides a quantitative measure of the information transmitted that is equivalent to the reduction of uncertainty. Applications of communication theory are discussed, before introducing several models of human processing. The latter discussion focuses on how human performance and decision making are influenced by selective attention, working memory limitations, the mode of processing, and other factors related to how people process information. Several models of human decision making are then introduced that can be used to evaluate how people make choices and make inferences. Such models can be used to make quantitative predictions in many different applications. Examples include models of customer preference and choice, models of fault detection and troubleshooting tasks performed by expert operators, and medical applications, such as diagnoses of illnesses by physicians.

Introduction

In many applications of human factors and ergonomics, the designers of products, processes, or systems need to understand, evaluate, and predict the results of mental activity followed by people when they perform tasks. This is especially the case for human–machine systems in which people interact with the system using a computer interface, but it is equally important when evaluating a wide variety of other issues, such as product usability, customer preferences, not to mention human error. Models of human information processing and decision making address this need of designers in several different ways.

First, such models provide helpful measures of task performance. For example, communication theory provides a quantitative measure of how much information is transmitted and lost when messages are sent to receivers using noisy communication channels. Signal detection theory (SDT) further describes this process in terms of receiver sensitivity and bias. Other predictive approaches, such as policy-capturing models, reveal which types of information are actually used when making decisions and how this information affects preferences, inferences, or judgments. Models of human information processing and decision making can also explain errors and biases, for particular design parameters and task conditions, in terms of user knowledge, goals, and information-processing limitations, such as the capacity of working memory and attention. Development of a good model can in some cases result in an expert system capable of providing assistance to operators or product users or even performing the task on its own.

Communication theory

Almost all human activities and tasks involve some form of communication in which information is exchanged between one or more participants. People communicate with each other by talking, writing, and various forms of nonverbal communication such as gestures and facial expressions. People's interactions with machines, tools, computers, consumer products, and other devices they use to perform tasks can also be viewed as a form of communication—that is, people *tell* machines or other devices what to do, using controls and input devices, and are *told* what to do and what not to do by displays, signs, signals, and other information-bearing cues in their environment.*

Communication failures are a common contributing factor to human error and accidents. Designing effective forms of communication is, consequently, one of the most important tasks faced by ergonomic designers.† Communication theory provides a starting point for addressing effectiveness-related issues, as discussed in the following section. One of the more fundamental contributions of this theory is that it describes the basic elements of the communication process, and shows how they fit together as a system. A second contribution of communication theory is that it provides quantitative measures of how much information is transmitted and lost in noisy communication channels. Communication theory can also be used to evaluate how much information is transmitted internally at particular information-processing stages within the human. Consequently, communication theory provides a method for evaluating why a communication is effective or ineffective and can suggest how much improvement is possible if sources of information are modified in certain ways.

A communication system (see Figure 10.1), as originally proposed by Shannon (1948), consists of

1. An information source that produces the message(s) to be communicated (i.e., a person talking)
2. A transmitter that converts or codes the message into a signal suitable for transmission (i.e., a telephone)
3. A channel that transmits the signal from transmitter to receiver (i.e., a telephone line)
4. A noise source that adds noise to the channel (i.e., static)
5. A receiver that reconstructs or decodes the message from the signal (i.e., a person listening on telephone)

These components of a communication system can vary greatly between settings. For example, the *source* of a message might be a regulatory agency, consumer group, educational institution, standard-making organization, manufacturer, or peer group. A *transmitter* could be the designer who develops a particular form of communication or a learned intermediary (such as a physician) who conveys or personalizes the meaning of a scientific study of drug effectiveness to a particular patient. The *signal*, or coded version of the message, could be a particular sound pattern, wording, or pictorial. The *channel* typically corresponds to the particular sensory channel (i.e., vision, hearing, touch, taste, or odor) or media (television, print, training course, on-product label, product cue or signal, auditory alarm, etc.) used

---

* The design of controls and displays is covered in Chapter 12.
† Note that employers are required by law in the United States to communicate information about workplace hazards to workers. For more about hazard communication, see Chapter 17.

Chapter ten: Modeling and evaluation of cognitive tasks

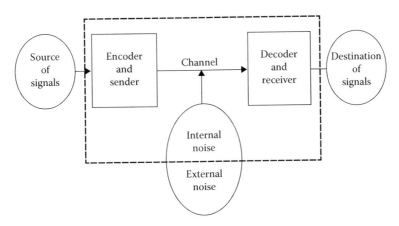

Figure 10.1 Model of a communication system.

to convey the signal. Sources of *noise* might include personal experience, peer groups, and various advocates who overstate or understate the significance of particular issues. The *receiver* might be the particular person who decodes the message.

The extent to which messages *from* a machine are correctly communicated *to* people is often important. From a design perspective, a machine or other device acts as an intermediary that transmits messages from the *mind of the designer* to the *device user*. The degree to which these messages are effectively communicated has a large impact on product quality* and usability. Poorly communicated messages will make it difficult to use a device as intended, and will result in errors, and in some cases, serious accidents.

Information transmission

To successfully communicate a message, the message must first be sent. For example, a speedometer might display the current speed of a vehicle.* The intended receiver must then notice, understand, and respond appropriately after receiving the message. The amount of information that is transmitted during the communication process depends upon both how much information is sent, and its effect on the receiver. In some cases the transmission is unsuccessful. Returning to the previous example, a broken speedometer that always says you are going 100 m/h provides no information about how fast you are going. An example of successful transmission is when you look at the speedometer and slow down because you realize you are speeding. If you do not look at the speedometer, the information sent by the display is lost. Information also is lost if you do not understand the message or simply ignore it. The point is that we can measure how much information is transmitted by considering how receiver responses change as a function of the information presented to them.

A way of calculating the transmitted information is given in *Shannon's Mathematical Theory of Communication* (Shannon, 1948). This theory provides a quantitative method of measuring the relationships between sets of categorical variables. To be more precise, suppose that we have one set of variables $X = \{x_1, x_2, ..., x_n\}$ corresponding to a set of inputs to the human operator and another set of variables $Y = \{y_1, y_2, ..., y_m\}$ corresponding to outputs of the human operator. How much will we know about Y if we know X? As pointed out by Shannon, the answer depends on how uncertain we are about Y before and

---

* See Chapter 12 for more information on the design of displays.

after knowing X. If we are completely certain about Y, before knowing X, then X cannot tell us anything about Y. Shannon defined the amount of uncertainty in Y before being told anything about X, using the following equation:

$$U(Y) = -\sum_{i=1}^{m} p(y_i) \log_2 p(y_i) \tag{10.1}$$

where
    $U(Y)$ is the uncertainty of Y
    $p(y_i)$ is the probability of the *i*th value of Y

Note that if we know Y, then $U(Y) = 0$. For example, suppose $Y = y_1$. Then $p(y_1) = 1$ and the probability of all other values of Y is 0. For this case it is easily verified using Equation 10.1 that $U(Y) = 0$. Consequently, when $U(Y) > 0$, we know that more than one value of Y has a nonzero probability of being true. The next step is to think about how uncertain we are about Y, if we know $X = x_j$. The conditional uncertainty of Y, given $X = x_j$, is simply

$$U(Y|X = x_j) = -\sum_{i \forall 1}^{m} p(y_i|x_j) \log_2 p(y_i|x_j) \tag{10.2}$$

where
    $U(Y|X = x_j)$ is the uncertainty of Y, given $X = x_j$
    $p(y_i|x_j)$ is the conditional probability of $Y = y_i$, given $X = x_j$

The average uncertainty in Y after observing X can be calculated by weighting $U(Y|X = x_j)$ by the probability of each possible value of X. This results in the following expression:

$$U(Y|X) = -\sum_{j \forall 1}^{n} p(x_j) U(Y|X = x_j) \tag{10.3}$$

where $p(x_j)$ is the probability of $X = x_j$. The fact that $U(Y|X)$ is an additive function of $U(Y|X = x_j)$ is one of the most useful features of Shannon's definition of uncertainty. Another important result is that we can use the same approach described here to describe the uncertainty of X, $U(X|Y = y_i)$, and $U(X|Y)$. That is,

$$U(X) = -\sum_{j=1}^{n} p(x_j) \log_2 p(x_j) \tag{10.4}$$

$$U(X|Y = y_i) = -\sum_{j \forall 1}^{n} p(x_j|y_i) \log_2 p(x_j|y_i) \tag{10.5}$$

$$U(X|Y) = -\sum_{i \forall 1}^{m} p(y_i) U(X|Y = y_i) \tag{10.6}$$

Note that $U(X)$ will often not be equal to $U(Y)$. This follows because there is no reason to expect that the uncertainty of X should equal that of Y. The mutual information in X and Y or, in other words, the average amount of information conveyed by X about Y is

$$I_t(Y:X) = U(Y) - U(Y|X) \qquad (10.7)$$

Intuitively, the above equation states that the amount of information transmitted by X about Y is simply how much more certain we are about Y after observing X. For example, if we view X as a stimulus (e.g., the presence or absence of a warning) and Y as the response (taking or not taking a precaution), $I_t(Y:X)$ tells us how much of the information in the warning is transmitted by the response. $U(Y|X)$ tells us how much of the information in the response is not explained by the warning. The latter quantity is normally referred to as *noise* but, as expanded upon later, can often be explained using some other set of variables along with X and Y. For example, a person might be obtaining information from product cues in addition to the warning. The effects of this additional information on the responses of a person would be part of the noise not explained by the presence or absence of the warning (Boxes 10.1 and 10.2).

It can also be easily shown that Equation 10.7 is equal to the average amount of information conveyed by Y about X, so we get a second relation of

$$\begin{aligned} I_t(X:Y) &= U(X) - U(X|Y) \\ &= U(Y) - U(Y|X) \end{aligned} \qquad (10.8)$$

Returning to the just described example, $I_t(X:Y)$ tells us how much of the information in the response is contained in the warning. $U(X|Y)$ tells us how much of the information in the warning is lost or not transmitted. The latter quantity is normally referred to as *equivocation*, or lost information. Noise and equivocation are very different, but easily confused measures. The key point is that noise measures the additional information in the response not explained by the sent signal, while equivocation measures how much of the sent information was not transmitted.

Rate of information transmission

All communication processes require time to be successfully completed. The maximum rate at which information can be transmitted is called the channel capacity, and can be measured as the number of bits transmitted per second. As pointed out by Shannon (1948), the rate at which information can be transmitted over a channel depends on how the messages are encoded. The basic idea is that each message is encoded as a particular combination of symbols. For example, we might code the number "8" with a binary code of "111." If we wrote this binary value down on a piece of paper, and looked at it sometime in the distant future, the time needed to transmit the number "8" would then be determined by how long it takes to read the three binary digits, plus the time needed to convert the binary representation into the number "8." The latter step might be done by simply retrieving the decimal equivalent from a table or long-term memory, or it might involve several additional decoding operations.

Optimal coding methods can maximize the rate of transmission by using shorter codes for more frequent messages. The key assumption is that the receiver must have the necessary knowledge to decode the message. One way of implementing this principle is to use symbols, icons, or abbreviations instead of more lengthy text for more frequently

BOX 10.1 EXAMPLE CALCULATION OF THE AMOUNT OF TRANSMITTED INFORMATION

Suppose an experiment was performed using two randomly presented stimuli x_1 and x_2, which were red and green lights. The responses y_A and y_B, respectively, were depressing keys A and B. One-half of the time the red light was the stimulus. In the remaining trials, the green light was the stimulus. Suppose that 16 trials were conducted, resulting in the data shown in the following matrix:

Stimulus	Response y_A	Response y_B	Row Sum
X_1	7	1	8
X_2	1	7	8
Column Sum	8	8	16

That is, the subject pressed key A seven times when the stimulus was a red light, and so on. The average uncertainty of the inputs, X, is

$$U(X) = -\frac{8}{16}\log_2\left[\frac{8}{16}\right] - \frac{8}{16}\log_2\left[\frac{8}{16}\right] = 1$$

The output or response uncertainty $U(Y)$ is computed similarly, and also equals 1. The noise or, in other words, the uncertainty in the response given we know the stimulus is calculated as

$$U(Y|X) = 0.5 * \left[\frac{7}{8}\log_2\left[\frac{7}{8}\right] + \frac{1}{8}\log_2\left[\frac{1}{8}\right]\right] - 0.5 *$$

$$-\left[\frac{7}{8}\log_2\left[\frac{7}{8}\right] + \frac{1}{8}\log_2\left[\frac{1}{8}\right]\right]$$

$$= 0.5435 \text{ bits}$$

The information transmitted is then calculated as

$$I_t(X:Y) = U(Y) - U(Y|X)$$

$$= 1.0 - 0.5435 = 0.4564 \text{ bits}$$

Some further calculations show that the equivocation or information lost from the response is equal to the noise in this example. Note that the transmitted information, plus the noise, plus the equivocation add up to the joint uncertainty. In most cases, the information lost will not be equal to the noise.

Now consider an experiment similar to the one shown above, where the results are as follows:

Stimulus	Response y_1	Response y_2	Row Sum
x_1	8	0	8
x_2	0	8	8
Column Sum	8	8	8

Note that the row and column sums are exactly the same as in the earlier example, so $U(X) = U(Y) = 1.0$. However, there is no uncertainty in the response given; we know the stimulus, as is intuitively obvious. This can be shown using Equation 10.3:

$$U(Y|X) = -0.5*[1*\log_2[1]+0*\log_2[0]]$$

$$-0.5*[1*\log_2[1]+0*\log_2[0]]$$

$$= 0 \text{ bits}$$

The information transmitted is then calculated as

$$I_t(X:Y) = U(Y) - U(Y|X) = 1.0 - 0$$

$$= 1.0 \text{ bit}$$

Note that all of the information in the stimulus is transmitted, that is, once we know the stimulus, we know the response with certainty. Consequently, this situation is referred to as perfect transmission of information.

provided messages. This principle is very frequently applied by information designers. For example, computer displays and web pages often use icons instead of text. The use of symbols in traffic signs is another example of this principle. Other examples of this principle include the use of abbreviations and signaling devices, such as traffic signals, safety markings, signal lights on vehicle instrument panels, and auditory alerts.

Another important issue is that people respond more slowly to surprising or unexpected stimuli. Unlikely events or stimuli convey more information as explained earlier in this chapter. The relationship between reaction time and the amount of information transmitted by a stimulus (S) about a necessary response (R) is known as the *Hick–Hyman law*.* That is,

$$\text{Reaction time}(s) = kI_t(S:R) + C \qquad (10.9)$$

Both the slope, or k, and the constant C, in this equation tend to be larger for novice subjects. With experience, both terms decrease significantly. One possible explanation of the Hick–Hyman law is that people have more opportunity to practice responses to familiar stimuli. Consequently, they respond more slowly to surprising or unexpected stimuli. Precision in positioning is another phenomenon where uncertainty appears to play an important role.† This relationship, called Fitts' law, shows that movement time is a simple linear function of a so-called index of difficulty, calculated as shown here:

$$\text{ID} = \log_2 \frac{2A}{W} \qquad (10.10)$$

---

* See Hick (1952) or Hyman (1953).
† Chapter 6 addresses Fitts law in reference to positioning tasks in more detail.

Some Applications of Information Theory

The ability of people to make *absolute judgments* is often analyzed using information theoretic measures. In an absolute judgment task, a person is asked to classify or assign a stimulus, such as a sound, into an appropriate category. For example, a subject might be asked to judge the brightness of a light or the length of a line on a 10-point scale. In the latter examples, people would be classifying a perceived quantity into 1 of 10 possible categories. When the number of categories (N) is small, people find it easy to make judgments and few errors result. For example, in a binary judgment task people might simply describe a light as being bright or dim, which obviously should not be too challenging in most situations. The task would still be quite easy if we used three categories. However, as N gets larger, the judgment task becomes increasingly difficult, and more errors occur. This phenomenon is nicely described in terms of the amount of information transmitted. Initial increases in N (e.g., going from two to three categories) increase the amount of information transmitted. The latter effect is illustrated by the following two matrices:

	S_1	S_2
R_1	0.5	0
R_2	0	0.5
Sum	0.5	0.5

$I_t(S:R) = 1$ bit

	S_1	S_2	S_3
R_1	0.33	0	0
R_2	0	0.33	0
R_3	0	0	0.34
Sum	0.33	0.33	0.34

$I_t(S:R) = 1.58$ bit

As shown earlier, more information is transmitted when three categories are used. On the other hand, suppose that going from two to four categories resulted in the following matrix:

	S_1	S_2	S_3	S_3
R_1	0.125	0.125	0	0
R_2	0.125	0.125	0	0
R_3	0	0	0.125	0.125
R_4	0	0	0.125	0.125
Sum	0.25	0.25	0.25	0.25

$I_t(S:R) = 1$ bit

As shown in the latter matrix, the amount of information transmitted is exactly the same as for the binary judgment task. Note that the amount of transmitted information would have doubled had it been transmitted perfectly. The reason this did not happen is that the subjects could not accurately distinguish S_1 from S_2, or S_3 from S_4. A number of studies have shown that maximum amount of information transmitted depends upon the sensory mode used and the number of dimensions involved. The upper limit for visual and auditory judgments on a single sensory dimension is somewhere around 2.8–3.1 bits—that is, people reach their limit for auditory and visual judgment tasks, on a single sensory dimension, when the number of potential categories is somewhere between

7 and 9. Lower values are normally obtained for other senses, such as touch. These studies have all found that convergence always tends toward a fixed amount. It is of interest that there is a fixed convergence for each different type of absolute judgment situation, but all such situations exhibit the same fundamental phenomena.

Another interesting application of information theory to inspection is given by Hancock (1966). In his analysis, Hancock started by assuming a human information-processing rate of 3 bits/s. He reasoned that when the incoming quality of a product is 98%, the uncertainty of good and bad items can be calculated as

$$U(\text{conforming}) = \log_2 \frac{1}{0.98} = 0.0291 \text{ bits}$$

$$U(\text{nonconforming}) = \log_2 \frac{1}{0.02} = 5.644 \text{ bits}$$

Given a human information-processing rate of 3 bits/s, the time required for inspection of conforming items is 0.0291/3 or 0.0097 s, and for nonconforming items, the required time is 5.644/3 or 1.881 s. It then follows that the expected time per item inspected is

$$E(\text{time}) = \frac{0.98(0.0291 \text{ bits})}{3 \text{ bits/s}} + \frac{0.02(5.664 \text{ bits})}{3 \text{ bits/s}}$$

or 0.0661 s per object inspected, without including the motor response time associated with handling the object. Hancock also used information theory to describe an optimum sequence for inspecting a collection of objects during online production. Other applications of information theory include efforts to measure the information content of assembly tasks (Goldman and Hart, 1965). The time required to monitor displays has also been shown to be related to uncertainty (Lehto and Buck, 1988).

The index of difficulty is measured in bits, and can be loosely interpreted as a measure of uncertainty. The applicability of this measure has been demonstrated for a wide variety of tasks, including eye tracking (Ware and Mikaelian, 1987), foot pedal activations (Drury, 1975), dual tasking (Shehab and Schlegel, 1993), and precision in actuating computer mouses, touch panels, and joysticks.

Another observation is that people tend to adapt their processing rates to task conditions. When stimuli are presented more frequently, responses become faster, but the number of errors also tends to increase. A quicker response rate increases the rate of information transmission, while the increased number of errors reduces the amount of information transmitted. One of the more remarkable things discovered about people is that they can often balance these two effects to obtain a consistent transmission rate. This constant processing rate is known as the *rate channel capacity* of a particular communication system. However, when stimuli are presented rarely, as in vigilance tasks, people have a tendency to miss or not notice the stimuli. This results in the so-called *vigilance decrement*.

In other words, the rate of transmission drops because people fail to process the incoming stimuli due to boredom or difficulty maintaining attention.

A final point is that the rate of information transmission can often be increased by providing information on multiple sensory channels or by increasing the number of distinguishing dimensions. For example, the channel capacity is higher for two-dimensional tracking than for one-dimensional tracking. A similar increase occurs when people judge the location of dots inside a square rather than judging location on a single dimension or when people judge the location of a dot within a cube instead of in a square. The bad news is that the effect of adding extra dimensions *is not simply additive*. Going from one to two dimensions normally results in a rate of transmission less than double the rate at which information is transmitted for a single dimension, and the incremental benefits tend to decrease as more dimensions are added.

Information partitioning and hypothesis testing

During task performance, intervening processes such as decision making and the use of other sources of information can influence the responses made by the receiver. In the previously described approaches these effects would be measured as noise (See Box 10.1). However, it turns out that the effects of other factors on the response can be directly measured by partitioning the transmitted information into multiple components. This follows because the additive properties of Shannon's measure of information allows the previous equations to be extended to multiple sets of variables (see DeWeese and Meister, 1999). That is, the average information transmitted about a variable Y by the combination of two variables X and Z can be described as

$$I_t(Y:X,Z) = I_t(Y:X) + I_t(Y:Z|X)$$
$$= I_t(Y:Z) + I_t(Y:X|Z) \tag{10.11}$$

where

$$I_t(Y:Z|X) = \sum_{j=1}^{n} p(x_j) I_t(Y:Z|x_j) \tag{10.12}$$

and

$$I_t(Y:Z|X=x_j) = \sum_{i=1}^{n} \sum_{k=1}^{d} p(y_i, z_k|x_j) \frac{p(y_i, z_k|x_j)}{p(y_i|x_j)p(z_k|x_j)} \tag{10.13}$$

Note that $I_t(Y:Z|X)$ specifies the average additional information about Y given by Z, after taking into consideration the information given by Y. $I_t(Y:Z|X=x_j)$ specifies the additional information about Y given by Z, after taking into consideration the information given by $X=x_j$. Returning to the simple warning example given previously, $I_t(Y:X,Z)$ might describe the total amount of information in the response explained by the warning combined with other cues in the task. This value would be the sum of the transmitted information explained by the warning $I_t(Y:X)$ and the other cues $I_t(Y:Z|X)$. The remaining unexplained information contained in the response would be

Chapter ten: Modeling and evaluation of cognitive tasks

$$I_n = U(Y) - I_t(Y:X) - I_t(Y:Z|X) \tag{10.14}$$

In other words, the original uncertainty in Y is partitioned or segmented into explained and unexplained components. Dividing each component by the maximum amount of information that could be transmitted about the variable of interest, in this case $U(Y)$, gives the proportion of uncertainty that is explained and unexplained. The explained (or transmitted) proportion is

$$\frac{I_t(Y:X,Z)}{U(Y)} = \frac{I_t(Y:X) + I_t(Y:Z|X)}{U(Y)} \tag{10.15}$$

and the unexplained proportion is

$$\frac{I_n}{U(Y)} = \frac{U(Y) - I_t(Y:X) - I_t(Y:Z|X)}{U(Y)} \tag{10.16}$$

The latter approach makes the results more easily interpretable, and can help guide further examination of other variables that might explain some of the uncertainty in the criterion variable in an analyzed data set. It also should be pointed out that each component $I(X:Y)$ of the transmitted information is χ^2 distributed with value

$$\chi^2 = 2N\log(2)I_t(X:Y) \tag{10.17}$$

and

$$df = \prod_i c(x_i) - \sum_i (c(x_i) - 1) - 1 \tag{10.18}$$

where
  N is the number of observations
  $c(x_i)$ is the number of categories or levels of each variable

Consequently, the transmitted information can be partitioned and tested for significance following a procedure similar to analysis of variance (ANOVA; see Hosseini et al., 1991). An application of this approach to analyze issues related to the effectiveness of warnings is shown in Box 10.2.

Signal detection theory

In its original form, SDT assumes that an ideal observer is attempting to determine whether a signal has been sent or not, based on observational evidence and knowing that noise is present in the transmission (Green and Swets, 1966). This situation breaks down into four decision-event-related outcomes shown in Table 10.3. In some cases, the observer decides the signal was sent or is present, even though it really was not. This corresponds to a false alarm. A miss occurs if they decide it is not present when it really is. A hit occurs if they decide it is present when it really is sent. A correct rejection occurs if they decide it is not present, when it really is not. Note that *hits* and *correct rejections* are correct responses, whereas *misses* and *false alarms* are errors.

BOX 10.2 EXAMPLE OF INFORMATION PARTITIONING

To illustrate how information partitioning might be used, consider the case where the analyst has data on how often a person complies with a warning label, located in one of two locations, and has also observed whether the person read or did not read the label. An example study fitting this description compared how often 60 subjects followed an instructional label placed on either the shipping container a file cabinet was contained within or on the cabinet (Frantz and Rhoades, 1993). The latter researchers also measured whether the subjects read the label. In terms of information analysis, the location of the warning would be a warning feature, and reading the label would be a warning process. Compliance with the warning would be the response variable or criterion. Equation 10.7 can then be used to calculate the information about the response transmitted by the warning feature alone and warning process alone. The results are shown in Table 10.1.

The next step is to use Equations 10.11 through 10.13 to develop a model that best explains how location and reading influence the response. An initial step in this analysis is to calculate the amount of information jointly transmitted by the warning feature and warning process. As shown in Table 10.2, $I_t(R{:}r,L) = 0.244$ bits. Since warning location logically influences the reading of the warning, the analyst might choose the following explanatory model:

$$I_t(R{:}r,L) = I_t(R:L) + I_t(R{:}r|L) = 0.244 \text{ bits}$$

$I_t(R{:}L) = 0.143$ bits (as shown in Table 10.1). Consequently, $I_t(R{:}r|L)$ can be calculated using the previous results as follows:

$$I_t(R{:}r|L) = I_t(R{:}r,L) - I_t(R{:}L)$$

$$= 0.101 \text{ bits}$$

Also, from Equation 10.14 we get

$$I_n(R{:}r,L) = U(R) - I_t(R:L) - I_t(R{:}r|L) = 0.615 \text{ bits}$$

Equations 10.15 and 10.16 can then be used to normalize the calculated values in terms of the proportion of uncertainty explained. Doing so shows that the model explains 28.4% of the uncertainty in the observed response. The location of the warning explains 16.7% of the uncertainty, and reading the warning explains an additional 11.7%. Both effects are statistically significant,* at the $\alpha = 0.05$ level: $\chi^2(r, df=1) = 8.4$ and $\chi^2(L, df=1) = 11.9$.

* The values of χ^2 and the df are obtained using Equations 10.17 and 10.18.

In SDT, the relative frequencies of events in which the signal was sent or *not* sent are noted respectively as $p(S)$ and $p(N)$. Those two relative frequencies add to unity because only one of those two events can happen on any particular observation. Now the four decision outcomes in the previous matrix are really defined in terms of the event. Accordingly, pairs of those outcomes are dependent on the signal being present or not. In the case that a signal was

Table 10.1 Compliance Information Transmitted by Warning Label Location and Reading of the Label

	Reading (r)					Location (L)		
	Yes	No	Sum			Package	Cabinet	Sum
R	14	3	17		R	0	17	17
~R	11	32	43		~R	15	28	43
Sum	25	35	60		Sum	15	45	60
	$H(R:) = 0.859953$ bits					$H(R:) = 0.859953$ bits		
	$H(r:) = 0.979869$ bits					$H(L) = 0.811278$ bits		
	$I_t(R:r) = 0.201458$ bits					$I_t(R:L) = 0.14261$ bits		

Source: Based on data from Frantz, J.P. and Rhoades T.P., *Hum Fact*, 35, 719, 1993.

Table 10.2 Compliance Information Jointly Transmitted by Warning Label Location and Reading of the Label

	Location = Read = Yes	Package Read = No	Location Read = Yes	Cabinet Read = No	Sum
R	0	0	14	3	17
~R	0	15	11	17	43
Sum	0	15	25	20	60
	$H(R:) = 0.859953$ bits				
	$H(r, L) = 1.554585$ bits				
	$I_t(R:r, L) = 0.244345$ bits				

Source: Based on data from Frantz, J.P. and Rhoades T.P., *Hum Fact*, 35, 719, 1993.

Table 10.3 Four Basic Outcomes in SDT

Signal Sent	Yes	No
Observer says signal is present	HIT	FALSE ALARM
Observer says signal is not present	MISS	CORRECT REJECTION

sent, the two outcomes of *hits* and *misses* are conditioned on that event. Therefore, the relative frequencies of a hit (or detection) and a miss, given that a signal was present, add to unity:

$$p(D) + p(M) = 1 \qquad (10.19)$$

It also follows that the conditional relative frequencies corresponding to false alarms and correct rejections also add to unity:

$$p(FA) + p(CR) = 1 \qquad (10.20)$$

To illustrate how this model works, assume that an electrical engineer wanted to analyze how well a signal was transmitted over a TV broadcast communication channel. This engineer would likely put a signal detector at one end of the channel and a signal generator at the other. After doing so, the engineer observed random electromagnetic

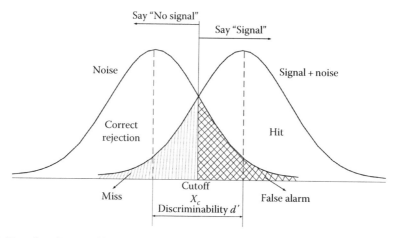

Figure 10.2 Signal and normally distributed noise distributions in signal detection theory.

disturbances similar to the signal that varied in strength. In SDT, those random disturbances are referred to as *noise*. Sometimes this noise would be sufficiently strong that it would give the impression of a signal. When this electrical engineering problem was formulated in statistical terms, it was assumed that noise varied in strength in a way that could be described with a normal distribution.* The mean of that noise distribution was arbitrarily set at zero and with a standard deviation of 1. Adding a constant signal to the noise results in another normal distribution with a standard deviation of 1 in normalized units, but the mean will be larger depending upon the signal strength, as shown in Figure 10.2. The difference between the mean of the noise distribution and the signal-plus-noise distribution is denoted as d', the *discriminability*. Since d' is measured in units of standard deviations, d' really is the signal-to-noise ratio.

The X axis in Figure 10.2 describes the measured or perceived strength of the signal. To decide whether a signal is present, the decision maker will check whether the observed signal strength is higher than some predetermined cutoff point. We will refer to that point as the cutoff point X_c. If the measured or perceived strength of the observation is equal to or greater than the cutoff point, the decision maker will decide that a signal is present. If not, the decision maker will decide that the signal is not there. In other words, the decision maker assumes that the measurement is noise alone. Now it is obvious that detection errors result from this operating procedure. The right-hand tail of the noise distribution that extends beyond X_c denotes the relative frequency of false alarms or $p(FA)$. Also the left-hand tail of the signal-plus-noise distribution below X_c denotes the relative frequency of misses or $p(M)$.

That is,

$$P(M) = \int_{-\infty}^{z_M} h(s)\,dz \qquad (10.21)$$

$$P(FA) = \int_{z_{FA}}^{\infty} g(z)\,dz \qquad (10.22)$$

---

* The terms normal and Guassian are used interchangeably to refer to this form of noise. See Noise and Signal-plus-Noise curves in Figure 10.2.

Chapter ten: Modeling and evaluation of cognitive tasks

$$d' = z_{FA} - z_M \tag{10.23}$$

where h(s) and g(s) are the probability density functions.

The SDT model also describes an effect called the *bias*. Consider the fact that the two normal distributions in Figure 10.2 have means that are exactly d' units apart. As such, the selected location of X_c specifies the size of one tail relative to the other tail. Note that a larger value of X_c in Figure 10.2 reduces false alarms but increases misses. The reverse changes occur with a smaller value of X_c. Therefore, it is presumed that a value of X_c is selected that appropriately reflects the relative cost of errors of each type. The exact setting of X_c determines the bias of the decision maker toward minimizing false alarms versus misses. The measure of bias in SDT is the parameter β defined as

$$\beta = \frac{p(X_c | \mu = d')}{p(X_c | \mu = 0)} \tag{10.24}$$

In effect, β is the ratio of the probability *ordinate* of misses to the probability *ordinate* of false alarms, both measured precisely at X_c. The real beauty of this theory is that it allows one to separate bias from sensitivity. A value of d' and a β value can also be easily calculated if P(FA) and P(M) are known (see Box 10.3 for example calculations of d' and β). These two measures can be used to evaluate human performance for a variety of applications, varying from inspection tasks (see Chapter 15) to the use of a vehicle collision warning system (Papastavrou and Lehto, 1995). The latter example extended the simple SDT model discussed earlier to a team decision where the collision avoidance system adjusted its value of β to reflect the fact that the driver was the primary decision maker and was also observing whether it was safe to pass.

The so-called receiver operating characteristic (ROC) curve is another important element of SDT. An ROC curve plots out the probability of a hit on one axis and the probability of a false alarm on the other, which will occur for a fixed value of d', at different levels of bias β. Figure 10.3 shows an example of an ROC. The curve traces out combinations of P(FA) and P(D) obtained for different values of β. In other words, a receiver with a particular sensitivity measured as d', will perform the task in a way that results in false alarm and detection rates between the two points [P(FA) = 1, P(D) = 1] to [P(FA) = 0, P(D) = 0].

Note that a different ROC curve is obtained for any value of d'. Higher values of d' correspond to better performance. If $d' = 0$, the ROC curve is a straight line between the points [P(FA) = 1, P(D) = 1] and [P(FA) = 0, P(D) = 0]. This situation corresponds to a signal strength or sensitivity value of 0. The curve shown in Figure 10.3, corresponds to a d' of around 2. Values of d' between 0 and 2 would trace out similarly shaped curves between the one shown and a straight line between the points [P(FA) = 1, P(D) = 1] to [P(FA) = 0, P(D) = 0]. Higher values of d' would trace out similar curves that lie above the curve shown in the figure.

If costs are assigned to false alarms, misses, hits, and correct rejections, it becomes possible to calculate an expected cost for each point on the ROC. An "ideal" operator will use a decision rule that selects the point on the ROC that optimizes performance. This approach can be applied to model performance without making the assumptions of the original version of SDT, such as normally distributed noise or a signal decision maker.

BOX 10.3 EXAMPLE CALCULATIONS OF d' AND β

Example 1

Suppose that 100 trials were done in an experiment in which an auditory alarm was given to subjects performing a task in a noisy environment. The alarm was given in half of the trials and not given in the others. At the end of each trial, the subject was asked if they heard the alarm. The results of the experiment are given as follows:

Response	Signal y_A	No Signal y_B	Row Sum
Yes	40	10	50
No	10	40	50
Column sum	50	50	100

To determine the value of d' and β for these data, it is necessary to first determine $P(FA)$ and $P(M)$ using the numbers just given. That is,

$P(FA) = $ (# of subject said yes when no signal was given)/(# of no signal trials)

$= 10/50 = 0.2$

$P(M) = $ (# of subject said no when a signal was given)/(# of signal trials)

$= 10/50 = 0.2$

The next step is to find the value of z_M. Note that Equation 10.21 corresponds to a left-tailed test, so if $P(M) < 0.5$, we know z_M is negative. Using the table in Section B.1.1 of Appendix B, and after switching signs, because the latter table is for a right-sided test, we see this corresponds to a z value of about −0.84 for z_M. We also find that z_{FA} is about 0.84. Then using Equation 10.23,

$$d' = z_{FA} - z_M = 0.84 - (-0.84) = 1.68$$

To determine β we need to find the ordinate values for a normal (Gaussian) distribution for each of the z scores that were used to determine d'. Using the table in Section B.1.2, we see that a z value of 0.84 corresponds to an ordinate value of about 0.28. Since both tails are the same, both ordinates are too; so using Equation 10.24 we get

$$\beta = 0.28/0.28 = 1.0$$

This result of $\beta = 1$ shows that both errors are treated as if the costs per miss and per false alarm are the same. In this situation the bias is zero; both errors are viewed equally.

Example 2

Consider another example where the frequency of false alarms is 0.10 and the frequency of misses is 0.08. Following the same steps as earlier, we find that these two frequencies yield standardized normal deviates, respectively, of 1.28 and −1.40, after interpolating the values in Section B.1.1. Consequently,

$$d' = 1.28 - (-1.4) = 2.68$$

Ordinates corresponding to those frequencies are 0.17 and 0.14 after interpolating the values in Table B.1.2. It follows using Equation 10.24 that

$$\beta = 0.1486/0.1754 = 0.8471$$

The result $\beta < 1$ indicates that this person viewed misses as more costly than false alarms.

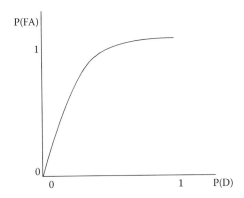

Figure 10.3 Example ROC curve.

Human information processing

At its most basic level, an information-processing model of the human will include the following stages: exposure to stimulus, message perception, storage and retrieval of information, decision making, and ultimately a response. Information is assumed to flow through these information-processing stages when the human being performs a task (Figure 10.4). The output of one stage becomes an input to a following stage. The rate at which information is processed at each stage is assumed to be limited. This means later stages of processing must wait until the necessary processing at earlier stages is completed. Interference occurs when two or more task elements require the same processing resources at the same time.

The following discussion will introduce some important elements of human information processing. Much of the emphasis here is on showing how and why task familiarity, receiver experience, and knowledge are critical determinants of human performance. Many such effects can be explained and/or predicted in terms of the way knowledge is represented within human memory and the mode of cognitive control. Simply put, a novice will normally behave very differently and know less than a more experienced person.

Models of attention

One finding of early research on human information processing was that people are limited in their ability to transmit information. Most of the incoming sensory data are filtered out at an early stage in the perceptual process (Welford, 1976). Welford (1976) states, "it is well known that far more data are transmitted by our sense organs to the

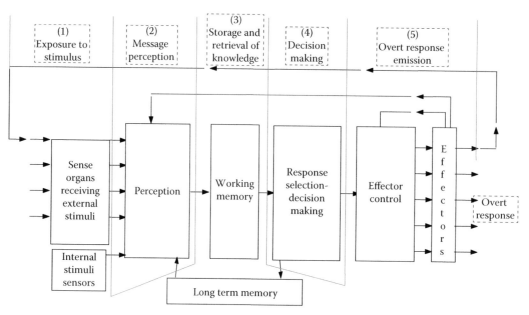

Figure 10.4 Sequential model of human information processing.

brain than we in fact perceive." For example, the transmission rate of the optic nerve could be said to be around 1 million bits/s, while the rate at which information enters short-term memory, in activities such as in reading, is probably around 100 bits/s (Kelley, 1968). Such effects are closely related to the allocation of attention. If the human could not selectively attend to individual aspects of the task, performance would be impossible. On the negative side, failures to notice and attend to displays, warnings, and environmental cues are a fundamental cause of human error and accidents. Attentional failures are related to factors that cause people to ignore relevant information or attend to irrelevant information. Such factors include attentional capacity and demands, expectations, and the stimulus salience.

Attentional capacity is perhaps the most fundamental constraint. The unitary view is that attention is a single, undifferentiated, but limited resource (Kahneman, 1973), which is allocated between tasks during task performance.* The multiple resource perspective of attention extends the unitary view by identifying separate resources associated with particular information-processing stages (Navon and Gopher, 1979). One of the most obvious conclusions suggested by either modeling perspective is that designers should try to use sensory channels that have *spare capacity*. For example, if the primary task performed by the product user places heavy demands on vision (in driving, for example), the designer might want to consider the use of auditory displays.

The unitary and multiple views of attention also lead to the conclusion that task-related *workload* is an important consideration. Under conditions of either extremely high or low workload, people become less likely to attend to potentially important sources of information. Attentional failures will also become more likely when there are many competing

---

* Certain models assume that thte total amount of attention that can be allocated is a function of arousal (Kahneman, 1973). Arousal refers to the activity of the autonomic nervous system. The general relationship between arousal level and performance is an inverted U. Greater levels of arousal increase performance to some optimal point, after which performance decreases.

sources of information or when noise is present in the task environment. Expectations are another important issue. People are less likely to notice unexpected stimuli. Perceptual confusions are also often related to expectations. Simply put, people have a tendency to see or hear what they expect to see or hear. One of the more common solutions proposed in such situations is to increase the *salience* of important information. Stimulus salience can be increased in many different ways. For example, lights can be made brighter, and tones or sounds can be made louder. Movement and changes in intensity also increase salience. The latter approach is illustrated by the use of flashing lights, frequency modulated tones, or waving flags.

The perceived value of attending to a source of information and the attentional effort required are other closely related issues that influence attention.* *Simply put, people will be less likely to attend to information displays if doing so is effortful or the information is not perceived to be important.* This tendency will be especially prevalent if there are competing demands for attention.

A number of generic strategies are available to the ergonomic designer for reducing the attentional effort required by an information display. One approach is to design a display to be consistent with existing *visual search patterns*. Information should be arranged so that the items are located where the typical user expects them to be, to help minimize the amount of time a person must spend searching for information. A second approach is to increase the *perceptual discriminability* of the items presented. One way to improve discriminability is to maximize the number of distinguishing features between individual elements of an information display. For example, two warning tones might differ in frequency, intensity, and modulation. Icons might differ in color, shape, and size. Another approach is to increase the *legibility* of the provided information.

Increasing the *conspicuity* of presented information is another related strategy. This approach is similar in many ways to increasing salience but focuses more on making an item *pop out* from its background. For example, warning symbols and icons might be designed to stand out from text in a warning sign by providing blank space around them and providing color contrast. Another strategy is to provide *redundant perceptual cues* on multiple sensory channels. For example, a visual display might be combined with a warning tone that is given when the monitored system is in an unusual state. Similarly, a tachometer might include a red line that indicates the point where the engine rpms become excessive.

Other explanations of attention and selective processing of information are related to working memory and the mode of processing, as expanded upon next.

Role of working memory

When people perform tasks, they are constantly shuffling information in and out of consciousness or, in other words, in and out of working memory. This process is nicely described by the production system model of human cognition (Newell and Simon, 1972; Anderson, 1983). From this perspective, task performance is described in terms of the changing contents of a working memory that has a limited capacity of five to nine items.

---

* Information overload has been cited as a reason for impaired decision making by consumers (Jacoby, 1977). As illustrated by the very lengthy, complex, patient package inserts provided by drug manufacturers in the United States, certain forms of product information impose a large workload on the intended audience. The degree to which normal consumers are willing to exert the effort needed to attend and process such information is obviously a very important issue.

The contents of the working memory change over time when new information entering working memory replaces or overwrites the existing contents. The new information can enter working memory when information from the environment is perceived or when facts are retrieved from long-term memory.

As pointed out by Lehto and Miller (1986), the production system model implies that attention will often be goal driven, making people less likely to pay attention to items unrelated to what they are doing. Research results consistent with this hypothesis are reported by Frantz (1992), who found that 85% of subjects using a wood preservative or drain cleaner stated they were looking primarily for directions on how to use the product. Placing the precautions in the directions resulted in an average compliance of 82% compared to 48% when they were in a separate precautions section.

The production system model also predicts that information perceived from displays or other sources of information will often not have any impact on behavior unless it is made available at the time it can be applied. This follows because items in working memory are constantly being replaced by new items. Numerous examples can be found where information is well integrated into a task, as, for example, when a lockout tag is placed next to an activated lockout, when switches are labeled on a control panel, or when a sign placed on a door warns against entering. Information that is well integrated into a task-specific context is also more likely to usefully exploit the human's knowledge. Well-integrated sources of information can act as cues that trigger the retrieval of additional, and hopefully relevant, information from long-term memory. Providing information only at the time it is relevant also reduces the chance of overloading working memory and minimizes the chance of distracting the operator.

Mode of processing

As stated by Anderson (1983)

> ... there are two modes of cognitive processing. One is automatic, less capacity-limited, possibly parallel, invoked directly by stimulus input. The second requires conscious control, has severe capacity limitations, is possibly serial, and is invoked in response to internal goals.

Cognitive psychology also makes a general distinction between procedural and declarative representations of knowledge stored in human long-term memory. Procedural knowledge corresponds to knowledge of *how to do something*, while declarative knowledge corresponds to knowledge of *facts*. The interactive roles of knowledge representation and the mode of processing are nicely captured by Rasmussen's skill-, rule-, and knowledge- (SRK)- based levels of task performance. As pointed out by Rasmussen, and expanded upon next, different forms of information are used at each level of performance (see Box 10.4).

Performance at the skill-based level involves the more or less automatic processing of *signals* from the task environment—that is, signal processing does not require a lot, if any, conscious control and is illustrated by a variety of sensory-motor tasks, such as tracking a moving object with your eyes or steering a vehicle. Performance at the rule-based level can be viewed as the sequential cuing of responses by consciously perceived cues or *signs*. For example, when a stoplight changes from red to green, most people decide to go. Skill-based and rule-based levels of performance are both procedural in nature. The primary difference is that skill-based behavior is a continuous procedure, in which each step is automatically executed, similar to the execution of a compiled computer program, or mental model,

> **BOX 10.4 SIGNALS, SIGNS, SYMBOLS, AND VALUES**
>
> The same stimulus might act as a different form of information, depending upon how it is processed. For example, consider the case where a person is observing a pressure gauge for a pressure vessel used in some industrial process. During a pressurization stage, an experienced operator might be making minor adjustments of a control value to fill the tank at a constant rate. During this stage of the task, the fluctuating movement of the display pointer would be acting as a *signal*.
>
> As the tank nears the proper pressure, the operator might then read the numerical value from the display, and decide to turn the valve off. At this point, the display is now acting as a *sign* or consciously named cue in working memory. Suppose the display shows that the pressure is continuing to build up after the operator shuts off the valve. In this situation, the operator might begin reasoning using a functional model of how the system works to determine why the pressure is still increasing. During this reasoning process, the information from the display might be represented as a *symbol* in some general rule (such as if the rate of pressure increase is greater than x, and the valve is turned off, and ... then component y might have failed). At the same time, the operator is likely to start reacting affectively to the situation. At the most basic level, the operator may become excited or highly aroused. They also might feel threatened. The display is now being processed in terms of *value* (in this case threat or a sense of potential danger), which both causes an affective reaction and alters goal priorities.

while rule-based behavior is a consciously executed step-by-step sequence. Performance at the knowledge-based level involves a much more laborious procedure in which people reason about the situation they face. This process involves the explicit formation of plans directed toward attaining some goal, and involves the meaningful manipulation of *symbols*.

A fourth level, called judgment-based performance (Lehto, 1991), focuses on how affective reactions to stimuli and *values* influence goals and objectives. As pointed out by Lehto (1991), the level of performance of the receiver should be carefully considered when designing products, warnings, and other forms of information. This process can be guided by attempting to match forms of information to the level of task performance. An initial observation is that the most important sources of information will often be elements of the product or environment, rather than supporting sources such as displays. Signals and cues from the product and environment are particularly important when performance is at the skill- or rule-based level. One of the key issues is that people are often unwilling to read instructions or consult complex, lengthy manuals. This tendency increases when people are highly familiar with the product or task they are performing.

Simply put, products and environments should be designed to provide signals and cues that are easily perceived, easily discriminated, and valid indicators of conditions, functions, hazards, and misuses. The latter requirements are closely related to the principles of operational visibility and consistency (Box 10.5). Signals and cues also can provide important forms of feedback during the use of a product or the operation of a control. Such feedback can communicate information about both correct and incorrect operation. For example, a control might emit an audible click when it is moved correctly into position, and a circular saw might vibrate and emit loud sounds when it is being used to cut an object it is not designed to cut.

BOX 10.5 OPERATION VISIBILITY AND CONSISTENCY

The principle of operational visibility states that easily perceived features of a device should communicate to the user how the device is to be used (Norman, 1988). Thus, when a person first encounters the device, its shape, size, texture, and other features should say to the potential user, "This is what I do and this is how I do it."

Many simple devices around us should but do not do that because architects and other designers often have a greater fascination for artistic beauty than knowledge of ergonomics or usability. For example, the doors leading out of public buildings often have a symmetrical panic bar running across the door and hide the hinges outside the door. Hiding the hinges makes it hard for the user to know which edge of the door opens and whether to push or pull. A simple flat plate on the moving side of the panic bar would communicate that the door should be pushed there, even without adding the word *push* to the plate. Note that a flat plate intuitively invites one to push it. Conversely, a vertical bar invites one to reach out and pull it. Adding these simple communication features greatly increases the usability of this simple device.

Consistency is an important related principle. If the features of a device or information display are consistently designed, these features can become valid, and in many cases easily learned, indicators to the user of when and how to perform necessary actions. This principle is illustrated by standards that prescribe the consistent use of particular words, colors, symbols, and shapes for the visual elements of traffic signs. Consistency is also one of the most emphasized elements of computer user interface design. Companies such as Microsoft, IBM, Apple Computer, and Hewlett Packard require that their software interfaces be reviewed for consistency with internal guidelines to be sure that particular elements are used in the same way throughout the interface. Web pages are another example where more sophisticated companies have established and carefully enforce consistency guidelines. Efforts are also made in many cases to promote consistency across products made by different companies. However, significant differences are sometimes observed. For example, the displays and controls used in automobiles will often differ between manufacturers. These differences will cause many drivers to spend some time learning, when they switch between vehicles. However, many drivers have strong preferences for particular configurations that they have become familiar with, so making changes to promote consistency across manufacturers is likely to create inconsistencies with the expectations of established customers, who are the majority after all, and may interfere with the fundamental principle of "providing current customers what they want."

Another issue is that various forms of information can encourage necessary shifts in the level of performance. One obvious application of this approach is the use of warnings or alerts in an effort to shift people upward from a routine skill- or rule-based level after they make errors or when unique or novel conditions are present. Educational forms of information also might be provided to help people develop or internalize procedures. When such information is procedural in nature, it helps people shift down from a knowledge-based level to the more routine rule-based task of simply following the instructions. For example, providing simple step-by-step instructions describing how to perform particular actions can be very helpful to inexperienced people who have not learned the proper procedure.

Declarative versus procedural knowledge representations

As noted earlier, the distinction between declarative and procedural knowledge is one of the most fundamental tenets of cognitive psychology. Procedural knowledge can be described in terms of schemas (Anderson, 1983), scripts (Schank and Abelson, 1977), production rules (Newell and Simon, 1972; Anderson, 1983), or goal hierarchies (Anderson, 1983). A schema or script might be viewed as a simple sequence of actions performed in a particular setting, such as the sequence of steps followed when getting into a car and preparing to drive. A rule might describe one or more steps to take when a certain condition is observed, such as deciding to stop when you see a stop sign. A goal hierarchy might describe the relationship between certain activities and the goals that the person is trying to attain. For example, to start a car, you need to find your keys, locate and insert the key into the lock, etc. All three representations can be used to describe the same set of procedures, as should be obvious from the preceding examples. Declarative knowledge is often described as propositions, encoded as linked nodes in an associative network, where nodes correspond to objects and links between nodes describe relations (Quilliam, 1968; Anderson and Bower, 1973). An example proposition is "asbestos causes cancer." In an associative network, "asbestos" and "cancer" would be nodes and "causes" would be a relation linking these two nodes.

The distinction between declarative and procedural knowledge representation made in the production system model, and others (including schema theory and scripts) has some very significant implications. For example, the ACT* production system model (Anderson, 1983) provides a detailed explanation of why it is difficult for people to perform tasks before they develop a good procedural knowledge representation. Part of the problem is that declarative knowledge and general rules must first be converted into an executable procedural form before they can be applied efficiently. For example, consider the relative difficulty of developing a solution from a general set of axioms compared to that of following a step-by-step algorithm. With greater experience, rules become more specialized or compiled to the extent that a sequence of actions can be performed without conscious processing once the specialized rule is triggered. An obvious implication is that simple step-by-step procedures describing how to perform a particular task may be helpful to inexperienced people who have not internalized the proper procedure. A second implication is that sources of information attended to when people are first learning how to do a task are less likely to be noticed after people become highly experienced and perform much of their task without conscious attention. Another important issue is that errors in task performance change depending on both the mode of processing and knowledge representation.

Error types and forms

As mentioned above, the types of errors people make during task performance tend to differ in form in a way that depends on the mode of processing. The Generic Error Modeling System (GEMS) developed by Reason (1990) extends Rasmussen's SRK framework to describe the tendency. Some common types of errors are discussed below from this perspective.

Skill-based errors
As noted by Reason (1990), errors at the skill-based level of performance tend to take the form of "strong but wrong" routines of behavior that occur when an attentional check

---

* ACT refers to, depending on the source, adaptive character of thought, adaptive control of thought, first 3 letters of activation, etc.

is omitted or mistimed. Specific examples include errors associated with inappropriate schema capture or schema fixation. Capture errors occur when an inappropriate schema is activated. A capture error can result in numerous observable consequences. These include perceptual failures, (i.e., distractions or, in other words, the failure to perceive an important condition because of an attentional capture resulting in either an incorrect or omitted response), and numerous response failures, that is, slips in which a person does something unintended when they react to a signal or a lapse in which a person forgets what they were supposed to do (i.e., overwriting of STM) because of an interruption, resulting in an omission or out-of-order responses.

Fixation errors occur when performing a particular task element consumes so much capacity that a critical attentional check needed elsewhere is omitted. As with capture errors, numerous different behavioral effects are associated with schema fixation. For example, a failure to perceive an important condition because of fixated attention can result in either an incorrect or omitted response. Errors at the skill-based level of performance can also occur because the task requirements are beyond the human's perceptual or motor skills or when people fail to shift up to rule-based or higher levels of performance at critical stages where conscious control is needed. A large set of factors come into play at this level that impact the possibility of error, including (1) variability in perceptual or motor skills between individuals as a function of the task, environment, and stressors; (2) signal strength and location; (3) the time available for a response; and (4) the required precision of responses.

Some common errors in skill-based behavior include scenarios where people follow scripts that are inappropriate because of changes in familiar settings or differences between products (e.g., a person walking might not change their gait before stepping on a wet spot, a driver might not notice a new stop sign at a familiar intersection, a person who previously used a nonflammable adhesive without ventilating their work area might do the same when using a flammable adhesive). Another common problem is that scripts are incomplete or contain unsafe steps that have been reinforced by past experience (i.e., entering, starting, and driving a vehicle without automatically fastening a seat belt).

To recover from or prevent an error from occurring at the skill-based level, the receiver must make a transition to a rule-based or higher level. There are theoretical reasons to expect shifts to a rule-based or higher level to become more likely at periods of transition between scripts or after interruptions in task performance (Reason, 1990). Some support for the theory that shifts occur at periods of transition between scripts is given by Frantz and Rhoades (1993) who found that 40% of their subjects noticed a warning placed on the top drawer of a file cabinet compared to 0% for a warning placed on the side of the package the cabinet came within. The exact moment when the warning label was noticed was not measured but probably occurred during a transition between the unpacking and filling scripts. Stronger support for the effect of interruptions is provided by their finding that the noticing of warning labels increased to 93% when (1) the top drawer was sealed shut with a warning or (2) a cardboard obstruction displaying the label was placed within the drawer. The latter conditions interrupted performance and concurrently showed a large increase in noticing of the warnings. However, the potentially positive role of task interruptions is limited because means of interrupting skill-based behavior are not always available and can lose effectiveness after initial use (e.g., in the study by Frantz and Rhoades, both methods of interrupting the task were disabled by filling the cabinet). Another problem is that interrupting the task can be annoying, and may cause serious user-acceptance problems, as illustrated by the public outcry against seat-belt interlock systems in the United States a few years ago. Interruptions are also a common cause of errors at the skill-based level of performance. After an interruption, people may fail to resume performance entirely or leave out steps in procedures (Reason, 1990).

Rule-based errors

At the rule-based level, errors occur when people apply faulty rules or fail to shift up to a knowledge-based level in unusual situations where the rules they normally use are no longer appropriate. This error mode can correspond to either misapplication of a good rule, or application of an inadequate rule (Reason, 1990). Inadequate rules may prescribe shortcuts or other experience-based behavior patterns that work most, but not all, of the time (Rasmussen, 1986). Intentional violations are another type of rule-based behavior and occur when people violate societal mores in routine ways, for instance, speeding except when police are present. The use of faulty rules leads to an important distinction between running and taking risks. Along these lines, Wagenaar (1992) discusses several case studies in which people following risky forms of behavior do not seem to be consciously evaluating the risk. Drivers, in particular, seem to habitually take risks. Wagenaar explains such behavior in terms of faulty rules derived on the basis of benign experience. In other words, drivers get away with providing small safety margins most of the time and, consequently, learn to run risks on a routine basis.

An interesting observation is that rule-based errors may be caused by skill-based errors. For example, schema capture or fixation errors can lead to perceptual failures, which in turn either prevent a needed transition up to the rule-based level or cause the incorrect rule to be applied. In the first case, the involved person fails to consciously perceive a condition so the behavior never reaches the rule-based level at the critical moment. In the second case, the person might fail to perceive or remember a contraindication against applying a particular rule. The observable consequences of this error mode include omitting task elements, performing task elements in the wrong order, or executing undesirable actions (as when counter-indications are forgotten). Efforts directed toward eliminating rule-based errors might focus on providing new rules or correcting incorrect rules that have been developed on the basis of experience (moved down the hierarchy from previous knowledge-based or judgment-based reasoning). This situation is undoubtedly a difficult one, because to remedy it, performance must be shifted to the knowledge based or higher level, during which the old rules containing unsafe actions (i.e., diving in shallow water, failure to wear seat belts, failure to wear helmets, failure to wear eye protection, etc.) are modified or new ones are learned. As emphasized by Rasmussen (1986), inducing upward shifts from the rule-based level is difficult, often because people are overconfident in the adequacy of their routine behavior.

Knowledge-based errors

Knowledge-based behavior is remarkably error prone and is indistinguishable from learning. Reason (1990) and other authors have provided rather extensive taxonomies of error types occurring when behavior is knowledge based. Such errors can be explained as either the inadequacies in the mental models that people use, errors made when manipulating mental models, or unintended consequences of exploratory testing of the system. Errors made when manipulating mental models are often related to the limited capacity of working memory.

Since people at a knowledge-based level of performance are likely to be seeking out information, it seems plausible that providing step-by-step instructions and other performance aids can help prevent errors when performance is at this level. One reason this approach may work is that people who are unfamiliar with a product seem least likely to have developed conflicting forms of skill or rule-based behavior that would have to be modified to attain compliance. In such situations, providing step-by-step instructions should allow people to perform at the rule-based level and, therefore, be highly useful.

On the other hand, knowledge-based behavior is less prevalent than skill- or rule-based behavior (Rasmussen, 1986), and the tendency to seek out information at the knowledge-based level of performance may also be outweighed by a profound lack of knowledge in certain instances. Johnson and Russo (1985) discuss research supporting the conclusion that simply providing information to inexperienced consumers may not be adequate to support intelligent choices between products. The extent to which the intended audience is able to correctly interpret the provided information is another important consideration. It will always be the case that some people will not understand some or all of the provided information. In some cases, this issue may be important enough to justify testing of comprehension.

Judgment-based errors

At the judgment-based levels, errors (or violations) can occur because of inappropriate affective reactions, such as anger or fear (Lehto, 1991). For example, frustration over being late for a flight might increase the desirability of speeding. Similarly, stress and anxiety over being behind schedule might lead an employee to take shortcuts or violate standard operating procedures. Generally, affective reactions are inappropriate when they are too large (i.e., rage over being cut off by a vehicle on the highway). However, the lack of an appropriate affective reaction (i.e., a failure to feel fear when driving at excessive speeds) could, of course, also contribute to violations. Both forms of inappropriate affective reactions can lead to inappropriate priorities and violations of social norms that are difficult to prevent. This leads us to the topic of human decision-making.

Decision making

Human decision making is a broad topic that is fundamentally related to the topic of information processing because people must gather, organize, and combine information from different sources to make many decisions. Decision making is often viewed as a stage of human information processing, but as decisions grow more complex, information processing actually becomes part of decision making, and methods of decision support that help decision makers process information become of growing importance. Decision making also overlaps with problem solving. The point where decision making becomes problem solving is fuzzy, but many decisions require problem solving and the opposite is true as well. Some of the earlier discussed cognitive models are, consequently, relevant for describing many aspects of human decision making. The following discussion will begin by briefly introducing the reader to classical decision theory and some of its prescriptions. This will include a brief discussion of decision rules and the subjective expected utility (SEU) model. Focus will then shift to some research results that compare human behavior to the prescriptions of classical decision theory, and some alternative modeling approaches that make less restrictive assumptions about how people should make decisions.

Classical decision theory

Classical decision theory (von Neumann and Morgenstern, 1947; Savage, 1954) frames the choices people make in terms of four basic elements:

Chapter ten: Modeling and evaluation of cognitive tasks

1. A set of potential actions or alternatives (A_i) to choose between
2. A set of events or world states (E_j)
3. A set of consequences (C_{ij}) obtained for each combination of action and event
4. A set of probabilities (P_{ij}) for each combination of action and event

From elementary probability theory, it can be shown that the return to a decision maker is maximized by selecting the alternative with the greatest expected value. The expected value of an action A_i is calculated by weighting its consequences C_{ik} over all events k, by the probability P_{ik} that the event will occur. The expected value of a given action A_i is, therefore,

$$EV[A_i] = \sum_k P_{ik} C_{ik} \qquad (10.25)$$

To illustrate how this approach might be applied, consider a case where a decision maker is deciding whether to wear a seat belt when traveling in an automobile. Wearing or not wearing a seat belt corresponds to two actions, A_1 and A_2. The expected consequence of either action depends upon whether an accident occurs. Having or not having an accident corresponds to two events, E_1 and E_2. Wearing a seat belt reduces the consequences (C_{11}) of having an accident. As the probability of having an accident increases, use of a belt should, therefore, become more attractive.

Classical decision theory has also specified a set of decision rules for choosing between alternatives once the choice has been framed in terms of the aforementioned elements. Frisch and Clemen propose that

> a good decision should (a) be based on the relevant consequences of the different options (consequentialism), (b) be based on an accurate assessment of the world and a consideration of all relevant consequences (thorough structuring), and (c) make trade-offs of some form (compensatory decision rule).

Consequentialism and the need for thorough structuring are both assumed by all normative decision rules. Most normative rules are also compensatory. However, because of cognitive limitations and the difficulty of obtaining information, it becomes unrealistic in many settings for the decision maker to consider all the options and possible consequences. To make a decision under such conditions, decision makers may limit the scope of the analysis by using simple heuristics or non-compensatory decision rules. The following discussion will briefly address some of these simple decision-making strategies and their implications.

Non-compensatory decision rules

Many non-compensatory decision rules have been identified. These include principles such as dominance, lexicographic ordering and elimination by aspects, and satisficing. *Dominance* is perhaps the most fundamental normative decision rule. An alternative is dominated by another, if it is never better than the other alternative, and is worse in at least one aspect. For example, a person choosing between products has an easy decision if one is both cheaper and safer. Dominance is obviously a normative decision rule, since an alternative is never better than an alternative that dominates it, and is often used by decision makers in naturalistic settings. The simplicity of this decision rule also makes it attractive.

The *lexicographic ordering principle* (see Fishburn, 1974) considers the case where alternatives have multiple consequences. For example, a purchasing decision might be based on both the cost and safety of the considered product. The different consequences are first ordered in terms of their importance. The decision maker then sequentially compares each alternative beginning with the most important consequence. If an alternative is found that is better than the others on the first consequence, it is immediately selected. If no alternative is best on the first dimension, the alternatives are compared for the next most important consequence. This process continues until an alternative is selected or all the consequences have been considered without making a choice. The *elimination by aspects* (EBA) rule (Tversky, 1972) is similar to the lexicographic decision rule. It differs in that the consequences used to compare the alternatives are selected in random order, where the probability of selecting a consequence dimension is proportional to its importance. Both EBA and lexicographic ordering are non-compensatory decision rules, since the decision is made using a single consequence dimension. Returning to the aforementioned example, if the decision maker feels safety is more important than cost, the lexicographic principle predicts that the customer would select a product that is slightly safer, even if it costs much more. EBA would select either product, depending on which consequence dimension the decision maker happened to focus upon first.

The *minimum aspiration level* or *satisficing* decision rule assumes that the decision maker sequentially screens the alternative actions until an action is found that is good enough. For example, a person considering the purchase of a car might stop looking once they found an attractive deal, instead of comparing every model on the market. More formally, the comparison of alternatives stops once a choice is found that exceeds a minimum aspiration level S_{ik} for each of its consequences C_{ik} over the possible events E_k. Satisficing can be a normative decision rule when (1) the expected benefit of exceeding the aspiration level is small, (2) the cost of evaluating alternatives is high, or (3) the cost of finding new alternatives is high. More often, however, it is viewed as an alternative to maximizing decision rules. From this view, people cope with incomplete or uncertain information and their limited rationality by satisficing in many settings instead of optimizing (Simon, 1955, 1983).

Subjective expected utility model and prospect theory

Classical decision theory proposes that a decision maker should select the alternative that maximizes expected value or utility (von Neumann and Morgenstern, 1947; Savage, 1954). While people do not always prefer alternatives that maximize expected utility (Winterfelt and Edwards, 1986; Stevenson, Busemeyer, and Naylor, 1991), the model does provide a base from which to analyze how consumers and others choose between alternatives when they are not sure of the outcome. Perhaps the most interesting aspect of the SEU model is that it takes into account differing degrees of risk aversion. This is modeled by fitting user preferences with a utility function $u(x)$ that assigns a utility to each of the possible consequences of choosing an action, where x is a quantity measured on a quantitative scale such as dollars, time, or severity of injury.

People who are risk averse have concave utility functions (See Figure 10.5), making the expected utility of a gamble less than the utility of its expected value. A linear utility function corresponds to risk neutrality, making the expected utility for a gamble equal to the utility of its expected value. A risk-seeking person, on the other hand, has a convex utility function, making the expected utility for a gamble greater than the utility of its expected value. In terms of the expected utility model, people's safety-related behavior is often risk seeking. In other words, they disproportionately emphasize the small loss associated with taking various safety precautions. Prospect theory (Kahneman and Tversky, 1979; Tversky

Chapter ten: Modeling and evaluation of cognitive tasks

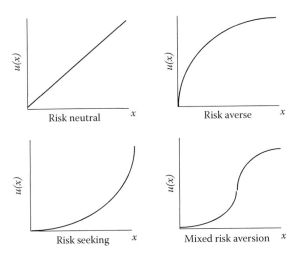

Figure 10.5 Utility functions for differing risk attitudes.

and Kahneman, 1981) is an extension of the SEU model that describes this and some other commonly observed effects using an S-shaped utility function. In the risk-seeking region of the utility function, a small, but certain, cost is disproportionately less attractive than a large, but uncertain, cost. In the risk-averse region of the utility function, a small, but certain, benefit may be disproportionately attractive compared to a large, but uncertain, benefit. The framing effect is another important part of the prospect theory. That is, framing a choice in terms of costs can cause a shift to the risk-taking portion of the utility function. Framing a choice in terms of benefits causes a shift to the risk-averse portion.

The best-known study of framing was conducted by Tversky and Kahneman (1981). Their data showed that preferences between medical intervention strategies changed dramatically, depending upon whether the outcomes were posed as losses or gains. Tversky and Kahneman concluded that this reversal illustrated a common pattern in which choices involving gains are risk averse, and choices involving losses are risk seeking. The interesting result was that the way the alternatives were worded caused a shift in preference for identical alternatives. More recently, it has been shown that such framing effects can be reduced or even be eliminated by changing the wording of problem statements (Kuhlberger, 1995). This latter study showed that standard framing effects could be reversed by certain problem wordings and eliminated by fully describing the problems.

Human judgment and inference

The ability of people to perceive, learn, and draw inferences accurately from uncertain sources of information has been a topic of much research. Research in the early 1960s tested the notion that people behave as intuitive statisticians who gather evidence and apply it in accordance with the Bayesian model of inference (Peterson and Beach, 1967). Much of the earlier work focused on how good people are at estimating statistical parameters, such as means, variances, and proportions. Other studies compared human inferences obtained from probabilistic evidence to the prescriptions of Bayes' rule.

Some example findings are given in Table 10.4. The research first shows that people can be fairly good at estimating means, variances, or proportions from sample data. Hertwig et al. (1999) point out that "there seems to be broad agreement with the conclusion" of

Table 10.4 Sample Findings on the Ability of People to Estimate Statistical Quantities

Accurate estimation of sample means	Peterson and Beach (1967)
Variance estimates correlated with mean	Lathrop (1967)
Variance biases not found	Levin (1975)
Variance estimates based on range	Pitz (1980)
Accurate estimation of event frequency	Hasher and Zacks (1989), Estes (1976), Jonides and Jones (1992)
Accurate estimates of sample proportions between 0.75 and 0.25	Edwards (1954)
Accurate detection of changes in relative frequencies	Robinson (1964)
Severe overestimates of high probabilities; severe underestimates of low proportions	Fischhoff et al. (1977); Lichtenstein et al. (1982)
Reluctance to report extreme events	Du Charme (1970)
Weather forecasters provided accurate probabilities	Winkler and Murphy (1973)
Poor estimates of expected severity	Dorris and Tabrizi (1977)
Correlation of 0.72 between subjective and objective measures of injury frequency	Rethans (1980)
Risk estimates lower for self than for others	Weinstein (1979); Weinstein and Klein, (1995)
Risk estimates related to catastrophic potential, degree of control, familiarity	Lichtenstein et al. (1978)
Evaluations of outcomes and probabilities are dependent	Weber (1994)

Jonides and Jones (1992) that people can give answers that reflect the actual relative frequencies of many kinds of events with great fidelity. However, as discussed by Winterfelt and Edwards (1986), subjective probability estimates are noisy like other psychophysical measures. Their accuracy will depend on how carefully they are elicited, and many other factors. Studies have shown that people are quite good at estimating probabilities between 0.25 and 0.75, when they have adequate exposure to the events they are asked to assess.

On the other hand, people have great trouble accurately estimating the probability of unlikely events, such as nuclear plant explosions. For example, when people were asked to estimate the risk associated with the use of consumer products (Dorris and Tabrizi, 1978; Rethans 1980) or various technologies (Lichtenstein et al., 1978), the obtained estimates were often weakly related to accident data. Weather forecasters are one of the few groups of people that have been documented as being able to estimate high and low probabilities accurately (Winkler and Murphy, 1973).

Part of the issue is that people will not be able to base their judgments on a representative sample of their own observations when events occur rarely. Most of the information they receive about unlikely events will come from secondary sources, such as media reports, rather than their own experience. This tendency might explain why risk estimates are often more strongly related to factors other than likelihood, such as catastrophic potential or familiarity (Lichtenstein et al., 1978; Slovic, 1978, 1987; Lehto et al. 1994). Media reporting focuses on newsworthy events that tend to be more catastrophic and unfamiliar. Consequently, judgments based on media reports might reflect the latter factors instead of likelihood.

Human inference

Inference is the procedure followed when a decision maker uses information to determine whether a hypothesis about the world is true. Hypotheses can specify past, present, or future states of the world, or causal relationships between variables. Diagnosis is concerned with determining past and present states of the world. Prediction is concerned with determining future states. Inference or diagnosis is required in many decision contexts. For example, before deciding on a treatment, a physician must first diagnose the illness.

From the classical perspective, the decision maker is concerned with determining the likelihood that a hypothesis (H_i) is true. Bayesian inference is perhaps the best-known normative method for statistical inference. Bayesian inference provides a well-defined procedure for inferring the probability (P_i) that a hypothesis (H_i) is true, from evidence (E_j) linking the hypothesis to other observed states of the world. The approach makes use of Bayes' rule to combine the various sources of evidence (Savage, 1954). Bayes' rule states that the posterior probability of hypothesis H_i given that evidence E_j is present, or $P(H_i | E_j)$, is given by the equation

$$P(H_i | E_j) = \frac{P(E_j | H_i)P(H_i)}{P(E_j)} \tag{10.26}$$

where

$P(H_i)$ is the probability of the hypothesis being true prior to obtaining the evidence E_j
$P(E_j|H_i)$ is the probability of obtaining the evidence E_j given that the hypothesis H_i is true

When the evidence E_j consists of multiple states $E_1, \ldots, E_n$, each of which is conditionally independent, Bayes' rule can be expanded into the expression

$$P(H_i | E_j) = \frac{P(H_i)\prod_{j=1}^{n} P(E_j | H_i)}{P(E_j)} \tag{10.27}$$

An example illustrating use of this equation is given in Box 10.6. Over the years, a large number of studies have compared human inference to the predictions of Bayes' rule obtained using Equation 10.27. The earlier studies found several significant deviations from the Bayesian model. These include the following:

1. Decision makers tend to be conservative in that they do not give as much weight to probabilistic evidence as Bayes's rule (Edwards, 1968).
2. They do not consider base rates or prior probabilities adequately (Tversky and Kahneman, 1974).
3. They tend to ignore the reliability of the evidence (Tversky and Kahneman, 1974).
4. They tend to overestimate the probability of conjunctive events and underestimate the probability of disjunctive events (Bar-Hillel, 1973).
5. They tend to seek out confirming evidence rather than disconfirming evidence and place more emphasis on confirming evidence when it is available (Einhorn and Hogarth, 1978; Baron, 1985). The order in which the evidence is presented has an influence on human judgments (Hogarth and Einhorn, 1992).
6. They are overconfident in their predictions (Fischhoff, Slovic, and Lichtenstein, 1977), especially in hindsight (Fischhoff, 1982; Christensen-Szalanski, 1991).
7. They show a tendency to infer illusionary causal relations (Tversky and Kahneman, 1973).

BOX 10.6 EXAMPLE OF BAYESIAN INFERENCE

Let us illustrate how this theorem handles information in an elementary situation. Suppose that there are two plain brown paper sacks with blue and white poker chips inside. You are not allowed to look in either sack, but you know that one of those two sacks has chips that are 20% blue and the other has 80% blue chips. Your decision information consists of two samples of three chips from each sack. So you pick one of the two sacks randomly. The probability that you select the sack with 80% blue chips is about 50–50 or 0.5. Now you sample three chips from that sack and find that *all three chips are blue*. What does your intuition tell you? You are right if it now seems *much more likely* that you have the sack with 80% blue chips. You are not certain because it is possible, although highly unlikely, that the sack with 20% blue chips could still yield a sample of three blue chips by random selection. Now say that you get another sample of three chips from the same sack and those are also all blue. Your intuition is now even stronger that you have the sack with 80% blue chips. Bayes' theorem works in a similar way, but it gives quantitative values. To start with, there are two hypotheses regarding which of the two sacks you are sampling from. Let us call sampling from the sack with 80% blue chips h_1 and sampling from the other sack h_2. Before any samples are taken, $P(h_1) = P(h_2) = 0.5$, corresponding to maximum uncertainty regarding which sack you are sampling from. Now if sampling from the sack is binomial in nature, the probability $P(e_1|h_1)$ of picking three blue chips from the bag containing 80% blue chips is given as

$$P[n = 3; x = 3] = \frac{3!}{3!0!} 0.8^3 (1 - 0.8)^0 = 0.512$$

while the probability $p(e_1|h_2)$ of picking three blue chips from the bag containing 20% blue chips is

$$P[n = 3; x = 3] = \frac{3!}{3!0!} 0.2^3 (1 - 0.2)^0 = 0.008$$

These two equations show the probability of obtaining the sample evidence for each of the bags. So by substituting these values into Equation 10.26 one obtains for the first bag:

$$P(h_1 | e_1) = \frac{0.5 * 0.512}{0.5 * 0.512 + 0.5 * 0.008} = 0.985$$

After taking the second sample of three blue chips, the preceding value becomes the prior probability, and the posterior probability you are sampling from the first bag becomes

$$P(h_1 | e_1, e_2) = \frac{0.985 * 0.512}{0.985 * 0.512 + 0.015 * 0.008}$$

$$= 0.9998$$

It follows that the probability associated with the other sack is the complement of this, or 0.0002. Thus, the information obtained from taking the two samples results in a change in the event probability from 0.5 to 0.9998.

Heuristics and biases

On the basis of such research, some psychologists have concluded that the prescriptions of classical decision theory have little to do with human decision making. One of the most influential contributions was made by Tversky and Kahneman who showed that many of the discrepancies or human biases could be explained by three heuristics (Tversky and Kahneman, 1973, 1974). The three heuristics they proposed were those of (1) representativeness, (2) availability, and (3) anchoring and adjustment. The representativeness heuristic holds that the probability of an item (A) belonging to some category (B) is judged by considering how representative A is of B. The availability heuristic holds that the probability of an event is determined by how easy it is to remember the event happening. Anchoring and adjustment holds that people start from some initial estimate and then adjust it to reach some final value.

Tversky and Kahneman state perceived probabilities will, therefore, depend on familiarity, salience, effectiveness of memory search, and imaginability. The implication is that people will judge events as more likely when the events are familiar, highly salient (such as an airplane crash), or easily imaginable. Events also will be judged more likely if there is a simple way to search memory. For example, it is much easier to search for words in memory by the first letter rather than the third letter. It is easy to see how each of the aforementioned items impacting the availability of information can result in biases. These biases should increase when people lack experience or when their experiences are too focused. On the other hand, it can be argued that these heuristics work remarkably well in a wide variety of naturalistic settings.

A lively literature has developed regarding these deviations and their significance[8] (Evans, 1989; Caverni et al., 1990; Wickens, 1992; Klein, Orasanu, Calderwood, and Zsambok, 1993; Doherty, 2003). From one perspective, these deviations demonstrate inadequacies of human reason and are a source of societal problems (Baron, 1998; and many others). From the opposite perspective, it has been held that the previous findings are more or less experimental artifacts that do not reflect the true complexity of the world (Cohen, 1993). A compelling argument for the latter point of view is given by Simon (1955, 1983). From this perspective, people do not use Bayes' rule to internally compute probabilities in their natural environments because it makes unrealistic assumptions about what is known or knowable. To compensate for their limitations, people use simple heuristics or decision rules that are adapted to particular environments. The use of such strategies does not mean that people will not be able to make accurate inferences, as emphasized by both Simon and researchers embracing the ecological[9] (i.e., Hammond, 1996; Gigerenzer et al., 1999) and naturalistic (i.e., Klein et al., 1993) models of decision making. In fact, the use of simple heuristics in rich environments can lead to inferences that are in many cases more accurate than those made using Naive Bayes or linear regression (Gigerenzer et al., 1999).

The finding that subjects are much better at integrating information when they are provided data in the form of natural frequencies instead of probabilities (Gigerenzer and Hoffrage, 1995; Krauss, Martignon, and Hoffrage, 1999) is particularly interesting. One conclusion that might be drawn from the latter work is that people are Bayesians after all, if they are provided adequate information in appropriate representations (Martignon and Krauss, 2003). Other support for the proposition that people are not as bad at inference as it once seemed includes Dawes and Mulford's (1996) review of the literature supporting the overconfidence effect or bias, in which they conclude that the methods used to measure this effect are logically flawed and that the empirical support is inadequate to conclude it really exists.

Part of the issue is that much of the psychological research on the overconfidence effect "over-represents those situations where cue-based inferences fail" (Juslin and Olsson, 1999). When people rate objects that are randomly selected from a natural environment, overconfidence is reduced. Koehler (1996) provides a similarly compelling reexamination of the base rate fallacy. He concludes that the literature does not support the conventional wisdom that people routinely ignore base rates. To the contrary, he states that base rates are almost always used and that their degree of use depends on task structure and representation, as well as their reliability compared to other sources of information.

Because such conflicting results can be obtained, depending upon the setting in which human decision making is observed, researchers embracing the ecological (i.e., Hammond, 1996; Gigerenzer et al., 1999) and naturalistic (Klein et al., 1993; Klein, 1998) perspectives on decision making strongly emphasize the need to conduct ecologically valid research in rich realistic decision environments.

Brunswik lens model of human judgment

A number of approaches have been developed for mathematically describing human judgments in natural environments. These approaches include the use of policy capturing models in social judgment theory, probabilistic mental models, and information integration theory.

Much of the work in this area builds upon the *Brunswik lens model* (Brunswik, 1952), originally developed to describe how people perceive their environment. The lens model is used in these approaches to describe human judgment on some criterion in terms of two symmetric concepts: (1) the ecological validity of probabilistic cues in the task environment, and (2) cue utilization. The *ecological validity* of a cue is defined in terms of the correlation or probabilistic relation between a cue and the criterion; *cue utilization* is defined in terms of the correlation or probabilistic relation between the cue and the judgment. The emphasis on ecological validity in this approach is one of its key contributions. The focus on ecological validity results in a clear, measurable definition of domain-specific expertise or knowledge and also can be used to specify limits in how well people can perform. More specifically, a perfectly calibrated person will know the ecological validities of each environmental cue. A second issue is that a good decision maker will utilize cues in a way that appropriately reflects ecological validity—that is, cues with higher ecological validity should be emphasized more heavily. The ultimate limit to performance is described in terms of the maximum performance possible, given a set of cues and their ecological validities. The latter quantity is often estimated in terms of the variance explained by a linear regression model that predicts the criterion using the cues (Hammond et al., 1975).

In *social judgment theory* (SJT) (Hammond et al., 1975; Brehmer and Joyce, 1988; Hammond, 1993), the focus is on developing *policy-capturing models* that describe how people use probabilistic environmental cues to make judgments. Linear or nonlinear forms of regression are used in this approach to relate judgments to environment cues. SJT has been applied to a wide number of real-world applications to describe expert judgments (Brehmer and Joyce, 1988). For example, policy-capturing models have been applied to describe software selection by management information system managers (Martocchio et al., 1993), medical decisions (Brehmer and Joyce, 1988), and highway safety (Hammond, 1993). Policy-capturing models provide surprisingly good fits to expert judgments. In fact, there is evidence and, consequently, much debate over whether the models can actually do better than experts on many judgment tasks (Slovic et al., 1977; Brehmer, 1981; Kleinmuntz, 1984).

Cognitive continuum theory (Hammond, 1980) builds upon social judgment theory by distinguishing judgments on a cognitive continuum varying from highly intuitive decisions to highly analytical decisions. Hammond (1993) summarizes earlier research showing that task characteristics cause decision makers to vary on this continuum. A tendency toward analysis increases, and reliance on intuition decreases, when (1) the number of cues increases, (2) cues are measured objectively instead of subjectively, (3) cues are of low redundancy, (4) decomposition of the task is high, (5) certainty is high, (6) cues are weighted unequally in the environmental model, (7) relations are nonlinear, (8) an organizing principle is available, (9) cues are displayed sequentially instead of simultaneously, and (10) the time period for evaluation is long. One of the conclusions developed from this work is that intuitive methods can be better than analytical methods in some situations (Hammond et al., 1987).

The theory of *probabilistic mental models* (Gigerenzer et al., 1991) is another Brunswikian model that has attracted a lot of recent attention. As in SJT, this approach holds that human knowledge can be described as a set of cues, their values, and their ecological validities. However, the *ecological validity* of a cue is defined as the relative frequency with which the cue correctly predicts how well an object does on some criterion measure, instead of as a correlation. Inference is assumed to be a cue-based process involving one or more pair-wise comparisons of the cue values associated with particular objects. Several different heuristics have been proposed within the framework of probabilistic mental models that describe how the inference process might be performed. One of the better performing heuristics is called the *take-the-best* heuristic. This simple heuristic begins by comparing a pair of objects using the most valid cue. If one of the objects has a positive cue value while the other does not, the object with the higher positive cue value is given a higher value on the criterion, and the inference process stops. Otherwise, the heuristic moves on to the next most valid cue. This process continues until a choice is made, or all cues have been evaluated.

Take the best is an example of what Gigerenzer et al. (1999) call a fast and frugal heuristic. This follows because the heuristic will generally make a choice without considering all of the cues. As such, take the best, differs in a major way from the multiple linear equations normally (but, not always, as emphasized by Hammond) used in the policy-capturing approach, and elsewhere. One of the most interesting results reported by Gigerenzer et al. (1999) was the finding that take the best was always as good and normally outperformed multiple linear regression, Dawes' rule (which sums the number of positively related cues and subtracts the number of negatively related cues), and Naive Bayes, when making predictions for several different data sets drawn from very different domains. The latter procedures used all the cues and involve more complex statistical operations, which makes the performance of take the best quite impressive. Gigerenzer et al. (1999) conclude that this performance demonstrates that fast and frugal heuristics can be highly effective.

Recognition-primed decision making and situation awareness

Klein (1989, 1998) developed the theory of recognition-primed decision making on the basis of observations of firefighters and other professionals in their naturalistic environments. He found that up to 80% of the decisions made by firefighters involved some sort of situation recognition, where the decision makers simply followed a past behavior pattern once they recognized the situation.

The model he developed distinguishes between three basic conditions. In the simplest case, the decision maker recognizes the situation and takes the obvious action. A second case occurs when the decision maker consciously simulates the action to check whether it should work before taking it. In the third and most complex case, the action is found to be deficient during the mental simulation and is consequently rejected. An important point of the model is that decision makers do not begin by comparing all the options. Instead, they begin with options that seem feasible based upon their experience. This tendency, of course, differs from the SEU approach but is comparable to applying the satisficing decision rule (Simon, 1955) discussed earlier.

Situation assessment is well recognized as an important element of decision making in naturalistic environments (Klein et al., 1993). Recent research by Klein and his colleagues has examined the possibility of enhancing situation awareness through training (Klein and Wolf, 1995). Klein and his colleagues have also applied methods of cognitive task analysis to naturalistic decision-making problems. In these efforts, they have focused on identifying (1) critical decisions, (2) the elements of situation awareness, (3) critical cues indicating changes in situations, and (4) alternative courses of action (Klein, 1995). Accordingly, practitioners of naturalistic decision making tend to focus on process-tracing methods and behavioral protocols (Ericsson and Simon, 1984) to document the processes people follow when they make decisions.

Orasanu and Salas (1993) discuss two closely related frameworks for describing the knowledge used by teams in naturalistic settings. These are referred to as shared mental models and the team mind. The common element of these two frameworks is that the members of teams hold knowledge in common and organize it in the same way. Orasanu and Salas claim that this improves and minimizes the need for communication between team members, enables team members to carry out their functions in a coordinated way, and minimizes negotiation over who should do what at what time. Under emergency conditions, Orasanu and Salas claim there is a critical need for members to develop a shared situation model. As evidence for the notion of shared mental models and the team mind, the authors cite research in which firefighting teams and individual firefighters developed the same solution strategies for situations typical of their jobs.

This notion of shared mental models and the team mind can be related to the notion discussed earlier of schemas containing problem-specific rules and facts (Cheng and Holyoak, 1985). It also might be reasonable to consider other team members as a form of external memory (Newell and Simon, 1972). This approach would have similarities to Wegner's (1987) concept of trans-active memory where people in a group know who has specialized information of one kind or another. Klein (1998) provides an interesting discussion of how this metaphor of the team mind corresponds to thinking by individuals. Teams, like people, have a working memory that contains information for a limited time, a long-term or permanent memory, and limited attention. Like people, they also filter out and process information and learn in many ways.

Models of time pressure and stress

Time pressure and stress are a defining characteristic of naturalistic decision making. Jobs requiring high levels of skill or expertise, such as firefighting, nursing, emergency care, and flying an airplane, are especially likely to involve high stakes, extreme time pressure, uncertainty, or risk to life. The effect of stressors, such as those mentioned here, on performance has traditionally been defined in terms of physiological arousal. The Yerkes–Dodson law (Yerkes and Dodson, 1908) states that the relation between performance and arousal is an inverted U; either too much or too little arousal causes performance to drop. Too little arousal makes it difficult for people to maintain focused attention. Too much arousal results in errors, more focused attention (and filtering of low-priority information), reduced working memory capacity, and shifts in decision strategies. One explanation of why performance drops when arousal levels are too high is that arousal consumes cognitive resources that could be allocated to task performance (Mandler, 1979).

Maule and Hockey (1993) note that people tend to filter out low-priority types of information, omit processing information, and accelerate mental activity when they are under time pressure. Time pressure also can cause shifts between the cognitive strategies used in judgment and decision-making situations (Edland and Svenson, 1993; Maule and Hockey, 1993; Payne et al., 1993). People show a strong tendency to shift to non-compensatory decision rules when they are under time pressure. This finding is consistent with contingency theories of strategy selection. In other words, this shift may be justified when little time is available, because a non-compensatory rule can be applied more quickly. Compensatory decision rules also require more analysis and cognitive effort. Intuitive decision strategies require much less effort because people can rely on their experience or knowledge, and can lead to better decisions in some situations (Hammond et al., 1987). As Klein (1998) points out, stress should impact performance if people use analytical choice procedures.

Novices and experts in novel, unexpected situations will lack domain experience and knowledge and, therefore, will have to rely on analytical choice procedures. Consequently, it is not surprising that time pressure and stress have a major negative impact on novice decision makers performing unfamiliar tasks. Interestingly, there is little evidence that stress or time pressure causes experienced personnel to make decision errors in real-world tasks (Klein, 1996; Orasanu, 1997). The latter finding is consistent with research indicating that experts rely on their experience and intuition when they are under stress and time pressure (Klein, 1998). The obvious implication is that training and experience are essential if people are to make good decisions under time pressure and stress.

Cognitive simulation

The modeling methods discussed earlier in this chapter can be used to simulate how people perform cognitive tasks. Cognitive simulations show how activity within the task is related to goals, knowledge, and information-processing activity. The approach can also describe how task sequences, depth of reasoning, critical types of information, knowledge requirements, memory demands, and other factors influence time requirements and result in errors. For example, errors might be traced to factors such as the similarity between the antecedent clauses of rules (Kieras, 1985), the forgetting of specific conditions, or the use of inappropriate heuristics (Johnson and Payne, 1985). Furthermore, such analyses can be done prior to the development of the system or product to be analyzed. The NGOMSL task description language, in particular, is specifically intended to play such a role and Kieras (1988) notes that the NGOMSL description of *how-to-do-it* knowledge can often be developed prior to system development as easily as after.

Example applications can be found pertaining to human–computer interaction, following of instructions, operation and learning of operating principles of electronic equipment, fault diagnosis, acquisition of flight skills, and air traffic control (Lehto, Sharit, and Salvendy, 1991). Several practical applications of this approach are discussed as follows.

GOMS: Cognitive work measurement of a skill-based task

Card et al. (1983) discuss research directed toward the practical analysis of a text-editing task that they consider to be representative of skilled cognitive task performance in general. During this analysis they seek answers to several general questions such as

Is it possible to describe the behavior of a user engaged in text editing as the repeated application of a small set of basic information-processing operators? Is it possible to predict the actual sequences of operators a person will use and the time required to do any specific task? ... How does the model's ability to describe and predict a person's behavior change as we vary the grain size of analysis?

As we shall see from the following discussion, the methods of analysis they chose to address these questions are practical extensions of the traditional forms of work measurement discussed earlier.

GOMS model of task performance

GOMS models human performance in terms of four basic components: (1) goals, (2) operators, (3) methods, and (4) selection rules for choosing among alternative methods. The basic structure of the GOMS model developed by Card et al. (1983) for use of a text editor is illustrated in Table 10.5. Note that the model hierarchically defines the goals of a person performing a text-editing task. More specifically, the goal EDIT-MANUSCRIPT is divided into multiple cases of the subgoal EDIT-UNIT-TASK. The subgoal EDIT-UNIT-TASK is further broken down into ACQUIRE-UNIT-TASK and EXECUTE-UNIT-TASK. Each of these goals is then further subdivided into subgoals.

Within the GOMS model, goals are symbolic structures that guide the sequencing of tasks during performance. Goals are satisfied by applying elementary perceptual, motor, and cognitive operators that relate inputs to outputs and last for fixed durations. Functional operators, shown in Table 10.5, include

GET-NEXT-PAGE,
GET-NEXT-TASK,
USE-QS-METHOD,
USE-LF-METHOD,
USE-S-COMMAND,
USE-M-COMMAND,
VERIFY-EDIT.

More detailed operators at argument and keystroke levels can be evaluated during experimentation, as will be expanded upon later. Methods relate operators and subgoals to goals and are conditionally activated. For example, one method in Table 10.5 is

GOAL:	ACQUIRE-UNIT-TASK
	GET-NEXT-PAGE
	GET-NEXT-TASK

Selection rules are used to conditionally select methods. For example, Card et al. note that Selection rules for the GOAL: LOCATE-LINE of the example model might read:

IF the number of lines to the next modification is less than 3,
THEN use the LF-METHOD; ELSE use the QS-METHOD.

Table 10.5 Structure of the GOMS
Model of Text Editing

GOAL:EDIT-MANUSCRIPT
. GOAL:EDIT-UNIT-TASK
. . *repeat until no more unit tasks*
. . GOAL:ACQUIRE-UNIT-TASK
. . . GET-NEXT-PAGE
. . . *if at end of manuscript page*
. . . GET-NEXT-TASK
. . GOAL:EXECUTE-UNIT-TASK
. . . GOAL:LOCATE-LINE
. . . . [select: USE-QS-METHOD
. . . . USE-LF-METHOD]
. . . GOAL:MODIFY-TEXT
. . . . [select: USE-S-COMMAND
. . . . USE-M-COMMAND]
. . . VERIFY-EDIT

Source: Card, S. et al., *The Psychology of Human-Computer Interaction*, Lawrence Erlbaum Associates, Hillsdale, NJ, 1983.

Experimental evaluation of the GOMS model

A series of nine experiments (Card, Moran, and Newell, 1983) was performed aimed at evaluating the accuracy of time predictions and task sequences generated by the GOMS model. Prior to performing this analysis, the elemental performance times and selection rules needed to implement the GOMS model were experimentally assessed, as were errors and their influence on task performance times. All such data were obtained during protocol analysis in which subjects were videotaped and verbal reports were recorded.

During evaluation, the grain of analysis used by the model was varied between UNIT-TASK, FUNCTIONAL, ARGUMENT, and KEYSTROKE levels. The UNIT-TASK grain of analysis corresponded to a single operator (EDIT-UNIT-TASK), which was repeatedly performed. The FUNCTIONAL level subdivided the UNIT-TASK into the goals and operators listed in Table 10.6. At the ARGUMENT level, operators were further subdivided into the commands and arguments used on the text-editing system. The KEYSTROKE level was defined in reference to the basic physical and mental activity of the user and included the operations of typing, looking, moving hand, and mental activity.

It was found that the GOMS model predicted unit-task performance times within 35% of mean observed times, which corresponded to an accuracy of 4% over a 20 min editing session. Errors occurred in 36% of the tasks and doubled performance times from 12.5 to 24.4 s. The UNIT-TASK model was somewhat less accurate in its predictions than the more detailed models, which were all approximately equivalent in accuracy. Task sequences were predicted with an accuracy of over 90% for all methods except the KEYSTROKE model. The latter model correctly predicted approximately 50% of the task transitions.

The results showed that the GOMS model was capable of accurately predicting performance time and that increasing the grain of analysis to a functional level was useful. However, further increases in detail did not bring about corresponding increases in the accuracy of time predictions because of decreased accuracy in predicting task sequences. Card et al. (1983) conclude that this latter effect took place because there was not enough

Table 10.6 Example Operators in NGOMSL

User actions pertaining to
  (1) Flow of control
      accomplish goal of ⟨goal description⟩
      report goal accomplished
      decide: if ⟨operator⟩ then ⟨operator⟩
         else ⟨operator⟩
      go to step ⟨step number⟩
  (2) Memory storage and retrieval
      recall that ⟨WM description⟩
      retrieve that ⟨WM description⟩
      forget that ⟨WM description⟩
      retrieve LTM that ⟨LTM description⟩
  (3) Primitive perceptual-motor acts
      hand movements (from/to)
      home-hand to mouse
      press-key ⟨keyname⟩
      type-in ⟨string of characters⟩
      move-cursor to ⟨target coordinate⟩
      find-cursor at ⟨cursor coordinates⟩
      find-menu-item ⟨menu-item-coordinates⟩
  (4) Analyst defined mental operators
      get-from-task ⟨name⟩
      verify-result
      get-next-edit-item

Source: Kieras, D.E., Towards a practical GOMS model methodology for user interface design, in Helander, M. (Ed.) *Handbook of Human-Computer Interaction*, Elsevier Science Publications B. V., North-Holland, Amsterdam, the Netherlands, 1988, pp. 135-157.

variation "within the editing tasks to trigger increased responsiveness from the finer grain models." Such results are encouraging for the practical analysis of cognitive tasks (about 60% of the time for this task was consumed during covert activity) because they indicate that fairly aggregated portions of skilled cognitive activity can be treated as unit tasks for the purposes of time estimation.

Natural GOMS language

Kieras (1988) discusses a GOMS methodology for human–computer interface design, which he calls natural GOMS language (NGOMSL). Use of the NGOMSL language results in a model of how the task is performed that can be simulated on the computer. More specifically, the model consists of a set of rules and methods that are executed to accomplish goals. The methods followed are selected in a Recognize-Act cycle where rules are triggered by perceived conditions.

To develop a model using NGOMSL, the analyst first specifies a set of top-level goals that the user of the system is expected to have. NGOMSL describes these as action–object

Chapter ten: *Modeling and evaluation of cognitive tasks* 419

pairs in the form: ⟨verbnoun⟩, such as activate system, setup system, etc. For each goal, the analyst describes a general method in terms of operators that are as high level (or general) as possible. The operators within this general method are then provided methods. This recursive procedure continues until each method is described in terms of analyst-defined or primitive operators.

The set of predefined primitive mental operators provided in NGOMSL describes user actions pertaining to

1. The flow of control
2. Memory storage and retrieval
3. Primitive perceptual-motor acts

Primitive operators for each of these categories are shown in Table 10.5. User-defined mental operators are also included that describe psychological processes too complex to be predefined in the NGOMSL language. Methods are described as a sequence of steps, as follows:

accomplish goal of ⟨goal description⟩

step 1. ⟨operator⟩
step *n*–1. ⟨operator⟩
step *n*. report goal accomplished.

Selection rules are described as:

IF ⟨condition⟩ THEN
accomplish goal of ⟨specific goal description⟩

The specific goal description is either associated with one specific method or triggers a new selection rule. Table 10.7 gives a partial listing of a method for text editing.

Kieras (1988) notes that developing the NGOMSL model allows interface designs to be evaluated prior to the development of a system in terms of several qualitative measures. These include

1. Naturalness of the design—Do the goals and subgoals make sense to a new user of the system?
2. Completeness of the design—Is there a method for each goal and subgoal?
3. Cleanliness of the design—If there is more than one method for attaining a goal, is there a clear and easily stated selection rule for choosing an appropriate method?
4. Consistency of design—Are methods for attaining similar goals consistent?
5. Efficiency of design—Are the most important goals accomplished by quick methods?

Quantitative predictions also become possible in NGOML, but only if the model is developed to the level of primitive operators. High-level operators, if left undefined, obviously consist of an indeterminate set of actions. If the model is developed to the level of primitive operators, performance and learning times can be inferred from the model. Along these lines, Kieras (1988) presents experimental data where learning times were well described as a function of the number of NGOMSL statements. Execution times were predicted in terms of the time required to execute the primitive operators required for a given method. The NGOMSL model also allows workload to be inferred in terms of the short-term memory load for specific tasks. Transfer of training can also be evaluated by comparing the methods associated with particular goals.

Table 10.7 Partial Listing of a Method in NGOMSL

1. *Method to accomplish goal of editing a file*
   Step 1. Retrieve—LTM that current item in command sequence is EDITING A FILE
   Step 2. Accomplish goal of performing file open
   Step 3. Accomplish goal of performing cut text
   Step 4. Accomplish goal of performing paste text
   Step 5. Accomplish goal of performing close file
   Step 6. Report goal accomplished
2. *Method to accomplish goal of performing File Open*
   Step 1. Accomplish goal of issuing file
   Step 2. Accomplish goal of issuing open file
   Step 3. Verify that correct file is chosen
   Step 4. Report goal accomplished
3. *Method to accomplish goal of issuing File*
   Step 1. Recall File and accomplish goal of locating File
   Step 2. Recall N and Press-Key N
   Step 3. Forget N and File
   Step 4. Report goal accomplished

Source: Reprinted from Lehto, M.R. et al., *Int. J. Prod. Res.*, 29(8), 1565, 1991. With permission.

Final remarks

The approaches for modeling the performance of cognitive tasks discussed in this chapter go well beyond those used in traditional forms of work measurement and methods analysis. One advantage of these approaches is that they explicitly show how activity within the task is related to goals, knowledge, and information processing. This can lead to a task description that is more detailed, and yet more flexible than traditionally developed in work measurement. More importantly, this approach can help the designer of tasks and products understand, evaluate, and predict the results of mental activity followed by people when they perform tasks. A good model can in some cases be implemented as a method of decision support for system operators or product users and, in some cases, can even perform the task on its own better than the experts that it attempts to describe.

Discussion question and exercises

10.1 What are the basic elements of a communication system? Explain the relation between entropy and mutual information.

10.2 What is an optimal coding method? What are the associated advantages?

10.3 An uncertainty analysis was made of how a person reacted to a loading signal while loading ships in a port. The results were as follows:

		Responses		
		a	b	Sum
Stimuli	a	0.51	0.10	0.61
	b	0.03	0.36	0.39
Sum		0.54	0.46	1.00

Chapter ten: Modeling and evaluation of cognitive tasks

What is the stimulus uncertainty in bits?

10.4 What are some ways of increasing the rate at which people transmit information? What are some basic limitations in human information processing?

10.5 Calculate U(R), U(r), and It(R:r) using the values in Table 10.1. Explain why the amount of information transmitted is not simply just It(R:L) + It(R:r).

10.6 In signal detection theory (SDT):
   a. What is d'?
   b. What is β?
   c. What role does the ROC serve?
   d. Determine the values of d' and β that would result in P(FA) = .1 and P(M) = .05.

10.7 What are some of the factors that influence attention?

10.8 What is the so-called Hick–Hyman law? Is it really a law of science?

10.9 Why is working memory important in task performance?

10.10 How and why is the mode of processing related to particular types of error? Explain how the mode of processing is related to signals, signs, symbols, and values.

10.11 How are the principles of operational visibility and consistency related to each other?

10.12 What is a cognitive simulation? Discuss the similarities and differences between GOMS and NGOML.

10.13 What is a value function? Why might it not necessarily be rational to always make decisions based on expected value? How does subjective expected utility differ from expected value? What is a multi-attribute utility function?

10.14 Consider the example of Bayesian inference shown in Box 10.6. Suppose that in both samples, two blue chips were selected. What would the probability now be that the samples were taken from the bag containing 80% blue chips?

10.15 What are some of the ways people's preferences for risky or uncertain alternatives violate basic axioms of SEU theory? How might prospect theory and the theory of contingent decision making explain some of these violations?

10.16 When people make intuitive estimates of probabilities and statistics based on sample data, how closely do the estimates by people compare to normative estimates? How does this compare to their ability to make statistical inferences?

10.17 What are some heuristics used by people to estimate probabilities? How well do these heuristics work? Why might the use of heuristics result in biases?

10.18 What are some alternative explanations of both biases and good performance on probability estimation and inference tasks? Why might it be unreasonable to draw conclusions from laboratory studies showing that people make inferences in unfamiliar settings that are inconsistent with the predictions of Bayes' Rule?

10.19 On what basis do the Brunswikian and naturalistic perspectives draw conclusions that people are in fact excellent decision makers when making choices in familiar natural settings or environments?

10.20 Discuss some of the similarities between the theory of contingent decision making, cognitive continuum theory, and the levels of task performance model. How and why might time pressure or stress impact the way people make decisions?

chapter eleven

Control tasks and systems

About the chapter
This chapter focuses on how people use control devices to control or guide the activity of machinery, vehicles, or processes. The chapter introduces some fundamental control concepts, statistics, and applications that provide perspective on how to model and measure human performance of control tasks, before briefly dicussing topics such as fuzzy-control logic and supervisory control. The latter topics are becoming important with the development of increasingly sophisticated equipment that can perform many of the functions done by people in the past, including communication.

Introduction
The word *control* has two meanings important to the field of ergonomics, depending on whether it is used as a verb or a noun. Used as a verb, control refers to the activities of one or more persons directly or indirectly guiding a system or process. For example, the crew of a ship might be taking steps to ensure they reach an intended destination, while maintaining a desired course and speed throughout the trip. Used as a noun, a control is a device used to control something. Steering wheels, key pads, knobs, push buttons, or switches are all good examples of controls. Both aspects of control will be introduced in this chapter.

Before moving into this discussion, it should be pointed out that control is an important aspect of many common tasks. At the most basic level, people control their limbs, hands, and other body parts in quite sophisticated ways when they make movements. Something as simple as reaching out to a light switch requires significant sensorimotor activity to keep track of the hand's current and projected position, and make necessary corrections in the path followed to avoid obstacles and stop at the target destination.* In most cases, people do much of this automatically, without being aware of how they actually make the movement or why certain movements, such as reaching to a small target some distance away, are more difficult and take longer to do. In somewhat more complex settings, people are consciously monitoring their environment, and using simple rules to decide what to do next to keep the system or process operating properly given what they have observed. For example, a driver looking ahead at the road will slow down when they observe that the roadway is blocked, or speed up if they notice they are traveling below the speed limit.

In even more complex settings, people will spend significant time and effort planning out responses to unusual conditions (such as an out-of control chemical process) as well as on learning from what happened to improve future performance. The most obvious common element to each of the previous settings is that something is being controlled or, in other words, kept within reasonable bounds, whether it is the position of your hand, the speed of your car, or the temperature of molten steel in a furnace. This implies that to control something, it is necessary to have (1) one or more goals or objectives, (2) observations

---

* There will be more on this topic later. Also, see the discussion in Chapter 6.

or measurements of how close the current situation is to meeting the goals or objectives, and (3) a means of adjusting performance to be better.*

Control systems

In many instances, the controlled system will have multiple controllers and system components. These different players in the system will often have differing goals and objectives that can often be organized as a hierarchy. One goal usually involves keeping the system on track by following some path through different states over time or by following some preconceived path. A simple example of such a control system is shown in Figure 11.1.

In the latter example, the captain wants to arrive at a specific location on a particular date. That is the highest goal in the hierarchy and is known as the *outer loop* of the control hierarchy. Next in line is the navigator who sets the specific course to be followed. The navigator makes a plan to fit the captain's orders and passes this plan down the control hierarchy. At the middle level is the officer of the deck, who specifies the average course heading in steering the ship, following the course set by the navigator but with correction for average sidewise ship movements occurring over his or her watch. The quartermaster who is controlling the wheel that moves the rudder takes orders from the officer of the deck. The quartermaster controls the rudder directly. He or she moves that wheel with various waves in order to keep the stated course. Thus, the inner hierarchical control provides error adjustment and correction. It follows that controllers at each of these hierarchical levels can anticipate some changes in the environment and some changes in the control hierarchy below them. It can also be argued that individual people exercise control by acting out all the roles aboard ship.

As Figure 11.1 describes, control starts out with goals and information in the mind of the person in charge of control, be it a person in direct control or a design engineer with an electromechanical device. However, let us hold our consideration to the individual person maintaining direct control of the system, and in this specific case, let us assume that the system is an automobile. In order to drive a car, one must turn the steering wheel in the intended direction, release the brake, direct the transmission to the proper general direction (forward or backward) and gear, and then apply pressure on the foot-operated throttle. All of those actions are *control signals* sent to a *control junction*, which converts that signal into some electromechanical action. The latter step results in the transfer of force from the engine, or *power source*, through different *control effectors* to the car's wheels.

These elements are all part of the control system shown in Figure 11.2. When the vehicle is operated, we have *control activities* and the associated *controlled variables* of speed and

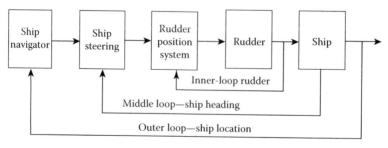

Figure 11.1 Control hierarchy in ship tending.

---

* Kelley (1968) devised this shipboard analogy many years ago. Kelley did a lot of work for the U.S. Navy and maybe that is why he came up with this analogy.

Chapter eleven: Control tasks and systems 425

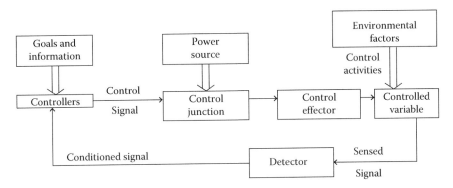

Figure 11.2 Example of control system.

direction of movement. The speed of the vehicle is sensed by a rotational detector that senses the revolutions of the wheels per minute. A speed gage converts that information into miles per hour or kilometers per hour, which the speedometer displays as a *conditioned signal* to the driver. The direction of the vehicle's travel is visually sent back to the driver through the windshield, revealing a travel path, current position, and velocity. All of these elements are part of the fundamental control system shown in Figure 11.2. In addition, there are environmental factors, such as rain falling on the pavement or a headwind plowing against the car, that alter the car's speed and direction.

When the conditioned signals indicate to the controller that the vehicle is not moving at the appropriate speed or direction, the controller takes this feedback information and alters the control signals sent to the control junction. A slight change of steering direction may be needed to move the car closer to the desired direction, and some throttle adjustment may be required to bring the car to the desired speed. In some cases, the friction between the tires and roadway may be low, so that the controller's added throttle causes the wheels to spin and the car's speed drops to near zero. To deal with this situation, the controller needs to be knowledgeable enough to realize that the cause of insufficient progress was slipperiness and that there is an appropriate action when that occurs.

Under more typical conditions where the pavement is dry and the friction is good, more power from the throttle would have produced speed increases, rather than a speed decrease. Human controllers learn how to control the vehicle through education, training, and experience. When the controller is a computer that the design engineer must program, that program is known as a *control model*. Such models vary from very fundamental to highly enriched, just as drivers vary from naive drivers, who are just starting out, to experienced drivers and, further, to professional race drivers who are very highly experienced. In part, control models are also used to describe features of human drivers as well as models to take the place of drivers. Also, some of these models are used to augment drivers, just as programs are used in aircraft to augment pilots' actions on the wing tab adjustments that need such rapid and complex corrections that people cannot perform them properly. Control models in the automatic braking systems in many new automobiles are another form of machine augmentation.

The immediate question posed is, "How would you start to devise a control model for even the simplest automobile situations?" You might say, "Let us start by constructing a simple model. We can embellish it more later." Most readers would agree with that strategy. You might also try to use introspection and imagine yourself as the controller. While there are clear differences between human and mechanistic controllers, most readers would agree that introspection provides a good start. A third approach might begin with an engineering model to start with, because engineers have been designing our cars in the past and they, more than many groups of people, would have a reasonable start on this subject.

Manual control

People often perform tasks in which they manually manipulate a control to meet some desired objective. In many cases, the primary objective is to move a controlled object from one location to another. For example, a forklift operator may be trying to lift a load up from the floor to a desired height. In this instance, the manual control task involves a sequence of control adjustments that are performed until the correct height is reached. After each adjustment, the operator looks to see how close they are to the correct height, and then uses this information to decide how much more to move the control. The information gathered provides feedback on how the task is progressing, and is essential to quick and efficient performance of the task.

BOX 11.1   ROLE OF VISIBILITY IN CONTROL APPLICATIONS

While it might seem trite or a little too simplistic, visibility is a very important aspect of many control tasks, particularly with vehicular control because vision is needed to guide the vehicle down the road or aisles between machine tools in the factory. In the most simplistic terms, designers should prevent visual obstructions from interfering with visibility. Improving visibility adds greatly to the safety and controllability of most vehicles. This fact is illustrated by Figure 11.3A through C, which shows the view from a windshield of a car at different eye positions. It is clear from this example that very important information can be lost with occluded vision. The child is unseen in Figure 11.3A and the traffic light is unseen in C.

Another example where visibility is important in design is crane pulpits. Figure 11.4 describes an old design of a crane pulpit used in the steel industry of about World War II vintage, and a newer design of about 1960 vintage where the operator has greatly improved visibility. Unfortunately, many crane operators in the steel industry still have obscured visibility due to fumes and smoke coming from hot metal; as a result, they need good supplemental voice communications with people on the ground level during some operations, particularly wen ladling molten metal.

In other cases, the objective is to maintain some amount of clearance between the controlled object and obstacles or unsafe areas. For example, consider a car moving through a congested construction zone or a forklift operating in a loading zone. Navigating through such areas involves a similar sequence of control adjustments, but the task is more complex than for the first example, because it requires keeping track of the distance from a changing set of environmental objects, some of which are moving. Speed and acceleration adjustments also might be needed, as in the case of a driver trying to maintain a constant distance behind a car in the construction zone that slows down at certain times. Having some type of advance knowledge of what the driver ahead is going to do clearly would be helpful in this situation. Such information is called feedforward or predictive information, and can be used to make control adjustments that anticipate what is going to happen next. Assuming that good predictive information is available, an anticipative strategy can greatly improve performance over the simpler control strategy of waiting and then reacting.

Chapter eleven: Control tasks and systems

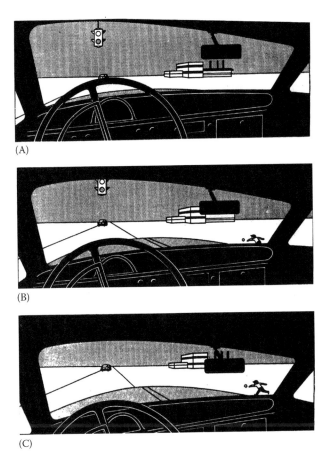

Figure 11.3 Visual fields from the eye position of persons whose height is (A) 5 percentile, (B) 50 percentile, and (C) 95 percentile when seated in a 1954-model automobile. (Adapted from King, B. and Sutro, P., *HRB Bull.*, 152, 1957.)

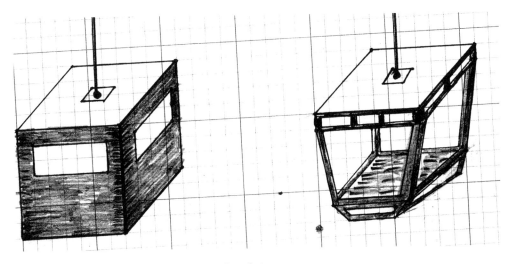

Figure 11.4 Old and newer crane-operator's pulpit.

Interestingly, everything that was mentioned in the previous discussion, and more, can be described with some relatively simple models of manual control, as expanded upon in the following sections. More advanced models have also been developed that can do as well or better than humans on certain control tasks. Such models are implemented in many different guises, including the advanced antilocking brake systems found in many modern automobiles and, not to mention, autopilots capable of flying airplanes, spacecraft, ships, or submarines.

Elementary model of manual control

An example of an elementary model of manual control is given in Figure 11.5. The basic idea is that the human operator uses information from a display, or simply by observing the environment, to make control responses that cause the controlled system to change its output in some desired way. More specifically, the objective of the human operator is make the output $o(t)$ of the system at some time t as close as possible to some input function $i(t)$.

For example, $o(t)$ might be the position of a car that the operator is driving, relative to the edge of the road, and $i(t)$ desired position. The difference between $i(t)$ and $o(t)$ at any particular time is the error $e(t)$. The operator observes the error $e(t)$ and, if the error is larger than acceptable, decides on a corrective response $r(t)$ proportional to the error. The response is made using a control device, after a time delay Δt. This results in an input $c(t+\Delta t)$ to the controlled system that causes the system output to change to a new value $o(t+\Delta t)$. The operator then observes the error $e(t+\Delta t)$ and chooses another corrective response. This process keeps the error at or very close to an acceptable level, assuming that the system is controllable and that the response corrections made by the human operator are appropriate and executed correctly (Box 11.2).

Tracking tasks

In many cases, the inputs to a system change over time. When this is true, the manual control task normally becomes a tracking task. Two different types of tracking tasks are very commonly performed.

In a *pursuit* tracking task, the operator tries to make the system follow some desired path through time and space. The operator can see both the input $i(t)$ (in other words, the moving target) and the output $o(t)$, and tries to make them stay close to each other. For example, suppose that your objective is to stay a fixed distance, let us say 100 ft, behind the car ahead of you as you drive down the highway. This would be a pursuit tracking task. In this case, you actually are pursuing the car ahead of you! Its position would be $i(t)$, and yours would be $o(t)$. The objective would be to keep $i(t) - o(t) = 100$ ft. Things get a little more abstract when there is no car ahead of you. Now, the objective is to follow the road instead of a car. The road is not

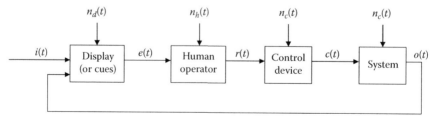

Figure 11.5 Elementary model of manual control.

Chapter eleven: Control tasks and systems

BOX 11.2   PERIODIC INPUTS, RESPONSE DELAYS, AND PHASE ANGLES

Effects of a response delay on a simple periodic function are illustrated in Figure 11.6. As shown in the examples, the output function is shifted back in time from the input by an amount equal to the time delay. For a periodic function, the amount of shift can be equivalently described in terms of what is called the *phase angle*. When the phase angle is 0 rad, the output is said to be perfectly in phase. As the phase angle increases, the match between input and output becomes increasingly out of phase. At a phase angle of π rad, the input and output are inverted—that is, the output reaches its maximum value when the input is at its minimum and reaches its minimum value when the input is at its maximum (see Figure 11.7).

Importantly, the phase angle for a response delay τ depends on the frequency ω of the periodic function. Frequency is normally expressed in Hz, where 1 Hz is equal to a frequency of 1 cycle/s. Also note that 1 Hz is equal to 2π rad/s. To illustrate these ideas, assume that we have a simple periodic function $f(t) = \sin(\omega t)$, where ω is frequency expressed in rad/s. This function after a phase shift is $f(t) = \sin(\omega t + \alpha)$, where α is the phase angle in rad/s. The phase angle α is equal to $\omega * \tau$. This means a given response delay results in a larger phase change, when the frequency ω of the periodic function is increased.

To illustrate this effect, suppose the response delay is 0.5 s. For an input function $i(t)$ with a frequency of 0.5 Hz, ω is π rad/s, and the phase angle is $\pi * -0.5 = -\pi/2$. So the input $i(t) = \sin(\omega t)$. The shifted function $o(t) = \sin(\pi t - \pi/2)$. Now consider an input function $i(t)$ with a frequency of 1 Hz. Here, ω is 2π rad/s, and the phase angle is $2\pi * -0.5 = -\pi$. So the input $i(t) = \sin(2\pi t)$. The shifted function $o(t) = \sin(2\pi t - \pi)$. It can be easily verified by putting in values for t, that $o(t) = -1 * i(t)$. The output reaches its maximum value when the input is at its minimum, and reaches its minimum value when the output is at its maximum.

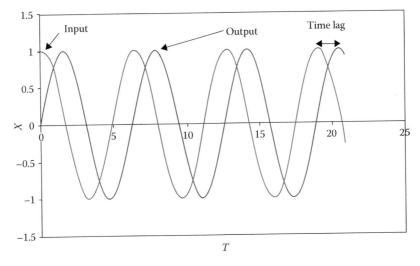

Figure 11.6 Example of phase shift due to response delay.

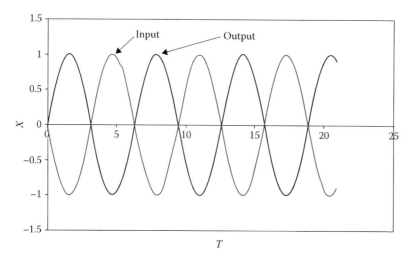

Figure 11.7 Example of out-of-phase input and output functions.

moving but you are, so you still are, in effect, chasing a moving target, that is, $i(t)$ will keep changing as you approach and pass curves, hills, cars, and a whole host of other objects.

In a *compensatory* tracking task, the operator is now trying to cancel out the effect of system disturbances or noise inputs. In some cases, this means that the operator is trying to keep the system $o(t)$ at some constant desired value. However, the operator is very often actually performing a *combined* pursuit and compensatory task. For example, a driver needs to make control corrections both to go around curves (a pursuit tracking element) and to adjust for gusts of wind pushing the car toward the side of the road (a compensatory tracking element). In many cases, the operator will not be able to predict or even see the noise or disturbance input itself. Instead, they will see its effect on the output. In pursuit tracking, on the other hand, the operator will often have various types of *preview* information available. For example, the driver can look ahead and see that they are approaching a curve several seconds before they reach it. On the other hand, they probably will not be able to tell that a wind gust is approaching, unless it happens to be kicking up dust or snow as it approaches. When predictive information is available, the operator can shift to a *feedforward* mode of operation where they make responses that anticipate what will happen. Doing so can reduce or even eliminate the reaction time delay.

The potential effect of system noise (n_s) on tracking performance is indicated in Figure 11.5 as an input to the system directly added to the system output $o(t)$. Several other noise sources are also indicated in the figure. Control noise (n_c) refers to errors added to the control output $c(t)$ caused by factors such as dead space, friction, or stickiness of the control device. For example, consider the situation where a sticky ball within a computer mouse makes it difficult to precisely move the cursor on a computer screen to a particular location. Operator noise (n_b) refers to errors added to the response during response selection and execution. Display (or perceptual) noise results in discrepancies between the actual and perceived value of the input $i(t)$, output $i(t)$, or error $e(t)$.

To better understand how these factors fit together, consider the case where a driver is currently in the center of their lane driving down a straight stretch of road. After being blown by a gust of wind or reaching a curve, the driver might observe that their car is too close to the edge of road (or equivalently too far from the desired position). The driver's response is then to turn the steering wheel some amount in a direction that causes the car to move away from the edge of road, closer to the desired position. In some cases, the initial correction may be adequate. However, additional corrections in the same or opposite direction are often required, depending upon whether the previous response was too small (undercorrecting) or too large (overcorrecting) to cancel out the error. Sometimes, the responses are unintentionally too large or too small because of errors related to display, operator, or control noise. In other cases, this effect is related to how much of the error the operator tries to cancel out with a single response.

Human controller

The elementary model of manual control discussed here corresponds to negative feedback control. In other words, the system output is fed back to the operator, as indicated by the loop on the lower part of Figure 11.5, and subtracted from the input function, resulting in an error signal:

$$e(t) = i(t) - o(t) \qquad (11.1)$$

If the operator response $r(t)$ is chosen in a way that causes the rate of change in system output $o'(t + \Delta t)$ at time $(t + \Delta t)$ to be proportional to $e(t)$, the output for a given input function can be expressed as a first-order differential equation.* That is,

$$o'(t + \Delta t) = Ke(t) = K[i(t) - o(t)] \qquad (11.2)$$

or equivalently,

$$o(t + \Delta t) = \int_0^t Ke(t)\,dt = K\int_0^t e(t)\,dt \qquad (11.3)$$

The latter equation shows that the output $o(t)$ is proportional to the integral of the error over time. The output function $o(t)$ can be determined by solving the differential equation

$$Ko(t) + o'(t + \Delta t) = Ki(t) \qquad (11.4)$$

Note that K and Δt are in many cases aggregated parameters of the particular system, control, and controller. This reflects the fact that the output $o(t)$ of the *controlled* system depends on both what type of control inputs are made by the operator and how the system responds to them.

For example, if we assume $i(t) = D$ for $t > 1$ and the initial conditions $o(t=0) = o'(t=0) = 0$ the solution to Equation 11.4 becomes

$$o(t) = 0 \quad t < 1 + \Delta t \quad \text{and} \quad o(t) = D(1 - e^{-K[t - \Delta t]}) \quad t \geq 1 + \Delta t \qquad (11.5)$$

where
  K is called the loop gain
  Δt is a time delay

---

* This assumption leads to the so-called crossover model of the human operator (McRuer and Jex, 1967). McRuer and Jex present an extensive set of studies showing that the crossover model closely describes the way people perform manual compensatory tracking tasks with random input. Such tasks are very similar to keeping a car in the center of its lane on a very windy day or keeping a boat on its proper course on rough water. For more on the crossover model and its application, see Hess (2007). Other models, such as the optimal control model provide similar predictions (again, see Hess [2007] or Jagacinski and Flach [2003] for a tutorial discussion of both models).

The output $o(t)$ starts after a fixed time delay Δt, and then increases over time, eventually becoming very close to the input value D. With larger values of K, the output increases more rapidly. The output function itself is called a first-order exponential lag, in this case having a time constant of $1/K$. Intuitively, the time constant describes how much the output lags behind the input after it first starts. As K becomes larger, $1/K$ becomes smaller, and the lag is less pronounced. Hence, a smaller time constant corresponds to less lag. A lag is quite different from a delay, due to the fact that a delay has a discontinuous effect. However, a very small lag will result in behavior similar to that of a delay. This can be seen by using a large value for K in Equation 11.6. Large values of K, however, can cause unstable behavior, especially when the input function is periodic.* For more on the issue of unstable behavior see Box 11.3.

BOX 11.3   TRACKING PERIODIC INPUTS

The inputs to a controlled system are often periodic. The ability of people to track or control the effects of periodic inputs is limited by how quickly they can respond (See Box 11.4). As a rough rule of thumb, human reaction times when controlling a single output process are typically 0.15–0.20 s (Rouse, 1980) but become as long as 0.40 to 0.50 s when people must generate a lead term in their output (McRuer and Jex, 1967). System response delays add to this value, causing the combined system–operator delay to become quite large in some situations.

The effects of a response delay of 0.5 s on the tracking of two simple periodic functions are illustrated here. Three examples are shown below. In the first example, the operator gain $K=1$, and ω is $\pi/8$ rad/s, which means the input function $i(t)$ repeats itself every 2 s. In the second and third examples, the operator gain K is, respectively, 1 or 2 and is 2 rad/s. In all three examples, it is assumed that the human controller changes the output function $o(t)$ every 0.5 s by an amount equal to $K * e(t - 0.5)$. This is a discrete approximation of the continuous behavior defined by Equation 11.2.

	$K=1$ $\omega=\pi/8$ rad/s $i(t)=\cos(\omega t)$				$K=1$ $\omega=2\pi$ rad/s $i(t)=\cos(\omega t)$				$K=2$ $\omega=2\pi$ rad/s $i(t)=\cos(\omega t)$		
Time	$i(t)$	$o(t)$	$e(t)$	Time	$i(t)$	$o(t)$	$e(t)$	Time	$i(t)$	$o(t)$	$e(t)$
0	1.00	0.00	1.00	0	1.00	0.00	1.00	0	1.00	0.00	1.00
0.5	0.98	1.00	−0.02	0.5	−1.00	1.00	−2.00	0.5	−1.00	2.00	−3.00
1	0.92	0.98	−0.06	1	1.00	−1.00	2.00	1	1.00	−4.00	5.00
1.5	0.83	0.92	−0.09	1.5	−1.00	1.00	−2.00	1.5	−1.00	6.00	−7.00
2	0.71	0.83	−0.12	2	1.00	−1.00	2.00	2	1.00	−8.00	9.00
2.5	0.56	0.71	−0.15	2.5	−1.00	1.00	−2.00	2.5	−1.00	10.00	−11.00
3	0.38	0.56	−0.17	3	1.00	−1.00	2.00	3	1.00	−12.00	13.00
3.5	0.20	0.38	−0.19	3.5	−1.00	1.00	−2.00	3.5	−1.00	14.00	−15.00
4	0.00	0.20	−0.20	4	1.00	−1.00	2.00	4	1.00	−16.00	17.00
4.5	−0.20	0.00	−0.20	4.5	−1.00	1.00	−2.00	4.5	−1.00	18.00	−19.00
5	−0.38	−0.20	−0.19	5	1.00	−1.00	2.00	5	1.00	−20.00	21.00

---

* For example, $i(t)=\sin(t)$ is a periodic function. Another example is $i(t)=(-1)^2 t$ when t is an integer.

In the first example, the output lags the input a little, but performance is quite good. Except for when $t=0$, which corresponds to the initial condition $o(t)=0$, the output at each of the discrete points in time shown is equal to $\cos(\pi t/8 - \pi/16)$, where $-\pi/16$ is the phase angle. Much poorer performance is shown in the second example! Here, the output lags the input by a phase angle equal to $-\pi$. Consequently, $o(t) = -1 * i(t)$ at each of the times shown, except for when $t=0$, which corresponded to the initial condition $o(t)=0$. In the third example, even worse performance occurs, since the output is unstable—that is, we see that the operator's output oscillates back and forth between increasingly large positive and negative values. These three examples are arguably oversimplified, but they do illustrate how response lag causes a phase shift in the output of a simple controller that increases with input frequency. They also show that the operator is better off not even trying to track inputs when the phase lag corresponding to their response delay approaches $-\pi$ (or equivalently a lag of 180°).

One of the more interesting aspects of the crossover model is that it assumes that the human operator adjusts his or her own gain so that the total system gain K results in a negligible response to frequencies above a value called the crossover frequency. The crossover frequency is equal to π/τ, where τ is equal to the effective delay (i.e., system plus operator). Note that the value π corresponds to a phase shift of $\alpha = \pi$ at the crossover frequency. In the example discussed here, τ would be 0.5, resulting in a crossover frequency value for of 2π rad/s. For more on this quite well-developed topic, see Hess (2006), Wickens (1984), Van Cott and Kincade (1972), or McRuer and Jex (1967).

BOX 11.4 REACTION TIMES IN CONTROL TASKS

As discussed in earlier chapters in this book, a number of factors have been shown to influence reaction time, including

1. The uncertainty associated with the events encountered.
2. The stimulus mode—reactions times to auditory stimuli are often 30–50 ms faster than reactions to visual stimuli, based on many laboratory studies.
3. Stimulus intensity—brighter and louder stimuli result in faster reaction times.
4. Stimulus discriminability.
5. Response complexity—greater.
6. Response precision.
7. Stimulus–response compatibility
8. *Location* compatibility or the degree to which the response movement is in the direction of the stimulus movement.
9. Practice or learning.

(continued)

BOX 11.4 (continued)

Most control situations in industry are rather complex, and, hence, longer reaction times should be expected. Another issue is that reaction times of people increase greatly when the need to respond is not anticipated. A frequently cited study by Johansson and Rumar (1971) measured braking reaction time data for a sample of 321 drivers. Figure 11.8 describes the distribution of reaction times in braking obtained in the study. The observed reaction times range from about 0.2 s to about 2.0 s. The mean of this distribution is around 0.7–0.8 s. The upper 95 percentile level is not clear but it approaches 1.6–2.0 s. While these values might not seem particularly high, an automobile traveling at 30 miles/h is moving at 44 ft/s. Consequently, the car moves 66 ft during a 1.5 s reaction time. This distance is obviously more than enough to make a difference between a near miss and having an accident in many foreseeable situations.

Naive theory of adaptive control

As mentioned earlier, the crossover model assumes that the operator response $r(t)$ is chosen in a way that causes the rate of change in system output $o'(t+\Delta t)$ at time $(t+\Delta t)$ to be proportional to $e(t)$. There are many ways this can be done. To illustrate the idea, let us consider a few simple examples.

To start with, consider the case where the output of the system is proportional to the control input. This means that

$$o(t) = K_c c(t) \quad \text{and} \quad o'(t) = K_c c'(t) \tag{11.6}$$

The latter situation is illustrated by the use of a pointing device such as a computer mouse or trackball to move a cursor on a computer screen. More specifically, movement of the mouse a distance x causes the cursor to move a distance $K_c * x$, where K_c is the control gain. Controls that act in this way are called *position controls*. Substituting the results of Equation 11.6 back into Equation 11.2, we get the result:

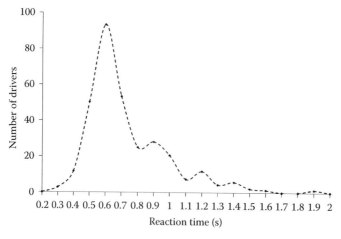

Figure 11.8 The distribution of driver breaking reaction time based on the results of the Johansson and Rumar study.

Chapter eleven: Control tasks and systems

$$o'(t + \Delta t) = K_c c'(t + \Delta t) = Ke(t)$$

which gives

$$c'(t + \Delta t) = \frac{K}{K_c} e(t) \quad \text{and} \quad c(t + \Delta t) = \frac{K}{K_c} \int_0^t e(t)\, dt \tag{11.7}$$

$$c(t + \Delta t) = \frac{K}{K_c} e(t) \tag{11.8}$$

The latter equation shows that the position of the input device is proportional to the integral of the error over time; that is, the operator changes the position of the input device at a rate proportional to the error. This has an effect on the output function equivalent to introducing a lag with time constant $1/K$. From a practical perspective, the time lag has the effect of smoothing out the response and results in the output function given in equation 11.5 when the input $i(t)$ is constant.

Let us now consider the case where the change in system output $o'(t)$ is proportional to the control input. This situation arises when *velocity controls* such as joysticks or steering wheels* are used to control the system. Following the same approach we used earlier to obtain Equation 11.7, we get the result

The position of the input device is set to a value proportional to the error time Δt units earlier. Now the human controller is acting as a simple gain! Another fairly common case arises in systems where the control input causes the system output $o''(t)$ to accelerate at some constant rate. Such behavior occurs when activation of a control causes a force to be exerted on the system, and corresponds to a second-order lag.† Controls that cause such effects are, consequently, called *acceleration controls*. For example, stepping on the gas pedal causes a vehicle to accelerate. Stepping on the brakes causes it to slow down. Another example is moving the steering wheel of a ship. Turning the steering wheel moves the rudder, which then causes a sideward force to be applied to the stern of the ship. This then causes a slow angular acceleration, which eventually causes the ship to turn (see Box 11.5 for a related discussion).

Let us now return to the idea that the operator is trying to respond in a way that causes the system output $o'(t + \Delta t)$ to be proportional to the error at the time t, as shown in equation 11.2. This leads to the result:

$$o'(t + \Delta t) = o'(t) + \int_t^{t+\Delta t} o''(t)\, dt = Ke(t)$$

Which gives

$$\int_t^{t+\Delta t} o''(t)\, dt = Ke(t) - o'(t)$$

---

* Setting the steering wheel to a given position causes a car to turn at a constant angular velocity.
† A second-order lag can be thought of as two first-order lags in series. The first lag describes how velocity lags the acceleration inputs. The second lag describes how position lags the velocity as a function of acceleration. In other words, the position is obtained by first integrating the acceleration to get velocity, and then integrating velocity to get position.

BOX 11.5 SLUGGISH SYSTEMS

Ships and other heavy objects have a lot of mass. Consequently, it takes a lot of force to accelerate them, which from a practical perspective means it takes a long time to get them moving, and, just as long or longer to slow them down. More than one ship has sailed into a pier with screws in full reverse! Mass is not the only issue. Any system with a lot of inertia has the potential to behave in a sluggish manner, when acceleration controls are used. For example, in several instances, the cores within nuclear power plants have overheated due to the difficulty operators have determining when to withdraw power to maintain the core heat at the proper level. To properly deal with systems with high inertia, the user must generate a response based on how quickly the system is changing, instead of its position alone. This corresponds to providing a response based on where the system will be sometime in the future. In the terminology of manual control, the operator is generating a lead that compensates for the system lag or sluggishness. Doing so correctly takes a lot of effort on the part of operators, especially when they are not highly familiar with how the system responds. In some cases, the system may have third- or higher-order dynamics that make it even more sluggish than a system that accelerates at a constant rate. Such systems are almost impossible for any but very highly skilled operators to control. The case of submarines is a good example of a system of this type.

It has now been known for many years that people have trouble controlling sluggish systems. In what seems to be the first formal study of the issue, Crossman and Cooke (1962) performed a now famous experiment that demonstrated this point quite unambiguously. This was done in an experiment that used the setup illustrated in Figure 11.9. Subjects in the experiment were instructed to operate the rheostat control in order to get the water to a prescribed temperature as fast as possible without overshooting it. Crossman merely changed the size of the waterbath to increase inertia and prove his point.

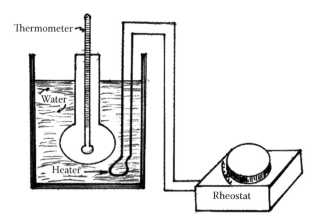

Figure 11.9 Crossman's waterbath experiment. (From Crossmn, E.R.F.W. and Cooke, J.E., Manual control of slow response systems, in *International Congress on Human Factors in Electronics*, Long Beach, CA, 1962.)

Chapter eleven: Control tasks and systems

If we assume that $o''(f)$ will be nearly constant in the interval t to $+\Delta t$, we can integrate the left side of the above equation to get:

$$C(t+\Delta t) = o''(t+\Delta t) = \frac{1}{\Delta t}[Ke(t) - o''(t)] \tag{11.9}$$

Now, the human operator is making a response proportional to a weighted function of both the error and rate of change of the system output observed, Δt time units earlier. It turns out that when the operator follows this strategy using an acceleration control, the system output is delayed in a way similar to that of a first-order lag (equation 11.5). The primary difference is that the initial response lags behind that of first order lag, and then catches up over time. This type of response to an input is referred to as a second-order lag. A second-order lag can be thought of as two first-order lags in series. The first lag corresponds to the effect of the parameter Δt in equation 11.9 and describes how velocity lags the acceleration inputs. The second lag corresponds to the effect of parameter K in equation 11.3 and describes how position lags velocity.

A second order lag can result in somewhat sluggish tracking performance, especially when the input $i(t)$ is changing rapidly overly time. To improve performance, the operator might also consider the rate of change of $i(t)$. That is, the operator might set the acceleration control at time $t+\Delta t$ to:

$$C(t+\Delta t) = \frac{1}{\Delta t}[K_2 e(t) + K_3 i'((t) - o'(t))] \tag{11.10}$$

That is, the position of the input device at time $t+\Delta t$ is set to a value proportional to both the error and the *rate of change of the error*, at a time t. In this case, the human controller is introducing a *lead* term in the output function that compensates for the lag in response of the system. We, consequently, see from the previous discussion that the human controller adapts in three different ways to how the system responds to control inputs. Another very important type of adaptive behavior, expanded upon later, is that people adjust the loop gain (K) to ensure good *error reduction* and system *stability* when tracking random periodic inputs of particular bandwidths (see Box 11.3).

However, it should be emphasized that all of these adaptive forms of behavior require extensive time and experience with the controlled system before they can be effectively used. Inexperienced operators are likely to control systems poorly, for a wide variety of reasons. A shortlist of some of the reasons inexperienced or even experienced operators may perform poorly includes (1) slow reactions, which add to the effective delay of the system plus operator; (2) a lack of understanding of the system dynamics, for example, causing the user to operate an acceleration control as if it were a position control; (3) trying to correct high-frequency inputs that are near or beyond the crossover frequency, resulting in operator-induced oscillations (see the third example in Box 11.3); (4) using too large a gain, resulting in jerky responses or even system instability; (5) not waiting for a response to take its effect before making another response; or (6) not making adequate use of preview information, for example, waiting until the vehicle gets into a curve before responding, rather than anticipating the arriving curve and timing the initiation of the steering response.

Positioning and movement time

A positioning task can be viewed as a type of control task in which the human operator is trying to move something to a fixed location. As such, it can be easily seen that the earlier example, resulting in Equation 11.5, corresponds to a simple positioning task. As expanded upon in Box 11.6, it can be shown that Equation 11.5 predicts movement or positioning times very similar to those predicted by Fitts' law. The latter results show how movement time can be predicted when *position controls* are used. Such controls include direct positioning controls such as touch screen or light pens as well as indirect positioning controls

BOX 11.6   RELATION OF FITTS' LAW TO THE FIRST-ORDER LAG*

Suppose that we view Equation 11.6 as giving the time needed to move close to a fixed location D distance units away from the current location. Now, let us assume that the movement stops once the nearest edge of a target of width W is reached. Recall from Chapter 6 (see Figure 6.6) that Fitts' law assumes that the center of the target is D units away. A little bit of algebra then leads us to the conclusion that the movement, therefore, stops when

$$o(t) = D - \frac{W}{2} = D\left(1 - e^{-K[t-\Delta t]}\right) \quad (11.11)$$

for $t > \Delta t$

or

$$\frac{W}{2D} = e^{-Kt} \quad (11.12)$$

After taking the natural logarithm of both sides of the preceding expression, we get

$$\ln\left[\frac{W}{2D}\right] = -K[t] \quad (11.13)$$

or

$$t = \frac{1}{K}\log\left[\frac{2D}{W}\right] \quad (11.14)$$

Recall that Equation 11.6 included a delay before movement started. The preceding equation gives the time needed to make the movement after the delay Δt. After taking the delay Δt into account, we get the following result:

$$t = \Delta t + \frac{1}{\log(2)K}\log_2\left[\frac{2D}{W}\right] \quad (11.15)$$

Recall that Fitts' law states that hand movement time (MT) is

---

* Jagacinski and Flach (2003) provide a similar derivation assuming a first-order lag without a delay. Wickens (1984) also discusses some of the similarities between the predictions of Fitts' law and a first-order lag.

$$MT = a + b\log_2 \frac{2D}{W} \qquad (11.16)$$

where
- W is the target width
- D is the distance from the target
- a and b are constants

It is easy to see that the first-order lag plus delay makes predictions very similar to Fitts' law. This similarity should gives us some confidence in the predictions of the control theory formulation, for movements within control tasks, since we know from an abundance of studies that Fitts' law is a good predictor of movement time.

such as the mouse, touch pad, or knobs, levers, or other devices that give a system output proportional to the input. Voice inputs* are also sometimes used as a method of position control. One advantage of voice input is that this frees up the hands and eyes for other activities. However, voice control continues to play a limited role as an input method due to the difficulty of developing adequately robust speech-recognition methods that can quickly and accurately recognize human speech, particularly in noisy environments (Wickenns, Lee, Liu, and Becker, 2004).

Velocity controls are also often used for positioning purposes. Some examples include joysticks on portable computers used to position a cursor on the computer screen, tuning buttons on radios that change the rate at which the tuner moves from station to station, not to mention a variety of buttons, levers, or switches that cause power lifts, cranes, hoists, or other devices to move in one or more directions. Many such controls cause the controlled device to operate on a binary basis, that is, either stopped or moving at some preset velocity. Others allow the velocity to be controlled. In our previous discussion we noted that the human operator using a velocity control ideally makes responses proportional to the error, in an effort to make the system behave like a first-order lag with a time delay. However, this strategy will result in significant overshooting of the target if too large a gain is used, for the given time delay Δt present. Many such systems also take some time to respond, which adds to the reaction time delay of the human operator.

This tendency for overshooting becomes even greater when the device operates on a binary basis, as illustrated by many electromechanical lifts. The problem is compounded when the preset velocity is quite high. Much of this effect is related to the fact that the velocity cannot be adjusted. Consequently, the operator is forced to predict when it will stop after hitting the off button, and time their response accordingly. Adding to the problem, such systems may coast for several seconds after the control is set to the stop position. It, consequently, may become very difficult to accurately position such systems, and especially so, for novice operators.

Control error statistics

The designers of control systems often create mockups of the system being designed and then test how well people perform. These mockups might vary from a simple setup where

---

* See Wickens et al. (2004) for more on the advantages and disadvantages of voice input and other control devices.

people use a joystick to track a moving object on a computer screen, to sophisticated flight or driving simulators.* Over the years, a number of methods have been developed for measuring how well people perform tracking or other control tasks. One of the main issues is that control errors can be either positive or negative. Consequently, the simple sum of control errors can be misleading, since the control errors might add up to zero, even if the operator is tracking poorly. To deal with this issue, one common practice is to compute squared error statistics. A commonly used measure of this type is called the root-mean-square (RMS) error. The latter measure is defined as

$$RMS = \sqrt{\frac{1}{T}\sum_{t=1}^{T} e_t^2} \quad (11.17)$$

where T is the time interval over which error statistics are being measured. Many people view this statistic as an average error amount, but that viewpoint is fallacious, for reasons explained later. The variance in error values relative to RMS is a less commonly used measure, defined as

$$VRMS = \frac{1}{T}\sum_{t=1}^{T}[e_t - RMS]^2 \quad (11.18)$$

Since the RMS is the mean of a squared variable and $VRMS$ is the variance about the latter mean, these statistics are really quadratic means and variances. Corresponding arithmetic means and variances of error are

$$A_e = \frac{1}{T}\sum_{t=1}^{T} e_t \quad (11.19)$$

$$S_e^2 = \frac{1}{T}\sum_{t=1}^{T}[e_t - A_e]^2 \quad (11.20)$$

Neither the quadratic statistics nor the arithmetic variance S_e^2 can be negative, but A_e can. It also follows that these four statistics are related to each other as shown here:

$$A_e = RMS - \frac{VRMS}{2RMS} \quad (11.21)$$

and

$$s_e^2 = RMS^2 - A_e^2 \quad (11.22)$$

Since RMS and $VRMS$ are both positive and the sign on the right-hand side in Equation 11.21 is negative, the arithmetic average A_e is never as great as the RMS. It further follows from Equation 11.22 that the square of RMS equals the sum of the arithmetic mean and variance. The RMS, therefore, includes both a mean and a variance measurement, rather

---

* For more on the topic of simulators, and some of their uses, see Chapter 9.

Chapter eleven: Control tasks and systems   441

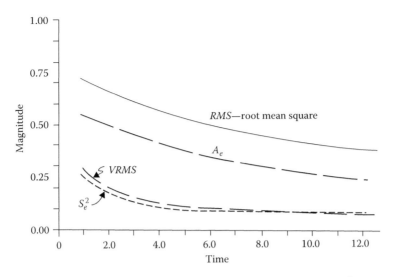

Figure 11.10 Example time plots of different tracking error statistics over time for response to a unit step function.

than just a mean, and that is why the *RMS* should not be considered a mean. Also, notice in Equation 11.22 that A_e^2 becomes positive when brought over to the left-hand side with S_e^2 so that an increase in either A_e or S_e^2 alone or in both causes an increase in *RMS*.

To illustrate the relation between these measures, consider the data plotted in Figure 11.10, showing the response of a simple control model to a unit step function.* Note the close similarities between the plots for S_e^2 and *VRMS*. The plots of the arithmetic and quadratic means are not nearly as similar in form, but the plot for A_e is clearly below that for the *RMS* error, as expected. The latter discrepancy would obviously become much larger for an example where the operator was tracking a periodic input, as some of the errors would become negative.

Partial means

It has been argued that the use of error squaring, while it takes care of the merging of positive and negative error, distorts the true error. For example, suppose someone had an error of –0.5 and +1.5. If these two errors were first squared, resulting in the values –0.25 and 2.25, and then summed, the average is 1.25. This value really does not make a lot of sense, because the average magnitude (or average absolute value) of the error is really 0.5. So what does *RMS* really mean? The problem is clearly the squaring transform of the error. This transform makes numbers below unity smaller and those above unity larger, all just to get rid of the sign problem. The argument goes further in saying that what one ought to do is to measure the negative-signed errors separately from the positive errors. Then the average negative errors can be compared to the average positive errors as a bias. In effect this suggestion is like computing the following:

$$E_{d(e_t,k)} = \frac{1}{T}\sum_{t=1}^{T} e_t \mid e_t < k \qquad (11.23)$$

---

* A unit step function jumps from 0 to 1 at time *t* and then stays constant.

$$E_{u(e_t,k')} = \frac{1}{T}\sum_{t=1}^{t=T} e_t | e_t > k' \qquad (11.24)$$

where k and/or k' are, respectively, the minimum and maximum acceptable magnitudes. That is, all magnitudes between k and k' are not considered errors. One ought to think of k and k' as thresholds of the start of error. In the special case where $k = k'$,

$$E_{d(e_t,k)} + E_{u(e_t,k')} = A_e \qquad (11.25)$$

Measurements $E_{d(e_t,k)}$ and $E_{u(e_t,k')}$ are formally defined to be partial means (See Box 11.7 for example calculation). Partial variances may be similarly defined as

$$V_{d(e_t,k)} = \sum_{t=1}^{T}[e_t - E_{d(e_t,k)}]^2 \qquad (11.26)$$

$$= E_{d(e_t^2,k)} - [E_{d(e_t,k)}]^2$$

BOX 11.7 EXAMPLE OF CALCULATION OF PARTIAL MEANS

Consider the following data where the average error is 3 units. Assume $k = k' = 3$.

Time t	Error e_t	Error Squared	$(e_t - A_e)$	$(e_t - A_e)^2$
1	1	1	−2	4
2	2	4	−1	1
3	3	9	0	0
4	4	16	+1	1
5	5	25	+2	4
Sum=	15	55	0	10

The partial means over time may be computed as

$$E_{d(e_t,k=3)} = 0.2(1) + 0.2(2) + 0.1(3)$$

$$= 0.9$$

Note that the threshold value 3 appears in both the upward and downward group, explaining why the value 3 appears in both expressions, but is weighted by a value of 0.1 instead of 0.2. The value of 0.2, corresponds to 1/5, for each observation since there were 5 time periods on which the error was observed.

$$E_{u(e_t,k'=3)} = 0.1(3) + 0.2(4) + 0.2(5)$$

$$= 2.1$$

Chapter eleven: Control tasks and systems

The sum of these two partial means is

$$0.9 + 2.1 = 3.0 = A_e$$

These errors are clearly biased upward, since the ratio of $E_{u(e_t, k'=3)}$ to $E_{d(e_t, k=3)}$ is equal to 2.33.

The partial means of the squared errors are

$$E_{d(e_t^2, k=3)} = 0.2(1) + 0.2(4) + 0.1(9) = 1.90$$

$$E_{u(e_t^2, k'=3)} = 0.1(9) + 0.2(16) + 0.2(25)$$
$$= 9.1$$

It follows that the arithmetic variance of the error is

$$V(e_t) = 9.1 + 9.1 - [0.9 + 2.1]^2$$
$$= 2.0$$

$$V_{u(e_t, k')} = \sum_{t=1}^{T} [e_t - E_{u(e_t, k')}]^2 \qquad (11.27)$$
$$= E_{u(e_t^2, k')} - [E_{u(e_t, k')}]^2$$

Note that the result of both Equation 11.26 and 11.27 is a function of the variable e_t^2, and e_t. Also, in the special case of $k = k'$, the arithmetic variance of error is

$$V_{e_t} = E_{u(e_t^2, k')} + E_{d(e_t^2, k)} + (E_{u(e_t, k')} + E_{d(e_t, k)})^2 \qquad (11.28)$$

and so the arithmetic mean and variance can be computed directly from partial means provided that the thresholds are equal. That is one of the advantages of using the partial means. Another is that partial means can be devised so that values of a variable that no one feels is an error can be left out of the basis of measuring errors.

As an illustration of this last point, consider a driver of an automobile who is traveling along a lane of traffic. We can assume that no reasonable driver feels that his or her driving is in error as long as the vehicle never gets closer than 15 cm from either side edge of the lane. In this way the mythical line 15 cm from the left side of the lane can be denoted as k, and the same line 15 cm inboard of the right edge can be denoted as k'. Now, there are no errors until part of the vehicle goes over either of those two mythical lines and the amount of downward error is the sum of all of those excursions to the left as measured as a constant interval series. The other partial mean consists of the excursions on the opposite side. In this case there are two measures of bias, the ratio of $E_{d(e,k)}/E_{u(e,k')}$ and the ratio of time spent in downward error to that in upward error. There are few if any drivers who care about a car drifting within the same middle of the lane, but drifts near or over the lane boundaries are another thing. If the particular lane one is driving in is the rightward lane of a two-lane road (i.e., two lanes going each way), the danger that the driver perceives in excursions onto the shoulder of the road may be vastly different from excursions into the faster lane of traffic. Hence, the bias between excursions left to excursions right may be substantial. On the other hand, if the driver was in the left lane of this same situation, one would expect a considerable difference in bias between excursions against slower traffic going the same way and fast traffic going the opposite way.

Numerical simulation of control activity

As discussed earlier in Chapter 9, computer simulations of human operators sometimes include a control task performed concurrently along with other types of activity. To illustrate how a control model might be developed and applied, let us assume that we can describe the position at any given time of a controlled element as the variable c_t. Returning to our earlier forklift example, c_t might correspond to how high the load is off the ground at some particular time t. In addition, let us assume that the target location at any given time is described by a forcing function f_t. In our previous forklift example, f_t would be the desired height the load should be lifted to, which in this case is a constant value. Since the function f_t specifies the intended goal at specific points in time, the control error e_t at a particular time t is simply

$$e_t = f_t - c_t \tag{11.29}$$

For the moment, let us assume that the human controller compares c_t to f_t every Δt time units. If c_t and f_t differ by no more than some acceptable value Δe, then no change is required. If the difference is greater than Δe, the human controller makes a response R_{t+d} at time $t+d$ of size

$$R_{t+d} = Ge_t \tag{11.30}$$

where
    d is the time needed to make the response
    G is the control gain

When d is greater than 0, the response by the human controller to changes in the forcing function is delayed by a constant time lag. Note that an additional time lag can be introduced by how often the controller observes the system (i.e., the sampling rate) and the time needed to take an observation. To keep things simple, we will, for now, assume that the human controller is continuously observing the system and that the observation time is included in the term d.

Gain is defined as the ratio of the amplitude of the control output to that of the control input. With more gain, a control requires less adjustment to make changes in the controlled system. For example, small movements of the steering wheel of a sports car will cause the car to turn quite sharply. In contrast, the steering system of a truck without power steering is geared very low. This gives the driver enough mechanical advantage to steer the truck during slow maneuvers, but large movements of the wheel are required to make small steering changes.

The response R_{t+d} causes the position of the controlled element to change to a new value:

$$c_{t+d} = c_t + \alpha e_t \tag{11.31}$$

where α is called the gain factor, and normally has a value between 0 and 1.* The error at time $t+d$, then becomes

$$e_{t+d} = f_{t+d} - c_{t+d} = f_{t+d} - [c_t + \alpha e_t] \tag{11.32}$$

Now if we assume f_t is a constant, as would be the case for our forklift example, where the human controller is attempting to lift a load to particular height, it can be shown that

$$e_n = (1-\alpha)^n e_0 \tag{11.33}$$

after n control responses are made. A little bit of algebra shows that n can be calculated as follows:

$$n = \frac{1}{\log(1-\alpha)} \log \frac{e_n}{e_0} = \frac{1}{\log(1-\alpha)} \log \frac{W}{D} \tag{11.34}$$

Note that W and D correspond, respectively, to the width of a target and the distance to the target. If we multiply n by the time per correction, it is interesting to observe that we have another expression for the time needed to reach a static, nonmoving target that is very similar to Fitts' law. This certainly seems reasonable, given that is exactly what Fitts' law is trying to predict. It also confirms that our simple discrete model behaves similarly to a first-order lag (see Box 11.6; Buck et al., 1994).

Overall performance of the model is shown in Table 11.1, assuming that f_t is a unit step function, a gain factor (α) of 0.5, and a sampling frequency and response delay both equal

Table 11.1 Performance of an Elementary Model of Manual Control for a Unit Step Forcing Function f_t, Given a Sampling Frequency of 1, Delay Lag of 1, and a Gain Factor of 0.5

Time, t	Forcing Function, f_t	Control Position, C_t	Control Error, e_t
0 Start	0	0	0
1	0	0	0
2	1.00	0	1
3	1.00	0.5	0.5
4	1.00	0.75	0.25
5	1.00	0.875	0.125
6	1.00	0.9375	0.0625
7	1.00	0.9688	0.0313

* Recommended values can be found in sources such as Murrell (1969), Morgan, Cook, Chapanis, and Lund, 1963; Boff and Lincoln, 1988; Kroemer, Kroemer, and Kroemer-Elbert, 1994; or MIL-HDBK-759C [DOD, 1995]).

to 1. This engineering model of control is admittedly naive compared to what people really do, but it is essentially similar to other models of feedback control. One advantage is that it is easily programmed on a spreadsheet, or could be included in a simulation model where people spend time on other activities. In such an analysis, performance of the model can easily be compared for different input functions, sampling rates, gains, and reaction times. This model can also be easily extended in a number of ways (see Fitts and Seeger, 1953).

Fuzzy control

The notion of *fuzzy sets* was put forth by Professor Zadeh (1965) of the University of California, Berkeley. His initial work has been applied in many diverse areas, including applications to control problems. Fuzzy set theory is interesting from a control perspective, because it seems better able to describe some of the subjective aspects of the human controller that are not directly addressed by traditional control theory. For example, fuzzy set theory provides a way to describe subjective reactions to a wide range of vehicle handling and control responses, not to mention driving *styles*. Along these lines, fuzzy rules might be developed that describe what makes particular people feel unsafe when they ride in certain vehicles driven by certain drivers. Other fuzzy rules and measurements might describe the handling characteristics that make certain vehicles more pleasant to drive for particular people. Such concepts are intuitive, but fuzzy, and can often be described using fuzzy linguistic variables and rules. Putting these fuzzy concepts into the mathematical language of traditional control theory would require considerably more data and relationships to be measured and inserted into the controller. The crisp equations used in traditional control theory also lead to controllers that may not be as robust or noise tolerant as fuzzy controllers.

Fuzzy measurements

Many concepts are imprecise and judgments are personal. Notions about comfort, beauty, goodness, safety, expensiveness, cloudiness, flavor, pleasant smell, etc. all fit into these categories of subjective evaluation and imprecise perceptions. Nevertheless, people think, speak, and act based upon their perceptions of these variables. Many times these perceptions shift with the individual's environment.

Consider the variable of automobile driving speeds over the range of 30–130 km/h. Now think about a particular kind of vehicle on an open stretch of an expressway. How would you rate degrees of comfort associated with various speeds? In terms of fuzzy set theory, the speeds are elements admitted to a set defined by the concept, and the speeds are rated according to compatibility with the concept. To be more precise, let X be a set of objects consisting of $\{x_1, x_2, \ldots, \text{and } x_n\}$. Fuzzy set A has a membership function denoted as $\mu_A(X)$ that admits elements of X to the set A by assigning a nonnegative real number. This operation yields a measurement function but with arbitrary compatibility values. When the largest of these arbitrary membership compatibility values is divided into every membership number, the measurement function is said to be *normalized*. Note that fuzziness exists without probabilistic notions and that fuzzy measurement is a very precise way to denote the degree of inexactness (Box 11.8).

Fuzzy sets follow certain logic. Two fuzzy sets are said to be equal, if for all x_i in X,

$$\mu_A(x_i) = \mu_B(x_i) \tag{11.35}$$

Chapter eleven: Control tasks and systems

BOX 11.8 EXAMPLE OF A FUZZY MEMBERSHIP FUNCTION

Let us assume that you had someone rate how comfortable they would be driving a Mercedes Benz 300D on a Sunday afternoon on the autobahn. Suppose that you found a person's ratings associated with 30, 40, ... , 110, and 120 km/h and normalized them as

km/h	30	40	50	60	70	80	90	100	110	120	130
$\mu_A(X)$	0.0	0.2	0.4	0.6	0.8	1.0	0.9	0.7	0.5	0.3	0.0

This result clearly shows that this person perceives very slow speeds and very high speeds on the Autobahn on Sunday to be uncomfortable. A function drawn through these values would be a fuzzy membership function.

A first reaction to this logic often is, "Hey, that is a very rigid requirement for something so fuzzy." Alternatively, a set similarity could be estimated by specifying a maximum discrepancy of the elements that belong in the sets, that is, two fuzzy sets are said to be similar if for all x_i in X there exists a constant k, such that

$$\mu_B(x_i) < [1+k]\,\mu_A(x_i) \tag{11.36}$$

and

$$\mu_B(x_i) > [1+k]\,\mu_A(x_i) \tag{11.37}$$

Another element of logic in fuzzy sets applies when one set is a subset of another. For example, fuzzy set B is a subset of fuzzy set A if for all x_i in X,

$$\mu_B(x_i) < \mu_A(x_i) \quad \text{or} \quad \mu_B(x_i) = \mu_A(x_i) \tag{11.38}$$

This subset case occurs when the range of set A only covers part of the range of set B. As another example, suppose that set B contains temperatures that are cold and set A contains temperatures that are *very* cold. It would follow that some temperature that seemed cold would not be perceived as very cold, but every temperature that was very cold would also be cold. Hence, A is a subset of B because membership function for B is contained in that for A.

Fuzzy logic

Another element of fuzzy set logic describes the *intersection* of two fuzzy sets. This intersection membership function is defined as:

$$\mu_{A\&B}(X) = \min\{\mu_A(X), \mu_B(X)\} \tag{11.39}$$

This intersection corresponds to the logical *and* so that the intersection membership is a fuzzy set that carries the properties of both fuzzy sets A and B. Note that when fuzzy set B is a subset of fuzzy set A, then the domain of the intersection is the subset. On the other hand, if fuzzy set *A* consists of those people who make an unreasonable number of errors and fuzzy set *B* includes those who are young, then the intersection of these fuzzy sets is young people who make unreasonable numbers of errors.

The *inclusive or* is another element of logic associated with fuzzy sets. The *union* of two fuzzy sets is

$$\mu_{A \cup B}(X) = \max\{\mu_A(X), \mu_B(X)\} \quad (11.40)$$

When fuzzy set *A* corresponds to middle-aged employees and fuzzy set *B* denotes good workers, the union of these two sets includes all employees who are middle-aged *or* good workers *or both*.

The logical complement of a normalized membership function is simply the normalized membership of the complement or

$$\mu_{\bar{C}}(X) = 1 - \mu_C(X) \quad (11.41)$$

where $\bar{C}$ means *not the set C* and that is equivalently the fuzzy set not-C. If fuzzy set *C* included those employees who made an unreasonable number of errors in their work, then the complement of set *C*, or $\bar{C}$, are those employees who do *not* make an unreasonable number of errors. Consider set *A* as those employees who are middle aged, and fuzzy set *B* as those who are good workers; then good workers who are middle-aged and do not make an unreasonable number of errors consists of $\mu_{A\&B\&\bar{C}}(X)$.

While the theory of fuzzy sets and fuzzy logic is interesting, can one really say that it is of any value in ergonomic design? The answer is a resounding, "Yes!" This theory has been used to develop a host of products and systems, primarily in Japan, and now the applications are developing in the United States and Europe. Those applications that extend to the operation of products and systems come about when the fuzzy logic of experienced users is coupled with heuristic *if–then* rules. A simple example of an if–then rule used in driving is: If the stop light has turned red; then stop. To obtain rules that might be used by a fuzzy controller, experienced human controllers are asked how they would control a system. The heuristic rules they use are then identified from their answers. The acquisition of such rules is no different than trying to develop an expert system. Rule acquisition for either case is often done in an operator-in-the-loop simulation.

Examples of fuzzy control

Williams (1992) provides an interesting and informative example of using fuzzy logic in control design. For this example, he selected a very famous process control problem that was originally used by Professor Ted Crossman many years ago. This process involves heating a vessel of water to a prescribed below-boiling temperature as quickly as possible using a single variable control electric rheostat. The vessel of water sits atop an electric heater. A human operator can turn the rheostat anywhere from the maximum power transmission to zero, at which point there is no power going to the electric heater. A thermometer is in the water vessel, and the human operator is free to alter the rheostat setting over time based on the thermometer reading. Note that there is no power cooling on this control system, so only the ambient room temperature and time can act as coolants. See Box 11.5 for more on Crossman's original study.

Chapter eleven: Control tasks and systems

If you were the human operator, you are likely to first observe that colder water in the vessel requires more heat, and so the initial control of the rheostat depends in part on the current temperature. However, it is unlikely that you would have a relational rule that said, "If the current temperature of the water is X degrees, then turn the rheostat to position Y." Rather, you would probably say, "If the water is cold, turn the rheostat on full. If the water is warm, then turn it on about three fourths of the way." Note that those two *if–then* rules are fuzzy because it is unclear what is meant by cold or warm water. Another variable that would affect one's perception of appropriate control is the size of the vessel. Since it will take a long time to heat a large vessel, you would likely say, "When the vessel is large, turn the rheostat to the top but only turn it to three fourths or half, respectively, for medium and small vessels."

Figure 11.11 gives examples of fuzzy rules corresponding to the previous statements. In particular, rule (a) relates water temperature and rule (b) container volume to how much the gas valve should be opened. Each of the shown fuzzy membership functions is assumed to be made up of straight-line segments. To see how we might interpret these functions, assume that the operator observes that the water temperature is 45° and the volume of the vessel is 70 gal. The temperature value corresponds to setting the rheostat to about 90% of capacity, according to the rule (a). Also, rule (b) indicates that a volume of 70 gal corresponds to setting the rheostat to about 70% of capacity. Since both conditions must be satisfied, the combination of the two rules (a) and (b) is min [90%; 70%] = 70%. Therefore, if the water temperature is 45° and the volume of the vessel is 70 gal, the rheostat will be set to about 70% of capacity.

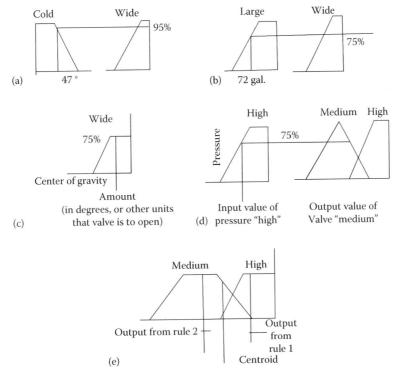

Figure 11.11 Fuzzy membership functions. (From Williams, T., *Comput. Des.*, 31(4), 113, 1992.)

Driving is another situation where fuzzy logic applies. Suppose that you were to set up rules that control the vehicle speed based on the distance to drive and the current driving speed on the highway. Following are four if–then rules about these two aspects of the driving:

Rule 1: IF the distance to be driven is *near* and the current vehicle speed is *slow*, THEN keep the speed constant.

Rule 2: IF the distance to be driven is *near* and the current vehicle speed is *fast*, THEN decelerate.

Rule 3: IF the distance to be driven is *far* and the current vehicle speed is *slow*, THEN accelerate.

Rule 4: IF the distance to be driven is *far* and the current vehicle speed is *fast*, THEN keep the speed constant.

Figure 11.12 shows membership functions for the fuzzy sets of *near* and *far* distances and for *slow* and *fast* car speeds. Note that these membership functions are all linear, for simplicity, which is a common practice. Use of such membership functions results in rules such as the following:

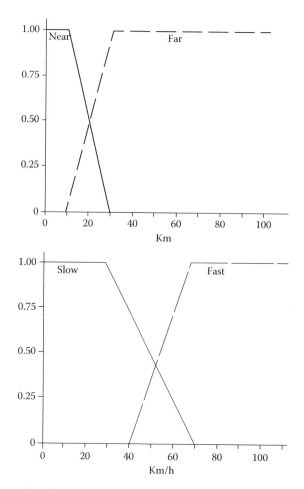

Figure 11.12 Fuzzy membership functions for an automobile driving example.

Chapter eleven: Control tasks and systems 451

1. If distance is less than 20 km and the speed is more than 52 k/h, then *decelerate*.
2. If distance is more than 20 km and the speed is less than 52 k/h, *accelerate*.
3. Otherwise, keep speed constant.

Supervisory control

Supervisory control is a relatively contemporary concept in ergonomics.* This subject is generally more relevant to process control than to vehicle control. The notion of supervisory control is that a human operator works directly with a computer and that computer, in turn, controls another computer that directly controls the process under question (Figure 11.13). Thus, the human operator works directly with a human interactive system (*HIS*). The *HIS* computer is connected to a slave computer called a task interactive system (*TIS*) and that slave system operates sensors and actuators associated with the process. This arrangement allows one to use all sorts of rules about running the process, but it puts a computer in charge of directly monitoring

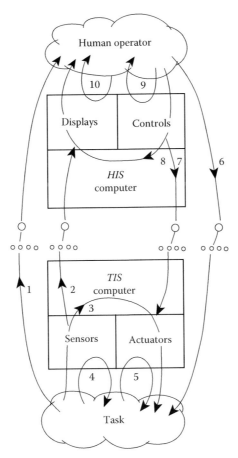

Figure 11.13 Illustration of Sheridan's supervisory control concept. (Adapted from Sheridan, T.B., Supervisory control, in Salvendy, G. (Ed.), *Handbook of Human Factors and Ergonomics*, 3rd edn., John Wiley, New York, 2006, pp. 1025–1052.)

---

* This concept was advanced by Sheridan (1983, 1984). For a more detailed discussion see Moray (1986) or Sheridan (2006).

the actual process, eliminating many of the shortcomings of human performance in monitoring that stem largely from the problem of human vigilance. Supervisory control also allows the human to perform at different hierarchical levels in the control process. For example, the operator might switch from simply monitoring the HIS, to operating the system manually in non-routine circumstances. Somewhat paradoxically, this makes the traditional ergonomics problem of *function allocation* easier in some cases, but much more complex in others.

In many applications, the supervisory control system can be used to reallocate tasks or functions to the human operator depending on the circumstances encountered. In situations, such as when the human operator is distracted or needs to focus their efforts on a particular task, it becomes obvious that the computer should take over some of the routine unrelated tasks normally performed by the human. For example, a pilot might put a plane on autopilot when they need to do some unrelated tasks. In other situations, such as emergencies, it may be not at all clear which functions should be done by machine versus human.

Moray (1986) describes some of the many functions of a supervisory control system as follows: "At one extreme the human performs the whole task up to the point where the computer carries out the instructions to complete the task." He goes on to show that the computer may

1. Help by suggesting options, or by both suggesting options and advising the operator as to which would be best to follow; at this level the human is free to accept or reject the advice
2. Select an action and leave it to the human to carry it out or not carry it out
3. Select an action and carry it out if the human approves it
4. Select and carry out a chosen action without human approval, unless the human intervenes to stop it (tacit approval)
5. Carry out the whole task and then inform the human what it has done
6. Complete the task and only tell the human if explicitly asked what it has done
7. Complete the task and then itself decide whether the human should be told what it has done
8. Decide whether to do the task, carry it out if it so decides, and then decide whether to tell the human

These functions mentioned by Moray show that the computer's role can range from that of a pure slave, to a compatriot, to the supervisor of the human.

Use of a supervisory control system can provide many benefits. Some of the more obvious are as follows:

1. System performance can be improved over what the human operator can do on their own.
2. Once the system has been taught by the operator on that operator's time scale, system performance is faster, smoother, and more nearly optimum. Automated braking systems on automobiles and high-speed controls of small flaps on the wings of aircraft perform tasks better than people could do alone. These examples are cases of control augmentation.
3. Operator's workload is greatly reduced. Consider the very simple case of a cruise control on a car or an autopilot on an aircraft. These devices allow the human controller to relax and intervene if it is warranted by a new situation.
4. Helps the operator to monitor, detect, and diagnose system failures. For example, a Kalman Filter might be used to detect whether the current state of a noisy process is outside its normal range.

Chapter eleven: Control tasks and systems

5. Performs tasks that the operator can specify but cannot do.
6. Provides a fail-safe capability when the operator's response time would not be adequate.
7. Improves the operator's task planning by providing online simulation capability, predictor displays, and other helping features. For example, the system might predict the future position of a ship over time as a function of the current control settings.
8. When necessary, it provides direct control and makes it easier and better by display and control aids.
9. Reduces cost and saves lives by obviating the need for the operator to be in hazardous environments and for life support required to send one there.

Unfortunately, there are very few studies of supervisory control that verify these advantages or pinpoint possible deficiencies of this mode of operation. Overall, it is sage to assume that use of a supervisory control system will introduce some of the same general problems found when operators use any form of automation. over reliance of the operator on the system is one major concern (Lehto et al., 2009). A related issues is that automated systems often are designed to be operated by relatively untrained personnel who function primarily in a passive role in which they simply observe the system. This increases the chance the operator will fail to notice when something goes wrong. The question also arises whether the operator will be able to operate the system under such conditions, despite these unanswered questions, the concept does provide an interesting design strategy for controlling industrial processes and other applications.

Relation to manual control

Manual control models, as briefly discussed earlier, assume the human operator is able to view the condition or state of a controlled element, such as a car, and make control adjustments to achieve desired effects. In contrast to the manual control model, supervisory control includes at least two additional control elements: the HIS computer and the TIS computer and process. This basic difference is illustrated in Figure 11.14 which describes manual control with a single feedback loop and supervisory control with multiple feedback loops. Note that the HIS system in a supervisory control application will often include a predictive display that allows the operator to preview the expected effect

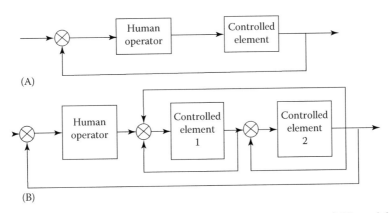

Figure 11.14 Comparison of the manual control (A) and supervisory control (B) models.

of control actions before they are downloaded into the TIS computer. Providing preview information has been shown to be a highly effective way of helping operators perform difficult control tasks (Kelley, 1968) and is just one of the functions the HIS system can perform to help the operator control a process or system.

Another important role of the HIS system is that it can help the human operator estimate the current and future state of a noisy process. In the simplest situation, a human observer monitors a signal over time and sees the signal y_j which consists of a real signal x_j and a noise value v_j where $j = 1, 2, 3, \ldots$ future time values. If the real signal is supposed to be the same all the time, the best estimate of the real signal is the average observed signal. An estimate of the real signal x'_j is

$$x'_j = \frac{1}{k}\left[\sum_{j=1}^{k} y_j\right] \tag{11.42}$$

Assuming that noise is a symmetrical distribution with a constant and finite noise variance, the greater the value of k, the better the estimate in the sense that there is lower error due to noise. It also follows that the estimate of the signal after one more time period is

$$x'_{k+1} = x'_k + \frac{1}{k+1}\left[y_{k+1} - x'_k\right] \tag{11.43}$$

If the next observed signal is the actual signal plus a noise value, and the expected value of the next signal is the current estimate of the signal, then any difference between the current and the next estimate of signal is

$$x'_{k+1} - x'_k = \frac{1}{k+1}\left[v_{k+1}\right] \tag{11.44}$$

where v is the noise for the next observation. It is easy to see that it may be helpful for the HIS system to do this estimation for the operator, using some form of an optimal model. If there is a cost for waiting longer in monitoring as well as a cost for estimation errors, the model of optimal estimation would say that one should delay until the marginal error of estimate equals the marginal cost for delay, assuming that one starts off with little or no basis for initially estimating the noise variance. Note that the previous discussion assumes that the signal is stationary (i.e., constant).

When the system is not a stationary process, the state of the system must be estimated by a process control monitor until a change in state is detected. After that change is detected, a control action must be made in the correct direction. Before making the decision of the correct control action, the operator must estimate the state of the system within the observed noise. To do that, scientists and control engineers have used the famous Kalman filter, which compares covariances between progressive errors and expected variable changes to help detect the system state from noise. A newer optimum control theory (OCT)* has been developed that combines optimum estimation theory using the Kalman filter with the assumption that people select optimum control actions, too. In so far as the OCT is an accurate model of human controllers, the actual operator must have an accurate model of the process. Second, the operator must be able to perform extensive computations

---

* Baron et al. (1970) describe an early version of this theory.

Chapter eleven: Control tasks and systems 455

in real time, unaided. Third, the operator has to have knowledge of the noise distribution on each variable that is being evaluated. These requirements have led to considerable controversies about the appropriateness of the model.

Final remarks

This chapter focuses on how people use control devices to control or guide the activity of machinery, vehicles, or processes. As should be apparent from the chapter, the topic is quite well developed, and a number of good approaches have been developed that can be used to model and measure human performance of control tasks. It also should be emphasized that adequate performance of control tasks depends greatly on whether appropriate controls are used and on how they are arranged. The discussion here, coupled with the related discussion on displays, provides a good introduction to the latter topic. Other issues addressed in this chapter, such as fuzzy control and supervisory control, are becoming especially important with the development of increasingly sophisticated equipment that can perform many of the functions done by people in the past, including communication.

Discussion questions and exercises

11.1 Why would the difficulty of Crossman's waterbath experiment increase with a larger quantity of water?
11.2 How does a pursuit tracking task differ from a compensatory tracking task?
11.3 Explain the difference between a first- and second-order lag.
11.4 What is the influence of introducing a longer time lag in a controller? Increasing the gain?
11.5 Suppose that someone said to you that the RMS error of tracking denoted the average error, but said nothing about the error variation. How should you correctly reply?
11.6 Why cannot probabilities be used in place of fuzzy sets? How does the fuzzy set membership function differ from probabilities?
11.7 Given the fuzzy membership function of Figure 11.12, what should be the fuzzy logic control action if the current distance is 20 km and the current speed is 50 km/h?
11.8 How does supervisory control differ from manual control?

chapter twelve

Design of displays and controls

About the chapter
This chapter focuses on the design of common types of displays and controls used in a wide variety of settings. The chapter begins by discussing the importance of designing an interface that provides a good mapping of the relationships between displays, controls, and their effects. Doing so improves usability and minimizes the need to provide additional displays that attempt to solve problems that could have been prevented in the first place by properly arranging the displays and controls. The remaining sections provide several design principles for both displays and controls. The latter discussion includes an explanation of each principle and examples illustrating how they might be applied.

Introduction
People's interactions with machines, tools, computers, consumer products, and other devices they use to perform tasks can be described as a form of communication where people determine what to do and what not to do using displays and tell machines or other devices what to do, using controls and input devices. In many cases, the provided displays perform a supportive role, by supplementing other sources of information that might be available to the person using the particular device. In others, the display is the sole source of information relied upon for task performance. The degree to which a display successfully performs its intended role of conveying information depends on many different factors. To design an effective display, the designer must first have a clear, well-defined objective—that is, the designer must know what information to provide and why it is needed. Only then can the designer move on and address the issue of how to design the display and evaluate its effectiveness.

The design of controls is a closely related topic. This follows, in part, because controls are often arranged in displays, such as control panels. The visual characteristics of a control will also often suggest both what it does and its current state. For example, a switch that is clearly in the on position tells users that the control has been activated. If the switch serves to turn on a lightbulb, the switch and the bulb are both display devices. A bright lightbulb tells the user that the switch is on and that the circuit is functioning properly. A dark lightbulb, on the other hand, indicates the switch could be off or that the bulb might have burned out. Another reason the two topics are closely related is that certain types of displays and controls normally are provided together in some type of an interface. For example, the driver interface in almost any vehicle will include both a speedometer and brake pedal. In some cases, the interface will have multiple displays and controls. For example, think of all the controls and displays provided in an aircraft cockpit. The arrangement of the displays and controls within the interface can greatly influence their effectiveness or usability, so the designer must take this factor into consideration when designing the interface. A closely related issue is that certain types of controls are much easier to use when particular forms of information are provided in a display. For example, a positioning task done using a joystick will normally be much easier to perform if the display simply shows the current position, rather than giving it in numerical coordinates.

The design of displays and controls is a particularly well developed topic in the field of human factors engineering or ergonomics. Much of the material available to the designer on this topic can be traced back to the so-called knobs and dials era of the field, which began in the 1940s when Paul Fitts and others first applied psychological principles to improve the display and control devices used by operators to manually control aircraft and vehicles (Hoffman and Militales, 2009). The latter perspective continues to be important, but has been broadened considerably over the years, to include applications that have little to do with manual control, such as designing the interface to a smartphone, or the warnings provided for a consumer product. In such applications, the designer will normally be focused on designing a device that combines display and control elements in clever ways to help ensure that the design will result in a product that performs its intended functions well, and is both usable and safe. To address these issues, the designer will normally have to carefully consider factors other than just the display or controls used, such as the nature of the task, who is doing it, and under what conditions. Models and methods introduced in earlier chapters have much to offer when addressing such issues. For example, as discussed in Chapter 11, the manual control model describes several important types of displays and controls, and provides a mathematical framework for analyzing control–display relationships. Communication theory and the models of human information processing, discussed in Chapter 10, provide a complementary framework for describing how people use displays and controls, and suggest methods of evaluating their effectiveness in terms of tasks and stages of performance. These, and other approaches covered elsewhere in this book will be referred to, along with supporting research, throughout the following sections to help explain and, in some cases, develop principles for designing displays and controls.

An important caveat is that the listed principles are by no means exhaustive and each could be discussed in much more detail. The intent is simply to provide enough material to convey some basic guidance on some of the more commonly recognized issues that come up in design applications. Subsequent chapters go into more detail on many of the other issues mentioned. The discussion in Chapter 13 is particularly relevant, and provides a more complete picture of the design process, including methods of user-centered design, as well as usability analysis and testing.

User-centered interface design

As mentioned earlier, displays and controls, and the way they are arranged, describe the interface between the operator and a device or system. The objective of user-centered interface design is to make interaction with the system or device as simple, effective, and efficient as possible. To satisfy this objective, the designer will normally conduct various forms of usability analysis, and testing, following methods described later in Chapter 13. In some applications such as when the user is driving a car, or operating a television set, the displays and controls are easily distinguishable designed elements normally called or named displays or controls. The design problem here is often primarily focused on selecting and arranging a set of displays and controls, appropriate for the particular task and potential users. However, for many consumer products it is not always obvious what the display is and what the control is. For example, many consumer products now on the marketplace will more or less seamlessly combine the display and control elements on a touch screen. Another issue is that the functional elements of many products,

Chapter twelve: Design of displays and controls 459

such as the burners on a stove, can also act as displays. This is important, because, as pointed out by Lehto (1991), the signals emitted by the products themselves can often be a more effective source of information than the designed displays (such as labels placed on the products).

Natural mappings

One way of addressing the preceding issues is to systematically consider the display-like properties of the functional and control elements of the device or product. Doing so can often lead to design solutions that eliminate the need to add display elements such as labels or instructions by creating an appearance that immediately communicates to the user important information, such as what the different parts of the device do or how to use the device. In practice, this often corresponds to treating certain surfaces of the product as a display and then studying how well it communicates certain messages, such as "how to turn the device on" or "which control activates which function" to the user. To illustrate how this principle might be applied, consider the two examples of burner and control layouts shown in Figure 12.1. In this case, the burners are the functional elements, the controls are dials, and a label provided by an imaginary designer underneath each control indicates which burner it activates. It is easy to see that both layouts provide a complicated mapping between each control and burner. To verify this, draw a line from each control to the burner it refers to. In the top example, the left most control addresses the top burner on the right and so on. The bottom example is a little better, in that the two controls on the left side and the two on the right refer to the two burners on the corresponding sides. However, the order is switched for the two burners on the left (back to front) compared to the two on the right (front to back). This switching of order introduces a new effect, in that the mapping is inconsistent from side to side. As might be expected, Chapanis (1959) found in his now classical study that

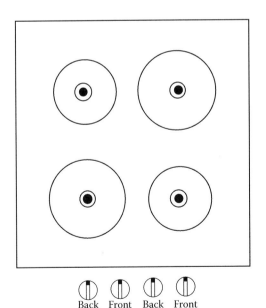 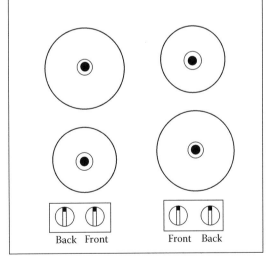

Figure 12.1 Stove with four top burners and corresponding control knobs, but poor mapping. (From Chapanis, A., *Research Techniques in Human Engineering*, The Johns Hopkins Press, Baltimore, MD, 1959.)

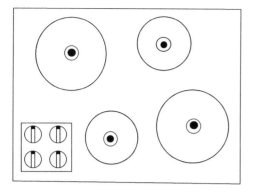 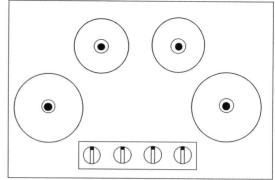

Figure 12.2 Alternative stove control–burner mappings. (From Chapanis, A., *Research Techniques in Human Engineering*, The Johns Hopkins Press, Baltimore, MD, 1959.)

many users had trouble understanding the association or mapping between the controls and burners for both of these layouts. For these reasons, our imaginary designer might have suggested we label the controls to improve usability. This solution is, however, what we might call a cognitive Band-Aid. Examples of much better solutions, in terms of conveying the association between the controls and burners, are shown in Figure 12.2. These solutions both provide a better mapping between each control and what it does, and eliminate the need to add the label. It is also easy to see that the top surface acts as a display that is not only functional, but also usable, in that it requires little cognitive effort to interpret.

As pointed out by Norman (1988, 1992), providing a good mapping of the relationships between controls and their effects, or between things, is an important way of improving communication between people and the devices they use. An example of a simple mapping is illustrated by vertically oriented light switches. In the United States, the up position maps to the on operation and the down position to the off operation.* Unfortunately, the reverse mapping is common in Europe, which leads to the topic of population stereotypes, discussed later. But first let us consider some other examples of mapping, both good and bad. The battery of switches that one typically sees immediately upon entering the front door of a home is an example of a poor mapping because it is unclear which light a switch activates. Even after living in a home for many years, most people still make the mistake of activating the wrong light. Norman (1988) shows the switch system used in his laboratory designed to address this issue. In this example, a plate holding several switches was placed in a nearly horizontal orientation and separate switches on the plate were arranged in positions corresponding to light fixtures about the room. In order to get the plate in a horizontal orientation, a small cove was made in a vertical section of the wall. That near-horizontal orientation readily corresponds to the layout of a room. The plate also showed a few other items about the room to better orient the user.

Some mappings are naturally obvious while others are merely conventional. Turning the steering wheel clockwise to turn right and counterclockwise for a left turn is the natural mapping of control activation. Pushing a throttle forward, or down and forward, to accelerate is another natural mapping. However, rotating a switch clockwise or counterclockwise to change velocity is not natural. Pushing a brake pedal forward or down and forward is not a natural mapping, but it is conventional. Mappings usually are improved

---

* When electrical switches are placed in a horizontal position, common practice is to rotate the switch so that *up* corresponds to *left* and *down* corresponds to *right*.

Chapter twelve: Design of displays and controls

by the use of displays next to the control to show the control to device association and the direction of control movement that corresponds to different desired effects.

Population stereotypes and stimulus–response compatibility

A good mapping will be consistent with population stereotypes. A population stereotype corresponds to a strong tendency for people from a particular group or culture to behave in similar ways in certain situations. Particular population stereotype can be measured by exposing people to a situation without explanation or instruction, and observing their first response. The strength of a particular stereotype can be expressed mathematically as:

$$\text{Stereotype strength} = \frac{\text{Highest relative use freq}}{1 - \text{Highest relative use freq}} \quad (12.1)$$

For example, suppose that 80% of the people who used a rotary switch without any instructions turned the switch clockwise to increase some quantity. The population stereotype strength is then 0.8/0.2 or 4. For proportions of 50%, 70%, 90%, and 95%, the ratios become 1.0, 7/3, 9, and 19, showing that the measured strength increases with relative frequency. Population stereotypes for light switches tend be strong, because people frequently operate switches, and most switches are designed consistently. However, the European and American stereotype behaviors are opposite, reflecting the fact that different populations may have different stereotypes. Turning on valves is another example of a strong stereotype. Normally, a valve is turned counterclockwise to open it and clockwise to close it. Cold and hot water valves are also usually located to the left and right, respectively. Population stereotypes can be measured for many other types of associations, such as which color of a traffic light means "it is safe to go."

In most cases, it makes sense to design the mapping between controls and displays to be consistent with any strong population stereotype that might be present. One of the reasons for doing so is that this helps assure strong stereotype behavior responses even during times of physical and emotional stress. However, this can create problems when different user groups have strong but conflicting population stereotypes (Box 12.1). An alternative approach is to allow users to customize features to be consistent with their own behavior, but unfortunately this may cause similar problems if other people use the device.

BOX 12.1 WHY ARE CERTAIN POPULATION STEREOTYPES SO STRONG?

Certain population stereotypes are especially strong because all people develop similar mental models for particular forms of behavior. For example, people from infancy learn that you can move an object in a certain direction by pushing it in that direction. Accordingly, it is easy to apply the same mental model to tell which direction to push a joystick to make a vehicle move in that direction. Other examples are more interesting. For example, why do most people automatically assume that turning a steering wheel clockwise will cause a vehicle to move to the right? Part of the answer is that when you are driving forward, the top of the steering wheel moves rightward against the horizon when you turn the wheel clockwise. When driving backwards, the relationship is reversed, making it harder to control the vehicle.

The notion of stimulus–response (S-R) compatibility (Fitts and Seeger, 1958; Proctor and Vu, 2009) refers to the fact that natural mappings of stimuli to responses result in quicker and more accurate performance. As pointed out by Proctor and Vu (2009), a large number of studies have been done over the past 60 years directed toward examining the issue of S-R compatibility, focusing on factors such as the consistency of displayed and control movements, spatial correspondence of display and control elements, sensory and response modalities, and how the information is coded. Much of this work has also focused on how the way visual information is represented or coded influences S-R compatibility.

The reason natural mappings help make displays and controls work well together can be explained in terms of the operator's mental model. That is, we can view the stimulus as the displayed information, and the response as behavior expressed either using a control, or in other ways, such as reaching to a fixed location with your hand, tapping a touch screen with your fingers, or saying something. The response is generated by the operator, using a mental model that maps the stimulus to appropriate responses. If the internal mapping described by a person's mental model is different from the mapping required by the task, performance will suffer, because people will have a strong tendency to respond consistently with their mental model. For example, anyone with years of driving experience in the United States will know that it is hard to keep from driving on the right side of the road when you visit countries where you are supposed to drive on the left side of the road, and especially so when the steering wheel is on the left side of the car.

Message meaning and comprehension

The content of a display and the extent to which it is correctly comprehended is an issue at the heart of display design. Several different modeling approaches address important aspects of these interrelated issues. The linguistic model provides a starting point by classifying factors that impact the meaning and comprehension of a message into three categories: semantic, syntactic, and context. The level of detail and breadth of coverage of a message is another important issue. Determining the appropriate balance between explicitly given and inferable elements of a message is complicated because the appropriate level of detail is primarily determined by what the user already knows or should know. The analyst facing this issue may find it helpful to parse the proposed messages into knowledge components and then systematically evaluate whether each element should be explicitly given. Knowledge components that seem easily inferable would be given a lower priority for inclusion in the message. Fuzzy set theory is another modeling approach that may be of value at this stage. The latter approaches are briefly addressed in the following section.

Linguistic modeling

For a message to be meaningful, the symbols used to convey the message must be meaningful to the receiver, as must their arrangement in the setting within which the message is given. For example, in the United States, a "stop" sign encodes its message using shape (an octagon), color (red), and the word "STOP". Each of these display elements should be understood by the targeted user population (drivers) to ensure that the message achieves its objective of convincing drivers to stop at intersections. These important factors that define the meaning of a message fall into the categories of semantics, syntactics, and context.

Semantics refers to the fundamental meaning the symbols themselves have for particular receivers. Verbal symbols are used in languages to express a wide variety of messages, and can also be distinguished as being either written or spoken. Nonverbal symbols include abstract symbols and pictographs, which can also be used to express a wide variety of messages. While written verbal symbols are abstract symbols, generally the term *abstract symbols* refers to nonverbal symbols that do not have an immediate, unlearned association with physical objects. In contrast, pictographs are symbols that explicitly resemble specific physical objects. Of interest is that certain pictographs can be generalized to become ideographs, or abstract symbols.

Syntax refers to the way an arrangement of symbols, as opposed to the symbols themselves, conveys meaning. The study of syntax considers how symbols can be arranged into particular patterns, each of which might have a different meaning. The dependence on syntax to convey meaning is directly proportional to the ratio of desired messages to the number of defined symbols. In other words, if a large number of messages are to be conveyed with a relatively small number of symbols, syntax becomes more important, or vice versa. Messages conveyed by verbal symbols are, therefore, highly dependent upon syntax, because a vast set of meanings can be conveyed by a relatively small set of words when they are combined into sentences. Importantly, the syntax of a given language follows standardized rules most receivers understand, provided they understand the language. This is not as true, in general, for nonverbal symbols.

The interpretation of abstract nonverbal symbols is also frequently dependent upon syntax, as in mathematics or other instances where the order of symbols conveys a particular meaning. Pictographs and some abstract symbols are the least dependent upon syntax because they are generally intended to convey very particular meanings independently of other symbols (i.e., there are nearly as many symbols as possible meanings, and usually exactly as many symbols as intended meanings). If abstract symbols or pictographs are combined to convey other than the simplest semantic information, syntax becomes important. This form of syntax might not follow a set of standardized rules understood by most receivers but might be easily learned. For example, a slash might be put across a picture of a fire to indicate no fires allowed.

The study of the context-specific meanings of symbols is frequently referred to as *pragmatics*. In other words, the meaning of individual symbols or strings may depend upon the exact environment within which they are introduced, upon a particular receiver, or even upon events that have taken place earlier. Such dependence occurs because most verbal symbols and many nonverbal symbols have multiple semantic meanings. The intended meaning of an isolated symbol with multiple meanings must be inferred from the context within which it appears. (For example, the word *fire* might mean one thing to someone working in an oil refinery and something entirely different to someone on a shooting range.) While syntactic processing of strings of symbols reduces the set of possible meanings, a syntactically correct string may still have multiple meanings, depending on the context within which it appears.

A large number of studies have shown that semantics, syntax, and context are all potentially important determinants of whether people will understand a particular message (for reviews see Lehto and Miller, 1986, 1988; Lehto and Papastavrou, 1993; Leonard, Otani, and Wogalter, 1999). Message comprehension can also vary greatly between user groups, due the many factors that can influence comprehension, it is often important to test how well the meaning of the different elements of a display are understood by the intended audience.

Testing the comprehension of symbols and display elements

Several standard-making organizations specify methods for testing how well symbols are comprehended. ANSI Z535.3 includes provisions for (1) obtaining a representative target audience of at least 50 people that includes subgroups with unique problems, and (2) evaluating the meanings the test audience associates with the symbol. A minimum of 85% correct responses and a maximum of 5% critical confusions (defined as the opposite of correct) are required before a symbol is deemed acceptable. Generic approaches for evaluating comprehension include recognition/matching, psychometric scales, and readability indexes, as briefly discussed below.

Symbol recognition/matching. Recognition/matching is a technique often used to measure symbol comprehension. In a recognition task, the subject provides an open-ended response describing the meaning of a symbol. In a matching task, the subject selects the most applicable meaning from a list. Recognition is less commonly studied than matching because classifying open-ended responses from a large group of subjects is difficult. The approach does provide the advantage of measuring the meaning of symbols independently from other symbols. A common deficiency in the application of both techniques is that they do not provide the contextual information commonly found within a task. One approach to alleviate this problem is to explicitly describe the context within which the symbol will be presented.

Message recall. The degree to which the message is correctly recalled sometime after viewing a sign, label, or other display provides an additional measure of comprehension. In evaluating recall, consideration should be given to the task-related context and the time passed since the display was viewed. Measures of recall in task-related settings more closely indicate the degree to which the message will be applied.

Psychometric scales. Subjective ratings of information displays by the target audience on psychometric scales are also used to measure comprehension. Application of this approach requires the development of appropriate measurement scales that denote relevant dimensions of message meaning. For example, subjects might rate the provided material on an overall understandability scale, and more specific items, such as the importance of taking certain actions, and the likelihood that certain events will occur. From these ratings various aspects of comprehension could then be inferred, using a variety of statistical methods. The semantic differential is a variant of this approach that has been used to evaluate the meanings associated with pictographs and symbols and seems to be of potential value in a wide variety of other contexts. For example, it has been used to evaluate associations between signal words and the level of perceived hazard. A limitation of the psychometric approaches is that they focus primarily on measuring associations (a measure of declarative knowledge). Behavioral observations, when feasible, are a better way of measuring how well procedures are comprehended or learned.

Testing the comprehension of text

Several computer tools have been developed that are intended to help writers improve the understandability of written text. These approaches include readability indexes (Klare, 1974) and grammar checkers (such as those implemented in Microsoft Word or other commercially available programs such as Correct Grammar). These approaches can be viewed as simple models of text understandability. A typical readability index describes the difficulty of written material in terms of word length and frequency of use, number of syllables, sentence length, or other similar variables. Examples include the Reading Ease, Flesch–Kincaid, and Gunning fog indexes. Commercially available software packages (such as Microsoft Word or Correct Grammar) also address grammatical issues such as

1. Core sentence grammar errors—that is, incomplete sentences, incorrect use of commas, subject–verb agreement, and commonly confused words
2. Context-sensitive grammar errors—that is, incorrect word order
3. Style errors—that is, colloquial word forms
4. Format errors—that is, incorrect punctuation with abbreviations and improper date formats
5. Punctuation errors—that is, placement of punctuation with regard to parentheses and quotations

The fact that such tools have some face validity, and are both available and easily applied, makes them attractive. However, not a lot of work has been done that addresses the effectiveness of these tools. Mixed results have been obtained in these studies (Box 12.2).

BOX 12.2 THE PREDICTIVE POWER OF READABILITY INDEXES

A few years ago, we performed a study of chemical workers in which we asked them to rate how well they understood a set of messages that had been proposed for inclusion in chemical labels and material data sheets. As part of the analysis, we examined how strongly these ratings were related to readability measures and other factors hypothesized to influence understanding of the phrases (Lehto and House, 1997). The Flesch ($r=0.16$) and Flesch–Kincaid ($r=-0.24$) readability indexes were both significantly correlated with the subject ratings, while the Gunning fog ($r=-0.12$) was not. All of the correlations were low, showing that the readability indexes were at best a weak predictor of worker ratings of comprehension. Word frequency, as measured in the Brown Corpus (Frances and Kucera, 1982) was also weakly correlated with worker ratings ($r=-0.1$).

Shorter sentences using short words scored better on the readability indexes, but this tendency alone was not a good predictor of worker understanding. Grammar checks were more useful, in that they flagged obvious problems, such as run-on sentences, use of the passive voice, or obscure acronyms (House and Lehto, 1999). Another useful predictor was a list of words that had been rated as being difficult by workers in a previous study. Interestingly, expert ratings of how much jargon a phrase contained ($r=-0.41$) were a much better predictor of worker comprehension than any of the other measures.

The *fuzziness* of the words commonly used in written messages given in signs, labels, tags, package inserts, material data sheets, and other information displays is another interesting issue. The use of fuzzy terms provides numerous advantages to the designers of such information displays. Simply put, writers can convey more information with fewer words, by relying on the ability of people to correctly interpret fuzzy terms. Such words can convey different meanings in different contexts, thereby allowing writers to avoid having to write out all of the context-specific meanings that might be relevant. This is important because the number of context specific meanings can be large in certain situations. Listing each one can result in long repetitive lists that might discourage people from reading the material. Long lists can increase the time required to find specific facts of interest (Lirtzman, 1984). Space limitations may also constrain the length of messages.

(continued)

> **BOX 12.2 (continued)**
>
> Not a lot of research has been done on how well people understand fuzzy phrases or whether the predictions of fuzzy set seem to hold. Some initial work has shown that there is significant variation in the meanings assigned to fuzzy phrases used to communicate hazards, by both consumers (Venema, 1989) and industrial workers (Lehto, House, and Papastavrou, 2000). In the latter study, 353 workers at nine locations were asked to interpret the meaning of hazard communication phrases containing linguistic variables and qualifiers (Lehto, House, and Papastavrou, 2000). Linguistic variables included the terms *toxic, acid, contact, ventilated, distance,* and *flush (eyes)*. Qualifiers associated with these terms included the terms *slightly, extremely, weak, strong, prolonged, poorly,* and *safe*. Contexts included *material contacted, health effect,* and *material burning*. Dependent variables included *duration, distance,* and *toxicity ratings* (on a 10-point scale). The results showed substantial fuzziness for certain terms, but, in general, the workers interpreted the phrases in meaningful ways, consistent with the predictions of fuzzy set theory.

The relatively weak predictive value of readability indexes is interesting, but perhaps not too surprising. Message understandability is obviously a function of knowledge and experience. Readability indexes and grammar checkers provide a superficial analysis of understandability that almost completely ignores the latter effects.

Due to the limited evidence that computer tools such as readability checkers really work, it is often necessary to conduct additional evaluations. One way of addressing this issue is to systematically evaluate which knowledge components are derivable from the rest of the message (Lehto and Miller, 1986). An example application of this approach is discussed by House and Lehto (1999). A total of 363 hazard communication phrases were parsed into different semantic categories. The six semantic elements considered were

1. Event—A statement describing an undesirable event
2. Cause—A statement describing the cause of the event
3. Consequence—A statement describing the results of the undesirable event
4. Action—A statement describing either actions to be taken to avoid an undesirable event or actions to take once an undesirable event has occurred
5. Likelihood of the consequence
6. Likelihood of the event

After parsing a phrase into the six semantic elements, each semantic element was then rated by the analyst as being either explicitly stated in the phrase or inferable. This process led to the identification of several changes and improvements of the messages. Numerous explicit knowledge components were added. In a few cases, explicit components were removed from the messages due to their obvious nature. Another approach is to provide the proposed messages to the target audience and ask them to interpret the intended meaning.

Types of displays

An incredibly wide variety of displays exist in the world around us. Visual displays are the most common type encountered, followed by auditory displays. Visual displays can be classified in several different ways. *Static* displays, such as signs, labels, or

Chapter twelve: Design of displays and controls

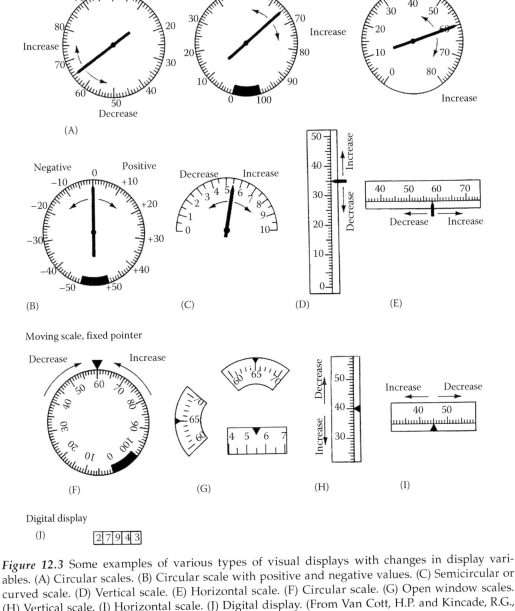

Figure 12.3 Some examples of various types of visual displays with changes in display variables. (A) Circular scales. (B) Circular scale with positive and negative values. (C) Semicircular or curved scale. (D) Vertical scale. (E) Horizontal scale. (F) Circular scale. (G) Open window scales. (H) Vertical scale. (I) Horizontal scale. (J) Digital display. (From Van Cott, H.P. and Kincade, R.G., *Human Engineering Guide to Equipment Design*, U.S. Government Printing Office, U.S. Department of Defense, Joint Army-Navy-Air Force Steering Committee, Washington, DC, 1972.)

books always present the same information. *Dynamic* displays, on the other hand, such as speedometers, fuel gauges, oil pressure or temperature indicators on automobiles display values that change in response to conditions (Figure 12.3). The latter category can be further divided into *analogue* and *digital* displays, depending on how the information is presented. For example, we might see a moving pointer on a dial versus the digits on an odometer. Displays can also be distinguished in terms of the function they play during task performance. *Status* displays, such as a speedometer, simply present information about current conditions. *Warning* or *alerting* displays indicate unusual or urgent conditions. *Predictive* or *historical* displays either project future states or present past trends or previous states. *Instructional*, *command*, or *advisory* displays explain, show, or suggest actions or procedures. Displays can also be distinguished in terms of how they code information. *Spatial* displays, such as diagrams, figures, and charts, show how items are related in time and space. *Symbolic* displays represent information with alphanumeric or nonverbal symbols. *Pictorial* displays show images, such as a flame or car. Other displays are distinguished by the *physical medium* by which they are provided. For example, a static display might be a piece of paper, a painted stair, a metal traffic sign, a stick-on label, or a stamped message on a part. A dynamic display might be a waving flag, a light, a cathode ray tube (CRT), a mechanical dial, or thermometer.

Auditory displays can be similarly classified. For the most part, auditory displays tend to be used for alerting or warning related purposes. This role is illustrated by a wide array of sirens, horns, bells, tones, beeps, or buzzers used in a many different settings to alert or warn people of the need to take some action. However, recent advances in technology have led to the increased use of displays that speak to people. This role is illustrated by navigation systems in cars that keep track of the vehicle's position using global positioning systems (GPS) and tell the driver where and when to turn to reach a particular destination. Synthesized speech chips are becoming quite economical, and even cheap toys are able to speak a limited set of messages quite clearly. This leads to the conclusion that auditory displays will become much more pervasive than they currently are. It is easy to see that auditory displays might be especially useful for visually impaired people who have trouble reading and following the directions on labels and other visual displays. For example, a bottle of medication might include a button that can be pressed to tell an older patient with poor vision how many pills to take at certain times of the day.

Other displays use sensory modalities such as touch or even people's ability to sense odors. For example, a cell phone might vibrate when a call arrives, or an odor might be added to a material that makes it easier to tell when it has leaked out and entered the air.

Principles of display design

As should be clear from the previous discussion, many different types of displays are available to the designer. The following discussion will present some general principles for display design* that can be applied in a variety of settings (Box 12.3). An effort will be made in this discussion to group related principles into somewhat nonoverlapping categories. Such groupings include sensory modality, display location and layout, legibility

---

* The topic of display design is quite well developed. As such, a huge amount of material on this topic can be found in a wide variety of sources, including nearly every textbook on human factors, handbooks (such as the *Handbook of Human Factors and Ergonomics*, edited by G. Salvendy), and last published in 2012. Journals and research publications, military and aviation standards for equipment design, not to mention ISO, ANSI, and government standards for signs, labels, tags, and markings. To avoid overwhelming the reader unfamiliar with the topic, no effort will be made in this chapter to address all of the design recommendations that can be found in these many sources. The focus, instead, is on identifying general rules or principles that can guide the designer. It also should be mentioned that some of the earlier discussion in preceding chapters is relevant here. The section on illumination in Chapter 4 is particularly relevant. Portions of Chapters 15, 16, and 11 also address aspects of display design relevant to fault identification, inspection, and control tasks, respectively.

BOX 12.3   USE OF MOCKUPS AND COMPUTER TOOLS

To help ensure effectiveness, Many ergonomics designers use mockups during preliminary testing of display prototypes and their layouts. Computer-aided design (CAD) tools are often used to draw and present alternative designs during this process. Designers have also applied computer programs used in plant layout. In particular, CRAFT and CORELAP have been used to specify ways of locating multiple displays or display elements (Barlett and Smith, 1973; Cullinane, 1977). The latter approaches attempt to minimize eye movements, subject to certain design constraints, such as keeping particular groups of displays in certain locations. Intuitively, computer solutions should be helpful when locations must be assigned to hundreds of different displays, such as in a control room of a nuclear reactor, especially if the constraints take human performance criteria into account. It turns out that the designs suggested by complex layout algorithms are often similar to those suggested by humans, but can be significantly better or worse, in some cases. The mixed conclusion is that computer solutions can provide some guidance but are clearly not a substitute for the expertise of a designer familiar with ergonomics.

---

of display elements, content, and coding. For the most part, this discussion will focus on visual displays.

Sensory modality

A key first step in design is to identify which sensory modality is most appropriate for the considered application. This typically comes down to a choice between auditory and visual modes of presentation, but other modes may be good in certain settings. This leads to our first principle of display design.

Principle 12.1

The most appropriate sensory modality of a display depends on its intended function, sensory demands of the background task, and sensory capabilities of its intended audience.

There is much more to this issue than can be addressed here without writing another chapter. Some general guidance is as follows. To start with, visual displays are pretty much the only good option for presenting a large amount of detailed instructional material or complicated relationships between items in time or space. This role is illustrated by instruction manuals, books, maps, charts, figures, or equations. When the intended function of the display can be satisfied by providing a small amount of information, the issue becomes less clear-cut. Status displays, warning lights, signs, or labels can all be a good option, if people will be able to see the display at the time the provided information is needed.

On the other hand, if people are busy doing other visual tasks, are visually impaired, or are often at locations where the display is not visible, it might be best to use an auditory or other type of display. Auditory or tactile buzzers are often good for attracting attention to changed, unusual, or urgent conditions. Auditory displays can also indicate location (as illustrated by a beeper on a moving forklift) and are good for giving short messages or instructions at the time they are relevant. On the other hand, auditory displays may be hard to hear in noisy conditions or may themselves be too noisy or distracting. Ringing cell phones, for example, are annoying to others. Tactile displays are often a good solution, when auditory displays are problematic. No one else hears the vibrating phone, and the vibration can be felt even in very noisy environments.

Principle 12.2

Displays that combine sensory modalities can be particularly effective when tasks are performed under changing conditions.

Visual and auditory elements can sometimes be effectively combined within a display, as illustrated by computer interfaces that beep when they display an urgent message. This approach is especially useful when people are monitoring several different displays as part of their task, because their attention might be distracted somewhere else. All of us also tend to drift off and daydream while doing such tasks. If an auditory alarm goes off when some display reaches a critical value, it, therefore, becomes more likely that people will react quickly before the problem becomes serious.

Display location and layout

Deciding where to place and how to place displays is one of the more important elements of designing a good user interface. Over the years, much has been learned on how to best arrange displays one of the more obvious principles that can guide ergonomic designers is as follows.

Principle 12.3

Locate visual displays where they can be seen and put more important visual displays in more central locations.

A common guideline is that displays should be located within 30° of the operator's typical line of sight. Clearly, displays that are not seen are not read. Placement in central locations helps assure that the more important displays are seen more easily, more frequently, and more precisely. Obstructions that block vision are another issue that should be considered. The latter issue is especially important when designing signs. From a practical point of view, rapidly growing vegetation can block traffic signs and other types of outdoor signs. Visual clutter is another issue that should be taken into consideration when locating displays and signs. Almost everyone has experienced the difficult situation where one is looking for a particular route sign or some identifying sign in a city full of neon lights and advertising lights. Another example where visual clutter might be an issue is when you are trying to navigate through a multiterminal airport to the gate your flight is leaving from in 15 min.

The principle of *temporal proximity* extends the previous principle by taking timing into consideration.

Principle 12.4

Displays should provide their information at the time it needs to be used.

Providing information at the time it is relevant reduces the need for people to remember it, and, consequently, can help eliminate errors due to forgetting or other reasons for the particular task that is being performed (Box 12.4). This is especially important if the display is providing warning or alerting information. The next set of principles addresses the issue of how displays or display elements should be grouped.

BOX 12.4 DISPLAY READING TASKS

People will use the information from a display in different ways depending on the task they are performing. For example, in a *check reading* an operator checks one or more displays to see if the operation is proceeding properly. In some cases, information from the display is used at a *qualitative* level. For example, "is the system operating normally?" In other cases, the operator might be seeking *quantitative* data from the display. Another type of display reading task occurs when the operator is contrasting the data from one display with another. For example, synchronizing motors or engines is a task that requires *comparison* of the data from multiple displays. In other cases, the operator may be interpreting a *warning* to decide if danger is actually present and what corrective or evasive action to take. A similar form of behavior occurs when an *advisory* message is received, such as when the computer beeps and tells you that you will lose your data if you do not back it up right away.

Principle 12.5

Displays and display elements should be grouped consistently with the sequences in which they are used by the operators.*

Operators must move their eyes from display to display in certain tasks. The amount of eye movement required can be reduced by placing displays that are normally viewed in sequence closer together. Grouping display elements consistently with viewing patterns can also reduce the time needed to find items within a display. Another issue is that ordering the elements within a display in a natural sequence reduces the need for the user to skip around to find the information they really need. This principle is especially important if each element corresponds to a particular step in a task, since it can reduce the chance of missing or failing to take a necessary action.

The *proximity compatibility* principle (Wickens and Carswell, 1995) provides a second grouping principle.

Principle 12.6

Tasks requiring information integration are better served by more integral object-like displays.

In other words, when the task requires someone to combine or integrate different pieces of information, the display elements or indicators should be grouped together in some way that makes it easier to perceive relationships and differences between the displayed items. Color coding is one commonly used strategy to help group-related items in a display, but other strategies can be followed. For example, suppose that a set of dials indicating the state of some industrial process is placed on the same panel in front of the operator. Also assume that the dials are designed consistently so that the pointer of each dial is in about the same position during normal operation. This allows an operator to easily tell if one of the pointers is at different position than the others, from a quick glance at the display. The deviation becomes an emergent feature that jumps out of the display without requiring the operator to carefully examine each dial.

---

* This principle came from the study by Fitts, Jones, and Melton (1950) where they studied eye movements of pilots between pairs of visual displays in aircraft cockpits during landings. This study was mentioned in Chapter 8 during the discussion of link diagrams.

In this example, the pointer deviation is an easily detected emergent feature because all of the displays are on the same scale and located close to each other. The latter observation leads to the principle of *spatial proximity*, which can be stated as follows.

Principle 12.7
Objects that are placed close together are more likely to be viewed as being related.
Thus, two adjacent displays may be assumed to be in some sense related. If they are, communications are correct; otherwise it is misleading. The effect of spatial proximity can also be increased by adding other organizing features to make the grouping more obvious. For example, a line might be drawn around related text in a manual, or a metal strip might surround a particular group of displays on an instrument panel.

Placing a label next to a control that identifies its function is another example illustrating the use of this principle.

Principle 12.8
Position displays or display elements so they have obvious spatial referents.
In many cases, a display or display element is used to convey information about where something is located. To do this, some element of the display must provide location cues. One way of doing this is to place the display on or close to the referred-to object. For example, a sign might be placed on a restroom door. The problem becomes more difficult when a sign or display has more than one possible referent. For example, suppose a navigation system tells you to turn right, at an intersection where three different streets intersect. Here, it would be important for the display to make it clear which of the possible right turns is correct. This might be done by showing both turns, while highlighting the correct right turn in a different color.

The arrangement of multiple signs is another important issue. People are likely to assume that each sign corresponds to a different object and use the arrangement as spatial cues. Such arrangements can be confusing, in some cases. For example, the horizontal arrangement of restroom signs in Figure 12.4 might be interpreted as implying there are three different restrooms, one on the left for males, one in the middle for disabled people, and one on the right for females. However, it might also imply that the door on the left leads to one unisex restroom, or to two gender-specific restrooms with capabilities for disabled persons. This ambiguity could easily be eliminated by placing an appropriate sign on each door. Another example where horizontal arrangement can be important is in the layout of highway exit signs. One of the more important issues here is to indicate which side of the highway the exit is on. Rather than relying on location alone to convey the directional cue, it might be more appropriate to include an *arrow* or phrase *exit left* to reduce the potential uncertainty of an unfamiliar driver.

Legibility of display elements

Legibility is one of the most heavily emphasized criteria in display design (see Box 12.5). Recommendations for ensuring that display components are legible can be found in many sources, including military standards for equipment design (i.e., MIL-STD 1472B) and consensus standards for product safety signs (ANSI Z535.4), safety signs and tags (ANSI Z535.2, ANSI Z535.5), and symbols (ANSI Z535.3; ISO 3641; ISO 9186). Human factors handbooks, texts, and research publications are another good source of legibility recommendations. It also should be mentioned that computer models have

Chapter twelve: Design of displays and controls 473

Figure 12.4 Are there two or three different restrooms?

BOX 12.5 EVALUATING LEGIBILITY AND OTHER PERCEPTUAL CHARACTERISTICS OF DISPLAY ELEMENTS

Several approaches have been developed for evaluating the perceptual characteristics of an information display. Such methods can be used to measure both perceptibility and the degree to which particular elements of the display attract attention.

Reaction time. Measures of reaction time have been used in many different contexts. The basic assumption is that more salient (or attention demanding) stimuli will result in quicker reaction times. Reaction time measures are particularly useful when quick reaction times are a requirement of the task (as in the driving of automobiles). Their value becomes less obvious when reaction time is not of essence to the task.

Tachistoscopic procedures and glance legibility. In this approach, a visual stimulus is presented to a subject for precisely timed intervals. After viewing the stimulus, subjects are then asked questions regarding its contents. The viewing times are generally very short, but longer viewing times may be used when evaluating displays containing several visual elements, such as a sign or label. By varying the presentation time, insight is obtained into how much time is required to correctly perceive different parts of the evaluated display. As such, these procedures provide a measure of how much attention is required to correctly perceive particular elements of the display. *Glance legibility* is a variation of this approach originally used to compare textual and symbolic traffic signs, which also involves very short viewing times. As for reaction time measures, these measures are most appropriate when quick perception is important. Several studies have found glance legibility to be unrelated to measures of comprehension or even other measures of legibility.

(*continued*)

BOX 12.5 (continued)

Legibility distance. Traffic and other signs are often evaluated in terms of the maximum distance at which they can be recognized by representative subjects. Some safety standards also recommend this approach. Legibility distance can be predicted accurately by computer programs that consider factors such as sign luminance, contrast, and visual angle. However, it traditionally has been directly measured by asking a subject to move toward the sign from a location at which the sign is not recognized. At the point where the sign is first recognized, the distance between the human subject and the sign is measured. Direct measurement can be inconvenient because the legibility distance of large objects is very long.

Eye movements. The tracking of eye movements during the viewing of visual displays has also been used to evaluate a variety of perceptual and attentional issues. This approach reveals how much attention people devote to specific elements of a sign or label, and can pinpoint areas of the sign or label that are hard to perceive or comprehend. It also illustrates the sequence in which items are viewed. A difficulty of this approach is that it generates a vast amount of data. Also, the equipment used to collect eye movement data can be quite intrusive (older systems restrained head movements).

Confusion matrices. In a confusion matrix, the various stimuli to be tested are listed in the same order on the x and y axes. Commonly, the x axis will correspond to the given stimuli, and the y axis will correspond to people's responses. Correct responses are on the diagonal line through the matrix described by cells for which the presented signal and elicited response are the same. All other cells in the matrix correspond to confusions. Much work has been performed in which auditory and visual confusions between phonemes and letters are tabulated in confusion matrices. The approach has also been used to document confusions between automotive pictographs.

---

been developed for predicting the legibility of traffic signs, road marking, and objects seen under a variety of driving conditions (see Olson and Bernstein, 1979, for an early example of a predictive model of the legibility distance of traffic signs). Among such models, TarVIP, a Windows©-based computer model, is available as freeware from the operator performance laboratory (OPL) of the University of Iowa (TarVip, 2004). The TarVIP model developed by Thomas Schnell and his colleagues at the University of Iowa allows the user to calculate the approximate detection distances for pavement markings and pedestrians, as well as the legibility distances of traffic signs under headlamp illumination. TarVIP includes a modeling capacity for determining the effects of factors such as (1) various levels of fog and oncoming glare conditions, (2) the type of headlamps used, (3) the use of retroreflective traffic materials and pedestrian clothing reflectivity, (4) ambient light levels, and (5) the type of roadway, among other factors introduced in the following sections.

The size of a display or display element is the most basic issue addressed in legibility standards. The required size depends on how far away a person is from the viewed object, how much light is present, and visual acuity. As discussed earlier in Chapter 2, the relationship between these factors is described by the size of the visual angle subtended by the viewed object. This leads us to Murrell's principle for designing legible displays.

Chapter twelve: Design of displays and controls

Principle 12.9

The minimum size of a critical detail should be 5 min of visual arc for novice observers and not less than 2 min of visual arc for experienced operators.

Murrell and Jones, in separate experiments (Murrell, 1969), found that the time required to read visual displays increased and the percentage of correct readings dropped off very rapidly when critical details were presented at small visual angles. This finding led Murrell to define this principle (see Box 12.6).

To put Murrell's principle into perspective, note that normal visual acuity is often defined as the ability to detect a detail subtending a visual angle of 1 min. For excellent viewing conditions, the average visual angle at which lines are detected can be as low as 2 s (or 1/30 min), and as low as 30 s (0.5 min) to distinguish the separation of lines. Normal vision on the Snellen eye chart, corresponding to the ability to make out the letter E, is

---

BOX 12.6   EXAMPLE APPLICATION OF MURELL'S PRINCIPLE

Consider a dial display that is to be read by novice observers located about 76 cm (about 30 in.) away from the display. The pointer must move over a scale divided into 100 equally spaced intervals. Murrell's principle tells us that each visible element of the scale must subtend a visual angle of at least 5 min. Since the tangent of a 5 min angle is 0.00145, we know by basic trigonometry that each element must have a minimum dimension of at least 0.1102 cm (i.e., 76 cm * 0.00145). We can use this information to determine how long the scale has to be, that is, since the scale must present 100 visual elements.

$$\text{Scale length} = 100 * 0.1102 \text{ cm}$$

$$= 11.02 \text{ cm}$$

If the scale goes all the way around the outer edge of the dial, the scale length will equal the circumference of the dial. We can use this information to calculate the dial diameter as shown here:

$$\text{Dial diameter} = \text{circumference}/\pi$$

$$= 11.02 \text{ cm}/3.1416 = 3.51 \text{ cm diameter}$$

The latter measurement of the diameter does not include any white space around the outer edge of the dial for placing numbers or other indicator values. Assume that we decided the alphanumeric characters need to subtend a visual angle of 20 min. Since the tangent of 20 min is 0.0058, this works out to a character height of 0.458 cm (76 cm * 0.0058). If we placed these characters around the outer edge of the dial, the diameter would have to be increased by about 0.92 cm (2 * 0.458 cm). However, a little more space would need to be provided, to separate them from the edges of the dial and scale indicators, so we would end up with a display diameter of around 5 cm.

about 5 min of arc. However, the legibility of alphanumeric characters varies greatly, depending on font, stroke width,* and character height-to-width ratio.

Design recommendations for alphanumeric characters vary anywhere from 10 to 30 min of arc, depending on the conditions addressed. The recommendations for text within books or similar printed materials fall at the lower end of this range. For signs it is often recommended that characters should subtend a visual angle of at least 25 min of arc at the intended viewing distance. Some good background information on how visual angle affects the legibility of alphanumeric characters is available from a study by Smith (1984) evaluating around 2000 subjects. He found that the average threshold in the population was 6 min; 90% of the people could make out characters at a visual angle of 9 min, 95% at 12 min, and 98% at 16 min. Smith's results and other sources lead us to the following design principle.

Principle 12.10

Symbols and alphanumeric characters should subtend a visual angle of at least 12 min of arc. When legibility is a primary concern, both should subtend visual angles of at least 16–25 min of arc. Characters should be in sans serif fonts, with character width-to-height ratios of 0.6:1–1:1.

To maximize legibility, stroke width-to-height ratios should also be around 1:5, if the alphanumeric characters are printed in black on a white background. The basic idea is that characters and symbols should be larger and bolder under poor viewing conditions or when good legibility is important. However, for highly luminous characters, much lower stroke width-to-height ratios should be used. A reasonable range for luminous characters is from 1:12 to 1:20.

Principle 12.11

Displays should provide an adequate contrast between visual elements and their background.

For printed materials, the brightness contrast between the characters and background should be at least 50%. This tends not to be a major concern, since most printed materials provide a contrast of 80% or higher. Contrast becomes more of a concern when a display has a layer of glass on its surface that separates the visual elements from the environment. For example, think of a CRT or LED display. Such displays are often evaluated using the following contrast ratio[†]:

$$\text{Contrast} = \frac{L - D}{L + D} \tag{12.2}$$

where
    L is the luminance of the brightest part of the display
    D is the luminance of the darkest part of the display

A general rule of thumb for such displays is that a contrast of 88% or more is needed to assure legibility of the critical details subtending a visual angle of less than 16 min of

---

* Text can be printed in many different fonts, such as Geneva, Helvetica, or Courier. Sans serif fonts are, in general, more readable. Stroke width refers to the width of the line segments in a character and is normally expressed as a ratio relative to character height. Bold fonts provide a higher stroke width-to-character height ratio and are more legible than normal fonts under low lighting conditions. Stroke width-to-character heights of fonts generally range from about 1:12 to 1:5, depending on the application.
† Contrast can be measured in different ways. The ratio defined by Equation 12.2 is often used during the evaluation of luminous visual displays such as CRTs, as expanded upon earlier in our discussion in Chapter 4 on veiling glare.

visual arc. More contrast is better, and larger displays should provide a contrast well above a minimum of 94%, as well as means of adjusting contrast. The fundamental problem is ambient light reflecting off the surface of such displays. As discussed in some detail in Chapter 4, this results in veiling glare that reduces the contrast.

Principle 12.12

Avoid crowding of display elements.

Crowding is a natural consequence of trying to put too much information in a limited space. To make everything fit, designers also are naturally tempted to make items on the display smaller. Unfortunately, crowding and the use of small characters can greatly reduce the legibility of the display. A variety of solutions might be feasible. These include reducing the number of visual elements, replacing text with pictorials or symbols, or simply making the display larger.

Principle 12.13

Take steps to deal with the effects of degraded legibility due to aging and adverse environmental conditions.

As pointed out by Lehto (1992), it may be important to consider aging effects and the potential effects of an adverse viewing environment, such as the presence of dirt, grease, and other contaminants, smoke or haze. It, consequently, may be important to test the legibility of a display under foreseeable conditions. It has been shown that the legibility of many symbols is less affected by dirt and contaminants than text is. Consequently, many traffic signs now convey their messages with symbols instead of text. Maintenance and replacement of damaged signs is another important issue in some environments.

Information content and coding

As discussed earlier in this chapter, the content of a display is perhaps the most important aspect of display design. Simply put, a display must provide information that the person needs to know, but does not know. The way this information is coded is a second, very important issue. Some codes make it a lot easier to correctly understand the intended message than others. Before addressing coding issues, let us start with a few principles on what types of information to provide.

One psychological theory is that negation of an action takes two steps: first the action and then the negation of that action. Whether this theory is exactly correct is of less concern in ergonomics than the fact that studies have shown that negative messages often take longer to react to than positive messages. The resulting principle is as follows.

Principle 12.14

Display instructions should, whenever possible, be stated in a positive manner.

In other words, it is better to say, "Do this," rather than, "Don't do that!" Our next very general principle is as follows.

Principle 12.15

Be selective.

Providing excessive amounts of information on a display increases the time and effort needed to find what is relevant. If the provided information is difficult to read quickly, many people will ignore most or all of it. The designer of a display consequently needs to focus on providing only what is needed to gain a particular behavioral objective. A second, closely related principle that becomes feasible in some cases is as follows.

Principle 12.16

Let the user control the amount and detail of information presented.

Both the amount and detail of information needed often change depending on who is the user of a display and what they are doing. For example, a pilot normally needs highly detailed information about an airport only when the aircraft is close to landing. Giving this information about the airport at other times, such as during takeoff, could actually interfere with the pilot's ability to do the task and cause a serious situation or even an accident.

The following principles go into a little more detail on the types of information that might be needed.

Principle 12.17

Information about the current or desired state or condition of a system, item, or environment is more likely to be useful if the state or condition is (a) unusual, (b) has recently changed, (c) is likely to change, (d) is expected not to be known by the user of the display, or (e) is not easily observed.

This principle follows from some of our earlier discussion in this chapter. Information about the rate of change of various parameters is also very important in some situations. For example, velocity is simply the rate of change in distance, and acceleration is the rate of change in velocity. It is easy to see that information about both velocity and acceleration is important to a driver of a car. The latter form of information is called *derivative* information. This leads to the next principle.

Principle 12.18

Displays that make derivative information easily available make it easier for people to predict what will happen.

It turns out that movement of a pointer on a circular dial is an excellent source of derivative information to a display user (Box 12.7). This follows because people can see both (1) how fast the tip of the pointer moves over the scale and (2) how quickly the angle of the pointer changes as it rotates. In fact, it is much harder to estimate either velocity or acceleration when a display shows only an object moving along a straight line, for the most part because the angular rotation cue is no longer there. *Strip-chart recorders* and other displays that plot out a variable over time also provide good derivative information. The slope of the plotted curve gives the first derivative of the plotted variable and its curvature gives the second derivative. Such plots are helpful in many different situations.

Displays, such as strip-chart recorders, that plot out or otherwise describe states of a process or system over time are called *historical* displays. Historical information can be used by a person to predict what will happen next. Some displays, called *predictive displays*, do this for the person. Predictive displays collect status and control information over time, and use various types of control logic to predict the future status of a controlled system. The user of a display can use these predictions to help decide whether corrections are needed to prevent problems or simply improve performance. Predictive displays have shown to greatly improve human performance in some settings. For example, several researchers* report fewer control errors and reduced fuel consumption for operators using predictive displays to control aircraft and spacecraft. Predictive displays are already being used and are likely to be used even more extensively in the future, because of their

---

* See C.R. Kelley (1968) for expanded descriptions of these experiments and some reported data.

Chapter twelve: Design of displays and controls 479

> **BOX 12.7 DIALS, POINTERS, AND MOVING LEGENDS**
>
> Many visual displays consist of a pointer moving around a dial. In some cases, as with a clock, multiple pointers are used. Other displays use a stationary pointer with a moving legend. Some general observations about such displays are as follows:
>
> 1. Dials with pointers are usually read faster than counters or digital displays that display a changing numerical value. The movement of the pointer also can be used to judge the rate at which a displayed variable changes. However, counters are usually read more precisely.
> 2. The displayed values on dials with fewer scale marks are read faster and more accurately than dials with more marks, but the estimated values are less precise. Speed and accuracy increase, because there are fewer items on the scale. However, precision drops, because there is a limit to how accurately people can make interpolations between adjacent scale markings.
> 3. Full open scales are preferred to dials with only partial scales exposed. Showing only part of the scale makes it harder to judge the overall significance of the displayed value, and increases the chance of making errors.
> 4. Moving pointers are preferred to moving legends. Most people do not like moving legends. Quickly moving legends can cause dizziness.
> 5. Multiple pointers should be avoided. Dials with multiple pointers are notorious for causing confusion. In particular, an altimeter with multiple pointers commonly used a few years ago became famous for how hard it was to use and the number of aircraft accidents that it caused. It is difficult to make a display with multiple pointers that are never confused. Even clocks can pose some confusion. When children are taught to tell time from the clock on the wall, most parents refer to the big hand and the little hand. In fact, there are adult jokes about this reference. But, many children have trouble telling whether the big hand is the short and fat one or the tall and skinny one.
> 6. Well-designed pointers will not make it hard to read dial indicators.

positive impact on human performance in areas such as process control. However, substantial progress still needs to be made on determining what to predict, to best help the human, and on how to do it accurately in real time.

Some additional principles of visual displays given next pertain to how the information is displayed and lead into some coding issues.

Principle 12.19

Be as realistic as possible in describing the variable you are trying to communicate.

If that variable is altitude, then try to show it on a vertical scale. As you might expect, this principle is called *pictorial realism*.[*] The principle of pictorial realism includes the use of internationally accepted symbols to mark the variable being described by a display and appropriate control devices. Another of Roscoe's principles is that of the *moving part*. This principle states the following.

---

[*] This and the next principle are discussed and described more fully in S.N. Roscoe (1980) relative to aircraft operations. Wickens (1992) also contains some further elaboration of these principles.

Principle 12.20

The direction of movement of an indicator on a display should be compatible with the direction of movement in the operator's mental model of the variable whose change is indicated.

An example of this principle is where the gasoline gage of an automobile shows the fuel level as a vertical dial with a moving legend pertaining to the level of the fuel in the tank. Another example of this principle is a thermostat where the vertical dial resembles an analogous thermometer.

Principle 12.21

Color coding is a good tool for conveying relationships between display elements. Color also conveys a limited set of meanings in certain contexts.

Color can be a strong organizing tool for designers. For example, Kopala discusses the use of color in a display intended to enable pilots to form a rapid appreciation of threats faced in a combat environment (Kopala, 1979). Hostile and nonhostile airborne threats were coded in red and green, respectively. Color coding might also be used on a chart to identify different departments, or on a map to identify countries. Several studies have verified experimentally that objects that appear different are viewed serially, but multiple attributes within a single object are usually perceived together (i.e., in parallel). The advantage of good color coding is that it enables people to process information from different parts of a display in a more parallel way that is much faster than serial processing. The latter effect is especially helpful for speeding the time taken to scan a display containing many items.

However, color has its limitations. One is that it may be difficult for people to distinguish certain colors. Chapanis and others have shown that colors that are farther from each other in the C.I.E. diagram (see Chapter 2) are more easily and correctly identified.* Another limitation is that no more than five or six colors should be used in the same display or collection of displays. Too many colors can be distracting, particularly if they are not serving any purpose other than the whims of the designer. Also, color stereotypes may differ, particularly between international populations. For example, one study showed that most European subjects thought that aircraft exits would be indicated with green lights, while most U.S. subjects selected red (Jentsch, 1994). The latter results mirror the convention of marking exits of buildings with green lights in Europe and red lights in the United States. In general, however, several studies show that people tend to associate the color red with danger, yellow with intermediate levels of hazard, and green or blue with safety.

Principle 12.22

Many coding schemes other than color, including shape, size, texturing, shading, and intensity, can be effective. These coding schemes are often not as good as color for speeding search performance. As with color, these coding schemes can convey a limited set of meanings in certain contexts.

Over the years, a huge number of studies have been done on the effectiveness and understandability of various coding schemes used in visual displays.† Differences in shape, size, texturing, shading, or intensity can all be effective ways of making differences in visual elements more obvious. Shapes combined with color often convey levels of hazard or functions of familiar signs. For example, almost all people recognize a red octagon as meaning *stop*.

---

* Ergonomic designers must be particularly careful in the use of colors for products intended for international markets because colors often carry social or religious meanings that differ throughout the world.
† Short summaries of the many studies performed can be found in the annotated bibliography published by Miller and Lehto (2001).

Chapter twelve: Design of displays and controls 481

These studies have also shown that people often have difficulty understanding the meaning of proposed symbols or pictographs. Consequently, safety critical symbols, one required in most standards to be tested first before they are adopted.

Principle 12.23

Verbal and numerical codes tend to be better understood and require less learning than symbols or other coding methods.

The basis for this principle has already been discussed much earlier in this chapter. One of the primary implications of this principle is that symbols should often be supplemented with supporting text, both, to aid learning and to improve understanding of people unfamiliar with the symbols.

Principle 12.24

The use of analogies and metaphors often can greatly improve the learnability and understandability of display elements.

This approach is illustrated by Apple Computer's pioneering development of the now common graphical user interface (GUI) used in computers. Their approach, later copied in "Windows," displays information and commands with symbols analogous to commonly used objects in the outside world. The use of the *garbage can* on the computer screen for a place to drag unwanted files is a typical example. *Pencils* for drawing, *paintbrushes* for filling in areas, and *folders* for storing files are others.

Principle 12.25

In most cases, line graphs are preferred slightly over vertical bar graphs and preferred strongly over horizontal bar graphs.

One study compared the speed and accuracy of people using line graphs and bar charts (Schultz, 1961). In this study it was found that people could predict the direction of trends and read probable scores most accurately and quickly for the *line* graph, and worst for the horizontal bar graph. Although missing data reduced accuracy and speed no matter which type of graph was used, *line* graphs seemed to be more resistant to that deficiency.

Shultz also compared performance for graphs with multiple lines to that observed when several graphs with a separate line were used. Subjects were asked to both read points off the graphs and make comparisons. Performance was almost always better when a single graph with multiple lines was used. The use of color improved performance greatly for the graph with multiple lines in some conditions. The resulting design principle is as follows.

Principle 12.26

Multiple lines on a single graph are preferred over multiple single-line graphs in point reading and trend comparing tasks. An exception is when there are many line crossovers to create confusion.

One conclusion that might be drawn from this principle is that putting the lines on a single graph makes it easy to integrate the information. A corollary to this principle is that one should use colored lines, when possible, for multiple-line graphs to reduce confusion when lines cross over. A final coding principle reflects the fact that is as follows.

Principle 12.27

Variations in pitch, amplitude, and modulation can be used to specify easily distinguished sounds. Some of these sounds convey stereotypical levels of perceived urgency.

Many sounds, other than speech, can convey a great amount of meaningful information. Loud, rapidly modulated sounds, such as some of the sirens used on modern fire trucks or ambulances, can convey a great amount of urgency compared to the quiet roar of the ocean. Over the years, a lot of work has been done on developing sounds that are associated with particular levels of urgency (see Edworthy and Adams, 1996). Significant progress has also been made in designing sounds that are easily localized.

As mentioned earlier, auditory displays are often used for warning purposes. The sirens used by police cars, ambulances, and fire trucks (see Box 12.8) immediately come to mind, but there are many other situations where auditory warnings are used. For example, in industrial settings, auditory warnings often indicate the presence of cranes or robots operating nearby, trucks that are backing up, press operations, ingot rolling in primary metal industries, and a host of other cases. A good auditory warning can be heard above the background noise level, demands attention, and has good noise penetration. A standard recommendation is that a warning should be at least 10 dB, preferably 15 dB, louder than the background noise. To make the warning sufficiently urgent, many sources recommend that the sound should be presented intermittently for durations of 100–150 ms. However, the latter topic is a subject of some debate.* Also, if more than one auditory warning is used in the same area, each warning should modulate the emitted sound differently to ensure that the warnings are easily distinguished from each other.†

BOX 12.8 SIRENS ON EMERGENCY VEHICLES

The sirens on emergency vehicles work quite well when people are traveling at low speeds, but not so well at high speeds in bright daylight conditions. In the latter situation, most people report seeing an emergency vehicle in their side-view or rear-view mirror before hearing it. To see why this happens, suppose a car is traveling at highway speeds, say at 90 km/h (i.e., 56 mph), and an emergency vehicle traveling at 120 km/h (about 90 mph) is coming up behind the automobile. When the vehicle is 50 ft away, it can be shown that sound of a standard siren is often only about 4 dBA above the ambient sound level within the car. This is well below the recommended level of a least 10 dB difference. Yet, the driver has only about 6 s to get out of the way, if the emergency vehicle does not slow down. The conclusion is that few people will hear the emergency vehicle in time to move over to the outside lane without obstructing the speed of the emergency vehicle. In all likelihood, the emergency vehicle will approach to a distance of 30 m or less before being heard by the driver ahead. That is why the flashing lights in front of the emergency vehicle are so important. Unfortunately, emergency lights can also be hard to see in bright sunlight. An emergency signaling system that triggers a warning inside of the vehicles in the path of emergency vehicles might work better than the system we now have.

---

* Some older recommendations from England called for single frequencies of about 3400 Hz. More recent recommendations studies by the highway safety people in the United States suggest the use of about 10 harmonically spaced components where at least four prominent components are in the range of 100–4000 Hz. Single-tone frequency sounds should not exceed 2000 Hz.
† Many years ago in Los Angeles, CA, several incidents occurred where fire trucks collided with each other. Part of the problem seemed to be that all the fire trucks were outfitted with the same continuously sounding sirens. The sounds were identical, and most of the sound heard by a driver was coming from their own truck. Consequently, the drivers had trouble hearing the other trucks.

Chapter twelve: Design of displays and controls

Table 12.1 Some Common Types of Control Devices

Control Categories	Types Within Each Category
Alphanumeric keyboards	Alternative alphabets
Balls	Track balls and computer mice as simply inverted
Cranks	Hand with and without offset handle, use for high-speed turning up to 200 rpm, max. rate decreases with the diameter
Joy sticks	Both discrete and continuous versions available, push–pull is usually more accurate than left–right or up–down
Knobs	Slide or rotary
Levers	All three classes but one-dimensional, hand operated
Plates	Pressure, force
Pedals	Rocker or linear movement, knee angle is critical for seated operator, maximum forces at about 160°, close to standing operator
Pushbuttons	Fixed position, automatic position return but one-dimensional operation
Rudders	Both hand and foot operated
Switches	Light-actuated, push–pull, rotary, rocker, sound-actuated, toggle
Treadles	Usually foot-operated but can be hand-operated, usually center hinged
Wheels	Hand, indent, steering

Design of controls

A wide variety of controls have been developed over the years. Some of the more commonly used types of controls are listed in Table 12.1 and examples are illustrated in Figure 12.5. Controls such as those shown here are generally available in versions that provide for either *discrete* or *continuous* movements. While continuous controls are operated faster than discrete controls, positioning is less accurate when there is no feedback. As mentioned in Chapter 10, controls can be distinguished in terms of their effect on the system. Recall that in that discussion, these included position controls, velocity controls, and acceleration controls. Many other devices which have control-like elements, for example handles are used to control certain types of hand tools, when people perform tasks like precision cutting with a knife as discussed earlier in chapter 5. Generally, discrete controls are preferred over continuous controls when

1. There are a limited number of control states (i.e., <25)
2. On–off or yes–no replies are required for the operator
3. Alphanumeric data are to be entered
4. Only a small mechanical force is needed to operate the control

Either type of control has a control dimension called bandwidth, which is the number of discrete positions or the length of movement in a continuous control.

General recommendations for selecting controls can be found in many sources. Among these sources, Murrell provided a good first-order basis for selecting controls based on task requirements for speed, accuracy, force, and range of movement. Table 12.2 describes the relative suitability of a particular type of control for each of those four criteria. Some additional principles are as follows. One of these is the principle of antagonistic forces, given below.

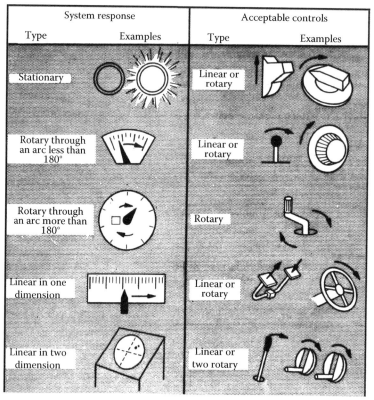

Figure 12.5 Some examples of different types of control devices. (From Van Cott, H.P. and Kincade, R.G., *Human Engineering Guide to Equipment Design*, U.S. Government Printing Office, U.S. Department of Defense, Joint Army-Navy-Air Force Steering Committee, Washington, DC, 1972.)

Principle 12.28

When a control requires the use of muscle bundles that are antagonistic to each other, the control actions are more accurate but usually slower.

The steering wheel of an automobile is usually operated using antagonistic forces because the left and right hand are pulling it in opposite directions. In fact, one-handed driving has been shown to be less accurate. Some people also believe that the two-handed basketball free throw is more accurate than one handed shots for the same reason. However, there are many good one-hand shooters.

Another principle of controls pertains to member movements.

Principle 12.29

The control should fit the body part used to actuate it in a way that allows natural movement.

For example, a smart-phone should be designed to fit comfortably in the user's hand, while allowing data input with natural easily performed finger movements. The basic idea is that the control should be designed to fit the human body in whatever positions the body might be rather than requiring the users to adjust to a poorly sized or located control. The control should be designed correctly in the first place — that is, be human-centered in the design of controls. At times, this means making the controls moveable to other locations and/or orientations.

The following principle addresses the important issue of control feedback.

Table 12.2 Characteristics of Different Types of Control Devices Ranked Relative to Four Criteria as an Aid to Designer Selection of an Appropriate Device

Type of Control	Speed	Accuracy	Force	Range of Movement	Loads
Cranks, small	1	3	4	1	Up to 40 in. lb
Cranks, large	3	4	1	1	Over 40 in. lb
Handwheels	3	4	2–3	2	Up to 150 in. lb
Knobs	4	2	1	2	Up to 13 in. lb
Lever-horizontal	1	3	3	3	Up to 30 lb
Lever-to-from body Vertical	1	2	Short 3 Long 1	3	
Lever across body Vertical	2	2	2	4	1 hand <20 lb 2 hands <30 lb
Joystick	1	2	3	3	Up to 5 lb
Pedals	1	3	1	4	30–200 lb
Push-buttons	1	4	4	4	Up to 2 lb
Rotary selector switches	1	1	4	4	Up to 10 in. lb
Joystick selector Switches	1	1	3	4	Up to 30 lb
Suitability = Standing operator uses body weight	1 is most suitable	4 is least suitable # depends on leg flexion			

Source: Reported by Murrell, K.F.H., *Ergonomics: Man in His Working Environment*, Chapman & Hall, London, U.K., pp. 372–378, 1969.

Principle 12.30

Controls should provide feedback to the activating person showing which control was touched and activated, when the activation occurred, and what control command was activated.

Feedback can be provided in several different forms by a control. Some of the more common forms include visual, auditory, and tactile feedback. It usually is better if a control provides more than one type of feedback. For example, shape coding and surface roughness on the different controls help the person recognize when the correct or incorrect control is grabbed. Force resistance is another form of feedback. Audible clicks with tactile sensations are still another form. Usually a visual check affirms the positioning, but it is better to use other senses for feedback and checking, too. The feel of the control should tell the operator if the correct control is activated. Tactile feedback is commonly advised to identify the control activated. For example, there are often dimples on the "d" and "k" keys of a QWERTY keyboard so that the operator knows that the hands are positioned on the home row and are in the correct lateral position. Surface shape patterns and surface roughness provide other forms of tactile and kinesthetic feedback. Laminated control labels are added forms of feedback.

Another principle of control design is that of discontinuous effort.

Principle 12.31

Controls should not require continued effort to be held in a null position or in the position of the last setting.

A famous control that violated this principle was the dead-man's throttle on a train. The train engineer had to hold the throttle to the desired speed setting. If he released the control, the train slowed to a stop. It was believed that if the engineer should die while driving the train, then this so-called *deadman's switch* would bring the train to a safe stop. It is probably not surprising to hear that some such controls have been wired in the operating orientation.

Several other principles address the arrangement of controls in control panels, such as the following.

Principle 12.32

If controls are usually activated in a definite sequence, arrange them to follow that sequence.

The notion here is analogous to silverware at a place setting on a dining table. The sequence of silverware is typically from the outside in as the courses of the meal are served. While control arrangements may follow other directional patterns, the idea is to aid the operator. A related principle is the following.

Principle 12.33

Arrange the controls to fit the operator's mental model of the process being controlled.

The simplest case of fitting mental models is drawing a line between controls that actuate valves along the same pipe. If controls have related functions, then arrange them in those perceived functional arrangements. That could be startup controls, process-operation controls, and shutdown controls, as a functional arrangement from left to right or top to bottom. Alternatively, controls could be arranged so that similar processes would have the same or similar controls in the same orientations in different control panels. In that way, the panels identify the different processes. Still another approach to this principle is to have the same function control on one panel in the same orientation for each process. This arrangement is particularly useful when the operator is running all processes simultaneously, and any changes in one process are typically reflected by changes in all processes. For example, two processes with agitation motors on each may need to be changed depending upon the viscosity of the chemicals flowing through both processes.

A third principle on control arrangements follows.

Principle 12.34

Arrange controls consistently from panel to panel, using identical or similar controls when the action is the same. When the action is different but the function is similar, arrange the same but use a different control as seen by the operator.

It is important to tell the operator when things are the same, and when they are not the same but are similar.

The final arrangement principle is as follows.

Principle 12.35

Keep control arrangements confined but not crowded.

Operators need to be able to activate the correct control quickly without fear of activating an adjacent control instead or in addition. If controls are to be activated by the hands and the operators may be wearing gloves, add extra space between adjacent controls to account for the space taken by the gloves.

A further control-selection principle is as follows.

Chapter twelve: Design of displays and controls 487

Figure 12.6 Multiple control positions of rotary switches. (From Van Cott, H.P. and Kincade, R.G., *Human Engineering Guide to Equipment Design*, U.S. Government Printing Office, U.S. Department of Defense, Joint Army-Navy-Air Force Steering Committee, Washington, DC, 1972.)

Principle 12.36

Controls should have labels that identify each switch action briefly but unambiguously (no abbreviations) and that show automatic control changes.

Seminara (1993) describes a study of controls used by operators of a nuclear reactor control room. The study identified operators of various degrees of experience and interviewed them about the function of the control, actions associated with control positions, and several other relevant control features. Only 21% of the questions were fully and correctly answered. The interviewed operators were especially uncertain in these interviews about the actions associated with intermediate control positions when there were three or more control positions of the rotary switches. Figure 12.6 provides examples of rotary switches with multiple positions.

Another principle regarding the type, orientation, and other features of controls is the following.

Principle 12.37

Select and orient controls to avoid accidental activations.

This principle is particularly important for critical controls, such as power activations of machine or power tools. Controls can be guarded against unintentional activation by recessing them or placing shields around them. A simple strategy for protecting controls from accidental activations is to make the controls moveable and give them to the person who is most likely to be hurt if they are accidentally activated. Another approach is to preselect minimum forces that are needed to push or turn a control so that accidental control contacts will not cause activation. When a control is very critical, a removable or fixed key lock can be used. When security is also important, the keys can be removable. An even better way to prevent accidental control activations is to use a portable control panel that the person who is in potential danger can carry.

A principle about locating controls and evaluating a good layout is the following.

Principle 12.38

Operators should be able to actuate controls from any reasonable posture during the task without loss of stability.

Some tasks, such as maintenance and repair, force the person to assume a wide variety of postures as they perform their jobs. However, if the posture becomes unstable due to the job, change the control locations in order to regain stability.

Some other principles for selecting appropriate controls in design follow from the principles of work methods, such as the following.

Principle 12.39

When one needs quick, fine-control adjustments, select those controls that employ body members of minimum mass.

The corresponding work-method principle is to use the member of least classification that is identical to the minimum mass body member. Another control principle that corresponds to the design of work methods is the use of antagonistic muscle groups for greater accuracy but slow action. That principle is still contested in both applications, but it seems to hold for larger controls such as steering wheels. Some other control principles that are not controversial include the following.

Principle 12.40

Single controls that move in two and three dimensions are better than two or three single-dimension controllers in terms of both speed and accuracy.

Orlansky (1949) conducted a study, several years of human operators demonstrating this result. Barnes (1937), from his studies of human movement provides the basis for another such principle, which is as follows.

Principle 12.41

Terminate control actions with a mechanical stop. Use auditory feedback to indicate a change in control activation state.

The mechanical stopping provides clear feedback, and the added sound has redundant features. At the time Ralph Barnes came up with this principle, feedback was not a well-known concept in the world.

Switches are probably the simplest type of controls. However, their activation is anything but simple. When operators must activate a number of different switches, they must first locate the appropriate switch before sending the hand to actuate it. If the switch is outside of the immediate peripheral vision, the operator usually visually locates the switch and starts to move the hand toward it. As the hand nears that switch, another visual glance often fine-tunes the hand movement onto the actual control device portion of the switch. With a simple two-position toggle switch, there is not much more to the activation except to throw the switch and move the activating hand back to the next position. However, with switches that have more settings, another visual glance must be made to see where the current switch is currently set so that an appropriate change in the switch can be made. After the switch has been actuated, another visual glance can verify that the correct control change was made. It is also known that performance times to reach to the switch, activate it, and move the hands back to either neutral positions or other controls have negative correlations within a control activation. This shows that people compensate in component actions in order to maintain normal and consistent performance times. Also, it is natural to recognize that the activation time is longer if there are more control changes to be made on a particular switch, but activations also take longer when there are more conditions possible on a switch even though only a few control changes need to be made. Oddly enough, there appears to be a small effect of increased performance time in reaching and hand returns when there are more possible conditions in switch changes.

Principles for better control–display relationships

Whenever a control is associated with a particular display, there is a control–display movement ratio and direction. That is, when control is moved x units in a particular direction, the display moves y units, but not always in the same direction as the control.

Chapter twelve: Design of displays and controls   489

When these movements are Δx and Δy, the ratio Δx/Δy is referred to as the *control–display* (C/D) *ratio*. Normally, there is a uniform linear relationship between the movement of the control and what is shown on the display. Note that a large C/D ratio requires a lot of control movement to move the display across its bandwidth. Therefore, large C/D ratios are undesirable when the operator must frequently move the dial large amounts. On the other hand, large C/D ratios are helpful for fine positioning of the display. Conversely, large dial movements and approximate tuning call for small C/D ratios. Since most tasks are composed of large movements and fine-tuning movement, a trade-off is most often the appropriate design action. This leads to the following principle.

Principle 12.42
Match the C/D ratio to appropriately reflect speed and accuracy needs of the operator.

Display–control compatibility is a term that refers to the relationship between what is shown on the display and control movements. Warwick performed many experiments in the late 1940s and developed the following principle.

Principle 12.43
The greatest compatibility is achieved with a rotary control when the part of the control nearest to the index of the display moves in the same direction as the index.

That is, the closer the display movement approximates the direction of the control, the better the compatibility. Also, the less the rotation of correspondence of control movements relative to movements in displays, the better the compatibility. Consequently, to gain better compatibility, keep the control near the corresponding display, moving in the same direction as that display, and on the same plane as that display. Figure 12.7 shows some of the C/D preference relationships that Warwick found in his experiments. The more leftward situations were more preferred, and those cases have the control and display on the same plane. Shifts across planes, as shown on the right, were least preferred. Warwick also found that when compatible and incompatible display–control combinations were mixed on the same operating panel, performance was worse than when all combinations were incompatible.

One of the most famous C/D incompatibility experiments was by done Fitts and Seeger.* In this experiment, three display panels and three control panels were used for a combination of nine display–control situations. Each display was geometrically distinct from the others and the control panels were identical to displays. This situation gave stimulus–response combinations that were clearly very compatible and other combinations

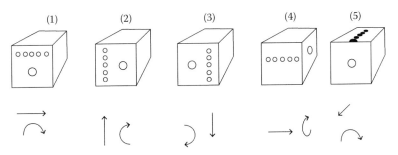

Figure 12.7 Arrangements of control devices and displays (lights) with the strongest preference relationships. (Adapted from Warrick, M.J., Direction of movement in the use of control knobs to position visual indicators, USAF AMc Report No. 694-4C, 1947.)

---
* See Kelley (1968) for expanded descriptions of these experiments and some reported data.

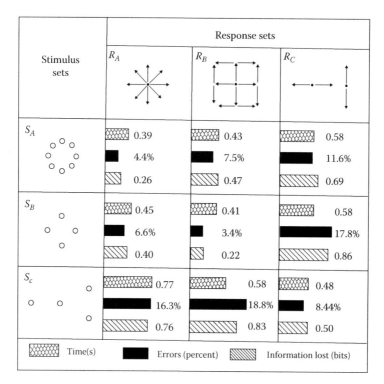

Figure 12.8 Results from the Fitts and Seeger (1953) experiment. (From Fitts, P.M. and Seegar, C.M., *J. Exp. Psychol.*, 46(3), 199, 1953.)

that were not. Fitts and Seeger also measured average response times, percentage errors, and bits of uncertainty lost per presentation. Figure 12.8 shows the nine stimulus–response situations along with the average performance measurements. Note that the major diagonal shows the three highly compatible combinations; the other six combinations show various degrees of incompatibility. The three highly compatible combinations provided the least performance time, fewest errors, and the least lost stimulus uncertainty. Hence, the message is clear: make the stimulus and response patterns as close as possible to each other in order to gain the greatest stimulus–response compatibility.

In addition to stimulus–response or display–control compatibility, ergonomists need to be concerned about the compatibility between successive responses. This form of *response–response compatibility* is particularly important when a signal triggers a series of required responses. Emergency shutdowns are a case in point. If a series of valves are to be closed, they should all close in the same rotational direction. Fortunately, valves all close in clockwise rotations. However, if the emergency shutdown procedure involves turning off electrical rheostats between valve closings, problems arise because the convention for rheostats is opposite to those of valves. Valves and rheostats should be separated by panels, sequence, and/or different operators for better compatibility.

Operator-in-the-loop simulation (OLS) can be helpful when evaluating the layout of controls and displays. Control errors are often related to display incompatibility or to faulty process models. OLS also offers a way to train people in the use of controls, not to mention data on how people learn. Even low-fidelity simulators can be useful for such purposes. Most people who have experience with a process can readily fill in missing elements of a simulation. However, naive subjects tend to need more realism than experienced subjects.

Final remarks

This chapter introduces some of the many issues that must often be addressed to design controls and displays. The first few sections provide a theoretical foundation for the design principles covered later in the chapter. Some effort is made to show how the presented principles follow from the many relevant studies that have been conducted over the years. However, most of these principles are based on a mixture of science and experience, and will continue to evolve as more is learned about the topics addressed.

Discussion questions and exercises

12.1 What are some basic methods for evaluating the perceptual elements of a display?

12.2 Give examples of good and poor mappings.

12.3 What are some methods that can be used to measure the comprehension of symbols and messages? How does semantic parsing differ from use of a readability index? What do we mean when we say some words are *fuzzy*?

12.4 Why is decision making an important consideration during display design?

12.5 What are the basic types of displays? Explain their differences and similarities.

12.6 If an operator needs to keep track of the status of a machine that needs immediate attention when it has a problem, and this operator must move away from the machine and do paperwork on occasion, then which of the following displays are appropriate for this message?
   a. Red flashing indicator light at 3 Hz
   b. Circular dial
   c. Strip-chart record
   d. Alphanumerical display (e.g., LED)
   e. Audio display

12.7 Explain the following display design principles: temporal proximity, proximity compatibility, and spatial proximity. When might it be useful to use CRAFT or CORELAP in display design?

12.8 What are some sources of information on display element legibility?

12.9 A visual display needs a horizontal dial of 100 spaces. Novice viewers are expected to see this dial from a viewing distance of 30 in. Based on the Murrell–Jones principle, what is the minimum length of the dial in centimeters?

12.10 For problem 12.9, now assume that the visual display will display the message WARNING. How much space is needed for this word?

12.11 Explain (mathematically) why veiling glare reduces the contrast of a CRT.

12.12 What are some advantages of dial displays with moving pointers?

12.13 Explain when and why derivative information is important. What types of displays provide such information?

12.14 What is the principle of pictorial realism?

12.15 What types of coding methods are commonly used in display design? Give an example for each.

12.16 What types of information are normally provided with an auditory display? Tactile display?

12.17 A hand-operated control is designed so that both hands operate on the device and one hand tends to pull in an opposite direction from the other. This device utilizes the principle of

  a. Antagonistic forces
  b. Member movements
  c. Discontinuous effort
  d. Minimum mass
  e. Postural freedom
  f. Helson's U-hypothesis
  g. Rhythm principle
  h. Population stereotype

12.18 When is a control device also a display device?

12.19 Steering wheels of most vehicles are held with the two hands pulling in opposite directions. Some of the older cars had a steering tiller, similar to boats, where direction control could be performed with a single hand. Which form of control is supported by one of the principles cited here, and which principle is it?

12.20 Since Fitts and Seeger introduced the concept of stimulus–response incompatibility, consider a similar concept that deals with sequences of control tasks, where the preceding control actuation might interfere with the control operation that follows. How might someone examine human-control situations to see if post-control actions could interfere with future actions?

12.21 A plating process has a control and display panel. Displays show the flow of chemicals into and out of tanks and the electrical power being used. Controls consist of fluid valves and direct current electrical rheostats. What is likely to be the ergonomic problem at the display–control panel? Some possibilities are

  1. The C/D ratio
  2. The crank design
  3. The treadle design
  4. Stimulus–stimulus incompatibility
  5. Stimulus–response incompatibility
  6. Response–response incompatibility
  7. Cumulative trauma syndrome
  8. None of the above

12.22 Consider a control associated with a particular display, with the display and control on the same plane. Suppose the selected display is placed such that the display movement is along an arc for an increase in the control, and the control rotates immediately below the display when it is increased. With what principle is this situation associated?

12.23 A knob on an emergency radio dial is used to monitor a very narrow band of radio frequencies with extremely high rejection. The control should have

  1. A low C/D ratio to minimize the travel time
  2. A low C/D ratio to minimize the adjustment time
  3. A moderate C/D ratio to minimize total time
  4. A high C/D ratio to minimize the travel time
  5. A high C/D ratio to minimize the adjustment time
  6. A high C/D ratio to minimize total time
  7. A joystick selector switch instead of a knob
  8. A mouse instead of a knob
  9. None of the above

chapter thirteen

Ergonomics of product quality and usability

About the chapter

This chapter focuses on the emerging role of ergonomics and industrial engineering in product design, especially in regard to rapid prototyping and usability testing at all stages of the product design cycle. Some of the specific topics addressed include quality function deployment (QFD), requirements analysis, expert assessment, usability and preference testing, and the use of designed experiments. The central theme of this discussion is that the design of products should be Human Central, to help ensure products satisfy critical customer requirements.

Introduction

Almost every company today insists that product quality is not only important, it is vital. Product sales depend on the product's perceived usefulness and utility compared to competitive products. When quality appears to be down, so are sales. That is the revenue side of the equation, but there are also cost considerations. Poor quality clearly costs more in terms of product liability, warrantees, and company reputation, and probably costs more in production, too.

Quality is defined by many of the gurus* of the quality movement as the ability of a product to meet the needs and desires of customers from various populations and subpopulations. To achieve this goal, designers must identify those people being served and understand the needs and desires they are seeking to satisfy when they use the product. Consequently, ergonomics specialists need to play an important role in determining what constitutes quality, how to design quality into products, and how to allocate design and manufacturing resources to achieve and maintain total quality (Box 13.1).

We have all seen a product in a showroom and said, "That's quality!" How do we really know? For example, suppose you are checking out a Rolls-Royce at an automobile dealer. The door opens easily without any play in it, like a light-weight bank vault door. You step on a rug that is several centimeters thick and the floor is as solid as a concrete patio. You see the hood ornament with the manufacturer's name and you know the legend of Rolls-Royce quality. The point of this very short story is that expensive luxury easily suggests quality, and companies who earn a reputation for quality products are perceived accordingly. But if the designers are clever enough, even inexpensive products can signify quality to potential buyers and owners who know what they want.

---

* For example, W.E. Deming, G. Taguchi, P. Crosby, and J.M. Juran.

> **BOX 13.1 WHO IS THE USER?**
>
> Buti (1995) extends the meaning of product or system user to include the following:
>
> 1. People who operate WITH the object (i.e., normal intended use).
> 2. People who operate ON the object (i.e., service, maintenance, or repair tasks).
> 3. People who HANDLE the object (i.e., shipping, packaging, and handling tasks).
> 4. People who ELIMINATE, REMOVE, or DESTROY the object (i.e., recycling and waste disposal activities).
>
> Buti's insight is important because it is altogether too easy to focus only on people who operate with the product during normal use. People who operate on, handle, or dispose the product are also obviously important users and are involved in a disproportionately large number of product-related accidents. Part of the issue might be that designers should better address the special needs of these users.

In the early phases of product design, the design team needs to identify consumer needs and wants. Juran,[*] another of the gurus of quality, refers to this identification and the development of products and processes to meet or exceed those needs as *quality planning*. That planning is not just about the current needs but also includes checking trends to determine future needs. At this phase of product design, ergonomic designers meet marketing personnel who are keenly interested in product sales as well, but they may not see eye to eye with ergonomics people. It is therefore important to understand some of these differences in perspective and why they arise. To start with, most ergonomic designers suggest that alternative product designs should be compared on the basis of *human performance during actual use*. Marketing people are moved more strongly by the subjective evaluations or preferences of large populations of people whose opinions are revealed by questionnaires and interviews. The argument of marketing people is that most users of commercial products neither know nor care about the product's performance, at least from the perspective of maximizing or optimizing performance.

The basis of this argument, and it is a good one, is that there is often no solid relationship between performance and preference.[†] On the other hand, satisfied users provide some of the best advertisements for new products, so it is essential to satisfy those people who first purchase the *new design* so that they will speak up for it. Satisfactory performance is therefore a necessary condition for continued sales, and on this point, there will be strong agreement between marketing and ergonomic perspectives. In some cases, improved ergonomics even becomes a potential selling point. However, the key question that first must be answered from the marketing perspective is, "Do products that are better from an ergonomic viewpoint sell better?" The answer is "not necessarily." The problem is that purchasing decisions normally are based on multiple criteria, and a particular ergonomic criterion may not be particularly important or might conflict with others. For example,

---

[*] J.M. Juran has been a strong advocate of quality in manufacturing for many decades. His consulting company has advised many manufacturers over this period. More recently, he established the Juran Institute for Quality Management.

[†] See Meyer and Seagull (1996) for greater detail regarding this argument and some case studies. These authors also show that preference can be validated empirically, a point that is addressed later in this chapter.

many ergonomically *improved* keyboards have been developed over the years that increase typing rates over the standard QWERTY layout. None of these sells particularly well. Most users are unwilling to put in the extra effort to learn the new layout. This is the type of result that could easily be discovered through a market survey!

To help avoid such problems, many ergonomic practitioners argue that product design should be *user centered*,* that is, satisfying user goals and objectives should be a focus at each stage of product design. From a practical perspective, this means that user inputs should be obtained at each stage of the design cycle, beginning with market analysis to identify user goals, followed by task analysis, user testing and evaluation, and postimplementation surveys (see Box 13.2). In other words, the design team should use preference ratings obtained from questionnaires and interviews concurrently with various kinds of performance testing. If the performance tests and measures are directly related to user goals, as they should be, preference will follow performance and conversely. However, empirical experimentation is often needed to evaluate the relationships between human perceptions of how their needs and wants are met and how these perceptions are affected by design changes that involve trade-offs between different needs and wants. This testing requires that actual performance be compared to customer perceptions. Quite obviously, action should be taken if the two differ.

Quality management and customer-driven design

Nearly all large manufacturers have adopted some form of quality management in their efforts to improve the quality of the products and services they provide (Edosomwan, 2001). These approaches focus on increasing customer satisfaction through continuous improvement of product quality. This approach is customer driven, as it focuses on identifying and satisfying customer needs and wants. The customer-driven design of products and services requires answers to many different questions, such as the following:

1. Who is the customer? Can customers be classified into fundamentally different groups? Which customers are the most important? Should certain customers be targeted? Should others be discouraged from purchasing the product or service?
2. What are the needs and wants of the customer? How should customer requirements be measured? Do all customers have the same requirements? Which requirements are the most important for each targeted group of customers? What is the relation of customer requirements to satisfaction?
3. What is the relation between customer requirements and product or service features or functionality? Can customer requirements be translated into product specifications?
4. How well are customer requirements satisfied by existing products? Which products meet particular requirements the best? Which requirements are poorly addressed?
5. What changes to the product or service would better satisfy poorly addressed requirements? Are these changes technically feasible? How would making certain changes affect satisfaction of other requirements? How expensive are such changes?
6. How effective is the product or service after making changes intended to improve its quality? Why were the changes effective or ineffective? If the changes were not effective, what should be done?

* The topic of user- or human-centered design is discussed in some detail in Chapter 1. Also see Flach and Dominguez (1995) for further discussion of this topic.

> **BOX 13.2   STAGES IN SOFTWARE DESIGN**
>
> A variety of systematic procedures are used in software engineering to help ensure that software is developed on time, within budget, and meets quality standards. The complexity of modern software, the wide range of customers, and the need to coordinate the effort of numerous programmers, engineers, technical writers, etc., have accelerated this trend. Several years ago, Mantei and Teory (1988), at that time both professors at the University of Michigan (Mantei in the School of Business and Teory in Computer Science), joined forces and proposed an approach for incorporating human factors analysis into the software development cycle. The traditional stages of software design and the new, human factor–focused stages proposed by Mantei and Teory are shown here. The human factor–focused stages are highlighted in bold:
>
> 1. *Market analysis*—use of focus groups to identify user needs and perceptions
> 2. Feasibility study—comparison of proposed software to similar products already in market
> 3. Requirements definition—what functions or features to provide in the software
> 4. *Acceptance analysis*—initial feedback from users obtained using mockups
> 5. *Task analysis*—evaluation of usability issues for selected subtasks
> 6. Global design—grouping of tasks to be performed by the software into modules
> 7. Prototype construction—rapid development of a preliminary version of software elements or modules (often done using rapid prototyping tools)
> 8. *User testing and evaluation*—observation of potential customers using the prototype. Evaluation of performance and customer ratings of product features
> 9. System implementation—a modified version that incorporates changes suggested during testing of the prototype
> 10. Product testing—testing of system performance
> 11. *User testing*—same as Stage 8, but involving the product instead of a prototype
> 12. Update and maintenance—correct problems and otherwise improve the system
> 13. *Product survey*—obtain inputs from customers who purchased the product to determine which features customers like, problems, overall satisfaction, etc.
>
> This view of the product design cycle is focused on software development, but it does not take much imagination to see its relevance in other settings. Just replace the word software with product. The important point is that human factors can play a critical role throughout the development cycle.

The answers to these questions can be obtained at varying levels of detail depending on the methods used to answer them and the particular stage of design in which they are applied. In the early stages of design, designers are often concerned about identifying problems and what customers want. Once designers have some idea of what the customer wants, they can then concentrate on analyzing how well particular designs satisfy customer requirements. Benchmarking comparisons against competing products are particularly useful at this stage. It then becomes necessary to generate and evaluate new designs or design changes. This process often involves the development and testing of design prototypes. Rapid prototyping and quick assessment of the impact of design changes (or redesigns) is especially critical in today's competitive environment. After the design is nearly finalized, usability tests, user trials, and field studies are often conducted to answer other questions. Before we consider the methods that might be used to answer these questions, let us first consider how we might systematically consider customer needs during design.

QFD: A framework for quality assessment and improvement

Several years ago, Hauser and Clausing (1988) proposed a structured approach for improving product quality called QFD. In the following years, QFD has been used in many different industries. For example, QFD has been heavily used within the United States automotive industry in their successful efforts to improve vehicle quality over the past 20 years. QFD involves the following steps:

1. Identify customer attributes (CAs)—characteristics of the product expressed in the customer's own words.
2. Identify engineering characteristics (ECs) of the product that are related to CAs and how the ECs affect CAs.
3. Determine how important the CAs are to particular customers.
4. Evaluate the CAs and ECs of the product both in isolation and in relation to competing products.
5. Set target values for each CA and EC.
6. Identify trade-offs between ECs.
7. Specify the technical difficulty and cost of implementing proposed improvements of ECs.
8. Implement or deploy improvements of ECs that are technically feasible and cost effective.

Hauser and Clausing organize the information generated while going through the previous steps in a structure called the *House of Quality*. The House of Quality provides a systematic way of organizing efforts to both assess and improve product quality. The basic idea is illustrated in Figure 13.1. As shown in the figure, the House of Quality consists of several matrices. Normally, these matrices are developed separately for different components or subsystems of the product to keep the analysis from becoming too complicated. For example, a House of Quality might be developed for the door of an automobile. Some of the different types of information captured within this structure are summarized next.

The analyses summarized in the House of Quality provide a concise and convenient way of determining *what* the customer wants, how well these wants or needs are satisfied by existing products, and *how* to best modify or improve product to satisfy the customer. The analysis begins with identifying CAs. CAs, as mentioned earlier, describe characteristics of the product in the customer's own words.* If the evaluated product was the door of an automobile, example CAs might be "easy to open from the outside, stays open on hill, doesn't leak in rain, protects against side impact." Note that the CAs are grouped into categories. As indicated in Figure 13.2, some logical groupings of CAs for an automobile's

---

* It is interesting to note that the meaning of most CAs (such as "easy to close") is a bit fuzzy. Fuzzy set theory provides a way of quantifying fuzziness and consequently may be helpful when interpreting or analyzing customer ratings CAs. For more on this topic, see Chapters 10 and 11.

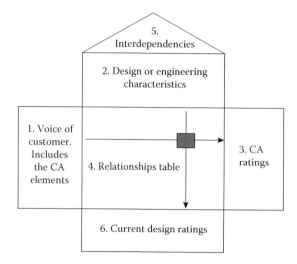

Figure 13.1 The House of Quality. (From Anoorineimi, A. and Lehto, M.R., Simulation of an auto transport facility, Unpublished technical report, Purdue University Technical Assistance Program, 2010.)

door are "easy to open and close," and "isolation." Each of these groupings corresponds to a general customer requirement (Box 13.3).

Once the CAs have been identified, the next step involves asking customers to rate both (1) the importance of each CA and (2) how well the product does compared to targeted competitors (again, on each CA). In many cases, the product is also compared to earlier versions that do not include proposed design changes. Example ratings are shown in Figure 13.2.* Such ratings are normally provided by more than one member of a targeted group of customers. In some cases, an effort is made to determine whether there are statistically significant differences between subgroups. This, of course, means that it is important to measure both the mean and the standard deviation of the obtained ratings.

Poor ratings of the product on CAs rated as important by customers indicate a need for improvement. The House of Quality guides the systematic identification, analysis, and comparison of different design solutions. The first step in this process is to identify the ECs of the product that are likely to affect the CAs. The identified ECs for the product are also grouped into categories, or general design requirements, directly corresponding to those used to organize CAs—that is, each design requirement corresponds to a customer requirement. For example, as shown in Figure 13.2, one grouping of ECs for the door of an automobile is "open-close effort" and another is "sealing insulation." The corresponding groupings of CAs are "easy to open and close" and "isolation." Note that House of Quality also shows how each EC impacts the design requirement it falls within. For example, as shown in Figure 13.2, "the energy to close the door" has a strong negative relation to minimizing "open-close effort"—that is, it is harder to open the door when more energy is required.

Once the ECs have been identified, it then becomes necessary to determine how they should be measured. Ideally, the measurement is a physical quantity as close as possible to what a customer would perceive. For example, the force required to close the door of a car

---

* Note that the importance ratings in this example are measured on a different scale than the performance ratings. Other measurement scales can obviously be used. For more on rating scales, see Chapter 14.

Chapter thirteen: Ergonomics of product quality and usability

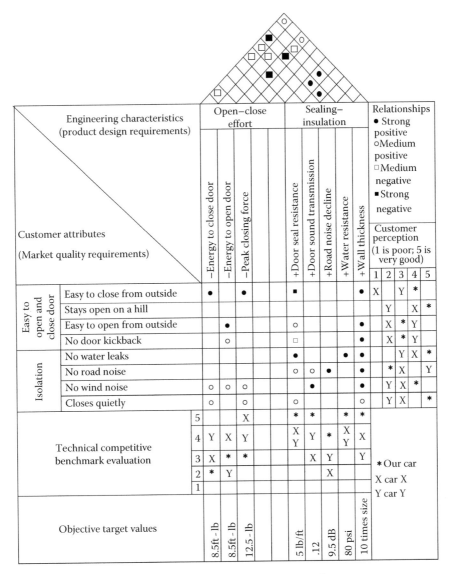

Figure 13.2 Example House of Quality for the door of an automobile.

could be measured directly with a telescoping spring balance mounted in the location of the average customer's hand. The device could then be pressed against an open car door to measure the force level at which the door begins to move. Similarly, to test the adequacy of sealing from outside rain, water could be sprayed at the car door. The water pressure could be raised until leaking occurs. The pressure at this point yields a numerical physical measure of sealing resistance. Using such techniques, measurements can be placed under each EC for the company's product and for targeted competitor products, as shown in Figure 13.2.

The next step in analysis is to specify how the ECs are related to CAs. When an EC has no effect on any CA, it is redundant and not included in the House of Quality. For example, "using more sealing insulation" is an EC that affects rain coming in and is therefore a relevant EC to the CA that the car door does not leak. However, the statement of "more sealing insulation" is too imprecise for a proper EC (i.e., "more" compared to what.) A better description would be, "door seal resistance." The latter description can be defined in engineering units and has a strong positive relationship with preventing leaks and a strong negative relationship with ease of closing the door.

The relation between CAs and ECs is shown in the *relationship matrix*. Customer requirements, expressed as CAs and their rated values, form one axis of the relationship. Design requirements, expressed as ECs, form the other axis (see Figures 13.1 and 13.2). The cells of the matrix contain values showing how strong the relationship is and whether it is positive or negative. In some cases, the relationships are obtained by surveying or studying potential customers (see Box 13.4). Often, however, they are simply assigned by the designer or others familiar with the product and potential customers. In any case, the relationship matrix shows how engineering design influences customer-perceived qualities in both a positive and negative sense.

Several other forms of analysis are summarized in the House of Quality. One of the more interesting involves setting performance targets for each EC. In some cases, the performance target is to be the *best in class*. In other words, the objective is to be at least as good as the best competitor on each EC. This approach is, however, rather simplistic. Another approach is to try to select the most cost-effective approach. One method along these lines is to calculate a cost-benefit ratio for proposed changes to ECs. Changes can be weighted in terms of their expected impact on particular CAs. The CAs themselves can be weighted in terms of the importance values assigned by the customer. The technical difficulty of changing the ECs may be an important consideration. Another issue is that ECs are often related to each other. Relationships between ECs are shown in the triangular matrix at the top of the House of Quality. The information in this matrix helps identify both design trade-offs and synergistic effects. Trade-offs are very common. For example, a larger door makes it easier to enter into a vehicle, but can make it harder to close. More door seal resistance may reduce rain leakage, but also may make the door harder to close. Synergistic relationships are less common, but do occur. For example, a smaller door may improve structural integrity, while also making it easier to open.

Another common form of analysis involves comparing customer perceptions of the product to objective physical measures of the ECs. Such comparison provides a basis for checking consistency. If the objective measurements are not consistent with customer perceptions, further analysis is warranted. The inconsistency could be due to improper measurement techniques or skewed perceptions. In either case, there is work to be done. If the objective measurements are going to drive engineering solutions, they need to be consistent with human perceptions. Ensuring that this is the case is another important challenge to ergonomic practitioners involved in product design.

As should be apparent from the preceding discussion, QFD and the House of Quality provide an excellent organizing structure for quality assessment and improvement. On the other hand, not a lot of detail is provided on how to generate the data needed to do the analyses. Some approaches for generating the needed data are covered in the following sections. These include customer requirement analysis, rapid prototyping, subjective assessment, preference testing, user trials, usability analysis, and usability testing.

Identifying customer requirements

Customers want products that are reliable, safe, and easy to use. Most importantly, the product must perform its intended function well. Aesthetic aspects of the product are also

> **BOX 13.3 ROOTS OF CUSTOMER-DRIVEN DESIGN**
>
> The quality movement in the United States occurred in response to trade deficits between the United States and some off-shore countries during the 1980s. Many people sought causes and tried to fix blame when we really should have been identifying and fixing the problem. What seems clear now is that
>
> 1. Companies do a lot more listening to current and potential future customers than previously
> 2. U.S. management cannot arbitrarily assume that it knows what is always best for customers
> 3. Industry must be more agile in adjusting to new messages that come directly and indirectly from the public
>
> But how do designers listen efficiently and how do they assume that they have heard correctly? Some of the answers are obtained during customer requirements analysis as expanded upon in the next.

often very important. Customer requirements can be identified in many different ways, and at many different levels of detail. As products grow more complex, the number of requirements grows accordingly. To further complicate the issue, not all customers have the same requirements. Consequently, identifying customer needs and wants can be a difficult task.

One way of dealing with this situation is to first determine which customers are the most important. Greater efforts are then devoted to identifying, and ultimately satisfying, the needs of the most important customers. The needs for each group, demographic factors, and other data are used to develop customer profiles. Marketing efforts and design features can then be tailored to fit particular profiles. Most marketing departments maintain customer databases, which can be used to determine different user groups and some of their needs.

Although much can be learned by analyzing existing customer data, it is normally essential to do a market survey at the start of the product design process to gain a better understanding of customer needs. Focus groups, customer surveys and questionnaires, and user trials are all potentially useful sources of information at this stage. The latter approaches are especially useful when developing new or unique products. In such situations, the product will not have a history. When products do have a history, substantial insight can often be obtained from earlier market research. For example, a House of Quality might be available for an earlier version of the product. The latter situation is definitely the case in certain settings, such as the automotive industry. Duplicating earlier research is obviously a waste of time and effort, unless there is reason to question its validity or relevance. Making this information available to those who need it is the main problem in many companies.

Ergonomists and other experts can often provide very useful input regarding customer needs and how to organize them in a meaningful way. For example, consider the organization of CAs for automobiles into categories and subcategories shown in Box 13.5. This particular structuring was developed by ergonomists at the University of Michigan Transportation Research Institute. Such structuring is very helpful even in the very early stages of design or for novel products that do not have a history. This follows because the analyst *needs* to know something about the topic of interest before conducting a focus

BOX 13.4 TESTING OF SEAT CUSHION QUALITY

John Versace discussed several efforts at Chrysler to determine customer desires and their relationship to ECs. Some of the tests that Chrysler performed were conducted at regional automobile shows in order to get nearly nationwide representation. Each test involved a designed experiment. One of these experiments involved testing of seat cushions under showroom feel conditions. Six different seating cushions were tested. The participants were classified as tall, medium, or short in height, and every participant rated all six cushions. The cushions were tested in six different sequences. The resulting analysis of variance (ANOVA) is shown in the table below.

The proper tests for statistical significance are determined by the expected mean squares (EMS), as indicated in the table below. For example, the type of cushion (factor C) has an EMS of $\sigma^2 + 432\sigma c^2$ so its significance is tested by dividing its mean square value (44.1) by that of the residual error term (mean square value of 5.7 and an EMS of σ^2). As can be easily verified from the below table, the ratio $[\sigma^2 + 432\sigma c^2]/[\sigma^2]$ is large for cushions, indicating that this factor significantly affected the ratings. This approach was followed for each of the sources of variance. Interpreting significant sources of variation is an important step of this process. For example, if the trials factor turns out to be significant, it is likely that the raters learn or become fatigued in rating over time. Also, taller people usually are heavier, and it could be that these people rate cushions with less support lower than shorter people do. In that case, the interaction of $C \times H$ (H stands for the person's height) should be large and so forth. The point is that the results of the ANOVA point out possible causes of differences in the ratings. This information can then be used to guide design-related decision making.

ANOVA Table for Cushion Rating Example

Source of VAR	Degrees of Freedom df	Mean Square	Expected Mean Square
Cushions (C)	5	44.1	$\sigma_2 + 432\sigma_c^2$
Sequences (S)	5	5.3	$\sigma_2 + 432\sigma_s^2$
Trials (T)	5	1.0	$\sigma_2 + 432\sigma_T^2$
Heights (H)	2	70.1	$\sigma_2 + 36\sigma_R^2 + 114\sigma_{S\times H}^2 + 864\sigma_H^2$
$C \times H$	10	8.0	$\sigma_2 + 432\sigma_{C\times H}^2$
$S \times H$	10	24.6	$\sigma_2 + 432\sigma_{T\times H}^2$
$T \times H$	10	12.2	$\sigma_2 + 36\sigma_R^2 + 114\sigma_{S\times H}^2$
Raters/$S \times H$	54	14.2	$\sigma_2 + 36\sigma_R^2$
Residual[a]	330	5.7	σ^2

Source: Versace, J., Subjective measurements in engineering. Society of Automotive Engineers. *Annual Meeting Proceedings*, Detroit, MI, 1959.

[a] Nonsignificant effects were pooled with this residual.

In a second experiment, tire thump ratings* were obtained from a fixed panel of four raters who drove several of 18 randomly selected production cars on selected road sections. Two ratings were obtained from each observer: one in the front seat and the second in the rear seat. This test formed an 18 × 4 × 2 factorial design. In another experiment involving a "tire roughness rating" test, Chrysler used different raters in differing numbers to rate 10 sets of 5 different tire types. These raters rated ride roughness at two different road speeds. Each rating was taken as the average of the individual ratings. A final example of Versace's experiments was one that yielded "shake ratings."[†] In this test, 4 engineers took 3 days to rate a total of 12 cars produced by targeted competitor. The results provided a good benchmark for Chrysler to judge the quality of their product and identify market opportunities.

---

* Automobile tires in the early part of the twentieth century were often asymmetrical due to manufacturing difficulties. This caused the tires to make a thumping sound and vibrate the car with each revolution. The so-called *thump rating* measured how people perceived the thumps. Better tires had lower thump ratings as they were quieter and gave a smoother ride.

† Shake ratings were similar to thump ratings. Vehicles that were tightly welded didn't rattle a lot under ordinary driving conditions. Better cars responded to a bump with a single sound rather than the sound of loosely connected parts hitting against each other for a longer period.

---

group or administering a survey to customers. Without some preliminary structuring, the market survey will deliver results that are difficult to interpret and, in many cases, completely useless.

As illustrated by the example in Box 13.5, customer requirements can normally be organized into meaningful generic categories. Without such organization, the sheer number of attributes potentially important to customers may cause the analysis of customer requirements to be overwhelming for all but the most simple products. Another point is that generic categories, such as those shown in Box 13.5, can be used as brainstorming tools to help identify customer requirements. Ergonomists and others with experience in the area will have little difficulty identifying such items. Along these lines, a generic framework for identifying customer requirements is given in Box 13.6. In this framework, generic product features that are likely to be important at particular stages of the product life cycle are mapped to generic quality requirements. Some of the product features overlap or fall into more than one requirement category. For the sake of conciseness, no attempt is made in this chapter to exhaustively map features to requirement categories.

Note that the product features shown correspond to CAs identified by the analyst. This is done by systematically considering each requirement category at each stage of the product use cycle and whether customers are likely to need the corresponding product features. For example, at the purchasing stage, many customers will want to purchase a product that meets their requirements. Consequently, the focus is on whether advertising and other forms of information are available that describe which product features are present. One relevant CA might be "Product has a warning label explaining the hazards of this product." Another might be "Product has a warranty."

At the installation, assembly, or setup stages, the customer is likely to be concerned about the effort involved. Example CAs might be "Product is easy to assemble" or "Product assembly instructions are easy to understand." At the use stage, the customer will probably be especially concerned about usability and functionality. Example CAs might

BOX 13.5 IMPORTANT CUSTOMER ATTRIBUTES FOR AUTOMOBILES

Category	Subcategory	Selected Issues
Occupant protection	Seat belts	Perceived level of protection, comfort, ease of fastening/unfastening
	Air bags	Perceived level of protection, perceived hazard, effect of occupant size and positioning
	Interior structure	Padding of instrument panel, steering wheel energy absorption
	Interior materials	Flammability
Vehicle control	Steering	Force levels, responsiveness, slack
	Braking	Force levels, response time, distance, pedal size
	Ride quality	Smoothness, effect of potholes, road surface
	Seat comfort	Accommodating widest range of user weights, user shapes, user heights; comfort of seat material, seat shape, seat hardness
Exterior design	Field of view	Vision blockages, driving forward vs. backward
	Headlights, taillights	Minimize glare to others, maximize illumination on road, maximize forward view
	Mirrors	Minimize glare to others, maximizes detection distance, minimizes blind spots
	Conspicuity	Vehicle detected at greatest distance, greatest aesthetic appeal
Interior design	Controls	Best switch type, location, shape, control/display ratio, force resistance, travel distance
	Displays	Best layout, conspicuity of visual elements, understandability of text, understandability of symbols, lighting, no reflections, not influenced by glare
	Thermal comfort	Best air flow, temperature control, radiant load, evenly distributed
	Air quality	Fewest fumes, dust
	Sound	Lowest exterior and interior noise, best audio quality, most intelligible speech (on phone), most audible warnings
Access, maintenance and repair	Ingress, egress	Best door size and swing angle, easiest to use handles and supports
	Trunk and under hood	Best access, fewest hot spots
	Maintenance and repair	Minimum time to replace tires, fluids, other commonly services items
	Documentation and manuals	Most understandable, satisfies legal requirements, ease of use

Source: Adapted from Waller, P.F. and Green, P.A., Human factors in transportation, in Salvendy, G. (Ed.), *Handbook of Industrial Engineering*, 3rd edn., John Wiley & Sons, New York, pp. 1972–2009, 1997.

BOX 13.6 GENERIC FRAMEWORK FOR IDENTIFYING CUSTOMER REQUIREMENTS*

Stage ×	Requirement Category =	Product Features/Attributes Commonly Related to Requirement Category
	Safety	Hazards: physical, chemical, thermal, biological
		Risk: severity, probability
		Precautions
		Informed consent
		Safety devices
		Brochures and advertising
		Point of purchase warnings, on-product warnings, warning labels, MSDSs
		Failure modes and effects, error modes and effects
Purchasing		MTBF, MTTR, failure rate
Installation		Product specifications: conditions of use, intended users
Assembly		Diagnostic equipment, special tools, parts list, parts delivery
Setup		
Use		Customer assistance (online, call center, other)
Malfunction	Reliability	Repair manuals, repair procedures, JPAs
Service		Service schedules
Repair		Service plan
		Warranty
	Functionality	Provided features/functions, performance measures and specifications
		Effort: mental, physical, sensorimotor
		Comfort: tactile, kinesthetic, muscular fatigue, thermal, etc.
	Usability	Displays and controls: location, conspicuity, C/D ratio, S/R compatibility, etc.
		Warnings: mode of presentation, content, format
		Product affordances, constraints
		User manuals, instructions, training: content, organization
	Aesthetics	Perceived attractiveness, quality, status, etc.

* Requierments fall within each of the stages.

be "Displays are easy to read" or "Seat is comfortable." Prior to malfunctions the customer is especially likely to want warnings and diagnostic help; after malfunctions, the customer is likely to be interested in getting assistance in resolving the problem. Example CAs might be "Product warns before a failure occurs," "The product tells you what the problem is," "A service technician will be sent out immediately after a problem occurs," etc.

Specifying design requirements

When a product is designed, it will have certain ECs that ultimately lead to a satisfied or dissatisfied customer (Box 13.7). In theory, the necessary ECs can be specified by first determining which CAs are important, and then specifying ECs in a way that maximizes desirable CAs. In practice, this approach can be applied in only a limited sense. The problem is that most products have far too many ECs for all of them to be formally specified by CAs. Yet many, if not the majority, of ECs are potentially important influences on product quality. The normal solution to this quandary is very simple. Designers select components that meet standards and guidelines established either in-house or by standard-making organizations, that is, most ECs of a product are set by complying with standards and guidelines.* Quality improvement efforts are normally focused on the much smaller set of ECs that is directly related to CAs customers feel are very important. Ergonomists are especially likely to be focused on ECs that impact usability, but it is easy to find applications of ergonomics for the other generic areas of CAs discussed in the previous section related to issues such as reliability, safety, and aesthetics.

For example, no automotive designer would specify the design of a seat belt to be installed within a vehicle sold in the United States without first examining the relevant portions of the U.S. Federal Motor Vehicle Safety Standard (FMVSS 208) and the Society of Automotive Engineers standard (SAEJ128) and making sure each component of the seat belt system satisfied the specified design requirements of each standard. However, the designer might be interested in increasing use of seat belts by improving their quality. Given that most customers rate comfort as a very important CA for seat belts, they might focus their efforts on first determining what ECs influence perceived comfort. Once this relationship is understood, they might then specify changes in seat belt design that improve comfort and also satisfy FMVSS 208 and SAE J128.

Ergonomists and industrial engineers are uniquely positioned to help answer such questions. The methods of usability analysis discussed later in this chapter provide one useful way of addressing this issue. Besides having a toolbox of methods for evaluating usability issues and task performance, ergonomic experts can often suggest improvements based on expert opinions during design reviews. They also are aware of useful standards, guidelines, handbooks, scientific publications, and other sources of information.† Ergonomists and industrial engineers can also guide empirical investigations of how ECs are related to customer perceptions and other subjective measures of product quality. This leads us to the issue of rapid prototyping and testing of products.

BOX 13.7 HOW DO CUSTOMERS CHOOSE BETWEEN PRODUCTS?

A fundamental notion in economics is the model of a consumer in a free economic society. According to this model, each consumer is rational and has limited monetary resources. These resources are exchanged in a marketplace for products that satisfy their needs. This foundation of economics is derived from utility theory, which was first articulated by Daniel Bernoulli and his brother during the early 1700s and was eventually formalized by John von Neumann and Oskar Morgenstern in 1944. Utility corresponds to the perceived value of something. Since utility is a perceived value rather than an objective value, different people will have different utility functions.

---

* Note that standard-making organizations, such as the American National Standards Institute (ANSI), as well as regulatory agencies, such as the U.S. Department of Transportation, consider issues such as safety, economics, functionality, and reliability when they develop their standards.
† Chapters in this book present numerous examples and references to sources of more detailed information on topics such as tool design, controls and displays, safety, hazard communication, keyboards, chairs, etc.

This theory assumes that people will both appraise the utility of the outcomes of their buying decisions and act to maximize their expected utility over time.

This perspective assumes that the perceived utility of a product is somehow related to its features. The unanswered question is: How do people do this? The famous psychologist E. Brunswik developed a model of how people make choices that shows how this might be done. A very simplified version of this model is shown in Figure 13.3. A person's preference for a particular product is a weighted function of the particular cues they perceive. Each cue corresponds to characteristics of the product or object in the world. In many cases, these preferences can be described with a simple additive function of two or three cues. The important point is that people do not consider all of the features of a product when making buying decisions, and some of the features are much more important than others.

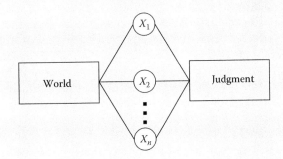

Figure 13.3 Simplified version of the Brunswik Lens Model.

Rapid prototyping and testing

Quickly getting the design from a concept to a product on the market is one of the more keenly sensed needs in product design today. While the work of each discipline in this process is important, management today recognizes that it cannot be the purely sequential process that it was in the past. The old practice of over-the-wall design is obsolete for much of industry. Today there is an urge to rapidly develop and test a prototype of the product to determine how well it will satisfy users. The development of a prototype allows the user interface of the product to be studied and refined, as expanded upon further in the next section. Such work is important because the product's interface usually has more to do with a user's impression of the product than any other factor.[*] Prototypes can also be used to speed up the manufacturing process once other design considerations have been addressed.

Rapidly generating prototypes is helpful in many design situations because doing so not only quickens design activities, but also creates a common reference point among the members of the design team to help in advancing further design improvements. Many of these new procedures for more rapidly producing a prototype involve the use

---

[*] Numerous people have attested to the importance of focusing on the user interface in design. Meister (1991) has been particularly strong in his advocacy of this approach.

of computer models that allow the product and rudimentary uses to be viewed in three dimensions. Unfortunately, most such systems cannot easily generate a physical model that can be handled and used as the real product would be, and this limitation reduces usefulness in ergonomic design. However, a number of computer-aided design (CAD) graphic modelers and those with wire-wrapped modeling characteristics are reported to be very useful for such purposes (see Snow et al., 1996). Even if the ability to make physical models is limited today, dramatic improvements have already occurred in other areas. For example, highly realistic interfaces for a wide variety of interactive systems can be quickly developed using markup languages and programming toolkits containing programming widgets (such as buttons, icons, windows, spreadsheets, menus, scroll bars, etc.).

While *high-tech* techniques offer advantages in quickly generating prototypes that can be tested early in the design process, a wide variety of *low-tech* methods have been used for a number of years and continue to be useful. At the earlier phases of product development, there may only be a paper description of the product. In this case, the best one can do is to identify important usability characteristics or have potential users think through how they might use the device. Another form of testing involves a partial prototype. These kinds of prototypes are often analog models of the device (i.e., they work in a manner similar to the device) but the physical appearance may be considerably different. In those cases, testing concentrates on the features that perform in a similar fashion, but there is little or no testing on other features. Partial prototypes are commonplace in software development when some new features are under consideration but other parts of the program are still being programmed. The next level up in form is the full prototype that simulates both the complete function and the full appearance of the device. Full prototypes look and act like the device in question, but may be made of different materials or differ in other minor ways from the final version of the product. Of course, the final version of a product is also subjected to testing of various types, including usability testing. At this point of development, it is not uncommon to consider field testing, but it is usually better to perform laboratory testing first. Most products are tested throughout their development, so many prototypes will often be developed for any given product.

The manner in which the testing results get embedded into the design depends to some extent upon the organization of the design team.* Organizations following concurrent or iterative design procedures are more likely to make good use of the testing results. If a serial design process is followed where the ergonomics expert simply does a design review often the product is developed, it is doubtful that there would be much adaptation to user needs and wants unless these needs are very well known, clearly delineated, and not subject to changes with time. On the other hand, organizations that implement rapid prototyping methods as part of a concurrent or iterative design process can be extremely effective at capturing user needs and wants well before the design is frozen and changes become too difficult to implement.

Customer ratings and preference testing

Listening to the customer is one of the most important elements of human central design. Customer ratings and preferences are subjective but still can be measured in ways that are helpful to product designers and engineers. This idea occasionally meets resistance from engineers and others who want "hard" data. One of the earlier voices

---

* See Chapter 19 for more on design teams and how the way they are organized can affect team performance.

heard in response to this resistance was provided by John Versace* who said that all measurements are somewhat subjective. This is particularly true for measures of comfort, attractiveness, effort, or other perceived quantities. There is also no single objective measure of less obviously subjective concepts such as safety, reliability, usability, or quality. Customer ratings and preferences are assessed using surveys, interviews, or focus groups. These approaches are addressed in Chapter 14, so will not be further discussed here. One advantage of these approaches is that they can be applied at any stage of the product life cycle to guide design decisions. Some experience based principles of preference testing are given in Box 13.8. To a large extent, Karlin's insights should be viewed as a warning that customer rating and or biased. Karlin's last point is particularly important, that is, ratings obtained in a group setting may be biased one way or another. To avoid contaminating the data, the people participating in the study should be kept separate from each other at the time the data are being collected. It is also essential that the people who have already provided their input do not speak to those who have not yet participated to avoid influencing their ratings.

To ensure that the results of the preference testing are valid, steps must be taken to eliminate these potential sources of bias. To a large extent, the issue faced and potential solutions, are the same as those faced in usability testing.

BOX 13.8 SOME PRINCIPLES OF PREFERENCE TESTING

Karlin (1957) experimented with users of various types of telephone equipment in order to determine what customers wanted. He provided many practical suggestions, including the following:

1. Preferences and opinions from people without actual use experience are unreliable.
2. Users should try out equipment under normal, real-life conditions in order to find out what their preferences really are.
3. Initial preference opinions based on brief experience with the product may be reversed by subsequent experience.
4. Many people have preconceived biases. Experience with a new device should be extensive enough to overcome those biases, enabling the user to evaluate all of its important properties.
5. Experience with a new device should duplicate the field situation as closely as possible.
6. People cannot artificially generate or imagine needs that are not real. Users should not be expected to ignore certain features in an experimental device that the tested experimental device does not possess but which would be present in the planned future device.
7. Users in an experiment should be allowed to use the device the way they want to use it.

---

* John Versace of Chrysler Corporation was one of the first people to perform usability testing. The statements accredited to him here are paraphrased from his article published by the Society of Automotive Engineers in 1959.

BOX 13.8 (continued)

Regarding the latter suggestion, experimenters should try to minimize contact with the user to help the user forget that he or she is participating in an experiment. Some further suggestions from Karlin* without added comment include the following:

1. Sudden changes in a device may produce unfavorable preference reactions. The same changes may be unnoticed if made gradually in a series of small steps.
2. Preference for a new device is more rapidly evaluated if the user alternates between his current device and the new one at suitable time intervals than if he or she uses the new device constantly.
3. The user is not conscious of some of the properties of a device that affect his or her preference.
4. Unless the user's experience is fresh in his or her mind, preference opinions may be unreliable.
5. The user's preference for a new device may be influenced by knowledge of how other users react.

* See Karlin (1957). Some recent papers may give readers the impression that usability testing was recently invented. Karlin's work shows that there is a bit of history here.

Usability analysis and testing

Usability analysis and testing focuses on how efficiently and effectively different groups of people use or operate products. Efficiency is often measured in terms of how much time and effort is required to use the product. Effectiveness normally refers to how well different users attain their goals or objectives when using the product. Satisfaction with the product and its performance is also often evaluated in usability studies. As such, it is easy to see that usability measures are closely related to product quality. Usability studies are, however, primarily focused on how well people operate (or use) the product rather than on its internal function. This focus is to a large extent on the interface used to activate and control product functions.

Part of the reason for focusing on the interface is that poorly designed interfaces cause usability problems even if the product functions well when operated correctly. Products that are unreliable or function poorly in other ways will also cause usability problems, so usability analysis and testing can help identify problems with the product's functionality as well as the interface. The focus on the product interface, or in other words, on how the person interacts with the product rather than on how the functions are implemented, also allows usability analysis and testing to be conducted in parallel with other design activities during product development. For example, Web-based and telephone interfaces to an electronic banking system might be developed in parallel with the database management system (DBMS) that processes the transactions. The designer of the interface would not need to know all the details of how the DBMS functions to evaluate the usability of the interface. Knowledge of the required inputs from the customer for each transaction would be sufficient to build and evaluate the usability of different interface prototypes.

Several different approaches for evaluating usability are followed by ergonomists, usability engineers, and others attempting to improve product usability. Many of these approaches involve some form of task analysis, where tasks and subtasks for each product function are first identified and then separately analyzed. Some of these approaches are analytical, while others are more oriented toward observation or empirical data collection.

Table 13.1 presents some of these techniques, their general areas of use, and their respective advantages and disadvantages. The following discussion will consider these techniques and their application in more detail.

Task analytic methods

The task analytic method focuses on the use of cognitive task analysis to identify potential usability problems. This approach involves a systematic decomposition of the task into subtasks and the sequences followed during task performance. The focus of cognitive task analysis in usability applications is usually on (1) documenting the knowledge and information required to do the task, (2) documenting how and when the needed information is provided, and (3) predicting the time required.

Several different ways of doing cognitive task analysis have been developed for use in product usability-related applications, including GOMS (goals, operators, methods, and selection rules), NGOMSL (natural GOMS language), and the MHP (Model Human Processor).* GOMS and NGOMSL both emphasize developing a hierarchical description of how users can attain their product use-related goals. This hierarchical description describes the knowledge required to do particular tasks and can be used to predict how much time it will take. GOMS and NGOMSL descriptions also provide a way to quantitatively predict the learning needed to operate a product and the transfer of learning between similar products. GOMS or NGOMSL analysis also allows interface designs to be evaluated prior to the development of a system in terms of several qualitative measures. These include the following:

1. Naturalness of the design—Do the goals and subgoals make sense to a new user of the system?
2. Completeness of the design—Is there a method for each goal and subgoal?
3. Cleanliness of the design—If there is more than one method for attaining a goal, is there a clear and easily stated selection rule for choosing an appropriate method?
4. Consistency of design—Are methods for attaining similar goals consistent?
5. Efficiency of design—Are the most important goals accomplished by quick methods?.

The MHP is a more theoretically grounded approach that provides time predictions and expected ranges for simple cognitive tasks. The task analytic methods have the potential advantage of providing predictions very early in the design of a system. This follows because task descriptions can be developed and predictions can be made without testing or a physical prototype of the system. On the other hand, a GOMS or NGOMSL description *does not directly show whether a problem exists or how to fix it*. This step requires substantial interpretation on the part of the analyst and perhaps some additional testing. The GOMS or NGOMSL descriptions often correspond to how a skilled user is expected to do the task. Expert evaluation, user testing, or other methods may end up being a much easier and efficient way to identify errors and other usability problems for less skilled users.

Expert evaluation

Expert evaluation is one of the most commonly used methods of product usability analysis. This approach has the obvious advantage of providing strongly diagnostic information,

---

* Recall that GOMS and NGOMSL, and the MHP were discussed in Chapter 10.

Table 13.1 Advantages and Disadvantages of Existing Usability Evaluation Techniques

Evaluation Method	Example Tools/Techniques	General Use	Advantages	Disadvantages
Analytic/theory based	Cognitive task analysis GOMS	Used early in usability design life cycle for prediction of expert user performance	Useful in making accurate design decisions early in the usability life cycle without the need for a prototype or costly user testing	Narrow in focus; lack of specific diagnostic output to guide design; broad assumption on user's experience (expert) and cognitive processes; results may differ based on the evaluator's interpretation of the task
Expert evaluation	Design walk-throughs Heuristic evaluations Process/system checklists Group evaluation	Used early in design life cycle to identify theoretical problems that may pose actual practical usability problems	Strongly diagnostic; can focus on entire system; high potential return in terms of number of usability issues identified; can assist in focusing observational evaluations	Even the best evaluators can miss significant usability issues; results are subject to evaluator bias; does not capture real user behavior
Observational evaluation	Direct observation Video Verbal protocols Computer logging Thinking aloud techniques Field evaluations Ethnographic studies Facilitated free play	Used in iterative design stage for problem identification	Quickly highlights usability issues; verbal protocols provide significant insights; provides rich qualitative data	Observation can affect user performance with the system; analysis of data can be time and resource consuming

Survey evaluation	Questionnaires Structured interviews Ergonomic checklists Focus groups	Used any time in the design life cycle to obtain information on user's preferences and perception of a system	Provides insights into user's options and understanding of the system; can be diagnostic; rating scales can provide quantitative data; can gather data from large subject pools	User experience important; possible user response bias (e.g., only dissatisfied users respond); response rates can be low; possible interviewer bias; analysis of data can be time and resource consuming; evaluator may not be using appropriate checklist to suit the situation
Experimental evaluation	Quantitative measures Alternative design comparisons	Used for competitive analysis in final testing of the system	Powerful and prescriptive method; provides quantitative data; can provide a comparison of alternatives; reliability and validity generally good	Experiment is generally time and resource consuming; focus can be narrow; tasks and evaluative environment can be contrived; results difficult to generalize

Source: Adapted from Stanney, K.M. et al., Human–computer interaction, in Salvendy, G. (Ed.), *Handbook of Industrial Engineering*, 3rd edn., John Wiley & Sons, New York, pp. 1192–1236, 2001.

quickly, and at almost any stage of design (Box 13.9). On the other hand, experts like all people can make mistakes, are subject to biases, and may not notice all the problems. There are many different ways of obtaining expert opinions on product usability. Following is a brief discussion of some of the more commonly applied methods.

> ### BOX 13.9 USE OF PROFESSIONAL RATERS
>
> Most experts on the topic of subjective human judgment agree that a fixed panel of professional raters will provide more consistent ratings. A professional panel will be able to detect subtle differences that will often go unrecognized by the typical randomly selected rater. Some principal differences are related to experience and the role of self-selection or rejection. As in wine tasting, some people may be naturally better at it, and those who are poor at it will often refuse to participate on a professional panel. But that same person may be delighted to be a randomly selected rater just for the brief experience of doing something unusual. On the other hand, a professional panel of raters may be strongly biased by their experience. For that reason, the average ratings of a professional panel may be considerably different from ratings obtained from the public at large.

Design walk-throughs and reviews are perhaps the most commonly followed method of usability analysis. These approaches involve asking the expert to step through the basic steps required to operate the product and identify problems. In some cases, the process is completely open-ended. In others, the expert is asked to consider particular types of users or to evaluate a particular feature identified elsewhere as potentially troublesome. For example, in cognitive walk-through analysis, the expert is asked to focus on potential problems faced by a first-time user. The assumption is that such users will be learning as they interact with the system. The expert is then asked to simulate the user's behavior as they interact with the system. Consequently, a fairly complete description of the user interface is needed before a cognitive walk-through can be conducted. However, such analysis can be conducted using a very primitive prototype, such as a paper version of the interface. The focus of the analysis is on the following:

1. How evident will each required action be to the user?
2. What incorrect actions might be taken?
3. Will the user be able to correctly interpret the response of the system to both correct and incorrect actions?
4. If an error occurs, will the user be able to determine what to do next to correct the problem?

Normally, the expert will be provided a form to record these types of information for each step of the interaction. Other information that might be recorded includes the type of user assumed, how serious the identified problems are, the types of knowledge needed, and suggested corrections.

Heuristic evaluation and the use of checklists are closely related techniques where an expert (or in some cases a user) rates the usability of the product on a number of criteria. Heuristic evaluation is done in many different ways, but the basic idea is to obtain a *quick and dirty* analysis from one or more experts on how well the interface complies with accepted usability guidelines. Sample usability heuristics are shown in Box 13.10. Note that these examples are for interactive systems (normally computer). In practice, the first step in a heuristic evaluation involves identifying a set of perhaps 10 usability guidelines considered to be most important for the evaluated product. In some cases these heuristics might be identified from an earlier analysis of CAs for the product. Published guidelines, checklists, standards, or other sources are also useful sources.

Once a set of heuristics has been identified, a small number of *usability experts* then use the guidelines to identify usability problems. In many cases, these experts are actually potential users rather than true experts in human factors or product usability. Typically, each of the experts participates in a 1–2 h session in which they first use the system and then use the heuristics to evaluate each feature of the product or step in performance. The result of the analysis is a list of the problems identified and suggested solutions. The descriptions of each problem should include an explanation of why it is a problem. This explanation should also indicate which heuristics were violated, the nature of the violation, and its severity. In some cases, the severity of each violation is rated on a numerical scale. Debriefing sessions are often conducted after the analysis to help determine how to fix the identified problems.

To help guide this process Nielsen (1993) suggests that each usability problem should be plotted on a two-dimensional graph where one axis of the graph corresponds to problem severity and the second, to the number of people in the targeted-user population who are likely to be affected by the problem. When each problem is plotted in this way, the problems furthest from the origin are the most important problems and are good candidates for design improvements. Comparing a design that has a very important deficiency to designs that do not have the particular problem can also help inspire solutions. Another issue pointed out by Nielson (1994) is that a single evaluator will never be able to identify all of the usability problems. One way around this issue is to use multiple evaluators. As shown in Figure 13.4, the proportion of identified problems increases with the number of

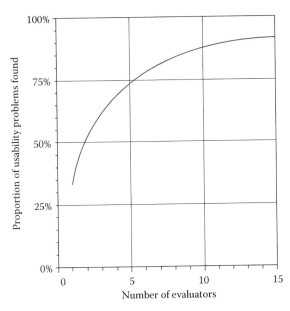

Figure 13.4 Proportion of usability problems found using various numbers of evaluators. (Reprinted from Nielsen, J., Heuristic evaluation, in Nielsen, J. and Mack, R. L. (Eds.), *Usability Inspection Methods*, John Wiley & Sons, New York, 1994. With permission.)

evaluators. However, there are diminishing returns. Typically, the first few evaluators find most of the problems. As a rough rule of thumb, Nielson estimates that five evaluators will find about 75% of the problems, but the exact amount will depend on how good the evaluators are and how complicated the product is. Various forms of cost-benefit analysis can be performed at this stage to help decide whether the expected benefit of adding one more rater is worth the cost of doing so.

Typical procedures in usability testing

Normally, usability testing begins by briefly telling the subject why the testing is being done. The product that is being tested is then described, and its use is demonstrated. At this stage, the experimenter may also explain to the subject how to record their preferences or choices during the usability test. Sometimes instructions are included that describe the product and how to use it. Determining the usability of the instructions is often part of the usability study. As part of this process, the subjects are normally asked to perform some simple task with the product in order to show that they understand the directions.

BOX 13.10   TEN USABILITY HEURISTICS FOR COMPUTER INTERFACES

1. Visibility of system status – Keep users informed about what is going on through appropriate feedback within reasonable time.
2. Match between system and the real world – Speak the user's language, with words, phrases, and concepts familiar to the user. Follow real-world conventions, making information appear in a natural and logical order.
3. User control and freedom – Provide a clearly marked *emergency exit* to leave the unwanted state. Support undo and redo.
4. Consistency and standards – Users should not have to wonder whether different words, situations, or actions mean the same thing.
5. Error prevention – Error prevention is better than good error messages.
6. Recognition rather than recall – Make objects, actions, and options visible. The user should not have to remember information from one part of the dialogue to another. Instructions should be visible or easily retrievable whenever appropriate.
7. Flexibility and efficiency of use – Accelerators—unseen by the novice user—may often speed up the interaction for the expert user such that the system can cater to both inexperienced and experienced users. Allow users to tailor frequent actions.
8. Aesthetic and minimalist design – Dialogues should not contain information that is irrelevant or rarely needed.
9. Help users recognize, diagnose, and recover from errors – Error messages should be expressed in plain language (no codes), precisely indicate the problem, and constructively suggest a solution.
10. Help and documentation – Documentation should be easy to search, focused on the user's task, list concrete steps to be carried out, and not be too large.

Source: Adapted from Nielsen, J., Heuristic evaluation, in Nielsen, J. and Mack, R. L. (Eds.), *Usability Inspection Methods*, John Wiley & Sons, New York, 1994.

These demonstrations are called walk-throughs. Once the subject is adequately oriented, the subject will perform a series of tasks. Performance of these task is observed and recorded. The final part of a typical testing procedure is a debriefing stage, where the subject's opinions are elicited. Many people who specialize in usability testing say that debriefing is often the most valuable part of usability testing. As part of this process, users like to be told how well they did. They also are often interested in finding out what effect the testing is expected to have on the ultimate product.

Some Methods of Collecting Data in Usability Studies

Hutchinson (1995) lists a variety of methods employed in usability studies in order of increasing usefulness:

1. Observation, directly or with TV cameras
2. Interviews
3. Dynamic mockups or prototypes
4. Questionnaires
5. Checklists
6. Rating Scales
7. Static mock-ups or prototypes
8. Time-motion studies

Many of these methods are part of the traditional assessment procedures one would naturally employ, but their ranked degrees of usefulness are somewhat surprising. While it is perfectly obvious that usability studies are crucial to product success because designers cannot possibly catch every important consideration associated with a product, more simple and inexpensive data collection methods should be employed first in the testing sequence. More sophisticated methods can follow when the design team approaches the final design phase.

It also should be mentioned that most large computer companies in the United States have invested in sophisticated usability laboratories that make it easy to observe people using interactive computer systems. Such facilities will often include one-way mirrors that allow the user to be nonintrusively observed, multiple video cameras, computers, and monitors, as well as a variety of other data collection devices (Figure 13.5). Somewhat paradoxically, the manufacturers of noncomputer consumer products may need more complicated setups to simulate the naturalistic settings in which their products are used. For example, a manufacturer of child products might have a playroom in which children are observed playing with the products they sell. While good usability studies can be performed without having a dedicated usability laboratory, making such a facility available certainly makes it easy to do good work.

The tasks in the sessions are often performed in several different orders by the subjects. This is done to control for possible learning effects. When learning is confounded with the task sequence, there is no way to completely separate out the effects of different variables from learning. Some experimenters advocate placing one or more common tasks throughout the sequence between the tasks of principal interest. Performance on the common tasks can then be fitted by a learning curve to the data of each individual subject, using the methods

Figure 13.5 Control room in a usability laboratory at Hewlett Packard Company. (Courtesy of Hewlett Packard Company, Palo Alto, CA.)

discussed in Chapter 5. The performance data for the principal tasks can then be adjusted using the learning curve. This approach assumes that objective data are being collected such as time or errors. Other (i.e., subjective) criteria cannot be as easily dealt with.

A principal purpose of usability testing is to determine the relationships between identified CAs and ECs. The testing usually does not seek a simple yes–no response. Instead, the usability testing measures how well specific ECs match CAs. To help answer such questions, it is often helpful to prepare a partial prototype of the product for use in these tests. While building a partial prototype may seem to be a formidable task, it usually is not difficult. Model builders have learned a number of tricks for making partial prototypes, including the use of modeling clay for molding small things, soft wood for larger objects, and foam core (i.e., a sheet with hard paper surfaces and a Styrofoam center) and a wooden frame for very large objects. Those materials can all be painted to resemble the final expected material. Model builders also use thick transparent plastic instead of glass and may insert flat pieces of lead into plastic, clay, and wood when the weight and feel of the product unit are important. Some model makers use heavy-weight aluminum foil to build up the shape of the partial prototype. A secret is to go to a local art shop and look around. A number of new plastic molding products (clays) are very malleable in their natural state but become fixed in a final form when heated a short time in an oven or microwave.

Usability testing also has a number of pitfalls to watch out for.* One of the most common is in the selection of subjects. It is easy to grab a few fellow workers to act as subjects in this testing. However, for a variety of reasons, conclusions based on the haphazard selection of a few subjects are tenuous at best. Fellow workers usually are not *representative* of the population of potential customers. If the subjects are *not* representative, the results cannot logically be generalized to that population. Another problem with using in-house subjects is that they tend to be strongly motivated to behave as they believe *good subjects* should act. Often they identify with the aims of the research and tend to behave

---

* Patrick A. Holleran (1991) provides an excellent discussion of many of these pitfalls in usability testing of computer software. See that paper for further details.

consistently with the hypotheses upon which the device or concept was developed. This case is often referred to as a self-fulfilling prophesy. Also, there is a danger in having the designer conduct the usability testing. Designers typically find it very difficult to present information in a positive but neutral manner and thus tend to bias the subjects' responses toward the newer design, or they overcompensate and bias subjects' responses away from the product being tested.

How Many Subjects?

Earlier in this chapter we discussed the issue of how many raters are needed in subjective trials and heuristic evaluations. This problem can be addressed analytically using approaches covered in Chapter 7. Such calculations will often reveal that a large number of subjects are needed to show statistically significant effects. From a practical perspective, the ergonomist may not have enough time to run a large number of subjects.

Fortunately, even a very small group of subjects is often enough. Ergonomists who work primarily in usability studies report anecdotal evidence that most of the severe problems associated with a product design are discovered by the first few subjects who try to use the design and that about 80% of the usability problems are detected with four or five subjects. Experiments involving a large group of subjects are normally needed only when the tested-for effects are less obvious.

Failure to avoid these pitfalls obviously reduces the validity of the experiment. In other words, the experiment becomes less likely to measure what it is supposed to measure. Valid results can be obtained only if the subjects of the experiment are adequately representative of the population of people for which the product is intended. Another requirement is that the tasks performed by the subjects in usability testing must be representative of the tasks the intended users would perform with the product being examined. Usability tests in laboratory settings typically focus on short, quickly performed tasks. Thus, laboratory tests usually leave the question of long-term usability unanswered and unevaluated. However, if there is a problem in short-term usage, severe problems are likely to occur in long-term usage. Consequently, laboratory tests are a good starting point, but they do not answer all the important questions.

Self-reporting[*] and *thinking aloud*[†] protocols pose a further threat to validity. Because people do not like to be thought of as inconsistent, subjects may *forget* initial attitudes to make them more consistent with their attitudes at the end of the study. Moreover, subjects may report effects of factors that do not appear to have any influence on their behavior (Nisbett and Wilson, 1977). For example, studies have shown that people's self-reports of how they use computers contrast sharply with the computers' internal records of their usage.

---

[*] Self-reporting techniques of information collection include questionnaires, check-off forms, and other schemes by which the subject directly collects the data. Nisbett and Wilson (1977) examined the question of validity in self-reporting and it has stirred a great deal of controversy.

[†] The technique of *thinking aloud* asks the subjects of an experiment to verbalize what they are thinking as they perform the tasks. L. Bainbridge (1990) describes the very effective use of this technique in studies of process control tasks. H.A. Simon has urged the use of the thinking aloud procedure, but many people question whether verbalization affects the use of the product and in turn affects the user's perception of the ease or effectiveness of use (Ericsson and Simon, 1984).

Many who perform ergonomic studies also worry that the thinking aloud protocol intrudes on other aspects of task performance. Several researchers* have reported that thinking aloud often affects, among other things, task performance time, the way a task is performed, and information retention. Verbal reporting during concurrent tasking tends to be less accurate when there is a heavy cognitive load imposed by those other tasks. That feature should not come as any surprise to you if you can remember driving and talking to someone else in the car and trying to keep on talking when the traffic suddenly gets very heavy. On the other hand, self-reporting protocols may be the only way to obtain an operator's knowledge of an operation. Perhaps, the primary lesson here is not that these self-reporting and thinking aloud protocols should never be used but rather that experimenters should avoid a total reliance on those techniques. Making a few auxiliary tests with the same subjects without the thinking aloud protocol to see what happens can be a good insurance policy. Other techniques can be embedded to provide consistency tests on those data. If other, more objective, sources of information are consistent with self-reports or verbal comments, the analyst obviously can be more confident in them.

Reliability is another important issue. A reliable test is one that provides a measurement on one occasion that correlates well with measurements gathered on other occasions. One way to test reliability in a questionnaire, for example, is to ask a second question later in the questionnaire that is redundant to one asked earlier. If the question is identical, the person responding will likely look back to maintain consistency, but a question asked differently but using the same logic as an earlier question can produce correlations that can be used to test reliability within persons. Another form of reliability mentioned previously is *interrater* consistency. When there is low agreement among, the data is likely to be inaccurate the tested people.

Interviews and postexperimental questionnaires

As mentioned above, Interviews or questionnaires are almost always part of the debriefing stage of usability testing. Some of the things to ask for in a debriefing interview include:

1. Any difficulties experienced during the trial testing with the product
2. A general evaluation of the product
3. An account of any improvements recommended for the product
4. Users' thoughts at the time they made an error
5. Users' own particular experiences with similar products

Some people prefer to be interviewed on a one-to-one basis because they feel embarrassed answering such questions in front of others. However, there are times when group interviews are useful. In group interviews, a statement by one interviewee can sometimes spark confirmation or extension by another person.

Some other useful suggestions on how to conduct an interview (see McClelland, 1990) are given:

1. Be courteous and exude a feeling of welcome.
2. Keep the users informed of what is to happen so they do not experience any surprises.
3. Take a positive lead in directing events but always make sure the user understands the questions.

---

* Ericsson and Simon (1980, 1984) and Russo, Johnson, and Stephens (1989) report such effects for *thinking aloud* protocols.

4. Ensure that users feel that you value their contributions.
5. Do not make users feel they are taking part in an inquisition.
6. Insure that users' judgments are actually based on their experiences during the testing rather than other experiences that might produce bias.
7. Be prepared to prompt and question the user in depth, but be careful not to badger the user. When a question is asked and answered, go on to something new. If necessary, return to that subject later.
8. Ask the user to make choices through rating or ranking scales.
9. When users are to make choices, employ contrasting design solutions to highlight differences.
10. Ask users to write down their views as a trial progresses.
11. Develop a clear structure for the interview; do not leave it open-ended.

The development of a good interview is similar to developing good questionnaire in many ways, as expanded upon in Chapter 14.

Designed experiments

In some cases, it is beneficial to design an experiment to systematically evaluate the factors that impact product quality and usability. A good example of a designed experiment directed towards such objectives was given earlier in Box 13.4. As shown there, a designed experiment provides a way to identify which of the studied factors have a significant effect on some variable of interest. The design of experiments involving human subjects differs from that of experiments on inanimate objects. People have intelligence and memories. They consequently react in negative ways when experiments appear to be too artificial or violate their expectancies of what should be happening. These complexities of human behavior must be kept in mind when planning usability experiments.

One of the most important issues is that it is necessary to conduct the experiment in a way that controls for the effects of factors not of primary interest to the study. Some principles for doing so are given in Box 13.11. At a more basic level, the designer must identify a set of independent variables and determine how to best manipulate them to measure their effects on one or more dependent variables, as expanded on below.

Independent variables

Experiments involving human subjects tend to use variables that fit within the Fitts' LIMET acronym. Recall from chapter 1 that LIMET stands for Learning, Individual differences, Motivation, Environment, and Task. Each element of LIMET identifies a general category of factors that might be considered in a usability study. Each of these categories can, of course, be divided into many subcategories. For example, factors falling within the category of individual differences that might be studied include the effect of age, experience, training, ethnic origin, gender, intelligence, visual and other sensory capabilities, time since last meal, general state of health, reaction time to audio signals, height, weight, and so forth. Many of these may be broken down into even finer categories. Generally, independent variables can be manipulated by the experimenter without affecting the other independent variables, but this is not always true. For example, two variables may be proposed as independent variables: the workload in station A and the workload in station B. If the input of station B is the output of

BOX 13.11   PRINCIPLES OF EXPERIMENTAL CONTROL

Principle 13.1

Make the instructions to the subjects very clear and consistent.

If there is a sensitive speed-accuracy trade-off effect between the tasks that are being studied, the subject must know exactly how to make that trade-off in order to render consistent data. Even minor voice modulations can send different messages to the subjects. For this reason, most ergonomic experimenters use one of the following methods:

Principle 13.2

Use written instructions or voice-taped instructions.

This technique may not prevent misinterpretations of instructions, but it will keep them consistent. Most human subjects in experiments try to follow the directions given to them, so it is important that the instructions are given consistently.

Principle 13.3

Avoid speaking to the subjects on any matters related to their conduct of the tasks.

Doing so will help avoid experimenter-induced shifts in subject attitudes.

Principle 13.4

Maintain within-subject variations of differences or, better yet, match subjects on as many features as possible.

In some scientific studies, the experimenters use identical twins to reduce variation, but this approach is usually restricted to medical research on family-related diseases. A major problem with attempts to reduce variability in the data by placing restrictions on subjects is that this approach reduces the population to which the results apply and otherwise makes the test results less believable. Within-subject contrasts of the new designs are good so long as the experience does not induce different behaviors on the part of the subjects.

Principle 13.5

Randomize the sequence of the experimental conditions tested in the experiment.

The old rule of thumb in experimental design is to randomize those factors that are not important to the testing so long as this does not affect subject behavior.

---

station A, then the workload in station B is nested into the workload of station A, and accordingly these two variables are not truly independent.

Another name given to independent experimental variables is *factors*. Each independent variable or factor has levels or values that are specified by the experimenter when planning the experiment. Perhaps the most obvious point is that the difference between the maximum and minimum levels of a factor should cover the range of interest in the experiment. Setting the maximum and minimum levels of a factor is not usually difficult because there typically is a natural range of operations. Setting the intermediate levels of a factor normally poses more difficulty. Every added level increases the experimental effort in an exponential manner, but, on the other hand, being too sparse limits the results that can be obtained experimentally. When a variable has two levels, one can only assume that the effect is linear. A third level would be needed to test linearity. A common practice is to use three levels per factor in the first experiment. Another common practice is to set the levels to be equally spaced unless there are compelling reasons to do otherwise. Sometimes unequal spacing allows one to view collection of data points as an embedded factor. In that way, constancies can be tested for the embedded factor.

But why not set a factor at a single constant level? On the plus side, setting a factor at a constant level and leaving it unchanged greatly reduces the amount of experimentation to be performed. However, one cannot logically infer anything at any other condition but that one. The point is, if you always operate that variable at the fixed condition and do not intend to change it, fix the level to only that condition. Otherwise, vary those conditions that are of interest and randomize the rest. Factors that are not of particular interest should be randomized unless randomization makes the experiment appear clearly artificial. For example, when presenting various conditions to a subject, one is not usually interested in the sequence of condition presentations. Therefore, choose the conditions in the sequence randomly. Since every subject gets a different sequence of conditions, any consistent differences in performance cannot be attributed to the sequence of conditions.

Dependent variables

In usability studies, the most common design goal is to attain a competitive advantage in the marketplace with the product design. Other objectives in usability studies include reducing training costs and/or customer support costs, raising productivity, improving user image through fewer errors, and enhancing the company's image. If any of these objectives are of concern to the design team, the dependent variables in the experiment should be designed accordingly.

Some typical dependent variables that might be measured in a usability study include:

1. Time to complete a task with alternative designs
2. Number of tasks completed in a fixed period of time
3. Number of user errors
4. Ratios of successes to errors
5. Time spent recovering from errors
6. Number of times the user was sidetracked
7. Number of features the user remembered after the study ended
8. Fraction of user statements that were positive toward the newer designs
9. Number of frustration exclamations used during the test
10. Number of times the user requested or needed help to perform a task with the product

It is easy to see that dependent variables such as those listed above are for the most part focused on performance. In other applications, the focus might be on measuring preferences, comfort, or aesthetic factors. Perhaps the most important consideration is that the dependent variable must be relevant to the usability issue that is being analyzed.

The sensitivity of the dependent variable is another important issue. Sensitivity describes how much the dependent variable changes when an independent variable is manipulated. For example, the dependent variable metabolism is quite sensitive to a change in the frequency of heavy-lifting tasks but quite insensitive to frequency changes in a cognitive task. The basic idea is that the dependent variable must not only provide relevant measure but also be sensitive enough to capture the effects of the independent variables, assuming they exist.

Basic experimental designs

Experiments involving human subjects differ depending on whether they manipulate the independent variables within the subjects or between the subjects. As a general rule of thumb variables with smaller effects should be manipulated within subjects because there are greater differences and more variability between subjects than within subjects. When all subjects experience the different levels of each primary independent variable, contrasts between those levels can be kept within each subject. Because there is less variability within subjects, the primary independent variables are more likely to show statistically significant effects in a within-subjects design. Between-subjects designs are often necessary when experiencing independent variable condition A influences how the person will behave in condition B. In this situation, the only way to cleanly determine the effect of each condition is to have one subject perform in condition A and another subject perform in condition B. The problem is that the potentially large differences between individual subjects must then be disentangled from the effect of condition A versus condition B. In other words, the error term becomes larger, reducing the power of the test in the between-subjects case.

To help decide between a within- or between-subjects design, ask yourself the question, "Will the person's performance change if one condition of the independent variable is performed before instead of after another condition?" If the answer is "yes" and the performance change is not what the experiment is trying to uncover, a between-subjects design should be used—that is, different subjects should be used in each condition. When the answer to the previous question is "no," then it makes sense to use the same subjects in multiple conditions. However, fatigue or boredom on the part of the subject may limit the number of conditions they can realistically participate in, forcing use of a between-subjects design. So in some cases, this is not an option.

Another important issue is that the experimental conditions must be adequately realistic in the mind of the subject. If they are not, the experiment will be perceived as a contrived situation. This sometimes creates a problem because it may be difficult to get large enough performance changes to show statistical significance without creating an artificial situation. A second strategy is to try to increase the observed differences in the dependent variables between experimental conditions. Some ways of doing this are:

1. Increase the range of the independent variables.
2. Increase the task difficulty.*
3. Use a more sensitive measure of the dependent variable.
4. Increase environmental stressors.

While all of these procedures can and do increase the chance of finding statistically significant effects, they also increase the chance that the results will have little practical significance. In particular, if the independent variables are outside their normal range,

---

* In physical tasks, one can increase the difficulty in a number of ways, such as by increasing the forces needed, increasing the frequency of exertions, or increasing the amount of torque needed to be overcome. For mental tasks, more things to process, more demands on memory, fuzzier discriminations, and more repetitions are all ways to increase task difficulty.

Chapter thirteen: Ergonomics of product quality and usability

the observed differences in performance differences are typically not of interest in applied ergonomics. Similarly, if the independent variable shows a significant effect only under extreme high-stress conditions, the finding will have little relevance to normal use conditions. From a practical perspective, this is a good reason to focus on good experimental control, as the latter approach focuses on reducing the noise in the data, rather than increasing the independent variables beyond their normal range.

In most ergonomic experiments, the question posed is whether mean performance differs between the two or more conditions being compared. In the simpler case where two means are being compared, say μ_1 and μ_2, the student t test measure, t, is simply

$$t = \frac{[u_1 - u_2]\sqrt{N}}{\sigma} \tag{13.1}$$

where
   N is the number of data points
   σ is the standard deviation of the dependent variable values

This student t-statistic is statistically significant with a sufficiently large t value.* Obviously, the t-statistic is shown by Equation 13.1 to be larger when there is a greater difference between means, a larger N, or a smaller σ. The size of the standard deviation σ depends upon within- and between-subject differences, variation in the measurement system, and other variations in the experiment. Better experimental control is one way of reducing the standard deviation of the dependent variable. Doing so, makes it easier to find statistical significance. That is, the difference $(u_1 - u_2)$ does not have to be as large to be statistically significant.

Final remarks

Product quality depends on how well the features and properties of the design match the needs and wants of the users. If a product fails to meet the customer's needs and desires, sales will fall and eventually the product dies. History is full of such cases. However, quality is multifaceted and so it is difficult to generate an overall strategy that will assure good product quality during the design process. Hauser and Clausing tried to develop such a strategy and succeeded in part. Their current strategy is limited in its ability to cope with complexity, a challenge to future research. In the meantime, complex products need to be partitioned into components so that a strategy similar to Hauser and Clausing's can be followed in which the interconnections of components are carefully managed throughout the design. No matter what strategy succeeds, usability testing conducted by ergonomists will remain an essential feature of quality assurance efforts in the foreseeable future.

Discussion questions and exercises

13.1 How do the well-known advocates of product quality define the term and how does their definition fit with usability studies and quality control as these terms are generally known in industry?

---

* A table of critical student t values is given in Appendix B.3.

13.2 Explain the columns and rows in the House of Quality. What is the relationship matrix? Explain how the correlation matrix can facilitate engineering creativity.

13.3 Critique the Hauser–Clausing House of Quality concept from positive and negative perspectives.

13.4 Who were some of the early people involved in the ergonomics of customer products and what did they do?

13.5 What are the advantages of using a wide variety of people as raters instead of a fixed panel of professional raters?

13.6 Why is it recommended that people who make ratings between alternative designs be given adequate time to fully experience the new device before being asked to rate it?

13.7 Why is it advocated that in usability studies one should (a) ask users to shift between the old and the new device in evaluation, (b) not ask users to state their preferences in the presence of other users in the study, and (c) sequence gradual changes between successive items to be rated?

13.8 What is the difference between a partial and a full prototype?

13.9 What do the terms *walk-throughs* and *debriefing* mean in usability studies?

13.10 Do better ergonomically designed products sell better?

13.11 Conduct a heuristic evaluation of a product you are highly familiar with. Explain why each heuristic was chosen.

13.12 Compare and contrast the cognitive walk-through method with the task analytic methods of usability analysis.

13.13 What are some evaluation methods used in usability studies? What are the advantages and disadvantages of each method?

13.14 How many subjects and/or raters are needed to develop valid and reliable conclusions? How does this change depending on the stage of the product design cycle? How does this change in designed experimental studies versus other situations?

13.15 What types of questions need to be answered in user-centered or customer-driven design? What procedures need to be performed? Discuss the role of ergonomists for each procedure.

13.16 Use the generic framework (Box 13.6) for identifying customer requirements to identify a complete set of CAs and ECs for a product you are very familiar with.

13.17 Why is better experimental control important? What are some ways of attaining this objective?

13.18 How do methods of identifying customer requirements differ from methods of specifying design requirements?

chapter fourteen

Questionnaires and interviews

About the chapter

The objective of this chapter is to introduce the reader to the problem of constructing questionnaires and designing interviews as information-gathering tools for ergonomic design. In many cases, the designer needs to determine the attitudes, concerns, objectives, and goals of various subpopulations of people on the use of a particular product. In most cases, the questionnaire can be used to gather such information economically. In other cases, questionnaires are used after experiments to collect subjective opinions of the experimental subjects during or subsequent to the experiment. Although questionnaire data or interviews of persons leaving an experiment may not be viewed as hard evidence, subjective appraisals are infinitely better than no evidence and sometimes are the only available evidence.

Introduction

A questionnaire is a list of questions presented to prospective respondents. The purpose is to collect information and sometimes to measure statistical accuracy. Since a large number of questionnaires can be distributed and the results can be easily summarized, the principal appeal of this technique is its relatively low cost in quickly gaining access to many people. However, the skill needed to construct a useful questionnaire can be substantial. A well-designed questionnaire requires careful planning and consideration of potential problems that may arise.

One of these concerns is the validity and reliability of the responses. Both of these features are difficult to check unless the questions relate to highly factual and noncontroversial topics. In the latter case, the correctness of factual responses can be tested to examine response validity. The developer of the questionnaire must also take steps to assure that the answers to the questions asked are consistent.* If the results are inconsistent, any conclusions based on the questionnaire data are questionable. In addition, the questions asked must not bias the respondents. The so-called *halo* and *horn* effects are a special concern. The halo effect occurs when people view a person or situation so positively that they give high ratings across the board, even for weaker attributes of the rated person or situation. For example, a supervisor who thinks one employee is a *star* might give that employee a higher performance rating than he or she deserves. The horn effect is an exactly opposite tendency to give poor ratings even when they are not deserved. Quite obviously, such bias must be minimal to develop valid conclusions.

Interviewing techniques are much more flexible than questionnaires. Questionnaires can be prepared that allow the respondents to skip irrelevant questions. However, that is not the same as the flexibility one has in an interview. During an interview, one can explore the unexpected in great detail. Also, it is difficult in questionnaires to assure that the

---

* Surveys and questionnaires often ask the same question more than once, in slightly different ways. Inconsistent answers to the related questions raise a flag, indicating that respondents are not responding carefully or are having trouble understanding the questions.

respondent understands the question. A question can include a *do not understand* response category, but there is always the possibility that subjects will be reluctant to admit they do not understand. During a face-to-face interview, it is relatively easy to determine if the respondent understands your question, so restating or explaining the question to the subject can largely avoid any problems.

Another issue is that questionnaires are obviously designed to be answered by a single person. Interviewing techniques allow more than one person to be studied in a group setting. Group-oriented interviewing methods include focus groups, brainstorming groups, the Delphi method, and the nominal group method. Focus groups and brainstorming groups can provide a lot of detailed information, in a single session, which saves time and money in data collection. Nominal group and Delphi methods provide some additional advantages in specialized applications. These relative advantages and disadvantages of questionnaires and interviewing techniques will become more apparent from the following discussion.

Questionnaire design

As noted earlier, a well-designed questionnaire requires careful planning and consideration of potential problems that may arise during its use. This means that a systematic procedure should be followed during the design and development of the questionnaire. Some of the steps normally followed in questionnaire design are as follows:

1. Planning the questionnaire
2. Developing the sampling procedure
3. Constructing the questionnaire
4. Pilot testing
5. Data collection and analysis

The following discussion will consider each of these steps in some detail.

Planning the questionnaire

There is much to consider in planning a questionnaire. The first principle in questionnaire planning and design was stated clearly by Sinclair[*]: *Know thy respondent!* Simply put, one should define clearly the population of respondents and their culture, language, and knowledge base. The investigator must be aware that population differences can alter the results. Any potential effects of this type must be investigated. If such effects appear to be a possibility, the questionnaire must be designed to minimize their influence. As discussed later, special steps may be taken to stratify the sample to either test the effect of population factors or simply ensure the sample is representative. In some cases, the questions may have to be reworded or modified for a special population. Also, recognize that the sampled population is probably not homogeneous. Even with excellent planning, the questionnaire may miss part of the population, but it will miss much more without adequate planning.

Another extremely important part of questionnaire planning and design is to define the objectives of the questionnaire in precise (ad nauseam) detail. The person writing the questions will probably have a rather narrow scope in mind for a particular question. For example, the analyst may be interested in how often the respondent wears their seat belt

---

[*] See Sinclair (1990) for a more comprehensive discussion regarding development of questionnaires and interviews.

when driving a car in the United States. However, the respondent will not know this unless told. In this example, if the respondent was a recent immigrant from Turkey (where people tend to not use seat belts), they might give a very different answer for what they do in Turkey, instead of in the United States. If the question did not specifically state it applies only to the United States, some of the subjects may interpret the question inappropriately, leading to noisy data.

Respondents cannot ask; they can only assume. If that assumption is wrong, there is usually no way for them to know. That way, bad data can infiltrate good data, and no one knows which is which. To help avoid such problems, the reason for each question needs to be identified, along with the possible responses, and the accuracy and precision needed. The analyst also should identify the relationships between particular questions and those used in other questionnaires, interviews, or other studies. About one-third to one-half of the questionnaire preparation time should be spent on detailing objectives. If the questions are more difficult to write than was expected, then more planning time is needed.

Sampling procedure

Having an appropriate sampling procedure is one of the most important elements of any study involving use of a questionnaire. To ensure this objective is met, the designer must first determine the appropriate population to sample from. Doing so helps determine what sampling method is most appropriate. Obviously, one does not want to sample subjects outside the defined population. In rare cases, the population is small enough that all members can be sampled; however, the normal practice is to sample a subset of the population. Totally random sampling is a desired goal, but it is difficult to achieve.* Some forms of nonrandom sampling can create a systematic bias in the results. This happens when greater proportions of some subsets of the population are sampled than other subsets. If that subset represents behavior that is distinctive, then the results will not accurately describe the total population. Some convenient methods of sampling can create significant biases. For example, social security numbers in the United States are often assigned to regions at different points in time. If a systematic sampling of these numbers is used, such as every 1/12 number, the results may be biased. It is important to check these possibilities. A classical case of biasing caused by the sampling method occurred in a U.S. presidential election about 50 years ago. Voters were asked over the telephone which presidential candidate they voted for. A winner was declared on the basis of the telephone survey.[†] Major newspapers across the country reported the results, as fact. Somewhat embarrassingly, this prediction turned out to be wrong. At that time, many less affluent voters did not have telephones, but they had votes.

Nonresponses are another important concern. Surveys sent to busy individuals are likely to end up in paper recycling. The respondent's first reaction to the survey usually determines what happens. As the advertisement says, you never have a second chance to make a first impression. Lost or missing responses often create systematic biases. For example, some people spend a lot of time traveling away from home or the office. When frequent travelers get their mail, they tend to go through their mail very quickly and efficiently. A survey that requires little, if any, time is likely to be thrown away, unless the recipient can see significant benefits from filling it out. *The questionnaire must not require more time than can be reasonably expected from the respondent*. It is also important that the respondent be shown the potential significance of the survey results or be rewarded in

---

* Chapter 7 discusses sampling methods in more detail.
† President H. S. Truman was the true winner.

some way for participating. If it is at all possible, the sampling plan should keep track of who responds and who does not respond. This practice helps determine if there is something special about the people who do or do not respond. Phone follow-ups are also common practice, for the same reason. Follow-up calls and reminder surveys are also used to increase the sample size.

Constructing the questionnaire

Questionnaires can be thought of as having four major segments. The *prologue* is to the questionnaire as the appetizer is to the meal. This part introduces the respondents to the subject of the questionnaire, motivates them to answer the questions, and provides any needed clarification. In some cases, a section may follow in which the respondent signs off or consents to use of the data provided. The next section is either the main body of information-seeking questions or questions about the respondent. The latter section typically inquires about the respondent's age, gender, experience, and other background data. The information gathered in this section is often called *respondent demographics*. If there is a need to direct some respondents into alternative parts of the information-seeking section, it is usually better to place this section before the main body of questions. For example, you may have seen filter questions of the type that say, "If you have been a victim in this kind of accident, go directly to question Y." It is particularly important to arrange questions in the information-seeking section in a logical manner from the *respondent's point of view*. The final section of a questionnaire is the *epilogue*. Here is where one thanks the respondent or states the reward to come following the response. If there are special instructions for returning the completed questionnaire, they can be included in the epilogue. In the questionnaire-meal analogy, the epilogue is the dessert.

The information-seeking section is clearly the *pièce de résistance* of the questionnaire. The questions in this section must be clear and short, and the number of questions must be kept within reasonable bounds. Factual questions with numerical or yes–no answers are relatively easy to construct and can normally be answered quickly. It is normally better to design the questions so that the responses are mutually exclusive (where one and only one answer can logically be given), but multiple nonexclusive (e.g., denote each applicable category) responses are sometimes allowed. If there is a large number of responders, the answers will probably be later entered into a computer for analysis. The data files for such analysis are much simpler if the responses are mutually exclusive.

When the answer to the information-seeking question is a subjective rating or a ranking, then there are several other points to consider in the construction of the questions. First, the question should not bias the respondent's answer. When a respondent is requested to rate agreement or disagreement to a statement such as, "The instructor is quite smart for a college professor," bias is injected into the question. The point is, the attitudes of the responder are of interest, not that of the questionnaire designer.

A further recommendation is that questions should be precise and unambiguous. Stay away from ambiguous questions or, as they say in industry, "Keep it simple!" In particular, keep the language similar in style to that found in daily newspapers. Keep the questions short! Also avoid imprecise words such as *frequently, typically, usually, somewhat,* or *approximately*. It is better to provide some examples for the responder rather than to get mired down in ambiguous words.

Still another suggestion is to avoid *prestige bias* or *leading questions*. The former occurs when respondents are given a chance to brag invisibly such as, "You clearly take the *Wall Street Journal*?" The respondent may view the negative answer as an admission of being uninformed. Leading questions push the respondent to a preconceived answer. An obvious example is, "Don't you think that the policies of President Frank G. Schwarzengrubber are unjust?" Finally, hypothetical questions should be avoided. Questions that start out, "If you were faced with the situation where Pauline was tied to the railroad track, a locomotive was heading in her direction, and you were the only person in sight." Even a reasonably stated hypothetical question invites an unreliable response.

Some experts recommend wording some of the questions positively and others negatively within the questionnaire. Proponents for this approach suggest that it controls some of the potential bias due to wording of a question. Wording the same question in opposing ways provides a means of testing whether responses are consistent. This approach may also force the respondent think more carefully about their answers. Opponents to this approach say that it makes the questionnaire more difficult to answer and it may cause the responder to drop out or to continue in a confused manner. Including multiple wordings of the same question also increases the number of questions, which will increase the time required and may discourage subjects from answering all the questions carefully. There obviously are not any perfect solutions to this dilemma. Perhaps the best advice is to evaluate positive and negative wordings of questions during pilot testing of the questionnaire. If concerns continue to be present, multiple questions with alternative wordings can be used in the final version of the questionnaire to provide consistency checks for important items.

Questionnaire response scales

The response part of a questionnaire is where the respondent answers the question posed. At the most general level, the questions can be either open-ended or structured. The answers to open-ended questions are usually unrestricted, except in length. Open-ended questions are very difficult to analyze, so they are normally used in the exploratory phase of a study. Structured questions limit the answers to alternatives presented to the respondent. The answers provided to the questions correspond to different types of measurement scales. These include what are called nominal, ordinal, interval, and ratio measures. A nominal measurement scale simply assigns a name to the response. For example, the subject might give their name or say "yes," "no," or "don't know" in response to the question. In an ordinal measurement scale, the possible answers form a ranking from a low to high value on some measurement criteria. For example, a rating of *strongly agree* is higher than *slightly agree* on an agreement scale (Table 14.1). In an interval scale, the distance between measurement units is constant. This type of measurement is illustrated by the Celsius temperature scale (i.e., a temperature of 20°C is 10° warmer than a temperature of 10°C and 10° cooler than 30°C). In a ratio scale, the ratio of measurement units is meaningful. For example, a probability of 1 is twice as much as a probability of 0.5. Note that this is not true for an interval scale (i.e., a temperature of 20°C is not twice as warm as a temperature of 10°C).

Responses to the questions on a questionnaire may be measured on any of these measurement scales. It is important to consider the measurement scale carefully during the early part of questionnaire development because the scales used constrain the types of analysis that become appropriate and the types of conclusions that can be drawn from the data. When the responses are on nominal scales, chi-square analysis becomes appropriate. Ordinal rankings can be evaluated with nonparametric tests such as the Wilcoxon sign test and Spearman rank correlation coefficient. Interval and ratio data are often evaluated with *t*-tests and analysis of variance (ANOVA), even though the latter practice is controversial when the data are on an interval scale. Some illustrations of these approaches will be given later in this chapter.

In the case of an open-ended question, space is provided to write in an answer. For example, the question might ask: "Do you have any suggestions for improving the

Table 14.1 Example of a 5-Point Response Form Plus a "Does Not Apply" Category

1. The agency helped me to find a job that matched my interests and abilities	Strongly agree	Agree	Neutral	Disagree	Strongly disagree	Does not apply
2. The agency helped me to deal with financial and transportation concerns	Strongly agree	Agree	Neutral	Disagree	Strongly disagree	Does not apply
3. The agency had resources available for me in non-traditional occupations	Strongly agree	Agree	Neutral	Disagree	Strongly disagree	Does not apply
4. My general experience with the agency was satisfactory	Strongly agree	Agree	Neutral	Disagree	Strongly disagree	Does not apply

process?" Quite often it is a good idea to give the respondent some indication of how long of an answer is appropriate. For structured questions, the possible answers are normally shown next to each question. In some cases, space is also provided for writing in an explanation of why the particular answer was given.

Many questionnaires are focused on measuring attitudes. Such questions often provide a statement followed by a measurement scale on which the respondent indicates how strongly they agree with the statement. A typical example is shown in Table 14.1. Note that respondents can also indicate that the question does not apply to them. When a higher degree of precision is needed than the 5-point system provides, it is easy to step up to a 7-point or a 9-point system. All that is needed is to add more shadings of agreement and disagreement, such as *absolutely disagree* and *agree*, *very strongly disagree* and *agree*, *disagree* and *agree a lot*, and *slightly disagree* and *agree*. Any of these ranked response systems provides a balanced response pattern. Also, these systems provide a center point where the responses on each side provide symmetry. However, as mentioned previously, the intervals between the indicated values on the scale may not be equal. This creates difficulty in analysis because it is not possible to calculate a mean value without knowing how far apart each of the values is. Also, different people may interpret the scales differently. To reduce response uncertainty and responder-to-responder variability, one can give some examples to define the ratings, especially extreme ratings. Anchors are statements at the end of a rating scale that specify the extreme condition envisioned for such a rating.

Another helpful approach is to assign numbers to each response. So for the example of Table 14.2, the *strongly agree rating* might be assigned the number 5. Table 14.2 illustrates part of a questionnaire showing a numbered 5-point response system. With this response form, the respondent is requested to darken the appropriate box to the right of the question. The responses are coded as integers. Note that a machine can be used to read and record the responses, which is an important consideration when a very large number of

Table 14.2 Example Response Form with Five Ratings

1. Do you follow diet in order to lose weight?	1	2	3	4	5
2. Do you avoid eating when you are hungry?	1	2	3	4	5
3. Do you eat diet foods?	1	2	3	4	5
4. Do you avoid foods with sugar in them?	1	2	3	4	5
5. Do you display self-control around food?	1	2	3	4	5

Chapter fourteen: Questionnaires and interviews

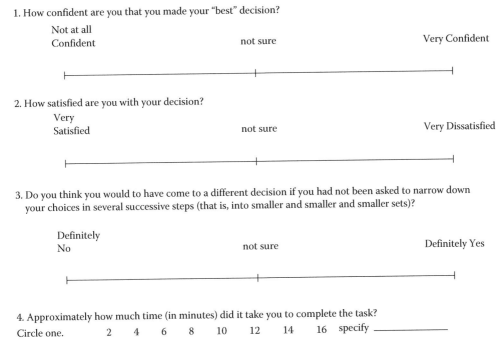

Figure 14.1 Illustration of a 10-level response form.

questionnaires are being considered. It is interesting to observe that the latter scale shows a number, without any verbal description.

A better approach is illustrated in Figure 14.1. As shown in the figure, the responses are marked on a horizontal line, which shows the midpoint and endpoints of the scale. If there are anchor conditions, key anchor words can be placed at the appropriate locations. Most people also suggest putting a vertical line in the middle with an appropriate neutral anchor word. When there are intermediate conditions, other vertical lines can be placed there to create proportional spacing. Respondents should be instructed to mark the line proportionately with their rating. Distance measurements from one end of the line provide a continuous measurement rating, which can be described with continuous statistical distributions (in some cases, the normal distribution) provided that the respondent was told to mark the line proportionately to their rating—that is, it is more reasonable to assume that the responses correspond to a ratio measurement scale. This approach gives the analyst a more reasonable basis for using parametric statistics (i.e., ANOVA and t-tests) to make statistical statements about the collection of responses made.* For example, means and confidence levels can be determined. Also, the responses from different subsets of the population can be compared to determine if those subpopulations differ with statistical significance.

---

* It should be stated here that many nonparametric statistics are excellent analytical devices, but they clearly are not as powerful as parametric statistical methods. The point here is try not to create a questionnaire that requires the use of nonparametric statistics during analysis.

Table 14.3 Rating Form with Numerous Response Categories

Rate the items below on the following 5-point scale				
Disagree strongly	Disagree a little	Neither agree nor disagree	Agree a little	Agree strongly
1	2	3	4	5

I see myself as someone who…

1. Is talkative _____
2. Tends to find fault with others _____
3. Does a thorough job _____
4. Is depressed, blue _____
5. Is original. Comes up with new ideas _____
6. Is reserved _____
7. Is helpful and unselfish with others _____
8. Can be somewhat careless _____
9. Is curious about many different things _____
10. Is full of energy _____

In practice, analyzing or interpreting the marks made by respondents on a line becomes quite tedious. So, in many cases, questionnaire designers compromise and ask their subjects to write in the numbers instead of marking them on a scale. An example along these lines is shown in Table 14.3. This form is also much more compact than the preceding examples. Another issue is that most experienced designers of questionnaires agree that attitude or subjective evaluation responses should be specified using an odd number of categories so that there is a distinct center point on the scale, where the scale is symmetrical about that center point. However, it is not hard to find examples where researchers violate this rule of thumb.

Not all requested information on a questionnaire is in the form of ratings or rankings. Much of the desired information will often be categorical in nature (i.e., gender, location, tool, or product used, etc.). Other data may be naturally described on a ratio scale (i.e., age, frequency of using equipment, etc.).

Demographic information Demographic information is almost always collected. An example form for collecting demographic data is shown in Table 14.4 that includes questions about the person's age, educational background, and home location. Note

Table 14.4 Form for Collecting Demographic Data

1. What country do you live in? _____
2. What is your age, in years? (check one)
    _____ a. Under 25
    _____ b. 25–40
    _____ c. Over 40
3. What is the highest level of education you have completed? (check one)
    _____ a. Less than grade 12
    _____ b. High school diploma or GED
    _____ c. Some college
    _____ d. Certificate
    _____ e. Associate's degree
    _____ f. Bachelor's degree
    _____ g. Graduate degree

that answers to some questions, such as home location, require simple, open-ended responses. Other answers are selected from categories specified in the form. In the latter case, the questionnaire designer must carefully consider the precision desired in the answers.

Collecting demographic information is sometimes vital because many questionnaires are focused on determining differences in ratings due to age, gender, or other background differences. This is especially true if the questionnaire pertains to a new product to be provided to the public or when a new public policy is being considered. In such cases, the ergonomic designer needs to first identify the reasons for negative or positive reactions within subpopulations concerning existing products or policies. Once this has been done, designers can work on correcting the reasons for dissatisfaction, while maintaining the features or conditions people desire. A variety of techniques are used to analyze demographic differences in rating or rankings. This issue will be addressed further after briefly discussing the role of pilot testing.

Pilot testing and data collection

Perhaps no part of the development of a questionnaire, except defining the objectives, is more vital than pilot testing. Untested questionnaires are bush league. Sinclair (1975, 1990) recommends three forms of pilot testing. First ask a colleague, or friend, to criticize and comment on the preliminary questionnaire. After the problems have been corrected, send the revised questionnaire to about 10 respondents. After these respondents have had a chance to examine the questionnaire, contact them. Then ask them to elaborate on how they perceived each question and to elaborate on the precise meaning of their answers. After corrections have been made to the revised questionnaire, send the revised version to a larger sample of respondents. When these questionnaires are returned, they should be checked for meaningless patterns of answers or invalid responses. This procedure should be repeated until there appear to be no further errors.

The last steps in the process will also indicate the expected degree of nonresponses. If one plans to follow up the nonresponses by telephone, the potential workload involved must first be estimated. Furthermore, one must know who responded and who did not. Of course, responders will likely suspect that there is no anonymity if they see that the questionnaires have an identification code. That probably will not cause any problems if the questionnaire is well designed and the sender has a reputation of being discrete, unless the subject matter is of a highly personal nature.

If the sampled population is very large, you will likely want to tabulate the responses on a computer as the returns come in. This practice has two advantages. Computerized tabulations can be quickly checked for out-of-range and illogical responses. Most importantly, corrections can be easily made. Respondents sometimes put responses in the wrong space or give an answer in incorrect units. For example, weights of 220 lb, 100 kg or 15.7 stones are all equal, but if these numbers are recorded incorrectly without accounting for the measurement units, huge discrepancies will occur in the data. The raw data consequently must be examined for consistency, and will have to be corrected where necessary.

Another issue is that some respondents may fill in the questionnaire without a lot of thought or may even give deceptive answers. Such data should obviously be removed. It requires some effort to give consistently misleading answers, so including a few redundant questions as consistency checks may be sufficient to identify lazy or careless respondents.

Data analysis

After correcting data errors or removing suspicious data, the next step is to analyze it. When one can anticipate the kinds of analyses that need to be done before tabulating the responses, those data can be placed in matrices to enable calculations without further manipulations. At this stage it may be useful to rescale all questions so all responses are directionally consistent. This consistency-adjusting process allows the analyst to generate statistics about the responses without the negative–positive shifts in the question construction affecting the results. While means and standard deviations of responses from individual questionnaire questions can be reported without this adjusting process, those statistics are inappropriate for groups of questions unless they are all currently in the same positive–negative orientation. This feature is important because questionnaires often contain sections of questions on specific topics. Reporting statistics without adjusting for directional effects increases the variances and tends to bring the means toward the neutral ranking.

The following discussion briefly introduces some of the more common forms of analysis performed on questionnaire data. Note that ANOVA is also commonly used to analyze questionnaire data. For example, categorical demographic variables might be independent variables in an ANOVA where subject rating scores are a dependent variable. However, since use of ANOVA is discussed elsewhere (see Chapter 9), it will not be covered in this chapter.

Chi-square analysis

Chi-square analysis is one of the most commonly used methods of analyzing data from questionnaires. This approach involves comparing the observed frequency of responses to the expected frequency for different stratifications of the data. For example, the ergonomic designer might be interested in comparing seat belt use of drivers and passengers. The Chi-square (χ^2) statistic is calculated for a sample by first finding the difference between the observed and expected frequencies for each category. The differences are then squared and divided by the expected frequency. The obtained value is then compared to tabled values of χ^2 to determine whether it is possible to reject the null hypothesis at a given level of statistical significance. The null hypothesis is that the observed frequency is equal to the expected frequency.

The value of χ^2 is calculated by summing values over the k categories as shown here:

$$\chi^2 = \sum_k \frac{(f_e - f_o)^2}{f_e} \tag{14.1}$$

where

f_e equals the expected frequency
f_o equals the observed frequency for the respective categories

The degrees of freedom are equal to $k - 1$.

To illustrate this approach, assume that a survey was done in a shopping mall involving 1000 people. The first question asked whether they drove or rode as passenger to the mall. A second question asked if they were wearing their seat belt. In 700 cases, the person filling out the survey said they were wearing their belt. A total of 700 people said they were the driver; 450 of these people said they were wearing their belts. A total of 300 people said they were riding in the vehicle as a passenger; 250 of them said they were wearing their belts. The null hypothesis is that drivers and passengers are just as likely to wear their seat belts. The expected frequency for drivers is consequently $0.7 \times 700 = 490$; for passengers, the expected frequency is $0.7 \times 300 = 210$. The calculated value of χ^2 is

$$\chi^2 = \frac{(490-450)^2}{490} + \frac{(210-250)^2}{210} = 10.88$$

Note that since we are comparing drivers to passengers, the number of degrees of freedom is 1. The null hypothesis can be rejected at the $\alpha = 0.01$ level, for a single degree of freedom, if the calculated value of $\chi^2 > 6.63$ (see Appendix B.4). Since we obtained a value of 10.88, the null hypothesis is rejected.

More advanced analysis involving categorical variables can be performed using statistical packages such as SAS. The latter programming environment includes a program called CATMOD that can be used to develop a logistic regression model, where all the variables are categorical. Returning to the previous example, assume a third question on the survey recorded the gender of the observed person. A logistic regression model could be used to predict the proportion of people wearing seat belts as a function of gender and whether they are driving or riding. For further information on CATMOD and related topics, the reader is encouraged to review the SAS manual.*

Spearman rank correlation coefficient

One important question that frequently arises is whether the answers to different questions in the questionnaire are related. If the answers are measured on an ordinal scale, standard measures of correlation may be inappropriate, and rank order measures should be used instead. One of the oldest procedures for calculating rank correlation involves calculating the Spearman rank correlation coefficient r_s within respondents. Using this computational method, let X be the rating on one question and Y be the rating of the other by the same respondent. Suppose that N respondents each had a score for each of those two questions as shown here in Table 14.5.

After all of the responses have been recorded, they can be ranked over the five subjects as shown in Table 14.5. Columns four and five show those rankings. Note that all subjects with equal X or Y values receive the same rank. For example, respondents 2, 3, and 5 all had X responses of 3, and so all three share the same rank on X. Also note that ties are assigned a rank by averaging the ranks of the tied items. Returning to this example, respondents 2, 3, and 5 each receive a ranking value of 2 on X. The latter value is the average of the

Table 14.5 Example Data for Calculating the Spearman Rank Correlation

Respondent	X Response	Y Response	X Rank	Y Rank	Difference
1	4	3	4.5	5	−0.5
2	3	1	2	1	+1.0
3	3	2	2	3	−1.0
4	4	2	4.5	3	+1.5
5	3	2	2	3	−1.0

---

* The SAS version 8 manual is available online from the SAS Institute Inc. at http://v8doc.sas.com/sashtml/

rankings 1, 2, and 3 that would have been assigned if the values were slightly different. Respondents 1 and 4 are also tied on X, and receive a ranking of 4.5, which is the average of 4 and 5. The next step is to find the rank difference as shown in the last column to the right in Table 14.5, where the convention is

$$X_i - Y_i = \text{the difference or } d_i$$

The Spearman rank correlation coefficient* is

$$r_s = 1 - \frac{6 \sum_{i=1}^{N} d_i^2}{N^3 - N} \quad (14.2)$$

when there are unique ranks in X and in Y, or at least very few rank duplications. Box 14.1 shows an example calculation of a Spearman rank correlation coefficient using Equation 14.2. If there are many duplications of ranks, adjustments need to be made. The correction for rank duplications in X and in Y is

$$T_k = \frac{1}{12} \sum_{i=1}^{k} (t_k^3 - t_k) \quad (14.3)$$

where t_k is the number of rank ties of a given rank in the X category. Note there may be more than one group of tied rank, and so the k index specifies each group. With the substitution of Y for the subscript of T, Equation 14.3 applies to category Y as well. Then the sum of squares for X and Y is

$$\sum_{i=1}^{N} x_i^2 = \frac{N^3 - N}{12} - \sum_{k=1}^{t_{i,k}} T_k \quad (14.4)$$

$$\sum_{i=1}^{N} y_i^2 = \frac{N^3 - N}{12} - \sum_{k=1}^{t_{i,k}} T_k \quad (14.5)$$

The Spearman rank correlation coefficient is

$$r_s = \frac{\sum_{i=1}^{N} x_i^2 + \sum_{i=1}^{N} y_i^2 - \sum_{i=1}^{N} d_i^2}{2 \sqrt{\left[\sum_{i=1}^{n} x_i^2 \right] \left[\sum_{i=1}^{N} y_i^2 \right]}} \quad (14.6)$$

With small sample sizes, tied ranks significantly inflate the apparent correlation, as the example calculation in Box 14.2 demonstrates. The statistical significance of a sample

---

* Sidney Siegel describes the Spearman correlation coefficient and provides numerical examples on pages 202–213 of his book (Siegel, 1956). Other related measures include Kendall's rank correlation, a partial rank coefficient, and a coefficient of concordance.

BOX 14.1 EXAMPLE OF CALCULATION OF THE SPEARMAN RANK CORRELATION COEFFICIENT

In the numerical example in Table 14.5, the rank differences between X and Y are given in the last column. The sum of those squared differences is

$$\sum_{i=1}^{N=5} d_i^2 = 0.5^2 + 1^2 + 1^2 + 1.5^2 + 1^2$$

$$= 0.25 + 1 + 1 + 2.25 + 1 = 5.50$$

and $N=5$ so the correlation is

$$r_s = 1 - \frac{6(5.50)}{5^3 - 5} = 1 - \frac{33.0}{120} = 0.725$$

BOX 14.2 EXAMPLE FROM BOX 14.1 WITH TIE CORRECTIONS

The sums of squares for X and Y, from the Box 14.1 example, corrected for ties, are

$$\sum_{i=1}^{N} X_i^2 = \frac{5^3 - 5}{12} - \frac{3^3 - 3}{12} - \frac{2^3 - 2}{12}$$

$$= 10 - 2 - 0.5 = 7.5$$

$$\sum_{i=1}^{N} Y_i^2 = \frac{5^3 - 5}{12} - \frac{3^3 - 3}{12} = 10 - 2 = 8.0$$

The corrected Spearman correlation coefficient is

$$r_s = \frac{7.5 + 8.0 - 5.5}{2\sqrt{7.5(8.0)}} = \frac{10.0}{15.49} = 0.645$$

The ties in ranking reduce the sums of squares and that in turn lowers the correlation coefficient from 0.725 without considering ties to 0.645 after ties are considered.

---

correlation can be determined for a particular sample size, N, and level of significance α. For larger sample sizes, smaller sample correlation coefficients become statistically significant. Table 14.6 shows the minimum statistically significant sample rank correlation coefficient* as a function of the sample size when $\alpha = 0.05$.

---

* See Siegel (1956) for an extended version of this table.

Table 14.6 Minimum Spearman Rank Correlation Coefficients that Are Significantly Different from Zero in a Statistical Sense for Different Sample Sizes N

$N =$	5	7	10	14	18	22	26	30
$r_s =$	0.9	0.714	0.564	0.456	0.399	0.359	0.329	0.306

Wilcoxon sign test

The Wilcoxon sign test is a nonparametric test that can be used to measure the statistical significance of differences in paired ordinal ratings of items A and B. The null hypothesis is that the $P(A > B) = P(B > A) = 0.5$. For example, n people might rate two different products (A and B) on an ordinal rating scale, such as ease of use. To apply the sign test, the first step is to count how many subjects rated A higher than B and how many rated B higher than A. Tied ratings are ignored. If the first quantity is referred to as x and the second as y, then the null hypothesis implies that x and y are binomially distributed with $p = 1/2$ and $N = x + y$. The cumulative binomial distribution can be used to estimate the statistical significance of the obtained results by summing the tail areas of the binomial probability density function discussed in Chapter 7 (Equation 7.1). The latter approach becomes quite tedious for larger samples.

A simpler approach is to check whether the observations fall within the Beta-1 confidence interval corresponding to the null hypothesis. The observation x corresponds to a sample estimate of q, where $q = x/N$. The Beta-1 confidence interval corresponding to the null hypothesis is described by the parameters a, b, and α where

$$a = b \qquad (14.7)$$

and

$$a + b = N \qquad (14.8)$$

and α is the desired level of confidence. For example, assume that 11 subjects rated two products (A and B) on ease of use. A total of 8 subjects rated $A > B$, 2 rated $B > A$, and one rated them the same. So $N = 8 + 2 = 10$. Using Equations 14.7 and 14.8, the parameters of the Beta-1 distribution for the null hypothesis are found to be $a = b = 5$. Recall that Beta-1 confidence intervals are given in Appendix B.2. If $\alpha = 0.05$, the upper and lower bounds of the confidence interval for the null hypothesis, corresponding to $a = b = 5$, can be found in Appendix B.2. These values are, respectively, 0.788 and 0.212. The observed value of q in the sample was $8/10 = 0.8$. This falls outside the confidence interval of the null hypothesis, so the results are statistically significant at the $\alpha = 0.05$ level. An exact calculation obtained by summing the binomial probabilities* reveals that $\alpha = 0.0434$.

It should be noted that Wilcoxon sign test does not consider how large the differences are between the ratings. It only considers how often the items are rated higher or lower. Consequently, it is more difficult to show significance with the Wilcoxon sign test than with parametric tests, such as t-tests, that consider how large the differences are. This again, is a reason for using interval or ratio measurement scales rather than ordinal measures. One advantage of the Wilcoxon sign test is that it is fairly simple to do and does not

---

* $p(x \leq 2) = p(x = 0) + p(x = 1) + p(x = 2) = 0.0217$. The quantities on the right side of this equation are each calculated using Equation 7.1, with $q = 0.5$ and $N = 10$. Note that $\alpha = 2^* p(x \leq 2) = 0.0434$ because the null hypothesis corresponds to a two-sided test. Since $q = 0.5$, the distribution is symmetrical, and the two tail areas are equal.

Chapter fourteen: Questionnaires and interviews

make a lot of statistical assumptions. It consequently provides a good conservative check of other statistical measures.

Reliability

Questionnaire reliability* is another issue that must be considered. In other words, how repeatable are the results obtained by the questionnaire? Reliability is an especially important issue when developing rating scales that will be used to measure subject attitudes or other quantities.[†] Reliability is also a major issue when trying to develop a simplified version of a questionnaire. Test–retest reliability is one common measure. In its simplest form, subjects might be asked the same question more than once. The correlation between answers by the same subject to the same question should be high. In other cases, responses for different versions of a questionnaire might be compared. For example, versions A and B of a questionnaire might be given to all of the subjects. Half of the people perform A before B; the others perform B before A. The correlation, $r_{x:x'}$ between scores on versions A and B within each person provides an estimate of reliability. Another traditional approach is to give half of the people version A and the other version B and then measure the correlation between test scores, hopefully for comparable people. Matched pairs of people are often used in the analysis to help ensure the correlations are based on comparable people.

The different correlations mentioned herein provide measures of test reliability. The standard error of measurement is another traditional measure of test reliability. This measure is calculated as

$$S_E = S_X \sqrt{1 - r_{x:x}} \tag{14.9}$$

where
 S_E is the standard error of measurement
 S_X is the standard deviation of the observed scores for the population of interest
 $r_{x:x}$ is the test reliability (or correlation) coefficient for that population

Smaller values of S_E correspond to more consistent ratings. Note that S_E can be thought of as the standard deviation of the test score for a single person who takes the test multiple times.

An important point is that S_E will differ between individuals in a test population. Researchers often assume that S_E is constant within a test population for the sake of convenience. However, this assumption is questionable. Several studies have shown that test reliability varies with the average ranking of the respondent. The observed S_E tends to be lower for respondents with very high or very low rankings than for those who fall in the middle (see Blixt and Shama, 1986; Cumming, 1997; Feldt et al., 1985; Jarjoura, 1986; Thorndike, 1951).

To overcome some of this criticism, some analysts suggest reporting conditional standard errors of measurement (CSEM), that is, standard errors of measurement are separately presented for particular ranking ranges. A practical concern is that each range must contain enough respondents to allow the conditional correlation coefficients

---

* This issue appears again in Chapter 18, with regard to employee testing and placement.
[†] The Cronbach α coefficient (Cronbach, 1951) is a commonly used measure of internal consistency when multiple measurement scales are used.

BOX 14.3 CONDITIONAL STANDARD ERROR OF MEASUREMENT

In a recent study, questionnaires were administered to 40 people in the Chicago area regarding a variety of automotive navigation systems. The subjects ranked each automotive navigation aid on several different scales. All of the rankings were on 5-point scales adjusted for consistency. Means of the individual questions ranged from 2.64 to 4.95, while standard deviations on individual questions among all respondents ranged from 0.22 to 1.26. Subject scores for the entire questionnaire were divided into six intervals. CSEM were calculated for each interval as shown in Table 14.7. Note that the standard errors of measurement were smallest for intervals 1 and 6. That is, the responses were more consistent for subjects who provided either the highest or lowest ratings. The largest CSEM (interval 4) is 1.51 times larger than the smallest (interval 1).

Table 14.7 Conditional Standard Errors of Responses to a Question

Interval	1	2	3	4	5	6
Score range	81 and below	82–84	85–87	88–90	91–93	94–96
CSEM	3.37	4.03	4.03	5.08	3.51	3.39

between the split questions in the questionnaire to be calculated. An example illustrating this approach* is given in Box 14.3.

Interviews

Interviews and questionnaires bear many similarities. In fact, interviews often involve questionnaires that are administered face to face. Much of the previous discussion is consequently also applicable to interviews. The primary difference is that interviews are much more personal and flexible than questionnaires. The personal nature and flexibility of an interview is its primary advantage over use of a questionnaire, but this can lead to problems. One of the great dangers of interviews is that many people do not adequately prepare for them and, as a result, they fail to get all the relevant information or they may bias the answers through carelessness. While interviewing has traditionally been considered an art form, some interesting ways of automating the interviewing process have been developed in knowledge engineering (Lehto, Boose, Sharit, and Salvendy, 1992).

There are many situations where interviews are much more appropriate than questionnaires. When the questions are complicated, a good interviewer can quickly tailor them to the interviewed person. Questionnaires are rather pedestrian in comparison because the questions do not normally change as a function of the previous answers. There are also times when the respondents need some help in responding. In this case the interviewers can help fill the gap. When there are widely divergent viewpoints, interviewers can often elicit those differences. A questionnaire would not distinguish differences unless those differences were well understood before it was developed. Many interviewers are also trained to observe body language and nonverbal cues.

---

* The questionnaire was developed by Michael Mollunhour and the reliability analysis was developed by Tammie L. Cumming, CCAD, of the University of Iowa.

BOX 14.4   CONTRIBUTIONS OF PSYCHIATRISTS TO ERGONOMICS

One school of thought in psychiatry focuses on interviewing people to talk through their problems. Personal construct theory is an interviewing technique developed many years ago by a psychiatrist named George Kelly. Personal construct theory provides a powerful technique for mapping or describing the way people categorize experiences and classify their environment. In this technique, subjects are asked to successively compare triads of objects and describe in their own words how one of the objects differs from the other two. Performing these comparisons results in the development of a set of distinguishing bipolar dimensions or constructs.

Each of the objects is then rated on the bipolar construct scales suggested by the subject, resulting in a repertory grid that describes the relationship between objects. The obtained constructs are often hierarchical in that many constructs have subconstructs, which themselves may have sub-subconstructs, and so on. This technique has the advantage that it can often uncover relationships that subjects are unable to express directly. This interviewing method is now used in many knowledge elicitation tools, intended to help knowledge engineers develop expert systems. For more about this topic, see Lehto, Boose, Sharit, and Salvendy (1992).

Although one-on-one interviews are often a good source of data, this process can be incredibly time consuming and expensive. Much greater efficiency can be attained by interviewing a group of people instead of individuals. Group-oriented interviewing methods include focus groups, brainstorming groups, the Delphi method, and the nominal group method.

Interviewing methods

There are different approaches in interviewing. Nondirected interviews are where the respondent leads and the interviewer just keeps the interview going. Some people liken this situation to that in psychiatry where the patient lying on the couch is urged to say whatever comes to mind (see Box 14.4). Directed interviewing techniques ask questions about a predetermined set of topics. In the most extreme case, the analyst simply reads questions to the subject and writes down their answers. Depending on the situation, either approach can be useful. Combined approaches are often the best. This offers the interviewer more flexibility to adapt the interview to the subject. Most successful interviewers are flexible and make adjustments depending on how the interviewed subject is reacting to the situation.

Perhaps the most important advantage of interviews over questionnaires is that interviewers can skip questions if desired. At the same time, interviewers must constantly be aware of the possible biasing that can occur when the interviewer participates in the discussions. Interviewing is definitely a very underrated talent. It must be recognized that the ultimate success of an interview will often depend upon the skill of the interviewer. The skill of the interviewer is particularly important when it comes to reading body language, and especially so when the interviewer is reaching across cultural and age gaps. If possible, the entire interview should be recorded. The interviewer will not always be able to write down everything important, so the recording can be a very useful backup.

Brainstorming and focus groups

A variety of approaches have been developed for obtaining information from groups of people. Brainstorming groups and focus groups can provide a lot of detailed information in a single session (see Box 14.5). This obviously saves time and money in data collection and can set the stage for further analysis. Brainstorming is a popular technique for quickly generating ideas. In this approach, a small group (no more than 10 individuals) is given a problem to solve. The members are asked to generate as many ideas as possible. Members are told that no idea is too wild and encouraged to build upon the ideas submitted by others. No evaluation or criticism of the ideas is allowed until after the brainstorming session is finished. Buzz group analysis is a similar approach, more appropriate for large groups. Here, a large group is first divided into small groups of four to six members.

BOX 14.5 STRUCTURED FOCUS GROUP ANALYSIS

A few years ago, Lehto and DeSalvo conducted a study to determine how well workers understood hazard communication phrases proposed for inclusion in the ANSI Z129.1 Chemical Labeling Standard. Part of their analysis involved a structured focus group analysis where workers rated the understandability of tested phrases as a whole, indicated difficult words, and provided general comments about phrases. The facilitator also asked the group to discuss particular phrases.

The focus group sessions lasted an average of 110 min. Questions regarding each phrase were both read aloud and included on an answer sheet provided to subjects. The procedure involved first reading the phrase to be evaluated. For the majority of the phrases, the subjects then rated their understanding of the phrases and indicated words they thought were difficult. After rating the phrases, the subjects were asked specific leading questions by the facilitator. Subjects were allowed to ask for clarification and to make oral and written comments regarding each of the specific questions. After addressing each of the specific questions, topics initiated by the subjects were discussed.

A total of 87 subjects were interviewed, with group sizes ranging from 2 to 10 people. The workers came from eight different chemical plants. Subjects tended to be male and were between 18 and 57 years old. The majority of the subjects had been employed over 5 years and tended to have over 10 years of experience working with chemicals. The majority of the subjects were classified as plant operators. Almost all subjects had received a training course on hazard communication.

The findings indicated that nearly all of the phrases were well understood by the workers. Those statements that were rated lower were often rated highly by experienced workers who were familiar with the situation referred to, implying that training should be used to supplement the statements with lower ratings. The workers thought that including additional information in the MSDS would be useful for many of the phrases. The desire for additional information often increased when a referred-to hazard condition was severe. The cross-validation revealed that neither (1) measures of readability, obtained using readability indexes and grammar checkers, nor (2) measures of understandability, obtained using word frequency, past citation of words as difficult, or labeling of words as abstract, were strongly related to self-reported understanding of the phrases by the workers. It seems likely that these findings were related to the high degree of experience of chemical workers. Further information on this study is given in Annex N of the ANSI Z129.1 Standard.

Each small group goes through a brainstorming-like process to generate ideas. They then present their best ideas to the entire group for discussion.

Focus groups are commonly used to provide feedback regarding consumer reactions to various product features in market research studies. The technique generally involves presentation of various products to a group of subjects. The subjects can be selected to closely correspond to some population of interest to the researcher. The subjects are then asked to provide comments in an open discussion format. Traditionally, a facilitator will first introduce a general introductory topic. The questions are then narrowed until the topic of interest is addressed. An advantage of this approach compared to traditional surveys and questionnaires is that subjects are less limited in their responses. This occurs because the subjects have some control over the flow of discussion. On the other hand, the open-ended responses may be difficult to interpret.

The use of brainstorming and focus groups will normally provide a substantial amount of creative suggestions, especially when participants build upon each other's ideas. However, personality factors and group dynamics can also lead to undesirable results. Simply put, some people are much more willing than others to participate in such exercises. Group discussion consequently tends to center around the ideas put forth by certain more forceful individuals. Group norms, such as deferring to participants with higher status and power, may also lead to undue emphasis on the opinions of certain members.

Nominal group and delphi technique The nominal group technique (NGT) and the Delphi technique attempt to alleviate some of the disadvantages of working in groups. The NGT consists of asking each member of a group to independently write down and think about their ideas. A group moderator then asks each member to present one or more of their ideas. Once all of the ideas have been posted, the moderator allows discussion to begin. After the discussion is finished, each participant rates or ranks the presented ideas. The subject ratings are then used to develop a score for each idea. The NGT is intended to increase participation by group members and is based on the idea that people will be more comfortable presenting their ideas if they have a chance to think about them first.

The Delphi technique allows participants to comment anonymously, at their leisure, on proposals made by other group members. Normally, the participants do not know who proposed the ideas they are commenting on. The first step is to send an open-ended questionnaire to members of the group. The results are then used to generate a series of follow-up questionnaires, in which more specific questions are asked. The anonymous nature of the Delphi process theoretically reduces the effect of participant status and power. Separating the participants also increases the chance that members will provide opinions independently of others.

Interrater consistency

In many cases, interviews involve ratings by more than one interviewer. The consistency of the obtained ratings may be an important issue in certain situations, such as when each interviewer ranks applicants for a position. Techniques are available for evaluating this issue. For example, with k interviewers and N applicants, the rankings of each interviewer go from 1 to N. Tied applicants are given a tied ranking (i.e., two ranked applicants below the second rank and above the fifth rank are both ranked 3.5, which is the average of rank 3 and rank 4). Note that if all k interviewers are perfectly consistent in their rankings, one applicant would be ranked all ones, another all twos, ... and another all Ns. On the other hand, zero consistency between interviewers would result in rankings that sum to $Nk/2$ for the k

interviewers. In most cases, rater consistency will be somewhere between these extremes. Kendall's coefficient of concordance (W) provides a measure of ranking consistency, useful in the latter situation. The coefficient of concordance is calculated as shown here:

$$W = \frac{s}{(1/12)k^2(N^3 - N)} \quad (14.10)$$

in which s is the *rank* sum of squares over the N people or objects being ranked, and k is the number of people doing the ranking.* Note that the denominator of Equation 14.10 is the maximum rank standard deviation when all k people doing the ranking are perfectly consistent. A numerical example showing how to calculate Kendall's coefficient of concordance is given in Box 14.6.

BOX 14.6 EXAMPLE OF CALCULATION OF CONCORDANCE

Consider the situation where three interviewers ranked six applicants for a job. The results of this ranking process are shown below in Table 14.8.

Table 14.8 Numerical Example of Three Interviewers Ranking Six Applicants

Interviewer	A	B	C	D	E	F
A	1	4	2	5	3	6
B	2	3	1	6	4	5
C	4	2	3	5	1	6
Ranked sum	7	9	6	16	8	17

To calculate Kendall's coefficient of concordance, the first step is to determine the rank sum of squares s. To determine s, we first need the average of the ranked sums, which is

$$\frac{1}{6}(7+9+6+16+8+17) = \frac{1}{6}(63) = 10.5$$

The rank sum of squares s is then calculated as

$$s = (7-10.5)^2 + (9-10.5)^2 + (6-10.5)^2$$
$$+ (16-10.5)^2 + (8-10.5)^2$$
$$+ (17-10.5)^2 = 113.5$$

Finally, Kendall's coefficient of concordance W is calculated, using Equation 14.10:

$$W = \frac{113.5}{(1/12)3^2(6^3 - 6)} = 0.721$$

---

* See Siegel (1956) for more description and additional examples.

Final remarks

The unfortunate thing about questionnaires and interviews is that too many of us do not give them the adequate review and preparation they deserve. Ergonomists frequently give exit interviews to subjects in experiments and they often write questionnaires administered to the public to try to ascertain what the public thinks about a particular issue or product. They are often, but wrongly, treated as fishing expeditions. While there is a lot of art involved in preparing either a questionnaire or an interview, there is some scientific basis for designing questionnaires and interviews in particular ways, and that is the emphasis of this chapter.

The analysis of data from questionnaires and interviews often involves ranking or other forms of nonparametric statistics. This means that Gaussian (i.e., normal) and parametric statistics cannot be used effectively to analyze the situation at hand, unless special steps are taken to ensure the data meet some of the associated statistical assumptions. Some nonparametric statistics are consequently introduced and illustrated. In fact, considerable amounts of statistics have been introduced in this book because they are important in guiding ergonomic designers. However, it is important to distinguish the different role of statistics in science in contrast to its role in design. Science creates attitudes and tests about bringing hypotheses as facts or laws and, in that compulsion, insists on hypotheses meeting adequate statistical confidence intervals or, equivalently, sufficiently small type I and type II errors as to be almost certain that the hypothesis is true. Designers, however, need to create an effective design, not necessarily optimal, and to do so within a time frame and budget. That objective requires designers to select a design alternative that appears better but may not pass the statistical tests of science. As Wickens (1998) states, designers are and should be more interested in criterial difference magnitudes than in meeting specified confidence intervals or error type probabilities. For that reason, the statistical focus here is to use statistics as a design tool and guide to augment the designer's common sense and design philosophy.

Discussion questions and exercises

14.1 What is your first reaction when receiving a letter requesting that you complete an enclosed questionnaire? What features cause you to (a) Fill out the questionnaire immediately or as soon as possible? (b) Put the questionnaire in your briefcase to be filled out when you feel like it? or (c) Toss it out straight away?

14.2 What is the *horn effect* in questionnaire design? How does it differ from the *halo effect*?

14.3 Why is it important for the person designing the questionnaire to define the audience very precisely before developing the questions and to define the objectives of the questionnaire with extreme precision?

14.4 What can a questionnaire constructor do to encourage a prompt and accurate response from respondents?

14.5 What does the term *respondent's demographics* mean? Provide an example.

14.6 What is the ending part of a questionnaire called and what should it contain?

14.7 What are the arguments for and against the *question shift* procedure in constructing questions?

14.8 Questionnaire responses typically cover N categories when the responder is asked to rank agreement or disagreement with a statement. What numbers are typically used for N and why? How should one characterize the extreme categories?

14.9 Suppose that respondents were instructed to make a mark on a line where the far left meant extreme disagreement, the far right meant extreme agreement, and the middle was a neutral position. The person making up the questionnaire planned to assume that the marks were made to fit a normal distribution. However, when the questionnaire responses were returned, a very large fraction was close to one end and another large fraction was near the other end. How would you interpret the meaning of this, and how could these results be analyzed for this question?

14.10 What is questionnaire reliability? How can it be measured? And what is its significance?

14.11 In what ways do interviews differ from questionnaires? What are some advantages of one over the other? What situations make interviews better than questionnaires?

14.12 Suppose that two interviewers ranked four interviewees as

Applicants =	A	B	C	D
Interviewer 1	2	4	1	3
Interviewer 2	4	3	2	1

What is the degree of ranking concordance between these interviewers?

14.13 Discuss each of the data analysis techniques covered. When is each approach appropriate? How does a parametric approach differ from a nonparametric approach?

14.14 What are the advantages of group interviewing techniques compared to one-on-one interviews? When might each approach covered in the chapter become appropriate?

chapter fifteen

Quality control and inspection

About the chapter

This chapter focuses on the important role of inspection in quality control and other applications. Some common ergonomic issues are addressed, and suggestions are given for how to analyze and improve the design of inspection tasks. The chapter also introduces the use of signal detection theory (SDT) as a quantitative framework for describing and evaluating inspector performance. The following sections focus on inspection economics and the topic of continuous improvement, emphasized so heavily by Deming.

Introduction

A number of steps can be taken during the design process to improve product quality, as discussed in Chapter 13. However, quality control is essential to ensure that products reach the customer in their intended form. Quality can be lost due to manufacturing defects, but this is only part of the issue. Another major concern is that products may be damaged during shipping and delivery, in some cases intentionally. Inspection is one of the most important quality control activities, but it is also an important part of activities such as security screening and preventative maintenance programs. In most modern production facilities, inspection is done by people designated as inspectors, as well as machinists, people who package the product, and all other persons whose close contact with the product or equipment enables them to identify faults and fix or withdraw product units. In some cases, assistance from the customer may also be needed to identify defects. This becomes especially true if the defect does not become apparent until long after the product was purchased and first operated. Quality control or assurance is consequently essential beyond the product design phase, not only to maintain quality but also to provide feedback to designers for improving quality.

Since about the 1960s, the role of quality control has changed considerably in the U.S. industry and service organizations. Prior to that, almost all companies separated the quality control functions from production operations. The assumption was that production control and quality control should occupy *adversarial* roles. Inspection was performed by inspectors who swore their allegiance to the quality control department and often regarded production personnel almost as an enemy.

Things began to change in the mid- to late 1960s when many U.S. manufacturers first noticed the effects of increasingly difficult overseas competition. Many U.S. consumers felt the competition offered high-quality products compared to those offered by the domestic manufacturers. The improved quality of the products offered by the overseas competition was largely attributed to the writings of Edward Deming, an applied statistician from the United States, who went over to Japan after World War II to help transform the Japanese industry from producing military goods to civilian products for worldwide competition. Many of Deming's ideas differed from current industrial practices in the United States. As a consequence, U.S. managers, in their typical precipitant style, dropped many practices that conflicted with those Deming advocated, based on the premise that what worked for Japan should work for the United States. Some companies exhibited almost religious fervor for Deming's ideas* and refused to consider practical alternatives. Recent years have brought about an increased willingness to modify Deming's ideas, in a variety of ways.

At about the same time, as U.S. customers started to buy overseas products because of higher quality, courts in the United States began shifting their attitudes about product liability. Many courts started finding manufacturers at fault if they could have reasonably foreseen an inherent danger in their product. This legal trend accelerated the quality movement by highlighting losses that are noticeable during design and manufacturing. As the economic equation shifted, industry accordingly moved away from separate inspection and began using the operators on the production line to assist in the inspection process. Some analysts (Taguchi (1978) and Deming (1986)) advocated getting rid of separate inspection departments and urged companies to scrap the adversarial roles between production and inspection and make them cooperative.

But even if the company abandons separate inspection activities, inspection must continue to be performed. The current tendency is to rely more on the workers themselves rather than designated inspectors. Part of the issue is that workers have always had some responsibility for assuring quality. For example, machine tool operators are inspectors of their own work and always have been. In earlier days, few managers gave workers credit for performing an inspection function and tended to pressure them into more production. Even now, people promoting tools such as value stream mapping tend to view inspection as a non-value-added step. In reality, an inspection does add value, but only if it increases quality. Accordingly, workers should be appropriately credited for taking the necessary steps to eliminate quality problems.

Today, there is general acknowledgment that inspection plays too valued a role in quality control and in other aspects of industrial and service operations to be sacrificed. Juran defines *quality control* as evaluating actual performance, comparing that performance to the goal, and taking action if the two differ. As such, it is easy to see that inspection is an essential part of the information collection activities that must take place in quality control. Inspection results allow manufacturing performance to be compared against goals or objectives and provide useful feedback to the people operating the processes. Tracking trends over time is especially useful. Any indications that product units are tending to go outside of acceptable bounds need to be brought to the attention of supervisory personnel immediately so that apparent problems can be diagnosed and corrected before any product units become defects. In many cases, quality losses can be traced back to product components that are not meeting specifications as they should. Dimensional and other measurements often obtained during inspections can be analyzed to determine process variability.† In any case, the numbers used in such analyses are often obtained during various forms of inspection. The quality of inspection data therefore becomes an important issue.

One last issue is that inspector performance normally is somewhere between perfection and complete neglect. The theory of signal detectability (SDT) provides an excellent framework for measuring inspector performance that addresses human abilities, task difficulty, and the costs associated with failing to detect bad items and rejecting good items. SDT helps ergonomists contribute to quality control by making it easier to compare

---

* Many of Deming's ideas are summarized in his "Fourteen Points" (Deming, 1986).
† Uncontrolled variability is a primary cause of quality loss in manufacturing.

different inspection situations and also suggests ways of making inspection tasks more effective and easier to perform. The later issue is the primary focus of this chapter.

Some common types of inspection

Several different types of inspection are done in industry and elsewhere. In most cases, inspections are done by people, sometimes with the assistance of various devices or even animals. The different types of inspection can broadly be divided into at least four somewhat overlapping categories:

1. Quality assurance inspections
2. Maintenance inspections of machinery, equipment, or facilities
3. Screening of potentially dangerous materials, people, or devices before they are transported or brought into places they should not be
4. Detection of tampering or damage to shipped or stored materials

Quality assurance inspections

Most manufacturers conduct several different types of inspections to help ensure the quality of the items they produce.* These include testing of incoming parts and materials to make sure they meet minimum quality standards, as well as testing and inspection of the produced items throughout the manufacturing process. In most cases, these tests are done by workers who look at, feel, or manually manipulate the item. Sometimes, they also use various types of tools and testing equipment appropriate for the particular application.

As mentioned earlier, quality assurance inspections are an important part of controlling production costs. Such inspections can flag defective parts before they are installed. Doing so can provide huge savings because it costs a lot of money and time to remove the part later and replace it with a good one. Another issue is that warranty and product liability costs can be huge if a significant number of defective items reach the customer. Simply put, items that do not meet specifications need to be caught as soon as possible, preferably well before they get out the door and reach the customer. Quality tracking systems are often also used to help identify and target the causes of defects, to help eliminate significant problems.† The principal question is how to accomplish this effectively and economically.

Maintenance and safety inspections

Maintenance and safety inspections are important to make sure products, systems, buildings, and processes are operating properly, safely, and reliably. Such inspections might be part of the normal start-up procedure followed by a machine operator, a routine periodic check by a mechanic, or an intensive audit conducted by a team of specialists. Maintenance inspections are often required by law. For example, the Federal Aviation Administration (FAA) specifies rules for required inspections on engines, aircraft structural components, and all sorts of other equipment. During this periodic

---

* Quality assurance inspections are also done in other sectors, such as in health care, food processing, and the restaurant industry. Such inspections are often required by law, and the results are made available to the public. For example, the sanitation conditions of restaurants observed by health inspectors are published in local newspapers.
† See Drury (1992) for some specific advice on how to achieve that goal.

inspection, deficiencies that are discovered need to be properly fixed before the aircraft can be returned to service. Failure to comply can lead to criminal sanctions. Other similar inspections occur throughout industry, particularly in industries where corrosive or dangerous chemicals can cause considerable danger to personnel, plant equipment, and the environment. In the United States, both the Environmental Protection Agency (EPA) and the Occupational Safety and Health Administration (OSHA) enforce laws that require frequent inspections and impose heavy fines on companies where accidents occur because no inspection program is in place.

Screening of people and materials

This approach is illustrated by the airline industry's inspection of passengers and their luggage for firearms, gas canisters, explosives, and other weapons. Such inspections are intended to prevent hostile passengers and weapons, biological agents, dangerous chemicals, or explosives from entering the aircraft. This form of inspection is aided by machines and contemporary technology, as bags are x-rayed and other forms of detectors are employed. Similar screening inspections are done in other transportation settings, courtrooms, border crossings between countries, mail rooms (think of anthrax detectors), and elsewhere aggressively hostile people and their weapons are a special concern.

Detection of tampering

Inspection of shipped or stored materials (i.e., off the shelf of a store) is another common activity. In recent years, significant efforts have been made to design products or packaging that make damage or tampering more obvious. The development of *tamper proof* bottles by the pharmaceutical industry is a classic example. This was done shortly after an unfortunate incident where several bottles of pain medication on the shelves of a retail store were poisoned with strychnine. The new bottle designs make it almost impossible to remove the cover without first tearing a plastic seal wrapped around the bottle top. Consequently, it becomes easy for the customer to see if the cover has been taken off by someone earlier. To reinforce the message, the packaging also includes the instruction *not to buy* if the seal is broken. In other situations, companies put moisture-sensitive devices in their shipping boxes that indicate if moisture has intruded. Other devices can indicate if the box has been placed in the wrong orientation or shaken too much during shipping. In those cases, the customer or an intermediate person can be the inspector. The critical issue is that the devices on or inside the boxes need to convey to the person at the end of the shipping process information about wrongful acts or damage that occurred during the shipping operation.

Human inspection

Most inspection tasks are performed by people using their sense of vision, but other senses are also used. For example, people often check if parts fit properly by taking the assembly apart and putting it back together. If something rubs or catches, the fit might not be good enough. In other cases, people run their hand or fingers along a surface, often with a thin paper between the surface and the finger, to check for surface imperfections. Quite often, people use several senses at the same time. For example, someone might feel and touch the inspected object while they look at it. At the same time they might be listening for odd sounds.

Chapter fifteen: Quality control and inspection

As implied by the previous comments, human inspection normally involves three steps:

1. Sensory activity—looking or scanning, touching or feeling, or listening
2. A measurement of sensory elements—perceived size, shape, color, brightness, roughness, tightness, force, loudness, or tone
3. A decision—identified features (is something there or not) or tolerances (is the measurement within specification)

The ability of a human inspector to successfully do these three steps depends on many factors. At the most general level, it is clear that performance is related to inspection strategies, viewing time, task pacing and time pressure, differences between individuals, and illumination levels. Before going into detail into these factors, the following discussion will briefly introduce SDT as a model of human inspection.

Signal detection theory

As mentioned earlier in Chapter 10, SDT is a special case of decision theory and can be viewed to be an abstract model of human inspection. As a descriptive inspection model, SDT allows the analyst to compare inspector performance for quite different inspection situations. When SDT is used to model human inspection tasks, it is assumed that a false alarm corresponds to what is called a type-I error in quality control, that is, it corresponds to *rejecting a good* or non-defective product. A miss corresponds to a type-II error or *accepting a bad* or defective product. Table 15.1 shows this relationship between these outcomes of an inspection task.

To continue the analogy, the difficulty of the inspection task, or equivalently the ability of the inspector to detect the defect, corresponds to d' in SDT. Inspector bias toward accepting bad parts or rejecting good parts corresponds to the parameter β. Figure 15.1 shows how these parameters of the basic SDT model can be mapped to an inspection task. To determine the values of d' and β, for an inspection task, all that is needed is the false alarm and detection rates for the inspection process (see Box 15.1). Ideally, this would be done at the level of an individual inspector. This is often not an issue, as many quality control programs will try to collect both type-I and type-II errors for their inspectors and inspection processes.

Over the years, several studies have been done evaluating how inspection task conditions influence some of the basic parameters used in SDT. One finding in these studies was that d' is influenced by viewing time. Figure 15.2 shows the results of two different studies where subjects viewed objects moving past them on a conveyer belt. As shown in the figure, d' tended to increase consistently with longer viewing times for nearly all of the tested targets. Many other studies have used d' measures to evaluate inspection task difficulty. For example, consider the results of the study shown in the top part of Figure 15.3.

Table 15.1 Relationship between the Four Basic Outcomes in SDT and Type-I and Type-II Errors in Inspection

Inspector Actions	Defect Is Present	No Defect
Reject part	HIT Correct rejection of part	FALSE ALARM Type-I error
Accept part	MISS Type-II error	Correct rejection Correct acceptance of part

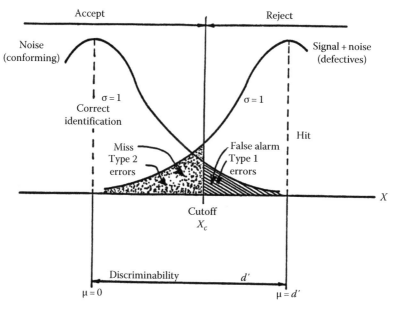

Figure 15.1 SDT model of visual inspection.

BOX 15.1 EXAMPLE CALCULATIONS OF D' AND β FOR AN INSPECTION TASK

Suppose that data were available showing the results of inspections conducted by the inspectors in a particular plant. The obtained data are given in the following:

Response	Product Defective	Product Not Defective
Reject	0.92	0.1
Accept	0.08	0.9
Column sum	1	1

That is, P(false alarm) $= P(FA) = 0.1$ and P(miss) $= P(M) = 0.08$.
Note that the detection rate, $P(D) = 1 - P(M) = 0.92$.

The next step is look up the right-tailed z-scores for $P(FA)$ and $P(M)$. Using the table in Appendix B.1.1, we find standardized normal deviates respectively of 1.28 and 1.40 by interpolating the values in Appendix B.1.1. d' is their sum so,

$$d' = 1.28 + 1.40 = 2.68$$

The next step is to look up the ordinates for the two z-scores. By interpolating the values in Appendix B.1.2, we find an ordinate value of 0.17 for the z-score of 1.28 and a value of 1.40 for the z-score of 0.14.

Chapter fifteen: Quality control and inspection

Recall from Chapter 10 that

$$\beta = \frac{P(X_c|\mu = d')}{P(X_c|\mu = 0)}$$

it follows that

$$\beta = \frac{0.1486}{0.1754} = 0.8471$$

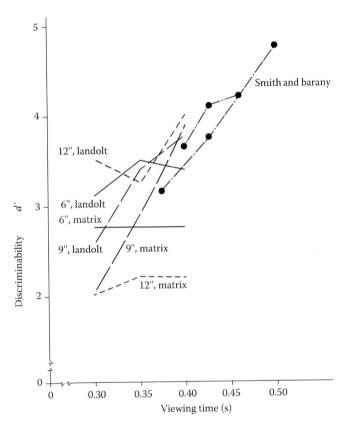

Figure 15.2 Discriminability d' for various types of targets and interspacings as a function of viewing time in seconds. (Adapted from Nelson, J.B. and Barany, J.W., *AIIE Trans.*, 1, 327, 1969; Smith, L.A. and Barany, J.W. *AIIE Trans.*, 4, 298, 1971.)

As shown there, d' consistently increased with longer exposure time for each of the three conveyer belt velocities tested. Decreases in belt velocities also caused d' to shift upward, that is, the highest d' occurred when the lowest belt velocity was combined with longest exposure time. Overall, these effects of the inspection conditions on d' certainly seem very reasonable. Somewhat more surprising effects were observed when β was plotted out for the same set of task conditions, as shown in the lower part of Figure 15.3, that is, β did not

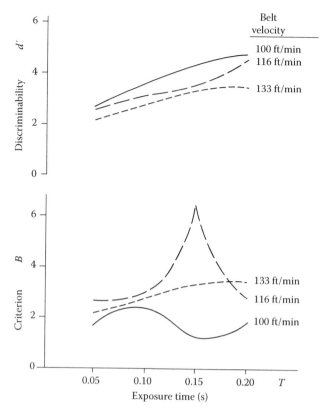

Figure 15.3 Discriminability d' and criterion effects as a function of exposure times in seconds and belt velocities in feet/minute. (Adapted from Rizzi, A. et al., *AIIE Trans.*, 11, 278, 1979.)

remain constant. It is not at all clear why subjects were shifting their bias, but it is interesting to observe that such large shifts were occurring. This, of course, is something that would be desirable to correct, because it will lead to overall poorer performance, by almost any objective standard of inspector performance.

A large number of other studies show similar trends to those mentioned already—that is, fairly logical effects of task conditions on d', but not really logical, and sometimes hard to interpret changes in β. As such, SDT offers a good conceptual framework for comparing results from various studies or, more importantly, for comparing the performance of particular inspectors in particular work circumstances. Since the model can specify exactly how the inspector *should* be performing, it provides a useful benchmark to compare different operators and inspection task designs against each other. These results also would seem to be useful feedback to the inspector to help them learn how to improve performance.

Nonparametric extensions of SDT

As noted in Chapter 10, the traditional version of SDT assumes the presence of normally distributed noise. This assumption allows d' to be conveniently measured, which can be quite helpful. In reality, however, the observations used to infer the presence of many types of defects are really measured or perceived as a small number of discrete states rather than on a continuous scale. For example, an inspector might be able to perceive two or three levels of brightness, roughness, or fit. In such cases, we can use Bayes' rule directly to judge the probability that the product is actually defective, using the conditional probabilities and prior belief that a defect is there.

For example, assume that the inspector can judge three levels of roughness. The probability that the product is defective $P(D|J_i)$, given a particular level of judged roughness, is given in the following:

$$P(D|J_i) = \frac{P(D) * P(J_i|D)}{P(J_i)} \tag{15.1}$$

where
- $P(D)$ is the prior probability of a defect being present
- $P(J_i|D)$ is the conditional probability of the observation J_i given the defect is present
- $P(J_i)$ is the marginal probability of the evidence being present

Note that the latter probability can be calculated as

$$P(J_i) = P(D)P(J_i|D) + P(\bar{D})P(J_i|\bar{D}) \tag{15.2}$$

where
- $P(\bar{D})$ is the prior probability of the defect not being present, or simply, $1 - P(D)$
- $P(J_i|\bar{D})$ is the conditional probability of the observation J_i given the defect is not present and remaining terms are as defined earlier

Note that we don't have to make any assumptions at all about noise or how it is distributed. This is a real advantage for a number of reasons. First, it doesn't make sense to assume that noise is normally distributed for a given inspection situation, unless there is some good evidence it really is. Many real-world situations seem to be better described as situations where the human judgments and their relation to whether the defect is present fall into three categories:

1. A region where the judged quantity is very low, for which the defect is very likely to be present or absent, depending on the situation
2. A region where the judged quantity is very high, where its relation to the presence or absence of the defect is the opposite of that when the judged quantity is very low
3. An intermediate region, where the relationship is ambiguous

For example, the probability of a product being defective might be close to 1 if the judged roughness is very low and close to 0 if the judged roughness is very high. However, if it is judged to be moderately rough, the probability of it being defective might be about 0.25.*

It is easy to see that an inspector performance tracking system might want to keep track of (1) the judged quantities measured on some simple scale and (2) the presence or absence of the defect. The conditional relationship between judged quantities given by particular inspectors and the actual presence or absence of a defect could then be kept track of in contingency tables. This information could then be used as a highly specific type of feedback to inspectors to help them improve performance. It also would provide

---

* The validity and benefit of assuming normally distributed noise, as is done in traditional SDT to model the latter situation, is questionable at best.

guidance on how to develop better decision rules for inspectors that should lead to better performance from the company's perspective.

A simple rule of this type might take the following form:
  If judged quantity x is ___, then accept or reject.

A slightly more complex rule might take the following form:
  If judged quantity x is ___, and judged quantity y is ___, then accept or reject.

It is easy to see that such decision rules provide highly specific information to the inspector that might serve as a good decision aid or training tool, especially when the relationship between judged quantities and the presence or absence of a defect is noisy or complicated. Also, the rules might be included in a computer decision support tool that takes judged quantities from the inspector and then suggests an optimal decision. The key requirement, and potential pitfall with this approach, is that the necessary data must be collected by a quality tracking system. Although many companies are or are starting to use quality tracking systems, not many yet collect the information needed to set up the necessary contingency tables to apply this approach at this time.

Inspection strategies

Inspection tasks differ greatly in complexity. In some cases, people are checking whether one or two elements are within tolerance. In others, they are checking through a long list of desired and undesired properties of a product. For obvious reasons, people find it harder to do a good job when they need to check for a lot of different things. This leads to a practical principle for designing inspection tasks that can be easily applied when more than one inspector is available:

Principle 15.1

Divide up the tasks so each inspector doesn't have to go through a long list of things to check on each of the inspected items.

Breaking up the larger inspection task into smaller groups of 7–10 attributes or features for each inspector greatly simplifies the task and often provides a number of other benefits. Breaking up the task into shorter lists for inspectors essentially means each inspector becomes a specialist rather than a generalist who inspects everything. Doing this should increase the rate of learning, leading to both quicker performance and lower error rates. Other advantages might also be observed. For example, suppose that we have a complicated product with many components that need to be checked. If we gave a single inspector a long list showing each component, and several things to check for each of the components, it would take the inspector a long time to go though the list. If the inspector finds a problem and sends the product back to be reworked, in many work settings they would end up having to go through the entire list once again, when the fixed product shows up again at their station for inspection.

Another issue is that certain inspection sequences can be more efficient than others. While the solution to this problem is not obvious, it follows common sense. Order the attributes or features to be inspected by their *decreasing probability of occurrence*. Or stated differently, we have the following principle:

Principle 15.2

Look first for rejection attributes or features that are most probable and inspect features from the most to the least probable.

Chapter fifteen: Quality control and inspection

This rule assumes that a product unit is rejected when the first defect is found. This assumption is normally reasonable because rejected items are usually sent to a repair center where not only the originally detected flaw can be fixed but also other attributes or features can be checked and repaired if necessary. This rule reduces needed inspection activity on the production floor, and those items that will be rejected anyway are rejected sooner.

A related issue is that defects are sometimes more likely to be in particular locations than others. It turns out that training people to follow certain visual search patterns can help people find things more quickly (Rizzi, Buck, and Anderson, 1979). This leads us to another principle:

Principle 15.3

Train people to follow more efficient visual search patterns in which they look first at areas where the defect or other attribute of interest is more likely to be.

An issue comes up in situations where the inspector has a batch of several items to be checked, each of which has several features that need to be checked. In some cases, inspectors check all the features of an item before looking at the next item. For example, a person might measure both the height and length of a part before inspecting the next part. In other cases, the inspector checks all of the items, one feature at a time, before moving on to the next feature. For example, the inspector might first check the height of all of the parts, before checking length. One study compared the two strategies and found fewer inspection errors when people followed the second strategy (Konz and Osman, 1978). The first strategy, however, for the previous example, is quicker, since the part doesn't have to be picked up and set down as often. This leads us to another principle.

Principle 15.4

One item at a time can be faster but one feature at a time gives better inspection accuracy.

Viewing time and movement

As mentioned in Chapter 2 and again in Chapter 4 during our discussion on illumination, visual acuity is related to viewing time and movement. People need more light and have greater difficulty making out details when viewing times are short or when the objects are moving. This issue has been extensively evaluated for a wide variety of inspection tasks, leading us to the following design principle:

Principle 15.5

Inspection performance drops with shorter viewing time and when the inspected items are moving.

Many investigators have reported that inspection errors actually decrease *exponentially* as a function of the available viewing time (for example, see Buck and Rizzi, 1974; Drury, 1973, 1975; Nelson and Barany, 1969). One such study (Rizzi, Buck, and Anderson, 1979) of visual inspection developed an equation showing that the probability of making a correct detection was as follows:

$$p(\text{correct detection}) = 1 - 2.58e^{-a} \qquad (15.3)$$

where

$$a = 10.58T - \frac{0.075}{V} \quad (15.4)$$

T is exposure time in seconds
V is conveyer belt velocity in feet per minute

Also note that for this study,

$$0.05 \leq T0.05 \leq 0.25 \text{ s}$$

The coefficient of determination for this equation was 0.79, showing that it gave a good fit to the data. The important aspect of this equation is its general trend and the fact it shows that viewing time and target velocity have *separate effects* on inspection performance.

Task pacing and time pressure

Task pacing refers to both how quickly the tasks arrive and the variability of the time between arriving tasks. Pacing determines the pressure on the inspector to inspect fewer or more items per unit of time. If tasks arrive quickly, the inspector will have less time for each task. This problem can be especially severe if the time between tasks is unevenly distributed. For example, suppose the items to be inspected on a conveyer belt arrive in random piles of 2–10 items, with wide separations between each pile. When a clump of ten arrives, the operator will obviously be under a lot more time pressure than when a clump of two items arrives. This time pressure tends to negatively affect inspection performance.*

The next principle directly follows from the previous comments:

Principle 15.6

Inspection performance will be impaired if pacing rates result in inadequate viewing times and time pressure.

A related variable is whether the inspection is *self-* or *externally paced*. In a self-paced task, people control the rate at which they do the task. Consequently, they can allow more time when the inspection is difficult and less when it is easy. This observation leads us to another principle.

Principle 15.7

Inspection performance will be better if people are self-paced instead of externally paced.

Some externally paced tasks still allow the worker a lot of freedom. For example, if the product units to be inspected are delivered to an inspector in tote pans that hold a specified number of product units, the pacing depends on the frequency of delivery and pickup as well as the size of the tote pans. When the tote pans are large and deliveries and pickups are infrequent, considerable freedom is given to the inspector, who can vary the time used in inspection among the items in the tote pan. As the tote pans get smaller and the frequency of delivery–pickup is greater, that freedom diminishes. At one extreme,

---

* Chapman and Sinclair (1975) show the effects of time pressure on performance when inspecting butchered chickens and jam tart pastries moving on belts.

the pace of deliveries and pickups of product units is strictly in the hands of the workers performing the pickup and delivery. Such a situation is analogous to inspection of product units coming in on a moving belt and going out at the same rate.

The belt-pacing example introduces another inspection issue—the distinction between *static* visual inspection of stopped items and *dynamic* inspection of moving items. The average amount of time available for inspecting an object on a moving belt decreases when the belt moves faster and when the spacing between successive objects is shorter. Shields and visual obstructions can further constrain the viewing time by forcing the observation to take place within a smaller viewing window. Some additional time is needed for the eyes to find and lock onto the moving target. After the eyes are visually locked onto the target, the target must be tracked while visual inspection occurs.

The latter observations lead to yet another important principle.

Principle 15.8

For dynamic visual inspection tasks, belt speeds must be adjusted to reflect the effect of visual obstructions on the available viewing time, the time needed to locate the target, and reduced visual acuity due to a moving target.

In static inspection, the inspector's eyes must simply move to a static object and then search it for a defect. Although a person's eyes can move very rapidly,* saccadic eye movements to a moving object take longer (typically from 300 to 500 ms) than movement to a static object. The latter situation often requires less than 200 ms. Based on this observation, one might wonder why dynamic inspection should ever be used. The answer is that dynamic inspection eliminates the need to move the inspected item on and off the belt during the inspection operation. In some cases, the associated time savings more than enough justify making people do a harder inspection task, especially if the inspection doesn't require a lot of visual acuity or the belt is moving quite slowly.

Another point is that it is not enough to talk about the number of product units inspected per hour, because this measure does not describe the variability of inspection times. Most studies of dynamic visual inspection have tested performance for equally spaced targets moving past an observer on a belt. This situation, of course, gives a fixed time between targets. In many practical situations, belt vibrations will cause the targets to vary in position so the distance between successive targets does not remain constant. This observation led Fisher, Moroze, and Morrison (1983) to model inspection of the moving targets as a queuing system.

One last observation is that visual inspection can be a difficult, tiring, and monotonous task. This leads to another important principle.

Principle 15.9

Rest breaks and job rotations may be necessary to maintain inspector performance at adequate levels.

Inspection can be a highly fatiguing task, especially if the necessary level of visual acuity is high and the decision rules for acceptance or rejection are complex. Rest breaks and employee rotations are both important strategies for avoiding losses in vigilance (Parasuraman, 1986).

---

* Yarbus (1967) reports saccadic eye movement velocities well in excess of 400 degrees/s.

Individual differences

The influence of gender, intelligence, personality, age, visual abilities, experience, and training on inspector performance have all been studied to varying degrees over the years.* Regarding the role of gender, there really isn't any strong evidence showing that one gender should be a better inspector than the other. However, many more women carry the title of inspector than men. Many people have hypothesized that the traditional social role of young women was to work on sewing and other projects that required a great deal of patience and tolerance for repetitive tasks. Certain researchers have suggested that women are consequently socially conditioned to be better inspectors.

Studies have also tested whether inspection performance is related to intelligence or personality. One study† revealed no correlation between inspection performance and whatever qualities intelligence tests measured at that time. Similar questions about inspection performance and personality variables as well as gender have also been investigated. In terms of the introvert–extrovert personality variable, conflicting evidence was reported, but there is some reason to believe that introverts might be better able to deal with the monotony of many inspection tasks. In general, the evidence doesn't seem strong enough to conclude that gender, personality, or intelligence will have much impact on inspector performance.

Some references in the literature suggest that younger people perform better in inspection than older people. Other references show that people with better visual acuity are better at visual inspection. Since more visual acuity is needed to detect fine details and that acuity deteriorates with age, it is not surprising that younger people would be better at visual inspection. Many companies now test visual acuity on an ongoing basis to make sure their inspectors can see well enough. In most cases, static visual acuity tests are used. This approach certainly seems reasonable when the inspected objects are stationary. However, if people are inspecting moving targets, the validity of the latter practice might be somewhat questionable. This follows because static visual acuity is only modestly correlated to dynamic visual acuity (DVA).‡ One study focusing specifically on visual inspection of moving objects found that a measure of people's ability to recognize blurred images was a better predictor of performance than static visual acuity (Nelson and Barany, 1969).§ It also has been argued that DVA should be a good measure of people's ability to inspect moving targets (Wentworth and Buck, 1982). However, the correlations between DVA tests and dynamic visual inspection performance were only slightly better than those for the blur test (about 0.7). DVA and the *blur test* both explain about 50% (i.e., $R^2 = 0.7 * 0.7 = 0.49$) of the variance in inspection performance, which is quite good, but still doesn't explain everything. These comments lead us to another basic principle.

Principle 15.10

Consider using measures of static and DVA to guide the selection of workers for inspection positions.

Other research has evaluated learning effects and whether training can improve inspection performance (Rizzi et al., 1979; Wentworth and Buck, 1982). These studies

---

* Wiener (1975) discusses this subject quite extensively.
† Wyatt and Langdon (1932) studied the relationship between intelligence and inspection accuracy.
‡ The relationship between static and DVA is not particularly strong. Some reported correlation coefficients are as high as 0.5, but Burg (1966) reported lower values. Even at the upper value, knowing the static visual acuity should really only reduce the variance in DVA by about 25%.
§ Nelson and Barany's *blur test* involved asking people to identify photographs that varied in how much they were out of focus.

showed that inspectors learned to deal with the limited time available when objects moved past them on a conveyer by changing their visual scanning patterns. When the belt speed increased or the available viewing time decreased, inspectors spent less time getting their eyes onto the object and spent most of the available time focused on the target. Both studies also showed that training can improve inspection performance. Research on the scanning patterns followed when people view radiological images also shows that people learn better visual search patterns with experience and further verifies that training can improve inspection (Kundel and LaFollette, 1972).

Illumination levels

The amount and quality of light falling on the inspected object is a very important job feature that strongly affects inspection performance. It is well documented in the ergonomics literature that greater illumination levels on targets improve human visual acuity. In more visually demanding tasks, greater illumination levels improve performance greatly. Many examples of this result can be found.* Thus, another principle of inspection is as follows.

Principle 15.11

If greater visual acuity is required in inspection, try adding more illumination.

It should be mentioned that too much illumination can be detrimental if the product units have high reflectivity. In that case, inspectors can be blinded by the reflected light in their eyes and may not distinguish defectives from nondefectives. As discussed in more detail in Chapter 4, the relationship between luminance (visual brightness), viewing time, and the size of an object at threshold detection has been studied quite extensively.† Simply put, people need more time and are less able to make out fine details when light levels are lower.

Our next principle reflects the fact that *more illumination* isn't necessarily the answer.

Principle 15.12

Try enhancing contrast by using a supplemental lighting system.

Many years ago, it was shown at Eastman Kodak that lighting systems could be designed to emphasize the visual differences between defects and nondefects (Faulkner and Murphy, 1975). Various experiments tested low- and high-angle lighting, spot versus diffused lighting, and different color light sources. In a number of examples, a little experimentation led to lighting arrangements that dramatically emphasized defects. Scratches on film under general illumination can be very difficult to see, for instance, but when low-level spotlights were used, scratches virtually leaped out at the observer. Figure 15.4 illustrates scratch enhancement by using a different viewing angle. Both the left and right figures are under the same light source, which was directly overhead, but the scratches were viewed at a low angle on the right. Similar enhancements are shown in Figure 15.5, where the surface-glazing illumination on the left brings out the seam in an article of clothing in contrast to the same seam under general illumination, as shown on the right.

---

* Blackwell (1970) shows many examples of this tendency. The most demanding task he studied was detecting black thread on black cloth.
† For example, Teichner and Krebs (1972) show that the logarithm of threshold luminance decreases linearly with the logarithm of added exposure time for fixed visual angles greater than 1 s of visual angle.

Figure 15.4 Scratches on a blackboard with two viewing angle under direct light source in overhead orientation.

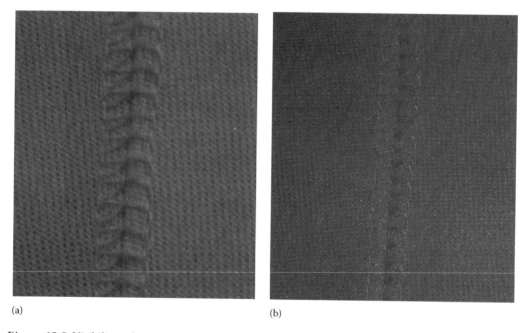

Figure 15.5 Visibility of seam in article of clothing under direct light source in overhead versus grazing orientation.

Other studies have shown that marks on steel plates left by rollers can be emphasized with low-angle lighting directed toward the inspector. Colored light and light with good color rendering properties often make it easier to see color differences. There are also reports of using polarized light and a black background to enhance detection of scratches on glass surfaces or burrs on blade edges.

In another example, a French razor blade company placed completed blades in boxes lined with black felt and then shone focused light across the sharpened edges in order to reveal imperfections on the blades. Moiré-pattern lenses have been used to enhance surface curvature, negative filters to enhance color defects, and stroboscopic lighting to enhance repetitive defects in rotating shafts or drums. There are many cases where rather simple tests with lighting can make defect detection so much easier.* Most of those alternative lighting schemes are easily implemented, so *do not rely solely on general overhead lighting* for inspection stations or for areas in or near machine tools where people perform this task.

Visual freedom and other issues

Many other variables also affect inspection: viewing constraints, for example. In one interesting study, targets were placed on a moving belt every 23 cm (9 in.). When the opening for viewing the moving targets was 46 cm, two targets were visible some of the time, but only one was visible with the 23 cm opening. A significant increase in inspection errors was observed for the 23 cm opening compared to the 46 cm opening. Note that regardless of belt speed, the average available viewing time per object was the same for either opening size. The available viewing time per object when two objects are viewed in a 46 cm opening (V_{46}) on a belt moving at a given speed is

$$V_{46} = \frac{46 \text{ cm}}{(2*\text{Speed})} = 23 \text{ cm/speed} \tag{15.5}$$

It is easy to see that the available viewing time per object above is the same as for one object in a 23 cm opening. Consequently, the essential difference between the conditions was that the larger opening gave the inspector more freedom to move their attention from one object to the other. A resulting principle is as follows.

Principle 15.13

Maintain visual freedom in inspection.

The greater visual freedom of the wider viewing window allowed the inspector to spend some extra time looking at the next item if they had time left over after looking at the first item. Even though the average time available was not increased, it is easy to see that this added flexibility should be helpful, because people are not robots who always need the same amount of time to do a task.

Another question is whether the direction the object is moving has an impact on inspection accuracy. One study found that for sheet inspection, inspectors were more accurate when the sheet moved toward the person (see Figure 15.6), rather than away from the person, or from side to side (Drury, 1975; Faulkner and Murphy, 1975). Inspection is also affected by attention and expectancies. Since full-time inspection tends to be a dull and monotonous task, maintaining attention can be a problem. Job rotation, breaks, and having other people around to talk to are all possible solutions to help break the monotony. Other strategies for improving performance include providing feedback on how well the

---

* See Eastman Kodak (1986) for an extensive table of special purpose lighting that several groups of Kodak ergonomists experimented with and some of their successes.

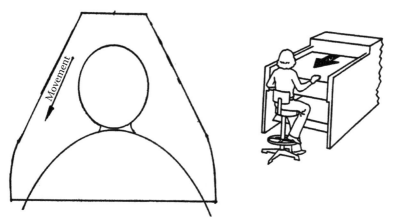

Figure 15.6 Direction of viewing a moving plastic sheet.

BOX 15.2   THE INFAMOUS *F* TEST

How many *F*'s are in the following sentence?

Finished files are the result of years of scientific study combined with the experience of many years.

This simple little exercise shows that most people do not look for targets in prepositions (e.g., *of*) or *F*'s buried in the middle of words (e.g., *scientific*) that do not have a strong *F* sound. Most people expect to find the *F*'s at the beginning of the words. Consequently, they find those *F*'s quite easily. Expectancy is a strong factor influencing our perceptions.

---

inspector is doing and feedforward (or predictive) information about defects.* Feedback and predictive information can help inspectors develop more accurate expectations of where defects are located. The strong role of expectations on inspection performance is illustrated by the infamous *F* test (see Box 15.2).

Analysis and improvement of inspection tasks

One of the better known strategies for analyzing and improving the performance of inspection tasks is a systematic checklist procedure devised by Drury (1992), based on maintainability work by Bond (1987). This procedure uses what are called factor/defect matrices. Basically the procedure starts by selecting an inspection task and identifying the factors associated with that task that potentially affect inspection. These factors can be organized into four principal categories:

1. Present—How the items are presented for inspection.
2. Search—How these items are visually searched to locate possible flaws.

---

* Sheehan and Drury (1971) and Drury and Addison (1973) show that feedforward and feedback effects can improve inspector performance.

3. Decision—Decide if each flaw is within or outside of an acceptable range.
4. Action—Take appropriate rejection or acceptance action.

The specific factors observed in any particular application vary greatly and do not always fit exactly into one of these categories. However, thinking about the categories while analyzing a task can help the analyst come up with potentially important factors. Many such factors have been discussed earlier, such as viewing time, movement, object size, contrast, color differences, illumination, task and decision complexity, and so on.

A key contribution of this approach is that the factors present in a given situation are organized in a matrix that also shows the defect the inspector is trying to find, along with relevant strategies for dealing with particular factors. This approach is best illustrated by looking at an example factor/defect matrix, such as the one shown in Table 15.2. As shown there, factors influencing performance are shown in the leftmost column of the matrix. The columns shifted over to the right list the defects that need to be detected by the inspector. The numbers assigned to possible combinations of defects and factors (i.e., the cells of the matrix) correspond to strategies listed at the bottom of the table. For example, Strategy 1, *Ensure that the component is visible for all inspector positions*, is a way of improving *accessibility* and, therefore, performance when the inspector is checking for missing, wrong, damaged, or reversed components. Strategy 12, *Provide comparison standards for all defects close to the line of sight*, is a way of making it easier to compare the object to a standard *good* or, in some cases, *bad* part. While the factor/defect matrix is not a quick fix, it does provide a systematic way to redesign or otherwise improve inspection tasks.

Inspection economics

Inspection should be principally an information-gathering process, directed toward eliminating *production* errors. To do that well, one needs to reduce errors in inspection because those errors misdirect management about the actual quality of their products. A secondary but important consideration is to minimize the costs of inspection. Economic considerations are particularly important when there is a chance of product liability problems.

Inspection costs

Inspection costs include those associated with setting up and operating the inspection station, the consequences of inspection errors, and production slowdowns caused by the inspection station. Inspection costs can be organized in a decision tree, as shown in Figure 15.7. In this figure, q is the proportion of defective items. As noted earlier, type-I inspection errors occur when the inspector rejects an item that meets specification, and type-II errors occur when the inspector accepts defective items. Accordingly, a type-I error only occurs when the item is good and so the probability of a type-I error (p_1) is a conditional probability. Thus, the joint probability of rejecting a good item is simply $(1 - q)p_1$. It similarly follows that the joint probability of accepting a bad item is qp_2, where p_2 is the probability of a type-II error. If we assume that N items are inspected, the results fall into four categories:

N_1 good items are rejected (type-I errors)
N_2 bad items are accepted (type-II errors)
N_3 good items are accepted
N_4 bad items are rejected

Table 15.2 Factor/Defect Matrix and Strategies to Improve Inspection

Factors Affecting Performance	Missing Component	Wrong Component	Damaged Component	Reversed Component	Missing Solder	Poor Solder	Excessive Solder
Accessibility	1	1	1	1			
Location of areas to inspect	2	2					
Handleability of circuit board	3	3	3	3			
Defect/field contrast	4	5		6			7
Field complexity	8	8	8	8	9		
Defect size	10	11		6	9		
Defect versus standard		12	12, 13	12		12	12
System noise		14			9	9	15
Action complexity	16	16	16	16	16	16	16

Strategies

1. Ensure that component is visible from a range of eye/board angles.
2. Subdivide board into visually logical areas to simplify recognition of area that needs inspection.
3. Use automated insertion that grips component to prevent falling off the board during inspection.
4. Place contrasting colored patch behind component to increase discriminability of missing part.
5. If possible, code components to match identifiers on the board.
6. Use obvious asymmetric component to detect reversals or mispositioned components.
7. Use a dark-colored undersurface of board to provide a good contrast with solder.
8. Subdivide board into visually logical areas so that patterns of parts are easily recognized.
9. Use a regular grid of solder joints to simplify detection of missing or inadequate solder.
10. Use a large colored patch to increase conspicuity of missing components.
11. Use components with lettering or identifiers in large printing.
12. Provide comparison standards for all defects close to the line of sight.
13. Use colored patches the same size and shape as components to simplify presence and orientation.
14. Reduce the number of different component types so that conspicuity of wrong component is better.
15. Keep the board undersurface as simple as possible to differentiate bridges and connectors.
16. Reduce the number of defect types searched for at a time by splitting the task to each board side.

Source: Reproduced from Drury, C.G., Product design for inspectability: A systematic procedure, in Helander, M. and Nagamachi, M. (Eds.), *Design for Manufacturability, A Systems Approach to Concurrent Engineering and Ergonomics*, Taylor & Francis, Ltd., London, U.K., Table 13.4, p. 213, 1992b. With permission.

Chapter fifteen: Quality control and inspection 569

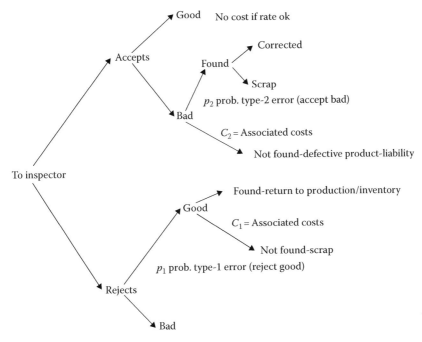

Figure 15.7 Tree graph of inspection error costs.

The expected number of cases observed for each of the four categories is as shown here:

$$N_1 = N(1-q)p_1$$
$$N_2 = Nqp_2$$
$$N_3 = N(1-q)(1-p_1);$$
$$N_4 = Nq(1-p_2)$$

Note that the cost of making a type-I error depends on whether the error is discovered. If a reinspection shows the rejected item is actually good, it can be brought back into production or inventory, which reduces the cost of the type-I error. If the error is not discovered, the company will lose the investment in materials and work that went into the item before that inspection, less any scrap value. Given that the probability of discovering the type-I error is p_d, the average cost of a type-I error is

$$C_1 = \text{cost of reinspecting} + p_d(\text{cost of returning a good part})$$
$$+ (1-p_d)(\text{materials} + \text{labor costs} - \text{salvage})$$

A similar analysis follows for type-II errors. Some inspection errors are discovered before the defective item leaves production. This allows the error to be corrected by either fixing the defective item or scrapping it. Other defective items will go out to the customer. Many of these defective items that get into the hands of the customer are simply returned to dealers who replace the items and return defective items to the company. Others may cause property damage or injury and may wind up in court as product liability cases. If the company loses the lawsuit, the cost of the type-II error can be extremely expensive, but the fraction of cases that gets this far is extremely low. Assuming that p_d is the fraction of type-II errors that is discovered prior to leaving the factory, the cost of a typical type-II error can be estimated as

C_2 = cost of reinspecting + p_d(cost of repairing or scrapping the product)

+ $(1-p_d)$(cost of returning product, unhappy customers, products liability, etc.)

The overall cost of inspection errors can be estimated as

$$\text{Insp. error costs} = N*[(1-q)p_1 C_1 + q\ p_2 C_2] \quad (15.6)$$

The latter equation does *not* include the cost of the inspector, effects of slow inspection in reducing production rates, and inspection station costs, all of which are additional costs. It is obvious from Equation 15.6 that reducing either p_1 or p_2 will reduce inspection error costs.

This point reemphasizes the importance of reducing inspection errors. We will return to this topic after addressing a couple other potentially important issues.

Location of the inspection station

The appropriate location of inspectors and inspection stations is another issue that can influence inspection costs. This issue can be approached in a number of ways but, due to its complexity, is sometimes analyzed using digital computer simulation.* An example of a situation where computer simulation was used is illustrated in Figure 15.8. The latter example evaluated the effects of locating an inspection station A in four different locations. The tested configurations all also included a second station B, always located at the end of the production sequence.

Five different production operations are also indicated in Figure 15.8 by numbered squares. Operation 5, indicated by a diamond shape, is an assembly operation where components from two different productions are joined together. The two different inspection operations are shown by circles. As noted earlier, four different configurations, each corresponding to different locations of inspection station A, were tested in the simulation.

For each configuration, numerous simulations were conducted using several different probabilities of type-I and type-II errors. The latter quantities were selected to be representative of false alarm and miss rates found in earlier studies of inspectors, such as those discussed earlier in this chapter.

---

* Computer simulation was discussed in Chapter 9. The case shown here is another example of that technique.

Chapter fifteen: Quality control and inspection   571

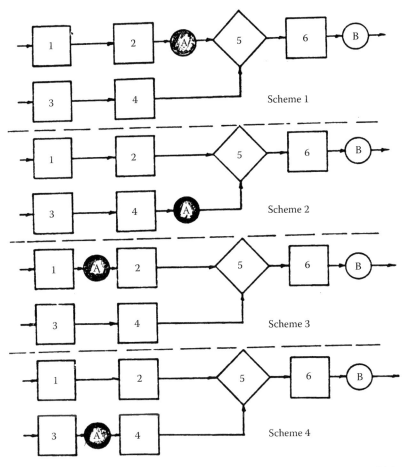

Figure 15.8 Four alternative configurations for locating an inspection station within production operations.

A summary of some of the results found in the simulations are given in Figure 15.9. As shown in the figure, the equivalent annual cost varied with the probability of type-I inspection error.* Note that in the cases shown here, no matter what the probability of type-I errors, Scheme 3 results in a lower cost than any other scheme. Also, Scheme 4 is almost as good as Scheme 3 and better than Schemes 1 and 2, where the inspection station A is located directly in front of the assembly operation. This result suggests that the advantage of particular schemes was primarily related to how much they starve the assembly operation of good parts. There is nothing special about this example other than that it shows how simulation can be used to evaluate the economic effects of where inspection stations are located. In other situations, the effect of inspection station locations may

---

* If a uniform series of costs at A dollars each period is equated to the actual present worth and over the actual duration, then the A parameter value of the series based on the average year gives the annual cost. In monthly time periods, for example, compute A per month and multiply by 12 for annual operations. It is that simple.

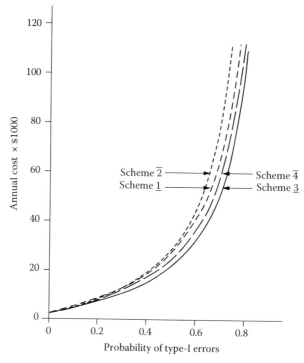

Figure 15.9 Equivalent annual costs of the four schemes of locating inspection station A as a function of the probability of type-I errors.

be very different. Predicting the significance of such effects can be difficult, which is why simulation can be helpful.

Learning and quality improvement

Quality improvement over time, due to organizational learning, is another issue with large economic implications. A focus on continuous improvement is one of the most important elements of the quality movement in the United States and directly corresponds to Deming's philosophy discussed in this chapter. Over the years, several studies have shown that a continued focus on the improvement of product quality does in fact lead to continuous improvement. Learning curves describing improvements in product quality over time have been described by Hedley (1976) for plastic production in the United Kingdom, electronic components in the United States, and some air frame deliveries from World War II on. Strata (1989) and others have verified this fact and developed learning curves to measure the rate of improvement that they observed. Choobineh (1986) also describes continuous quality improvements over time for programmable manufacturing cells.

Strata (1989) points out that the slope of the learning curve is related to the time required to identify, prioritize, and eliminate the causes of quality problems. Numerous learning curves and functional forms have been proposed showing how quality improves with experience. For example, Compton, Dunlap, and Heim (1992) developed an index

Chapter fifteen: Quality control and inspection

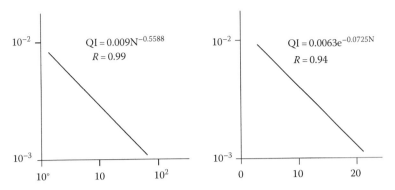

Figure 15.10 Failure rates as a function of the operations quantity.

showing how quality improvement is related to the number (n) of units produced. Two functional forms were developed, as given here:

$$Q_n = Q_1 n^m \tag{15.7}$$

$$Q_n = Q^* + Q_0 e^{-n} \tag{15.8}$$

respectively, for the power form and exponential learning models. In both equations, Q_n refers to the failure rate after n units are produced. Figure 15.10 describes the results reported for both models.

The key point is that changes in quality over time or with greater experience occur and can be described with simple learning curves. These equations can guide economic analyses by providing an estimate of how much learning can be expected and of how long it will take.

Final remarks

This chapter has analyzed some of the ergonomic aspects of quality control efforts intended to prevent quality from being lost or badly decreased before products reach the consumer. Inspection during manufacturing is a key element of quality control, serving an information-finding function as well as a means of eliminating defective components before products are assembled. However, when the first function of inspection is successfully achieved, the second is not needed.

Discussion questions and exercises

15.1 Inspectors who accept defective items are making errors of the _____ type. Inspectors who reject a non-defective item are making an error of the _____ type.

15.2 Inspection errors are known to change with the time T allowed to view the average item being inspected. The change in these errors as a function of T is best described as (check one)
An accelerating increase____
A linear increase____
A decelerating increase____
A decelerating decrease____
A linear decrease____
An accelerating decrease____

15.3 Recall the equation of inspection performance as a function of viewing time and other variables in dynamic visual inspection. Now suppose that your current inspection situation involves a belt velocity of 100 ft/min and the viewing time per item is 0.2 s. What is the probability of correct detections?

15.4 An inspector went through a batch of 200 parts with and without defects, accepting parts and rejecting parts as shown here:

	With Defects	Without Defects	Total
Accepted	8	83	91
Rejected	92	17	109
Totals	100	100	200

In terms of SDT, what is the detectability measure d' for this inspector? What is the SDT measure of beta and what does the value mean?

15.5 What human abilities or individual differences seem to be justifiable for employee selection testing purposes?

15.6. In SDT,
   a. What do d' values tell you about type-I and type-II errors?
   b. What does β tell you about these two types of error?
   c. What role does the ROC serve?
   d. Why has SDT been used in psychophysics?

15.7 It was discovered that inspection errors were being made at an inspection station; the number of type-I and type-II errors amounted to 10% and 5%, respectively, of those presented to the inspector. In terms of the theory of signal detection, what are the values of d' and b?

15.8 An inspector was observed to inspect 10,000 items and made 250 type-II errors and 100 type-I errors. Based on SDT and its conventions, what are the d' and b values associated with this situation if it is known that 85% of the items are good?

15.9 How, when, and why might increasing the operator's visual freedom improve performance?

15.10 Develop a simple factor/defect matrix for improving your ability to find mistakes in the printed (on paper) version of your last long term paper.

15.11 What are some of the problems associated with using SDT for describing dynamic visual inspection?

15.12 When and why might nonparametric forms of SDT be more appropriate?

15.13 Why should one expect errors in the inspection of factory products to change as the fraction (lot) defectives q increases from zero to greater fractions?

15.14 What is the likely benefit for an inspector to visually scan first for the most probable defectives? Please explain why.

15.15 What was the normal relationship between production and quality departments prior to the introduction of Just-in-Time (JIT) assembly methods? Why?

15.16 How could airline preboarding inspection be improved, and what benefits would be expected as a result? (For discussion.)

15.17 If you worked for the FAA, how could you help improve aircraft inspection for maintenance and replacement? (For discussion, assume that the agency and the airline would cooperate regarding your suggestion.)

Chapter fifteen: Quality control and inspection

15.18 Inspection records show that an inspector's record on a product was

	Conforming	Defectives	Total
Accepted	845	115	960
Rejected	15	25	40
Total	860	140	1000

What is the sensitivity d' of this inspector?
What is the β value and what does it mean?

15.19 Look up Deming's points for continuing improvement on the internet. Which ones do you most agree with? Disagree with? Why?

chapter sixteen

System reliability and maintenance

About the chapter

Industrial engineers and ergonomists can contribute to a maintenance and repair program in many different ways. Maintenance and repair activity can be reduced and in some cases eliminated by designing more reliable products and processes. System downtime after a failure occurs can be reduced by providing better means of detecting, diagnosing, and correcting faults. The timely delivery of technical information, tools, parts, and materials is another particularly important consideration. This chapter presents some of these strategies and design principles for increasing the effectiveness of maintenance and repair.

Introduction

Maintenance is not a glamorous activity, but it is extremely important to management and others who recognize it as one of the fastest accelerating costs in industry. Accordingly, maintenance is a challenge that ergonomics designers need to tackle.

Reliability is a closely related issue. Less reliable products obviously need to be repaired more often. This increases costs and reduces productivity. The complexity of many new production machines increases the chance that malfunctions and errors will occur and can make it difficult to detect and correct them before they cause system failures or accidents. Some of the more easily analyzed failures and error modes can be directly mapped to particular components of the system. More complex error modes involve interactions between system components and activities and may force difficult trade-offs to be made. For example, performing maintenance may increase reliability but, at the same time, may expose operators to additional hazards.

Part of the problem is that many of the newer forms of production equipment have not been around long enough to be standardized, so their designs vary greatly between manufacturers. Even within a particular manufacturer's product line, very significant differences are often present. To help deal with this issue, some manufacturers offer service programs and training courses to their customers. The delivery of technical information in appropriate form has also become a major concern. Because of competitive pressure, some manufacturers rush new products out before developing adequate instructions for operators and maintenance personnel. Other companies embrace a throwaway philosophy and assume the product will be replaced rather than repaired after it fails.

Designers can do many things to make repairs and other maintenance easy, or at least easier. Designers can also take steps to enhance reliability and continued usability so that maintenance and the cost of ownership become less problematic. Solutions to reliability and maintenance problems can be found at each stage of the product's life cycle. Implementing solutions in the design and development stages increases the chance of completely eliminating problems. Such solutions include design redundancy, modular components, fault detection, and error correction algorithms. Solutions applicable in the production phase are drawn from the field of quality assurance and include establishing specifications and tolerances, use of high-reliability components, online inspection and control procedures, testing and component burn-in, and employee training and motivation programs. In the use stage, well-designed maintenance programs play a particularly important role.

The following discussion will first introduce the topic of system reliability and availability. The basic elements of a maintenance program will then be outlined in some detail. Improving reliability is, of course, a primary objective of a maintenance program. Attention will then shift to approaches for reducing maintenance effort and means of support. Although these topics will be addressed in separate sections, an effort will be made to show the strong linkage between these topics throughout this discussion.

Reliability and availability

The reliability of a system is normally defined as the probability that the system will give satisfactory performance (i.e., not fail) over a given time interval. The frequency at which failures occur over time is the failure rate, $\lambda(t)$, and is measured as the number of failures occurring per unit time.* The mean time between failures (MTBF) is another commonly used measure of reliability. For a constant failure rate, λ, the MTBF is equal to $1/\lambda$.

In practice, production equipment may not be operational even if the equipment itself has not failed. Shipping, handling, storage, installation, operation, maintenance, and other activities can all reduce the proportion of time production equipment is operational. Operational readiness is the probability that a system is operating or can be operated satisfactorily and takes into account the effects of such factors.

Availability is the probability that, at any point in time, the system will be ready to operate at a specified level of performance. Availability depends on both the MTBF and the mean time to restore (MTTR) or repair the system after a failure occurs. This relationship is expressed as

$$\text{Availability} = \text{MTBF}/(\text{MTBF} + \text{MTTR})$$

$$= 1/(1 + \text{MTTR}/\text{MTBF}) \quad (16.1)$$

Equation 16.1 shows that the availability of a system improves when the MTBF is increased or the MTTR is reduced.† Preventative maintenance is performed to increase the MTBF. The MTTR will depend on how quickly, easily, and accurately a malfunction can be diagnosed and corrected.

Reliability analysis

Reliability analysis is an important function in many organizations. This is especially true in health care or other areas where there are safety critical systems or for factories using forms of automation, such as robotics or computer-integrated production systems, in which the failure of a single subsystem can be incredibly expensive because it stops all

---

* The reliability of a system generally changes over its life cycle. When a system is first put into service, the failure rate is often elevated. This so-called *infant mortality* phase often reflects quality problems or manufacturing defects. Toward the end of a system's useful life, the failure rate again increases as components of the system begin to wear out.
† If a system is required to have a certain availability factor and its MTBF is known, Equation 16.1 can be used to calculate an acceptable MTTR. Note, however, that the availability factor alone may be insufficient. In production circumstances, equipment having an MTBF of 500 h and a restoration time of 4 h (MTTR/MTBF ratio of 0.008) may be much more acceptable than equipment having a 5000 h MTBF but a 40 h MTTR (also a ratio of 0.008).

production. The movement toward lean manufacturing has compounded the problem by its focus on eliminating work in progress (WIP), because doing so reduces the buffering capability of having WIP between work stations.

Analytical tools are available that can be used to help identify the critical failure modes of a product or process. Such approaches guide the systematic identification of reliability problems, including the impact of human error, and help develop countermeasures. Following such procedures documents what has been done, helps prevent critical omissions, and in some cases, provides quantitative estimates of system reliability that can be used to measure the expected costs and benefits of particular countermeasures. It should also be emphasized that these approaches can be applied to evaluate reliability, safety, and productivity, alike.

Failure modes and effects analysis (FMEA) is a technique often used to analyze the effects of system malfunctions on reliability and safety (Hammer, 1993). Variants of this approach include preliminary hazard analysis (PHA) and failure modes effects and criticality analysis (FMECA). In all of these approaches, worksheets are prepared that list the components of a system, their potential failure modes, the likelihood and effects of each failure, and both the implemented and the potential countermeasures that might be taken to prevent the failure or its effects. Methods of assessing human reliability have also been developed. All of such approaches involve task analysis. They diverge significantly, however, in the degree to which they attempt to develop quantitative measures of error.

The technique for human error rate prediction (THERP—see Box 16.1) (Swain and Guttman, 1983) provides tabled estimates of human error probabilities (HEPs) for a limited set of tasks originally designed for use in the nuclear power industry. HEART is another well-known system for assessing human reliability that is similar in many ways to THERP (Williams, 1999). Application of THERP begins with human-error-related events in the system fault tree. The associated task is divided into a sequence of elemental subtasks that are organized into an event tree. HEPs are then assigned to the outcomes of each task from a tabled set of values. (The tabled values include provisions for the influence of performance-shaping factors (PSFs) and dependencies between subtask failures.) The probability of task failure is then calculated from the event tree. While advantageous in that it provides a quantitative measure of human error, the overall applicability of THERP to industrial tasks is limited by the limited set of HEPs available in its database.

The subjective likelihood index methodology (SLIM-MAUD—See Box 16.2) was developed to address the latter issue. SLIM-MAUD uses expert judgment as a means of estimating the probability of an accident or human error (Embrey et al., 1984). The approach follows a structured format in which human factors and domain experts estimate (both) the importance of PSFs as causes of accidents or human error for the analyzed task (Table 16.1) and the degree to which each PSF is present in the evaluated scenarios. This information is then used to develop a mathematical ordering of the likelihood of the analyzed errors.

Application of these approaches has become a well-known specialty area in the field of human factors called human reliability analysis. This approach is also applied by practitioners in the field of systems safety and reliability (Moriarty, 1988).

Fault avoidance and fault tolerance

Fault avoidance and fault tolerance are two other heavily emphasized strategies for improving the reliability of products and production equipment. Methods of fault avoidance focus on preventing or reducing system failures by improving the reliability of system components. Methods of fault tolerance, on the other hand, focus on reducing the effect on the system if one or more components fail.

BOX 16.1 ESTIMATING THE PROBABILITY OF HUMAN ERROR USING THERP

The technique for human error and reliability prediction (THERP) is one of the better-known methods of assessing the probability of human error and has been applied in the field of risk analysis to develop quantitative measures of the risk of accidents. The application of THERP involves the following steps:

1. Task analysis
   a. Break down task into elements.
   b. Analyze each task for possible errors.
   c. Talk through real scenarios.

2. Develop HRA event trees and assign HEPs
   a. HEPs tabulated for simple tasks such as monitoring, display reading, or adjusting control
   b. Tables are based on empirical data and assume that

   - The plant is operating normally
   - The operator not wearing protective equipment
   - The administrative control at industry one
   - The tasks are performed by licensed personnel
   - The working environment is adequate to normal

3. Assessment of PSFs and how they modify HEP
4. Calculations of human reliability:
   a. Effects of recovery factors
   b. Sensitivity analysis

To illustrate this approach, consider the following simple example where we are interested in assessing the probability of human error when monitoring a PRESSURE display in control room:

Step 1 is to divide the task into the four subtasks listed in the following:

1. Perform monitoring.
2. Read display correctly.
3. Initiate cooldown procedure.
4. Perform cooldown correctly.

Step 2 is to determine HEPs for each subtask. For example, let us assume that two operators are monitoring a pressure display. The check list is long, with no checkoff procedure. It turns out that one of the human error tables in the THERP system gives a HEP = 0.01 for this situation. Assume now that a PSF is that the personnel are experienced and task is discrete. A second table indicates that the HEP should be multiplied by 2 for this situation, which increases the HEP to 0.02. A similar analysis of the next subtask (i.e., of reading the display correctly) results in a HEP of 0.03, when it is assumed the display is chart recorder, read by experienced personnel.

In Step 3, the HEPs for subtasks are combined with failure rates of other system components to calculate overall system reliabilities.

Chapter sixteen: System reliability and maintenance

BOX 16.2 ESTIMATING THE PROBABILITY OF HUMAN ERROR USING SLIM-MAUD

SLIM-MAUD is a risk-assessment method in which expert judgments are used to develop quantitative estimates of the probability of human error as a function of performance-shaping functions, such as workload, stress, training, or environmental factors. The first step in this approach is to specify a subjective likelihood index (SLI) that measures the presence and importance of PSFs (see Table 17.7) expected to influence the probability of human error. The next step is to translate the SLI into a HEP $P(E)$. Normally, the latter step is done by assuming

$$\log P(E) = a(\text{SLI}) + b$$

For example, let us assume we are interested in predicting the probability of operator error $P(E)$ in a variety of driving situations. For the moment, let us assume that $P(E)$ is potentially a function of three PSFs, where

PSF1 → Reaction time required
PSF2 → Bad environment
PSF3 → Distractions

Also, let us assume we have three driving tasks of interest:

Task 1 → Ordinary driving on freeway, with a known probability of driver error $p(E) = 10e^{-4}$
Task 2 → Skidding on ice—with a known probability of driver error $p(E) = 0.1$
Task 3 → Driving in construction area—where $p(E)$ is unknown

The first step in SLIM-MAUD is to rate each of the tasks on each PSF, resulting in the following rating matrix:

	Task 1	Task 2	Task 3
PSF1 → RT	1	9	5
PSF2 → E	1	9	9
PSF3 → D	1	5	6

These ratings indicate that

- Task 2 is worst on RT and tied for worst on environment
- Task 3 is worst on distractions and tied for worst on environment
- Task 1 is best on all PSFs

The second step is to rate the PSFs in terms of importance. There are many possible rating methods that might be used. Perhaps the simplest is Saaty's pairwise method rating used in AHP. Along these lines, suppose your expert feels that PSF1 is twice as important as PSF2 and four times as important that PSF3. These pairwise comparisons result in the following matrix:

	PSF1	PSF2	PSF3
PSF1 →	1	2	4
PSF2 →	1/2	1	2
PSF3 →	1/4	1/2	1

(continued)

BOX 16.2 (continued)

The third step is to develop a weight vector (W) from the pairwise ratings that describes the importance of each PSF. In AHP, this is normally done by calculating the eigenvectors of the pairwise matrix. However, if the pairwise ratings are quite consistent, the eigenvector solution is closely approximated by a weight vector obtained by first normalizing each column and then calculating an average weight from the vectors obtained from each normalized column. To normalize a particular column within the pairwise ratings matrix, first sum up the values within a column. The normalized column vector is then obtained by dividing each element of a column vector by the column sum. For example, the first column in our example is

	PSF1
PSF1 →	1
PSF2 →	1/2
PSF3 →	1/4

The column sum $= 1 + 1/2 + 1/4 = 1.75$. Dividing each element in the column by the column sum gives a normalized vector for the first column of

$$W = \begin{matrix} 1/1.75 \\ 0.5/1.75 \\ 0.25/1.75 \end{matrix} = \begin{matrix} 0.58 \\ 0.28 \\ 0.14 \end{matrix} = \begin{matrix} w_1 \\ w_2 \\ w_3 \end{matrix}$$

where w_1, w_2, and w_3 correspond to the importance weights assigned to PSF1, PSF2, and PSF3, respectively. It turns out that all of the pairwise ratings in the example are perfectly consistent, so exactly the same result is obtained for columns 2 and 3. The next step is to use the ratings of each task on the PSFs along with the importance vector W to calculate a SLI for each task. This results in

$$SLI[Task1] = w_1 PSF1(Task1) + w_2 PSF2(Task1)$$
$$+ w_3 PSF3(Task1) = 0.58(1) + 0.28(1) + 0.14(1) = 1$$

$$SLI[Task2] = 0.58(9) + 0.28(9) + 0.14(5) = 8.44$$

$$SLI[Task3] = 0.58(5) + 0.28(9) + 0.14(6) = 6.26$$

We then solve for a and b in the equation $\log P(E)$ $a(SLI) + b$. This is possible because we know the error probabilities for Tasks 1 and 2. For Task 1, after doing some algebra, we determine that

$$\log(P(E)) = \log(10^{-4}) = a(1) + b \text{ and therefore } a + b = -4$$

BOX 16.2 (continued)

Similar analysis for Task 2 gives the following results: log (0.1) = a(8.44) + b, and therefore 8.44a + b = −1. A little more algebra gives the result:

$$a = 0.4032 \quad b = -4.403$$

The final step in analysis is then to solve for the unknown HEP for Task 3 using the equation

$$\log P(E) = a(\text{SLI}) + b$$

after plugging in the values of a, b, and SLI for Task 3, we get the result:

$$\log P(E) = 0.4032(6.26) - 4.403 = -1.879$$

A little more algebra then gives us the answer we wanted:

$$P(E) = 0.013$$

Table 16.1 Examples of PSFs Considered in THERP and SLIM-MAUD

Situational characteristics	*Psychological stressors*
Quality of environment	Stress level, duration
Supervision	Distractions
Work hours	Inconsistent cueing
Job and task instructions	*Physiological stressors*
Procedures required: written	Duration
Work methods	Extreme energy, material, etc., levels
Cautions and warnings	Movement constriction
Task and equipment characteristics	*Person-specific factors*
Demands: perceptual, motor, cognitive	Training/experience/knowledge/practice
Frequency and repetitiveness	Motivation
Interface	Physical condition
Group versus individual performance	

The most basic strategy for fault avoidance is to select highly reliable components. Safety margins can be provided by designing components to withstand stresses well above those expected in normal operation. Probability theory can then be applied to predict the probability of component failure, for particular safety margins and conditions (Kapur, 1992). In this approach, probability density functions $f(x)$ and $g(y)$ are used to, respectively, describe the stresses incurred, x, and the ability to withstand particular levels of stress, y. The probability of component failure is then calculated as

$$P(F) = P(y > x) \tag{16.2}$$

$$= \int_{-\infty}^{\infty} g(y) \int_{-\infty}^{y} f(x) dx \, dy \tag{16.3}$$

Preventing misuse of the equipment through training, supervision, and provision of technical information is another strategy for fault avoidance. Ergonomists are especially likely to play an important role here. Other commonly used strategies include burn-in, timed replacement, component screening, software screening and debugging, and accelerated life testing (Hammer, 1993). Burn-in methods involve subjecting the component to expected stress levels for a limited time prior to installation. This approach is useful when a component has a high *infant mortality* failure rate. Timed replacement involves the replacement of components prior to entering the *wearout period* in which failure rates are increased. Component screening involves the establishment of tolerances, or upper and lower limits for component attributes. Components outside of the tolerable range are rejected. Software screening and debugging involves testing the software under a wide variety of foreseeable conditions, as well as applying methods used in software engineering to eliminate errors in code. Accelerated life testing involves testing components under more adverse conditions than those normally present.

Fault tolerance takes a different approach. Ideally, when components fail, systems should lose functionality gradually rather than catastrophically. A variation of this idea is that the system should shut down or go into a neutral state when critical components fail. Fault tolerance is a necessary strategy, because highly reliable components can be very expensive. Furthermore, the complexity and large number of components within some systems make hardware and software failures inevitable. Other methods for providing fault tolerance include providing redundancy, error correction, and error recovery.

Redundancy is said to be present when a function of the system is performed by two or more system components. This situation corresponds to a configuration in which system failure can theoretically occur only when all of the parallel components fail. The use of standby systems illustrates a similar form of hardware redundancy. Other approaches that can reduce the time needed to correct malfunctions are discussed next.

Fault Detection Algorithms

Fault detection algorithms play a general role in both improving and maintaining the reliability of production systems and are a critical element of real-time fault tolerant computing. A wide variety of fault detection algorithms have been developed. Analytical redundancy is a concept often exploited by these algorithms—that is, knowledge of the structural and functional redundancy of the system is used to guide the comparison of data obtained from different sensors. System faults cause sensors to provide data that deviate by more than the fault detection threshold from values calculated or obtained from other sources. For example, one sensor might indicate a part is still in the machine, while another indicates the ejection cycle has been completed.

Determining the appropriate fault detection threshold can be difficult, because both sensor noise and the criticality of particular errors change, depending on the operation performed, component failures, and other factors. Fault detection algorithms can be quite good at detecting internal malfunctions, but symptoms of inferior performance are not always easily identified without human inputs. For example, people can detect many problems in the quality of color laser printouts that are very difficult to detect using sensors.

> **Fault Tolerant Computing**
>
> Fault tolerant computing schemes that provide for error correction and recovery are used extensively in advanced production systems. Such methods include parallel processing of information, use of modular code, voting schemes, and a variety of system rollback schemes. Parallel processing and the use of modular code are both forms of redundancy. Voting schemes and probabilistic methods provide a means of error correction by comparing information received from multiple redundant sources. System rollback schemes store system-state information. This theoretically allows the system to return to an earlier correct state and resume operation after an error occurs. In the simplest implementations, the system will retry the operation after a brief pause. More complex schemes will respond to the error by appropriately reconfiguring the system before attempting the operation again. Expert systems provide another means of implementing fault tolerance.

Maintenance programs

Modern maintenance programs focus on predicting and preventing failures rather than simply fixing them after they occur. Well in advance of installation and start-up of production equipment, the maintenance manager will want to establish a program of preventive maintenance, coordinate it with the production manager, and provide for its support. Formally establishing a maintenance program is an essential first step in making sure production equipment will be properly maintained. Some of the tasks coordinated in a maintenance program are to

1. Specify, assign, and coordinate maintenance activities
2. Establish maintenance priorities and standards
3. Set maintenance schedules
4. Store and provide necessary technical information
5. Provide special tools, diagnostic equipment, and service kits
6. Control spare parts inventory
7. Provide training to maintenance personnel and operators
8. Measure and document program effectiveness

Industrial engineers and ergonomists can make many important contributions to a maintenance program as briefly discussed next.

Assigning and coordinating maintenance activities

Maintenance and production have traditionally been viewed as independent activities. However, coordination of maintenance activities with production is essential, as is developing a mutual understanding of the needs and purposes. Production must allot time for maintenance. If a one-shift operation is involved, scheduling should be no problem. If it is a three-shift operation, then either weekend maintenance must be planned for or downtime must be scheduled on one of the shifts. Some typical maintenance activities or tasks are as follows: repair, service, inspections, calibration, overhauls, testing, adjusting, aligning, and replacing components.

In some companies, the particular activities are performed by different classes of skilled people (e.g., electricians, millwrights, electronic technicians, hydraulic trade persons) who have separate and divided responsibilities. Other companies use personnel with integrated skills. There are merits to both approaches and the solution is what works best for the specific installation. However, for the benefit of efficiency, proficiency, and fast response time, the trend is toward integrated skills and responsibilities. The latter trend, inspired much by developments in Japan, both emphasizes the critical role the operator must play in maintenance and focuses on cooperation between production and maintenance departments as the key to increasing equipment reliability (Nakajima, Yamashina, Kumagai, and Toyota, 1992).

Rather than assigning all responsibility for maintenance to the maintenance department, operators are now being assigned many routine maintenance tasks. Tasks assigned to operators include (Nakajima et al., 1992) the following:

1. Maintaining basic equipment conditions (cleaning and lubrication of equipment, tightening bolts)
2. Maintaining operating conditions through proper operation and inspection of equipment
3. Detecting deterioration of equipment through visual inspection
4. Early identification of symptoms indicating abnormal equipment operation

Maintenance personnel are expected to concentrate their efforts on less routine tasks. Such tasks include the following:

1. Providing technical support for the production department's maintenance activities
2. Restoring deterioration of equipment through inspections, condition monitoring, and repair activity
3. Improving operating standards by tracing design weaknesses and making appropriate modifications

In larger companies, maintenance and repair personnel often work in crews. Sometimes these crews are organized by craft, such as plumbers, electricians, and carpenters, who go out as individuals or pairs to make specialized repairs. When one crew has completed its work, another crew is called in. Ordinarily, someone precedes the crews to estimate the amount of work to be done and to identify which crews are to go and in what sequence. Although such a plan is normally followed, something inevitably goes askew and revisions must be made. Revisions create problems in transportation when one member of a crew goes on to another job while the other remains to finish up the work at the previous job. These problems are minimal, of course, when the plant size is relatively small. Another set of difficulties ensues when crafts need to be sequenced back and forth on a job. Of course, the intelligent way is to revise the crew so that its members are multicraft personnel. In contrast to production operations, maintenance activities must be ultraflexible to accommodate jobs that arise unexpectedly.

Sampling Maintenance Activities

As suggested earlier in chapter 8, activity sampling is a useful way to evaluate maintenance activities. Time spent on particular tasks and specific causes of unplanned personnel idleness or loss of productivity, including idleness due to various forms of interference, can be identified. Maintenance personnel spread out all over the work site can carry devices, which page them at random times. After being paged, the worker presses a button on the device corresponding to their current activity or otherwise records what they are doing. Special equipment can be purchased for this purpose. Portable pagers with a numerical response pad are also suitable. When space is not a problem, an individual can be assigned to make the observations and identify the activities observed.

Figure 16.1 shows some of the types of information that can be obtained with this approach.

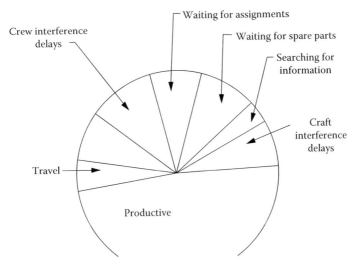

Figure 16.1 Lost productivity of maintenance personnel identified by activity sampling.

Maintenance and repair work can be planned just like production work, but it is necessary to know the exact work needed and to predict performance times for it. Thus, some type of performance time standards is needed for these operations. Work standards can be set using standard data systems (see Chapter 6)* for various families of maintenance tasks as a function of variables that affect the amount of work to be done. For example, floor-cleaning times can be estimated on an area basis for task requirements (i.e., sweeping, mopping, and waxing) as a function of factors that make the task more difficult (e.g., kinds of floor surfaces, moving many chairs and tables, or causes of floor staining). Many organizations have such standards, and some consulting firms often have access to government or industry standards on such matters. A special form of setting indirect work standards is a technique called *slotting*, using Universal Indirect Labor Standards. The general idea is to group jobs of a given kind into small collections, called slots. All the jobs in the same slot are assigned the same time standard.

---

* Recall that chapter 6 includes a section on the use of standard data systems to predict performance time. As expanded in there, such systems are normally developed from historical data using methods such as regression analysis.

Setting maintenance priorities and standards

Maintenance efforts must be prioritized. This follows because maintenance programs are expensive, and the resources allocated to maintenance must consequently be allocated efficiently. For example, the annual cost of robot maintenance has been estimated to be about 10% of the initial procurement cost (Dhillon, 1991), but obviously these costs will vary depending upon the specific setting.

Maintenance priorities must reflect the cost to the company of equipment and process failures. Identification of failure modes and potential countermeasures is guided by past experience, standards, checklists, and other sources. This process can be further organized by separately considering each step in the production of a product. For example, the effects of a power supply failure for a welding robot might be separately considered for each step performed in the welding process. The priority assigned to particular equipment should reflect the degree that failures disrupt production, waste expensive materials, reduce quality, increase downtime, and impact safety.

Once priorities are established, maintenance standards and procedures, as well as maintenance plans, should then be established that reflect these priorities. Maintenance priorities should also be reviewed on a periodic basis because they are likely to change over time for a variety of reasons. Typical high-priority activities include inspection and test procedures, lubrication of effectors, change of internal filters in hydraulic and pneumatic robots, test of hydraulic fluids for contamination, check of water separators in pneumatic equipment, change of air filters for cooling systems, and functional checks of controls and control settings. Maintenance standards and procedures will include specifications, tolerances, and methods to be followed when performing maintenance tasks. Checklists, maintenance manuals, and training programs are common means of making sure this information is made available to maintenance personnel.

Maintenance schedules

Maintenance activity is generally scheduled to take place either at regular intervals, after breakdowns, or when certain conditions are met. *Preventive* maintenance involves a number of different activities performed on a regular basis before a problem occurs, such as changing the oil in an automobile after every 5000 miles of operation. These and other regularly recurring maintenance activities typically follow a schedule suggested by the manufacturer. Inspections are especially likely to be performed on a periodic basis. Daily inspections are normally performed by the operator prior to beginning operations. Less routine inspections are often performed on weekly, monthly, biannual, and annual schedules by maintenance personnel. After sufficient running time and performance results have been accumulated, a company may choose to deviate from the original schedule provided by the manufacturer. Replacement analysis provides one means of formally analyzing this problem by comparing the expected value of replacing versus not replacing parts at particular points in time.

Corrective maintenance is called for when a machine or device fails to work properly and someone must diagnose and correct the problems. *Breakdown-based* maintenance policies schedule the replacement of parts and other repair activity after failures occur. This approach can maximize the life of equipment components that might otherwise be replaced prior to completing their useful life. On the other hand, a breakdown-based policy can result in disruptive interruptions of production and serious damage to equipment that far outweighs the cost of replacing components before they begin to wear out.

Condition-based, or predictive, maintenance policies schedule part replacements and repairs after the equipment shows signs of deterioration. This approach theoretically results in less frequent maintenance activity than interval-based policies and fewer disruptions of production than a breakdown-based policy. Condition-based maintenance policies typically schedule inspections at regular intervals by the operator or maintenance personnel. Control charts and other methods used in the field of statistical quality control (Brumbaugh and Heikes, 1992; Speitel, 1992) are also used to determine deterioration over time in production quality and a consequent need for equipment maintenance activity, including nonroutine inspections, adjustments, recalibration, part replacement, or other repairs.

Other techniques used to detect deterioration include vibration analysis, wear debris (contaminant) monitoring of lubricants and hydraulic fluids, behavior monitoring, and structural integrity monitoring (Nakajima et al., 1992). More complex equipment, such as industrial robots, contain internal diagnostic functions that are useful during troubleshooting and can also signal the need for conducting maintenance on the machine.

Storing and delivering technical information as needed

In today's industry, it is increasingly important to keep technical information available when needed. Production equipment is becoming more and more sophisticated because of national and overseas competition and because higher labor costs in the United States must be offset with greater effectiveness and efficiency by U.S. companies. These facts lead industry to buy more flexible and smarter equipment that, in turn, has created substantially greater need to make technical information about repair and maintenance available to the people responsible for these functions. Good networked computer systems have helped. Once the technical information is in the system, it can be called up at any time and delivered immediately to the individuals who need it. In fact, e-mail routes appear to be well suited for some of these functions.

Tools and Diagnostic Equipment

Maintenance normally requires that maintenance personnel be provided special tools and diagnostic equipment. Manufacturers can provide help and guidance regarding this issue and generally provide such information as part of the documentation package. Ordinarily, the investment in special tools will be minimal, but the cost without them can be significant. Among the most common special tools are circuit card pullers, torque wrenches, seal compressors, accumulator charging adapters, and alignment fixtures. The list will be, to some extent, peculiar to the specific equipment involved.

In addition to special tools, the maintenance department must be equipped with proper diagnostic equipment. In some cases, this will only involve oscilloscopes, multimeters, gauges, and similar devices probably already available in the department. In other cases, special diagnostic tools must be purchased. Specialized equipment that is infrequently used by multiple crafts persons normally does not need to be duplicated for multiple crews.

Specialized tools, fixtures, and equipment used in production operations are often held in tool cribs. Tools and equipment used by maintenance personnel can be held in the same cribs used for production operations. It is sometimes better to provide a separate tool crib for maintenance operations to allow quicker and more flexible access. Maintenance tools are often needed at the most unexpected times.

Many companies are now putting out maintenance and repair instructions on compact disks that are hooked to or are part of the manufacturer's or the user's computers. That arrangement allows repair–maintenance personnel to interact with the instructions directly to find the relevant materials. Video images can explicitly show conditions under which the maintenance worker must check on the machine. The more reliable a manufacturing machine, the longer it functions before needing repair, and the less the maintenance and repair people remember about maintenance procedures. Hence, the technical information system is especially important when equipment is reliable.

Tools, component parts, and materials delivery

Maintenance personnel often will not know which tools, component parts, or materials they need until they start doing the task. This is particulary true for corrective maintenance. For this reason, maintenance personnel may need help in tracking down unexpected parts requirements and getting them delivered to the appropriate personnel. One issue is that maintenance personnel will need part numbers and other technical information before component parts in inventory can be issued and/or ordered with appropriate priority. Another issue is that the person performing the repair on the shop floor will often need to contact appropriate personnel in the company that made the production equipment for assistance. This is especially likely when interpreting system diagnostics. If such interactions indicate the need to get a new component part as soon as possible, so be it. Many companies simply cannot afford to have complex production equipment unavailable.

An adequate selection of spare parts can greatly reduce downtime after equipment failures occur, especially when the company uses a breakdown-based part replacement policy. Most manufacturers will provide a list of parts they recommend keeping on site, based on the number of machines in the facility and the probability they will be needed. The cost of keeping spare parts available is often small compared to the cost of downtime. Manufacturers also often provide preventive maintenance kits. When the necessary components are included ahead of time in kits, a long list of items need not be pulled from inventory before going to the job site.

Spare parts must be in secure areas and stocked in an organized manner to reduce the time required to find them. A well-designed inventory control system will also specify reorder points and order quantities that appropriately balance the cost of keeping parts in inventory against the expected cost of running out of needed parts. Inexpensive parts or parts that frequently fail normally should be kept in stock. On the other hand, it may be cost effective to return expensive parts, or parts that rarely fail, to the supplier for repair or replacement rather than keeping them in inventory.

Training for maintenance and repair

A skilled workforce is a prerequisite of any well-planned and well-executed preventive maintenance program. Personnel need to be trained on how to maintain, troubleshoot, and repair production equipment. They also need to know how the equipment works and how to operate it. Although manufacturers will usually have a field service organization

Chapter sixteen: System reliability and maintenance

of highly trained technicians, it is essential that these skills reside in-house so that the user can be virtually self-supporting and defer to others only when confronted with unusual problems. The timeliest use of manufacturers' technicians is during installation and start-up phases, but even then, it should be a team effort during which the customer personnel receive on-the-job training.

Aircraft Maintenance Errors and Safety

Aircraft maintenance is one of the best managed maintenance operations. However, despite the oversight of the Federal Aviation Administration, maintenance errors creep in. A study by Boeing attributed 39 of 264 (or 15%) major aircraft accidents to maintenance during 1982–1991. Of those 39 accidents, 23% involved incorrect removal or installation of components, 28% were due to manufacture or vendor maintenance/inspection error, and 49% were design related.

Another study of airplane maintenance disclosed maintenance error and some estimated costs of those errors: 20%–30% of in-flight engine shutdowns were caused by maintenance errors at an estimated cost of $500,000 each, 50% of flight delays were due to maintenance errors at an estimated cost of $10,000 per hour of delay, and 50% of flight cancellations were due to maintenance errors, with average cancellation costs of about $50,000.

Engine shutdown rates due to maintenance failures differed between air carriers by a factor of 10, showing that some companies much better manage maintenance problems than others. What is clear about maintenance errors is that technicians do not make errors on purpose. Errors result from many contributing factors, but a large portion of those factors can be better managed.

Although manufacturers of production equipment will often provide training to their customers, there still may be reasons to establish in-house maintenance training courses. Prior to starting this effort, management must understand the need for training and be committed to meeting that need. Once management is committed, the next step is to meet the people who will be directly involved to describe and discuss what is being planned and why it is needed. A series of *working* meetings should then follow that involve personnel at all levels. During these meetings, the concerned parties need to express their concerns, discuss potential problems, and develop plans of action. In a cooperative working environment, the often-neglected maintenance department will ideally become an integral part of the team, ready and able to fulfill its function when and as needed.

During this process, it is essential that the group determines what skills are needed, what skills are lacking, who will be assigned what tasks, and how they will be trained. After the initial training program, it may be useful to train and upgrade personnel on a continuing basis to maintain skills and make sure that personnel learn how to maintain newly acquired equipment, perform new maintenance procedures, and other important tasks.

Record keeping and measurement of maintenance program effectiveness

A final consideration is that the effectiveness of the maintenance program should be periodically reviewed. This, of course, means that all maintenance work has to be carefully documented, as does the cost of the maintenance program and its influence on system performance. Accomplishing this latter goal requires that a failure reporting and documentation system be established. Inspection and repair records must also be kept. At a minimum, it is necessary to describe the failure and the repair, when it happened, how long the equipment was operated since the last repair, the repair person's name, the place of repair, the model and serial numbers of repaired equipment, the repair starting date and time, and the type of inspection or event that triggered the repair.

In the short term, documentation of what has been done provides invaluable information for people working on different shifts. In the long term, documentation of what has been done, and of the effectiveness of doing so, provides a database for establishing trends and fine-tuning the maintenance program. It can also provide the manufacturer with a history should there be a chronic problem. To help ensure that accurate and complete data are collected, repair personnel and operators must be made aware that documenting failures is as important as repairing them. Forms used to report failures, and record inspections and repairs, must be made conveniently available, simple, and clear. Simply put, filling out the necessary documentation forms should be quick and easy.

Ergonomics personnel can use a number of strategies to evaluate the effectiveness of the maintenance program. As noted earlier, one of these strategies is to capture information on the maintenance operations through activity analysis or work sampling (see Chapter 8). Figure 16.2 describes some possible results of an activity sampling study conducted over a longer time period to measure effectiveness. This diagram illustrates the proportion of effort over time spent in different types of maintenance activities, such as emergency repairs, unscheduled repairs, scheduled repairs, inspections, and preventive maintenance. Both emergency and unscheduled repairs are clear signs of problems when they are large or frequent. Overtime is also a high-cost item, which needs to be tracked. Sometimes breakdowns result through no fault of anyone, but once the problem is known, maintenance management can address it and make policies, which can better preclude these difficulties in the future.

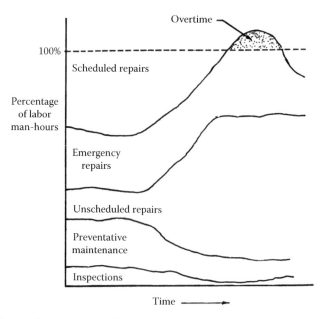

Figure 16.2 Tracking maintenance activity over time.

Reducing maintenance effort

Several approaches can be followed to reduce a company's maintenance- and repair-related costs. Some maintenance and repair activity can be eliminated entirely through product design. More reliable and maintenance-free products are often available, and service contracts can be purchased from manufacturers. If maintenance and repair activity can't be eliminated, it can be often reduced or simplified through ergonomic design. The following discussion describes several methods and principles for eliminating, simplifying, and otherwise reducing maintenance effort.

Fault and malfunction detection

Malfunctions of machines and equipment used in industry today can, for the most part, be grouped into three basic categories: mechanical, electrical, and software. For each type of malfunction, there is normally an associated set of observable symptoms and less readily observable causes. The symptoms are generally physical or mechanical in nature and remain observable as long as the malfunction is present. Causes, on the other hand, often correspond to transient events that are no longer observable at the time troubleshooting takes place.

The symptoms associated with particular malfunctions vary greatly in their degree of specificity. Sometimes the equipment does not function at all. More often, the equipment shows symptoms of inferior performance or of an impending breakdown that must be detected before any troubleshooting or repair activity can take place. As noted earlier, many companies schedule inspections by workers or maintenance staff to identify such symptoms. Unfortunately, human inspectors make a lot of mistakes (Drury, 1992). NASA's aviation reporting system recently showed 90 maintenance-related incidents, of which 60% were due to failure to follow and complete operating maintenance procedures. In regard to the visual detection and location of faults, the mean probability of detection has been reported to be as low as 0.46–0.50, with standard deviations of 0.13–0.15.

Ergonomists can take steps to improve fault detection by workers and inspectors. A fundamental design principle that can help ensure faults will be detected in a timely manner is as follows.

Principle 16.1
Make faults or conditions leading to a fault obvious to the operator.

A wide variety of approaches can be followed to attain this goal. One approach is to include fail-safe features that stop the equipment from operating after a fault occurs. Needless to say, if the machine or equipment stops operating, it becomes obvious that something has gone wrong. *Alerts* and *status displays* of various types can also be provided that keep the operator informed.* In some cases, the status information can be sent directly from the machine to the maintenance department or the manufacturer via intranets, Internets, or telephone lines. The latter approach is especially useful for triggering needed maintenance activities before a malfunction or, in extreme cases, an accident occurs. In many cases, the information required to detect a potential malfunction is similar to that needed to start up or run the system. The primary difference is that certain variables have values distinctly different from normal operations. Fault detection algorithms play a critical role in identifying critical deviations of this type.

Alerts and status displays of various types are a common feature in modern machinery and equipment. Examples include warning lights, buzzers, and tones, or

---
* The design of displays is addressed in Chapter 12.

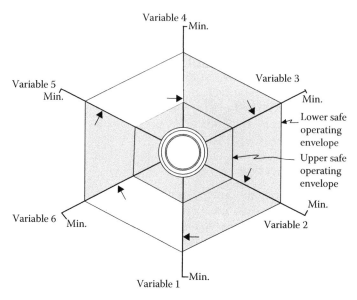

Figure 16.3 System safety parameter display for nuclear power reactor controls used by Westinghouse.

LED displays such as those on printers and copiers. A more complex status display developed by Westinghouse designers for use in nuclear reactor control rooms is shown in Figure 16.3. The status of eight different variables is shown on this particular display. Each variable is displayed on one of the lines stretching outward from the center of the display. The two hexagons describe upper and lower operating limits of the variables. The operator consequently can monitor operations using a single display and easily tell if something important has gone wrong in the system. This leads us to the next principle:

Principle 16.2

Help users identify symptoms of inferior performance by providing benchmarks that allow current performance to be easily evaluated.

People can identify much smaller differences when they can directly compare items against each other. Consequently, they should be much better at detecting problems in current performance if they have something to compare it against. There are many different ways of providing benchmarks. One common approach is to give examples of defects to the user or operator. For example, the manuals provided with printers will often include examples of print defects that can be compared to current printer outputs. One problem with this approach is that users can start detecting problems when the product or equipment is still within tolerance! This is a good reason to also give the user examples of normal, or within tolerance, performance.

A more ambitious approach is to simulate and display normal behavior of the system to the user in real time, as illustrated in Figure 16.4. The system might compare actual and desired levels of production and notify the operation when there is a discrepancy. In other configurations, it might just display the actual level to the operator.

The next closely related principle applies to situations where system performance is changing over time.

Chapter sixteen: System reliability and maintenance 595

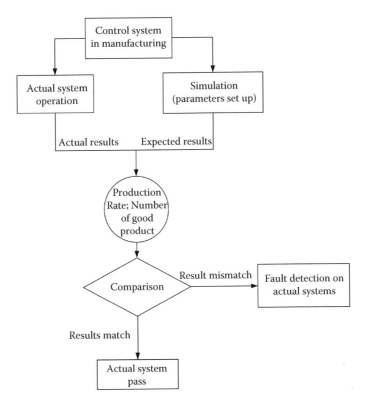

Figure 16.4 Use of system simulation to detect faults.

Principle 16.3

Help users identify trends and changes by making it easy to compare current performance to past performance.

It can be very difficult for people to detect changes in performance. This is especially true when the changes occur slowly over time. For this reason, it is often useful to provide plots of performance over time. Control charts, as mentioned earlier, are one very useful method for identifying changes in performance, such as quality problems, that should trigger maintenance activity.

Many types of plots can be used. For example, Anyakora and Lees (1972) found that control-room operators in chemical plants associated particular faults with specific patterns of fluid flow out of valves and pumps. Figure 16.5 illustrates some of the patterns identified by operators.

Principle 16.4

Provide Job Performance Aids (JPAs) and training to ensure inspection procedures are properly followed.

It is essential that ergonomists evaluate the perceptual and cognitive demands of inspection tasks.* Before better inspection procedures can be developed, the designer must know what symptoms the inspector should be looking for, how difficult they are to identify,

---

* The topic of visual inspection is covered in more detail in Chapter 15. Since inspection is almost always part of maintanance, much of the discussion in chapter 15 is relevant here.

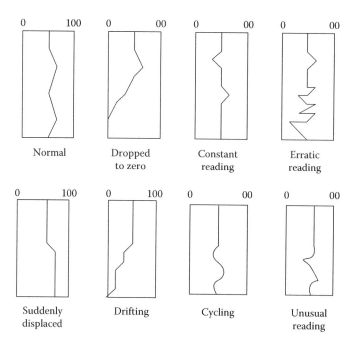

Figure 16.5 Strip chart recording patterns corresponding to faults in chemical operations. (From Anyakora, S.N. and Lees, F.P., *Chem. Eng.*, 264, 304, 1972.)

and why they might be missed. These issues can then be focused upon in employee training. In many cases, the problem is that inspectors forget to perform important steps in the process. Too often this happens because the inspector does not have the necessary technical information with them. In other cases, the information is available but buried in a large service manual.

JPAs and training are both potential solutions to the latter problem. JPAs often are formatted as checklists or step-by-step instructions tailored to the specific inspection task. JPAs also may contain tables, graphs, illustrations, flow charts, or other information. The essential point is that JPAs focus on providing the information most relevant to the particular task being performed, as opposed to a service manual that contains a lot of irrelevant information. The use of checklists and step-by-step instructions within JPAs helps prevent inspectors from forgetting to do certain steps and is especially appropriate for nonroutine activity. Checklists can also require inspectors to indicate the presence or absence of critical symptoms. Some companies are now using computer systems to print out JPAs for maintenance staff before they go out to inspect particular equipment. Another approach is for the inspector to bring a handheld computer with them that contains the step-by-step instructions and/or other checklist items.

Diagnostic support

Once a fault is known to exist, its cause must be located and rectified. Symptoms of a malfunction can be classified into three basic categories: alarms, inferior performance, and stall (or inoperability). Alarms are messages displayed by the machine or equipment when it detects a malfunction. Alarms provide information regarding the type of malfunction that has taken place, but it must be recognized that any given alarm can be triggered in many different situations. Inferring the cause of an alarm signal can therefore be difficult and time consuming but, in most situations, less so than for the categories of stalls and inferior performance.

JPAs versus Training

Smillie (1985) provides a set of rules for choosing between JPAs or training. These rules, though not intended as absolute principles, appear to be quite useful when they are considered in the context of a particular task to be performed. They include the following:

1. Ease of communication. Put into training those tasks that are hard to communicate through words. Put into JPAs those tasks that would benefit from the inclusions of illustrations, tables, graphs, flow charts, etc.
2. Criticality of the tasks. Put into training those tasks for which the consequences of error are serious. Put into JPAs those tasks that require verification of readings and tolerances.
3. Complexity of the task. Put into training those tasks with difficult adjustments and procedures that can only be achieved through practice. Put into JPAs those tasks that require long and complex behavioral sequences that are extremely costly to teach.
4. Time required to perform the task. Put into training those tasks with a required response rate that does not permit reference to a printed instruction. Put into JPAs those tasks that are long and require attention to detail.
5. Frequency of the task or similar task. Put into training those tasks that are easy to learn through experience. Put into JPAs those tasks that are rarely performed.
6. Psychomotor component of the task. Put into training those tasks that require extensive practice for acceptable performance. Put into JPAs those tasks in which references to printed instructions are not disruptive to task performance.
7. Cognitive component of the task. Put into training those tasks that require evaluation of numerous existing conditions prior to making a decision. Put into JPAs those tasks in which binary fault trees can be developed into a decision aid.
8. Equipment complexity and accessibility. Put into training those tasks in which equipment is easily accessed. Put into JPAs those tasks that provide detailed procedures to properly access equipment.
9. Personnel constraints. Put into training those tasks that require a team effort. Put into JPAs one- or two-man tasks.
10. Consequences of improper task performance. Put into training those tasks in which an occasional error will not damage equipment. Put into JPAs those tasks that require branching, such as a diagnostic decision aid that lists failure mode symptoms.

Inferior performance is a less specific category of malfunction-related symptoms that cannot be detected automatically by the equipment. Accordingly, the worker must first detect such symptoms before any troubleshooting can take place. Detecting such symptoms of malfunction can be difficult. Any given symptom of inferior performance can also be caused in several different ways, as was also true for the alarm category. This latter tendency can make troubleshooting particularly difficult when the inferior performance is caused by multiple interacting factors.

Stall refers to the situation in which the equipment does not function at all and is generally caused by hardware or software failure. Software malfunctions occur primarily in the early stages of machine design and can only be debugged by using sophisticated diagnostic tools such as logic analyzers, making them very difficult to analyze.

Troubleshooting is one of the most time-consuming and difficult maintenance and repair activities. System downtime related to troubleshooting is a major concern for complex production equipment, not to mention aircraft and transportation systems. Even small reductions in downtime can yield large economic benefits. Quite often, finding the fault is a matter of isolating and testing portions of the system. This divide-and-conquer procedure requires some way to determine when an isolated part of a system is properly operating or not. The following design principle is one of the most basic methods of ensuring this can be done efficiently.

Principle 16.5

Provide isolation points and other design features for isolating faults in equipment and machines.

Equipment and machines can be designed to make it easier to isolate and test separate subsystems. One way of doing this is to provide test points that allow people to progressively isolate systems, subsystems, and ultimately components to locate the fault. Design modularity is another feature that helps isolate faults and reduce downtime. Components can be grouped into modules that can be separately tested and quickly replaced. Replacing the entire module reduces downtime. The faulty elements of the module can be diagnosed and repaired later in shop or by the manufacturer.

Expert System for Troubleshooting Malfunctions of Production Equipment

Naruo, Lehto, and Salvendy (1990) discuss the development of an expert system for diagnosing the malfunctions of a machine used by the NEC Corporation to mount chips on integrated circuit boards. Development of the expert system was justified by the inability of operators to efficiently diagnose many malfunctions of the chip-mounting machine, the associated cost of production delays, and the disruption incurred when experts were forced to leave unrelated tasks to help operators troubleshoot malfunctions.

The first step in development of the expert system was to elicit and organize the machine designer's knowledge. This process resulted in a hierarchical classification of malfunction symptoms and causes, a set of 15 flow diagrams documenting the designer's troubleshooting procedures for particular malfunction symptoms, and a matrix documenting design information. The flow diagrams were translated into a large logic network diagram, which was directly translated into a set of 94 rules. An additional set of 270 rules were derived from the design matrix. The resulting 364 rules were then implemented in an expert system. Onsite validation revealed that 92% of the chip-mounting machine's malfunctions occurring in 1988 and 1989 were successfully diagnosed by the expert system.

In many cases, equipment will contain self-diagnostic routines that use network logic to isolate the fault to varying levels of specificity (see Box 16.3). Figure 16.6 illustrates a network of connections where input nodes 1, 2, 3, and 4 lead to output nodes 25, 26, 27, and 28. Box 16.9 shows how faults can be isolated for this example network by checking

Chapter sixteen: System reliability and maintenance

**BOX 16.3 ILLUSTRATION OF NETWORK
LOGIC TO ISOLATE FAULTY NODES**

Consider the network shown in Figure 16.6. One can identify the following paths through the network from the input nodes to output nodes as follows.

Path	Input	Nodes along the Path					Output
A	1	5	9	13	18	22	25
B	1	5	10	13	18	21	26
C	1	5	10	14	17	23	27
D	2	6	11	15	18	21	26
E	2	6	11	15	18	22	25
F	2	6	11	15	20	23	27
G	3	6	11	15	18	21	26
H	3	6	11	15	18	22	25
I	3	6	11	15	20	23	27
J	4	7	11	15	18	21	26
K	4	7	11	15	18	22	25
L	4	7	11	15	20	23	27
M	4	8	11	15	18	21	26
N	4	8	11	15	18	22	25
O	4	8	11	15	20	23	27
P	4	8	12	14	17	23	27
Q	4	8	12	16	19	24	28

Each path shown in the table begins with an input node and ends with an output node. The number of paths connecting input nodes to output nodes is shown in the following table:

	Outputs			
Inputs	25	26	27	28
1	1	1	1	0
2	1	1	1	0
3	1	1	1	0
4	2	2	3	1

Paths from input nodes 1–3 are unique to output nodes 25, 26, and 27. However, interior nodes are often on multiple paths. For example, node 5 is on paths ABC, node 6 is on paths DEFGHI, node 7 is on paths JKL, etc.

To illustrate how a fault might be isolated, consider the case where a test of node 1 results in an output on nodes 26 and 27, but no output on node 25. Since node 25 was not one of the outputs, path A must contain a failure. Note that nodes 9, 22, and 25 are on path A but not on paths B and C. If a test of node 2 results in an output on node 25, path E is okay, so nodes 22 and 25 are operative. The only node remaining is node 9, so this is where the failure must be.

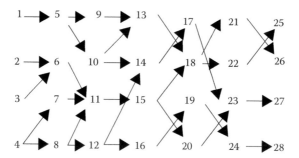

Figure 16.6 Network diagram showing logical connections between components.

which nodes are activated for each input signal. This process might be followed by a self-diagnostic routine or by a person who checks each node using diagnostic tools. The next design principle applies to the latter situation:

Principle 16.6

Provide troubleshooting tools and strategies.

Many different approaches are available for making it easier for people to troubleshoot complex systems. Some generic methods that might be implemented in a variety of different ways include decision tables, logic diagrams, flow diagrams, step-by-step procedures, and Bayesian inference. Decision tables are quite useful when there is a direct, one-to-one, mapping between symptoms and malfunctions. Symptoms can be listed in one column of the table, with diagnoses and suggested solutions in other columns. For example, if printouts from a laser printer are covered with black blotches, the malfunction is almost certainly a toner leak, and the solution is to replace the toner cartridge and vacuum out the inside of the printer.

When there is a one-to-many mapping between symptoms and malfunctions, the problem becomes more difficult—that is, more than one malfunction is possible for the observed symptoms, so it becomes necessary to obtain additional information to diagnose the problem correctly. This is where some of the other methods become especially useful. It costs time and effort to collect information. Each observation taken or test conducted has a cost, so it is important to guide or even optimize the troubleshooting process. Flow diagrams, decision trees, and step-by-step procedures provide one way of guiding the steps followed during troubleshooting. With the benefit of past experience, an efficient sequence of steps can be embedded into these methods. These procedures can also be implemented in very different media, such as service manuals, JPAs, or computer help systems. Note that it is becoming increasingly common to put such information on the computer. This allows free text searches, as well as the use of hypertext links to help users obtain information.

Analytical methods can also be used to provide decision support to troubleshooters. This includes the use of Boolean algebra and logic to infer what diagnoses are possible given the known symptoms.* Bayesian inference is another approach that helps the

---

* The earlier example shown in Box 16.3 should give the reader some feel for how logical reasoning can be used to constrain the possible diagnoses. A number of additional techniques that will not be covered here include (a) techniques used in propositional and predicate calculus to prove that particular logical expressions are true or false and (b) search methods, such as forward and backward chaining through a set of logical propositions or rules.

troubleshooter develop inferences that best fit the available data. One interesting aspect of the latter approaches is that they are nonprocedural. Given the information available at some particular time in the troubleshooting process, they can be used to calculate a good *best guess*. It also should be pointed out that these techniques can be implemented in a variety of different ways. One of the more common approaches is to use them as the embedded engine of an expert system.

Bayesian inference is based on Bayes' rule, which describes how the probability a particular diagnosis is true changes when sample information is available. This relationship is given in the following equation:

$$P(D_j|S_k) = \frac{P(S_k|D_j)P(D_j)}{\sum_{j=1}^{m} P(S_k|D_j)P(D_j)} \qquad (16.4)$$

where
D_j is a particular diagnosis (or fault) j
m is the number of possible diagnoses
S_k is a symptom (an observable situation) k
$P(D_j|S_k)$ is the probability of a particular diagnosis (or fault) j, given the particular symptom k, and is known as the posterior probability
$P(S_k|D_j)$ is the probability of the symptom k, given a particular diagnosis (or fault) j, and is known as the evidence
$P(D_j)$ is the prior probability of a particular diagnosis (or fault) j before obtaining any evidence

Note that the numerator of the right-hand side of Equation 16.4 is the product of the prior belief and evidence for a particular diagnosis. The denominator is a normalization constant. The left-hand side of the equation gives the posterior probability calculated after obtaining the evidence.

Numerical Example of Diagnosis Using Bayes' Rule

Suppose that 100 malfunctions in the past were diagnosed into categories D_1 and D_2 and D_3. Three symptom patterns were identified as S_1, S_2, and S_3. The observed combinations of symptoms and diseases are shown in the following:

	D_1	D_2	D_3	Sum
S_1	20	5	0	25
S_2	5	40	5	50
S_3	0	5	20	25
Sum	25	50	25	100

Each value in the 3×3 matrix divided by 100 is an estimate of the joint probability of a particular combination of symptom and diagnosis. The conditional probabilities can be calculated by dividing the number of times each combination occurs by the corresponding row or column sum. For example,

$$P(S_1|D_1) = 20/25; P(S_2|D_1) = 5/25;$$

$$\text{and } P(S_3|D_1) = 0/25. \ P(D_1|S_1)$$

$$= 20/25; P(D_1|S_2) = 5/50;$$

$$\text{and } P(D_1|S_3) = 0/25$$

Suppose that the relative frequencies of D_1, D_2, and D_3 are, respectively, 0.2, 0.5, and 0.3 based on years of data. Also, a recent study of the joint occurrence of symptoms and diagnoses resulted in the data given earlier. These two sources of information can be combined using Bayes' rule. The prior probabilities are given by the earlier relative frequencies. The conditional probabilities are given by the recent study. For example,

$$P(D_2|S_2) = \frac{P(S_2|D_2)P(D_2)}{P(S_2|D_1)P(D_1) + P(S_2|D_2)P(D_2) + P(S_2|D_3)P(D_3)}$$

$$= \frac{(0.8)(0.5)}{(0.2)(0.2) + (0.8)(0.5) + (0.2)(0.3)}$$

$$= \frac{0.4}{0.04 + 0.4 + 0.06} = 0.8$$

Verifying the fault and the correction

Once there is reason to believe that the fault is in a particular location or that the fault is a malfunctioning component, mechanics, electronic technicians, or other specialists often go into the system and remove that component for repair or replacement. Since most of the pressure is on fixing the malfunctioning machine, the need to determine whether or not that component was malfunctioning and why is frequently overlooked. The latter activity is important, because some of the replaced parts may have been working properly. Once a new part is installed, technicians rarely remove them even if they don't fix the problem. This leads us to another important principle:

Principle 16.7

Document the malfunction symptoms and the parts that fixed the problem.

Documenting which parts actually corrected the problem for the observed symptoms can help reduce the time needed to fix future malfunctions. Decision support tools need these data to predict and diagnose malfunctions. Such data also can be used to validate existing means of diagnostic support and provide a measure of how effective the existing repair procedures are.

Design for disassembly

Almost all maintenance requires some *disassembly*, the efficiency of which depends upon the component parts of the machine, the mating of these parts, and connector devices.

Chapter sixteen: System reliability and maintenance 603

When these parts are properly designed, maintenance can be performed faster and with more assurance that it is done properly. Finding some leftover parts or connectors after reassembly is not a pleasant outcome. Missing parts or connectors at some stage of reassembly is another common problem. One of the principles of component part design to improve maintenance is as follows.

Principle 16.8

Minimize the number of component parts and connectors that must be disassembled to do the maintenance task.

When fewer parts and connectors must be taken apart, maintenance and repair can be done more easily. This reduces both the time needed and the chance of making mistakes. Products can and should be designed to be easy to take apart and put back together for all foreseeable types of maintenance activity. To ensure this goal is satisfied, it is often helpful to first build a model of the product. Such models can be created on the computer or from simple, rough materials such as Lincoln logs, Erector-set components, or Fischertechnik®* assembly pieces. An example devised by Baggett (1992) is shown in Figure 16.7. The lift shown in this figure includes part numbers for each component.

It is often helpful to prepare a graph showing how the parts of a product connect with each other. For example, Figure 16.8 shows how the parts in Figure 16.7 connect. It is possible to tell from Figure 16.7 that part 71 connects with parts 77 and 74 and maybe with others as well. Figure 16.8 shows a node-branch graph of the lift where part 71 is on the right-hand side, near the bottom, with branches connecting to parts 77, 76, and 69. Part 69 is hidden from view in Figure 16.7 and connects directly to part 74. The point is that it is difficult to see all of the connections even from a photograph, just as it is difficult for the maintenance worker or the designer to see the connection.

Another important point is that Figure 16.8 can be easily partitioned into three distinct major parts, the left grouping, the center grouping, and the right grouping. Those three major groupings are separated by a few physical connections. The left grouping has part 27 connecting to parts 21 and 22 in the center grouping. Note that part 27 is shown in Figure 16.7 as the base plate under the shaft that connects the lift. Parts 33 and 50 of the left grouping connect physically with part 80 of the center grouping. Part 80 of the center grouping physically connects to part 64 of the right grouping. Accordingly, the center grouping consists of the base plate, shaft, and top piece on the shaft. The mechanism behind the base plant corresponds to the left grouping, and the platform-motor assembly, which goes up and down on the shaft (shown in the middle of Figure 16.7), corresponds to the right grouping.

One reason for partitioning off the three major groupings is to identify clusters of nodes such that the connections between nodes are mostly within the cluster, with only a few connections between clusters. It further follows that within a major grouping one can develop subclusters following the same approach. While this logic is not always apparent in Figure 16.9, it is approximated there. It also follows that pieces of the Fischertechnik in the inner clusters could be permanently joined to make a component part with a small number of mating connections. Designers could next consider each cluster within a grouping to see if the connected clusters should be joined or not, and so forth. In this way,

---

* Fischertechnik® is a product made by a German company that consists of plastic blocks that connect to one another. Special pieces with pulleys and a small electric motor are also included. Semifunctional models of machine tools and other devices can be built and operated using this product.

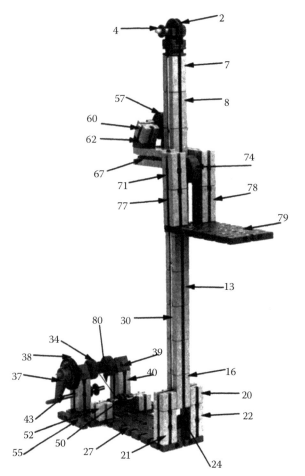

Figure 16.7 Lift made from 80 pieces of a Fischertechnik® assembly kit. (Reprinted from Baggett, P., Putting it together with cognitive engineering, in Helander, M. and Nagtamachi, M. (Eds.), *The International Ergonomics Society, Design for Manufacturability—A Systems Approach to Concurrent Engineering and Ergonomics*, Taylor & Francis, Ltd, London, U.K., pp. 160–170, 1992, Figure 10.1, p. 165. With permission.)

the number of parts can be minimized subject to other needs, such as the parts becoming too heavy or cumbersome to disassemble and reassemble.

This leads us to a related design principle.

Principle 16.9

Group components by function into modules that can be separately removed or installed.

Ideally, no more than one disassembly step is required for each maintenance task. Although this goal is often not achievable, minimizing the number of steps is especially important for modules that are frequently replaced. Grouping by function also helps locate faults and simplifies the process of fault diagnosis.

Some additional considerations are described next. One is that the physical connections between two component parts may be symmetrical, semisymmetrical, or asymmetrical

Chapter sixteen: System reliability and maintenance 605

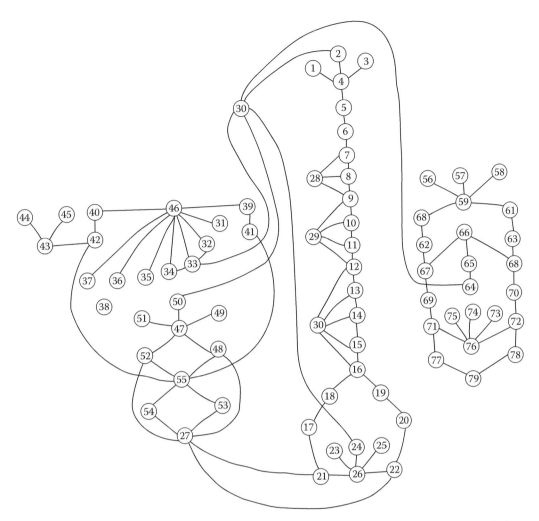

Figure 16.8 Graph of nodes (representing component parts) and branches. (Reprinted from Baggett, P., Putting it together with cognitive engineering, in Helander, M. and Nagtamachi, M. (Eds.), *The International Ergonomics Society, Design for Manufacturability—A Systems Approach to Concurrent Engineering and Ergonomics*, Taylor & Francis, Ltd, London, U.K., pp. 160–170, 1992, Figure 10.2, p. 166. With permission.)

in character. There is a clear and obvious advantage of symmetrical parts, as they can be joined in many orientations. Symmetrical parts reassemble faster than semisymmetrical or asymmetrical parts, as MTM tables show.* On the other hand, if operation of the machine requires a particular orientation of the component parts, a connection that seems to be symmetrical could mislead the maintenance person to reassemble the part incorrectly, which would necessitate another disassembly and reassembly. This leads us to the next principle.

* See Chapter 6 for a discussion of the MTM system and examples. The position and apply pressure subtable used in MTM shows the result referenced to here.

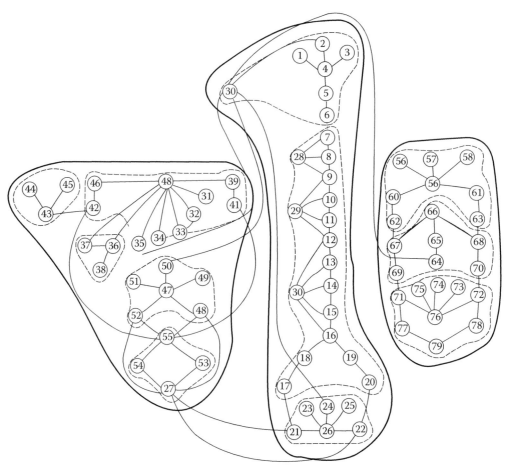

Figure 16.9 Typical conceptualization of major assemblies and minor component parts (containing one or more pieces of Fischertechnik® kit) for the lift. (Reprinted from Baggett, P., Putting it together with cognitive engineering, in Helander, M. and Nagtamachi, M. (Eds.), *The International Ergonomics Society, Design for Manufacturability—A Systems Approach to Concurrent Engineering and Ergonomics*, Taylor & Francis, Ltd, London, U.K., pp. 160–170, 1992, Figure 10.3, p. 167. With permission.)

Principle 16.10

Symmetrical part mating is preferred when the assembly operates properly when so assembled, but otherwise go to semisymmetrical or asymmetrical mating.

It also follows that it is easy during part disassembly to forget how the part was oriented. So a second principle that helps reduce maintenance costs is to include markings that communicate how the part should be oriented. Writing across a part can indicate the part's orientation. Another example is marking parts with arrows that are aligned when two mating parts are correctly oriented. So another principle is as follows.

Principle 16.11

Design the part in such a way that it is obvious how it is to be oriented during reassembly.

It is better if mating parts can be fit together without requiring vision, but providing visual means of orientation is far better than providing nothing.

Parts should be designed for easy manipulation during manual handling. The MTM tables show that certain types of grasps* are more difficult and time consuming than others. So the MTM grasp table provides useful information on part manipulation. Part size and weight are also important considerations. Both factors limit the members of the population who can perform the service as maintenance people come in multiple sizes and strengths. Ideally, the parts will be small enough for most workers to handle. Problems exist at the other extreme, too. Small parts are often easily bent or warped during disassembly or reassembly. Bent or warped parts must be replaced, and this further increases the time and effort required to finish the task. Sometimes it becomes necessary to assign particular maintenance personnel to tasks that require great strength or small but dexterous limbs. Most would agree, however, that such requirements should be avoided if possible.

Some parts have an obvious sequence of disassembly and reassembly. It also follows that the person who performed the disassembly does not always perform the reassembly. Hence, memory cannot be counted on as a reliable guide to reassembly. While the reassembly sequence typically is the opposite of the disassembly sequence, at times that may not be the case. In such cases, some form of warning should be provided to help prevent mistakes. One way to communicate a sequence for the component parts is to have parts that increase gradually in size with disassembly.

Another design consideration is given in the following.

Principle 16.12

Provide tactile feedback when the part is properly and fully joined, such as an audible snap and a noticeable vibration that occurs when parts are joined and unjoined.

Such feedback tells the assembler or disassembler that the part is properly and effectively mated or free from the connective part. In either case, the person disassembling components or reassembling them does not have to use vision or shake the part to determine that the connection has been adequately made.

Fasteners and tools

Connective fasteners are another important maintenance consideration. Almost everyone who has had to fix a flat tire on an automobile on the road understands the basic problem. In that situation, one of the first things one has to do is disassemble the flat tire by removing the lug bolts, and a typical ploy is to put them in the hubcap as they are disassembled. However, in many forms of equipment, fasteners do not need to be removed from the outside part during disassembly. Often the end thread of the connector can be stopped so that the connector is stored with the disassembled part and the connectors are immediately ready for reconnecting during reassembly.

Principle 16.13

Provide cues that make it easier to locate and identify fasteners.

Locating all of the fasteners is difficult for many products. Simple signs can be useful that direct the maintenance person's attention to the locations of fasteners. It is also useful to show the maintenance person the type of fastener. A few years ago there were only two principal types of heads for screw fasteners: slot heads and Phillips heads. Today

---

* Note that the MTM system includes time predictions for various forms of grasps. Here again, see Chapter 6 for more information about the MTM system.

there is a much wider variety that includes the square head and several other kinds. Consequently, it is helpful to indicate which tool is needed to disconnect the fastener.

Many maintenance situations require a variety of tools. If maintenance workers must walk some distance to the machine requiring maintenance, they will normally want to carry only the tools they will need. Some readers will recognize this situation as the *classical knapsack* problem that is analogous to a hiker filling a knapsack before a trip and attempting to prepare for all needs and desires without exceeding the weight limitations of his own carrying capacity. This problem can be addressed by *making the tools more multipurpose*, even at a loss of efficiency. Otherwise, travel time between the task and the tool-storage location may become excessive. For the case of multiple types of fastener heads, one could elect a screwdriver with interchangeable bits for each type of head. On the other hand, the best design follows the following principle.

Principle 16.14

Minimize the size and numbers of and the types of connective fasteners in the machine design, subject to adequate strength.

Unfortunately, many companies overlook this principle when they order component parts from vendors without controlling the number and types of connective fasteners. Many of the principles of maintenance listed previously follow those in assembly production, but there are some departures. When it comes to hand-tool design, most of the principles still hold, except that most maintenance operations are not sufficient in numbers of people or in time durations to justify much special development in tool design.

Work envelope

A work envelope is merely a defined spatial region that the parts of a person's body and/or hand tools occupy at some time while performing a particular task (see Dempster, 1955). It has been known for many years that providing adequate space to both see and manually access parts makes it much easier to perform maintenance and repair tasks. Some newly developed technologies have made it much easier to determine work envelopes for a wide variety of settings using the computer.*

The associated design principle is as follows.

Principle 16.15

Maintain open space in the design to accommodate at least the work envelope and vision requirements within this envelope.

Some of the older studies on work envelopes used photography to establish the envelope's spatial bounds. Long-exposure, still and strobe light photography as well as both regular and slow-motion videotape have all been used for this purpose. The use of videotape poses some advantages by allowing images to be viewed immediately. Digital videotape makes it particularly convenient to record images that can be processed and ultimately used in computer-aided design (CAD) systems.

Along these lines, Figure 16.10 shows a work envelope developed several years ago using photographic techniques. This particular example shows the *swept volume* for a person using a screwdriver to turn in a screw in the horizontal plane. Note the low level

---

* See discussion in Chapter 3 on computer modeling methods, such as SANTOS.

Chapter sixteen: System reliability and maintenance 609

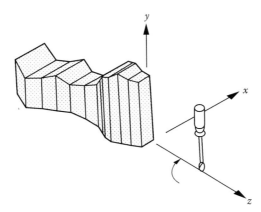

Figure 16.10 Drawing of the "swept volume" for a screwdriver used to turn a screw in the horizontal plane. (From Altman, J.W., Guide to design of mechanical equipment for maintainability, Report ASD-61-381, U.S. Air Force, Air Force Systems Command, Wright-Patterson Air Force Base, OH, 1961.)

of resolution provided, which is the natural consequence of generating the envelope dimensions manually from a sequence of photographs, a tedious task to say the least. Stereophotographic means were typically used in devising these work envelopes with measurement techniques similar to those used in aerophotography.

Modern approaches, such as use of the DataGlove™,* allow three-dimensional work envelopes to be generated in real time. The glove records the (time-variant) shape of the hand and is intended to be used with a tracker system that records hand location on the x, y, and z planes. The glove and tracker data are processed to compute the outer edges of the hand in space either in real time or nearly so. While this approach greatly reduces the effort needed by ergonomic specialists to develop work envelopes, it does not have the capacity to easily track the extreme reach of any tool held in the hand. Some tracker systems are also affected by nearby metal enclosures.

Many companies are now implementing CAD-based approach for generating work envelopes for maintenance and repair tasks. For example, General Electric has developed a system called Product Vision that uses a path-generating algorithm to predict the route of a component when it is removed, including all turns and wiggles. The dimensions of the component as it moves along this path define the work envelope (see Schroeder et al., 1994).

Final remarks

As should be apparent from the preceding discussion, maintenance is a broad and complicated topic. This chapter has focused primarily on methods of reducing maintenance effort and delivering the maximum value for the cost. Attaining this goal begins with the design of more reliable products and processes. Numerous strategies can then be followed to prevent failures, including routine or preventive maintenance procedures. Since failures will inevitably occur, it also becomes essential to develop better means of detecting,

---

* Majoros and Taylor (1997) describe the DataGlove™, a product made by VPL Research, Inc. that consists of a latex glove with optical fibers extending down the fingers and thumb used to measure the flexion of joints in the hand. Gloves of this type are also used in *virtual reality* studies.

diagnosing, and correcting faults to minimize system downtime. The timely delivery of technical information, tools, parts, and materials is a particularly important consideration. It is also critical to document corrective actions, and whether they were correct, to improve future performance. Industrial engineers and ergonomists can contribute throughout this process in many different ways.

Discussion questions and exercises

16.1 Given a MTBF of 40 h and a MTTR of 1 h, calculate system availability.

16.2 Consider some routine maintenance activities that are part of janitorial activities. Identify some of the variables that affect performance time in that task and others belonging to the same family. Also identify some devices or machines that one could consider buying in order to reduce the costs of performing that activity. How could you determine if the purchase of the device or machine is economically feasible?

16.3 What is a *part-removal envelope* or a *work envelope*? Think of the maintenance task of checking the oil level and coolant level in an automobile, and describe how you would identify the work envelope using two video cameras. How would you do it using a single video camera?

16.4 How would you devise a kit of tools that maintenance workers of a specialized shop could use? How would you test its effectiveness? How would you determine if there are too few or too many tools in the kit?

16.5 What are some of the advantages and disadvantages of having operators do routine maintenance?

16.6 What methods are commonly used in industry to schedule maintenance? What are some of the advantages and disadvantages of each method?

16.7 Why is it important to deliver technical information, tools, and parts in a timely manner to maintenance and repair personnel? How can this goal be attained?

16.8 Discuss the role of record keeping in maintenance programs.

16.9 How does fault avoidance differ from fault tolerance? How might these features be attained in the design of production equipment?

16.10 Why do people have trouble detecting and diagnosing faults? What methods might be followed to improve their performance?

16.11 Select a common item around you, preferably one that is not too complex. Identify the parts by number and draw a graph with nodes representing each part and branches (or links) representing physical connections. Suppose that you want to make subassemblies of these parts, how would you determine those parts that should be combined into subassemblies?

16.12 Data collected over the past year relating three diagnoses to three different symptoms are given here:

	D_1	D_2	D_3	Row Total
S_1	8	20	2	30
S_2	35	3	2	40
S_3	1	2	32	35
Column total	44	25	36	105

a. Find the probability of each diagnosis given symptom 2, based only on the data set for the past year.
b. Suppose that the three diagnoses are known to have a probability of occurrence of 0.42, 0.20, and 0.38 respectively for D_1, D_2, and D_3. Based on Bayes' theorem, find the probability of each diagnosis given symptom 2.
c. Compare and contrast your answers in parts (a) and (b) of this question. Why are they different?

16.13 For the network diagram shown in Figure 16.6, suppose an input to node 2 activates nodes 25, 26, and 27. An input to node 3 results in no output. Where is the fault located?

16.14 List and discuss five principles for the disassembly and design of fasteners.

chapter seventeen

Occupational safety and health management

About the chapter

This chapter covers one of the most important aspects of macroergonomics—the management of occupational safety and health. The chapter begins by presenting a brief history of workplace safety and health issues, including the evolution of legal responsibilities of the employer. Attention then shifts to important elements of occupational safety and health, including methods of classifying causes of injury, illness, and accident prevention. Next, contemporary methods for managing occupational safety and health are discussed. These methods are primarily instituted through a workplace safety program, which carries out activities such as ensuring compliance with safety and health standards, accident and illness reporting and monitoring, identification and management of workplace hazards, and administering methods of hazard control.

Introduction

Occupational health and safety is an interdisciplinary field that focuses on preventing occupational illnesses and injuries. Government agencies, universities, insurance companies, trade associations, professional organizations, manufacturers, and service organizations all play important roles in reducing the burden of occupational illnesses and injuries to society.

Prevention efforts have paid off dramatically in industrialized countries. The annual number of accidental work deaths per 100,000 workers in the United States, as of 2008, has dropped to about one-eighth of what it was 60 years ago. Nevertheless, the accident toll continues to be high. Single events, such as the British Petroleum oil rig fire and explosion in 2010, can cause multiple deaths and injuries, and disrupt entire sectors of the economy, with costs mounting in tens of billions of dollars. Overall, in the United States alone, the National Safety Council (NSC) estimates that there were about 4,300 accidental work deaths and 3,200,000 disabling injuries in 2008; with an associated cost of 183 billion dollars (NSC—Injury Facts, 2010). Major components of this cost estimate include insurance administration costs, wage and productivity losses, medical expenses, and uninsured costs. Worldwide, the direct costs of accidents are estimated to exceed several hundred billion dollars. If indirect costs are considered, such as lost productivity, uninsured damage to facilities, equipment, products, and materials, and the cost of social welfare programs for injured workers and their dependents, the bill raises to several trillion dollars annually. Consideration of pain and suffering raises the cost to even a greater level.

It should also be emphasized that despite the fact that industrial accident rates are decreasing in most industrialized countries, economic estimates of the annual costs of accidents show the opposite trend. For example, the 2009 Liberty Mutual Workplace Safety Index (WSI) shows an increase of 42.8% in direct U.S. workers compensation costs for the most disabling workplace injuries over the 10 year period from 1998 to 2007 (Liberty Mutual, 2009). Worker injuries due to overexertion, slipping, tripping, or falling alone accounted for nearly two-thirds of these costs.

The root causes of these and many other types of worker injuries and illnesses can often be addressed through better workplace ergonomics as part of an occupational safety and health program. The application of ergonomic principles to reduce workplace injuries and illnesses began years ago and continues to be a critical element of safety and health management. However, better ergonomics is only part of the solution. Many other important strategies are available that can be used by management to keep risks at an acceptable level. An important key to success is that instead of simply reacting to accidents, injuries, or illnesses, management should be proactively taking steps to prevent them from occurring in the first place.

Some historical background

Before the nineteenth century, the responsibility for occupational safety and health was placed on the individual employee. The attitude of most companies was, "Let the worker beware!"

With the advent of the industrial revolution in the nineteenth century, workplace injuries became more prevalent. Dangerous machinery caused many accidents. As manufacturing moved to larger plants, there were increasing numbers of workers per plant and an increased need for supervision. There was little theory available to guide management in reducing workplace dangers. This lack of management knowledge, and an unfortunate indifference to social concerns, caused many safety problems during this era.

During the mid-nineteenth century, the labor movement began as a response to worker concerns about issues such as wages, hours, and working conditions. Unions began to address safety issues. However, industrial safety did not measurably improve. Unions had limited bargaining power as well as limited knowledge about how to effectively reduce workplace hazards. Furthermore, employers did not have a strong legal incentive to improve workplace safety. During this period, the law was focused on the "master–servant rule" whereby workers looked to common law to win redress. The employer was not legally obligated to provide a reasonably safe and healthful environment. Employees' actions could void employer liability if

1. The risks were apparent and the employee continued to work.
2. There was some negligence by the employee.
3. Fellow workers contributed to the injury.

Public acceptance of this situation eventually began to waiver. During the late nineteenth and early twentieth centuries there were numerous changes in the law. One was the federal Employers' Liability Act of 1908, which provided for railroad employees to receive damages for injuries due to negligence of the company or its managers, even when there was some contributory negligence on the part of the injured employee. However, the burden remained for the employee to prove the company was negligent through litigation. During this time, there were also increasing efforts by unions, companies, and trade associations to promote industrial safety. Later came "Workman's Compensation" laws enacted by individual states. These laws provided for the compensation of all injured employees at fixed monetary levels, for the most part eliminating the question of fault and the need for litigation. Most states used a casualty insurance company or a self-insurance fund to compensate the employee for an accident under these laws. Also during this time, there were some federal government departmental efforts by the Bureau of Mines, the Bureau of Labor, and the National Bureau of Standards to improve worker safety.

Progress since the late 1930s has been impressive. During the early part of that decade, U.S. federal government actions forced management to accept unions as a bargaining agent of employees. Unions became more widespread, and safety conditions were one aspect of the contracts negotiated between employers and unions. During World War II (1939–1946), there were widespread labor shortages, which resulted in increased effort by employers to recruit and retain personnel, including a greater attention to safety for workers.

The Environmental Protection Act of 1969 secured many sources of employee protection against chemical and radiation sources in industry. In 1970 the Occupational Safety and Health Act (Williams-Steiger Act) was passed which placed an obligation on employers to provide a workplace free from recognized hazards. This act also established the Occupational Safety and Health Administration (OSHA), a federal agency whose objective is to develop and enforce workplace safety and health regulations. As shown in Table 17.1 a large number of laws were added over this period that directly affect worker safety and employers responsibility toward workers. For most hazards, the enactment of laws and regulations has evolved over time with growing recognition and understanding of their nature. This tendency is illustrated in Table 17.2, which shows how laws, regulations, and standards limiting exposure to lead have evolved over time in the United States.

Table 17.1 Timeline of Selected Laws and Acts Related to Occupational Safety in the United States

Year	Legislation
1938	Fair Labor Standards Act
1938	Federal Product Safety and Health Legislation: Food, Drug, & Cosmetic Act
1953	Flammable Fabrics Act
1958	Longshore Safety Act
1959	Radiation Hazards Act
1960	Federal Hazardous Substances Act
1966	Federal Metal & Nonmetallic Mine Safety Act
1966	National Traffic & Motor Vehicle Safety Act
1966	Highway Safety Act
1967	Flammable Fabrics Act Amendment
1968	Radiation Control for Health & Safety Act
1968	Fire Research & Safety Act
1969	Contract Work Hours & Safety Standards Act
1969	Federal Coal Mine Health & Safety Act
1969	Environmental Protection Act
1970	Williams-Steiger Occupational Safety & Health Act
1970	Poison Prevention Packaging Act
1970	Lead-Based Paint Poison Prevention Act
1970	Other Federal Safety and Health Legislation: Clean Air Amendments
1975	Hazardous Materials Transportation Act
1976	Toxic Substances Control Act

Table 17.2 Timeline of Lead-Related Regulations, Standards, and Laws in the United States

1880–1930	Lead, especially from paint, is suspected by health care professionals of causing deaths to children with previous symptoms of seizure, drop wrist, drop foot, etc.
1910	Marion Rhodes, a congressman from Missouri, wants to extend labeling provision of the Food and Drug Act to cans of lead paint. His bill is rejected
1930	*White House Conference on Child Health* alerts participants about lead paint on toys
1930s	Paint manufacturers advise consumers of availability of lead-free paints for toys and cribs
1933	Massachusetts—"Revised Rules, Regulations and Recommendations Pertaining to Structural Painting" states that "toys, cribs, furniture and other objects with which infants may come in contact should not be painted with lead colors"
1935	Baltimore Health Department becomes first in nation to offer blood test for lead as a diagnostic test
1949	Maryland state legislature passes Toxic Finishes Law, making it "unlawful for any person to manufacture, sell or offer to sell, any toy or plaything, including children's furniture, decorated or covered with paint or any other material containing lead or other substance of a poisonous nature, from contact with which children may be injuriously affected." The law was repealed a year later because of ambiguities of its definitions ("injuriously"), lack of enforceability, and failure to define acceptable levels of lead for paint
1951	Baltimore passes ordinance banning the use of paints containing lead pigments, June 29, 1951
1954	New York City Health Department approves following label on October 29: "Contains lead. Harmful if eaten. Do not apply on toys, furniture, or interior surfaces which might be chewed by children"
1955	ASA standard Z66.1 is adopted. The standard limits lead in interior paint to 1%
1958	Baltimore mayor signs ordinance mandating warning label on paint cans: "WARNING—Contains Lead. Harmful if Eaten. Do not apply on any interior surfaces of a dwelling, or of a place used for the care of children, or on window sills, toys, cribs, or other furniture"
1963	Congress passes Clean Air Act
1970	Public Law 91-695 is enacted. It provided federal funds for mass screening, treatment, education, and research on how to lessen lead hazard. Also mandated that federally funded public housing meet the 1% paint standard
1972	U.S. Food and Drug Administration bans paint with an excess of 0.5% lead from interstate commerce under provisions of the Federal Hazardous Substances Act
1974	Congress asks the Consumer Product Safety Commission (CPSC) to assess danger of multiple layers of paint and to determine a "safe" amount of lead in paint. CPSC arrives at 0.5% as a "safe" level
1977	CPSC adopts as a "safe" level recommendation of the AAP and National Academy of Sciences to lower the "safe" level to .0670
1978	National consensus is reached that lowers lead content of all household paints—interior and exterior—from 1% to 0.06%
1978	Tetraethyl lead is removed from gasoline

Safety and health achievements and future promises

A number of notable successes and a few failures have been observed since the passage of the Occupational Safety and Health Act of 1970 and the resulting establishment of the OSHA and its sister agency, the National Institute for Occupational Safety and Health

(NIOSH). One of the most notable successes is the huge reduction in the incidence of occupational poisoning. Previously, many miners were poisoned with phosphorus and many painters and printers were poisoned with lead. There has also been a decline in silicosis, silio-tuberculosis, black lung, and other associated lung diseases that principally occurred in mining and also in the manufacture of asbestos products. Some success has occurred in reducing accidents from physical fatigue and in machine safeguarding; however, several forms of guarding have created losses in productivity. Also some very unsafe factories have been closed down, but those are special cases rather than general cases of success.

Several failures have also been noted. These failures include continuing incidence of chronic bronchitis, occupational dermatitis (which is the most common industrial occupational disability), diseases from vibration, chemical poisonings other than lead and phosphorus, lesions to bones and joints, chronic vascular impairment, neuroses, and mental disorders. Several of these failures are the focus of current research.

Several other ongoing changes in industry are influencing safety and health. One is that during the past two decades the U.S. industry has become more capital intensive through automation. One of the positive side effects of automation is that fewer workers are exposed to some of the more hazardous occupational environments associated with tasks such as painting and welding.

Another phenomenon that is affecting safety and health is the enactment of equal employment opportunity laws by the federal government prohibiting discrimination in employment based on gender, age, race, origin of birth, and religion. Employee screening, such as strength testing, conducted to ensure workers are able to perform certain jobs safely, may disproportionately eliminate older and female applicants. The Equal Employment Opportunity Commission requires evidence that such selection criteria are in fact necessary. Obtaining such evidence can be difficult; so many firms prefer to reduce job requirements that require abilities that tend to differ among people by age, gender, and other bases. The Americans with Disabilities Act of 1990 further states that industry must provide access to jobs for qualified individuals with disabilities, through the design of processes and workspaces to accommodate their needs, when it is not economically infeasible to do so. This law puts the burden of proof for economic infeasibility on the company.

Fundamental elements of occupational safety and health

Much of the improvement in occupational safety and health that has been attained over the past 100 years can be attributed to two complementary, sometimes overlapping, approaches. The first approach focuses on preventing accidents and mitigating their effects by designing safer systems and taking steps to ensure they operate as intended. The second focuses on controlling the exposure of workers to harmful substances, energy sources, or other environmental or work-related stressors that cause illnesses or cumulative trauma. The two approaches reflect the traditional dichotomy between safety and health functions still present in some organizations, but have much in common. In recent years, the trend has been toward combined approaches where safety and health professionals (see Box 17.1) work together to identify, predict, control, and correct safety and health problems (Goetsch, 2008).

The successful application of either approach requires knowledge and understanding of

1. The types of hazards encountered by workers and how often they result in injuries, illnesses, or death
2. The causes of accidents and other forms of exposure to hazards
3. The strategies for preventing or controlling accidents and exposure to hazards, and their effectiveness

> **BOX 17.1 SAFETY AND HEALTH PROFESSIONALS**
>
> Many safety and health professionals are certified in particular technical specialties. The Board of Certified Safety Professionals administers the Certified Safety Professional (CSP) while the Board of Certification in Professional Ergonomics offers the Certified Professional Ergonomist (CPE) and Certified Human Factors Professional (CHFP). Certification in the practice of Industrial Hygiene is available from the American Board of Industrial Hygiene. Also, all of the individual states in the U.S. license engineers as Registered Professional Engineers (PE). Although specific licensing in safety is not available, one of the principles of licensure is safety. The National Society of Professional Engineers has as one of its Fundamental Canons that engineers, in the fulfillment of their professional duties, shall hold paramount the safety, health, and welfare of the public. Those working in the safety area have several professional organizations that they may join including the American Society of Safety Engineers (ASSE) and the Human Factors and Ergonomics Society (HFES).

As expanded upon in the following sections, classifications of occupational injuries and illnesses published by organizations such as the Bureau of Labor Statistics (BLS), National Institute for Occupational Safety and Health (NIOSH), and Centers for Disease Control (CDC) in the United States provide a good starting point for developing understanding and perspective of particular hazards by identifying patterns of injury and illnesses, risk factors, and levels of exposure. Theories and models of accident causation build upon this perspective by identifying why accidents occur and suggesting methods of hazard control.

Classifications of occupational injuries and illnesses

Organizations such as the BLS and NSC in the United States routinely collect and disseminate information regarding the incidence of occupational injuries and illnesses for particular occupations or industries. The rationale for following this approach is that some occupations have higher rates of certain injuries and disorders than others. Identifying such trends can guide prevention efforts by government regulators, management, and others by focusing attention on the industries and occupations with elevated incident rates. Example statistics of this type collected by the BLS are given in Table 17.3 for the occupations in the United States with the highest rates of injuries between 2005 and 2008.

More detailed classifications focus on particular elements of hazards, accidents, and injuries. One of the better known coding schemes of this type is the Occupational Injury and Illness Classification System (OIICS) used by the BLS in the United States to code characteristics of injuries, illnesses, and fatalities. This coding scheme is used by the BLS in their annual Survey of Occupational Injuries and Illnesses and Census of Fatal Occupational Injuries programs. In the OIICS system, occupational injuries or illnesses are classified in terms of the nature of injury or illness, body parts affected, primary and secondary sources of injury or illness, and the event or exposure type using a hierarchical coding scheme specified in the OIICS coding manual (http://www.bls.gov/iif/oshoiics.htm). The manual also provides selection rules and coding instructions. At the top level of the coding hierarchy, each major category (or section) is divided into 1-digit codes (Table 17.4). The 1-digit codes are then further divided into 2-digit codes, and so on. For example, event code "313 - Contact with overhead power lines" is a division of the

Table 17.3 U.S. Occupational Injury and Illness Statistics Reported in 2008 for Occupations with High Incidence of Injuries

Occupations	Number of Injuries and Illness	Incident Rate per 100,000 Full-Time Workers
Laborers and freight, stock, and material movers	79,590	440
Heavy and tractor-trailer truck drivers	57,700	362
Nursing aids, orderlies, and attendants	44,610	449
Construction laborers	31,310	383
Retail salespersons	28,900	90
Janitors and cleaners	28,110	243
Light or delivery service truck drivers	28,040	324
General maintenance and repair workers	20,800	213
Registered nurses	19,070	114
Maids and housekeeping cleaners	18,650	278
Carpenters	18,160	236
Stock clerks and order fillers	18,020	130

Source: Bureau of Labor Statistics, Washington, DC.

broader code "31 - Contact with electric current," which is a division of the even broader event code "3 - Exposure to harmful substances or environments."

The source of injury or illness, with well over a thousand four-digit codes, is the largest single category of codes in the OIICS classification. The nature of injury or illness, with several hundred codes, is the second largest category, followed by event or exposure, and part of body affected, both with around 300 codes. One of the advantages of the OIICS classification system is that its hierarchical structure provides a systematic well-developed way of organizing the huge number of ways occupational injuries or illnesses can occur. For example, hundreds of chemicals or chemical products are systematically grouped into different subcategories within the source of injury or illness category. Patterns of occupational injury and illnesses can also be examined by examining combinations of categories, such as cross referencing the part of body affected for a particular source of injury.

Organizations such as BLS, NIOSH, and NSC in the United States routinely follow this approach to describe the nature of injury or illness, body parts affected, source of injury, event or exposure type, etc. for particular occupations or industries. Other examples include the earlier-mentioned Liberty Mutual WSI, which breaks out direct U.S. workers compensation costs for the most disabling workplace injuries using two-digit OIICS event or exposure codes (Table 17.5).

Classifications of this type guide injury prevention efforts by identifying particular groups of events (such as overexertion, falls, etc.) that can be focused upon with specific control strategies. However, simply identifying the event or exposure that resulted in the injury or illness tells us little about why it occurred. The latter issue has traditionally been approached by studying the causes of accidents, as expanded upon in the following section.

Accident causation

Over the years, much has been learned about the causes of occupational injuries and illnesses, by studying accidents. This work has led to numerous models and theories of

Table 17.4 Single-Digit Codes Used in the BLS Occupational Injury and Illness Classification System

Section, division, and title

Nature of injury or illness
0 Traumatic injuries and disorders
1 Systemic diseases or disorders
2 Infectious and parasitic diseases
3 Neoplasms, tumors, and cancer
4 Symptoms, signs, and ill-defined conditions
5 Other conditions or disorders
8 Multiple diseases, conditions, or disorders
9999 Nonclassifiable systemic diseases or disorders

Part of body affected
0 Head
1 Neck, including throat
2 Trunk
3 Upper extremities
4 Lower extremities
5 Body systems
8 Multiple body parts
9 Other body parts
9999 Nonclassifiable

Source of injury or illness
0 Chemicals and chemical products
1 Containers
2 Furniture and fixtures
3 Machinery
4 Parts and materials
5 Persons, plants, animals, and minerals
6 Structures and surfaces
7 Tools, instruments, and equipment
8 Vehicles
9 Other sources
9999 Nonclassifiable

Event or exposure
0 Contact with objects and equipment
1 Falls
2 Bodily reaction and exertion
3 Exposure to harmful substances or environments
4 Transportation accidents
5 Fires and explosions
6 Assaults and violent acts
9 Other events or exposures
9999 Nonclassifiable

Source: http://www.bls.gov/iif/oshoiics.htm

Table 17.5 Top 10 Causes of Disabling Injuries in 2007

Cause	Cost ($ Billions)	Percentage
Overexertion	12.7	24.0
Falls on same level	7.7	14.6
Falls to lower level	6.2	11.7
Bodily reaction (after slipping or tripping)	5.4	10.2
Struck by object	4.7	9.0
Highway incident	2.5	4.7
Caught in/compressed by	2.1	3.9
Repetitive motion	2.0	3.8
Struck against object	2.0	3.8
Assaults/violent acts	0.6	1.1

Source: Liberty Mutual, *Liberty Mutual Workplace Safety Index*, National Safety Council (2010), Hopkinton, MA., *Injury Facts*, National Safety Council, Chicago, IL, 2009.

accident causation that explain why accidents occur and also suggest generic strategies for accident prevention (Lehto and Salvendy, 1991; Goetsch, 2008).

Some of the earliest research on accident causation was done by Heinrich in the 1920s. Based on his analysis of 75,000 industrial accidents, Heinrich concluded that 88% of the accidents were caused by unsafe acts, 10% by unsafe conditions, and 2% by unpreventable causes. His conclusion that 98% of accidents are potentially preventable by eliminating unsafe acts and conditions was a major departure from the prevailing opinion that industrial accidents were an unavoidable cost of progress and set the stage for a whole host of approaches for accident prevention.

Heinrich also developed what is now known as the Heinrich accident triangle, which showed that for every accident resulting in serious injury, 29 resulted in minor injuries, and 100 in no injury. Since serious accidents are rare events, many incidents resulting in minor or no injury are likely to occur before particular unsafe acts and conditions cause serious accidents to happen. Consequently, minor accidents and near misses can act as an early warning to an organization that serious problems are present. The victims of serious accidents may also be unable or unwilling to tell analysts much about what caused the accident. For these and other reasons, it is now well accepted that it is important to study near misses to learn more about the causes of accidents, and many organizations, such as the Nuclear Regulatory Commission, have established policies of tracking all incidents, regardless of whether there is an injury. Other examples include the Aviation Safety Reporting System, administered by NASA, which collects voluntary safety related reports submitted by pilots, air traffic controllers, flight attendants, mechanics, and dispatchers. Another example is the National Fire Fighter Near-Miss Reporting System, funded by the U.S. Department of Homeland Security, which collects voluntary reports submitted by fire service professionals.

Multifactor theories

Heinrich went on to develop domino theory, which organized the sequence of events leading to an injury in terms of five factors (or dominos): social environment and ancestry, fault of person, unsafe act or condition, accident, and finally injury. Domino theory proposed that an injury could be prevented by removing any single factor from the accident sequence. Over the years, a large number of other models and theories have been developed that are built upon this basic contribution of domino theory.

Extensions of Heinrich's early focus on the accident sequence include the development of multifactor models that show how multiple chains of events converge to cause most, if not all, accidents (Lehto and Salvendy, 1991). This approach often involves the use of event and fault trees during probabilistic risk assessments to calculate the criticality of particular event sequences leading to accidents. Other approaches including network models, such as Benner's multilinear sequencing model, flowcharts, or PERT/CPM networks, place the events and unsafe acts that lead to unsafe conditions, and ultimately, an accident, on time lines to show both temporal and logical relationships between events, and are especially useful during accident investigation to explain how and why accidents occurred.

The epidemiological model provides another framework for organizing the multiple factors that may play a role in accident causation. As first proposed by Gordon (1949), factors influencing accidents may be subdivided in those associated with the *host* (accident victim), *agent* (deliverer of the injury), and *environment* (the accident setting). The epidemiological model has guided an immense amount of accident research over the years. Extensions of this model include the Industrial Accident Model (Johnson, 1973), which shows how characteristics of the victim, environment, and accident agent fit into the accident sequence as background factors, initiating factors, intermediate factors, immediate factors, and measurable results. By organizing a large number of potential accident causes in a meaningful way, the model seems to have significant potential for guiding managerial efforts for reducing accidents. Another example is the Haddon Matrix (Haddon, 1975), which organizes causes of traffic accidents and methods of improving traffic safety, related to the driver, car, and road environment into pre-accident, accident, and post-accident stages. This framework has been used for years to guide safety programs of the National Highway Traffic Safety Administration.

Haddon (1975) also proposed 10 generic countermeasure strategies that focus on the role of energy as a cause of injury. Each countermeasure falls within different stages of the accident sequence and focuses on different elements of the epidemiological model. The strategies can be paraphrased as (1) prevent the initial buildup of energy, (2) reduce the potential energy, (3) prevent the release of the energy, (4) reduce the rate of release, (5) separate the host from the energy source, (6) place a barrier between the host and energy source, (7) absorb the energy, (8) strengthen the susceptible host, (9) move rapidly to detect and counter the release, and (10) take procedures to ameliorate the damage. The energy model serves a useful purpose in focusing attention on energy as a potential cause of accidents and can be applied in a wide variety of ways in industrial settings. The model can also be extended to generic categories of accidents involving undesired transfers or blockage of material flows. For example, the exposure to toxic materials can be viewed as a flow from some source to a susceptible host.

Role of human error and unsafe behavior

Human error and intentionally unsafe behavior are a common cause of accidents (Lehto and Salvendy, 1991; Reason, 1991; Lehto, 2006). The role and significance of human error can be viewed from many different perspectives. At the most general level, accidents are often the predictable consequence of design errors and management oversights or omissions. Such failures include poorly designed facilities, inadequate risk assessments, safety policies, supervision, and less than adequate operating, inspection, and maintenance procedures. At the operational level, errors are often distinguished as either (1) errors of omission, such as skipping steps in critical procedures or failing to take precautions, or (2) errors of commission, such as operating a machine at the wrong speed. Errors or unsafe acts are also commonly classified in terms of their consequences, their revocability, and their detectability (Altman, 1964). Errors that have delayed consequences are often called latent errors (Reason, 1991). Latent errors and violations, such as failing to reactivate an alarm system after performing maintenance on it, are a particularly important cause of accidents, as they often are not detected or corrected until after an accident happens.

Much effort has been directed toward determining why errors and violations occur. The overall conclusion is that a large number of performance shaping factors can play a role, including task demands, social norms, incentives and rewards, operator objectives, environmental factors, operator skill, and the presence or absence of feedback. One school of thought is that errors and violations are largely due to the lack of proper incentives to behave safely. This perspective is the cornerstone of Behavior Based Safety (Geller, 1996), a management system that focuses on developing a supportive culture in which workers receive positive reinforcement for behaving safely. A corollary to this view is that workers are sometimes rewarded for taking shortcuts or failing to follow precautions. This can result in the development of routine, or habitual, patterns of unsafe behavior such as not using personal protective equipment, or not using designated pedestrian walkways. Once such patterns of behavior become engrained, it becomes difficult to eliminate them.

Risk compensation is another issue. For example, the benefit of modifying a forklift to make it more stable might be reduced because the operator starts driving faster. People also sometimes show an overreliance on safety devices (Lehto et al., 2009), and technology, in general, (e.g., a user might rely too heavily on an alerting system to detect a hazard). The latter tendencies are arguably rational reactions in that they are predicted by economic theory, if we assume people adjust their choices to maintain an acceptable level of risk.

Risk management and systems safety

In any organization it is essential that risks be kept to an acceptable level. Rather than simply reacting to accidents, management should be proactively taking steps to prevent them from occurring. To do this properly, management must balance the severity and likelihood of the hazards faced against the effectiveness and cost of control measures. The first step in this process is to systematically identify which hazards are potentially present. The next step is to assess the criticality of each hazard on some type of risk index that takes into account both severity and likelihood. Management can then use this information to prioritize hazards and decide upon control measures that keep risks at an acceptable level.

Practitioners in the field of systems safety apply this proactive approach throughout the life cycle of a product, program, or activity (Roland and Moriarty, 1990). This process involves hazard analyses, design reviews, and specification of safety requirements at each stage of development as the design moves forward from an initial concept to production and deployment in the field. One advantage of this approach is that it is usually much easier and cheaper to eliminate hazards by taking steps early in the design process. Control measures added in response to accidents that occur after a design has been launched are also often less effective. The systems safety approach also requires hazard analyses and design reviews for proposed changes. This is important, as changes in systems, processes, or products can often create hazards. Change analysis (Johnson, 1973, 1980) is also a useful technique for identifying root causes of system failures and accidents after they occur.

Applications of the systems model often involve various forms of probabilistic risk assessment to evaluate the reliability of safety-critical subsystems and determine the effect of design changes and other control measures. In this approach, fault trees are used to determine which combinations of component failures and human errors of omission or commission can cause a subsystem to fail and how different subsystem failures can come together to cause unsafe conditions, accidents, and ultimately injuries. The probability of system and subsystem failures as well as accidents can then be calculated after assigning probabilities to component failures and human errors of omission or commission.

A major strength of following the systems safety approach is that it provides a "divide-and-conquer" strategy for systematically identifying management oversights and omissions that cause accidents (Johnson, 1975, 1980). Johnson used the systems approach to organize a large number of factors contributing to accidents in the management oversight and risk tree (MORT). In its original form, MORT was a method for accident investigation (Johnson, 1975) but eventually evolved into a model of safety management that follows a systems approach to organize and address a large number of factors contributing to management oversights and omissions, including (1) failures to adequately assess risk and (2) less than adequate safety policies, supervision, engineering controls, and standard operating, inspection, and maintenance procedures.

Occupational health and safety management programs

As stated in the Occupational Health and Safety Act, employers are required to provide a workplace free from recognized hazards. This is normally done by establishing a health and safety management program. Guidance on how to do this is available from a large set of sources. For example, the American National Standards Institute (ANSI) has developed a standard for occupational health and safety management, ANSI/AIHA Z10-2005: American National Standard for Occupational Health and Safety Management Systems. The seven sections of the standard include the following: management leadership and employee participation, planning, implementation and operation, evaluation and corrective action, and management review. The standard outlines what has to be accomplished by the organization but is a "performance" standard and leaves the specific tasks to be determined by each organization for their unique circumstances.

Some of the features of successful safety management programs that have been identified in studies of program effectiveness are summarized in Table 17.6. Typical activities conducted in a health and safety management program include (1) ensuring compliance with safety standards and codes, (2) accident and illness monitoring, (3) identifying and analyzing workplace hazards, (4) implementing controls to reduce or eliminate hazards, and (5) hazard communication.

Compliance with standards and codes

An essential aspect of a health and safety management program is to determine which standards, codes, and regulations are relevant, and then ensure compliance. Many standards and codes are developed by consensual organizations and cover such topics as chemical labeling, personal protective equipment, and workplace warning signs. In addition, states develop regulations and standards that pertain to workplace safety and health.

The broadest applicable regulation in the United States is that of the OSHA general industry standard. In addition to the OSHA general duty clause, which places an obligation on employers to provide a workplace free from recognized hazards, OSHA also specifies a large set of detailed and mandatory health and safety design standards. OSHA also requires that many employers implement comprehensive hazard communication programs for workers involving labeling, material safety data sheets, and employee

Table 17.6 Elements of Good Safety Management

Management commitment

1. Safety officer holds high staff rank
2. Top officials are personally involved in safety activities; e.g., they make personal plant safety tours and give personal attention to accidental injury reports
3. High priority is given to safety in company meetings and in decisions on work operations
4. Management sets clear safety policy and goals

Safety committee and safety rules

1. The safety committee holds regular, frequent meetings
2. Safety rules are regularly reviewed and updated in light of accident experience
3. There is evidence of management and staff compliance with rules

Hazard control

1. There is a high level of housekeeping
2. There is orderly design/layout of work processes
3. There are good environmental qualities (ventilation, lighting, and noise control)
4. There is a greater number and variety of safety devices of on operating machinery

Inspections and communications

1. There are daily worker–supervisor contacts on safety or other job matters
2. Formal inspections are made at regular, frequent intervals
3. There is a smaller span of supervisor control
4. There are numerous informal contacts between workers and top officials

Accident investigations

1. Investigations and records are kept both on disabling (lost time) and record keeping injuries and nondisabling injuries
2. Investigations are made of property accidents and "near misses"
3. There is regular use of reports for prompting hazard control measures

Employee support

1. There are well-established procedures for job placement and advancement
2. There are personal counseling services
3. There are recreational facilities and programs for off-job hours

Safety motivation

1. A humanistic approach is used in disciplining safety violators
2. Worker families are enlisted in safety promotions
3. Specially designed posters or displays are used for hazard recognition
4. Individual praise and recognition are given for safe job performance

Safety training

1. Safety is included in new worker orientation
2. Workers are given initial and follow-up training in safe job procedures
3. Supervisors are given special safety training
4. A variety of safety training techniques (lectures, films, group discussions, simulations) are used

(continued)

Table 17.6 (continued) Elements of Good Safety Management

Makeup of workforce
1. Workers are generally older
2. Workers generally have longer experience in their jobs
3. There are more married workers
4. There is less turnover and absenteeism in the workforce

Source: Zimolong, B., Occupational risk management, Salvendy, G. (Ed.), *Handbook of Human Factors and Ergonomics*, 2nd edn, Wiley, New York, pp. 989–1020, Chapter 31, 1997.

training. OSHA enforces these standards by conducting inspections and imposing fines if the standards are not met. In addition, employers are required to maintain records of work-related injuries and illnesses, and to prominently post both an annual summary of injury and illness, as well as notices of noncompliance with standards.

Accident and illness monitoring

The recording and reporting of work-related injuries and illnesses is another critical component of any safety management program. In most cases, this process involves some form of accident investigation. Another requirement is to post and maintain statistical records. These activities are mandated by the Occupational Health and Safety Act and help the employer evaluate the extent and severity of work-related incidents and identify patterns of occurrence that should be the focus of safety management efforts.

Accident reporting

OSHA requires employers to submit a report of each individual incident of work-related injury or illness that meets certain criteria as well as maintain a log of all such incidents at each establishment or work site. Reportable incidents are those work-related injuries and illnesses that result in death, loss of consciousness, restricted work activity or job transfer, days away from work, or medical treatment beyond first aid. Detailed instructions on how to decide if an incident is reportable and then report it are given in the *OSHA Recordkeeping Handbook* (http://www.osha.gov/recordkeepinghandbook).

OSHA form 301, "Injury and Illness Incident Report," is used to report a recordable workplace injury or illness to OSHA. Information that must be provided on this form includes:

- Employee identification
- Medical treatment received
- Time of event and how long the employee had been at work
- What the employee was doing just before the incident occurred (such as "climbing a ladder while carrying roofing materials")
- What happened (such as "When ladder slipped on wet floor, worker fell 20 feet")
- What the specific injury or illness was (such as "strained back" or "broken wrist")
- What object/substance directly harmed the employee (such as "concrete floor")
- Whether the employee died

In addition, the same injury or illness must be recorded on OSHA form 300, "Log of Work-Related Injuries and Illnesses," which must be maintained at each work site. Each entry in the log identifies the employee and their job title, and gives the date of the injury or onset of illness. Each entry also describes where the event occurred, and gives a short description of the injury or illness, the parts of the body affected, and the object or substance that injured or made the person ill. The case is then classified into one of the following four categories based on outcome:

1. Death
2. Days away from work
3. Job transfer or restriction
4. Other recordable case

If applicable, the number of days away from work must be recorded as well as the number of days on job transfer or restriction. Finally, the case is classified as either an injury or one of five major types of illnesses.

Data from the form 300 log must be summarized at the end of each year on OSHA form 300A, "Summary of Work-Related Injuries and Illnesses." OSHA mandates that these summary statistics be posted at the work site, accessible for viewing by all employees, from February 1 to April 30 of the following year.

OSHA periodically visits plants to determine if they are in compliance with the law. When OSHA inspectors call, they have the right to see all of the aforementioned reports/logs. Also, certain events will automatically trigger an OSHA visit, such as an accident resulting in a fatality.*

Calculation of incidence rates

Data submitted to OSHA can be used to calculate incidence rates for various types of injuries or illnesses. An important component of an effective safety management program is to calculate and track these rates. Incidence rates should be tracked over time, and also compared to industry rates as a whole, in order to identify new problems in the workplace and/or progress made in preventing work-related injuries and illnesses.

An incidence rate is defined as the number of recordable injuries and illness occurring among a given standard number of full-time workers (usually 100 full-time workers) over a given period of time (usually 1 year). For example, the incidence rate of total nonfatal injury and illness cases in meatpacking plants in 2002 was 14.9 per 100 workers. This means that an average of 14.9 injuries and illnesses would occur for every 100 workers over the course of a year in a meatpacking plant.

The formula for calculating the incidence rate of injuries and illnesses per 100 workers is as follows:

$$\text{Incidence rate per 100 workers per year} = N \times \left(\frac{200,000}{EH} \right) \qquad (17.1)$$

where
  N is the number of reportable injuries and illnesses at an establishment during a 1 year period
  EH is the annual total number of hours worked by all employees at the establishment in a 50 week year. Note that the value of 200,000 in the formula corresponds to

$$\text{Annual total number of hours worked by 100 workers} =$$

$$(100 \text{ employees} \times 40 \text{ h/week} \times 50 \text{ weeks/year}) = 200,000$$

---

* OSHA will always visit when a fatality occurs or more than six persons are injured in a single accident. At other times, OSHA targets certain industries because of an industry-wide problem and this can also trigger visits.

The following example illustrates how to use this formula. Note that it is necessary in this example to annualize the number of accidents and number of hours worked by all employees at the establishment.

Example 17.1

If a plant experiences four accidents during a 26 week interval for their 80 employees who worked 40 h weekly, then the incidence rate is calculated as follows:

$$N = 2 \times 4 \text{ accidents in a 26 week period}$$

$$= 8 \text{ reportable accidents in an annual period}$$

$$EH = 80 \text{ employees} \times 40 \text{ h/week} \times 50 \text{ weeks}$$

$$= 160,000 \text{ total employee hours worked in 50 weeks}$$

$$\text{Incidence rate per 100 employees per year} = N \times \left(\frac{200,000}{EH}\right)$$

$$= 8 \times \left(\frac{200,000}{160,000}\right)$$

$$= 10$$

Another important statistic is the *lost workdays rate*. The lost workdays rate is defined as the number of lost workdays resulting from all recordable injuries and illness occurring among a given number of full-time workers over a given period of time. This statistic gives a measure of the severity of the injury and the illness experience of a given establishment. For example, two different establishments may have the same incidence rate of injury and illness. However, one establishment could have a higher lost workday rate than the other for the same number of injuries/illnesses. In this way, the lost workday rate tells us that the former establishment has a higher severity of injury/illness.

The formula for calculating the lost workday rate per 100 workers is

$$\text{Lost workday rate per 100 workers per year} = M \times \left(\frac{200,000}{EH}\right) \tag{17.2}$$

where

M = Number of lost workdays of employees due to reportable injuries and illnesses during a 1 year period

$= N \times (\text{average number of workdays lost per injury/illness})$

Example 17.2

In the case of the previous example, if the four cases average 5 days lost time each, then

$$\text{Lost workday rate per 100 workers per year} = M \times \left(\frac{200,000}{EH}\right)$$

$M = N \times (5 \text{ average workdays lost per injury/illness})$

$= 8 \times 5$

$= 40 \text{ lost workdays due to reportable injuries and illnesses during a 1 year period}$

Each year the BLS reports accident statistics for each major industry. An example of such statistical data was given earlier in Table 17.3, which showed the average incidence rate of certain industries.

It can be very beneficial to compare incidence and lost workday rates for a particular company to industry averages. Note, however, that some caution must be observed when making these comparisons. While industry-wide rates are usually reported per 100 employees, in a few cases the rates are reported *per 10,000 employees*. Rates are typically reported per 10,000 employees only for incidents of low occurrence, such as certain types of illnesses. *When comparing company incidence rates to industry-wide statistics*, it is necessary to be sure whether the industry rates are reported per 100 employees or per 10,000 employees.

Hazard and task analysis

The process of identifying and classifying hazards has been recognized for many years as an important first step in injury and accident prevention. For example, over 50 years ago, Heinrich identified a generalized procedure for improved safety, which he called, "the hazard through-track." This through-track starts with a hazard recognition phase (Heinrich, 1959). This phase includes developing knowledge about probable hazards and their relative importance. The second phase is to find and name hazards explicitly. Identifying the particular hazards involves analyses, surveys, inspections, and inquiries. Supervisory and investigative reports usually form the basis of this identification. Selection of a remedy forms the third phase of this through-track. Some of the options here include engineering revisions of product or process, personnel adjustments or reassignments, and training, persuasion, and/or discipline. The last phase of this through-track consists of implementing the remedy. An important step here is to verify that the remedy is an improvement. Sometimes a postaudit is needed to verify that a remedy works.

Critical-incident analysis

One of the challenges of hazard and task analysis is the combinational explosion of conditions and events that occur in a reasonably complex work environment. Whereas accident investigation can focus on "what happened" using deductive reasoning and physical evidence, hazard analysis must usually rely on inductive reasoning to determine "what can happen." The classic problem in accident analysis is that the analyst must consider how a possibly very large number of subsets of conditions and events might interact to produce unusual and undesirable results. Consequently, often there is not enough accident data to determine how likely certain types of accidents are to occur.

The critical-incident technique (CIT) is a type of analysis that is helpful in bridging the gap between accident investigation and pure prospective hazard analysis (Flanagan, 1954). The CIT is used to gather useful information from "near misses," or accidents that almost happened. Since the causes of a near miss could have resulted in an accident, their analysis can be as useful as investigating an actual accident.

Work safety analysis

A number of hazard identification techniques have also been developed that focus on dividing work into sequences of subtasks, which are then individually evaluated. Work safety analysis is typical of such approaches and involves the development of a matrix in which hazards are identified for each subtask and then described in terms of causative factors and corrective actions (Suokas and Rouhiainen, 1984). To guide this process, checklists of generic accidents and their causes are provided to the analyst. The analyst also rates the probability and consequence of each hazard on a five-point scale. The hazards are then ranked on a risk index obtained by multiplying the probability score by the consequence score, both before and after the corrective action.

Action error analysis (AEA) is a similar approach that also involves the division of work procedures into sequential stages (Suokas and Pyy, 1988). A matrix is then developed in which potential errors are identified for each stage and described in terms of primary consequences, secondary consequences, means of detection, and measures for prevention. As such, performing AEA is practically identical to developing a detectability/revocability/consequence (D/R/C) matrix (Altman 1964, 1967) as a way of prioritizing, in terms of importance, the human errors found during task analysis.

Control strategies

Strategies for preventing or controlling accidents and occupational exposure to hazards can be distinguished in several different ways. The most effective approaches often are focused on organizational objectives other than injury or illness prevention, such as increased reliability, quality, or productivity. Such solutions create a "win–win" situation making it easier to gain support across the organization for their implementation. Simply put, few people would argue with the fact that well designed physical facilities, equipment, processes, and jobs combined with a well trained, motivated workforce will result in higher productivity, better quality products and services, and safety.

The basic point is that efforts to improve productivity or other objectives of the organization can also lead to safety improvements. However, in many situations, additional hazard control measures are needed to attain a reasonable level of safety, such as changes in the design of facilities, equipment, processes, and jobs that in some cases conflict with productivity or other objectives of the organization. Many control measures are highly specific to particular hazards, as expanded upon earlier in Chapter 4 for common hazards found in the workplace. The following discussion will introduce some general control strategies, beginning with job and process design. Attention will then shift to the so-called hierarchy of hazard control that provides a way of organizing control measures.

Job and process design

Good job and process design can eliminate the root causes of accidents and occupational exposure to hazards in many different ways. To a large extent, this approach involves the application of techniques used in the fields of industrial engineering and ergonomics (Salvendy, 2001). Some of these approaches are as follows:

1. Principles of facility and plant layout focus on the separation of people from hazardous operations, efficient patterns of material flow that minimize the need for people to cross traffic areas, properly located and spaced aisles and walkways, and appropriate locations for exits and emergency egress.

2. Value stream mapping focuses on eliminating unnecessary movements of material from location to location, reducing forklift traffic and the need for potentially hazardous exertion while moving items.
3. Methods of production control schedule operations to help avoid ebbs and surges of activities that place excessive demands on operators to rush operations.
4. Methods of workplace layout and design help reduce unnecessary reaching, lifting, or unsafe postures.
5. Methods of task analysis can be used to systematically assess task requirements and develop appropriate solutions when they are excessive. Such solutions include providing power tools to reduce excessive exertion, training, checklists and instructions, modifying the task to make it less demanding, adding additional staff, scheduling rest breaks, and job rotation.
6. Standardization of operations, tools, and parts to reduce critical confusions and errors.
7. Methods of inventory control help ensure that necessary parts, tools, and equipment are available at the time they are needed, allowing activities to be performed promptly and correctly.
8. Technical information systems help ensure that accurate updated maintenance and repair instructions, material data sheets, and other important information are made available in a timely manner when it is needed, and which can help prevent critical errors.
9. Quality control and preventative maintenance programs help ensure that equipment and tools remain in a good state of operational readiness.

All of the aforementioned approaches address root causes of accidents. In recent years, these approaches have been repackaged in various ways. Examples include concepts such as total quality management, Six Sigma, 5S (see Chapter 4), the Visual Factory, Factory Physics, or lean manufacturing. Companies applying these approaches have often observed large considerable improvements in both safety and productivity. It also should be mentioned that these approaches, involving organizations such as the Purdue Regenstrief Center for Healthcare Engineering (http://www.purdue.edu/discoverypark/rche/), are currently driving major efforts in health care settings to apply methods of industrial engineering to improve patient care and safety through better delivery of critical services.

Hierarchy of hazard control

The so-called hierarchy of hazard control can be thought of as a simple model that prioritizes control methods from most to least effective. The basic idea is that designers should first consider design solutions that completely eliminate the hazard. If such solutions are technically or economically infeasible, solutions that reduce but do not eliminate the hazard should then be considered. Behavioral controls, such as training, education, employee selection, and supervision, fall in this latter category for obvious reasons. Simply put, these behaviorally oriented approaches will never completely eliminate human errors and violations.

An important point is that few design solutions completely eliminate the possibility of human errors and violations. Behavioral solutions can supplement and, in some cases, be preferable to design solutions. For example, a behavioral solution that reduces the number of automobile collisions, such as enforcement of speed limits, obviously supplements design solutions such as better seat belts and is arguably acting at a more fundamental level by helping prevent the accident as well as reducing its consequences. The obvious conclusion is that designers and others need more guidance than the hierarchy of hazard control provides to select appropriate intervention strategies.

The emphasis on nonbehavioral solutions also seems to conflict with the traditional view of accident researchers that human error and intentionally unsafe behavior are the predominant cause of accidents (e.g., Cooper, 1961; Kowalsky, Masters, Stone, Babcock, and Rypka, 1974; Ramsden, 1976; Heinrich, Peterson, and Roos, 1980). Part of the blame for human error can be given to poor product and equipment design (Norman, 1992). However, many other factors play a role. For example, one analysis of accidents at sea found that habits, incorrect diagnoses, lack of attention, lack of training, and unsuitable personality contributed to 93 of the 100 accidents studied (Wagenaar and Groeneweg, 1988). The authors of the latter study conclude that preventing human error is the most promising approach to reducing accidents. Some of their proposed solutions include better training, working conditions, behavioral controls, and incentives. This again supports the conclusion that behavioral solutions are an important tool in preventing accidents and deserve more consideration than the hierarchy of hazard control would suggest.

It should be emphasized that there are many ways of eliminating or reducing undesired behavior that can be quite effective. Some of these alternative solutions are as follows:

1. Design for usability and understandability
2. Behavioral constraints—elements of product design that make the undesired behavior difficult or impossible
3. User selection—making the product available only to selected, qualified, and responsible users
4. Supervision, enforcement, and incentives

As pointed out by Norman (1988), human errors can often be eliminated by designing products to be more usable. Some of his suggested solutions include the use of affordances and constraints, visible and natural mappings, and the provision of feedback. Such features make correct and incorrect uses of the product more obvious to the user, and reduce the need for instruction manuals, warning labels, and other types of product information. Behavioral constraints are features of the product that make it hard or impossible to perform certain behaviors. Examples include features such as interlocks, lock-ins, lock-outs, guards, or barriers. Other behavioral constraints require that the user have certain knowledge (such as a password) to operate the product and are often targeted to prevent use of a product by unqualified users. Some related strategies include screening out employees with alcohol or drug dependence, bad driving records, or who have not taken training courses. Supervision- and enforcement-related strategies focus on detecting and stopping the behavior, as illustrated by need for supervisors to detect and enforce willful violations of safety rules.

Behavioral incentives include methods of rewarding safe behavior, such as reduced insurance premiums to nonsmokers, or to drivers who use seatbelts. Other uses of incentives include punishment, such as issuing tickets for failing to wear a seatbelt. At this point, it should almost be unnecessary to state that there are many fundamentally different approaches to modifying human behavior that will often be effective. Behavioral controls, such as safety information supplement product design by making certain hazards more obvious. They can also supplement supervision, enforcement, and use of behavioral incentives by reminding or informing people such programs are in place.

A warning sign that informs drivers that they have entered a radar speed control zone or that speeding penalties are doubled in highway construction work zones illustrates this role. A closely related role of a warning is that of providing feedback that informs the user when they make errors. The importance of the alerting and feedback roles of a warning implies that warning systems that detect and selectively respond to intermittent hazards are especially desirable.

Warnings and other forms of safety information, such as safety precautions, can also serve as performance aids that help people decide what to do. In the latter role, the information sources often are serving as concise forms of external memory that help people remember and apply what they already know. This can happen in at least three different ways. That is, safety information can identify or describe the hazard, describe actions that should be taken to reduce the hazard or its effects, or direct the person's attention to other sources of information. For example, a warning label might inform the user that a product contains hydrochloric acid, briefly describe what to do if it enters a person's eye, and direct the user to a material safety datasheet (MSDS) and other sources of more detailed information. Such information can be a useful supplement to training, instruction manuals, education, and experience. This role is especially likely to be important when people do not know or have forgotten how to perform certain safety-critical tasks.

Hazard communication

In most industrialized countries, governmental regulations require that certain forms of safety information be provided to workers. For example, in the United States, the EPA has developed several labeling requirements for toxic chemicals. The DOT makes specific provisions regarding the labeling of transported hazardous materials. OSHA has promulgated a hazard communication standard that applies to workplaces where toxic or hazardous materials are in use. Training, container labeling, and MSDS (Box 17.2) are all required elements of the OSHA hazard communication standard.

BOX 17.2   MATERIAL SAFETY DATA SHEETS

The OSHA hazard communication standard specifies that employers must have a material safety data sheet (MSDS) in the workplace for each hazardous chemical used. The standard requires that each sheet be written in English, list its date of preparation, and provide the chemical and common name of hazardous chemicals contained. It also requires the MSDS to describe (1) physical and chemical characteristics of the hazardous chemical; (2) physical hazards, including potential for fire, explosion, and reactivity; (3) health hazards, including signs and symptoms of exposure, and health conditions potentially aggravated by the chemical; (4) the primary route of entry; (5) the OSHA permissible exposure limit, the ACGIH Threshold Limit Value, or other recommended limits; (6) carcinogenic properties; (7) generally applicable precautions; (8) generally applicable control measures; (9) emergency and first aid procedures; and (10) the name, address, and telephone number of a party able to provide, if necessary, additional information on the hazardous chemical and emergency procedures.

In the United States, the failure to warn also can be grounds for litigation holding manufacturers and others liable for injuries incurred by workers. In establishing liability, the theory of negligence considers whether the failure to adequately warn is unreasonable conduct based on (1) the foreseeability of the danger to the manufacturer, (2) the reasonableness of the assumption that a user would realize the danger, and (3) the degree of care that the manufacturer took to inform the user of the danger. The theory of strict liability only requires that the failure to warn caused the injury or loss.

From an ethical and legal perspective, employees have a right to know about hazards in the workplace so they can make informed decisions about how to respond to them. Providing safety information to workers can also be thought of from the control perspective as either a way of preventing unsafe acts by alerting, reminding, or instructing people what to do or as a strategy for ensuring safe behavior by building safety awareness and motivating people to behave safely.

Sources of safety information

Manufacturers and employers throughout the world provide a vast amount of safety information to workers. The many sources of safety information made available to workers include materials provided in training courses, material safety data sheets, written procedures, safety signs, product labels, and instruction manuals. Information provided by each of these sources varies in its behavioral objectives, intended audience, content, level of detail, format, and mode of presentation. Each source also provides its information at different stages in a way that corresponds to behavioral objectives and when it is provided (Table 17.7).

Table 17.7 Objectives and Example Sources of Safety Information Mapped to Accident Sequence

	Task Stage in Accident Sequence			
	Prior to Task	Routine Task Performance	Abnormal Task Conditions	Accident Conditions
Objectives (behavioral)	Educate and persuade worker of the nature and level of risk, precautions, remedial measures, and emergency procedures	Instruct or remind worker to follow safe procedures or take precautions	Alert worker of abnormal conditions; specify needed actions	Indicate locations of safety and first aid equipment, exits, and emergency procedures; specify remedial and emergency procedures
Example sources	Training manuals, videos, or programs; hazard communication programs; MSDSs; safety propaganda safety feedback	Instruction manuals; job performance aids; checklists; written procedures; warning signs and labels	Warning signals: visual, auditory, or olfactory; temporary tags, signs, barriers, or lock-outs	Safety information signs, labels, and markings; MSDSs

Source: Reprinted from Lehto, M.R. and Miller, J.M., Principles of prevention: Safety information, in Stellman, J.M. (Ed.), *ILO Encyclopedia of Occupational Health and Safety*, 4th edn., International Labour Office, Geneva, Switzerland, pp. 56.33–56.38, 1997. With permission.

BOX 17.3 WARNING SIGNS, LABELS, AND TAGS

ANSI and other standards provide very specific recommendations for how to design warning signs, labels, and tags. These include, among other factors, particular signal words and text, color coding schemes, typography, symbols, arrangement, and hazard identification (Table 17.8).

Among the most popular signal words recommended are *danger*, to indicate the highest level of hazard; *warning*, to represent an intermediate hazard; and *caution*, to indicate the lowest level of hazard. Color-coding methods, also referred to as a *color system*, consistently associate colors with particular levels of hazard. For example, red is used in all of the standards in Table 17.8 to represent the highest level of danger. Explicit recommendations regarding typography are given in nearly all the systems. The most general commonality between the systems is the recommended use of sans serif typefaces. Varied recommendations are given regarding the use of symbols and pictographs.

The FMC and the Westinghouse systems advocate the use of symbols to define the hazard and to convey the level of hazard. Other standards recommend symbols only as a supplement to words. Another area of substantial variation shown in Table 17.8 pertains to the recommended label arrangements. The proposed arrangements generally include elements from the previous discussion and specify the *image*—graphic content, color; *background*—shape, color; *enclosure*—shape, color; and *surround*—shape, color. Many of the systems also precisely describe the arrangement of the written text and provide guidance regarding methods of hazard identification.

Certain standards also specify the content and wording of warning signs or labels in some detail. For example, ANSI Z129.1 specifies that chemical warning labels include (1) identification of the chemical product or its hazardous component(s), (2) signal word, (3) statement of hazard(s), (4) precautionary measures, (5) instructions in case of contact or exposure, (6) antidotes, (7) notes to physicians, (8) instructions in case of fire and spill or leak, and (9) instructions for container handling and storage. This standard also specifies a general format for chemical labels that incorporate these items. The standard also provides extensive and specific recommended wordings for particular messages.

Design specifications, such as those just discussed, can be useful to developers of safety information. However, many products and situations are not directly addressed by standards or regulations. Certain design specifications are also scientifically unproven. In extreme cases, conforming to standards and regulations can reduce the effectiveness of safety information. To ensure effectiveness, developers of safety information consequently may need to go beyond safety standards. Recognizing this issue, the International Ergonomics Association (IEA) and International Foundation for Industrial Ergonomics and Safety Research (IFIESR) supported an effort several years ago to develop guidelines for warning signs and labels (Lehto, 1992) that reflect published and unpublished studies on effectiveness and have implications regarding the design of nearly all forms of safety information.

Six of these guidelines, presented in slightly modified form, are as follows: (1) match sources of safety information to the level of performance at which critical

(continued)

BOX 17.3 (continued)

errors occur for the relevant population, (2) integrate safety information into the task and hazard-related context, (3) be selective; (4) make sure that the cost of complying with safety information is within a reasonable level, (5) make symbols and text as concrete as possible, and (6) simplify the syntax of text and combinations of symbols. Satisfying these guidelines requires consideration of a substantial number of detailed issues as addressed in the earlier parts of this chapter.

Sources of information, such as safety training materials, hazard communication programs, and various forms of safety propaganda, including safety posters and campaigns, are used to educate workers about risks and persuade them to behave safely. Such information is often provided away from the job in the classroom or safety meetings. Occasional safety and health meetings are especially appropriate when things are changing in manufacturing areas that may have safety and health implications. Meetings called to inform operators of some expected changes and to elicit employee suggestions about details of those changes are also very valuable, especially if they affect safety and health. Inexperienced workers are often the target audience, and the information provided is often quite detailed and focused on building safety awareness and motivating people to behave safely.

Other sources of information, such as written procedures, checklists, instructions, warning signs, and product labels, often provide critical safety information to the operator during routine task performance. This information usually consists of brief statements that either instruct less skilled workers or remind skilled workers to take necessary precautions. Following this approach can help prevent workers from omitting precautions or other critical steps in a task. Statements providing such information are often embedded at the appropriate stage within step-by-step instructions describing how to perform a task. Warning signs at appropriate locations can play a similar role. For example, a warning sign located at the entrance to a workplace might state that hard hats must be worn before entering.

At the next stage in the accident sequence, highly conspicuous and easily perceived sources of safety information alert workers of abnormal or unusually hazardous conditions. Examples include warning signals, safety markings, tags, signs, barriers, or lockouts. Warning signals can be visual (flashing lights, movements, etc.), auditory (buzzers, horns, tones, etc.), olfactory (odors), tactile (vibrations), or kinesthetic. Certain warning signals are inherent to products when they are in hazardous states (i.e., the odor released upon opening a container of acetone). Others are designed into machinery or work environments. Safety markings refer to methods of nonverbally identifying or highlighting potentially hazardous elements of the environment (i.e., by painting step edges yellow or emergency stops red). Safety tags, barriers, signs, or lock-outs are placed at the point of hazard and are often used to prevent workers from entering areas or activating equipment during maintenance, repair, or other abnormal conditions.

At the final stage in the accident sequence, the focus is on expediting worker performance of emergency procedures at the time an accident is occurring or performance of remedial measures shortly after the accident. Safety information signs and markings conspicuously indicate facts critical to adequate performance of emergency procedures (e.g., the locations of exits, fire extinguishers, first-aid stations, emergency showers, eye wash stations, emergency releases). Product safety labels and material safety data sheets may specify remedial and emergency procedures to be followed.

Table 17.8 Summary of Recommendations in Selected Warning Systems

System	Signal Words	Color Coding	Typography	Symbols	Arrangement
ANSI Z129.1 precautionary labeling of hazardous chemicals	Danger Warning Caution Poison optional words for "delayed" hazards	Not specified	Not specified	Skull and crossbones as supplement to words. Acceptable symbols for three other hazard types	Label arrangement not specified; examples given
ANSI Z535.2 environmental and facility safety signs	Danger Warning Caution Notice (general safety) (arrows)	Red Orange Yellow Blue Green as above; B&W otherwise per ANSI Z535.1	Sans serif, upper case, acceptable typefaces, letter heights	Symbols and pictographs per ANSI Z535.3	Defines signal word, word message, symbol panels in one to three panel designs. Four shapes for special use. Can use ANSI Z535.4 for uniformity
ANSI Z535.4 product safety signs and labels	Danger Warning Caution	Red Orange Yellow per ANSI Z535.1	Sans serif, upper case, suggested typefaces, letter heights	Symbols and pictographs per ANSI Z535.3; also SAE J284 Safety Alert Symbol	Defines signal word, message, pictorial panels in order of general to specific. Can use ANSI Z535.2 for uniformity. Use ANSI Z129.1 for chemical hazards
NEMA Guidelines: NEMA 260	Danger Warning	Red Red	Not specified	Electric shock symbol	Defines signal word, hazard, consequences, instructions, symbol. Does not specify order
SAE J115 Safety signs	Danger Warning Caution	Red Yellow Yellow	Sans serif typeface, upper case	Layout to accommodate symbols; specific symbols/pictographs not prescribed	Defines three areas: signal word panel, pictorial panel, message panel. Arrange in order of general to specific

(continued)

Table 17.8 (continued) Summary of Recommendations in Selected Warning Systems

System	Signal Words	Color Coding	Typography	Symbols	Arrangement
ISO Standard: ISO R557, 3864	None. Three kinds of labels Stop/prohibition Mandatory action Warning	Red Blue Yellow	Message panel is added below if necessary	Symbols and pictographs	Pictograph or symbol is placed inside appropriate shape with message panel below if necessary
OSHA 1910.145 Specification for accident prevention Signs and tags	Danger Warning (tags only) Caution Biological Hazard, BIOHAZARD, or symbol (safety instruction) (slow-moving Vehicle)	Red Yellow Yellow Fluorescent Orange/Orange–Red Green Fluorescent Yellow–Orange and Dark Red per ANSI Z535.1	Readable at 5 ft or as required by task	Biological hazard symbol. Major message can be supplied by pictograph (Tags only). Slow-Moving Vehicle (SAE J943)	Signal word and major message (tags only)
OSHA 1910.1200 (Chemical) Hazard Communication	Per applicable requirements of EPA, FDA, BATF, and CPSC		In English		Only as MSDS
Westinghouse Handbook; FMC Guidelines	Danger Warning Caution Notice	Red Orange Yellow Blue	Helvetica bold and regular weights, upper/lower case	Symbols and pictographs	Recommends five components: signal word, symbol/pictograph, hazard, result of ignoring warning, avoiding hazard

Source: Adapted from Lehto, M.R. and Miller, J.M., *Warnings Volume I: Fundamentals, Design, and Evaluation Methodologies*, Fuller Technical Publications, Ann Arbor, MI, 1986; Lehto, M.R. and Clark, D.R., Warning signs and labels in the workplace, in Karwowski, W. and Mital, A. (Eds.), *Workspace, Equipment and Tool Design*, Elsevier, Amsterdam, the Netherlands, pp. 303–344, 1990.

Before safety information can be effective, at any stage in the accident sequence, it must first be noticed and understood and, if the information is not immediately applicable, also be remembered. Then, the worker must both decide to comply with the provided message and be physically able to do so. Strategies for attaining these objectives have been discussed in earlier chapters of this book.

Final remarks

Much more information is available on each of the topics noted here, and many issues are not addressed. However, the presented material should be enough to familiarize the reader with the importance of safety and health, types of occupational hazards, causes of accidents, control strategies, and important functions of occupational safety and health management. There are numerous books on the subject of occupational safety and health management and engineering as well as specialty books on subjects such as toxicology, human error, system engineering, reliability, and human factors engineering. The Web is another important source of information. Most safety organizations, such as the NSC, Board of Certified Safety Professionals, OSHA, NIOSH, and BLS, have website offering large amounts of helpful information. Contact information for many of these sources of additional information is provided in the following.

Discussion questions and exercises

17.1 What does the Heinrich accident triangle show?
17.2 What OSHA form must be posted at the work site summarizing the injuries and illnesses that occurred? What government organization reports accident statistics collected from OSHA (and other sources) for each major industry in the United States?
17.3 What objectives of safety information naturally fit into particular stages of the accident sequence? Explain why particular sources of information fit particular objectives better than others.
17.4 What are some of the legal requirements associated with hazard communication? What standards apply? What kind of recommendations are given in the standards? How consistent are they?
17.5 What are some features of good safety management?
17.6 What is OIICS? The BLS? Is the cost of accidents increasing or decreasing in the United States? How about accident rates?
17.7 What were some common law defenses used by employers to avoid responsibility for employee injuries?
17.8 Why are behavioral solutions to safety problems often important?
17.9 Discuss the Hierarchy of Hazard Control.

Resources

American National Standards Institute (ANSI), 1819 L Street, NW, Suite 600, Washington, DC 20036, http://www.ansi.org
American Society of Safety Engineers (ASSE), 1800 E Oakton St, Des Plaines, IL 60018, http://www.asse.org
ASTM International (originally known as the American Society for Testing and Materials), 100 Barr Harbor Drive, West Conshohocken, Pennsylvania, http://www.astm.org

Board of Certification in Professional Ergonomics, PO Box 2811, Bellingham WA 98227–2811, http://www.bcpe.org

Board of Certified Safety Professionals, 208 Burwash Avenue, Savoy, Illinois USA 61874, http://www.bcsp.org

Department of Labor, 29 Code of Federal Regulation (CFR) 1910.95: Occupational Noise Exposure, http://www.osha.gov

Department of Labor, 29 Code of Federal Regulation (CFR) 1910 subpart D: Walking-Working Surfaces, http://www.osha.gov

Department of Labor, 29 Code of Federal Regulation (CFR) 1910 subpart O: Machine Guarding, Part 212: General Requirements for all machines, http://www.osha.gov

Department of Labor, 29 Code of Federal Regulation (CFR) 1910 subpart Z: Toxic and Hazardous Substances, Part 1200: Hazard Communication, http://www.osha.gov

Department of Labor, 29 Code of Federal Regulation (CFR) 1910 Subpart Z: Toxic and Hazardous Substances, http://www.osha.gov

Human Factors and Ergonomics Society (HFES), P.O. Box 1369, Santa Monica, CA 90406-1369, http://www.hfes.org

National Safety Council (NSC), 1121 Spring Lake Drive Itasca, IL 60143-3201, http://www.nsc.org

National Society of Professional Engineers, 1420 King Street, Alexandria, VA 22314, http://www.nspe.org

chapter eighteen

Personnel selection, placement, and training

About the chapter

This chapter addresses some of the ergonomic issues related to personnel selection, placement and training that arise in many organizations. The first part of the chapter introduces these topics and provide some background material on why they are important. Focus then shifts to testing methods often used in personnel selection and placement issues. One objective of the latter discussion is to introduce readers to the concepts of test *validity*, *reliability*, *power*, *efficiency*, *selection ratios*, *skill level*, and *Z-scores*, and their importance in ability testing. The following sections then focus on training. The later discussion briefly introduces some commonly used training methods, along with methods of measuring and improving the effectiveness of a training program. The use of job aids of various forms as a substitute for training is then briefly discussed.

Introduction

As the title implies, this chapter deals with personnel selection, placement, and training. Since engineers and ergonomic specialists design systems that involve people, this subject is clearly a relevant ergonomic issue in system design. Perhaps the most obvious reason this issues is relevant is that these approaches provide an opportunity to modify the human element present in the system in a way that will improve performance. Simply put, it is essential in all organization to bring in the right people, put them in appropriate jobs, and train them are needed. To achieve this goal, it is necessary to develop methods of selecting, placing, and training personnel that are both economically and legally defendable. This follows because people are both a major resource and a major cost in manufacturing and many other applications. Further, the Equal Employment Opportunity Commission (EEOC) takes a great deal of interest in the way those decisions are handled. With the recent Americans with Disabilities Act (ADA), hiring, training, and placing personnel will come under even greater scrutiny in the future. Such scrutiny necessitates a scientific basis for making personnel decisions. Some recommendations exist for incorporating people into system design, but unfortunately there are also knowledge gaps today.

Personnel selection and placement

Personnel selection and placement involves many decisions. In fact, a job application evokes an interrelated decision series that includes the following options:

1. Should the applicant be selected as an employee of this company?
2. If selected, what type of job should be recommended?
3. When an applicant is selected for a particular class of jobs, what type of training is required?

To make these decisions, one must collect data from interviews, application forms, and knowledge and skill testing. The immediate question here is how this data can be used to help in the decision-making process. First, let us look into some fundamental concepts of ability testing.

Personnel systems

All companies have a personnel system. In smaller companies, personnel management is the duty of each individual supervisor. As companies grow, these activities become the responsibility of the Human Resources Department, which receives employment applications, screens applicants, arranges for interviews and other tests, maintains non-payroll personnel records, operates training programs, and generally assists the company in personnel selection, placement, and training. This department is usually involved in determining wage structures, wage administration, and job evaluation.* It also collects information about the needs of various production departments relative to personnel requirements and ability patterns. Such information helps them advertise for job applicants.

A flow of job applications precipitates a flow of applicants and personnel through the department and the company. Figure 18.1 illustrates this flow and shows applicants undergoing tests and interviews, providing biographic data, and submitting personal references. Based on the results of this testing and other activities, some applicants are hired and others

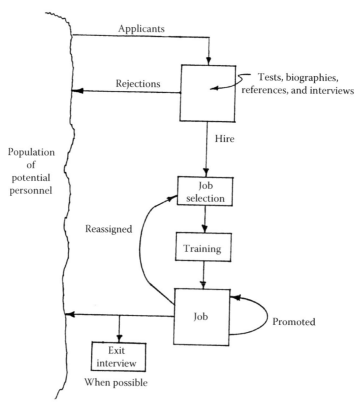

Figure 18.1 Flows of personnel through the company.

* The subjects of wage administration, job analysis and job evaluation are discussed later in Chapter 20.

BOX 18.1   EQUAL EMPLOYMENT OPPORTUNITY LAW*

Federal Laws in the United States prohibiting unfair job discrimination include the following:

- The Equal Pay Act of 1963 (EPA)—It applies to virtually all employers and requires equal pay for men and women who perform substantially equal work in the same establishment.
- Title VII of the Civil Rights Act of 1964 (Title VII)—It prohibits discrimination based on race, color, religion, sex, or national origin by private employers, state and local governments, and education institutions that employ 15 or more individuals. Title VII also applies to employment agencies and labor organizations.
- The Age Discrimination in Employment Act of 1967 (ADEA)—It protects individuals who are 40 years of age or older from age-related discrimination. The ADEA applies to all private employers with 20 or more employees, state and local governments (including school districts), employment agencies, and labor organizations.
- Title I and Title V of the Americans with Disabilities Act of 1990 (ADA)—It prohibits employment discrimination against qualified individuals with disabilities by private employers with 20 or more employees, state and local governments (including school districts), employment agencies, and labor organizations.
- Sections 501 and 505 of the Rehabilitation Act of 1973—It prohibits discrimination against qualified individuals with disabilities who work in the federal government.
- The Civil Service Reform Act of 1978—It prohibits certain personnel practices concerning employees of the federal government.

The U.S. Equal Employment Opportunity Commission (EEOC) enforces all of these laws. These laws make it illegal to unfairly discriminate in any aspect of employment, including the following:

- Hiring and firing; layoff or recall; assignment or classification of employees
- Compensation; pay, fringe benefits, retirement plans
- Transfer, promotion, or disability leave
- Job advertisements, recruitment, or testing
- Training and apprenticeship programs; use of company facilities
- Other terms and conditions of employment

Many states and localities also have antidiscrimination laws. The agencies responsible for enforcing those laws coordinate their efforts with the EEOC to avoid duplication of effort. The EEOC was authorized by Title VII of the Civil Rights Act of 1964 to endeavor to eliminate unlawful discrimination in employment by conciliation and persuasion. All laws enforced by EEOC, except the EPA, require filing a charge with EEOC before a private lawsuit may be filed in court. There are also strict time limits

---

* This box is a condensed and edited compilation of material published by the EEOC (http://www.eeoc.gov/facts/qanda.html, May 24, 2002 and http://www.eeoc.gov/ada/adahandbook.html#dodonts, February 4, 2004). Additional details on Equal Employment Opportunity Law can be obtained from the EEOC website (http://www.eeoc.gov).

(continued)

BOX 18.1 (continued)

within which charges must be filed. The EEOC provides a mediation service to help resolve disputes in which employment-related discrimination is alleged to have occurred. The EEOC also works with the parties in such disputes to reach mutually agreeable settlements.

Anyone may file a charge of discrimination with EEOC if they feel that their employment rights have been violated. The EEOC will then investigate whether there is sufficient evidence to establish that discrimination occurred. If not, a notice is issued that closes the case and gives the charging party 90 days in which to file a lawsuit on their own behalf. If the EEOC finds reasonable cause to believe discrimination has occurred, the charged organization (employer, employment agency, labor union, or other charged entity) and the charging party will be informed of this in a letter of determination that explains the finding. The EEOC will then attempt to develop a conciliation agreement in which the charged organization agrees to cease the unlawful discrimination and provide relief and remedies to the affected individuals. If the case is successfully conciliated, or if a case has earlier been successfully mediated or settled, neither EEOC nor the charging party may go to court unless the conciliation, mediation, or settlement agreement is not honored. If EEOC is unable to successfully conciliate the case, the agency will decide whether to bring suit in federal court. If EEOC decides not to sue, it will issue a notice closing the case and giving the charging party 90 days in which to file a lawsuit on his or her own behalf.

BOX 18.2   RELIEF AND REMEDIES FOR UNLAWFUL DISCRIMINATION*

Relief or remedies for unlawful discrimination may include financial compensation for back pay, front pay (i.e., pay lost due to not being promoted), fees paid for expert witnesses, attorney's fees, court costs, and so on. Punitive damages may also be included if an employer acted with malice or reckless indifference. Other common remedies include promotion, hiring or reinstatement of the employee, and providing reasonable forms of accommodation. Affirmative action may also be required to comply with Title VII Civil Rights Act of 1964 when past employment practices have caused the percentage of minority workers in the workforce to be significantly lower than the percentage of minority workers in the local population.

Some of the criteria that may be used (by the EEOC or judge or jury in a lawsuit) to determine whether unlawful discrimination has occurred and decide upon appropriate remedies are as follows:

- Was race, color, sex, age, national origin, religious preference, or handicapped status used as an occupational qualification for hiring, promotion, training, pay, or setting other work-related conditions? Doing so is almost always unlawful. Even asking questions about any of these items without making it clear that

---

* This box is a condensed and edited compilation of material published by the EEOC (http://www.eeoc.gov/facts/qanda.html, May 24, 2002 and http://www.eeoc.gov/ada/adahandbook.html#dodonts, February 4, 2004). Additional details on Equal Employment Opportunity law can be obtained from the EEOC website (http://www.eeoc.gov).

response is entirely voluntary and will not be used as an occupational qualification may be unlawful. However, in a very limited set of circumstances, sex, age, national origin, or religious preference may be used as a bona fide occupational qualification (BFOQ). For example, sex and age are both BFOQs for certain jobs requiring physical characteristics possessed by only one sex, such as actors or models; national origin may be a BFOQ for certain groups promoting the interest of a particular nationality; and religious preference may be a BFOQ for religious organizations.

- Is each job qualification job related? More specifically, is each job qualification necessary for successful and safe performance of the job? Has this been adequately documented? An employer must be prepared to prove that any physical or mental requirements for a job are due to "business necessity" and the safe performance of the job. This is one area where ergonomists, engineers, and other experts can play a very important role in both eliminating unlawful discrimination in the workplace and protecting the legitimate interests of the employer.
- *What is the effect of each job requirement on potentially affected groups?* For example, what percentage of female workers would be screened out by requiring them to be able to lift a weight of 40 lb from the floor to the work surface? What would be the effect of reducing this requirement to 20 lb? What would be the effect on task performance of reducing the requirement? Careful analysis of such questions is strongly recommended, as Title VII and other law may require remedies even when employment practices unintentionally discriminate against individuals because of their handicaps, race, color, national origin, religion, or sex. It is easy to see that employment practices such as making selection or placement decisions based on criteria such as education levels, verbal and mathematical skills, physical strengths, sensory abilities, and other capabilities may lead to unintentional discrimination. This follows because such criteria may differ systematically between potentially affected groups. For example, males tend to be stronger and larger than females.
- Do testing methods used to measure qualifications result in a higher rejection rate for minority, female, handicapped, older people or other protected groups? Are the testing methods used reliable and valid? Are these tests administered in an equitable manner? If a test has a higher rejection rate for protected groups, it must be validated, and shown to be an accurate, reliable, predictor of job performance. All applicants should be required to take the test (including medical exams). For example, requiring females or handicapped people, but not able-bodied males, to take a strength test would be discriminatory. The ADA also requires reasonable modifications of testing procedures, such as providing readers larger text or extra time to complete a written test for people with certain handicaps.
- Can job requirements be reduced or eliminated through training, job aids, job redesign, or other reasonable forms of accommodation? Reasonable accommodations recognized by the EEOC may include, but are not limited to, making existing facilities used by employees readily accessible to and usable by

(continued)

BOX 18.2 (continued)

persons with disabilities; job restructuring; modification of work schedules; providing additional unpaid leave; reassignment to a vacant position; acquiring or modifying equipment or devices; adjusting or modifying examinations, training materials, or policies; and providing qualified readers or interpreters. However, an employer generally is not obligated to provide personal use items such as eyeglasses or hearing aids.

- *Will the accommodation cause "undue hardship" to the employer?* An employer is required by the EEOC to make a reasonable accommodation to a qualified individual with a disability unless doing so would impose an undue hardship on the operation of the employer's business. Undue hardship means an action that requires significant difficulty or expense when considered in relation to factors such as a business' size, financial resources, and the nature and structure of its operation. Also, an employer is not required to lower production standards to make an accommodation.

are rejected. After hired personnel are trained formally or on-the-job, they are assigned to specific jobs in the company in this process of *job placement* or *job selection*. Over time they are either retained at the entry level or promoted. Some personnel are reassigned to other jobs as the job requirements change and as they become skilled. Since errors occur occasionally in matching people to jobs, or employees find and select alternatives to working for the company, there is an outflow of people who quit or who are fired. Most companies conduct an exit interview, when possible, to determine the reasons for resignations. For example, when people quit because of more competitive wages at other companies, the exit interview provides information that can be used to modify the local wage structure data.

The personnel department also keeps track of personnel turnover or attrition. High attrition rates, particularly from a particular production department, are symptoms of personnel problems; higher attrition can drain company resources through increased recruiting and training costs.

In addition, personnel departments need to keep track of the population of potential employees. Companies compete with each other for the supply of potential employees. When the supply is not adequate, the company must either increase wages (or benefits) or find other ways to meet production needs. This could mean reassigning production to other company plants, automating or otherwise improving productivity, searching for vendors to reduce the personnel needs, or relocating the existing plant. Some companies that have relocated have chosen rural locations where there is a smaller supply of skilled personnel. Generally this strategy has resulted in much larger investments in training.

Concepts of ability testing

Many organizations use test scores, along with other data, to select employees and place them into appropriate job positions. Many different types of tests are used for this purpose that measure abilities, skills, and other employee characteristics. For legal and economic reasons, it is essential that the tests used are both valid and reliable (McCormick, 1979; Swezey and Pearlstein, 2001). As outlined in Box 18.1, one reason this follows is that selection and placement criteria must be based on a bonafide occupational requirements to meet the legal requirements. If a test fails to meet this standard, it will not only lead to potential legal sanctions, but also will have little value as basis for selection or placement purposes.

The *validity* of a test is defined as its ability to measure what it is supposed to measure. One way of examining validity is to administer a developed test to a group of workers whose job performance has been previously assessed. These performance quality measurements might be shop ratings, records of production rate, and/or quality work or opinions of shop supervisors. The next task is to find the correlation between the performance measurements and test results. Higher correlations indicate greater validity, and a correlation of less than 0.3 is usually considered insufficient to qualify as a valid test. Of course, the more objective the basis for making these correlation tests, the more the validity measurement will be accepted. This is often referred to as *face validity*.

The *reliability* of the test used is another important issue. A reliable test gives the same measurements consistently in the same situations. Reliability is sometimes called *test repeatability* or *consistency*. Reliability is typically measured by creating two forms of the same test. If the same group of people is given both forms of this test, the *correlation of scores within-persons* provides an appropriate measurement reliability. Generally, reliability correlations must be above 0.7 to be considered reasonable.

Tests also should be designed in such a way that test scores vary enough to discriminate poor, adequate, and excellent performers from each other. Otherwise, the test will have little practical value. A poorly designed test may also lead to unexplained fluctuations in test scores (called *instabilities*) The latter effect is obviously undesirable and management will soon lose confidence in testing if the results show such tendencies.

Two related measurements in testing are *power* and *efficiency*. *Power* is the test's ability to detect small ability differences. *Efficiency* is the power/cost ratio. Most people believe that simulators have much more power than written tests, but they also cost much more. When a battery of tests is assembled for personnel selection, low correlations within-persons between different parts of the same test battery usually indicate greater test power. A low correlation indicates minimal redundancy within a test, meaning that the different parts of the test are measuring different things. Note that this is a necessary but clearly not a sufficient condition for adequate test power. Obviously, power is difficult to measure.

Three other measures have important implications in personnel testing. One is the *selection ratio* or in other words, the fraction of applicants who are hired. Another measure with important implications is the *skill level* needed to perform satisfactorily. A high skill level means that only a few persons in the population can perform the needed tasks. A third is the so-called Z-score. The latter measure is the proportion of hired people who provide acceptable performance. Figure 18.2 shows the statistical relationships between skill level, selection ratio, and the Z-score. Note that the Z-score is equal to area A in this figure, divided by the sum of areas A and B. As with almost all statistical techniques, there are two types of errors in personnel selection. First, some people are selected who cannot perform the job adequately, and that error is shown in Figure 18.2 as area B. The other type of error consists of those people who are rejected by the test but who could perform adequately on the job, and that error corresponds to area D.

Figure 18.3 shows the effect of differences in validity and selection ratios on Z-scores, assuming a population skill level of 50%. The different curves describe the Z-scores (indicated on the y-axis) as function of the selection ratio for nine different values of the validity coefficient. Note that a validity coefficient of zero indicates that there is no relationship between the test scores and ability in the job. In the latter situation, the Z-score is equal to the skill level regardless of the population skill level. This follows because a test validity of zero means selection is purely by chance, and so the expected fraction

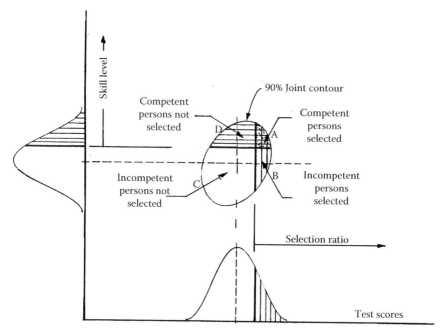

Figure 18.2 Statistical relationships among test scores, skill level, and selection ratio.

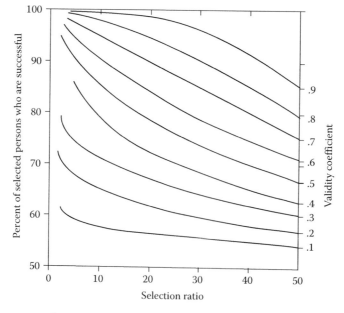

Figure 18.3 Z-scores as a function of the selection ratio and validity where skill level is 50%.

of selected personnel should be the ratio of personnel who can perform the job satisfactorily. If the validity increases to 0.5, we now see that the Z-score increases when the selection ratio decreases from a little less than 70% for a selection ratio of 50% to about 8% for a selection ratio of 10%. This result demonstrates that fewer people at a lower skill level are selected on the basis of this test when we use a smaller selection ratio.

Chapter eighteen: Personnel selection, placement, and training 649

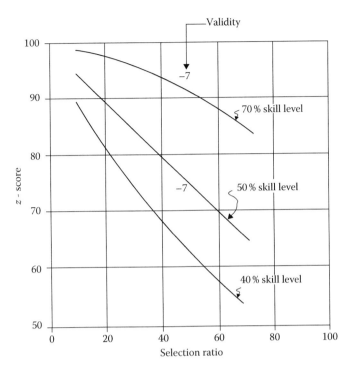

Figure 18.4 Z-scores as a function of the selection ratio, skill level, and test validity.

Similar conclusions can be drawn from Figure 18.4, which shows Z-scores as a function of the selection ratio, skill levels, and test validity. Note that with an increase in the *selection ratio* from 30% to 60%, and skill level and validity remaining at 50% and 0.7%, the Z-score drops from around 84% to about 70%. On the other hand, had there been a simultaneous increase in skill level to 70%, the Z-score would be about 87%, for a selection ratio of 60%. Figure 18.4 clearly shows that increases in skill levels when other factors are constant results in an increased Z-score. Figures 18.3 and 18.4 demonstrate that increases in the selection ratio decrease the Z-scores.

To describe a larger range of Z-scores and other important aspects of personnel selection statistics, tables of correlated statistics are included in Section B.1.3 of Appendix B of this book. These tables show decile percentages of bivariate normal distributions, given that both specified variables occurred. Seven of the tables in this appendix denote those statistics for correlations of 0.9, 0.8, …, 0.4, and 0.3 respectively. In the top table of Section B.1.3 corresponding to a correlation of 0.9, the top decile jointly on both variables has a percentage of 68.8%—that is, for a test with a 0.9 validity, 68.8% of the people who have the highest 10% in test scores would be expected to also exhibit the top 10% in job performance. In comparison, a correlation of 0.5 would be expected to result in a lower percentage of people who would fit the previous situation. Accordingly, if we move down in Section B.1.3 to the table corresponding to $r = 0.5$, we see that the percentage of people who are jointly at the top 10% of both test scores and in job performance is expected to be 32.3%; this is only about half the number of people we found for a correlation coefficient of 0.9. When those correlations correspond to test validity values, Z-scores and other statistics can be computed.

To illustrate how this works, let us return to Figure 18.2. To determine the Z score, for a given combination of selection ratio, population skill level, and test validity, we need to first determine areas A and B indicated in the figure. Appendix B.1.3 can be used to do this. For example, let us assume a value of $r=0.5$, corresponding to the test validity, a selection ratio of 30%, and a population skill level of 50% level. The next step is to find the table in Appendix B.1.3 that corresponds to $r=0.5$. Then, since we are assuming a selection ratio of 30% and a population skill level of 50%, to determine areas A and B we need to use the part of this table corresponding to columns 8–10 (i.e., the 30% for selection ratio) and for rows 6–10. This part of the table and the associated values for areas A and B is given below:

Deciles	8	9	10	Sum	
10	13.9	19.1	32.3	65.3	
9	14.2	16.5	19.1	49.8	
8	13.4	14.2	13.9	41.5	
7	12.4	12.2	10.5	35.1	
6	11.3	10.5	8.0	29.8	
Sum	65.2	72.5	83.8	221.5	Area A
Complement	34.8	27.5	16.2	78.5	Area B
				300.0	Area A+B

In this example, we obtained the complement for each column by subtracting each sum from 100. Note that we can do this because the total sum of each column or row is always 100, for each of the tables in Appendix B.1.3. In other words, the complement in this particular example corresponds to sum of the respective values for rows 5–1 the sum of rows 1–5 in the table in Appendix B.1.3. After determining the complement, we then added the sums across the columns to get area A, and added the complements to get area B. The resulting Z-score is area A divided by the sum of area A and area B, or $221.5/300 = 0.738$. If the test validity was 0.9, instead of 0.5, the table and corresponding values for areas A and B would be as follows:

Deciles	8	9	10	Sum	
10	6.5	22.4	68.98	97.8	
9	23.6	36.2	22.4	83.1	
8	27.6	23.6	6.5	56.2	
7	21.3	11.5	1.7	34.4	
6	12.7	4.5	0.4	18.0	
Sum	91.5	98.2	99.9	289.6	Area A
Complement	8.5	1.8	0.1	10.4	Area B
				300.0	Areas A+B

The Z-score is area A divided by the sum of area A and area B, or $289.5/300 = 0.965$. Note that the Z-score increases from 74% to 96.5% with that change in validity.

Individual differences on the job

Large individual differences are often observed in job performance between people of similar background and training. Test scores, even of experienced workers, will often vary accordingly. For example, Figure 18.5 shows the percentage of experienced shop personnel with particular test scores on a shop test. The test scores in this example certainly show a very wide distribution, raising question about the usefulness of the test in the

Chapter eighteen: Personnel selection, placement, and training 651

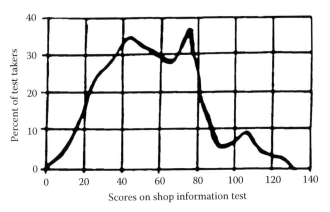

Figure 18.5 Example distribution of test scores on a shop test by experienced shop personnel.

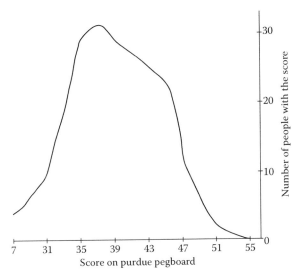

Figure 18.6 Number of aircraft electricians and their test scores on the Purdue Pegboard out of 227 who took the test (measuring manual dexterity).

applicant selection decision process, assuming all of the tested personnel were performing adequately on the job. On the other hand, if further analysis revealed that the test scores were in fact correlated with objective measures of job performance, or perhaps training time, the test might serve a useful role in selecting new applicants.

Another example of scores on a proficiency test is shown in Figure 18.6. This proficiency test was the Purdue pegboard test of physical dexterity. The test consists of placing small metal pins into holes drilled into a board and then putting washers and collars over the pins. Performance is timed for right hand only, left hand only, and both hands, and the composite result provides the score. The small size of these hand-manipulated parts gives this test reasonable face validity. In the example shown in Figure 18.6, the test was performed by 227 aircraft electricians. The wide range of test scores observed here are consistent with the conclusion that the previous selection criteria for these particular electricians did not include the forms of manual dexterity measured by the Purdue pegboard test. The distribution of test scores is reasonably close to normal (i.e., Gaussian distributed).

The test results shown earlier are fairly typical in that they reveal a great deal of variability in human performance. Some variability clearly is acceptable in current work situations. However, too much variability in performance between workers can lead to obvious problems. Reducing such variability is one of the objectives and potential benefits of a good personnel selection program.

Economic considerations in personnel selection

A good method of personnel selection can potentially provide many benefits to an organization, such as improved productivity, lower training costs, and less employee turnover. However, these potential benefits must be balanced against the administrative costs incurred during the personnel selection process. As a general rule of thumb, the validity of the selection criteria or testing methods used can be thought of as quality measure. This follows because, as discussed earlier, the Z-score (or probability of a successful hire) increases with greater test validity. However, this increased quality may come with a price. That is, the use of a simple selection test (with a lower validity) is likely to cost less than the use of a more sophisticated test (with high validity). An example of economic calculations illustrating this trade-off is illustrated in Table 18.1. As shown in the table, the cost associated with unsatisfactory hires drops when test validity increases. The lowest unsatisfactory hiring costs occur for a test validity of 0.9 and the highest occurs for a test validity of 0.3. However, the opposite trend is shown for the (administrative) cost of the test. The total cost (sum of both types of cost) turns out to be the lowest for a test with a validity of 0.8. The costs with r equal to 0.9 or 0.7 are only 7.7% and 3.7% higher than the estimated lowest cost. Validity coefficients below 0.6 result in very substantial cost increases.

Overall, these results are consistent with the Helson U-Hypothesis mentioned earlier in Chapter 1 that states to be precisely optimal is not very important, but being approximately so is critical. To see how some of the numbers in Table 18.1 might be calculated, suppose a company observed that new hires who did not perform well at the start simply quit during the first 2 weeks. The cost of hiring a person who could not perform adequately might be estimated as about 2 weeks' pay and benefits, plus the test and any training costs. If new hires were paid about \$14/h with initial employee benefits, 2 weeks' pay amounts to 2 weeks × 40 h/week × \$14/h or \$1120.00. Now, let us assume that information from the Human Resources Department reveals that most jobs can be performed satisfactorily by 50% of the population, the typical selection ratio is 20%, and there are usually 50 applicants. This information can then be used along with the cost of 2 weeks pay to estimate the expected cost of hiring personnel who cannot perform the jobs as \$1120 × (1 − z) s where s is the number of people hired. The value of s is simply the product of the selection ratio and the number of applicants. That is, $s = 0.2 \times 50$ applicants, or 10 people hired.

Table 18.1 Summary of Economic Calculations on Personnel Selection Costs as a Function of Test Validity

Validity r =	0.9	0.8	0.7	0.6	0.5	0.4	0.3
Unsat. hiring cost	\$100.80	\$582.40	\$1176.00	\$1747.20	\$2447.20	\$3080.00	\$3712.80
Test cost	\$2603.95	\$1929.06	\$1429.08	\$1058.69	\$784.30	\$581.02	\$430.43
Total cost	\$2704.75	\$2511.46	\$2605.08	\$2805.89	\$3231.50	\$3661.02	\$4143.23
percentage > least	7.7	0.0	3.7	11.7	28.7	45.8	65.0

The next step in the analysis is to determine z, which is simply the Z-score, for each value of the validity coefficient r, a selection ratio of 20%, and population skill level of 50%. To do this, we follow the same procedure discussed in the earlier examples of calculating the Z-score. The only difference is that since the selection ratio is now 20%, we use columns 9 and 10 from the appropriate table in Section B.1.3 instead of columns 8–10. Along these lines, as can be easily verified by eliminating column 8 from the table used in the previous example, and recalculating the sums, we obtain a Z-score of 0.991 for $r = 0.9$. The expected cost is then $\$1120 \times (1 - 0.991) \times 10 = \$100.80$, as indicated in Table 18.1 for a test validity of 0.9.

Training

Training is an alternative strategy that can supplement personnel selection to help ensure a skilled and knowledgeable work force. A wide variety of training methods are used in industry. One of these methods is lecturing. For example, the lecture method is often used in the introductory safety training programs because it can be adapted easily to different plant practices. Another method that is commonly employed in industrial settings is demonstrations by supervisors, including on-the-job training. Perhaps one of the oldest methods is sometimes referred to as "go sit by Nellie," with the implication that the trainee is to watch an experienced person, such as Nellie, and imitate that person. Unfortunately, sometimes that person is not performing the job correctly and the trainee learns incorrect procedures. That is why the "go sit by Nellie" method is used frequently as a negative contrast. A more recent emphasis on programmed learning has meant that ergonomists with training expertise put together a program of training. Some firms buy television tapes that are prepared by specialists and have trainees sit through a short training program and then discuss applications or problems in their own plant. Programmed training can be done using a computer interactively, but such training usually entails very elaborate planning and testing to assure that trainees do not get caught in some loops of logic. Also, training devices and simulators that are used in programmed training can be sufficiently elaborate and realistic to be used in team training, particularly to instruct workers about emergency procedures or the handling of dangerous situations.

Designing a Training Program

A brief but interesting summary of the steps to follow when designing a training program was given by Rouse (1991). These steps are briefly summarized in the following:

1. Define the tasks and duties of personnel, particularly those that need to be certified through a simulation.
2. Identify existing knowledge and skills in the target population that are needed to operate the system.
3. Define the training requirements by determining the additional training that should be included in the training program.
4. Define the methods for developing the required knowledge and skills.
5. Identify the training equipment and simulation required to support the training curriculum.
6. Prepare the course material.
7. Evaluate the course by presenting it formally.

Some training principles

In designing training systems, a number of principles are important. One of the first principles is *motivation*. Highly motivated people learn faster and better than people with lower motivation. Before beginning a training program, trainees should be told the purposes of the training program and the benefits to the individual and the company in terms of advancement, security, satisfaction, and/or pay. Most people desire these incentives and, in exchange, cooperate with the training program.

Another requirement is *freedom from distractions and peer pressures*. Distractions mostly reduce the efficiency of the training program. However, if the distraction is not recognized and the part of the training program in which the distraction occurred is not repeated, the training can be incomplete. Since all forms of distraction cannot be totally avoided, it is important to be redundant in key points of the program. Also, peer pressures can reduce the amount of learning or the motivation of the learner. If a trainee senses peer envy or disapproval, even if unjustified, the trainee is likely to be distracted.

If you, as a child, had the experience of learning to play a musical instrument, you will recognize that a key to training is *practice*. In fact you probably are thinking of the cliché "Practice makes perfect!" Indeed, improvement with practice on sensorimotor tasks can go on for years. The question of how much practice is enough is a very difficult one to answer because there almost always is room for improvement. One well-known training principle is that learning tends to improve when trainees receive feedback or knowledge of results (KOR) regarding how well they are doing during the training sessions. The sessions, themselves, should be short and frequent to allow trainees to reflect upon their performance between sessions and devise ways of improving.

In many applications, KOR is provided on a *reinforcement* schedule. In the initial stages of training, it can be helpful to provide immediate feedback giving text results and encouragement on most of the trials. If performance on a trial is good, KOR alone is reinforcing. Later, reinforcement can be done on a less frequently basis such as every three to five trials on the average providing feedback more frequently to start with and the gradually reducing its frequency trainee learns how to generate KOR on their own. That is, they, become aware of how well they are doing and start setting performance targets on their own. In fact, this concept of having people provide their own reinforcement and motivation is probably the single most important idea in *skill retention*.

Practice is a rehearsal of a task. One way to rehearse is to simply recite some words over and over. This type of recitation is known as *rote rehearsal* and it is known to help maintain information in a short-term working memory. However, it is not known to be very effective in long-term memory (LTM) retention. On the other hand, *elaborative rehearsal* requires one to describe the *meaning* of the actions as well as the process actions. To do this, ask the people being trained to go over the session as if they were explaining it to a new trainee. Thus, elaborative rehearsals strengthen semantic LTM.

Reduction of concurrent task loads is another important condition for training. Learning, during the training program, is a task in itself. More rapid learning rates tend to occur when the job people are training to perform is at first simplified to reduce the concurrent task load. After the simplified tasks can be performed without error, the job can be made more complex. It is particularly beneficial if the addition of complications toward reality can be geared to the individual learner. In this way, individuals can graduate from the formal training program as soon as they reach an acceptable level of proficiency. Others who need more skill training can then be brought back for additional training. This is the underlying notion behind adaptive training.

Another effective training method is training through an *analogy* or a metaphor.* Of course, to be effective, the selected analogy must be familiar to the trainee or much easier to describe than the object of the training. One of the best uses for analogy and metaphors in training is in imparting a correct mental model of systems, processes, or operations. An example of an analogy in training people in nuclear reactors is that the system works similarly to a tea kettle, rod control is analogous to turning the heat up or down under the kettle and that it is very important to anticipate the rate of change in the water temperature before there is too much steam created so that it cannot be easily handled. Another example of an analogy in training is in computer operations where the filing cabinet is used as an analogy to saving copy and storing information for easy retrieval. Proper mental models provide the trainee with knowledge that can be called upon when other procedures are not sufficient. Often this knowledge involves intermediate variables and an understanding of how those variables depend upon the controlled variables. Understanding these relationships between variables is the foundation of knowledge-based behavior in contrast to skill-based or rule-based behavior. In addition, learned analogies can provide a way to retain procedural knowledge; this class of knowledge tends to be rapidly forgotten when not rehearsed.

Another principle of training is *accuracy in the criteria* and *similarity in the job situation*. Trainees usually behave in a manner reflecting the criteria indicated by the training when the program is over. If they discover that the criterial stress in the training program is incorrect for the job, trainees are likely to question the accuracy of the procedures taught. Similarity between the job during the training and the actual job will greatly enhance the *transfer of training* from the program to the job. The fidelity of the training devices and procedures *usually* helps improve the transfer of training; the general exception occurs when fidelity requires overwhelming complexity in procedures so that the training program itself is slowed down. Another case that speaks against higher fidelity is one in which a new system requires the training of quite different responses to similar stimuli.* An example of a situation with similar stimuli but with different responses is the changing of color coding. It is better to disassociate the job where the color code is being changed so that the new code can be learned without old cues to confuse the trainee. Intermediate training must alter the similarity of appearance until the new responses are much stronger than the older ones.

Transfer of training

The objective of a training program is to improve employee performance in order to justify training employee, the skills developed in the training must therefore transfer over to the job. To measure the amount of transfer of training, one must first know how long it takes a population of people to acquire the needed skills on the job without training. In some cases, this can be done by monitoring a group of employees with no training overtime until their performance time on a collection of tasks has reached a standard level. For example, suppose that the company uses methods-time measurement (MTM) as a standard of performance. New personnel may be assigned directly to the job and their performance can be monitored over time.

---

* To remind the reader, an analogy is something else that is similar in likeness or in operation, whereas a metaphor is a figure of speech that is physically wrong but suggestive of a resemblance.

These times can be fit to a learning curve* that can be used to estimate the time required to reach a standard of proficiency. An example of a standard of proficiency is the cycle time as estimated by MTM.† Suppose that a group of persons without training took 100 working hours (12.5 days with 8 h shifts) to reach that standard. This time to criterion (TOC) is 100 h. Another group of personnel, who were thought to be equivalent to the first, were then given G hours of training before going on the job. Afterward, the persons who were first trained were observed and a learning curve was fit to their performance. If this second group with training achieved standard performance more quickly, then the savings in performance time could be attributed to the transfer of training to the learning situation. Suppose the average person in the group with training achieved standard time performance within 80 working hours. The typical method for computing the percentage transfer of training is to take the difference in time to reach criterion between the control group and the training group relative to the control group. Thus, the percentage of transfer is

$$\% \text{ transfer} = \frac{\text{control group TOC} - \text{training group TOC}}{\text{control group TOC}} \quad (18.1)$$

In this example, the percentage of transfer‡ is

$$\% \text{ transfer} = \frac{100\,\text{h} - 80\,\text{h}}{100\,\text{h}} = 20\%$$

It follows that a training group that takes longer than the control group exhibits *negative transfer*, and the actual training is inhibiting the person's skill or knowledge learning. For example, if the training group had a TOC of 150 h, then the % transfer is −50%.

Another measure of training transfer is the transfer effectiveness ratio (*TER*).§ This measure is defined as follows:

$$TER = \frac{\text{control group TOC} - \text{training group TOC}}{\text{control group TOC}} \quad (18.2)$$

Where TOC corresponds to how long it tasks to reach a desired level of proficiency.

If the training group took 80 working hours to reach proficiency and that group had been given 10 h of training in contrast to the control group that took 100 h to reach proficiency, then the TER is

---

* Singley and Andersen (1989) shows that negative transfer occurs when equipment or procedural similarity (or identity) creates a positive transfer of wrong responses.
† Recall that Chapter 5 includes a discussion with learning curves and how those curves can be fit to empirical performance data.
‡ Note that the control group averaged 100 h each to reach the performance criterion of MTM, while the group with G hours of training averaged only 80 h to reach the performance criterion. This example shows a 20 h savings due to the training program. Relative to the control group of personnel, the percentage of transfer is (100 − 80 h)/100 h or 0.20 or 20% transfer of training. Note that the G hours of training did not enter into this percentage of transfer calculation.
§ Povenmire and Roscoe (1973) developed this measure of training transfer.

Chapter eighteen: Personnel selection, placement, and training

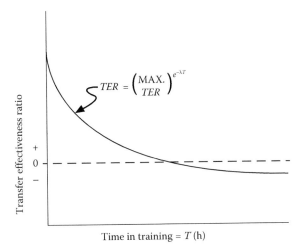

Figure 18.7 A typical transfer effectiveness ratio function with time in training.

$$TER = \frac{100\,h - 80\,h}{10\,h \text{ in the training program}} = 2$$

The ratio in this example shows that for every hour spent in training, the transfer of training effect provided a savings of 2 h.*

The relationship between the time in a particular training program and the resulting *TER* is usually a negative exponential function such as the one in Figure 18.7. This figure shows diminishing TERs with greater times in training. When the training involves a complex concept, the *TER* curve can first rise and then decay as the time in training increases. In such situations, the training program should continue past the point in time where the TER begins to decay. This follows because not enough learning has occurred to improve performance until we reach this point where *TER* is maximum.

*TER*s describe how effective the training program is. As indicated by Figure 18.7, continuing the training past a certain point will no longer result in time savings, compared to the alternative of learning on the job. This point is shown in the figure where the two lines cross. In some applications, it makes sense to use this crossing point as a basis for deciding how long to train the employee. This follows because based purely on the basis of time. The training is no longer cost effective when the TER is less than 1. However, training can be justified for other reasons even if the TER is less than 1. For example, if training is done on a simulator, accidents will not result in dangerous situations during the training. Another example would be where product losses due to inexperienced handling result in considerably larger costs on the job than during training, The overall conclusion is that training can be very cost effective even for *TER* values < 1.

Other training strategies

Many different types of training strategies have been followed by organizations. In some cases, the focus is on increasing the proficiency of crews or teams rather than individuals.

---

* In contrast to the percentage of transfer measure, TER does account for the amount of time spent in training.

Crew *proficiency training* usually occurs in the later learning stages after individuals have essentially learned their specific jobs. This training usually stresses coordination among different crew members. Proficiency training is often done on simulators where realism is high. Some companies specialize in training workers to use a particular type of equipment or particular operating procedures. Crews from nuclear reactor control rooms, for instance, usually go to specialty schools for training with new systems and updated display devices. Companies that made the reactor, such as General Electric, Westinghouse, Combustion Engineering, and Babcock and Wilcox, provide specialty training with their reactor systems. Often, on-site simulators enable control crews to practice special shutdown and other emergency procedures. Airlines also have specialty simulator schools for teaching emergency procedures with specialized types of aircraft, and many of the larger airlines run schools for training aircraft crews in coordinated efforts in emergency situations.

Segmented and *part-task training* are commonly used training strategies. Both of these approaches divide the entire job into tasks, so that training can be done separately on the component tasks. When the job entails performing sequential tasks A, B, and C, then segmentation is the training in task A first, B second, A and B together third, C fourth, B and C fifth, and then finally A, B, and C together.* If the job involves tasks A and B which are done together or serially but not in any particular sequence, then part-task training involves an alternation of training separately on tasks A, then B, then A, then B, and finally A and B together. Both of these training strategies reduce the concurrent workload to make learning more efficient. This training strategy also reduces the needs for task coordination until the individual tasks are well learned. Finally, some tasks require a great deal more skill learning than others, so these segmentation and fractionization strategies allow more time to be spent on those tasks that require the most learning. In that way a person can be equally proficient on all parts of the job.

Adaptive training allows trainees to perform the tasks first under the easiest condition. As the trainee's proficiency on the easier task conditions becomes acceptable, task complexity is increased. This training strategy speeds the program for faster learners and slows it for slower learners.† One of the benefits of this training strategy is that trainees have nearly homogeneous proficiency at the end of the program, even though the program time differs greatly between individuals.

Another strategy is *guided training*, a method that gives corrective feedback only when performance is outside of an acceptable level. The idea is that feedback on small details can interfere with learning the target task. Use of training wheels in learning to ride a bicycle is analogous to guided training because the training wheels only come into play as feedback when the bicycle tips too much. This concept leads to rapid learning, but not always to good transfer of training.‡

Many training programs will also include *group discussions* and *role playing* by the participants. Both of these techniques keep motivation high during training. Role playing is used more for *attitude development*. Having people play a role unlike the one they play on the job can make them aware of the problems faced by their coworkers. Loss of cooperation between working groups can be detrimental to everyone without people in either group realizing it.

---

* The particular grouping of these tasks is arbitrarily stated here. Some sequences of grouping may very well decrease the total learning time to a criterion level on the entire job.
† See Gopher et al. (1975) for details on adaptive training concepts and techniques.
‡ See Lintern and Roscoe (1980) for further details.

Even formal university courses tend to include almost all of these elements. However, formal university courses are usually focused on knowledge about the systems under study rather than skills, particularly physical skills.

Job aids

A job aid is something that enables a person to perform a task without training, which would otherwise be unachievable.* In a sense, a job aid is an alternative to training. Some examples of what people have in mind when they speak about job aids include instructions for assembling a bicycle that is shipped partially assembled, a tape recording of instructions for repairing a specific piece of equipment, computer diagnostics, job setup instructions, pictorial presentations of oscilloscope readouts that are used in maintenance activities, or a diagram of locations on an automobile chassis that require lubrication. In the case of very infrequent jobs where training is of questionable value, these job aids are very important. Manufacturers can ship partially assembled products to save money in shipping and still have the product assembled by the recipient. With untrained consumers, there is no economical way to impart the final assembly skills except through job aids. But even with employees, if a job is performed infrequently and at random times, training is not economical because skill retention is likely to be poor. In other cases, job aids serve mostly as a memory supplement. That capability is particularly useful when there is significant variability in the job, such as the station attendant who lubricates various kinds of automobiles or a computer repair person who deals with numerous kinds of personal computers. Job aids can also reduce the complexity of a task when they take the form of a transformation table or a graph that assists in computations.

Some of the economics of job aids are obvious. Training programs are expensive and time consuming. If a job aid will do as well or well enough, then adequate skills are developed with lower costs. Supplementing a job with job aids also reduces the skill level required of people, resulting in lower pay rates. Also, training investments are lost when the trainee leaves the company, but the job aid remains. Another benefit of job aids is the potential improvement in speed and reliability of persons using job aids when tasks require high degrees of memory recall.

The objective of a job aid is to elicit the correct behavior reliably on the first try. In order to achieve that objective, the designer of a job aid must know the desired behavior and minimum acceptable performance. Then the designer must create the stimulus to create correct behavioral changes. That stimulus must also fit the performers in terms of their level of understanding. Most job aids are visual stimuli because most need to exhibit spatial information. Traditional job aids are written instructions with diagrams. More recently, television tapes are a preferred form of job aid for certain applications. Television is particularly good for showing a person how to perform specific motor actions. The important requirement in designing a job aid is to test the aid over a large cross section of people to verify that it produces the correct performance on the first try.

Final remarks

This chapter focuses on the important topics of personnel selection, placement of personnel within the company, and training. These elements are highly related. When skill

---

* See Wulff and Berry (1962) for some examples and applications. For more on job aids versus training, see discussion in Chapter 16.

requirements are high and little training is planned, the only way to acquire proficiency is to find people who already have the needed skills. If more training is planned or is in consideration, these requirements are lessened, but steps must still be taken to ensure the potential skills are there. Many companies move newly hired personnel through several divisions of a company as part of their training and orientation process. This allows employees to learn how various parts of the company interface with each other and understand that cooperation between those divisions is important. In fact, some supervisors have new forklift operators first work at some of the stations supplied by the forklifts before operating those machines.

Formal personnel selection procedures have varied in popularity over the decades. Regardless of the current philosophy, some means must be used for selecting personnel. Ways of evaluating the effectiveness of any particular philosophy and procedures are the principal focus of the discussions in this chapter. Obviously substantial individual differences exist in skills and proficiency in various jobs. Procedures that help designers and managers do a better job of matching the right person to the right job deserve consideration.

Another ongoing activity in personnel management is training. Personnel turnover means that people are coming into the company who need to be trained on the skills that others had when they left the company. There are also new technologies to learn in order to do any job better, so even without turnover there is need for ongoing training. Since training is needed, the logical question is, "How is this training best delivered?" A number of different strategies in training are discussed. All have their advantages and disadvantages, but the key is to learn which works best for which situation. Regardless of the mode of delivering the training, the objective is to get the training to transfer to the job effectively. As discussed previously, there are some straightforward approaches to measuring transfer provided that one can measure the degree of proficiency equivalence. Therein lies a principal difficulty. Another issue is that analyst may want to consider providing job aids in addition to or in lieu of training.

Discussion questions and exercises

18.1 Identify four typical methods of industrial training and criticize each.

18.2 List at least three important principles of training.

18.3 What is a job aid? What does it have to do with training? What are at least two principal advantages of job aids over alternatives?

18.4 Describe differences between test validity and reliability.

18.5 If a selection test was more powerful than another, why is it not necessarily also more efficient?

18.6 A control group of new employees was placed on the job and achieved criterion performance in terms of the work factor synthetic system in 200 working hours. Another group of employees, which appeared matched in all respects to the control group, was given 4 h of special training before being placed on the job. Afterward their performance was monitored and the average person in the training group reached criterion performance in 160 working hours. What is the percentage of transfer? What is the TER?

18.7 When a computer is used for training and the computer screen employs a ticket window to imply data input, a filing cabinet to imply putting data into memory, and a garbage can to represent a place to trash unwanted files, what training principle is being employed?

Chapter eighteen: Personnel selection, placement, and training 661

18.8 A person is taught to perform car management activities without requiring the person to simultaneously steer the car or control the speed of the vehicle. Later these tasks will be combined. What principle of training is applied in this initial training?

18.9 What advantage does part-task training have over full-task training?

18.10 When using a simulator to help train a person to do a particular job, better fidelity usually contributes to better training. Why is this and when does better fidelity interfere?

18.11 Suppose that a control group of employees required 10 working hours to reach an acceptable level of proficiency, but another group who experienced 5 h of training required 13 working hours to obtain proficiency. What is the percentage of transfer and the *TER*?

18.12 Assume that 20% of people in a general application pool are qualified to perform a job. If a selection test has a validity of 50%, what percentage of the people scoring in the top 20% of the test would be expected to be successful on the job?

18.13 What is the meaning of the term Z-score?

18.14 What technical terms denote the ability of a test to measure very small ability differences and the ratio of this ability to cost?

18.15 Personnel of a company are selected by a formula that combined a test score and the rating of an interviewer. It was found that scores from this formula correlated with job performance rating with the coefficient of 0.9. When a selection ratio of 20% was used for a job where the top 30% could perform it satisfactorily, what is the expected Z-score of this procedure for the intended job?

chapter nineteen

Work groups and teams

About the chapter

Teams, crews, or other groups of people perform many operations in manufacturing or service industries. Examples include aircrews in an airline, control operators in chemical plants and nuclear reactors, and work groups in direct assembly-machining operations. This chapter reviews some of the studies that have been done on crews, teams, and other groups that are relevant to industry or the service sector, including product design teams. Ergonomics specialists and engineers are expected to play important roles in design teams and need to be aware of ergonomic issues that arise in such settings. Accordingly, one objective of this chapter is to introduce readers to some of the issues that arise when teams and crews are formed, mature, and eventually dissolve. The chapter also suggests computer simulation as a technique for studying teams in those settings and discusses some issues that come up when measuring team or crew proficiency.

Introduction

The mental and/or physical demands of a job or task will often exceed the capabilities of a single individual. If those demands do not lend themselves to automation, the task will need to be done by two or more people. In many situations, the members of this group must interact over time and work both interdependently and adaptively toward common goals. For example, in many manufacturing applications, people operate machine tools using parts and materials provided by forklift drivers who deliver these items to the point of operation. Coordinating these tasks performed by the team is often critical, because in most modern lean manufacturing systems, the parts and materials must arrive just-in-time (JIT) to avoid delays. Good teamwork is critical in an immense variety of other applications. For example, think of the activity taking place in the control room of a nuclear reactor, the cockpit of a commercial aircraft, a busy hospital, or an industrial research and development (R&D) laboratory.

An interesting story about the president of Lincoln Electric and his son illustrates the importance of teamwork in industry. The son was in college where he played varsity football with a passion, but his studies had little to do with business, industry, or Lincoln Electric. One day his father was struck by a heart attack and died very suddenly. After the funeral, the son went to Lincoln Electric and reorganized it according to the principles he learned from football. Not only did the company survive but also it became a case study used in business schools, such as Harvard.

When we hear the word *team,* most of us first think of sports, just as did the fellow from Lincoln Electric. However, when one reflects more on the subject, it is clear that the three major sports in the United States differ widely in many ways as will the analogies.

When it comes to movement and activity coupling among the team players, actions in basketball unfold in a fluid manner and encompass all the team members far more so than in football or baseball. In the latter two sports, actions mostly occur during brief time periods and are coupled between but a few team members. Baseball, far more than football, is loosely coupled, and it is mostly tactical in nature, except for the batting lineup and pitcher selection. In fact, Pete Rose once said, "Baseball is a team game, but nine men who reach their individual goals make a nice team." Football distinguishes itself among these three sports as the most strategic. Game plans are critical because of the strategic nature of the game and because fewer football games are played each year compared to baseball or basketball games.

It appears appropriate to use sports analogies in business situations, because both are competitive, both require cooperation, and both regard human resources in a strategic manner. But an important point is that analogies have a limited range of fit, and different situations are apt to fit one sport better than another. A manager of pharmaceutical R&D made this point on business–sports analogies clear. He said that the first stage in R&D is like baseball as it depends largely on the independent actions of scientists and engineers who are loosely focused. The second stage in R&D is like football as it involves many players who must work in a highly coordinated manner to get proper dosages, define manufacturing plans, test for toxic levels, and perform clinical tests. The third R&D stage is like basketball, as it requires various kinds of specialists who must work together almost synchronously to gain approval for the drug from the government.

To create an effective team like those found in professional sports, which are typically well structured and contain highly paid professional athletes, the work group must be carefully structured with specific goals in mind. Depending on the specific situation, this process may involve determining answers to questions such as how large should the group be, what types of personnel are needed, who does what, and what types of training are needed. In other situations, it may be important to measure how well particular individuals or the group as a whole is performing. Many of these questions can be addressed using methods covered elsewhere in other chapters of this book, and especially so in cases where people perform their tasks, more or less independently. For example, the jobs assigned to particular people can be analyzed using the methods of task analysis discussed in Chapter 8, tools and layouts can be analyzed using the methods covered in Chapter 5, personnel selection and training needs can be analyzed using the approaches covered in Chapter 18, the effect of including additional workers can be estimated using simulation methods covered in Chapter 11, appropriate compensation levels can be determined using methods covered in Chapter 20, and so on. There are, however, a number of important issues that arise when people work together not addressed in these other chapters. The following discussion will first introduce some of the many factors that impact the effectiveness of groups and teams. Attention will then be given in the remaining sections to issues such as measuring group performance and research regarding the effectiveness of different types of industrial work groups such as quality circles, semi-autonomous groups, and research, design, and development teams.

Simple model of group effectiveness

Several years ago, Goldstein (1984) developed a model of group effectiveness that provides a nice summary of some of the many factors found in previous studies that influence the effectiveness of groups. The elements of the model can be organized into four categories: (1) group composition and structure, (2) organizational resources and structure, (3) task demands, and (4) group processes. The most fundamental set of requirements is associated with the composition and structure of the group. The basic idea is that an effective group will be composed of members with adequate skills and experience, clear goals and roles, and a behavioral norm of high expected levels of performance within the group. The members of the group should also have control over the tasks they perform and clear leadership.

The second set of factors influencing group performance is related to how well the organization the group is within supports and encourages good performance. Some of the ways the organization can help support high levels of group performance include providing adequate training and technical resources. The organization can also encourage better group performance by providing both adequate rewards for good performance of the group and adequate supervision and monitoring of group performance. The third set of factors describe how demanding the tasks performed by the group are. Task demands increase with task complexity, task interdependence, and environmental uncertainty. The degree to which the group can adequately cope with more demanding tasks depends, of course, on group composition and structure, the organizational resources and structure, and group processes. The latter, fourth, set of factors include the adequacy of both within-group communications and the operational strategies followed. Conflicts may also arise within the group that can interfere with performance, as most readers can probably well remember from past experiences working in a group.

Conflict, despite its negative connotations, is a normal, expected aspect of group decision making and can in fact serve a positive role (Ellis and Fisher, 1994). One of the more classic models (Tuckman, 1965) describes this process with four words: *forming, storming, norming,* and *performing*. Forming corresponds to initial orientation, storming to conflict, norming to developing group cohesion and expressing opinions, and performing to obtaining solutions. As implied by Tuckman's choice of terms, there is a continual interplay between socio-emotive factors and rational, task-oriented behavior throughout the group decision-making process. A lack of conflict-settling procedures and separation or lack of contact between group members can also contribute to conflict. Conflict becomes especially likely during a crisis and often escalates when the issues are perceived to be important, or after resistance or retaliation occurs.

Polarization, lack of trust, and cultural and socioeconomic factors are often contributing factors to conflict and conflict escalation. Ellis and Fisher (1994) distinguish between affective and substantive forms of conflict. Affective conflict corresponds to emotional clashes between individuals or groups, while substantive conflict involves opposition at the intellectual level. Substantive conflict is especially likely to have positive effects on group decisions by promoting better understanding of the issues involved. Affective conflict can also improve group decisions by increasing interest, involvement, and motivation among group members and, in some cases, cohesiveness. On the other hand, affective conflict may cause significant ill will, reduced cohesiveness, and withdrawal by some members from the group process, making it important to establish good ways of resolving conflict to help ensure the group performs well (Box 19.1).

Life cycle of teams and crews

Changes taking place within a group over time, as they form, mature, and eventually dissolve, are a particularly important influence on group performance. No group lasts forever. Some do last for many years, but most do not. Over the limited life span of a group,

BOX 19.1   CONFLICT RESOLUTION AND MANAGEMENT

Members of groups can resolve their conflicts in many different ways. Discussion and argument, voting, negotiation, arbitration, and other forms of third-party intervention are all methods of resolving disputes. Discussion and argument is clearly the most common method followed. Other methods of conflict resolution normally play a complementary, rather than primary, role—that is, the latter methods are relied upon when groups fail to reach consensus after discussion and argument, or they simply serve as the final step in the process.

Group discussion and argument is often a less than rational process. For example, Brashers et al. (1994) state that "…the literature suggests that argument in groups is a social activity, constructed and maintained in interaction, and guided perhaps by different rules and norms than those that govern the practice of ideal or rational argument. Subgroups speaking with a single voice appear to be a significant force.… Displays of support, repetitive agreement, and persistence all appear to function as influence mechanisms in consort with, or perhaps in place of, the quality or rationality of the arguments offered." Brashers et al. also suggest that many members of groups do not participate in group discussions and arguments, because they are following group norms or rules of behavior, such as "…such as: (a) submission to higher status individuals, (b) experts' opinions are accepted as facts on all matters, (c) the majority should be allowed to rule, (d) conflict and confrontation are to be avoided whenever possible." A number of approaches for conflict management have been suggested that attempt to address many of the issues raised by Brashers et al. These approaches include seeking consensus rather than allowing decisions to be posed as win–lose propositions, encouraging and training group members to be supportive listeners, de-emphasizing status, depersonalizing decision making, and using facilitators (Likert and Likert, 1976). Other approaches that have been proposed include directing discussion toward clarifying the issues, promoting an open and positive climate for discussion, facilitating face-saving communications, and promoting the development of common goals (Ellis and Fisher, 1994).

Conflicts can also be resolved through voting and negotiation. Negotiation becomes especially important when the involved people have competing goals and some form of compromise is required. A typical example would be a dispute over pay between a labor union and management. Strategic concerns play a major role in negotiation and bargaining (Schelling, 1960). Self-interest on the part of the involved parties is the driving force throughout a process involving threats and promises, proposals and counterproposals, and attempts to discern how the opposing party will respond. Threats and promises are a means of signaling what the response will be to actions taken by an opponent and consequently become rational elements of a decision strategy (Raiffa, 1982). Establishing the credibility of signals sent to an opponent becomes important because if they are not believed, they will not have any influence. Methods of attaining credibility include establishing a reputation, the use of contracts, cutting off communication, burning bridges, leaving an outcome beyond control, moving in small steps, and using negotiating agents (Dixit and Nalebuff, 1991). Given the fundamentally adversarial nature of negotiation, conflict may move from a substantive basis to an affective, highly emotional state. At this stage, arbitration and other forms of third-party intervention may become appropriate due to a corresponding tendency for the negotiating parties to take extreme, inflexible positions.

its activities, responsibilities, and organization and member goals, knowledge, and behavioral norms tend to change in many ways. Morgan et al. (1986) modeled the typical life cycle of teams, crews, and other similar groups as a process of evolution, maturation, and eventual disintegration. Their model is called the team evolution and maturation (TEAM) model. This model provides a framework for describing important aspects of crews and teams as they develop and mature over time. One story goes that the ability of crews to alter their work procedures resulted in their eventual capacity to handle, with the same degree of efficiency, loads up to four times greater than the workloads they could barely handle at the outset. The most important reason for that improvement is not the improvement of the individual team members but rather the development of new working procedures that greatly improve group performance.

According to Morgan et al., when a team is first formed, the members spend their time learning task requirements, operating procedures, task-specific information, and other things that improve their individual skills. With more learning, teamwork develops due to better coordination of effort, and compensatory behavior emerges, in which team members adapt to each other's strong and weak points. Much of the improvement is due to *feedback* obtained from observing and communicating with other members of the group. Improvements in individual skills that occur during this process are, of course, often a necessary condition for adequate group performance, but the real payoff is when the members develop an ability to work effectively together in a coordinated way and develop a strong sense of cohesion or belonging to the team. Conflict within a group or the presence of outside disruption forces, such as lack of support by the larger organization, can interfere with this process and, in some cases, the disintegration of the team. Team *dysfunctions* may also occur, which either do not contribute to the goals or detract from the goals. As the team moves toward disintegration, power struggles and the formation of coalitions are commonly observed. Coalitions are especially disruptive when certain members of the group can gain by following a common course of action at the expense of the long-run objectives of the group as a whole. A typical example is when a subgroup of technical employees leaves a corporation to form their own small company producing a product similar to the one they had been working on.

Other group characteristics that tend to change over time and are important to teamwork include *group cohesion* and its *authority structure*. Group cohesion describes the extent to which the group members wish to remain in the group. Organizations, such as the U.S. military, have focused for years on developing high levels of group cohesion in terms of a *esprit de corps* because they believe better group cohesion results in better performance. A large number of studies have evaluated the effects of group cohesion on performance (Fleishman and Zaccaro, 1992). Somewhat surprisingly, mixed results have been found in these studies. One finding is that group cohesion seems to have positive effects when people in the group have high expectations and invest energy in the group (i.e., high group drive), but, if not, the relationship between group cohesion and cohesion is negative. A negative relationship is particularly likely when a cohesive group does not share the same goals as the larger organization they are within (think of rebels). In extreme cases, such individuals may form disruptive coalitions that struggle against other subgroups they interact with. Leadership style is another important consideration. Highly cohesive groups that perform well often are often led by people that are both focused on the task and people oriented. For less cohesive groups, people-oriented leadership styles are often especially desirable.

A related issue is that teams that experience time stress tend to rely heavily on their leader. This tendency corresponds to use of a more centralized *authority structure* and often results in a quicker team response than distributed forms of decision making where each member has a greater role in deciding what to do. However, while a centralized authority structure may result in faster decisions, it is not always better in terms of the quality of the team performance, the qualities of the decisions made, or team member satisfaction. Allowing various crew members to generate problem solutions in an open, uncritical manner in a group setting tends to result in greater variety and creativity of the responses. Hence, differences in the authority structure create differences in the types and timing of responses. It is true that military personnel who deal directly with an enemy force must be prepared for quick action when needed. That is not typically the situation in industry. In fact, many of the efforts to achieve continuous quality improvements in the United States stress the need for involvement (Denton, 1992). As they say, quality is everybody's business. Also, the notion of quality circles where the operating personnel get together occasionally to help identify those factors that cause quality to deteriorate is another example of distributed decision making.

Group performance and biases

There is little doubt that teams made up of highly experienced, well trained members, working together as true team toward a common objective, are often highly effective in many real-world applications (see Klein and Wolf, 1995). For example, think of firefighters, airplane crews, or the physicians and staff in a hospital intensive care ward. For less well-coupled groups, such as advisory panels, organizational management, legislators, or society as a whole, the quality of the decisions made is often much more questionable. Part of the issue here is the so-called phenomenon of groupthink, which has been blamed for several disastrous public policy decisions (Janis, 1972; Hart et al., 1997). Janis and Mann (1977) proposed that the results of groupthink include failure to consider all the objectives and alternatives, failure to reexamine choices and rejected alternatives, incomplete or poor search for information, failure to adequately consider negative information, and failure to develop contingency plans. Groupthink is one of the most cited explanations of how group decision processes can go wrong. Given the prominence of groupthink as an explanation of group behavior, it is somewhat surprising that only a few studies have empirically evaluated this theory. Empirical evaluation of the groupthink effect and the development of alternative modeling approaches continue to be active areas of research (Hart et al., 1997).

Other research has attempted to measure the quality of group decisions in the real world against rational, or normative, standards. Viscusi (1991) cites several examples of apparent regulatory complacency and regulatory excess in government safety standards in the United States. He also discusses a variety of inconsistencies in the amounts awarded in product liability cases. Baron (1998) provides a long list of what he views as errors in public decision making and their very serious effects on society. These examples include collective decisions resulting in the destruction of natural resources and overpopulation, strong opposition to useful products such as vaccines, violent conflict between groups, and overzealous regulations, such as the Delaney clause. He attributes these problems to commonly held, and at first glance innocuous, intuitions such as "Do no harm", "Nature knows best", "Be loyal to your own group", the need for retribution ("an eye for an eye"), and a desire for fairness.

A significant amount of laboratory research is also available that compares the performance of groups to that of individual decision makers (Davis, 1992; Kerr et al., 1996). Much of the early work showed that groups were better than individuals on some tasks. Later research indicated that group performance is less than the sum of its parts. Groups tend to be better than individuals on tasks where the solution is obvious once it is advocated by a single member of the group (Davis, 1992; Kerr et al., 1996). Another commonly cited finding is that groups tend to be more willing to select risky alternatives than individuals, but in some cases the opposite is true. One explanation is that group interactions cause people within the group to adopt more polarized opinions (Moscovici, 1976). Large groups seem especially likely to reach polarized, or extreme, conclusions (Isenberg, 1986). Groups also tend to overemphasize the common knowledge of members, at the expense of underemphasizing the unique knowledge certain members have (Stasser and Titus, 1985; Gruenfeld et al., 1996). A more recent finding indicates that groups were more rational than individuals when playing the ultimatum game (Bornstein and Yaniv, 1998).

Duffy (1993) notes that team biases and errors can be related to information-processing limitations and the use of heuristics, such as framing. Topics such as mediation and negotiation, jury decision making, and public policy are now being evaluated from the latter perspective (Heath et al., 1994). Much of this research has focused on whether groups use the same types of heuristics as individuals to make decisions (see Chapter 10), and whether they are subject to the same biases. This research has shown (1) framing effects and preference reversals (Paese et al., 1993), (2) overconfidence (Sniezek, 1992), (3) the use of heuristics in negotiation (Bazerman and Neale, 1983), and (4) increased performance with cognitive feedback (Harmon and Rohrbaugh, 1990). One study indicated that biasing effects of the representativeness heuristic were greater for groups than for individuals (Argote et al., 1986). The overall conclusion is that group decisions may be better than those of individuals in some situations but are subject to many of the same problems.

Measuring team–crew work performance

Team performance can be measured in a number of different ways depending upon the purposes served by the measurement. If the team is doing traditional physical work in industry, traditional measurements may be employed (e.g., the number of units assembled or tons of meal produced). These methods may also be employed in making comparisons among different teams of personnel who are performing similar jobs and/or operations. Such measurements are macro in nature because they deal with the whole team as a unit.

When measurements are made on individual team members, measurements are molecular in nature and they often consist of frequency counts of the particular behaviors that are observed. Communication frequency between pairs of members is one such measure. In some cases, the nature of the communications is also important, especially when assessing the cohesiveness of teams or crews. In the case of assessing various skill dimensions that are inherent in team operations, one needs to specify the dimensions and distinguishable behaviors that illustrate or negate the dimension. Some example dimensions that might be focused upon are given in Table 19.1.

These illustrations focus on the things one wants a team or crew to do as a group. However, the dimensions shown are not always clear and distinct. For example, is a team member's prompt of another a case of cooperation, coordination, or acceptance of suggestions/criticism?

Another approach to measuring team capability is to have observers rate the team on each dimension; typically, these ratings are on a 5-, 7-, or 9-point scale. However, observer reliability differs markedly among studies. One study of ratings made from videotapes showed rather low reliability ranging from 0.02 to 0.37. Other studies have found

Table 19.1 Skill Dimensions in Team Operations

Dimensions	Behavioral Examples
Giving suggestions or criticism	When asked if a procedure was right or not a reply was stated as unsure
Cooperation	Another team member prompts on what to do next
Team spirit and morale	A discussion occurs on improving team performance
Adaptability	Member changed the way a task was performed when asked by another
Coordination	Provided information to another on what to do next
Acceptance of suggestion and criticism	Thanked another member for catching a mistake

Source: Morgan, B.B. et al., Measurement of team behaviors in a Navy environment, Technical Report No. NTSC TR-86-014, Naval Training Systems Center, Orlando, FL, 1986.

Table 19.2 Reliability of Observer Ratings

Observer Group	Low Reliability	High Reliability
Self-raters	0.21 cooperation	0.60 team spirit
On-site raters	0.57 accept suggestions	0.81 coordination
Off-site raters	0.36 accept suggestions	0.74 cooperation

Source: Brannick, M.T. et al., *Hum. Perform.*, 6, 287, 1993.

reliabilities among judges from 0.56 to 0.86. The latter results show high repeatability levels with unsophisticated judges on abstract dimensions. Some additional results are from a study by Brannick, Roach, and Salas (1993), where different people rated themselves and other team members on several different criteria. This study also compared on-site ratings to those of off-site observers. The reliabilities of observer ratings found in this study are given in Table 19.2.

These results make one wonder about team member self-rating capability. Also, on-site observers appear to be much better at ratings than off-site observers, perhaps due to nonverbal communications unavailable to the off-site observers. The difficulty in rating team members is a particular concern when determining merit pay. Simply put, performance measurement is necessary for merit pay. One problem with merit pay is that it is a necessary inducement for high performance, but in itself is clearly not sufficient to assure that goal (McGinty and Hanke, 1989). What appears to be even more important is that the method of implementation of merit pay is perceived to be properly performed and that the magnitude of merit pay is substantially different from base pay.

The measurement of team–crew performance is still far from being a science, so work in this area suffers from many things including what was once exclaimed: there is nothing more practical than a good theory. As the science develops, measurement systems become more robust and reliable. Part of the team behavioral measurement problem is related to the difficulty of measuring improvements in team work overtime and perhaps learning curves would help. Another source of measurement difficulty is the diversity of applications. Most of the team–crew research has been in the military involving aircrews performing tactical decision making, which is quite different than most industrial or even sports applications. A third problem is the interrater reliabilities. Unlike in traditional work sampling the (activity analysis), the behaviors under observation are more complex and fuzzy in nature.

Industrial work teams

Industrial work teams are relatively small groups of people in industry who are placed together for some specific purpose. They are organized in a variety of ways and their longevity varies from very brief to long term. Members may be assigned, but voluntary teams appear to be more successful.* Since the late 1970s, there has been a management shift toward work teams and experimenting on their organization and operation. Early in this renaissance time period, the quality movement brought in the *quality circle* as an advisory council to identify and correct problems that led either directly or indirectly to losses in product quality. Later the work team notion gravitated to production-service applications in manufacturing, mining, data processing, and maintenance. Work teams were also used in research, development, design, and as project teams or task forces. The latter applications have been around for many years, although organized in a very different manner than today, but this application is quite different and is discussed separately later in this chapter.

A very wide variety of teams operate in industry under the names of quality circles, management teams, committees, product development groups, cockpit crews, task forces, and many others. The reasons for these teams are many and varied. An obvious reason for some teams is that one person alone cannot perform the job. Another typical reason is the need for a variety of expertise readily available, such as with air crews, or research and development teams.† An additional rationale frequently given for the use of teams or crews is that more heads are better than one or that more people provide redundancy to improve the safety of the system as in nuclear control room crews. Still another reason for multiperson teams is the need to achieve *personnel involvement* as a humanistic input to management.

The concept of quality circles was imported from Japan to the United States during the Deming-Juran‡ period in the late 1970s and early 1980s when the term *quality* dominated management thinking in the United States. In Japan, quality circles focused on how to improve product quality and productivity of the company. After the concept's importation to the United States, the focus on productivity was greatly reduced and the focus on the quality of working life became much stronger. One study on the effectiveness of 33 different quality circles in the United States revealed that 16 groups appeared to yield positive results, 9 yielded mixed or insignificant effects on their objectives, and 8 exhibited negative findings (Barrick and Alexander, 1987). Another study§ of over 46 quality circles noted attitudinal improvements along with increased productivity and decreased absenteeism. Not to refute these findings, it should be noted, however, that social studies, after the fact, tend to exhibit a positive bias because many ineffective situations tend to be forgotten or are submerged from view. Still other studies found immediate improvement effects due to quality circles, but then exhibited a fallback to original levels, which have been referred to as the *honeymoon effect*. Another effect reported¶ was differences between employee-initiated quality circles and management-initiated groups. The latter groups were reportedly smaller in size with higher management attendance. Those groups were also reported to solve more work-related problems and to solve them faster than employee-initiated groups.

---

* See Jin (1993) who studied numerous work crews in China.
† Discussed later in this chapter.
‡ Edward Deming and Juran were gurus of the Quality Movement which occurred when the U.S. manufacturers found out that many of their products were inferior in quality to those manufactured overseas and as a result sales were slipping badly to the offshore companies.
§ Marks, Mirvis, Hackett, and Grady (1986).
¶ Tang, Tollison, and Whiteside (1989).

The other type of industrial work group is the semi-autonomous group (SAG) in which jobs and work methods or modified internal technology, such as computer networks, have been redesigned so that a team of workers can perform the needed tasks without external supervision. These groups were envisioned in the 1950s as a way to facilitate the integration of technology into the corporate structure. SAGs have also been used for maintenance functions, a single unit of work (e.g., machining), or a single product line. European companies have utilized this form of group more than the U.S. companies.* One of the more noteworthy experiments has involved the autonomous assembly work groups in Sweden where the people in the team decide totally on their organization and operating plans.† Some such work groups have internally determined their rewards and recognition as well. Proponents of SAGs state that their use leads to enhanced productivity, quality, and internal positive attitudes toward themselves, their jobs, and the company.

An older study of mining operations provides a longitudinal and cross-sectional comparison of performance in parts of the company with and without SAGs.‡ Safety rule violations, accidents, and lost time all decreased with SAGs, but productivity decreased as well. The authors felt the latter effect was related to new safety rules that slowed operations. Workers in SAGs tended to be more autonomous and independent. Also, supervisors were called upon to make fewer decisions. However, jealousy between groups and perceived inequalities among workers, and perceived attempts of union busting caused the SAGs to be discontinued in this company. Others have reported higher turnovers in employees with SAG organizations and little or no differences in productivity were observed. Other studies§ have shown improved productivity, reduced absenteeism, lower turnover, more deadlines met, and fewer customer complaints in SAGs.

A study of the features of SAG that affected their performance¶ showed that effectiveness ratings were strongly associated with encouragement of self-observations and self-evaluation. This indicates that SAG leadership appears to operate differently than traditional management where influence flows from the top downward and self-management is a bottom-up effect. It was also reported that high-status workers in SAGs tend to benefit from their existence but low-status workers do not. Another finding was that group size was negatively related to the connectedness of the group and that group effectiveness was significantly related to information accuracy and openness in communications (O'Reilly and Roberts, 1977). Also, many SAGs recognized the advantages of job switching, and so they provided a path for those employees who wanted to switch jobs.

Some work teams were started as a management experiment, others because of a new technology. An example of the latter was in coal mining with the advent of the contemporary coal cutter and conveyor. With the older methods, coal miners worked independently of each other to a large degree. The new technology forced miners to perform various specialized tasks concurrently and to perform those tasks in a synchronized fashion (Trist, Higgins, Murry, and Pollack, 1963). Some applications in harvesting, and fast food operations are other examples of technology dictating the nature of the work team.

The wide diversity in work teams is evident from the information shown in Table 19.3. This table shows application categories and differentiates among them in terms of internal operations, external connection (i.e., integrations of the team and company), typical work

---

* Dainty and Kakabadse (1992) looked into the commonalties of 36 successful European work teams.
† Kelly (1982) describes the case.
‡ Trist, Higgins, and Brown (1977).
§ Archer (1975), Poza and Markus (1980), or Walton (1977).
¶ Manz and Sims (1987).

Table 19.3 Categories of Industrial-Service Work Teams and Their Similarities and Differences

Application Examples	Work-Team Differentiation	External Integration	Work Cycles	Typical Outputs
Advice Quality circles Advisory groups Panels	Low differentiation Representation, often short lived	Low integration Usually few demands for synchronized performance with other work units	Work cycles may be brief or long	Decisions Selections Suggestions Proposals Recommendations
Production Assembly Manufacturing Mining Aircrews Data processing Maintenance	Low differentiation Variable membership, high turnover, variable life of team	High integration externally paced inside and outside	Repeated or continuous cycles	Food Chemicals Components Assemblies Retail sales Service Repair activities
Projects Research Planning Architecture Engineering Task Forces	High differentiation expert specialists, specialized equipment	Low integration little external pressure, much external communications	Typically one cycle	Plans Designs Investigations Prototypes Reports
Actions Sports teams Expeditions Negotiating teams Surgery teams Cockpit crews	High differentiation expert specialists	High integration closely synchronized with counterparts	Brief performance often repeated	Combat missions Court lawsuits Concerts Surgical operations

cycles, and the outputs from the work team. It is easy to recognize the diversity of industrial work teams and why it is difficult to predict their effectiveness.

Principles of team effectiveness

Some general principles have evolved for improving effectiveness from the study of many work teams. One of these is that providing feedback will normally improve team performance, as mentioned previously. However, providing performance feedback often does not result in increased satisfaction among team members. In fact, the opposite is likely to be true for team members receiving low performance ratings. Performance feedback does encourage team members to set goals, and some positive things have been published regarding this issue. For example, one study showed that providing performance related incentives to members of aviation maintenance teams encouraged them to set higher goals (Pritchard, Jones, Roth, Stuebing, and Ekeberg, 1988). However, it is often difficult to determine how to give performance feedback to individuals in teams that produce a single output, such as design teams. Leadership and authority structure is another frequently cited issue, as mentioned earlier in this chapter. One principle that follows is given below:

Principle 19.1

Teams faced with unpredictable inputs or uncertain outcomes perform best with decentralized communication and flexible internal coordination (Susman, 1970; Tushman 1977; Argote, 1982; Campbell, 1988).

Another principle that appears to have research consensus is discussed in Principle 19.2.

Principle 19.2

The smallest number of people who can do the required tasks in the available time is the best team size (Hackman, 1987).

One of the effects identified in the literature is *social loafing*, describing a tendency for some individuals to decrease their effort when additional members are added to the crew. That effect appears to be eliminated when teams or crews are as small as possible or when individual member outputs are made identifiable. Hence, an important consideration in crew ergonomics is to find ways to identify individual crew member contributions to crew performance. One exception in minimizing crew sizes, perhaps, is teams or crews where some members perform heavy physical tasks. If those tasks come up too frequently, extra team members are needed or else the whole teams will be required to take time breaks.

A further principle in team operations deals with the *norms* and *rules of behavior*, that tend to be agreed upon by the team members. Teams that most researchers describe as effective have norms and rules consistent with high performance. Also, a property of team cohesiveness is as follows:

Principle 19.3

Team cohesion is highly correlated with communication and conformity to group norms (McGrath, 1984).

Effective work teams tend to be highly cohesive, but strong cohesion can also lead to ineffective teams as mentioned earlier in this chapter. Two correlated features of successful teams are *self-empowerment* and *acceptance of accountability*. The former term means that teams do not have to go back to management on all issues and this freedom to decide breeds accountability (see Logan, 1993).

Research, development, and design teams

Traditionally, industry has organized research and development teams as an isolated entity which functions separately from the product design teams. Conceptual ideas about new products being designed by a company were started by a research group that often was located away from the main company. Even now, in many organisations research groups often develop pie-in-the-sky concepts and hand off their concepts to a practical designer from the mechanical or electrical engineering department, depending on the principal nature of the product. The designer then attempts to create an almost finished product before sharing the design elsewhere. When the chief designer from that department feels that the time is right, he or she takes the preliminary plans to another department and seeks additional input about the design. At this point, some aspects of the design concept are frozen or fixed. Other aspects of the design may be changed depending upon the nature of the problem discovered in these later activities. After the second department adds their recommendations, the design concept goes to the third department and so on. During this process of design review and modification, the personnel reviewing the design must negotiate with the principal designer to bring their ideas into the design. This serial form of design organization and this procedure is known as *over the wall* design because the process is analogous to each department in turn doing their work and then tossing the plan over the wall to the next department. It is obvious that this procedure limits more and more those departments that get the plans later in the process. This process is changing in many firms, and some of the following sections describe how it is changing (Box 19.2).

BOX 19.2 PROBLEMS WITH THE TRADITIONAL SERIAL DESIGN PROCESS

One of the problems with the traditional serial design process is that it takes a great deal of time to come up with a completed design when this approach is followed. A study of several technology-based companies illustrating this effect was done several years ago by McKinsey & Co. (Gupta & Wilemon, 1990). They reported that the serial design process typically resulted in a six-month delay in reaching the market, causing profit losses due to competition amounting to around 33% over the next 3 to 5 years. In automobile manufacturing, the design time span has been traditionally 2 or more years. Another difficulty with the serial design concept is that about 70% to 80% of the ultimate product unit cost is often committed during the first 15% to 25% of the design process. The farther down a department is in the design sequence hierarchy, the less impact that department has on the design. If a department that is well downstream in the process sees a simple but important improvement, the incorporation of that idea depends heavily on which features of the design are strongly frozen and which are still fluid. Generally, the farther downstream, the more features are frozen. There are numerous anecdotes of designs that could not be manufactured without very costly retooling that could have been avoided (Stoll, 1988; Boothroyd, Dewhurst & Knight, 1994). An associated problem with the traditional serial design process is that top management often is not heavily involved during the study, design, and development phases of design projects where their potential influence is the greatest (Roberts, 1976). A common mistake is that management often tends to try to influence design actions during early production and marketing phases when there is little the design team can do without starting over. Another issue is that many of the new globally marketed products require multidisciplinary design efforts because of different conditions, cultures, and requirements in foreign markets. Names and typical patterns of use can differ between cultures and companies have made potentially embarrassing mistakes when they fail to adequately consider how language and cultural differences can lead to unintended perceptions. For example, the French automobile named *citroën* made some Americans think of *lemon*, a colloquial expression for a poorly designed automobile. Other examples include the Chevrolet *nova*, which was translated unfavorably in Spanish, by some people as a vehicle that doesn't go, or the Mazda *LaPuta*, the Nissan *Moco*, or the Mitsubishi *Pajero*, which all have unfortunate translations in Spanish.

(continued)

BOX 19.2 (continued)

Recognition of these and other difficulties with the traditional serial design process has caused industry to change in many ways. Companies making such changes recognize that global products need to be innovative, very-high quality, customer-tailored, long lasting, and flexible enough to accommodate language and cultural differences. To achieve new products with these features, many companies dispensed with management hierarchies and went to a much flatter management organization that interacts early with design teams. Flexible design teams bringing together multiple disciplines are another innovation. These companies recognize the tremendous importance of rapidly generating prototype models so that all the team members can communicate internally about design changes and effective tradeoffs. They have also emphasized to their design teams that they need to consider and address all elements of the design over the life cycle of the product, starting with research and development, through prototyping, early sales, mature sales, and phasing out the product. Emphasis is also placed on the design elements fitting into the product life cycle from the consumer's viewpoint including training, technical support development, field testing, general use, and disposal of the product. See Fabrycky and Blanchard (1991) for additional information on this subject.

Team roles

Team members of research, development, and design teams tend to fit certain roles that have been categorized into eight types. Sometimes a person will take one primary role and perform in a lesser capacity in another. Numerous studies have found team members to be consistent in the roles they take. The roles chosen by particular people are predictable from a psychometric test battery. These roles are as follows:

1. *Chairperson* displays powers of control and coordinates the resources within the group. Although chairpersons tend to be democratic and operate on a participative basis, they are ready to exercise direct control when they perceive it to be necessary. These persons are dominant but fairly relaxed, with enthusiasm and often charisma, but they are also usually detached, reserved, and objective. Although not usually creative, they are concerned primarily with what is practical, and they tend to have strong egos.
2. *Shaper* is a more traditional leader who is at the front with all guns blazing. Shapers prefer to act on decisions directly and personally. They like quick results and willing followers. Typically they have dual motives of getting the job done and satisfying their own need to be in charge. They may not be popular, nor do they get the best out of the team, but they get results. Shapers are usually assertive extroverts fueled by nervous energy rather than self-assurance, and they frequently display insecurity. They dislike rules and other constraints. Often they are skeptics, quick to criticize, but unduly sensitive to criticism. Typically they are competitive and compulsive, but when things go well they generate enthusiasm and make things happen. Shapers are best operating with informal teams. In traditional teams, shapers need to use self-discipline and be more coordinating, rather than leading.

3. *Innovators* are creative thinkers or idea people. They offer new ideas and strategies because of fertile and fluent minds that generate original ideas. When innovators are properly made part of a team, they can contribute greatly to team success. However, their personality profile is not geared to orthodox corporate structures, and when they are in those situations they are apt to be exiled to remote offices, given fringe projects to work on, and generally kept out of management's hair. They show concern for ideas over people, an underemphasis of pragmatic concerns, resistance to persuasion, undiplomatic behavior, and uninhibited self-expression. When the team or management rejects their idea, they tend to opt out of the team.
4. *Monitor-evaluator* persons tend to complement innovators by analyzing ideas and suggestions from both the team's viewpoint and viewpoints external to the team. They are often shrewd, cautious, objective, and astute, serving as strategists and judges, but not idea generators. Monitor-evaluators are often overcritical, negative, and short on persuasive and motivational powers. They need to exercise care to keep from lowering team morale. However, when properly used, these people can keep teams away from ill-advised ideas and notions and to not get too pumped up over impractical ideas. One of their abilities is in interpreting complex ambivalent data that confronts most managers. They are intellectually competitive, often to the point of conflict with colleagues. As a team leader, there is a danger of overdominating colleagues and stifling their contributions. As junior people, monitor-evaluators are overly cynical and skeptical in nature.
5. *Company workers* are people who accept the cultures and conventions of a company or organization and buckle down to the everyday work of producing the goods and other jobs within the firm. Their strengths are in translating general plans into a practical working brief for their team's participation. New ideas and innovations do not excite them as they seek practical routines and tangible results. They are uncomfortable in situations requiring flexibility, adaptability, and expediency. Their high scores are on sincerity, integrity, general character, and self-discipline. Their faith is more in people than ideas and they have low tolerance for change, experimentation, or quick results. Company workers tend to grow to intermediate or higher management but rarely to the top. They tend to be the team's backbone rather than its head.
6. *Team workers* are perceptive of the feelings, needs, and concerns of other people in the team, and, because of this they are the glue that often holds the team together. They exude team spirit, and improve within-team communications and cooperation. These people are persuasive, but in a low-key manner, and they tend to smooth any friction within the team. Team workers are stable extroverts but without a strong competitive spirit or a desire for power. Often they go unrewarded in career progression because they have a reputation for lacking decisiveness and resilience, characteristics that are frequently over touted in the corporate culture. As junior people, team workers are apt to be the behind-the-scene helpers where credit is often missed. When team workers grow into senior roles, they are often adept delegators and conscientious developers of a well-coordinated staff. The danger in this role is that they will likely seek coziness and harmony when some tension and strain would be more useful.
7. *Resource investigators* are people-oriented team members, who are moderately dominant extroverts with a spirit for life. They are hardly ever in the room with others very long and are frequently on the phone. There is a drive in them to explore resources

outside the team and maintain a useful set of outside contacts. When with the team, they maintain good relationships and encourage their colleagues to use their talents and try out their ideas. Generally, resource investigators are cheerful people, whose sources of satisfaction are related to their team contribution and are usually so recognized. They have a tendency to lack self-discipline, be impulsive, shift interests, and are often ready to drop a project. They need variety, challenge, and constant stimulation to be happy and productive. Unlike the innovator, they are not a source of original ideas, but they have the ability to stimulate others on and off the team, which is a principal mark of their contribution. They have a tendency to focus on irrelevancies and trivialities, but teams without resource investigators tend to be focused much too inwardly, be too defensive, and out of touch with the outside world.*

8. *Completers* are stable extroverts who remind everyone about the underlying philosophy of the project, the positive contributions each team member needs to make, and that everyone has weaknesses, but good teams balance them with each other's strengths. Completers are not necessarily those team members who dominate or manage the group. They are highly anxious, often compulsive, sometimes introverted, and tense. Frequently they develop a lot of nervous energy that they harness as a compulsion to finish the job at a high standard and on time, hence the name completers. Teams without a completer are prone to both errors of commission and omission, putting-off tasks, and complacency. Completers prevent these things from happening by showing concern, naggingly communicating a sense of urgency, and threatening any casual spirit. They have self-control, strength of character, and a sense of purpose. While not the easiest people to live with, they fight carelessness, overconfidence, and procrastination. But as such, they irritate fellow team members and are apt to lower team morale.

Most people are composites of these *pure team roles*. Also, certain combinations of roles are more typical than others. Two roles that are limited strongly by the person's talent are those of *innovators*, who must be creative, and *monitor-evaluators*, who must be capable of critical thinking. Some role combinations entail compatible personality traits such as those of *resource investigators* and *completers*. The *chairperson* role is most versatile, and so it is often seen in combination with another type, however, the other role can affect the behavior extensively. A principal point is that combined roles tend to have a prominent main role and a subsidiary role.

The team composition that results in the most-productive teams tends to contain people playing different roles. All teams need a *chairperson*, so that is a given. Also an *innovator* is usually critical to success in research, development, and design. *Completers* are not necessarily needed in a team unless it has productivity problems. The *monitor-evaluator* is another important role in most teams. Roles of *company workers, team workers,* and *resource investigator* may be duplicated as needed, but the latter role usually fills a vital need. *Shapers* are best not sharing the *chairperson* role. While good teams are balanced across roles, most have gaps in their personality coverage. Some teams consciously try to fill the gaps internally as they recognize deficiencies. In fact, situations have been reported where team members assessed their colleagues, pooled the results, and even had self-assessments of the pool, have succeeded in building a better team than by trying to mechanically slot individuals into teams (Belbin, Aston, and Mottram, 1976; Mottram, 1982).

A team role that is even better known than those discussed here is the *gatekeeper*, who serves as an intermediary between the average crew or team member and sources of

---

* The *gatekeeper* concept bears strong similarities with this *resource investigator* role, as discussed later.

technical information. While this role may be a composite of some already stated, it is closely linked to the role of a *resource investigator*. Generally gatekeepers receive a wide variety of external information sources and they maintain a high level of contacts outside the organization. They also tend to be well read in technical journals and other technical information sources. Team or crew members regard gatekeepers as valued resources because they save hours of labor for other team members by finding the needed technical information.

Another team-role differing from the roles noted above is the *bridge person*. This type of person is an agent of a team who links the team to the rest of the organization and plays the role of bringing the team or crew members together. The roles of *chairperson, company worker*, or *team worker* often fits the description of a bridge person who advocates the team purpose and protects and defends them from other specialists and management, particularly on budgetary issues, and they help back up the technical team leader.

Communications within research, development, and design teams

The importance of communications within Research, Development, and Design (RDD) teams has been known for some time. One study by Rubenstein, Barth, and Douds (1971) investigated variables that affected effective communications within and between 60 RDD groups and found that differences in work-related values did not impede communications, but that groups who perceived themselves to be mutually interdependent tended to have fewer communications problems. Also, they identified a number of features that contributed positively to communication effectiveness. These features include the level of technical respect, preference for participate decision making, and communication frequency. At about the same time, the concept of *gatekeepers* became known as those persons in these RDD groups who serve principally to link a group with sources of external technical information (Allen and Cohen, 1969; Tushman, 1977). Gatekeepers appeared to be very important in technical service projects, but not in more fundamental research. However, not all researchers found that communications frequency facilitated effective development; part of the confusion appears to be linked with the type of communications. More communication associated with problem solving tended to predict better research performance, but more communication with management tended to be negatively linked with performance (Katz and Tushman, 1979). Part of the issue is the uncertainty that often surrounds complex research projects where teams are usually decentralized. In contrast, service projects contain far less uncertainty, rely less on intrateam communications, and tend to be more centrally managed. Accordingly, optimal communications within and between RDD teams is task specific, where team individuals communicate more with each other and with outside sources when the tasks are complex and new ground is being uncovered. Loose organizational structures are more apt to occur with participative decision making and decentralized leadership. But when the tasks are less uncertain and more routine, external communications tend to funnel through gatekeepers and internal communications are less needed. Also, organizational structure tends to tighten and become more traditional.

Group cohesiveness is another important issue in RDD teams, and is personified by the characters in *The Three Musketeers*.* This characteristic of groups varies in degrees from this extreme to total independence of one another and appears to be related to RDD team effectiveness. A study by Keller (1986) of 32 RDD teams measured team performance in a number

---

* *The Three Musketeers*, by Alexander Dumas, is an historical novel set during the reign of Louis XIII (1610–1643). The characters in the book and in motion picture versions of the book maintained the "One for all, and all for one" motto rendered by the musketeers.

of ways including: meeting deadlines, rated value to the company, rated group performance, and rated budget/cost achievements. Group cohesiveness was positively related with every performance measurement.* Other factors positively related to some of those performance measurements included job satisfaction (positively), innovative orientation (positively), and physical distance between persons (negatively). However, management style did not have much influence. It was also found that more successful RDD groups, in terms of the number of technical papers written and patents registered, had greater longevity. Also, heterogeneously composed RDD teams tended to perform better than homogeneous teams, especially if the members were young. RDD teams functioned better when administrative managers restricted themselves to administrative and organizational affairs and technical managers had influence over technical aspects of the project. While those situations appear to be simply turf problems, those restrictions tend to simplify to all concerned who is handling what.

The prevailing trend in most companies is to move in the direction of concurrent design team, in order to provide an opportunity for people from multiple disciplines to work nearly simultaneously on the design (Box 19.2). Recent developments in computer networks, and computer-aided design graphics, have facilitated this process. As one group suggests a change, another can laud or criticize alternative plans from their specific points of view. Here again, communications plays a critical role in how individual members of the design team can cooperate with each other.

Another approach for improving team effectiveness is to involve multiple disciplines in a cyclical manner. At the first stage, each discipline provides their initial concerns. Second-, third-, and higher-level interactions then take place as the design becomes more refined during the design process.†

Some observations from NASA

Some recent changes in NASA led to some organizational freedoms in some of their task groups. One group, known as the MARS PATHFINDER group, experienced some remarkable successes, and they set out to tell the public what they did that they believed made them successful. Their story illustrates some breakthrough thinking and some notable leadership concepts. A few of the principles they expressed include the following:

1. *Set goals that make you stretch.* They believe that there is power in self-generated expectations. Starting a project with a dramatic destination fuels the creative process and aspirations of the team members.
2. *Let limitations guide you to breakthroughs.* NASA's viewpoint was that constraints are invitations to rise above the past. Limitations should therefore be viewed as blessings rather than curses. Take advantage of their recognition but do not expect each one to be overcome.
3. *Discipline creativity.* While many technical groups violate norms and break traditions, successful groups are highly demanding and downright exacting on individuals about meeting their commitments. Everyone on the team has their deliverable role and each person is expected to meet it.
4. *Invite different perspectives; embrace eccentricity.* Not everybody sees things alike and often a combination of different perspectives is enough to create a great solution. Bounce ideas off one another and let the inside renegade speak out.

---

* O'Keefe, Kernaghan, and Rubenstein (1975) found similar results and reported that cohesive groups were more responsive to change than noncohesive groups.
† Also see Figure 1.4 on software design.

5. *Develop robust solutions.* While risk taking is acceptable, failure is not. Build in margins that provide an acceptable safety zone. Learn through early mistakes and critiques how to make a better and, if possible, simpler solution.
6. *Take personal responsibility for fully trustworthy communications.* First maintain a climate of trust as a clear code of conduct and personify it through effective communications. "Information hoarding is a cardinal sin and openness is a key virtue." Another dimension to this parable is to be hard on issues, soft on people, or said differently, "Fix the problems, not the blame." These and other ideas highlight NASA's notion to design what is needed and make it "faster-better-cheaper."*

Final remarks

Work groups and teams play an essential role in many industries and service organizations. A large set of factors related to group structure and other issues influence their performance. Addressing their issues presents a difficult challenge for designers, responsible for selecting team members, measuring heir performance, taking steps to ensure the teams will successfully perform their tasks. Because teams are so essential in many industrial operations their functions need nurturing.

Discussion questions and exercises

19.1 In what development phase do crew members become proficient at communications, while operations strategies are planned, and conflict begins to arise?

19.2 What type of crew authority structure tends to make the crew quicker to respond? More creative?

19.3 When crew members express a strong desire to be members of a particular crew, as opposed to not being members, what crew feature is described?

19.4 In designing crew operations for a new manufacturing process that requires multiple people, what are some of the important factors to consider?

19.5 Many people try to use analogies from sports to describe industrial team–crew operations as though any sport is appropriate to the situation at hand. What is wrong and right with this approach?

19.6 Indicate whether the following statements concerning the use of computer simulation in studies of crew performance are true or false and explain why this use is important for designing systems for crew operations:
  a. Finding the active time fractions of individual operators in crews.
  b. Estimating the shop, mill, or operation performance with the crew of operators.
  c. Approximating the sensitivity of more or fewer resources (e.g., crew size) to system performance.
  d. Determining the impact of new technologies to system and crew member performance.
  e. Calculating the effective performance time of an individual crew member on various tasks.
  f. Verifying learning rates of operators on different tasks.
  g. Specifying task accuracy in cognitive task performance.
  h. Allocating specific tasks to jobs for better system performance and balanced workloads between the crew members.

---

* See Pritchett and Muirhead, 1998.

19.7 What are some uses of computer simulation for ergonomic design of crews in the past literature?

19.8 What are some of the problems of doing field studies in studying the ergonomic implications of increased automation in contrast to using computer simulation or operator-in-the-loop simulation?

19.9 If a team is said to have high cohesion, what does this mean?

19.10 What are *quality circles* and why were they popular in the United States at one time? Do you think that they are a good thing and worth the cost of operating? Why?

19.11 Who are gatekeepers and what do they do?

19.12 What are some primary difficulties in measuring team–crew performance as a group?

19.13 What appears to be the effect of group cohesiveness on RDD group performance? What about homogeneously versus heterogeneously composed RDD teams?

19.14 What are some principal difficulties in measuring team–crew performance as a group?

19.15 What are the roles and characteristics of team–crew members known as *shapers* and *monitor-evaluators* in research, development, and design teams?

chapter twenty

Job evaluation and compensation

About the chapter

Setting wages in a fair and consistent manner is one of the most important issues faced by organizations. Perceived failures to meet this objective can lead to labor strife, grievances, losses of good employees, and charges of unlawful discrimination. Job evaluation is a process in which jobs are systematically evaluated to determine their relative contribution to the organization. This information is then used along with performance measures and other information to establish or set wages. The emphasis of this chapter is on describing several commonly used or traditional methods of job evaluation, and some of their respective advantages and disadvantages. Other issues including methods of administering wages are also briefly discussed.

Introduction

Jobs can be evaluated in a number of ways and for a number of different purposes as discussed in earlier chapters of this book. The term "job evaluation," however, has traditionally referred to the systematic process followed by industrial engineers and other analysts to determine the relative worth of jobs performed within organizations. Traditional methods of job evaluation result in a ranking (or rating, depending on the evaluation method) of the evaluated jobs in terms of their value to the organization. Some methods of job evaluation, such as the so-called factor method, directly rate the jobs in terms of their monetary value to the organization. Most evaluation methods, however, require a second step in which the rankings or ratings of particular jobs are converted into monetary value using a wage curve or schedule. For a variety of reasons, jobs of comparable worth are often grouped during the latter step into "grades." When the latter approach is used, a range of pay rates is specified for each grade rather than for separate jobs. After pay rates have been determined for jobs or job grades, they are then adjusted for individual merit and other factors, to set wages and salaries for each employee.

Job evaluation serves several important purposes. The primary purpose is to provide an objective basis for wage administration by the company. The need to set wages and salaries in a fair and equitable manner that accurately reflects the contribution of the jobs performed within an organization is obviously a very compelling reason to consider implementing a formal job evaluation plan. Part of the issue is that employee satisfaction and job performance are both strongly related to how much people are paid, and whether people feel the amount is fair and equitable (see Box 20.1). A good job evaluation system clearly and unambiguously describes the basis for setting wages and salaries, which helps eliminate problems related to lack of understanding of the evaluation process by affected employees. In some cases, employees, unions, and other affected parties are provided the opportunity to participate in the job evaluation process, which improves understanding of the process and helps gain acceptance of the results.

BOX 20.1 COMPENSATION LEVELS AND PERCEIVED EQUITY

Anyone with practical experience in the field of labor relations knows that employees who file grievances, quit, or otherwise express their dissatisfaction often state in rather strong terms that they feel they are underpaid and that pay rates are not set in an equitable, fair, or consistent way. It therefore should not be particularly surprising to find a large amount of research exists that confirms wages and the way they are administered can have significant effects on productivity. One of the earliest findings was that financial incentives can lead to large increases in the productivity of workers (Wyatt, 1934). However, subsequent studies showed that productivity sometimes reverted back to levels close to those observed prior to introducing the incentives (Viteles, 1953). Employee fears that productivity increases will lead to reductions in the workforce or not be fairly shared are both possible explanations of the latter tendency.

Numerous other studies led researchers long ago to conclude that the perceived fairness and equity of pay has a significant impact on employee morale, performance, and satisfaction (Jacques, 1962). Equity theory (Adams, 1963) provides an interesting explanation of how perceptions of equity might be formed, and why such perceptions can have effects on performance that are independent of how much the worker is paid. The basic idea is that people develop perceptions of equity by comparing what they put into the job (inputs) and receive (outcomes) from a job against the inputs and outputs of other workers. Inputs include items such as effort, skill, experience, or education. Outcomes include items such as pay, status, recognition, or other benefits. Adams suggested that the result of this comparison could be mathematically expressed as two ratios:

$$\frac{\text{My outcome}}{\text{My inputs}} \quad \text{vs.} \quad \frac{\text{Other person's outcome}}{\text{Other person's input}}$$

Worker perceptions of equity are then defined in terms of the two ratios. If the ratios are about the same, the worker judges their pay to be equitable. Workers that perceive their own outcome to input ratio to be low compared to other workers, compensate by putting in less effort into a job. Conversely, workers that perceive their outcome to input ratio to be high, compensate by putting in more effort into a job.

To test the predictions of equity theory, Adams conducted an experiment in which he led some hourly-paid personnel to believe that they were *overpaid* relative to others doing the same job. He found that those persons who thought they were overpaid significantly *outperformed* the others even though the actual wages were the same.

Job evaluation is also done to help ensure compliance with laws prohibiting discrimination in the workplace, and provide a basis for defending employers against charges of discrimination. Several federal and state laws* in the United States currently require equal pay for equal work. For example, the Equal Pay Act of 1963 (EPA), which applies to

---

* See Boxes 18.1 and 18.2 in Chapter 18 for more information regarding unlawful discrimination in the workplace.

virtually all employers, requires equal pay for men and women who perform substantially equal work in the same establishment. The Age Discrimination in Employment Act of 1967 (ADEA) contains similar provisions regarding equal pay to older workers. Other provisions can be found in state laws. For example, Minnesota Rule 5200.0040 Equal Pay for Handicapped Workers specifies that a handicapped person performing equal work "shall be paid the same wage as a non-handicapped person with similar experiences." In addition to helping employers comply with legal requirements associated with laws such as those mentioned earlier, a good job evaluation system can help companies by reducing grievances, lawsuits, and labor disputes. The use of ad hoc, poorly defined, hiring and pay practices opens the door to criticism, and can also make it more difficult to negotiate agreements with employees and labor unions. Regarding the latter point, a good evaluation system provides factual evidence that strengthens the negotiator's stand, and helps focus the negotiations away from irrelevant points and onto more important issues.

It should be clear from the earlier discussion that job evaluation offers many potential benefits. However, it must be recognized that conducting a good job evaluation will require a very significant commitment of time from many different people within the organization. Furthermore, this time and effort must be invested on a continual basis to keep the evaluations up to date. Part of the issue is that new processes, machinery, and equipment will result in changes in job content, which creates a need to reevaluate the affected jobs. Another issue is that market forces will cause the going rates for people with certain skills to change over time. This may create a need to reassess or adjust the wages paid to particular employees. As pointed out by Hannon et al. (2001), to minimize equity concerns, pay should be higher for the higher ranked jobs than for the lower ranked jobs. However, this may not always be possible, due to market forces. Simply put, in some cases, a company may have to pay more than suggested by the job evaluation system to fill particular jobs, when people with the necessary skills are in short supply. Such issues may create significant tension within an organization, but that is to be expected when dealing with a topic as important as compensation.

Job evaluation systems

Over the years, several different systems or methods of job evaluation have been developed (McCormick, 1979; Dantico and Greene, 2001; Hannon et al., 2001). Job evaluation has traditionally been done by members of a job evaluation committee who systematically compare job descriptions to develop a ranking (or rating, depending on the evaluation method) of the evaluated jobs in terms of their value to the organization. These ranking or ratings are then converted into monetary value. Some controversy exists regarding the use of these so-called traditional systems, but there is little doubt that such systems play an essential role in most larger organizations with significant human resources. Other systems of job evaluation, such as the Position Aptitude Questionnaire (PAQ) developed by McCormick or market value–based compensation schemes, have also been used (see Hannon et al., 2001), but will not be discussed in this chapter.

The four different methods most commonly used in a job evaluation to compare jobs are as follows:

1. Job ranking methods
2. Job classification methods
3. Point systems
4. Job factor comparisons

These methods differ in complexity. The job ranking and job classification methods are the simplest methods. Point systems and job factor comparison are more complex and often are developed with the assistance of consultants. Regardless of the method used to compare the jobs, the job evaluation process normally involves three sometimes overlapping steps:

1. Developing job descriptions by doing a job analysis
2. Comparison of the jobs using one of the four evaluation methods mentioned earlier
3. Converting the results of the job evaluation into a wage schedule used to set base pay rates and other forms of compensation

The last step will be covered later, in the section Wage Administration.

Job analysis

Job evaluation starts with an activity called *job analysis*.* Job analysis is the process followed to collect information about the job. This information is documented in a written document called a *job description*. The job description will normally give the *job title* (in other words, name the job performed), the date of the last analysis, the name of the analyst, and other information such as

1. Tasks performed within the job
2. Methods of performing these tasks
3. Levels of acceptable performance
4. Knowledge and skill requirements (including certificates, licenses, and degrees)
5. Training requirements
6. Responsibilities
7. Working conditions

An example of a task description is given in Table 20.1. Almost anyone would agree that the information listed in this example should affect the rate of pay received. Hence, those features are included in the job description. A cautionary note is that requirements listed in the job description should reflect characteristics of the *job rather than the individual* who happens to be holding the job. For example, just because the person currently holding the job has a degree in engineering does not necessarily mean the job requires an engineering degree. Another point is that it is critical to date these forms and list the name of analyst preparing the report. The less recent the date, the more likely there have been some job changes. To help ensure the task descriptions are kept up to date, some periodic auditing and updating of these records is necessary. When job descriptions are put on computers, the dated job descriptions and evaluation forms can be listed in terms of the last date an audit or update was made. In that way, the job descriptions most likely to need updating can be identified easily.

Job ranking methods

Two types of job ranking methods are commonly used to compare jobs. The simplest approach is to order the job descriptions for the evaluated jobs from highest to lowest overall

---

* Task analysis was discussed at some length earlier in Chapter 8. In some cases the job analysis will require a task analysis to develop the job description, as expanded upon later.

Table 20.1 Example of a Written Job Description

Computer Programmer 3

As a fully qualified computer programmer, applies standard programming procedures and detailed knowledge of pertinent subject matter (e.g., work processes, governing rules, clerical procedures, etc.) in a programming area such as a recordkeeping operation (supply, personnel and payroll, inventory, purchasing, insurance payments, depositor accounts, etc.); a well-defined statistical or scientific problem; or other standardized operation or problem. Works according to approved statements of requirements and detailed specifications. While the data are clear-cut, related, and equally available, there may be substantial interrelationships of a variety of records and several varied sequences of formats are usually produced. The programs developed or modified typically are linked to several other programs in that the output of one becomes the input for another. Recognizes probable interactions of other related programs with the assigned program(s) and is familiar with related system software and computer equipment. Solves conventional programming problems.

Performs such duties as develops, modifies, and maintains assigned programs; designs and implements modifications to the interrelation of files and records within programs in consultation with higher level staff; monitors the operation of assigned programs and responds to problems by diagnosing and correcting errors in logic and coding; and implements and/or maintains assigned portions of a scientific programming project, applying established scientific programming techniques to well-defined mathematical, statistical, engineering, or other scientific problems usually requiring the translation of mathematical notation into processing logic and code. (Scientific programming includes assignments such as using predetermined physical laws expressed in mathematical terms to relate one set of data to another; the routine storage and retrieval of field test data; and using procedures for real-time command and control, scientific data reduction, signal processing, or similar areas.) Tests and documents work and writes and maintains operator instructions for assigned programs. Confers with other EDP personnel to obtain or provide factual data.

In addition, may carry out fact-finding and programming analysis of a single activity or routine problem, applying established procedures where the nature of the program, feasibility, computer equipment, and programming language have already been decided. May analyze present performance of the program and take action to correct deficiencies based on discussion with the user and consultation with and approval of the supervisor or higher level staff. May assist in the review and analysis of detailed program specifications and in program design to meet changes in work processes.

Works independently under specified objectives; applies judgment in devising program logic and in selecting and adapting standard programming procedures; resolves problems and deviations according to established practices; and obtains advice where precedents are unclear or not available. Completed work is reviewed for conformance to standards, timeliness, and efficiency. May guide or instruct lower level programmers; may supervise technicians and others who assist in specific assignments.

Source: U.S. Department of Commerce, *Standard Occupational Classification Manual (SOC)*, Appendix B, Occupational descriptions, National Technical Information Service, Washington, DC, 1991.

value to the organization. Since evaluators may differ in their judgment of the value of particular jobs, most companies will ask several people with knowledge of the job to participate in the ranking process. Pairwise comparison of jobs is also sometimes done. In this approach, the evaluator compares each job against every other job. In most cases, the analyst simply indicates which of the two compared jobs is more valuable. An overall ranking for each job can then be tallied by counting how often it was rated higher than another job by the raters.

The ordering developed in the ranking process is used to assign jobs into particular job grades. This requires the analyst to identify which jobs in the rank order bracket the grades to be assigned. For example, assume the analyst determined that the Job A falls at the boundary between Grades 1 and 2; and Job B falls at the boundary between Grades 2 and 3. Then all jobs ranked below Job A might be assigned to Grade 1. Job A and all other jobs ranked below Job B might be assigned to Grade 2, and so on. A wage range is then specified for jobs falling into a particular grade using methods expanded upon later in this chapter. The number of job grades used varies between organizations. Keeping the number of grades small helps reduce unnecessary administrative costs. Most organizations use somewhere around 10 different job grades.

Pairwise *ranking* systems are normally used only when a small number of jobs will be evaluated, because the number of required comparisons quickly becomes large when more jobs are evaluated. That is, if n jobs must be compared, the number of pairwise comparisons (N) becomes

$$N = \frac{n(n-1)}{2} \quad (20.1)$$

If 40 jobs must be evaluated, formula (20.1) shows that the number of comparisons would be 780, which is obviously asking a lot from the raters. Another issue is that ranking methods do not show how large the difference is between ranked jobs. Simply put, a rank order only tells the analyst which jobs are rated higher. Consequently, the analyst will need to consider information besides the rank ordering to decide how job grades should be assigned. The subjective nature of the ranking process and its dependence on the raters being sufficiently familiar with every one of the rated jobs are other potential concerns (Hannon et al., 2001). A more sophisticated approach might ask the raters to indicate how much better each job is on one or more carefully specified rating scales. A value ranking could then be calculated in the same way done for the SLIM-MAUD example discussed in Chapter 16 (See Box 16.2). The latter approach would be very similar to the job factor comparison method discussed a little later in this chapter.

Job classification methods

Job *classification* methods are commonly used and are arguably the simplest way of evaluating jobs. To apply this approach, the analyst starts with a list of predefined job categories or grades. Existing jobs are then sorted into particular job grades based on how closely the job description matches the grade description.

One of the best known job classification methods is the *General Schedule (GS) Federal Classification and Job Grading System* used to classify white-collar employees of the U.S. federal government. The GS schedule is separated into 15 grades (GS-1, GS-2, ..., GS-15). Entry-level positions usually are classified as GS-1 to GS-7, midlevel are GS-8 to GS-12, and top-level are GS-13 to GS-15. Each grade has 10 steps. New employees are normally classified to the first step of their assigned GS grade. Within each grade, employees move up steps after serving a prescribed period of service in at least a satisfactory manner. Employees can also move up steps within a grade, or to higher grades, when additional responsibilities are added to their jobs, or after they develop new job-related skills, or perform exceptional service. A separate *Federal Wage System (FWS)*, also containing 15 grades, is used to classify "blue-collar" employees of the U.S. federal government.* Both

---

* Very extensive information on the GS and FWS systems is currently provided on the web by the U.S. Office of Personnel Management (see http://www.opm.gov/fedclass/html/fwseries.asp).

systems contain job-specific grading standards describing how to assign the proper GS or FWS grades to different jobs falling within particular occupational groups (such as the Accounting and Budget Group, GS-500; the Engineering and Architecture Group, GS-800; or the General Administrative, Clerical, and Office Services Group, GS-300).

The latter standards spell out in considerable detail how a wide variety of jobs are to be classified to a particular grade. For some jobs, the standard is more or less a written description of a particular type of job at different grades. For example, *Federal Wage System Job Grading Standard for Machining, 3414*, provides separate descriptions for Machinists at Grades FWS-10 and FWS-11. To determine which grade to assign, the analyst compares the job being analyzed to the written descriptions in the standard. For other jobs, the grade is determined by applying a *Factor Evaluation System* to assign points to the job for different factors, and then converting the point score into a GS grade using a matrix defined in the particular standard applicable. Point systems are expanded in the following section.

Point systems

Point methods are the most commonly used method of job evaluation. Point systems are clearly a step above both job ranking and classification systems in sophistication, and yet relatively easy to use once developed. All point systems contain three basic elements:

1. A set of compensable factors that describe various attributes of the job such as the abilities required, job demands, managerial or supervisory needs of the job, and unpleasant or undesirable features of the job. It is very important that the compensatory factors used in a point system do not overlap. If the factors overlap significantly, the point system will in effect be counting certain items more than once, which is obviously a systematic source of bias that might lead to unfair evaluations. The system also must not leave out important compensable factors.*
2. A rating scale for each factor that measures the degree to which the particular factor is present for a particular job. Usually this is done using a 5-point rating scale. It is critical that clear unambiguous definitions be given for deciding on how the ratings should be assigned on the rating scales. Analysts should be carefully trained to ensure they can properly interpret these definitions. Analysts also must be familiar with the job to help ensure they provide valid ratings.
3. The number of points that should be assigned for each possible rating of the degree to which compensable factors are present. The way points are assigned to ratings reflects the importance of each compensable factor used in the system (see Box 20.2). It is essential that such assignments accurately reflect the value of each compensable factor to the company. From both a theoretical and practical perspective, it is obvious that the value of each compensable factor might vary between companies in a way that reflects their specific goals and objectives.

To use a point system, the analyst first identifies the degree to which each particular compensable factor is present for the evaluated job. In many cases, the analyst will have an earlier job description that gives specific details summarizing the results of an earlier

---

* A survey of 21 companies using point plans conducted many years ago, showed that the number of compensable factors and subfactors used in these plans varied from a low of 3 to a high of 35 (Patton et al., 1964). As pointed out by Hannon et al. (2001), five to seven factors are usually enough to capture the essence of most jobs in an organization. In practice, companies often use more factors to help show that the plan covers everything. This helps gain acceptance of the point plan by employees, which is clearly an important objective.

> **BOX 20.2   EXAMPLE MAPPING OF POINTS TO DEGREES**
>
> To illustrate how the number of points might be assigned when developing a point system, consider the following example. To keep things simple, suppose that two compensable factors will be considered for each job and that each factor is assumed to be equally important. Also, the degree of each factor is measured on a 5-point rating scale with equally spaced intervals. If we further specify that the maximum possible number of points that can be assigned to a job is 100, we know that the points assigned to each factor must always be between 10 and 50. This follows since
>
> Maximum possible total points for job = 100 = Maximum possible points on factor 1
>
> + Maximum possible points on factor 2
>
> Since each factor is assumed to be equally important, we also know that
>
> Maximum possible points on factor 1 = Maximum possible points on factor 2
>
> This gives two equations and two unknowns, and we therefore know that
>
> Maximum possible points on factor 1 = Maximum possible points on factor 2 = 50
>
> Furthermore, since each rating scale used to measure the degree of each factor has five equally spaced intervals, we now know the point values to assign to each possible rating. That is, for each factor, the possible point values allocated to each possible degree rating are {10, 20, 30, 40, 50}.

evaluation done using the existing point system. An example task description of this type is shown in Table 20.2 for a toolmaker class A. Note that this task description contains many specific details about the job which should be checked to see whether anything has changed since the last evaluation. After completing the aforementioned steps, the analyst looks up how many points to allocate for each factor. The total point value for the job is then calculated by adding up the point values assigned to each factor.

NEMA point system

The earlier discussion assumed the analyst will be using an existing point system. When this is not the case, a very significant effort will need to be invested to develop each of the three elements of a point system listed earlier. To save time and money, many companies adopt or modify an existing point system to their specific circumstances, often with the assistance of consultants. One of the earliest point systems was devised by the National Electric Manufacturers Association (NEMA) in the 1930s. Over the years, this system has been used and modified in various ways by many different companies and organizations. Well-known variations of the NEMA system include the point systems developed by the National Metal Trades Association (NMTA) and the American Association of Industrial Management (AAIM).* Use of point systems has a long history

---

* The developers of the AAIM system provide instruction manuals, training courses, and other resources to help practitioners better understand and apply their version of the NEMA system. The address of AAIM is 7425 Old York Road, Melrose Park (Philadelphia), PA 19126.

Table 20.2 Example Task Description Showing the Results of an Earlier Job Evaluation

Job Name: Tool Maker	Job Rating Specifications (Shop)	Class: A	Dept: Tool Room Grade: 2 Total Points: 340		
Factors	Substantiating Data			Degree	Points
Education	Use shop mathematics, charts, tables, handbook formulas, and trigonometry. Work from complicated drawings, sketches, and samples. Use all types of precision measuring instruments. Thorough knowledge of machine shop practices. Equivalent to a complete accredited apprenticeship			4	56
Experience	Over 4 years up to and including 5 years			4	88
Initiative and ingenuity	Diversified work, perform a wide range of tool room operations, considerable judgment to interpret drawings and samples, plan work, perform difficult machine and bench operation, maintain close tolerances and interrelationship			4	56
Physical demand	Moderate physical effort required for setup and operation of machine tools. Equivalent to occasionally handling heavy weight material			3	30
Mental or visual demand	Must concentrate mental and visual attention closely to a number of details to interpret drawings, plan work, make difficult tool setups, and perform close bench work			4	20
Responsibility for equipment or process	Improper setup or operation may result in jamming of tools and work causing tool breakage or damage to machine mechanisms. Probable damage seldom over $150			3	15
Responsibility for material or product	Failure to interpret drawings, improper machine tool or bench operations may result in spoilage of work or excessive down time, probable loss seldom over $200			3	15
Responsibility for safety of others	Thoughtlessness or inattention in handling tools or material, improper setup or operation, broken tools, flying particles, may result in lost-time injuries to others			3	15
Responsibility for work of others	Responsible for two and up to five persons for special project as assigned			2	10
Working conditions	Generally good working conditions, exposed to various shop elements, but with none continuously present to the extent of being disagreeable			2	20
Hazards	May crush hands or feet handling tools and materials. Exposed to hand, finger or eye injuries from machine tool operations			3	15

that can be traced back 50 or more years for some companies. The NEMA, NMTA, and AAIM point systems all specify the following four major compensable factors related to job demands and requirements:

1. *Skill*—education, experience, and other talents required by the job
2. *Effort*—physical and mental demands of the job

3. *Responsibility*—the number of different kinds of things the employee is responsible for, and the magnitude of significance of this responsibility
4. *Working conditions*—both undesirable and hazardous conditions at the workplace

As shown in Table 20.3, and expanded upon in Box 20.3, each factor in these point systems is divided into two to four sub-factors. Table 20.3 also shows how points are assigned for particular ratings. For example, as can be seen by checking the table, a degree 1 rating on the education sub-factor is given 14 points, while a degree 5 rating on the experience sub-factor receives 110 points. As mentioned earlier, the points assigned for each sub-factor are summed up to determine an overall score for the job. The overall scores are then often used to assign a grade to the job. In Table 20.4, we can see, for example, that a Grade 1 job has an overall (summed) score falling in the range of 0–139. Consequently, if the summed point values of the sub-factor ratings happened to be 120 for a particular job, it would be assigned a grade of 1. Somewhere between 10 and 20, labor grades are usually used.

Note that the points are mapped to ratings in the NEMA system in a way intended to reflect the importance of the factors and sub-factors. This can be seen by comparing the points assigned in Table 20.3 for the different sub-factors. For example, for the experience sub-factor, the difference between a degree 1 and degree 5 rating is 88 points (i.e., 110 points minus 22 points). For the job hazards factor, the corresponding difference is only 20 points (i.e., 25 points minus 5 points)! Such analysis shows that the NEMA system weights the skill factor much more highly than the other factors. In fact, the skill factor accounts for 50% of the points of the maximum number of points possibly assigned. This follows, because the total number of points assignable in the system is 500, and 250 of them are associated with skill sub-factors. That is, the sub-factors of education (14% of the maximum total), experience (22%), and initiative and ingenuity (14%) add up to 50%. The remaining factors such as effort, responsibility, and working conditions factors, respectively, account for up to 15%, 20%, and 15% of the total possible points.

BOX 20.3 COMPENSABLE FACTORS AND SUBFACTORS

NEMA and related point systems specify four major compensable factors. Each factor can be further divided and measured in different ways, as expanded upon in the following.

Skill is divided into three different sub-factors:

1. *Education*—measures the basic training, knowledge, and scholastic or academic background needed to *start* on the job, *prior to learning the job duties*. Note that the term "education" does not necessarily mean formal education.
2. *Experience*—measures the minimum length of time it would take a normally qualified person working under normal supervision to attain quality and quantity performance standards.
3. *Initiative* and *ingenuity*—measures the degree of independence of the worker on the job. This sub-factor varies between job settings where the employee is told what to do and has little or no choice to situations, where the person is nearly an independent operator.

BOX 20.3 (continued)

Effort is divided into two sub-factors:

1. *Physical demand*—measures the effort associated with lifting or moving materials. This sub-factor includes force and frequency considerations, as well as the physical position of the operator during the lifting or moving. The levels of this sub-factor vary from a low frequency of moving very light objects in easy positions to frequent lifting or moving heavy objects in difficult positions. Other considerations include the proportion of time that the physical effort is sustained.
2. *Mental or visual demand*—measures the mental or visual attention required by a job. The NEMA point systems system were originally developed for production shop work where there is little mental processing in contrast to some office work. The AAIM system includes specific provisions for clerical work involving more mental processing and visual activity.

Responsibility is described in terms of four sub-factors. These subfactors correspond to responsibility for

1. *Equipment or process*—defined in dollar amounts.
2. *Material or product*—also defined in dollar amounts.
3. *Safety of others*—defined in terms of the degree of potential injury and attention required to protect the safety of others.
4. *Work of others*—measures the number of persons supervised and the approximate fraction of time this responsibility holds.

Working conditions are divided into two sub-factors:

1. *Jobconditions*—measures physical conditions and their undesirability. Undesirable conditions involve exposure to large amounts of fumes, cold, heat, dirt, dust, grease, noise, or vibration. For jobs that require human operators to move from one region of a factory to another, a time-weighted average (TWA) should be calculated as described in Chapter 17.
2. *Job hazards*—measures the severity and likelihood of potential injuries and the relative likelihood of that exposure. Levels vary from a situation where an accident involving an injury is unlikely, to intermediate situations with significant likelihood of serious injury, such as a loss of body parts, to situations where death is a significant possibility.

Job factor comparison method

The job factor comparison method is similar in many ways to use of a point system. As for point systems, the jobs are quantitatively evaluated using compensable factors. The factors used in the evaluation also are similar to those commonly used in point systems. However, a fundamentally different process is followed that involves separate comparisons of the evaluated jobs against previously evaluated benchmark jobs on each factor. The job factor method also directly specifies the monetary value of a job. Each of the earlier discussed methods requires a separate step in which the results of the job evaluation are converted into monetary value using a wage curve or related method.

The first step in the job factor comparison method is to select a set of 15–20 benchmark jobs which are accepted to have fair rates of compensation. Ideally, this set of benchmark

Table 20.3 Compensable Factors and Subfactors Used in the NEMA Point System

	Degree				
	1	2	3	4	5
Skill					
Education	14	28	42	56	70
Experience	22	44	66	88	110
Initiative and ingenuity	14	28	42	56	70
Effort					
Physical demand	10	20	30	40	50
Mental/visual demand	5	10	15	20	25
Responsibility					
Equipment or process	5	10	15	20	25
Materials or product	5	10	15	20	25
Safety of others	5	10	15	20	25
Work of others	5	10	15	20	25
Working conditions					
Job conditions	10	20	30	40	50
Job hazards	5	10	15	20	25

Table 20.4 Example Assignment of Job Grades in the NEMA Point System

Score Range	Grade	Score Range	Grade	Score Range	Grade
0–139	1	206–227	5	294–315	9
140–161	2	228–249	6	316–337	10
162–183	3	250–271	7	338–359	11
184–205	4	272–293	8	360–381	12

jobs is representative of all jobs performed in the organization. Each of the jobs is then ranked on five commonly accepted compensable factors[*]: mental requirements, skill requirements, physical requirements, responsibility, and working conditions. Table 20.5 shows an example of such a ranking for five benchmark jobs.

The second step in analysis is to specify how much of the salary for each benchmark job should be allocated to each compensable factor. This process might be done by a compensation committee or an individual analyst. As for the other methods of job evaluation, it is important that people doing the evaluation are unbiased and familiar with jobs. Table 20.6 shows an illustrative example summarizing the result of this step for the same set of benchmark jobs shown in Table 20.5. After completing these two steps, the analyst should then check to make sure the results of each step are consistent. That is, for each compensable factor, the job with the highest ranking in Table 20.5 should have the highest monetary allocation in Table 20.6, and so on. In some cases, steps will then

---

[*] Other factors can also be used in this method. The five listed here were viewed by the original developers of the factor comparison method (Benge et al., 1941) as universal factors applicable to all jobs.

Table 20.5 Ranking of Benchmark Jobs on Compensable Factors

Job Title	Mental Requirements	Skill Requirements	Physical Requirements	Responsibility	Work Conditions
A. Machine operator	1	1	4	1	2
B. Janitor	3	4	1	3	1
C. Forklift operator	2	2	2	2	3
D. Assembler	4	3	3	4	4

Table 20.6 Example Allocation of the Contribution of Compensable Factors to Wages for Benchmark Jobs

Job Title	Wage Rate ($)	Mental Requirements ($)	Skill Requirements ($)	Physical Requirements ($)	Responsibility ($)	Work Conditions ($)
A. Machine operator	20.00	5.50	8.00	1.50	3.00	2.00
B. Janitor	10.00	1.50	1.00	2.50	1.50	3.50
C. Forklift operator	15.00	3.00	6.00	2.00	2.50	1.50
D. Assembler	10.00	1.00	5.50	1.50	1.00	1.00

have to be taken to make the ratings consistent, such as repeating the earlier steps or removing benchmark jobs that are inconsistently rated. Once the analyst is satisfied that the results are reasonably consistent, they are used to evaluate jobs not included in the original benchmarking process.

The third step is to compare each of the jobs to be evaluated against the benchmark jobs. This is done separately for each of the compensable factors. For example, suppose we wanted to use the results of the benchmarking process (Table 20.6) to evaluate a new job. We would then pick one of the compensable factors and determine which of the benchmark jobs are the most similar on this factor to the job we are evaluating. In most cases, an attempt then is made to bracket the analyzed job by finding one benchmark job that is slightly greater than the evaluated job on the considered factor and a second benchmark job that is slightly lower. The results of following such a procedure are shown in Table 20.7. The benchmark jobs bracketing the example job on each factor are indicated in the latter table with an "*."

The fourth step is to allocate a monetary value to the evaluated job that falls somewhere between the amount allocated to each of the two benchmark jobs bracketing the evaluated job. This is done separately for each factor, often by averaging the bracketing values or

Table 20.7 Ordering of Example Benchmark Jobs by Compensable Factors

Mental Requirements	Skill Requirements	Physical Requirements	Responsibility	Work Conditions
A*	A	B	A	B
C*	C*	C	C*	A
B	D*	D*	B*	C*
D	B	A*	D	D*

* Benchmark jobs bracketing an evaluated nonbenchmark job.

Table 20.8 Allocation of the Contribution of Compensable Factors to Wages for a New Job Corresponding to the Example Nonbenchmark Job in Table 20.7

Job Title	Wage Rate ($)	Mental Requirements ($)	Skill Requirements ($)	Physical Requirements ($)	Responsibility ($)	Work Conditions ($)
A. Machine operator	20.00	5.50	8.00	1.50	3.00	2.00
B. Janitor	10.00	1.50	1.00	2.50	1.50	3.50
C. Forklift operator	15.00	3.00	6.00	2.00	2.50	1.50
D. Assembler	10.00	1.00	5.50	1.50	1.00	1.00
E. New job	14.50	4.25	5.75	1.50	1.75	1.25

through interpolation. An even simpler approach is to assign the value of the closest benchmark job. The total amount allocated to the evaluated job is then calculated by adding up the separate wage allocations for each factor. Example results obtained by performing these steps for the previous example job are illustrated in Table 20.8.

The results of each earlier-discussed job evaluation method are used to set wages, following the methods expanded upon in the following section. In many cases, this task is performed by members of a human resources department within the organization that uses the results of the job evaluation along with other information, such as job performance measures, and prevailing labor rates to determine wages, salaries, and benefits.

Wage administration

Most organizations have a human resources department that administers a compensation program, policy, or plan that reflects its values and objectives. The compensation program, policy, or plan will usually specify the following:

1. The forms of compensation provided, and whether they will be above, below, or at the prevailing market rates
2. The criteria used to determine wages, benefits, merit increases, and other forms of compensation
3. How these criteria are applied
4. Formal appeal and grievance procedures
5. Steps taken to comply with Federal Wage and Hour Laws, laws against discrimination, and other applicable governmental guidelines

These elements of a compensation program, policy, or plan will vary between organizations and reflect the emphasis a particular organization places on attaining goals, such as

1. Rewarding employees for their past and current performance
2. Attracting and retaining good employees
3. Maintaining wage and salary equity
4. Motivating employees to improve and seek out promotions
5. Encouraging employees to participate in training and education programs
6. Keeping compensation costs within the budget
7. Avoiding labor disputes and charges of discrimination

Chapter twenty: Job evaluation and compensation

In some cases, the elements of a compensation program or plan are spelled in detail in a labor-management agreement. The following discussion will explore these elements and some of the ways they are specified in organizations. The latter process often relies heavily on the results of a job evaluation, but will involve several additional steps, as expanded upon in the following sections.

Forms of compensation

Many different forms of compensation are provided to employees by employers. Some of the different forms of compensation that might be provided by a particular organization are summarized in Table 20.9. As indicated there, the total compensation package will often include much more than payment for job performance. The different forms of compensation provided can be broadly classified as being either pay or benefits. For the most part, each form of compensation can be a negotiated element of a labor-management agreement. Laws such as U.S. Fair Labor Standards Act (FLSA) require employers to bargain in "good faith" when negotiating forms and levels of compensation for many categories

Table 20.9 Forms of Compensation Provided by Organizations[a]

Forms of Pay

1. Base pay, which is the fixed compensation paid to an employee for performing specific job responsibilities. Base pay is typically paid at an hourly rate or as a salary. Minimum hourly wage rates are specified by federal and state laws
2. Variable pay, which is compensation contingent on performance or results (of an individual worker, a group of workers, or of the organization as a whole). Examples include piecework, incentives, merit pay, bonuses, recognition awards, and distributions from profit sharing plans
3. Differential pay, which is a nonperformance-based pay usually given to accommodate a specific working condition. Such pay can be given for working overtime, working on night shifts, being on-call, hazardous duty, or as a replacement for certain benefits (such as social security in a limited set of cases). Employees who are classified as nonexempt under the U.S. Fair Labor Standards Act (FLSA) must be paid at a rate of 1 1/2 times their regular pay rate for time worked in excess of 40 h in their workweek. Differential pay for overtime is not required for managers, supervisors, and many other white-collar employees classified as exempt under the FLSA
4. Pay for nonproductive time at work, including breaks, training sessions, and special events (such as kaizen events, sensitivity awareness sessions, or team-building activities)

Benefits

1. Pay for nonproductive time away from work, including vacations, holidays, sick leave, jury duty, administrative leave, family leave, and emergency leave
2. Mandated income protection programs, required by law, including worker's compensation insurance, unemployment insurance, social security and Medicare
3. Voluntary income protection programs, including medical benefits and insurance, life insurance, pensions, retirement plans, and deferred compensation plans
4. Other financial benefits, such as price discounts for buying products made by an employer, tuition reimbursements, or stock options
5. Nonfinancial compensation, including on-the-job training, flexible work schedules, and relaxed dress codes

[a] The total compensation package provided to an employee will include a mixture of pay and benefits.

of workers. Federal and state laws also mandate* a minimum hourly wage, minimum differential pay for overtime for certain classes of workers, and income protection programs such as worker's compensation insurance, unemployment insurance, and social security.

Pay is classified in several ways, including its basis and intent. Much of the work performed in most organizations is compensated on an hourly basis. This is especially true when people perform the so-called blue-collar jobs. Employees compensated on an hourly basis are classified as hourly employees, or wage earners. Hourly employees are normally paid only for the time they work. Salaried employees receive pay computed on the basis of weekly, biweekly, or monthly pay periods. Normally, a salaried employee receives the same pay for each pay period, regardless of how many hours were worked. In some cases, employees receive pay contingent on how well they or, in some cases, the organization as a whole performed. In most cases, such pay is an increment or supplement to the base pay level, thought of as a bonus for good performance of the job. In direct incentive systems, the bonus is based on output measures, such as the number of units produced by the employee. Indirect incentive systems provide a bonus not directly tied to the output of the particular worker, such as a share of the profits when the company has a good year. Other forms of pay include pay differentials or additional compensation. Typical examples include compensation given for working longer hours than ordinary (overtime pay), working over a holiday weekend, or for working in unpleasant or hazardous conditions.

Benefits are similarly classified into categories. Importantly, pay provided for nonproductive time away from work, such as a paid vacation, is classified as a benefit by the U.S. Fair Labor Standards Act and most organizations. Many other types of benefits are provided by organizations to their employees. Some are required by law, as mentioned earlier. An often underappreciated truth is that the benefits provided to employees can be a very significant cost to an organization. These costs are often incurred long after a commitment is made to provide them, and may be much larger than expected due to the accumulated effect of cost of living increases and rapidly increasing costs of medical care. A focus on short-term objectives can lead organizations to unwisely make a long-term commitment to benefit plans they can ill afford in their efforts to control the costs of wages and salaries.

Specifying how wages, benefits, merit increases, and other forms of compensation will be determined is clearly the most important element of a compensation program or plan. One part of this process is to develop a wage structure or schedule that specifies the base pay (and in many cases benefits) that will be paid for the particular jobs performed within the organization. A second part is to develop a schedule of pay incentives, and a plan or method of implementing it, which rewards employees for good performance. These two aspects of wage administration will be briefly expanded upon in the following.

Establishing a wage structure

The process of establishing a fair, acceptable, and effective wage structure requires a careful trade-off between two, sometimes, conflicting objectives. The first is to maintain what is called external equity, or in simpler terms, payment of wages and salaries that are consistent with those paid by other organizations in similar areas and circumstances. The second is to maintain internal equity, or in other words, equal pay for equal work on all of the jobs performed within the organization. These two objectives can be addressed by developing a wage structure or schedule based on both job evaluations and analysis of the

---

* That is, these elements are required by law, and are not subject to negotiation.

prevailing labor rate (or market value) of a job. One advantage of this approach is that a job evaluation system provides an internally consistent* measure of the worth of particular jobs that can be adjusted to take into account the prevailing labor rates.

Assuming the results of the job evaluation are already available, this overall process often involves four basic steps:

1. Conducting a *wage and salary survey* to determine the prevailing labor rates
2. Measuring *external consistency* using a *wage calibration curve* that compares wages paid within the organization to prevailing labor rates
3. Measuring *internal consistency* with a *wage curve* that relates the results of the job evaluation to wages paid within the organization
4. Setting *rate ranges* for jobs within labor grades to accommodate differences between individual employees

Wage and salary surveys

Generally the first step in developing a wage structure is to survey the wages, salaries, fringe benefits, and pay differentials paid by employers in the labor market from which the organization recruits its employees. The labor market for some employees such as office personnel is local, whereas the labor market for other employees such as engineers might be national. Conducting the wage and salary survey allows an organization to determine how much it must pay to be competitive with other establishments.

Some organizations conduct their own wage and salary surveys. However, information obtained from surveys conducted by the U.S. Bureau of Labor Statistics (BLS) or other sources will often be sufficient to establish prevailing labor rates. In particular, the National Compensation Survey (NCS) conducted by the BLS provides comprehensive measures of occupational earnings; compensation cost trends, benefit incidence, and detailed plan provisions. Detailed occupational earnings are available from the NCS for metropolitan and nonmetropolitan areas, broad geographic regions, and on a national basis. The BLS makes this information conveniently available at no cost on the web (see http://www.bls.gov). Information is also available from wage surveys conducted by different states in the United States, and from trade groups such as the Society for Human Resource Management, the American Management Association, and the National Society of Professional Engineers. Unfortunately, existing surveys do not always adequately address the specific jobs of concern to an organization. In such situations, the organization may decide to do their own survey. Doing so can be quite expensive.

Wage calibration curves

The easiest way to test external consistency is to develop a wage calibration curve. A wage calibration curve simply plots the wages paid by the organization against the prevailing labor rate, or, in other words, the wages paid for similar jobs by competing organizations (see Figure 20.1). If wages paid by the organization perfectly match the prevailing labor rate, each of the plotted jobs should lie on a straight line with a slope equal to 1. The latter line is called the wage calibration line. Any job above the calibration line is paid at a higher rate than the competition. Jobs below the calibration line are paid less.

A quick analysis of the example in Figure 20.1 shows that most of the jobs lie above the calibration line. Also, if we draw a straight line through the plotted points, (the best-fit

---

* The obtained results are internally consistent, because all of the jobs in the organization are evaluated on the same criteria.

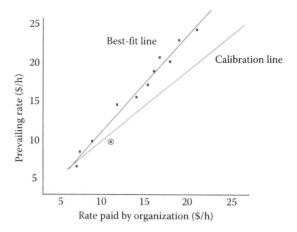

Figure 20.1 Example wage calibration curve. Points on the curve correspond to 12 different jobs.

line in the figure) we see that the two lines have very different slopes. The best-fit line is close to the calibration line for low paying jobs. As the rate of pay increases, the best-fit line deviates more and more. This difference in slope of the lines shows that there is a systematic trend for this organization to pay less than the prevailing labor rate for the higher paying jobs. This trend might reflect the desire or the need of the organization to pay less than its competition for the higher paying jobs. If not, the organization in this example might decide to raise its compensation rates for the higher paying jobs, in order to become more competitive. Also, the circled job seems to be an outlier because it is well below both the best-fit and calibration lines. The organization therefore might want to flag this particular job for further analysis.

Wage curves

A wage curve shows the relation between the results of job evaluation and wages paid. In many cases, separate wage curves are developed for wages paid by the organization, and wages paid by competing organizations for similar jobs.* An internal wage curve can be used to test the internal consistency of wages currently paid by the organization. Either internal or prevailing rates can be used to help set appropriate ranges of compensation for particular jobs. These uses of wage curves are illustrated in the following.

To illustrate how an internal wage curve can be used to test internal consistency, consider the example wage curve shown in Figure 20.2, which relates the wages currently paid by the organization to job evaluation scores obtained using a point system.† With the exception of the two circled points, the relationship is well described by a straight line. To improve internal consistency, the rate paid for the circled point below the line might be adjusted upward, while the point above the line might be adjusted downward. Rates too far above the line are often called *red circle rates*, and rates too far below the lines are called *green circle* rates. From a practical perspective, it is important to recognize that adjusting

---

* The latter curve is also used to test external consistency. For examples showing how this can be done, the reader is encouraged to consult McCormick (1979).
† A similar analysis would be followed using the results of the other job evaluation systems. A common approach is to plot base pay versus the labor grade determined by the job evaluation. This would allow the results of the classification and ranking methods to be plotted. The monetary worth obtained using the job factor comparison method can also be similarly plotted.

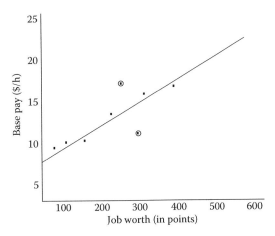

Figure 20.2 Example wage curve. Plotted points correspond to wages paid by an organization for jobs with different point values assigned by a job evaluation system.

wages downward is likely to create problems with the affected workers. Consequently some organizations resort to approaches such as holding wages constant for people currently performing the affected jobs until rates paid for other jobs catch up, offering buyouts, or transferring affected employees to another job that pays their current rate.

It is important to realize that the relation between wages and job evaluation measures doesn't necessarily have to be linear to be consistent. For example, the relation might be well described by a simple curve. Statistical approaches can also be used, such as fitting the relationship with linear or nonlinear regression models. One advantage of the latter approach is that internal consistency can then be measured mathematically using a correlation co-efficient. In addition, outliers can be formally defined in terms of the number of standard deviations away from the best-fit line. In practice, however, many organizations simply plot out the wage curve and visually determine necessary adjustments.

Rate ranges

The preceding discussion has assumed a single pay rate for each job or job grade. In practice, the pay given to different people for performing the same job may vary significantly, depending on the seniority or merit of the employee. It therefore usually becomes necessary to determine a reasonable range of pay increments for each job or job grade. For administrative reasons, this range is almost always established for a job grade, or grouping of jobs determined to be of similar worth in the job evaluation, rather than separately for each job. Jobs are already grouped into grades as part of the evaluation process when the job ranking or job classification methods are used. Point systems or the job factor method, however, provides a separate score for each job. Consequently, the latter method requires an extra step in which each job grade is specified as a range of point values or monetary worth. This step was discussed earlier, for the NEMA point system. A similar process is followed for the job factor comparison method.

An example wage curve showing the range of payment allocated to different job grades is given in Figure 20.3. Job grades are designated in the figure by boxes. The width of each box describes the range of points used to assign jobs to particular grades, and the height describes the range of payment possible within a grade. The slope of the line drawn through the boxes indicates how rapidly wages increase from the lowest to

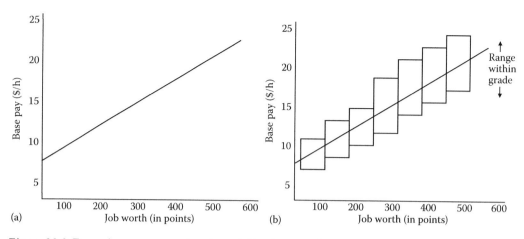

Figure 20.3 Example wage curve showing relation between base pay (in $) and job evaluation score (in points). Particular jobs fall within job grades (indicated by boxes). The vertical range within a grade (height of box) describes allowed adjustments to pay for individual experience or merit.

highest job grade. The bottom of the range for each job grade often is the starting wage of inexperienced workers. The top of each range is the wage assigned to the most qualified, senior, or productive employees in the job grade.* In many cases, the range of payment is greater in the higher job grades to allow more scope for merit rating adjustments, and encourage people to seek promotion. The ranges between adjacent job grades also often overlap, which allows highly qualified workers at lower grades to sometimes earn more than people just starting out in a higher grade.

Incentive plans

To raise productivity and lower labor costs in today's competitive economic environment, organizations are increasingly turning to incentives, as a way to motivate employees to work harder for both the company and themselves. A wide range of compensation options are currently used by organizations to reward outstanding performers for their contributions such as merit pay, cash bonuses, incentive pay, and various gain-sharing or profit-sharing plans. Some insight into the immense current interest in this topic can be obtained by browsing the web for companies marketing different incentive plans.†

Part of the reason organizations are interested in incentives is that strong evidence shows that the use of incentives can lead to average improvements of 20% or more (Condly, Clark, and Stolovitch, 2003). Unfortunately, designing a good pay-for-performance system is not easy. Careful consideration must be given to how employee performance will be measured, types (and amount) of the rewards provided, which employees to cover, and the payout method. The size of monetary bonuses for performance and their perceived value

---

* Wage compression is a common problem in organizations. Cost of living adjustments and seniority adjustments granted to union workers in labor agreements sometimes cause most of the employees in lower job grades to make more than the average worker in a higher grade not within a union. Market forces also often cause organizations to grant high starting salaries to new employees with little experience, well above the bottom rate originally assigned to the job grade. This may create a situation in which employees with many years of experience receive about the same pay as newly hired, and less qualified, workers.
† A recent search on Google (http://www.google.com) by one of the authors resulted in 11,800,000 hits for the phrase "incentive system." Obviously, not all of these hits are for companies marketing solutions, but many were.

to employees is a particularly important issue. Some desirable features of a good incentive plan include the following:

1. *Clarity*—Each element of the plan must be clearly defined and systematically documented to cover all contingencies. Poorly specified plans contain ambiguous elements and sometimes can be manipulated by either workers or employers.
2. *Simplicity*—The pay should be easily understood by the affected employees. Ideally, the plan also provides a straightforward way for employees to calculate their pay.
3. *Significance and incentive*—The potential rewards provided by the plan should be sufficiently large to justify the increased effort employees will need to exert to obtain them. Rewards must be perceived to be significant before they can realistically be expected to impact behavior.
4. *Equity and fairness*—Participants should receive a fair minimum pay rate. The benefits of improved performance should be fairly shared in a way that accurately reflects the contribution of each participant. Careful consideration of the potentially disruptive effects of rewarding particular individuals within a team is more highly than others.
5. *Pay for performance*—The measures used to judge employee rewards must be clearly linked to profitability and other important objectives of the organization.
6. *Measurability, consistency, and accuracy*—Sound and objective standards must be used to measure worker contributions and establish reward levels.
7. *Stability*—The plan should be set up for the long term. Changes to the plan will cause resentment when employees lose benefits they have grown accustomed to receiving. Careful consideration must be given on how to properly adjust performance targets over time.
8. *Maintainability*—Elements of the plan used to judge worker performance, such as time standards, must be kept up-to-date.

Incentive plans can be divided in terms of whether they are designed to reward performance at the individual, group, or organizational level.

Individual level plans

Plans at the individual level are the most popular form of incentive pay plan. Such plans directly reward participants based on their job performance. Examples include piecework systems, performance bonuses, performance-based merit increases, and sales commissions. Piecework and bonus plans traditionally have been used in manufacturing settings. However, the modern movement toward "pay for performance" focuses on a wide variety of jobs beyond those in traditional manufacturing settings.

In all such plans, the worker's output or performance is measured in some way. The measures used can be divided into three overlapping categories:

1. *Efficiency measurements* such as acceptable product units produced per operator hour, transactions conducted per operator hour, defective product units per hour, or sales dollars per expense dollar
2. *Effectiveness measurements* such as average time to respond to a customer's inquiry, fraction of orders tardy, rework hours per product unit, number of escalations requiring management intervention per customer contact, or complaints per transaction processed
3. *Inferential measurements* such as overtime hours per work period, attendance per total hours worked, machine downtime per scheduled hours, and a wide variety of subjective performance ratings of employees by supervisors

Efficiency and effectiveness measures, such as those listed earlier, are directly linked to production rates and quality, and consequently are easily seen to be valid measures of employee performance, assuming the employee has some control over them. Efficiency and effectiveness measures also tend to be more objective than inferential measures. Inferential measurements, such as reduced equipment downtime, good housekeeping, reliable employee attendance, and cooperative attitudes, do not directly measure production, but clearly have an impact that can be easily inferred. For example, machine downtime can slow production elsewhere, and such effects can be estimated. The validity of basing incentives on inferential measures, especially for subjective performance ratings, is obviously likely to be questioned by workers in some cases. Despite their lower validity, the later types of measures are often used to help set merit increases, because it is too expensive or difficult to obtain a better measure of performance.

A second aspect of plans implemented at the individual level is that rewards are given to the worker using various formulas that compare worker performance to desired levels of production or quality. In rare instances, the level of compensation is entirely determined by employee performance. This approach is illustrated by piecework systems where the employee is paid a set amount for each unit of production or by sales incentive plans where workers are paid a commission for the number of units sold. More commonly, the worker receives some set minimum rate that is supplemented by some percentage of performance beyond a preset goal. For example, a sales worker might receive a minimum salary supplemented by a bonus proportional to the number of units sold over a predetermined quota. Other plans reward employees based on quality measures, such as the fraction of non-defective parts produced. In the latter case, the bonus might be a proportion of the cost saving to the company associated with less rework and scrapped units. It also should be mentioned that bonuses often are calculated by combining multiple measures. For example, the total bonus might be determined as the combined cost savings associated with producing more good items and fewer defective items than the average employee.

One assumption of all of these individual level plans is that the employee can carry out all the activities required to achieve the performance measure. If performance is controlled by factors beyond the worker's control, the worker can be unfairly penalized or rewarded. Another point is that the performance standard must be carefully specified, clearly defined, and measurable before such a plan can provide fair results.

Group level plans

Group level plans are often used to reward the members of a team who must work together to reach the performance goal. Some jobs require cooperation and close coordination of activities within the team. For example, fire fighters must work closely together while fighting a fire. In such settings, rewarding the productivity of the team, as a whole, with perhaps some refinements for the contributions of each individual, obviously makes sense. Group level plans are also used when it is difficult to determine which individual is responsible for the over-all performance of a group or when objective performance standards are not available. One major advantage of group level plans is that they are much easier to administer than plans at the individual level. Group plans also reduce potential conflicts within a group, encourage cooperation, and increase group cohesion (see Chapter 19 for more detail on these issues). A disadvantage, however, is that group incentive plans provide an opportunity for some employees free-load. A group plan also reduces the opportunity for the true high performers to gain the rewards they deserve.

On the whole, there is evidence that group level plans can improve team performance significantly (Condly, Clark, and Stolovitch, 2003). In their meta-analysis of research on incentive effectiveness, the latter authors found an average increase of 48% in performance over baseline rates, when team incentives were used.

Organization level plans

Gainsharing and profit-sharing plans reward employees based on the overall costs and/or profits of an organization. Many organizations use such plans to help encourage a less adversarial relationship between employees and management. Gainsharing plans focus on sharing with employees the effects of reducing costs through productivity improvements. Profit-sharing, as implied by the name, focuses on sharing profits with employees.

The *Scanlon plan*, developed about 70 years ago, is one of the best known and still very popular gainsharing plans. Under this plan, a committee assesses the productive efficiency of each department of the company. Periodic reviews are conducted, typically at three month intervals. Everyone in a department is then given bonuses determined by comparing the payroll costs actually incurred against those expected for the amount produced. Thus, everyone shares the gain when the overall performance efficiency is improved. Another feature of the Scanlon plan is to award differential wages within the departments according to an efficiency index. The Rucker Share-of-Production and Kaiser plans are other well-known gainsharing plans. Well-known profit-sharing plans include the Nunn-Bush and Lincoln plans. In the Nunn-Bush plan, the company sets aside a *proportion* of its wholesale receipts. The specific proportion set aside each year is negotiated with the union, with an upper and lower limit set on the wage fund. Employees are paid from this fund, based on the hours worked, an amount such that the fund stays within the fund limits. In the *Lincoln plan*, employees are paid a piece rate plus stock in the company. The intent of this approach was to instill a sense of ownership in the company.

Final remarks

Job evaluation is performed by many organizations to help ensure that wages are set in a fair and consistent manner. The four commonly used methods of job evaluation discussed in this chapter are the job ranking method, job classification method, point systems, and job factor comparison. The results of job evaluation are used along with wage and salary surveys to help set wages that are both consistent with the wages paid to other jobs within an organization and competitive with prevailing rates paid by competing organizations. Developing appropriate worker incentive plans that "pay for performance" is another important issue when administering wages and salaries.

Discussion questions and exercises

20.1 What is the role of job analysis in job evaluation?

20.2 If one is using a ranking method of job evaluation, then
   a. How long would it take to make pairwise comparisons between 10 different jobs when a single comparison decision required 30 s on the average?
   b. Suppose that in 12 different jobs it was found that 3 pairs of those jobs were essentially equivalent. How many comparisons were saved by the recognition of these group equivalences compared to no equivalent groups, assuming pairwise comparisons?

20.3 Give an example of a currently used job classification system. How are job grades assigned in the system?

20.4 How are points assigned in the AAIM or a similar system?

20.5 In what way can the AAIM or a similar system be used to resolve issues of alleged gender or age discrimination? To answer, consider the case where one kind of job pays less and is populated almost entirely by females and a second kind of job pays more but is populated almost entirely by males. Describe how you could use this job evaluation system to investigate whether this situation is discriminatory or not.

20.6 What laws are relevant to issues of wage discrimination? What types of wage discrimination are illegal?

20.7 What forms of compensation are commonly provided by organizations? How do benefits differ from pay?

20.8 What is a wage curve? How does a wage curve differ from a wage calibration curve?

20.9 Give some examples of the types of wage data that can be obtained from the BLS.

20.10 How do red circle and green circle rates differ from each other?

20.11 Why are job grades assigned a range of compensation rates?

20.12 What are some desirable aspects of an incentive plan? How might rewards be assigned in individual level, group level, and organizational level plans? How effective are incentive plans? When might they fail?

Appendix A: Some probability distributions

Table A.1 Some Additional Probability Distributions

Density Function		Statistics	Comments
Continuous variable forms			
Gaussian (normal)	$\dfrac{1}{\sigma\sqrt{2\pi}} e^{-(x-\mu)^2/2\sigma^2}$ Parameters μ, σ Range $-\infty$ to $+\infty$	$E(x) = \mu$ $V(x) = \sigma^2$ mode $\tilde{m}(x) = \mu$	Symmetrical
Gamma	$\dfrac{x^{(a-1)} e^{-x/b}}{\Gamma(a) b^a}$ Parameters a, b Range 0 to $+\infty$	$E(x) = ab$ $V(x) = ab^2$ $\tilde{m}(x) = \begin{cases} (a-1)b & \text{if } a > 1 \\ 0 & \text{if } a \leq 1 \end{cases}$	Positively skewed, $(a-1)! = \Gamma(a)$ when a is integer
Negative exponential	$\dfrac{e^{-x/b}}{b} = \lambda e^{-\lambda x}$ Parameter $\lambda = 1/b$ Range 0 to $+\infty$	$E(x) = b = 1/\lambda$ $V(x) = b^2 = 1/\lambda^2$ $\tilde{m}(x) = 0$	Same as a gamma when $a = 1$
Chi square	$\dfrac{(x^2)^{(a-1)} e^{-x^2/2}}{(a-1)! 2^a}$ Parameter $a \geq 1$ Range 0 to $+\infty$	$E(x^2) = 2a$ $V(x^2) = 4a$ $\tilde{m}(x^2) = (a-1)/2 \; a > 1$	A special case of the gamma for $b = 2$ and where the variable is x^2
Beta-1 or β_1	$\dfrac{\Gamma(a+b)}{\Gamma(a)\Gamma(b)} x^{a-1}(1-x)^{b-1}$ Parameters $a \geq 1$, $b \geq 1$ Range 0–1	$E(x) = a/(a+b)$ $V(x) = \dfrac{ab}{(a+b)^2(a+b+1)} = \dfrac{E(x)[1-E(x)]}{a+b+1}$ $\tilde{m}(x) = \dfrac{a-1}{a+b-2}$	Either positive or negative skewness

Appendix A: Some probability distributions

Beta-2 or β_2	$\dfrac{\Gamma(a+b)}{\Gamma(a)\Gamma(b)}\dfrac{z^{a-1}}{(1+z)^{a+b}}$	$E(z) = a/(b-1)$	Either positive or negative skewness—$z = x/(1-x)$ if x is Beta-1—F distribution is a special case
	Parameters $a > 1, b > 1$	$V(z) = \dfrac{a(a+b-2)}{(b-1)^2(b-2)}$	
	Range 0–1		
Student	$\dfrac{((k-1)/2)!}{\sqrt{k\pi}\,((k-2)/2)!}\left(1+\dfrac{t^2}{k}\right)^{-(k+1)/2}$	$\tilde{m}(z) = \mu = \tilde{m}(x)$ $E(t^2) = (a-1)/(b-1)$ $V(t^2) = \sigma^2 k/(k-2)$ $\dfrac{((k-1)/2)!}{\left[((k-2)/2)!\right]\sqrt{k\pi}} = 0.3989 - \dfrac{0.1009}{k} + \dfrac{0.0203}{k^2}$	For a large k, the student-t is Gaussian—student t is a Beta-2 when $a = 1/2$, $b = k/2$ for $z = t^2/k$—ratio of a normal and a chi-square
	$t = (x - \mu)/\sigma$ Parameters $-\infty < \mu < +\infty$ $0 < \sigma < +\infty$ $V \geq 1$ Range $-\infty$ to $+\infty$		
Rectangular	$\dfrac{1}{b-a}$	$E(x) = \dfrac{(a+b)}{2}$	Symmetrical
	Parameters a, b	$V(x) = \dfrac{(b-c)^2}{12}$	
	Range a to b	No mode	

Appendix B: Tables of statistical distributions

This appendix contains a variety of tabled values of statistical distribution that are used in various locations in this textbook. These tables are compiled here for easy reference.

B.1 Normal (Gaussian) distribution tables

B.1.1 Cumulative normal (Gaussian) distribution

Standard deviates $z = (x - \mu)/\sigma$ from the mean

z	0	1	2	3	4	5	6	7	8	9
0.0	0.0000	0.0040	0.0080	0.0120	0.0160	0.0199	0.0239	0.0279	0.0319	0.0359
0.1	0.0398	0.0438	0.0478	0.0517	0.0567	0.0596	0.0636	0.0675	0.0714	0.0754
0.2	0.0793	0.0832	0.0871	0.0910	0.0948	0.0987	0.1026	0.1064	0.1103	0.1141
0.3	0.1179	0.1217	0.1255	0.1293	0.1331	0.1368	0.1406	0.1443	0.1480	0.1517
0.4	0.1554	0.1591	0.1623	0.1664	0.1700	0.1736	0.1772	0.1808	0.1844	0.1879
0.5	0.1915	0.1950	0.1985	0.2019	0.2054	0.2088	0.2123	0.2157	0.2190	0.2224
0.6	0.2238	0.2291	0.2324	0.2357	0.2389	0.2422	0.2454	0.2486	0.2518	0.2549
0.7	0.2580	0.2812	0.2642	0.2673	0.2704	0.2734	0.2784	0.2794	0.2823	0.2852
0.8	0.2881	0.2910	0.2939	0.2957	0.2996	0.3023	0.3051	0.3078	0.3106	0.3133
0.9	0.3159	0.3186	0.3212	0.3238	0.3264	0.3289	0.3315	0.3340	0.3365	0.3389
1.0	0.3413	0.3438	0.3461	0.3485	0.3508	0.3531	0.3554	0.3577	0.3599	0.3621
1.1	0.3643	0.3665	0.3686	0.3708	0.3729	0.3748	0.3770	0.3790	0.3810	0.3380
1.2	0.3549	0.3869	0.3888	0.3907	0.3925	0.3944	0.3952	0.3980	0.3997	0.4015
1.3	0.4032	0.4049	0.4066	0.4032	0.4099	0.4115	0.4131	0.4147	0.4162	0.4177
1.4	0.4192	0.4207	0.4222	0.4236	0.4251	0.4265	0.4279	0.4292	0.4306	0.4319
1.5	0.4332	0.4345	0.4357	0.4370	0.4382	0.4394	0.4406	0.4418	0.4429	0.4441
1.6	0.4452	0.4453	0.4474	0.4484	0.4495	0.4505	0.4513	0.4525	0.4535	0.4545
1.7	0.4554	0.4564	0.4573	0.4582	0.4591	0.4599	0.4608	0.4616	0.4625	0.4633
1.8	0.4641	0.4649	0.4656	0.4664	0.4671	0.4676	0.4666	0.4693	0.4699	0.4706
1.9	0.4713	0.4719	0.4726	0.4732	0.4738	0.4744	0.4750	0.4756	0.4761	0.4767
2.0	0.4772	0.4773	0.4763	0.4788	0.4793	0.4798	0.4803	0.4808	0.4812	0.4817
2.1	0.4821	0.4826	0.4830	0.4834	0.4888	0.4548	0.4848	0.4850	0.4854	0.4857
2.2	0.4861	0.4864	0.4868	0.4871	0.4875	0.4878	0.4881	0.4884	0.4887	0.4890
2.3	0.4893	0.4896	0.4898	0.4901	0.4904	0.4906	0.4909	0.4911	0.4912	0.4916
2.4	0.4918	0.4920	0.4922	0.4925	0.4927	0.4929	0.4931	0.4932	0.4934	0.4936

(*continued*)

z	0	1	2	3	4	5	6	7	8	9
2.5	0.4938	0.4940	0.4941	0.4943	0.4946	0.4946	0.4948	0.4949	0.4951	0.4952
2.6	0.4953	0.4956	0.4956	0.4957	0.4959	0.4960	0.4951	0.4962	0.4963	0.4964
2.7	0.4965	0.4966	0.4967	0.4968	0.4969	0.4970	0.4971	0.4972	0.4973	0.4974
2.8	0.4974	0.4975	0.4976	0.4977	0.4977	0.4978	0.4979	0.4979	0.4980	0.4981
2.9	0.4981	0.4982	0.4982	0.4983	0.4984	0.4984	0.4985	0.4986	0.4986	0.4986
3.0	0.4987	0.4987	0.4987	0.4983	0.4988	0.4989	0.4989	0.4989	0.4990	0.4990
3.1	0.4990	0.4991	0.4991	0.4991	0.4992	0.4992	0.4992	0.4992	0.4993	0.4993
3.2	0.4993	0.4993	0.4994	0.4994	0.4994	0.4994	0.4994	0.4995	0.4995	0.4995
3.3	0.4995	0.4995	0.4995	0.4996	0.4996	0.4996	0.4996	0.4996	0.4996	0.4997
3.4	0.4997	0.4997	0.4997	0.4997	0.4997	0.4997	0.4997	0.4997	0.4997	0.4998
3.5	0.4993	0.4993	0.4998	0.4998	0.4998	0.4998	0.4998	0.4998	0.4998	0.4998
3.6	0.4998	0.4998	0.4999	0.4999	0.4999	0.4999	0.4999	0.4999	0.4999	0.4999
3.7	0.4999	0.4999	0.4999	0.4999	0.4999	0.4999	0.4999	0.4999	0.4999	0.4999
3.8	0.4999	0.4999	0.4999	0.4999	0.4999	0.4999	0.4999	0.4999	0.4999	0.4999
3.9	0.5000	0.5000	0.5000	0.5000	0.5000	0.5000	0.5000	0.5000	0.5000	0.5000

B.1.2 Ordinates of the normal (Gaussian) distribution

Ordinates y at z standard deviates from the mean

z	0	1	2	3	4	5	6	7	8	9
0.0	0.3989	0.3989	0.3989	0.3988	0.3986	0.3984	0.3982	0.3980	0.3977	0.3973
0.1	0.3970	0.3955	0.3961	0.3956	0.3951	0.3945	0.3939	0.3932	0.3925	0.3918
0.2	0.3910	0.3902	0.3894	0.3885	0.3876	0.3867	0.3857	0.3847	0.3836	0.3825
0.3	0.3814	0.3802	0.3790	0.3778	0.3765	0.3752	0.3739	0.3725	0.3712	0.3697
0.4	0.3683	0.3668	0.3653	0.3637	0.3621	0.3605	0.3589	0.3572	0.3555	0.3538
0.5	0.3521	0.3503	0.3485	0.3467	0.3448	0.3429	0.3410	0.3391	0.3372	0.3352
0.6	0.3332	0.3312	0.3292	0.3271	0.3251	0.3230	0.3209	0.3187	0.3165	0.3144
0.7	0.3123	0.3101	0.3079	0.3056	0.3034	0.3011	0.2989	0.2966	0.2943	0.2920
0.8	0.2897	0.2874	0.2850	0.2827	0.2803	0.2780	0.2756	0.2732	0.2709	0.2685
0.9	0.2561	0.2637	0.2613	0.2589	0.2565	0.2541	0.2516	0.2492	0.2468	0.2444
1.0	0.2420	0.2396	0.2371	0.2347	0.2323	0.2299	0.2275	0.2251	0.2227	0.2203
1.1	0.2179	0.2155	0.2131	0.2107	0.2083	0.2059	0.2036	0.2012	0.1989	0.1965
1.2	0.1842	0.1919	0.1895	0.1872	0.1846	0.1826	0.1804	0.1781	0.1758	0.1736
1.3	0.1714	0.1691	0.1669	0.1647	0.1626	0.1604	0.1582	0.1561	0.1539	0.1518
1.4	0.1497	0.1475	0.1456	0.1435	0.1415	0.1394	0.1374	0.1354	0.1334	0.1315
1.5	0.1295	0.1276	0.1257	0.1238	0.1219	0.1200	0.1182	0.1153	0.1145	0.1127
1.6	0.1109	0.1092	0.1074	0.1057	0.1040	0.1023	0.1006	0.0989	0.0973	0.0957
1.7	0.0940	0.0925	0.0909	0.0893	0.0878	0.0863	0.0848	0.0833	0.0818	0.0804
1.8	0.0790	0.0775	0.0761	0.0748	0.0734	0.0721	0.0707	0.0694	0.0681	0.0669
1.9	0.0656	0.0644	0.0632	0.0520	0.0608	0.0595	0.0584	0.0573	0.0562	0.0531
2.0	0.0540	0.0529	0.0519	0.0508	0.0498	0.0488	0.0478	0.0468	0.0459	0.0449
2.1	0.0440	0.0431	0.0423	0.0413	0.0404	0.0396	0.0387	0.0379	0.0371	0.0363
2.2	0.0355	0.0347	0.0339	0.0332	0.0325	0.0317	0.0310	0.0303	0.0297	0.0290
2.3	0.0283	0.0277	0.0270	0.0264	0.0258	0.0252	0.0246	0.0241	0.0235	0.0220

Appendix B: Tables of statistical distributions

z	0	1	2	3	4	5	6	7	8	9
2.4	0.0221	0.0219	0.0213	0.0208	0.0203	0.0198	0.0194	0.0189	0.0184	0.0180
2.5	0.0175	0.0171	0.0157	0.0163	0.0158	0.0154	0.0151	0.0147	0.0143	0.0139
2.6	0.0135	0.0132	0.0129	0.0126	0.0122	0.0119	0.0116	0.0113	0.0110	0.0107
2.7	0.0104	0.0101	0.0099	0.0096	0.0093	0.0091	0.0088	0.0086	0.0084	0.0081
2.8	0.0079	0.0077	0.0075	0.0073	0.0071	0.0069	0.0067	0.0065	0.0063	0.0061
2.9	0.0060	0.0058	0.0056	0.0055	0.0053	0.0051	0.0050	0.0048	0.0047	0.0040
3.0	0.0044	0.0043	0.0042	0.0040	0.0039	0.0038	0.0037	0.0036	0.0035	0.0034
3.1	0.0033	0.0032	0.0031	0.0030	0.0029	0.0023	0.0027	0.0026	0.0025	0.0025
3.2	0.0024	0.0023	0.0022	0.0022	0.0021	0.0020	0.0020	0.0019	0.0018	0.0018
3.3	0.0017	0.0017	0.0016	0.0016	0.0015	0.0015	0.0014	0.0014	0.0013	0.0013
3.4	0.0012	0.0012	0.0012	0.0011	0.0011	0.0010	0.0010	0.0010	0.0009	0.0009
3.5	0.0009	0.0008	0.0008	0.0008	0.0008	0.0007	0.0007	0.0007	0.0007	0.0006
3.6	0.0006	0.0006	0.0006	0.0005	0.0005	0.0005	0.0005	0.0005	0.0005	0.0004
3.7	0.0004	0.0004	0.0004	0.0004	0.0004	0.0004	0.0003	0.0003	0.0003	0.0003
3.8	0.0003	0.0003	0.0003	0.0003	0.0003	0.0002	0.0002	0.0002	0.0002	0.0002
3.9	0.0002	0.0002	0.0002	0.0002	0.0002	0.0002	0.0002	0.0002	0.0001	0.0001

B.1.3 Correlated bivariate normal tables

$r: = 0.9$

	1	2	3	4	5	6	7	8	9	10
10	0.00	0.00	0.00	0.01	0.08	0.40	1.72	6.49	22.43	68.86
9	0.00	0.00	0.05	0.34	1.44	4.53	11.45	23.55	36.21	22.43
8	0.00	0.05	0.47	2.06	5.88	12.63	21.29	27.56	23.55	6.49
7	0.01	0.34	2.06	6.26	12.88	19.99	24.01	21.29	11.45	1.72
6	0.08	1.44	5.88	12.88	19.58	22.60	19.99	12.63	4.53	0.40
5	0.40	4.53	12.63	19.99	22.60	19.58	12.88	5.88	1.44	0.08
4	1.72	11.45	21.29	24.01	19.99	12.88	6.26	2.06	0.34	0.01
3	6.49	23.55	27.56	21.29	12.63	5.88	2.06	0.47	0.05	0.00
2	22.43	36.21	23.55	11.45	4.53	1.44	0.34	0.05	0.00	0.00
1	68.86	22.43	6.49	1.72	0.40	0.08	0.01	0.00	0.00	0.00

$r: = 0.8$

	1	2	3	4	5	6	7	8	9	10
10	0.00	0.02	0.13	0.42	1.09	2.51	5.35	11.05	23.18	56.24
9	0.02	0.27	0.98	2.40	4.81	8.45	13.52	19.94	26.42	23.18
8	0.13	0.98	2.78	5.53	9.09	13.16	17.19	20.15	19.94	11.05
7	0.42	2.40	5.53	9.20	12.86	15.90	17.62	17.19	13.52	5.35
6	1.09	4.81	9.09	12.86	15.52	16.63	15.90	13.16	8.45	2.51
5	2.51	8.45	13.16	15.90	16.63	15.52	12.86	9.09	4.81	1.09
4	5.35	13.52	17.19	17.62	15.90	12.86	9.20	5.53	2.40	0.42
3	11.05	19.94	20.15	17.19	13.16	9.09	5.53	2.78	0.98	0.13
2	23.18	26.42	19.94	13.52	8.45	4.81	2.40	0.98	0.27	0.02
1	56.24	23.18	11.05	5.35	2.51	1.09	0.42	0.13	0.02	0.00

(*continued*)

r: = 0.7

	1	2	3	4	5	6	7	8	9	10
10	0.04	0.26	0.72	1.51	2.79	4.78	7.90	13.01	22.22	46.78
9	0.26	1.18	2.61	4.51	6.92	9.86	13.36	17.41	21.68	22.22
8	0.72	2.61	4.86	7.33	9.91	12.49	14.89	16.79	17.41	13.01
7	1.51	4.51	7.33	9.87	12.06	13.76	14.81	14.89	13.36	7.90
6	2.79	6.92	9.91	12.06	13.43	14.02	13.76	12.49	9.86	4.78
5	4.78	9.86	12.49	13.76	14.02	13.43	12.06	9.91	6.92	2.79
4	7.90	13.36	14.89	14.81	13.76	12.06	9.87	7.33	4.51	1.51
3	13.01	17.41	16.79	14.89	12.49	9.91	7.33	4.86	2.61	0.72
2	22.22	21.68	17.41	13.36	9.86	6.92	4.51	2.61	1.18	0.26
1	46.78	22.22	13.01	7.90	4.78	2.79	1.51	0.72	0.26	0.04

r: = 0.6

	1	2	3	4	5	6	7	8	9	10
10	0.24	0.88	1.77	2.95	4.51	6.61	9.51	13.75	20.76	39.02
9	0.88	2.50	4.22	6.07	8.10	10.34	12.82	15.61	18.70	20.76
8	1.77	4.22	6.31	8.25	10.08	11.82	13.42	14.79	15.61	13.75
7	2.95	6.07	8.25	9.98	11.37	12.45	13.18	13.42	12.82	9.51
6	4.51	8.10	10.08	11.37	12.17	12.54	12.45	11.82	10.34	6.61
5	6.61	10.34	11.82	12.45	12.54	12.17	11.37	10.08	8.10	4.51
4	9.51	12.82	13.42	13.18	12.45	11.37	9.98	8.25	6.07	2.95
3	13.75	15.61	14.79	13.42	11.82	10.08	8.25	6.31	4.22	1.77
2	20.76	18.70	15.61	12.82	10.34	8.10	6.07	4.22	2.50	0.88
1	39.02	20.76	13.75	9.51	6.61	4.51	2.95	1.77	0.88	0.24

r: = 0.5

	1	2	3	4	5	6	7	8	9	10
10	0.74	1.89	3.09	4.45	6.04	7.99	10.47	13.85	19.10	32.40
9	1.89	3.92	5.59	7.18	8.79	10.46	12.26	14.25	16.56	19.10
8	3.09	5.59	7.31	8.76	10.07	11.27	12.40	13.42	14.25	13.85
7	4.45	7.18	8.76	9.93	10.86	11.58	12.11	12.40	12.26	10.47
6	6.04	8.79	10.07	10.86	11.35	11.59	11.58	11.27	10.46	7.99
5	7.99	10.46	11.27	11.58	11.59	11.35	10.86	10.07	8.76	6.04
4	10.47	12.26	12.40	12.11	11.58	10.86	9.93	8.76	7.18	4.45
3	13.85	14.25	13.42	12.40	11.27	10.07	8.76	7.31	5.59	3.09
2	19.10	16.56	14.25	12.26	10.46	8.79	7.18	5.59	3.92	1.89
1	32.40	19.10	13.85	10.47	7.99	6.04	4.45	3.09	1.89	0.74

r: = 0.4

	1	2	3	4	5	6	7	8	9	10
10	1.64	3.21	4.53	5.87	7.32	8.97	10.96	13.53	17.34	26.65
9	3.21	5.30	6.73	8.00	9.22	10.44	11.73	13.16	14.88	17.34
8	4.53	6.73	8.04	9.09	10.01	10.85	11.65	12.42	13.16	13.53
7	5.87	8.00	9.09	9.87	10.48	10.98	11.37	11.65	11.73	10.96
6	7.32	9.22	10.01	10.48	10.79	10.95	10.98	10.85	10.44	8.97

Appendix B: Tables of statistical distributions

	1	2	3	4	5	6	7	8	9	10
5	8.97	10.44	10.85	10.98	10.95	10.79	10.48	10.01	9.22	7.32
4	10.96	11.73	11.65	11.37	10.98	10.48	9.87	9.09	8.00	5.87
3	13.53	13.16	12.42	11.65	10.85	10.01	9.09	8.04	6.73	4.53
2	17.34	14.88	13.16	11.73	10.44	9.22	8.00	6.73	5.30	3.21
1	26.65	17.34	13.53	10.96	8.97	7.32	5.87	4.53	3.21	1.64

$r: = 0.3$

	1	2	3	4	5	6	7	8	9	10
10	3.00	4.74	5.99	7.15	8.34	9.62	11.09	12.92	15.53	21.62
9	4.74	6.58	7.69	8.63	9.50	10.36	11.25	12.25	13.48	15.53
8	5.99	7.69	8.61	9.33	9.95	10.53	11.08	11.65	12.25	12.92
7	7.15	8.63	9.33	9.83	10.23	10.56	10.84	11.08	11.25	11.09
6	8.34	9.50	9.95	10.23	10.41	10.52	10.56	10.53	10.36	9.62
5	9.62	10.36	10.53	10.56	10.52	10.41	10.23	9.95	9.50	8.34
4	11.09	11.25	11.08	10.84	10.56	10.23	9.83	9.33	8.63	7.15
3	12.92	12.25	11.65	11.08	10.53	9.95	9.33	8.61	7.69	5.99
2	15.53	13.48	12.25	11.25	10.36	9.50	8.63	7.69	6.58	4.74
1	21.62	15.53	12.92	11.09	9.62	8.34	7.15	5.99	4.74	3.00

B.2 Beta (type 1) Bayesian (high density)

		Confidence Intervals							Confidence Intervals				
a	b	0.90		0.95		0.99		a	b	0.90		0.95		0.99
1	1	(Flat distribution)						4	1	0.062	1.000	0.473	1.000	0.316	1.000
	2	0.000	0.684	0.000	0.776	0.000	0.900		2	0.395	0.957	0.330	0.974	0.220	0.991
	3	0.000	0.536	0.000	0.632	0.000	0.785		3	0.289	0.862	0.239	0.895	0.156	0.943
	4	0.000	0.438	0.000	0.527	0.000	0.684		4	0.235	0.775	0.184	0.816	0.118	0.882
	5	0.000	0.369	0.000	0.451	0.000	0.602		5	0.184	0.701	0.149	0.746	0.093	0.823
	6	0.000	0.319	0.000	0.393	0.000	0.536		6	0.155	0.639	0.124	0.685	0.077	0.768
	7	0.000	0.280	0.000	0.348	0.000	0.482		7	0.134	0.586	0.107	0.633	0.065	0.718
	8	0.000	0.250	0.000	0.312	0.000	0.438		8	0.117	0.542	0.093	0.588	0.056	0.674
	9	0.000	0.226	0.000	0.283	0.000	0.401		9	0.105	0.503	0.083	0.548	0.050	0.634
	11	0.000	0.189	0.000	0.238	0.000	0.342		11	0.086	0.440	0.068	0.483	0.040	0.567
	21	0.000	0.104	0.000	0.133	0.000	0.197		21	0.056	0.270	0.035	0.302	0.020	0.366
	51	0.000	0.044	0.000	0.057	0.000	0.086		51	0.018	0.125	0.014	0.141	0.008	0.176
2	1	0.316	1.000	0.224	1.000	0.100	1.000	5	1	0.631	1.000	0.549	1.000	0.398	1.000
	2	0.135	0.865	0.094	0.906	0.041	0.959		2	0.475	0.970	0.409	0.982	0.292	0.995
	3	0.068	0.712	0.044	0.772	0.016	0.867		3	0.368	0.892	0.315	0.919	0.223	0.958
	4	0.043	0.605	0.026	0.670	0.009	0.780		4	0.299	0.816	0.254	0.851	0.177	0.907
	5	0.030	0.525	0.018	0.591	0.005	0.708		5	0.251	0.749	0.212	0.788	0.146	0.854
	6	0.023	0.464	0.013	0.527	0.004	0.647		6	0.216	0.690	0.182	0.732	0.124	0.804
	7	0.019	0.416	0.011	0.476	0.003	0.593		7	0.190	0.639	0.159	0.682	0.107	0.758
	8	0.015	0.376	0.009	0.433	0.002	0.548		8	0.169	0.595	0.141	0.638	0.094	0.716
	9	0.013	0.343	0.007	0.390	0.002	0.508		9	0.152	0.556	0.126	0.599	0.084	0.677
	11	0.010	0.293	0.006	0.341	0.001	0.443		11	0.127	0.492	0.105	0.533	0.069	0.611
	21	0.005	0.168	0.002	0.199	0.001	0.268		21	0.069	0.111	0.056	0.342	0.036	0.405
	51	0.002	0.074	0.001	0.088	0.000	0.122		51	0.029	0.147	0.024	0.164	0.015	0.200

3	1	0.464	1.000	0.368	1.000	0.215	1.000		6	1	0.681	1.000	0.607	1.000	0.464	1.000
	2	0.288	0.932	0.228	0.956	0.133	0.984			2	0.536	0.977	0.473	0.987	0.353	0.996
	3	0.189	0.811	0.147	0.853	0.083	0.917			3	0.432	0.912	0.379	0.935	0.283	0.967
	4	0.138	0.711	0.105	0.761	0.057	0.844			4	0.361	0.845	0.315	0.876	0.232	0.923
	5	0.108	0.632	0.081	0.685	0.042	0.777			5	0.310	0.784	0.268	0.818	0.196	0.876
	6	0.088	0.568	0.065	0.621	0.033	0.717			6	0.271	0.729	0.234	0.766	0.169	0.831
	7	0.074	0.515	0.054	0.568	0.027	0.665			7	0.241	0.680	0.207	0.719	0.149	0.788
	8	0.064	0.471	0.046	0.522	0.023	0.619			8	0.217	0.637	0.186	0.677	0.133	0.748
	9	0.056	0.434	0.041	0.484	0.020	0.579			9	0.197	0.599	0.168	0.639	0.120	0.711
	11	0.045	0.375	0.032	0.421	0.015	0.511			11	0.167	0.535	0.142	0.574	0.100	0.646
	21	0.022	0.223	0.016	0.255	0.007	0.320			21	0.094	0.346	0.079	0.377	0.054	0.439
	51	0.009	0.100	0.006	0.116	0.003	0.150			51	0.040	0.168	0.034	0.185	0.023	0.222
7	1	0.720	1.000	0.652	1.000	0.518	1.000		11	1	0.811	1.000	0.762	1.0000	0.658	1.000
	2	0.584	0.981	0.524	0.989	0.407	0.997			2	0.707	0.990	0.659	0.984	0.557	0.999
	3	0.485	0.926	0.432	0.946	0.335	0.973			3	0.625	0.955	0.579	0.968	0.489	0.985
	4	0.414	0.866	0.367	0.893	0.282	0.935			4	0.560	0.914	0.517	0.932	0.433	0.960
	5	0.361	0.810	0.318	0.841	0.242	0.893			5	0.508	0.873	0.467	0.895	0.389	0.931
	6	0.320	0.759	0.281	0.793	0.212	0.851			6	0.465	0.833	0.426	0.858	0.354	0.900
	7	0.287	0.713	0.251	0.749	0.189	0.811			7	0.429	0.796	0.392	0.823	0.324	0.870
	8	0.260	0.672	0.227	0.708	0.170	0.774			8	0.398	0.762	0.363	0.790	0.299	0.840
	9	0.238	0.635	0.208	0.672	0.154	0.739			9	0.372	0.730	0.339	0.759	0.277	0.811
	11	0.204	0.571	0.177	0.608	0.130	0.676			11	0.328	0.672	0.298	0.702	0.242	0.758
	21	0.118	0.378	0.101	0.409	0.073	0.469			21	0.207	0.478	0.186	0.507	0.149	0.562
	51	0.052	0.187	0.044	0.205	0.031	0.242			51	0.098	0.254	0.088	0.273	0.069	0.311

(continued)

a	b	Confidence Intervals							a	b	Confidence Intervals						
		0.90		0.95		0.99					0.90		0.95		0.99		
8	1	0.750	1.000	0.688	1.000	0.562	1.000	21	1	0.896	1.000	0.867	1.000	0.803	1.000
	2	0.624	0.985	0.567	0.991	0.452	0.998		2	0.832	0.995	0.801	0.998	0.732	0.999
	3	0.529	0.936	0.478	0.954	0.381	0.977		3	0.777	0.978	0.745	0.984	0.680	0.993
	4	0.458	0.883	0.412	0.907	0.326	0.944		4	0.730	0.955	0.698	0.965	0.634	0.980
	5	0.405	0.831	0.362	0.859	0.284	0.906		5	0.689	0.931	0.658	0.944	0.595	0.964
	6	0.363	0.783	0.323	0.814	0.252	0.867		6	0.654	0.906	0.623	0.921	0.561	0.946
	7	0.328	0.740	0.292	0.773	0.226	0.830		7	0.622	0.882	0.591	0.899	0.531	0.927
	8	0.300	0.700	0.266	0.734	0.205	0.795		8	0.593	0.859	0.563	0.877	0.504	0.908
	9	0.276	0.664	0.244	0.699	0.188	0.762		9	0.567	0.836	0.538	0.855	0.480	0.889
	11	0.238	0.602	0.210	0.637	0.160	0.701		11	0.522	0.793	0.493	0.814	0.438	0.851
	21	0.141	0.407	0.123	0.437	0.092	0.496		21	0.374	0.626	0.351	0.649	0.308	0.692
	51	0.064	0.205	0.055	0.223	0.040	0.261		51	0.204	0.378	0.190	0.397	0.164	0.433
9	1	0.774	1.000	0.717	1.000	0.599	1.000	51	1	0.956	1.000	0.943	1.000	0.914	1.000
	2	0.657	0.987	0.602	0.993	0.492	0.998		2	0.926	0.998	0.912	0.999	0.878	1.000
	3	0.566	0.944	0.516	0.959	0.421	0.980		3	0.900	0.991	0.884	0.994	0.850	0.997
	4	0.497	0.895	0.452	0.917	0.366	0.950		4	0.875	0.982	0.859	0.986	0.824	0.992
	5	0.444	0.848	0.401	0.874	0.323	0.916		5	0.853	0.971	0.836	0.976	0.800	0.985
	6	0.401	0.803	0.361	0.832	0.289	0.880		6	0.832	0.960	0.815	0.966	0.778	0.977
	7	0.365	0.762	0.328	0.792	0.261	0.846		7	0.813	0.948	0.795	0.956	0.758	0.969
	8	0.336	0.724	0.301	0.756	0.238	0.812		8	0.795	0.936	0.777	0.945	0.739	0.960
	9	0.311	0.689	0.278	0.722	0.219	0.781		9	0.778	0.925	0.759	0.934	0.722	0.950
	11	0.270	0.628	0.241	0.661	0.189	0.723		11	0.746	0.902	0.727	0.912	0.689	0.931
	21	0.164	0.433	0.145	0.462	0.111	0.520		21	0.622	0.796	0.603	0.810	0.567	0.836
	51	0.075	0.222	0.066	0.241	0.050	0.278		51	0.419	0.581	0.404	0.596	0.375	0.625

Appendix B: Tables of statistical distributions

B.3 Cumulative student t-distribution table

df	$t_{0.995}$	$t_{0.99}$	$t_{0.975}$	$t_{0.95}$	$t_{0.90}$	$t_{0.80}$	$t_{0.75}$	$t_{0.70}$	$t_{0.60}$
1	63.66	31.82	12.71	6.31	3.08	1.376	1.000	0.727	0.325
2	9.92	6.96	4.30	2.92	1.89	1.061	0.816	0.617	0.289
3	5.84	4.54	3.18	2.35	1.64	0.978	0.765	0.584	0.277
4	4.60	3.75	2.78	2.13	1.53	0.941	0.741	0.569	0.271
5	4.03	3.36	2.57	2.02	1.48	0.920	0.727	0.559	0.267
6	3.71	3.14	2.45	1.94	1.44	.0906	0.718	0.558	0.265
7	3.50	3.00	2.36	1.90	1.42	0.896	0.711	0.549	0.263
8	3.36	2.90	2.31	1.86	1.40	0.889	0.706	0.546	0.262
9	3.25	2.82	2.26	1.83	1.38	0.883	0.703	0.543	0.261
10	3.17	2.76	2.28	1.81	1.37	0.879	0.700	0.542	0.260
11	3.11	2.72	2.20	1.80	1.36	0.876	0.697	0.540	0.260
12	3.06	2.68	2.18	1.78	1.35	0.873	0.695	0.539	0.259
13	3.01	2.65	2.16	1.77	1.35	0.870	0.694	0.538	0.259
14	2.98	2.62	2.14	1.76	1.34	.0868	0.692	0.537	0.258
15	2.95	2.60	2.13	1.75	1.34	0.866	0.691	0.536	0.258
16	2.92	2.58	2.12	1.75	1.34	0.865	0.690	0.535	0.258
17	2.90	2.57	2.11	1.74	1.33	0.863	0.689	0.534	0.257
18	2.88	2.55	2.10	1.73	1.33	0.862	0.688	0.534	0.257
19	2.86	2.54	2.09	1.73	1.33	0.861	0.688	0.533	0.257
20	2.84	2.53	2.09	1.72	1.32	0.860	0.687	0.533	0.257
21	2.83	2.52	2.08	1.72	1.32	0.859	0.686	0.532	0.257
22	2.82	2.51	2.07	1.72	1.32	0.858	0.686	0.532	0.256
23	2.81	2.50	2.07	1.71	1.32	0.858	0.685	0.532	0.256
24	2.80	2.49	2.06	1.71	1.32	0.857	0.685	0.531	0.256
25	2.79	2.48	2.06	1.71	1.32	0.856	0.684	0.531	0.256
26	2.78	2.48	2.06	1.71	1.32	0.856	0.684	0.531	0.256
27	2.77	2.47	2.05	1.70	1.31	0.855	0.684	0.531	0.256
28	2.76	2.47	2.05	1.70	1.31	0.855	0.683	0.530	0.256
29	2.76	2.46	2.04	1.70	1.31	0.854	0.683	0.530	0.256
30	2.75	2.46	2.04	1.70	1.31	0.854	0.683	0.530	0.256
40	2.70	2.42	2.02	1.68	1.30	0.851	0.681	0.529	0.255
60	2.66	2.39	2.00	1.67	1.30	0.848	0.679	0.527	0.254
120	2.62	2.36	1.98	1.66	1.29	0.845	0.677	0.526	0.254
∞	2.68	2.33	1.96	1.645	1.28	0.842	0.674	0.524	0.253

B.4 Cumulative chi-square table

Percentage Points, Chi-Square Distribution

$$F(\chi^2) = \int_0^{\chi^2} \frac{1}{2^{n/2}\Gamma(n/2)} x^{(n-2)/2} e^{-x/2} \, dx$$

df = n	0.100	0.250	0.500	0.750	0.900	0.950	0.975	0.990	0.995
1	0.16	0.10	0.46	1.32	2.71	3.84	5.02	6.63	7.9
2	0.21	0.58	1.39	2.77	4.61	5.99	7.38	9.21	10.6
3	0.58	1.21	2.37	4.11	6.25	7.81	9.35	11.3	12.8
4	1.06	1.92	3.36	5.39	7.78	9.49	11.1	13.3	14.9
5	1.61	2.67	4.35	6.63	9.24	11.1	12.8	15.1	16.7
6	2.20	3.45	5.35	7.84	10.6	12.6	14.4	16.8	18.5
7	2.83	4.25	6.35	9.04	12.0	14.1	16.0	18.5	20.3
8	3.49	5.07	7.34	10.2	13.4	15.5	17.5	20.1	22.0
9	4.17	5.90	8.34	11.4	14.7	16.9	19.0	21.7	23.6
10	4.87	6.74	9.34	12.5	16.0	18.3	20.5	23.2	25.2
11	5.58	7.58	10.3	13.7	17.3	19.7	21.9	24.7	26.8
12	6.30	8.44	11.3	14.8	18.5	21.0	23.3	26.2	28.3
13	7.04	9.30	12.3	16.0	19.8	22.4	24.7	27.7	29.8
14	7.79	10.2	13.3	17.1	21.1	23.7	26.1	29.1	31.3
15	8.55	11.0	14.3	18.2	22.3	25.0	27.5	30.6	32.8
16	9.31	11.9	15.3	19.4	23.5	26.3	28.8	32.0	34.3
17	10.1	12.8	16.3	20.5	24.8	27.6	30.2	33.4	35.7
18	10.9	13.7	17.3	21.6	26.0	28.9	31.5	34.8	37.2
19	11.7	14.6	18.3	22.7	27.2	30.1	32.9	36.2	38.6
20	12.4	15.5	19.3	23.8	28.4	31.4	34.2	37.6	40.0
21	13.2	16.3	20.3	24.9	29.6	32.7	35.5	38.9	41.4
22	14.0	17.2	21.3	26.0	30.8	33.9	36.8	40.3	42.8
23	14.8	18.1	22.3	27.1	32.0	35.2	38.1	41.6	44.2
24	15.7	19.0	23.3	28.2	33.2	36.4	39.4	43.0	45.6
25	16.5	19.9	24.3	29.3	34.4	37.7	40.6	44.3	46.0
26	17.3	20.8	25.3	30.4	35.6	38.9	41.9	45.6	48.3
27	18.1	21.7	26.3	31.5	36.7	40.1	43.2	47.0	49.6
28	18.9	22.7	27.3	32.6	37.9	41.3	44.5	48.3	51.0
29	19.8	23.6	28.3	33.7	39.1	42.6	45.7	49.6	52.3
30	20.6	24.5	29.3	34.8	40.3	43.8	47.0	50.9	53.7

Appendix B: Tables of statistical distributions

B.5 F Distribution table

$F_{\alpha;V_1,V_2}$ values are shown below where the first column corresponds to the value of V_2 the second to $1-\alpha$ and the remaining columns to values of V_1. Note that $F_{\alpha;V_1,V_2} = 1/F_{1-\alpha;V_2,V_1}$

V_2	$1-\alpha$	1	2	3	4	5...	10...	15...	20...	30...	60...	120...	∞
1	0.90	39.9	49.5	53.6	55.8	57.2	60.2	61.2	61.7	62.3	62.8	63.1	63.3
	0.95	161	200	216	225	230	242	246	248	250	252	253	254
	0.99	4050	5000	5400	5620	5760	6060	6160	6210	6260	6310	6340	6370
2	0.90	8.53	9.00	9.16	9.24	9.29	9.39	9.42	9.44	9.46	9.47	9.48	9.49
	0.95	18.5	19.0	19.2	19.2	19.3	19.4	19.4	19.5	19.5	19.5	19.5	19.5
	0.99	98.5	99.0	99.2	99.2	99.3	99.4	99.4	99.4	99.5	99.5	99.5	99.5
3	0.90	5.54	5.46	5.39	5.34	5.31	5.23	5.20	5.18	5.17	5.15	5.14	5.13
	0.95	10.1	9.55	9.28	9.12	9.01	8.79	8.70	8.66	8.62	8.57	8.55	8.53
	0.99	34.1	30.8	29.5	28.7	28.2	27.2	26.9	26.7	26.5	26.3	26.2	26.1
4	0.90	4.54	4.32	4.19	4.11	4.05	3.92	3.87	3.84	3.82	3.79	3.78	3.76
	0.95	7.71	6.94	6.59	6.39	6.26	5.96	5.86	5.80	5.75	5.69	5.66	5.63
	0.99	21.2	18.0	16.7	16.0	15.5	14.5	14.2	14.0	13.8	13.7	13.6	13.5
5	0.90	4.06	3.78	3.62	3.52	3.45	3.30	3.24	3.21	3.17	3.14	3.12	3.11
	0.95	6.61	5.79	5.41	5.19	5.05	4.74	4.62	4.56	4.50	4.43	4.40	4.37
	0.99	16.3	13.3	12.1	11.4	11.0	10.1	9.72	9.55	9.38	9.20	9.11	9.02
10	0.90	3.29	2.92	2.73	2.61	2.52	2.32	2.24	2.20	2.15	2.11	2.08	2.06
	0.95	4.96	4.10	3.71	3.48	3.33	2.98	2.84	2.77	2.70	2.62	2.58	2.54
	0.99	10.0	7.56	6.55	5.99	5.64	4.85	4.56	4.41	4.25	4.08	4.00	3.91
15	0.90	3.07	2.70	2.49	2.36	2.27	2.06	1.97	1.92	1.87	1.82	1.79	1.76
	0.95	4.54	3.68	3.29	3.06	2.90	2.54	2.40	2.33	2.25	2.16	2.11	2.07
	0.99	8.68	6.36	5.42	4.89	4.56	3.80	3.52	3.37	3.21	3.05	2.96	2.87
20	0.90	2.97	2.59	2.38	2.25	2.16	1.94	1.84	1.79	1.74	1.68	1.64	1.61
	0.95	4.35	3.49	3.10	2.87	2.71	2.35	2.20	2.12	2.04	1.95	1.90	1.84
	0.99	8.10	5.85	4.94	4.43	4.10	3.37	3.09	2.94	2.78	2.61	2.52	2.42
30	0.90	2.88	2.49	2.28	2.14	2.05	1.82	1.72	1.67	1.61	1.54	1.50	1.46
	0.95	4.17	3.32	2.92	2.69	2.53	2.16	2.01	1.93	1.84	1.74	1.68	1.62
	0.99	7.56	5.39	4.51	4.02	3.70	2.98	2.70	2.55	2.39	2.21	2.11	2.01
60	0.90	2.79	2.39	2.18	2.04	1.95	1.71	1.60	1.54	1.48	1.40	1.35	1.29
	0.95	4.00	3.15	2.76	2.53	2.37	1.99	1.84	1.75	1.65	1.53	1.47	1.39
	0.99	7.08	4.98	4.13	3.65	3.34	2.63	2.35	2.20	2.03	1.84	1.73	1.60
120	0.90	2.75	2.35	2.13	1.99	1.90	1.65	1.54	1.48	1.41	1.32	1.26	1.19
	0.95	3.92	3.07	2.68	2.45	2.29	1.91	1.75	1.66	1.55	1.43	1.35	1.25
	0.99	6.85	4.79	3.95	3.48	3.17	2.47	2.19	2.03	1.86	1.66	1.53	1.38
∞	0.90	2.71	2.30	2.08	1.94	1.85	1.60	1.49	1.42	1.34	1.24	1.17	1.00
	0.95	3.84	3.00	2.60	2.37	2.21	1.83	1.67	1.57	1.46	1.32	1.22	1.00
	0.99	6.63	4.61	3.78	3.32	3.02	2.32	2.04	1.88	1.70	1.47	1.32	1.00

References

Abdel-Malek, K., Yang, J., Kim, J., Marler, T., Beck, S., and Nebel, K. (2004) Santos: A virtual human environment for human factors assessment. Paper presented at the *24th Army Science Conference: Transformational Science and Technology for the Current and Future Force*, Orlando, FL, November 29–December 2, 2004.

ACGIH. (1971) Notice of intent to establish threshold limit values—Heat stress. In *Threshold Limit Values of Airborne Contaminants and Physical Agents*, American Conference of Governmental Industrial Hygienists, Cincinnati, OH, pp. 71–81.

Adams, J. S. (1963) Wage inequities, productivity, and work quality, *Ind Relat*, 3, 9–16.

Adrian, W. (2003) The effect of observation time and contrast on visual acuity, *Clin Exp Opt*, 86(3), 179–182.

Adrian, E. D. and Bronk, D. W. (1929) The discharge of impulses in motor nerve fibers: Part II. The frequency of discharge in reflex and voluntary contractions, *J Physiol*, 67, 119–151.

Aft, L. S. (1997) The need for work measurement: Some observations on the current state of affairs in the business world, *IIE Solut*, 29, 16–19.

Aghazadeh, F. and Waly, S. M. (1998) A design and selection guide for hand held tools. In Karwowski, W. and Salvendy, G. (Eds.) *Ergonomics for Plant Managers*. Society of Manufacturing Engineers, Dearborn, MI, pp. 65–86.

Agur, A. M. R. (1991) *Grant's Atlas of Anatomy*, 9th edn. Williams & Wilkins, Baltimore, MD.

Ainslie, G. (1975) Specious reward: A behavioral theory of impulsiveness and impulse control, *Psychol Bull*, 82, 463–509.

Alexander, D. C. (1986) *The Practice and Management of Industrial Ergonomics*. Prentice Hall, Englewood Cliffs, NJ.

Aliev, R. A., Fazlollahi, B., and Vahidov, R. M. (2000) Soft computing based multi-agent marketing decision support system, *J Intell Fuzzy Syst*, 9(1–2), 1–9.

Allais, M. (1953) Le comportement de l'homme rationel devant le risque: Critique des postulateset axioms de l'ecole americaine, *Econometrica*, 21, 503–546.

Allen, T. J. and Cohen, D. I. (1969) Information flow in research and development laboratories, *Admin Sci Q*, 14, 12–25.

Allen, D. and Corn, R. (1978) Comparing productive activities as different times, *Ind Eng*, 13(3), 40–43.

Altman, J. W. (1964) A central store of human performance data, *Proceedings of Symposium on Quantification of Human Performance*, Albuquerque, NM, August 17–19, 1964, pp. 97–108.

Altman, J. W. (1967) Classification of human error. In Askgren, W. B. (Ed.) *Symposium on Reliability of Human Performance in Work*. Aerospace Medical Labs, Technical Report No. 67-88, Wright-Patterson Air Force Base, OH, May.

Altman, J. W., Marchese, A. C., and Marchiando, B. W. (1961) Guide to design of mechanical equipment for maintainability, Report ASD-61-381. U.S. Air Force, Air Force Systems Command, Wright-Patterson Air Force Base, OH.

American Conference of Governmental Industrial Hygienists. (2004) *Industrial Ventilation: A Manual of Recommended Practice*, 25th edn. American Conference of Governmental Industrial Hygienists, Cincinnati, OH.

American National Standards Institute. (2007) *Human Factors Engineering of Computer Workstations* (Standard # ANSI/HFES 100-2007). American National Standards Institute, Washington, DC.

Anderson, N. H. (1981) *Foundations of Information Integration Theory*. Academic Press, New York.

Anderson, J. R. (1983) *The Architecture of Cognition*. Harvard University Press, Cambridge, MA.

Anderson, D. (1990) *Design for Manufacturability*. CIM Press, Cambria, CA.

Andre, A. D. and Segal, L. D. (1994, January) Design functions, *Ergon Des*, 1, 5–7.

Andress, F. (1954) The learning curve as a production tool. *Harv Bus Rev*, 32, 87–97.

Ang, C. H. and Gray, R. (1993) IDEF modeling for project risk assessment, *Comput Ind Eng*, 22, 3–45.

Ankrum, D. R. and Nemeth, K. J. (1995, April) Posture, comfort, and monitor placement, *Ergonomics in Design*, 3, 7–9.

Annett, J. and Duncan, K. D. (1967) Task analysis and training design, *Occup Psychol*, 41, 211–221.

Annis, J. (1996) Men and women: Anthropometric comparisons for the ergonomist. In Mital, A., Kumar, S., Menozzi, M., and Fernandez, J. (Eds.) *Advances in Occupational Ergonomics and Safety*. International Society for Occupational Ergonomics and Safety, Cincinnati, OH, pp. 60–65.

Anoorineimi, A. and Lehto, M. R. (2010) Simulation of an auto transport facility, Unpublished technical report, Purdue University Technical Assistance Program.

Anyakora, S. N. and Lees, F. P. (1972) Detection of instrument malfunction by process operator, *Chem Eng*, 264, 304–309.

Archer, J. T. (1975) Achieving joint organizational, technical, and personal needs: The case of the sheltered experiments of aluminum casting team. In Davis, E. and Cherns, A. B. (Eds.) *The Quality of Working Life: Cases and Commentary*. Free Press, New York.

Argote, L. (1982) Input uncertainty and organizational coordination in hospital emergency units, *Admin Sci Q*, 27, 420–434.

Argote, L., Seabright, M. A., and Dyer, L. (1986) Individual versus group: Use of base-rate and individuating information, *Organ Behav Hum Dec*, 38, 65–75.

Arkes, H. R. and Blumer, C. (1985) The psychology of sunk cost, *Organ Behav Hum Dec*, 35, 124–140.

Arkes, H. R. and Hutzel, L. (2000) The role of probability of success estimates in the sunk cost effect. *J Behav Dec*, 13(3), 295–306.

Arzi, Y. and Shatub, A. (1996) Learning and forgetting in high cognitive and low cognitive tasks. In Ozok, A. F. and Salvendy, G. (Eds.) *Advance in Applied Ergonomics—Proceeding of the First International Conference (ICAE'96)*, Istanbul, Turkey, Ozok, A. F. and Salvendy, G. (Eds.). USA Publishing, Istanbul-West Lafayette, IN, pp. 672–675.

Asfahl, C. R. (1990) *Industrial Safety and Health Management*, 2nd edn. Prentice Hall, Englewood Cliffs, NJ.

Asfahl, C. R. (2004) *Industrial Safety and Health Management*, 5th edn. Prentice Hall, Englewood Cliffs, NJ.

ASHRAE. (1977) *ASHRAE Handbook and Product Directory*. American Society of Heating, Refrigerating and Air Conditioning Engineers, New York.

Asmussen, E. and Heeboll-Nielsen, K. (1962) Isometric muscle strength in relation to age in men and women, *Ergonomics*, 5, 167–169.

Astrand, P. O. and Rodahl, K. (1977) *Textbook of Work Physiology. Physiological Bases of Exercise*, 2nd edn. McGraw-Hill, New York.

Aume, N. M. (1963) The effect of reach distance on reach time in a button pushing task. AMRL Memo. Systems Research Branch. Human Engineering Division, Behavioral Sciences Laboratory Aerospace Medical Division, Wright-Patterson Air Force Base, OH, p. 58.

Ayoub, M. M. and El-Bassoussi, M. M. (1978) Dynamic biomechanical model for sagittal plane lifting activities. In Drury, C. G. (Ed.) *Safety in Manual Materials Handling*. DHEW (NIOSH), Cincinnati, OH, pp. 88–95.

Badarinath, N. B. (1985) Development of cognitive task metrics. Unpublished master's thesis, The University of Iowa, Iowa City, IAO.

Baggett, P. (1992) Putting it together with cognitive engineering. In Helander, M. and Nagtamachi, M. (Eds.) and the International Ergonomics Society, *Design for Manufacturability—A Systems Approach to Concurrent Engineering and Ergonomics*. Taylor & Francis, Ltd., London, U.K., pp. 160–170.

Bailey, G. B. (1961) Comments on An experimental evaluation of the validity of predetermined time studies, *J Ind Eng*, 12(5), 328–330.
Bailey, R. W. (1989) *Human Performance Engineering*. Prentice Hall, Englewood Cliffs, NJ.
Bainbridge, L. (1990) Verbal protocol analysis. In Wilson, J. R. and Corlett, E. N. (Eds.) *Evaluation of Human Work: A Practical Ergonomics Methodology*. Taylor & Francis, London, U.K., pp. 161–179, Chapter 6.
Balzer, W. K., Doherty, M. E., and O'Connor, R. O., Jr. (1989) Effects of cognitive feedback on performance, *Psychol Bull*, 106, 41–433.
Bar-Hillel, M. (1973) On the subjective probability of compound events, *Organ Behav Hum Perform*, 9, 396–406.
Barlett, N. W. and Smith, L. A. (1973) Design of control and display panels using computer algorithms, *Hum Factors*, 15(1), 1–7.
Barlow, R. E. and Lambert, H. E. (1975) *Introduction to Fault Tree Analysis*. SIAM, Philadelphia, PA.
Barnes, R. M. (1937) *Motion and Time Study*. John Wiley & Sons, New York.
Barnes, R. M. (1963) *Motion and Time Study: Design and Measurement of Work*, 7th edn. John Wiley & Sons, New York.
Barnes, R. M. (1968) *Motion and Time Study: Design and Measurement of Work*, 6th edn. John Wiley & Sons, New York.
Barnes, R. M. (1980) An improved professional product resulting from development work. Report to Eastman Kodak Company. In Barnes, R. M. (Ed.) *Motion and Time Study Design and Measurement of Work*, 7th edn. John Wiley & Sons,. New York, pp. 113–115.
Barnes, R. and Amrine, H. (1942) The effect of practice on various elements used in screw-driver work, *J Appl Psychol*, 26, 197–209.
Baron, J. (1985) *Rationality and Intelligence*. Cambridge University Press, Cambridge, U.K.
Baron, J. (1998) *Judgment Misguided: Intuition and Error in Public Decision Making*. Oxford University Press, New York.
Baron, S., Kleinman, D. L., and Levison, W. H. (1970) An optimal control model of human response (Part I and II), *Automatica*, 6, 357–383.
Barrick, M. R. and Alexander, R. A. (1987) A review of quality circle efficacy and the existence of the positive-findings bias, *Personnel Psychol*, 40, 579–591.
Basak, I. and Saaty, T. (1993) Group decision making using the analytic hierarchy process, *Math Comput Model*, 17, 101–109.
Bazermen, M. (1998) *Judgment in Managerial Decision Making*, 4th edn. John Wiley & Sons, New York.
Bazerman, M. H. and Neale, M. A. (1983) Heuristics in negotiation: Limitations to effective dispute resolution. In Bazerman, M. H. and Lewicki, R. (Eds.) *Negotiating in Organizations*, Sage, Beverly Hills, CA.
Beach, L. R. (1990) *Image Theory: Decision Making in Personal and Organizational Contexts*. John Wiley & Sons, Chichester, U.K.
Beach, L. R. (1993) Four revolutions in behavioral decision theory. In Chemers, M. M. and Ayman, R. (Eds.) *Leadership Theory and Research*. Academic Press, San Diego, CA.
Beck, D. J., Chaffin, D. B., and Joseph, B. S. (1993) Lightening the load, *Ergonomics in Design*, 4, 22–25.
Bedford, T. (1953) Thermal factors in the environment which influence fatigue. In Floyd, W. F. and Welford, A. T. (Eds.) *Symposium of Fatigue*. H. R. Lewis Co, London, U.K.
Belbin, R. M., Aston, B. R., and Mottram, R. D. (1976) Building effective management teams, *J Gen Manage*, 3(3), 23–29.
Belding, H. S. and Hatch, T. F. (1955) Index for evaluating heat stress in terms of resulting physiological strains, *Heat Piping Air Cond*, 27, 129–136.
Belecheanu, R., Pawar, K. S., Barson, R. J., and Bredehorst, B. (2003) The application of case-based reasoning to decision support in new product development, *Integrated Manuf Syst*, 14(1), 36–45.
Bell, D. E. (1982) Regret in decision making under uncertainty. *Oper Res*, 30, 961–981.
Benge, E. J., Burk, S. L. H., and Hay, E. N. (1941) *Manual of Job Evaluation*. Harper & Row, New York.
Berliner, D. C., Angell, D., and Shearer, J. W. (1964) Behaviors, measures, and instruments for performance evaluation in simulated environments. Paper presented at the *Symposium and Workshop on the Quantification of Human Performance*, Albuquerque, NM.

Bernoulli, D. (1954) Exposition of a new theory of the measurement of risk. L. Sommer (trans.), *Econometrica*, 22, 23–36. (Original work published 1738.)

Beshir, M. Y., Ramsey, J. D., and Burford, C. L. (1982) Threshold values for the Botsball: A field study of occupational heat, *Ergonomics*, 25(3), 247–254.

Best's Safety & Security Directory Oldwick, A.M. Best Co, NJ.

Bevis, F. W., Finniear, C., and Towill, D. R. (1970) Prediction of operator performance during learning of repetitive tasks, *Int J Prod Res*, 8, 293–305.

Biel, W. C. (1962) Training programs and devices. In Gagné, R. M. (Ed.) Psychological *Principles in System Development*. Holt, Rinehart and Winston, New York, Chapter 10.

Birnbaum, M. H., Coffey, G., Mellers, B. A., and Weiss, R. (1992) Utility measurement: Configural weight theory and the judge's point of view. *J Exp Psychol [Hum Percept]*, 18, 331–346.

Blackwell, H. R. (1959) Development and use of a quantitative method for specification of interior illuminating levels on the basis of performance data, *Illum Eng*, 54, 317–353.

Blackwell, H. R. (1970) Development of procedures and instruments for visual task evaluation, *Illum Eng*, 65, 267–291.

Blixt, S. L. and Shama, D. D. (1986) An empirical investigation of the standard error of measurement at different ability levels, *Educ Psychol Meas*, 46, 545–550.

Boehm, B. (1988) A spiral model of software development and enhancement, *IEEE Comput*, 21(5), 61–72.

Boff, K. R. and Lincoln, J. E. (Eds.) (1988) *Engineering Data Compendium: Human Perceptions and Performance*. Armstrong Aerospace Medical Research Laboratory, Wright-Patterson Air Force Base, OH.

Bohen, G. A. and Barany, J. W. (1978) Predicting learning curves and the cost of training for individual operators performing self-paced assembly operations. School of Industrial Engineering, Purdue University, West Lafayette, IN.

Bolton, G. E. and Chatterjee, K. (1996) Coalition formation, communication, and coordination: An exploratory experiment. In Zeckhauser, R. J. (Ed.) *Wise Choices: Decisions, Games, and Negotiations*.

Bond, N. A. (1987) Maintainability. In Salvendy, G. (Ed.) *Handbook of Industrial Engineering*. John Wiley & Sons, New York, pp. 1328–1356.

Boothroyd, G. and Dewhurst, P. (1987) *Product Design for Assembly*. Boothroyd Dewhurst, Inc., Wakefield, RI.

Boothroyd, G., Dewhurst, P., and Knight, W. (1994) *Product Design for Manufacturing and Assembly*. Marcel Dekker, Inc., New York.

Borg, G. A. V. (1962) *Physical Performance and Perceived Exertion*. Gleerups, Lund, Sweden.

Borg, G. A. V. (1982) Psychophysical bases of perceived exertion, *Med Sci Sports Exer*, 14(5), 377–381.

Bornstein, G. and Yaniv, I. (1998) Individual and group behavior in the ultimatum game: Are groups more rational players? *Exp Econ*, 1, 101–108.

Bracken, J. (1966) Percentage points of the beta distribution for use in Bayesian analysis of Bernoulli processes, *Technometrics*, 8, 687–694.

Brainbridge, L. (1990) Verbal protocol analysis. In Wilson, J. R. and Corlett, E. N. (Eds.) *Evaluation of Human Work: A Practical Ergonomics Methodology*. Taylor & Francis, London, U.K., Chapter 6, pp. 161–179.

Brandt, J. R. (1959) Influence of air conditioning on work performance, *Res. Div. Tech. Bull.*, 161(January 12), 1–9.

Brannick, M. T., Roach, R. M., and Salas, E. (1993) Understanding team performance: A multimethod study, *Hum Perform*, 6, 287–308.

Brashers, D. E., Adkins, M., and Meyers, R. A. (1994) Argumentation and computer-mediated group decision making. In Frey, L. R. (Ed.) *Group Communication in Context*, Lawrence Erlbaum Associates, Hillsdale, NJ, p. 12.

Brauer, R. L. (2006) *Safety and Health for Engineers*, 2nd edn. John Wiley & Sons, Hoboken, NJ.

Brehmer, B. (1981) Models of diagnostic judgment. In Rasmussen, J. and Rouse, W. (Eds.) *Human Detection and Diagnosis of System Failures*. Plenum Press, New York.

Brehmer, B. and Joyce, C. R. B. (1988) *Human Judgment: The SJT View*. North-Holland, Amsterdam, the Netherlands.

Bridger, R. S. (2003) *Introduction to Ergonomics*, 2nd edn. Lawrence Erlbaum Associates, Hoboken, NJ.

Brookhouse, J. K., Guion, R. M., and Doherty, M. E. (1986) Social desirability response bias as one source of the discrepancy between subjective weights and regression weights, *Organ Behav Hum Dec*, 37, 316–328.

Brown, J. S. and Slater-Hammel, A. T. (1949) Discrete movements in the horizontal plane as a function of their length and direction, *J Exp Psychol*, 38, 84–95.

Brumbaugh, P. S. and Heikes, R. G. (1992) Statistical quality control. In Salvendy, G. (Ed.) *Handbook of Industrial Engineering*, 2nd edn. John Wiley & Sons, New York, Chapter 87.

Brunswick, E. (1952) *The Conceptual Framework of Psychology*. University of Chicago Press, Chicago, IL.

Buck, J. R. (1977) New strategies in ergonomic research and design. *Proceedings of the 28th Conference of the American Institute of Industrial Engineers*, Dallas, TX, pp. 95–99.

Buck, J. R. (1989) *Economic Risk Decision in Engineering and Management*. Iowa State University Press, Ames, IA.

Buck, J. R. (1996) Ergonomic issues with learning and forgetting curves. In Ozok, A. F. and Salvendy, G. (Eds.) *Advance in Applied Ergonomics—Proceedings of the 1st. International Conference (ICAE'96)*, Istanbul, Turkey, A. F. and Salvendy, G. (Eds.) USA Publishing, Istanbul-West Lafayette, IN, pp. 689–694.

Buck, J. R., Badarinath, N. B., and Kachitvichyanukul, V. (1988) Cognitive task performance time and accuracy in supervisory process, *IIE Trans*, 20(2), 122–131.

Buck, J. R. and Cheng, S. W. J. (1993) Instructions and feedback effects on speed and accuracy with different learning curve functions. *IIE Trans*, 25(6), 34–47.

Buck, J. R., Deisenroth, M. P., and Alford, E. C. (1978) Man-in-the-loop simulation, *SIMULATION*, 30(5), 137–144.

Buck, J. R. and Ings, D. M. (1990) *Human Performance Times of Elementary Cognitive Tasks Concurrently with Tracking*. Dept. of Industrial & Engineering, The University of Iowa, Iowa City, IA.

Buck, J. R. and Lyu, J. (1992) Human performance in cognitive tasks in industry: Sequences and other stream effects on human performance. In Mattila, M. and Karwowski, W. (Eds.) *Computer Applications in Ergonomics, Occupational Safety and Health. Proceedings of the International Conference on Computer-Aided Ergonomics and Safety ('92-CAES)*, Tampere, Finland, May 18–20. North-Holland Publishing, Amsterdam, the Netherlands.

Buck, J. R. and Maltas, K. L. (1979) Simulation of industrial man-machine systems, *Ergonomics*, 22(7), 785–797.

Buck, J. R., Payne, D. R., and Barany, J. W. (1994) Human performance in actuating switches during tracking, *Int J Aviat Psychol*, 4(2), 119–140.

Buck, J. R., Pignietello, J. J., Jr., and Raz, T. (1995) Bayesian Poisson rate estimation using parameter maps, *IIE Trans*, 27, 405–412.

Buck, J. R. and Rizzi, A. M. (1974) Viewing strategy and window effects on dynamic visual inspection accuracy, *AIIE Transactions*, 6, 196–205.

Buck, J. R., Stoner, J. W., Bloomfield, J., and Plocher, T. (1993) Driving research and the Iowa driving simulator. In Lovesey, E. J. (Ed.) *Contemporary Ergonomics, Proceedings of the Ergonomics Society's 1993 Annual Conference*, Edinburgh, Scotland, April 13–16, 1993, pp. 392–396.

Buck, J. R., Sullivan, G. H., and Nelson, P. E. (1978) Operational impact and economic benefits of a new processing technology, *Eng Econ*, 23(2), 71–92.

Buck, J. R., Tanchoco, J. M. A., and Sweet, A. L. (1976) Parameter estimation methods for discrete exponential learning curves, *AIIE Trans*, 8(2), 184–194.

Buck, J. R. and Wang, J. H. (1994) Is the whole equal to the sum of its parts? *Proceedings of the International Ergonomics Association Conference*, Toronto, Ontario, Canada, p. 5.

Buck, J. R. and Wang, M. I. (1997) Simulating micromethods of information-seeking tasks. *Proceedings of the 13th Triennial Congress of the International Ergonomics Association*, Tampere, Finland, Vol. 3, pp. 28–30.

Buck, J. R. and Yenamendra, A. (1996) Ergonomic issues on entering the automated highway system. In Noy, I. (Ed.) *Human Factors in Intelligent Vehicle Highway Systems*. Lawrence Erlbaum Associates, Hillsdale, NJ, pp. 309–328.

Buck, J. R., Zellers, S. M., and Opar, M. E. (1998) *Bayesian Rated Sequential Occurrence Sampling*. Department of Industrial Engineering, The University of Iowa, Iowa, City, IAO.

Budescu, D. and Weiss, W. (1987) Reflection of transitive and intransitive preferences: A test of prospect theory, *Organ Behav Hum Perform*, 39, 184–202.

Bui, T. X., Jelassi, T. M., and Shakun, M. F. (1990) Group decision and negotiation support systems, *Eur J Oper Res*, 46(2), 141–142.

Burg, A. (1966) Visual acuity as measured by dynamic and static tests: A comparative evaluation. *J Appl Psychol*, 50, 460–466.

Burgess, J. H. (1986) *Designing for humans: The Human Factor in Engineering*. Petrocelli Books, Princeton, NJ.

Bursill, A. E. (1958) The restriction of peripheral vision during exposure to hot and humid conditions, *Q J Exp Psychol*, 10, 113–129.

Buti, L. B. (1995) Ergonomic industrial design principles for product ergonomics, *Ergonomia Prodotti, Lavoro, Ricerca*, (5 September), 13–17.

Campbell, D. J. (1988) Task complexity: A review and analysis, *Acad Manage Rev*, 13(1), 40–52.

Carbonell, J. R. (1966) A queueing model for many-instrument visual sampling, *IEEE Trans Hum Fact Electron*, 7, 157–164.

Carbonnell, J. F., Ward, J. L., and Senders, J. W. (1968) A queuing model of visual sampling: Experimental validation, *IEEE Trans Man Mach Syst*, 9, 82–87.

Card, S. K., English, W. K., and Burr, B. J. (1978) Evaluation of mouse, ratecontrolled isometric joystick, step keys, and text keys of text selection on a CRT, *Ergonomics*, 21, 601–613.

Card, S., Moran, T., and Newell, A. (1983) *The Psychology of Human-Computer Interaction*. Lawrence Erlbaum Associates, Hillsdale, NJ.

Card, S., Moran, T., and Newell, A. (1986) The model human processor: An engineering model of human performance. In Boff, K. R., Kaufman, L., and Thomas, J. P. (Eds.) *Handbook of Perceptional and Human Performance*, pp. 1–35.

Carlson, J. G. (1973) Cubic learning curves: Precision tool for estimating, *Manuf Eng Manage*, 71(5), 22–25.

Caverni, J. P., Fabre, J. M., and Gonzalez, M. (1990) *Cognitive Biases*. North-Holland, Amsterdam, the Netherlands.

Centreno, M. A. and Strandridge, C. R. (1993) Databases: Designing and developing integrated simulation modeling environments. *Proceedings of the 25th Conference on Winter Simulation*, Los Angeles, CA, December 12–15, 1993, pp. 526–534.

Chaffin, D. B. (1974) Human strength capability and low-back pain, *J Occup Med*, 16(4), 248–254.

Chapanis, A. (1959) *Research Techniques in Human Engineering*. The Johns Hopkins Press, Baltimore, MD.

Chapanis, A. (1961) Men, machines, and models. *Am Psychol*, 16(3), 113–131.

Chapanis, A. (1995) Ergonomics in product development: A personal view, *Ergonomics*, 38(8), 1625–1638.

Chapman, D. E. and Sinclair, M. A. (1975) Ergonomics in inspection tasks in the food industry. In Drury, C. G. and Fox, J. G. (Eds.) *Human Reliability in Quality Control*. Taylor & Francis, Ltd., London, U.K., pp. 231–252.

Cheng, P. E. and Holyoak, K. J. (1985) Pragmatic reasoning schemas, *Cogn Psych*, 17, 391–416.

Chhabra, S. L. and Ahluwalia, R. S. (1990) Rules and guidelines for ease of assembly. *Proceedings of the International Ergonomics Association Conference on Human Factors in Design for Manufacturability and Process Planning, Human Factors Society*, Santa Monica, CA, pp. 93–99.

Choobineh, F. (1986) FMS: A totally programmable manufacturing cell. In Choobineh, F. and Suri, R. (Eds.) *Conference Proceedings in Flexible Manufacturing System: Current Issues and Models*. Industrial Engineering and Management Press, Atlanta, GA.

Christensen, J. M. and Mills, R. G. (1967) What does the operator do in complex systems? *Hum Fact*, 4(9), 329–340.

Christensen-Szalanski, J. J. and Willham, C. F. (1991) The hindsight bias: A meta-analysis, *Organ Behav Hum Dec*, 48, 147–168.

Chubb, G. P. (1971) The use of Monte Carlo simulation to reflect the impact human factors can have on systems performance. *Proceedings of the 5th Conference on Winter Simulation*, New York, December 8–10, 1971, pp. 63–70.

Chubb, G. P., Stokolsky, N., Fleming, W. D., and Hassound, J. A. (1987) STALL: A simple model of workload analysis in early system development. *Proceedings of the 31st Annual Meeting of the Human Factors Society Meeting*, Santa Monica, CA.

CIE. (1971; 1988; 2004) *CIE Publication 15 Colorimetry.* Commission Internationale de l'Eclairage (CIE), Vienna, Austria.
Clauser, C. E., McConville, J. T., and Young, J. W. (1969) weight, volume, and center of mass of segments of the human body (Report # AMRL-TR-69-70). Aerospace Medical Research Laboratory, Wright-Patterson Air Force Base, OH.
Clemen, R. T. (1996) *Making Hard Decisions: An Introduction to Decision Analysis*, 2nd edn. Duxbury Press, Belmont, CA.
Cohen, M. S. (1993) The naturalistic basis of decision biases. In Klein, G. A., Orasanu, J., Calderwood, R., and Zsambok, E. (Eds.) *Decision Making in Action: Models and Methods.* Ablex, Norwood, NJ, pp. 51–99.
Collan, M. and Liu, S. (2003) Fuzzy logic and intelligent agents: Towards the next step of capital budgeting decision support, *Ind Manage Data Syst*, 103(6), 410–422.
Coma, O., Mascle, O., and Balazinski, M. (2004) Application of a fuzzy decision support system in a design for assembly methodology, *Int J Comput Integ Manuf*, 17(1), 83–94.
Compton, W. D., Dunlap, M. D., and Heim, J. A. (1992) Improving quality through the concept of learning curves. In Heim, J. A. and Compton, W. D. (Eds.) *Manufacturing Systems: Foundations of World-Class Practice.* National Academy Press, Washington, DC, pp. 100–106.
Condly, S. J., Clark, R. E., and Stolovitch, H. D. (2003) The effects of incentives on workplace performance: A meta-analytic review of research studies, *Perform Improv Q*, 16(3), 46–63
Conley, P. (1970) Experience curves as a planning tool, *IEEE Spect*, June 7, 63–68.
Connolly, T., Ordonez, L. D., and Coughlan, R. (1997) Regret and responsibility in the evaluation of decision outcomes, *Organ Behav Hum Dec*, 70, 73–85.
Conway, R. and Schultz, A. (1959) The manufacturing process function, *J Ind Eng*, 10, 39–53.
Cook, B. (1998) High-efficiency lighting in industry and commercial buildings, *IEEE Power Eng J*, 12(5), 197–206.
Cooper, J. (1961) Human initiated failures and malfunction reporting, *IRE Trans Hum Fact Electron*, 2, 104–109.
Corlett, E. N. (1988) Where have we come from and where are we going? In Megaw, E. D. (Ed.) *Contemporary Ergonomics.* Taylor & Francis, Ltd. London, U.K., pp. 2–18.
Cronbach, L. J. (1951) Coefficient alpha and the internal structure of tests. In Mehrens, W. A. and Ebel, R. L. (Eds.) *Principles of Educational and Psychological Measurement.* Rand McNally, Chicago, IL, pp. 132–167.
Crosbie, M. J. and Watson, D. (2004) *Time-Saver Standards for Architectural Design*, 8th edn. McGraw-Hill Professional, New York.
Crosby, P. B. (1979) *Quality is Free.* McGraw-Hill Book Co., Inc., New York.
Crossman, E. R. F. W. (1956) Perceptual activity in manual work, *Res J Sci Appl*, 9, 42–49.
Crossman, E. R. F. W. (1959) A theory of the acquisition of speed-skill, *Ergonomics*, 2(2), 153–166.
Crossman, E. R. F. W. and Cooke, J. E. (1962a) *Manual Control of Slow Response Systems.* International Congress on Human Factors in Electronics. Long Beach, CA.
Crossman, E. R. F. W. and Cooke, J. E. (1962b) Manual control of slow response systems. In Edwards, E. and Less, F. P. (Eds.) *The Human Operator and Process Control.* Taylor & Francis, London, U.K., pp. 51–66.
Crossman, E. R. F. W. and Goodeve, P. J. (1983) Feedback control of hand-movement and Fitts' law: Communication to the experimental society, *J Exp Psych*, 35A, 251–278.
Cui, G. and Wong, M. L. (2004) Implementing neural networks for decision support in direct marketing, *Int J Market Res*, 46(2), 235–254.
Cullinane, T. P. (1977) Minimizing cost and effort in performing a link analysis, *Hum Fact*, 19(2), 155.
Cumming, T. L. (1997) Conditional standard errors of measurement for questionnaire data. Center for Computer Aided Design, University of Iowa, Iowa, IA.
Dainty, P. and Kakabadse, A. (1992) Brittle, blocked, blended, and blind: Top team characteristics that lead to business success and failure, *J Manage Psych*, 7(2), 4–17.
Damon, F. A. (1965) The use of biomechanics in manufacturing operations, *Western Electr Eng*, 9(4), 11–20.
Dantico, J. A. and Greene, R. (2001) Compensation administration. In Zandin, K. B. and Maynard, H. B. (Eds.) *Maynard's Industrial Engineering Handbook.* McGraw-Hill, New York, pp. 7.79–7.96.

Dar-El, E. and Vollichman, R. (1996) Speed vs. accuracy under continuous and intermittent learning. *Advance in Applied Ergonomics Proceedings of the First International Conference (ICAE'96)*, Istanbul, Turkey, Ozok, A. F. and Salvendy, G. (Ed.). USA Publishing-Istanbul, West Lafayette, IN, pp. 689–694.

Das, B. and Behara, D. N. (1995) Determination of the normal horizontal working area: A new model and method, *Ergonomics*, 38, 734-748.

Das, B. and Sengupta, A. K. (1995) Computer-aided human modeling programs for workstation design, *Ergonomics*, 38(9), 1958–1972.

Datta, S. and Ramanathan, N. (1971) Ergonomic comparison of seven modes of carrying loads on the horizontal plane, *Ergonomics*, 14(2), 269–278.

Davenport, T. H. and Prusak, L. (2000) *Working Knowledge: How Organizations Manage What They Know*. Harvard Business School Press, Boston, MA.

Davis, J. H. (1992) Some compelling intuitions about group consensus decisions, theoretical and empirical research, and interpersonal aggregation phenomena: Selected examples, 1950–1990, *Organ Behav Hum Dec*, 52, 3–38.

Davis, G. I. and Buck, J. R. (1981, October) Simulation-aided design of man/machine inter-faces in automated industries. *Winter Simulation Conference Proceedings*, December 9–11, Atlanta, GA, pp. 407–416.

Dawes, R. M., van de Kragt, A. J. C., and Orbell, J. M. (1988) Not me or thee but we: The importance of group identity in eliciting cooperation in dilemma situations: Experimental manipulations, *Acta Psychol*, 68, 83–97.

Dawes, R. M. and Mulford, M. (1996) The false consensus effect and over-confidence: Flaws in judgement or flaws in how we study judgement? *Organ Behav Hum Dec*, 65, 201–211.

DeGreene, K. B. (Ed.) (1970) *Systems Psychology*. McGraw-Hill, New York.

De Jong, J. R. (1957) The effect of increasing skill on cycle time and its consequences for time standards, *Ergonomics*, 1, 51–60.

Delbecq, A. L., Van de Ven, A. H., and Gustafson, D. H. (1975) *Group Techniques for Program Planning*. Scott, Foresman, Glenview, IL.

Deming, W. E. (1986) *Out of the Crisis*. MIT Press, Cambridge, MA.

Dempsey, P. G. and Leamon, T. B. (1995) Bending the tool and the effect on productivity: An investigation of a simulated wire twisting task, *Am Ind Hyg Assoc J*, 56(7), 686–692.

Dempster, W. T. (1955) Space requirements of the seated operator: Geometrical, kinematic, and mechanical aspects of the body with special reference to the limbs. Report WADC TR 55-15. U.S. Air Force, Wright Air Development Center, Wright-Patterson Air Force Base, OH.

Dennis, A. R., Haley, B. J., and Vandenberg, R. J. (1996) A meta-analysis of effectiveness, efficiency, and participant satisfaction in group support systems research. *Proceedings of the International Conference on Information Systems*, pp. 278–289.

Dennis, A. R. and Wixom, B. H. (2002) Investigators the moderators of the group support systems use with meta-analysis, *J Manage Inform Syst*, 18(3), 235–258.

Dennis, A. R., Wixom, B. H., and Vandenberg, R. J. (2001) Understanding fit and appropriation effects in group support systems via meta-analysis, *MIS Q*, 25(2), 167–193.

Denton, D. K. (1992) Building a team, *Qual Prog*, 25(10), 87–91.

DeSanctis, G. and Gallupe, R. B. (1987) A foundation for the study of group decision support systems, *Manage Sci*, 33(5), 589–609.

DeWeese, M. R. and Meister, M. (1999) How to measure the information gained from one symbol, *Net: Comp Neural Syst*, 10, 325–340.

Dhillon, B. S. (1991) *Robot Reliability and Safety*. Springer-Verlag, New York.

Dixit, A. and Nalebuff, B. (1991) Making strategies credible. In Zechhauser, R. J. (Ed.) *Strategy and Choice*. MIT Press, Cambridge, MA, pp. 161–184.

Doherty, M. E. (2003) Optimists, pessimists, and realists. In Schnieder, S. and Shanteau, J. (Eds.) *Emerging Perspectives on Judgment and Decision Research*, Cambridge University Press, Cambridge, NY, pp. 643–679.

Dooley, M. (November, 1982) Anthropometric modeling programs—A survey, *IEEE Comp Graph Appl*, 2(9), 17–25.

Dorris, A. L. and Tabrizi, J. L. (1978) An empirical investigation of consumer perception of product safety, *J Prod Liab*, 2, 155–163.

Dougherty, M. R. P., Gronlund, S. D., and Gettys, C. F. (2003) Memory as a fundamental heuristic for decision making. In Schnieder, S. and Shanteau, J. (Eds.) *Emerging Perspectives on Judgment and Decision Research*. Cambridge University Press, Cambridge NY, pp. 125–164.

Drucker, P. F. (1985) *The Effective Executive*. Harper Row, New York.

Drucker, P. (1995) The information executives truly need, *Harv Bus Rev*, 73(1), 54–62.

Drury, C. G. (1973) The effect of speed of working on industrial inspection accuracy, *Appl Ergon*, 4, 2–7.

Drury, C. G. (1975a) Application of Fitts' law to foot pedal design, *Hum Fact*, 17, 368–373.

Drury, C. G. (1975b) Inspection of sheet material-model and data, *Hum Fact*, 17, 257–265.

Drury, C. G. (1988) The human as optimizer. In Megaw, E. D. (Ed.) *Contemporary Ergonomics*. Taylor & Francis Ltd., London, U.K., pp. 19–29.

Drury, C. G. (1992a) Inspection performance. In Salvendy, G. (Ed.) *Handbook of Industrial Engineering*, 2nd edn. John Wiley & Sons, New York, Chapter 88.

Drury, C. G. (1992b) Product design for inspectability: A systematic procedure. In Helander, M. and Nagamachi, M. (Eds.) *Design for Manufacturability, A Systems Approach to Concurrent Engineering and Ergonomics*. Taylor & Francis, Ltd., London, U.K.

Drury, C. G. and Addison, J. L. (1973) An industrial study of the effects of feedback and fault density on inspection performance, *Ergonomics*, 16, 159–169.

Du Charme, W. (1970) Response bias explanation of conservative human inference, *J Exp Psych*, 85, 66–74.

Duffy, L. (1993) Team decision making biases: An information processing perspective. In Klein, G. A., Orasanu, J., Calderwood, R., and Zsambok, C. E. (Eds.) *Decision Making in Action: Models and Methods*. Ablex, Norwood, NJ.

Eastman, M. C. (1983) *Ergonomic Design for People at Work (Vol. 1)*. Van Nostrand Reinhold. (see Rodgers), New York

Eastman, M. C. (1986) *Ergonomics Design for People at Work (Vol. 2)*. Nostrand Reinhold, Van New York.

Eastman, M. C. and Kamon, E. (1976) Posture and subjective evaluation at flat and slanted desks, *Hum Fact*, 18, 15–26.

Edland, E. and Svenson, O. (1993) Judgment and decision making under time pressure. In Svenson, O. and Maule, A. J. (Eds.) *Time Pressure and Stress in Human Judgment and Decision Making*, Plenum Press, New York.

Edosomwan, J. A. (2001) Total quality leadership. In Salvendy, G. (Ed.) *Handbook of Industrial Engineering*, 3rd edn. John Wiley & Sons, New York, pp. 1793–1807.

Edwards, W. (1954) The theory of decision making, *Psycho Bull*, 41, 380–417.

Edwards, W. (1968) Conservatism in human information processing. In Kleinmuntz, B. (Ed.) *Formal Representation of Human Judgment*. John Wiley & Sons, New York, pp. 17–52.

Edwards, E. (1993) What is in a name? *The Ergonomist*, 275, 1–4.

Edworthy, J. and Adams, A. (1996) *Warning Design: A Research Prospective*. Taylor & Francis, London, U.K.

Einhorn, H. J. and Hogarth, R. M. (1978) Confidence in judgment: Persistence of the illusion of validity. *Psychol Rev*, 70, 193–242.

Einhorn, H. J. and Hogarth, R. M. (1981) Behavioral decision theory: Processes of judgment and choice, *Annu Rev Psychol*, 32, 53–88.

Elkin, E. H. (1962) Target velocity, exposure time, and anticipatory tracking as determinants of dynamic visual acuity (DVA), *J Eng Psychol*, 1, 26–33.

Ellis, D. G. and Fisher, B. A. (1994) *Small Group Decision Making: Communication and the Group Process*, 4th edn. McGraw-Hill, New York.

Elster, J. (Ed.) (1986) *The Multiple Self*. Cambridge University Press, Cambridge, NY

Embrey, D. E., Humphreys, P., Rosa, E. A., Kirwin, B, and Rea, K. (1984) SLIM-MAUD: An approach to assessing human error probabilities using structural expert judgement. Vol. 1, Overview of SLIM-MAUD. NUREG/CR-3518, US Nuclear Regulatory Commission, Washington, DC, March. 1984.

Epp, S. and Konz, S. (1975) Appliance noise: Annoyance and speech interference, *Home Econ Res J*, 3(3), 205–209.

Epps, B. W. (1986) Comparison of six cursor control devices based on Fitts' law models. *Proceedings of the Human Factors Society 30th Annual Meeting*, pp. 327–331.

Ericsson, K. A. and Simon, H. A. (1980) Verbal reports as data, *Psychol Rev*, 87, 215–251.

Ericsson, K. A. and Simon, H. A. (1984) *Protocol Analysis: Verbal Reports as Data*. MIT Press, Cambridge, MA.

Espinasse, B., Picolet, G., and Chouraqui, E. (1997) Negotiation support systems: A multi-criteria and multi-agent approach, *Eur J Oper Res*, 103(2), 389–409.

Estes, W. (1970) The cognitive side of probability learning, *Psych Rev*, 83, 37–64.

Etzioni, A. (1988) Normative-affective factors: Toward a new decision-making mode, *J Econ Psychol*, 9, 125–150.

Evans, J. B. T. (1989) *Bias in Human Reasoning: Causes and Consequences*. Erlbaum, London, U.K.

Fabrycky, W. J. and Blanchard, B. S. (1991) *Life-Cycle Cost and Economic Analysis*. Prentice-Hall, Englewood Cliffs, NJ.

Fallesen, J. J. and Pounds, J. (2001) Identifying and testing a naturalistic approach for cognitive skill training. In Salas, E. and Klien, G. (Eds.) *Linking Expertise and Naturalistic Decision Making*. Lawrence Erlbaum Associates Inc., Mahwah, NJ, pp. 55–70.

Farina, A. J. and Wheaton, G. R. (1973) Development of a taxonomy of human performance: The task characteristics approach to performance prediction. JSAS Catalog of Selected Documents in Psychology, 3(Ms No. 323).

Farley, R. R. (1955) Some principles of methods and motion study as used in development work, *Gen Motors Eng J*, 2, 20–25.

Faulkner, W. (1983) *Ergonomic Design for People at Work*. Van Nostrand Reinhold, New York.

Faulkner, T. W. and Murphy, T. J. (1975) Lighting for difficult visual tasks, *Hum Fact*, 15(2), 149–162.

Fazlollahi, B. and Vahidov, R. (2001) Extending the effectiveness of simulation-based DSS through genetic algorithms, *Inform Manage*, 39(1), 53–64.

Feather, N. T. (1966) Effects of prior success and failure on expectations of success and failure, *J Pers Soc Psychol*, 3, 287–298.

Feldt, L. S., Steffen, M., and Gupta, N. C. (1985) A comparison of five methods for estimating the standard error of measurement at specific score levels, *Appl Psychol Measure*, 9, 351–361.

Feller, W. (1940) On the logistic law of growth and its empirical verifications in biology. *Acta Biotheoret*, 5, 51–66.

Fischhoff, B. (1982) For those condemned to study the past: Heuristics and biases in hindsight. In Kahneman, D., Slovic, P., and Tversky, A. (Eds.) *Judgment under Uncertainty: Heuristics and Biases*. Cambridge University Press, Cambridge, NY.

Fischhoff, B. and MacGregor, D. (1982) Subjective confidence in forecasts, *J Forecast*, 1, 155–172.

Fischhoff, B., Slovic, P., and Lichtenstein, S. (1977) Knowing with certainty: The appropriateness of extreme confidence, *J Exp Psych Hum Percep Perform*, 3, 552–564.

Fishburn, P. C. (1974) Lexicographic orders, utilities, and decision rules: A survey, *Manage Sci*, 20, 1442–1471.

Fisher, E. L., Moroze, M. L., and Morrison, R. E. (1983) Dynamic visual inspection: Queuing models. *Proceedings of the Human Factors Society, 27th Annual Meeting*, pp. 212–215.

Fitts, P. M. (1954) The information capacity of the human motor system in controlling the amplitude of movement, *J Exp Psychol*, 47, 381–391.

Fitts, P. M. (1966) Cognitive aspects of information processing III: Set for speed versus accuracy, *J Exp Psychol*, 71, 849–857.

Fitts, P. M., Jones, R. E., and Milton, J. L. (1950) Eye movements of aircraft pilots during instrument landing approaches, *Aeronaut Eng Rev*, 9(Feb), 24–29.

Fitts, P. M. and Peterson, J. R. (1964) Information capacity of discrete motor responses, *J Exp Psychol*, 67, 103–112.

Fitts, P. M. and Seegar, C. M. (1953) S–R compatibility: Spatial characteristics of stimulus and response codes, *J Exp Psychol*, 46(3), 199–210.

Flach, J. M. and Dominguez, E. O. (1995, July) Use-centered design: Integrating the user, instrument, and goal, *Ergonom Des* 19–24.

Flanagan, J. C. (1954) The critical incident technique, *Psychol Bull*, 51(4), 327–359.

Fleishman, E. A. and Quaintance, M. K. (1984) *Taxonomies of Human Performance*. Academic Press, Orlando, FL.

Fleishman, E. A. and Zaccaro, S. J. (1992) Toward a taxonomy of team performance functions. In Sweney, R. W. and Salas, E. (Eds.) *Teams: Their Training and Performance*. Ablex, Norwood, NJ, pp. 31–56.

Flin, R., Salas, E., Strub, M., and Martin, L. (Eds.) (1997) *Decision Making under Stress: Emerging Themes and Applications*. Ashgate, Aldershot, U.K.

Fogel, L. J. (1963) *Human Information Processing*. Prentice-Hall, Englewood Cliffs, NJ.

Fong, G. T., Krantz, D. H., and Nisbett, R. E. (1986) The effects of statistical training on thinking about everyday problems, *Cogn Psychol*, 18, 253–292.

Foroughi, A. (1998) Minimizing negotiation process losses with computerized negotiation support systems, *J Appl Bus Res*, 14(4), 15–26.

Fox, W. (1967) Human performance in the cold, *Hum Fact*, 9(3), 203–220.

Francis, W. N. and Kucera, H. (1982) *Frequency Analysis of English Usage*. Houghton Mifflin, Boston, MA.

Frantz, J. P. (1992) Effect of location, procedural explicitness, and presentation format on user processing of and compliance with product warnings and instructions. Unpublished doctoral dissertation, University of Michigan, Ann Arbor, MI.

Frantz, J. P. and Rhoades, T. P. (1993) A task analytic approach to the temporal and spatial placement of product warnings, *Hum Fact*, 35, 719–730.

Friedman, J. W. (1990) *Game Theory with Applications to Economics*. Oxford University Press, New York.

Fuld, R. B. (1993, January) The fiction of function allocation, *Ergonom Des* (Human Factors and Ergonomics Society) 20–24.

Gagge, A. P., Stolwijk, J. A. J., and Nish, Y. (1971) An effective temperature scale based on a simple model of human physiological regulatory response, *ASHRAE Trans*, 77(part 1), 247–267.

Gagné, R. M. (1954) Training devices and simulators: Some research issues, *Am Psychol*, 9, 95–107.

Gagné, R. M. (1959) Problem solving and thinking, *Annu Rev Psychol*, 10, 147–172.

Gagné, R. M. (1962) Human functions in Systems. In Gagné, R. M. (Ed.) *Psychological Principles in System Development*. Holt, Rinehart, and Winston, New York, Chapter 2.

Gagné, R. M. and Melton, A. W. (Eds.) (1962) *Psychological Principles in System Development*. Holt, Rinehart, and Winston, New York.

Gann, K. C. and Hoffman, E. R. (1988) Geometrical conditions for ballistic and visually controlled movements, *Ergonomics*, 31(5), 829–839.

Garavelli, A. C. and Gorgoglione, M. (1999) Fuzzy logic to improve the robustness of decision support systems under uncertainty, *Comp Ind Eng*, 27(1–2), 477–480.

Garg, A. and Milliman, P. (1961, January–February) The aircraft progress curve—Modified for design changes, *J Ind Eng*, 12, 23–28.

Garrett, J. (1971) The adult human hand: Some anthropometric and biomechanical considerations, *Hum Fact*, 13(2), 117–131.

Gelberman, R. H., Szabo, R. M., and Mortensen, W. W. (1984) Carpal tunnel pressures and wrist position in patients with Colles' fractures, *J Trauma*, 24, 747–749.

Geller, S. E. (1996) *The Psychology of Safety: How to Improve Behaviors and Attitudes on the Job*. Clinton Book Company, Radnor, PA, p. 390.

Geller, E. S. and Williams, J. H. (2001) *Keys to Behavior-Based Safety: From Safety Performance Solutions*. ABS Consulting, Rockville, MD, p. 484.

Genaidy, A. M., Duggai, J. S., and Mital, A. (1990) A comparison of robot and human performances for simple assembly tasks, *Int J Ind Ergon*, 5, 73–81.

Gershoni, H. (1971) Motivation and micro-method when learning manual tasks, *Work Stud Manage Serv*, 15(9), 585–595.

Gertman, D. I. and Blackman, H. S. (1994) *Human Reliability and Safety Analysis Data Handbook*. John Wiley & Sons, New York.

Gibson, A. K. S. and Tyndall, E. P. T. (1923) The visibility of radiant energy, *Sci Pap Bur Stand*, 19, 131–191.

Gigerenzer, G. and Hoffrage, U. (1995) How to improve Bayesian reasoning without instruction: Frequency formats, *Psychol Rev*, 102, 684–704.

Gigerenzer, G., Hoffrage, U., and Kleinbolting, H. (1991) Probabilistic mental models: A Brunswikian theory of confidence, *Psychol Rev*, 98, 506–528.

Gigerenzer, G., Todd, P., and the ABC Research Group. (1999) *Simple Heuristics That Make Us Smart*. Oxford University Press, New York.

Gitlow, H., Gitlow, S., Oppenheim, A., and Oppenheim, R. (1989) *Tools and Methods for the Improvement of Quality*. Irwin, Homewood, IL.

Gladstein, D. L. (1984) Groups in context: A model of task group effectiveness, *Admin Sci Q*, 29, 499–517.

Glover, J. (1966) Manufacturing process functions II: Selection of trainees and control of their progress, *Int J Prod Res*, 8, 293–305.

Goel, V. and Rim, K. (1987) Role of gloves in reducing vibration: An analysis for pneumatic chipping hammer, *Am Ind Hyg Assoc J*, 48(1), 9–14.

Goetsch, D. L., (2008) *Occupational Safety and Health for Technologists, Engineers, and Managers*, 7th edn. Prentice-Hall Inc., Englewood Cliffs, NJ.

Goetsch, D. L. (2010) *Occupational Safety and Health for Technologists, Engineers, and Managers*, 7th edn. Prentice Hall, Upper Saddle River, NJ.

Goldberg, S. (1961) *Introduction to Difference Equations*. John Wiley & Sons, Inc., New York.

Goldman, R. F. (1977) Establishment of the boundaries to comfort by analyzing discomfort. In Mangum, B. W. and Hill, J. E. (Eds.) *Thermal Analysis Human Comfort—Indoor Environments, Proceedings of a Conference*. NBS Special Publication 491.

Goldman, J. and Hart, L. W., Jr. (1965, September–October) Information theory and industrial engineering, *J Ind Eng*, 16, 306–313.

Goldstein, W. M. and Hogarth, R. M. (1997) Judgment and decision research: Some historical context. In Goldstein, W. M. and Hogarth, R. M. (Eds.) *Research on Judgment and Decision Making: Currents, Connections, and Controversies*. Cambridge University Press, Cambridge, NY, pp. 3–65.

Gopher, D., Williges, B. H., Williges, R. L., and Damos, D. (1975) Varying the type and number of adaptive variables in continuous tracking, *J Motor Behav*, 7(3), 159–170.

Gordon, J. (1949) The epidemiology of accidents, *Am J Publ Health*, 39, 504–515.

Grandjean, E. (1988) *Fitting the Task to the Man: A Textbook of Occupational Ergonomics*, 4th edn. Taylor & Francis, London, U.K.

Grandjean, E., Hunting, W., and Pidermann, M. (1983) VDT workstation design: Preferred settings and their effects, *Hum Fact*, 25, 161–175.

Grandjean, E., Hunting, W., Wotzka, G., and Scharer, R. (1973) An ergonomic investigation of multi-purpose chairs, *Hum Fact*, 15(3), 247–255.

Green, G. H. (1974) The effect of indoor relative humidity on absenteeism and colds in schools, *ASHRAE Trans*, 80(part 2), 131–141.

Greenburg, L. and Chaffin, D. (1977) *Workers and Their Tools*. Pendall Publishing, Midland, MI.

Greenstein, J. and Rouse, W. B. (1980) A model of human decision making in multiple process monitoring situations. *Proceedings of the 16th Annual Conference on Manual Control*, Cambridge, MA, pp. 465–487.

Green, D. M. and Swets, J. A. (1966) *Signal Detection Theory and Psychophysics*. John Wiley & Sons, New York.

Groover, M. P. (2007) *Work Systems and the Methods, Measurement, and Management of Work*. Pearson/Prentice Hall, Upper Saddle River, NJ.

Grose, J. (1979) A computer simulation of an aircraft pilot during mid-flight. Unpublished master's project, School of Industrial Engineering, Purdue University. West Lafayette, IN.

Gruenfeld, D. H., Mannix, E. A., Williams, K. Y., and Neale, M. A. (1996) Group composition & decision making: How member familiarity and information distribution affect process and performance, *Organ Behav Hum Dec*, 67(1), 1–15.

Gupta, S. C. and Happ, W. W. (1965) A flowgraph approach to the La Place transform, *J Franklin I*, 280(August), 150–163.

Gupta, A. K. and Wilemon, D. L. (1990) Accelerating the development of technology-based new products, *Calif Manage Rev*, 32(2), 24–44.

Hackman, J. R. (1987) The design of work teams. In Lorsch, J. (Ed.) *Handbook of Organizational Behavior.* Prentice Hall, Englewood Cliffs, NJ, pp. 315–342.

Haddon, W., (1975) Reducing the damage of motor vehicle use, *Technol Rev, 77*(8), 52–59.

Hadley, G. (1967) *Introduction to Probability and Statistical Decision Theory.* Holden-Day, Inc., San Francisco, CA.

Hamilton, V. and Warburton, D. M. (Eds.) (1979) *Human Stress and Cognition.* Wiley, New York.

Hammer, W. (1972) *Handbook of System and Product Safety.* Prentice-Hall, Englewood Cliffs, NJ.

Hammer, W. (1993) *Product Safety Management and Engineering,* 2nd edn. American Society of Safety Engineers (ASSE), Des Plaines, IL.

Hammer, W. and Price, D. (2000) *Occupational Safety Management and Engineering,* 5th edn. Prentice Hall, Upper Saddle River, NJ.

Hammer, J. M. and Rouse, W. B. (1979) Analysis and modeling of freeform text editing behavior. *Proceedings of International Conference on Cybernetics and Society,* Denver, CO, October 1979, pp. 659–664.

Hammond, K. R. (1980) Introduction to Brunswikian theory and methods. In Hammond, K. R. and Wascoe, N. E. (Eds.) *Realizations of Brunswick's Experimental Design.* Jossey-Bass, San Francisco, CA.

Hammond, K. R. (1993) Naturalistic decision making from a Brunswikian viewpoint: Its past, present, future. In Klein, G. A., Orasanu, J., Calderwood, R., and Zsambok, E. (Eds.) *Decision Making in Action: Models and Methods.* Ablex, Norwood, NJ, pp. 205–227.

Hammond, K. R. (1996) *Human Judgment and Social Policy: Irreducible Uncertain, Inevitable Error, Unavoidable in Justice.* Oxford University Press, New York.

Hammond, K. R., Hamm, R. M., Grassia, J., and Pearson, T. (1987) Direct comparison of the efficacy of intuitive and analytical cognition in expert judgment, *IEEE Trans Syst Man Cyber, SMC-17,* 753–770.

Hammond, K. R. Stewart, T. R., Brehmer, B., and Steinmann, D. O. (1975) Social judgment theory. In Kaplan, M. F. and Schwartz, S. (Eds.) *Human Judgment and Decision Processes.* Academic Press, New York, pp. 271–312.

Hancock, W. M. (1966) The design of inspection operations using the concepts of information theory. *University of Michigan Summer Conf. Notes—New Developments in the Prediction of Human Performance on Industrial Operations,* Ann Arbor, MI, July 25–29.

Hancock, W. M. (1967) The prediction of learning rates for manual operations, *J Ind Eng, XVIII*(1), 42–47.

Hancock, W. M. and Foulke, J. A. (1966) Computation of learning curves, *MTM J, XL*(3), July–August, 5–7.

Hannon, J. M., Newman, J. M., Milkovich, G. T., and Brakefield, J. T. (2001) Job evaluation in organizations. In Salvendy, G. (Ed.) *Handbook of Industrial Engineering.* Wiley Interscience, New York.

Hardin, G. (1968) The tragedy of the commons, *Science, 162,* 1243–1248.

Harmon, J. and Rohrbaugh, J. (1990) Social judgment analysis and small group decision making: Cognitive feedback effects on individual and collective performance, *Organ Behav Hum Dec, 46,* 34–54.

Harris, D. H. (1987) *Human Factors Success Stories [Videotape].* The Human Factors Society, Santa Monica, CA.

Hart, P., Stern, E. K., and Sundelius, B. (1997) *Beyond Groupthink: Political Group Dynamics and Foreign Policy-Making.* University of Michigan Press, Ann Arbor, MI.

Hasher, L. and Zacks, R. T. (1984) Automatic processing of fundamental information: The case of frequency of occurrence, *Am Psychol, 39,* 1372–1388.

Haslegrave, C. M., Baker, A., and Dillon, S. (1992) Use of ergonomic workspace modeling in vehicle design. In Mattila, M. and Darwowski, W. (Eds.) *Computer Applications in Ergonomics, Occupational Safety, and Health.* North-Holland Publishing, Amsterdam, the Netherlands, pp. 145–152.

Hauser, J. R. and Clausing, D. (1988, May–June) The house of quality, *Harv Bus Rev, 66*(3), 63–73.

Hax, A. C. and Majluf, N. S. (1982) Competitive cost dynamics: The experience curve, *Interfaces, 12,* 5.

Heath, C. (1995) Escalation and de-escalation of commitment in response to sunk costs: The role of budgeting in mental accounting, *Organ Behav Hum Dec, 62,* 38–54.

Heath, L., Tindale, R. S., Edwards, J., Posavac, E. J., Bryant, F. B., Henderson-King, E., Suarez-Balcazar, Y., and Myers, J. (1994) *Applications of Heuristics and Biases to Social Issues*. Plenum Press, New York.

Hedge, A., McCrobie, D., Morimoto, S., Rodriguez, S., and Land, B. (1996, January) Toward pain-free computing, *Ergonom Des* 4–10.

Hedley, B. (1976) A fundamental approach to strategy development, *Long Range Planning*, December 2–11.

Heinrich, H. W. (1959) *Industrial Accident Prevention—A Scientific Approach*. McGraw-Hill, New York.

Heinrich, H. W., Petersen, D., and Roos, N. (1980a) *Industrial Accident Prevention: A Safety Management Approach*. McGraw-Hill, New York.

Heinrich, H. W., Peterson, D., and Roos, N. (1980b) *Industrial Accident Prevention*, 5th edn. McGraw-Hill, New York.

Helander, M. G. and Furtado, D. (1992) Product design of manual assembly. In Helander, M, Agamachi, M. N. and the International Ergonomics Society (Eds.) *Design for Manufacturability—A System Approach to Concurrent Engineering and Ergonomics*. Taylor & Francis Ltd., London, U.K., pp. 171–188, Chapter 11.

Helander, M. and Nagamachi, M. (Eds.) (1992) *Design for Manufacturability: A Systems Approach to Concurrent Engineering and Ergonomics*. Taylor & Francis Ltd., London, U.K.

Helson, H. (1949) Design of equipment and optimal human factors, *Am J Psychol*, 62, 473–497.

Hendrick, H. (1996) The ergonomics of economics is the economics of ergonomics. *Human Factors and Ergonomics Society (HFES) Presidential Address. Proceedings of the 40th Annual Conference*, Philadelphia, PA, pp. 1–10.

Hertwig, R., Hoffrage, U., and Martignon, L. (1999) Quick estimation: Letting the environment do the work. In Gigerenzer, G., Todd, P., and the ABC Research Group (Eds.) *Simple Heuristics That Make us Smart*. Oxford University Press, New York, pp. 209–234.

Hess, R. A. (2007) Feedback control models—Manual control and tracking. In Salvendy, G. (Ed.) *Handbook of Human Factors and Ergonomics*, 2nd edn. Wiley, New York.

Hick, W. E. (1952) On the rate of gain of information, *Q J Exp Psychol*, 4, 11–26.

Hirano, H. (1996) *5S for Operators: 5 Pillars of the Visual Workplace (For Your Organization!)*. Productivity Press, Inc., Portland, OR.

Hirsch, W. (1952) Text, *Int J Prod Res*, 8, 293–305.

Hitchings, B. and Towill, D. R. (1975) An error analysis of the time constraint learning curve model, *Int J Prod Res*, 13(2), 105–135.

Hoffman, R. and Militales, L. (2009) *Perspectives on Cognitive Task Analysis: Historical Origins and Modern Communities of Practice*. Taylor and Francis, CRC Press, New York, p. 516.

Hogarth, R. M. and Einhorn, H. J. (1992) Order effects in belief updating: The belief-adjustment model, *Cogn Psychol*, 24, 1–55.

Holleran, P. A. (1991) Pitfalls in usability testing, *Behav Inform Technol*, 10(5), 345–358.

Hollnagel, E. (1993) *Human Reliability Analysis: Context and Control*. Academic Press, London, U.K.

Holsapple, C. W. and Whinston, A. B. (1996) *Decision Support Systems: A Knowledge-Based Approach*. West Publishing, St. Paul, MN.

Honeycutt, J. M., Jr. (1962) Comments on An experimental evaluation of the validity of predetermined elemental time systems, *J Ind Eng*, 13(3), 171–179.

Holsapple, C. W., Lai, H., and Whinston, A. B. (1998) A formal basis for negotiation support system research, *Group Dec Negot*, 7(3), 203–227.

Hosseini, J. C., Harmon, R. R., and Zwick, M. (1991) An information theoretic framework for exploratory market segmentation research, *Dec Sci*, 22, 663–677.

Houghton, F. C., and Yagloglou, C. P. (1923) Determining equal comfort lines, *J Am Soc Heat Ventilat Eng*, 29, 165–176.

House, T. and Lehto, M. R. (1999, May 19–21) Computer aided analysis of hazard communication phrases. Paper presented at *International Conference on Computer—Aided Ergonomics and Safety*, Barcelona, Spain.

Howard, R. A. (1968) The foundations of decision analysis, *IEEE Trans Syst Sci Cyber*, SSC-4, 211–219.

Howard, R. A. (1988) Decision analysis: Practice and promise, *Manage Sci*, 34, 679–695.

Hsiao, H. and Keyserling, W. (1991) Evaluating posture behaviour during seated tasks, *Int J Ind Ergon*, 8, 313–334.
Huber, J., Wittink, D. R., Fiedler, J. A., and Miller, R. (1993) The effectiveness of alternative preference elicitation procedures in predicting choice, *J Market Res*, 30, 105–114.
Huchingson, R. D. (1981) *New Horizons for Human Factors in Design*. McGraw-Hill, New York.
Huff, E. M. and Nagel, D. C. (1975) Psychological aspects of aeronautical flight simulation, *Am Psychol*, 30, 426–439.
Hunting, W., Laubli, T., and Grandjean, E. (1981) Constrained postures, *Ergonomics*, 24, 917–931.
Hutchinson, R. D. (1981) *New Horizons for Human Factors in Design*. McGraw-Hill, New York.
Hutchinson, G. (1995) Taking the guesswork out of medical device design, *Ergonomics in Design*, April, pp. 21–26.
Hyman, R. (1953) Stimulus information as a determinant of reaction time, *J Exp Psychol*, 45, 188–196.
IESNA (1993) *Lighting Handbook: Reference and Application*, 8th edn. Illuminating Engineering Society of North America, New York.
ILO. (1979) *Introduction to Work Study*. International Labour Office, Geneva, Switzerland.
Imker, F. R. (1994, April) The back support myth, *Ergonom Des* 2(2), 9–12.
Industrial Ventilation. 2004 *A Manual of Recommended Practice*, 25th edn. American Conference of Governmental Industrial Hygienists, Inc (ACGIH), Cincinnati, OH.
Isen, A. M (1993) Positive affect and decision making. In Lewis, M., Haviland, J. M. (Eds.) *Handbook of Emotions*. Guilford Press, New York, pp. 261–277.
Isenberg, D. J. (1986) Group polarization: A critical review and meta analysis, *J Personal Soc Psychol*, 50, 1141–1151.
Jacoby, J. (1977) Information load and decision quality: Some contested issues, *J Market Res*, 14, 569–573.
Jacques, E. (1961) *Equitable Payment*. John Wiley & Sons, New York.
Jagacinski, R. J. and Flach, J. M. (2003) *Control Theory for Humans: Quantitative Approaches to Modeling Performance*. Erlbaum, Mahwah, NJ.
Jagacinski, R. J. and Monk, D. L. (1985) Fitts' law in two dimensions with hand and head movements, *J Motor Behav*, 17, 77–95.
Jagacinski, R. J., Repperger, D. W., Ward, S. L., and Moran, M. S. (1980) A test of Fitts' law with moving targets, *Hum Fact*, 22(2), 225–233.
Janis, I. L. (1972) *Victims of Groupthink*. Houghton-Mifflin, Boston, MA.
Janis, I. L. and Mann, L. (1977) *Decision Making: A Psychological Analysis of Conflict, Choice, and Commitment*. Free Press, New York.
Jarjoura, E. (1986) An estimator of examinee-level measurement error variance that considers test form difficulty adjustments, *Appl Psychol Meas*, 10, 175–186.
Jasic, T. and Wood, D. (2004) The profitability of daily stock market indices trades based on neural network predictions: Case study for the S&P 5000, the DAX, the TOPIX and the FTSE in the period 1965–1999, *Appl Fin Econ*, 14(4), 285–297.
Jasper, H. H. and Ballem, G. (1949) Unipolar electro-myograms of normal and denervated human muscle, *J Neurophysiol*, 12, 231–244.
Jentsch, F. G. (1994) Color coding of aircraft emergency exit lighting: Stereotypes among subjects from four language groups, *Proc Hum Fact Ergonomics Soc 38th Annu Meet*, Vol. 2, p. 969.
Jin, P. (1993) Work motivation and productivity in voluntary formed work teams: A field study in China, *Organ Behav Hum Dec*, 54, 133–155.
Johansson, G. and Rumar, K. (1971) Driver's brake reaction times, *Hum Fact*, 12(1), 23–27.
Johnson, W. G. (1973) Sequences in accident causation, Guest Editorial in *J Saf Res*, 5(2), 54–57.
Johnson, W. G. (1975) MORT: The management oversight and risk tree, *J Saf Res*, 5(2), 54–57.
Johnson, W. G. (1980) *MORT Safety Assurance Systems*. National Safety Council, Chicago, IL.
Johnson, C. A. and Casson, E. J. (1995) Effects of luminance, contrast, and blur on visual acuity, *Opt Vis Sci*, 72, 864–869.
Johnson, E. J. and Payne, J. W. (1985) Effort and accuracy in choice, *Manage Sci*, 31(4), 395–414.
Jones, L. A. (1917) The fundamental scale of pure hue and retinal sensibility to hue differences, *J Opt Soc Am*, 1, 63–77.

Jonides, J. and Jones, C. M. (1992) Direct coding for frequency of occurrence, *J Exp Psychol Learn Mem Cogn*, 18, 368–378.

Jordan, N. (1963, June) Allocation of functions between man and machines in automated systems, *J Appl Psychol*, 47(3), 161–165.

Jordan, R. (1965) *How to Use the Learning Curve*. Materials Management Institute, Boston, MA.

Juslin, P. and Olsson, H. (1999) Computational models of subjective probability calibration. In *Judgment and Decision Making Neo-Brunswikian and Process-Tracing Approaches*. Lawrence Erlbaum Associates, Hillsdale, NJ, pp. 67–95, Chapter 5.

Just, M. A. and Carpenter, P. A. (1987) *The Psychology of Reading and Language Comprehension*. Allyn & Bacon, Inc., Boston, MA.

Kachitvichyanukul, V., Buck, J. R., and Ong, C.-S. (1987) A simulation model for ergonomic design of industrial processes, *SIMULATION*, 49(5), 199–206.

Kahneman, D. (1973) *Attention and Effort*. Prentice-Hall, Englewood Cliffs, NJ.

Kahneman, D., Slovic, P., and Tversky, A. (1982) *Judgment under Uncertainty: Heuristics and Biases*. Cambridge University Press, Cambridge, NY.

Kahneman, D. and Tversky, A. (1979) Prospect theory: An analysis of decision under risk, *Econometrica*, 47, 263–291.

Kantowitz, B. H. and Elvers, G. C. (1988) Fitts' law with an isometric controller: Effects of order of control and control-display gain, *J Motor Behav*, 20, 53–66.

Kantowitz, B. H. and Sorkin, R. D. (1983) *Human Factors: Understanding People-System Relationships*. John Wiley & Sons, New York.

Kapur, K. C. (1992) Reliability and maintainability. In Salvendy, G. (Ed.) *Hand-Book of Industrial Engineering*, 2nd edn. John Wiley & Sons, New York, Chapter 89.

Karger, D. W. and Hancock, W. (1982) *Advanced Work Measurement*. Industrial Press, New York.

Karlin, J. E. (1957, November) Consideration of the user in telephone research, *Ergonomics*, 1(1), 77–83.

Katz, R. and Tushman, M. L. (1979) Communications patterns, project performance, and task characteristics: An empirical evaluation and integration in an R&D setting, *Organ Behav Hum Perform*, 23, 139–162.

Kaufman, L. and Thomas, J. P. (Eds.) *Handbook of Perception and Human Performance*. John Wiley & Sons, New York, Chapter 43.

Keen, P. G. W. and Scott-Morton, M. S. (1978) *Decision Support Systems: An Organizational Perspective*. Addison-Wesley, Reading, MA.

Keeney, R. L. and Raiffa, H. (1976) *Decisions with Multiple Objectives: Preferences and Value Tradeoffs*. John Wiley & Sons, New York.

Keller, R. T. (1986) Predictors of the performance of project groups in R&D organizations. *Acad Manage J*, 29, 715–726.

Kelley, C. R. (1968) *Manual and Automatic Control: A Theory of Manual Control and Its Applications to Manual and to Automatic Systems*. John Wiley & Sons, New York.

Kelly, G. A. (1955) *The Psychology of Personal Constructs*. Norton, New York.

Kelly, J. E. (1982) *Scientific Management, Job Redesign and Work Performance*. Academic Press, London, U.K.

Kemsley, W. F. F. (1957) *Women's Measurements and Sizes*. H.M.S.O. Published by the Board of Trade for Joint Clothing Council, Ltd., London, U.K.

Keren, G. (1990) Cognitive aids and debiasing methods: Can cognitive pills cure cognitive ills? In Caverni, J. P., Fabre, J. M., and Gonzalez, M. (Eds.) *Cognitive Biases*. North Holland, Amsterdam, the Netherlands.

Kerkhof, G. (1985) Individual differences and circadian rhythms. In Folkard, S. and Monk, T. H. (Eds.) *Hours of Work*. John Wiley & Sons, Chichester, U.K., Chapter 3.

Kerr, L. N., MacCoun, R. J., and Kramer, G. P. (1996) Bias in judgment: Comparing individuals and groups, *Psychol Rev*, 103, 687–719.

Kersten, G. E. and Noronha, S. J. (1999) WWW-based negotiation support systems: Design, implementation, and use, *Dec Supp Syst*, 25(2), 135–154.

Keyserling, W., Herrin, G., and Chaffin, D. (1980) Isometric strength testing as a means of controlling medical incidents on strenuous jobs, *J Occup Med*, 22(5), 332–336.

Kieras, D. E. (1985) The role of prior knowledge in operating equipment from written instructions. Report no. 19 (FR-85/ONR-19), University of Michigan, Ann Arbor, MI, February 20.
Kieras, D. E. (1988) Towards a practical GOMS model methodology for user interface design. In Helander, M. (Ed.) *Handbook of Human-Computer Interaction*. Elsevier Science Publications B. V., North-Holland, Amsterdam, the Netherlands, pp. 135–157.
King, B. and Sutro, P. (1957) Dynamic visual fields, *HRB Bulletin* 152.
Klare, G. R. (1974) Assessing readability, *Read Res Q*, 10, 62–102.
Klein, G. A. (1989) Recognition-primed decisions. In Rouse, W. (Ed.) *Advances in Man-Machine System Research*. JAI Press, Greenwich, CT.
Klein, G. A. (1995) The value added by cognitive analysis. *Proceedings of the Human Factors and Ergonomics Society—39th Annual Meeting*, San Diego, CA, pp. 530–533.
Klein, G. A. (1996) The effect of acute stressors on decision making. In Driskell, J. E. and Salas, E. (Eds.) *Stress and Human Performance*. Erlbaum, Hillsdale, NJ.
Klein, G. A. (1998) *Sources of Power: How People Make Decisions*. MIT Press, Cambridge, MA.
Klein, R. W., Dittus, R. S., Roberts, S. D., and Wilson, J. R. (1993) Simulation modeling and health-care decision making, *Med Dec Making*, 13(4), 347–354.
Klein, G. A., Orasanu, J., Calderwood, R., and Zsambok, E., (Eds.) (1993) *Decision Making in Action: Models and Methods*. Ablex, Norwood, NJ.
Klein, G. A. and Wolf, S. (1995) Decision-centered training. *Proceedings of the Human Factors and Ergonomics Society—39th Annual Meeting*, San Diego, CA, pp. 1249–1252.
Kleinmuntz, B. (1984) The scientific study of clinical judgment in psychology and medicine, *Clin Psychol Rev*, 4, 111–126.
Knowles, W. B. (1967) Aerospace simulation and human performance research, *Hum Fact*, 9(2), 149–159.
Knowles, A. and Bekk, L. (1950) Learning curves will tell you who's worth training and who isn't, *Fact Manage*, 108, 114–115, June.
Kochar, D. S. and Wills, B. L. (1971) Simulation of two-man interaction system. *Proceedings of the Winter Simulation Conference*, New York, pp. 56–62.
Koehler, J. J. (1996) The base rate fallacy reconsidered: Descriptive, normative, and methodological challenges, *Behav Brain Sci*, 19, 1–53.
Konz, S. (1975) Design of food scoops, *Appl Ergon*, 6(1), 32.
Konz, S. (1983) *Work Design: Industrial Ergonomics*, 2nd edn. Grid Publishing, Columbus, OH.
Konz, S. (1990) *Work Design: Industrial Ergonomics*, 3rd edn. Grid Publishing, Columbus, OH.
Konz, S. and Goel, S. C. (1969) The shape of the normal work area in the horizontal plane, *AIIE Trans*, 1(1), 70–74
Konz, S. and Johnson, S. (2000) *Work Design: Industrial Ergonomics*. Holcomb Hathaway, Publishers, Inc., Scottsdale, AZ.
Konz, S. and Johnson, S. (2004) *Work Design: Industrial Ergonomics*, 6th edn. Holcomb Hathaway, Scottsdale, AZ.
Konz, S. and Nentwich, H. (1969) A cooling hood in hot-humid environments. Kansas State University Engineering Experiment Station Report 81.
Konz, S. and Osman, K. (1978) Team efficiencies on a paced visual inspection task, *J Hum Ergol (Jpn)*, 6, 113–121.
Kopala, C. (1979) The use of color-coded symbols in a highly dense situation display. *Proceedings of the 23rd Annual Meeting of the Human Factors Society*, Human Factors Society, Santa Monica, CA, HFS397–HFS401.
Koriat, A., Lichtenstein, S., and Fischhoff, B. (1980) Reasons for confidence, *J Exp Psychol Hum Learn Mem*, 6, 107–118.
Kornblum, S. (1969) Sequential determinants of information processing in serial and discrete choice reaction time, *Psychol Rev*, 76, 113–131.
Kowalsky, N., Masters, R., Stone, R., Babcock, G., and Rypka, E. (1974) An analysis of pilot error-related aircraft accidents. (NASA, Report No. CR-2444), NASA, Washington, DC.
Kraus, N. N. and Slovic, P. (1988) Taxonomic analysis of perceived risk: Modeling individual and group perceptions within homogenous hazards domains, *Risk Anal*, 8, 435–455.

Krauss, S., Martignon, L., and Hoffrage, U. (1999) Simplifying Bayesian inference: The general case. In Magnani, L., Nersessian, N., and Thagard, P. (Eds.) *Model-Based Reasoning in Scientific Discover*. Plenum Press, New York, pp. 165–179.

Krishnaswamy, S. (1983) Determination of synthetic times for simple arithmetic tasks. Unpublished master's thesis, The University of Iowa, Iowa City, IA.

Kroemer, K. H. E. (1972) Human engineering the keyboard, *Hum Fact*, 14, 51–63.

Kroemer, K. H. E. (1986) Coupling the hand with the handle: An improved notation of touch, grip and grasp, *Hum Fact*, 28, 337–339.

Kroemer, K. H. E., Kroemer, H. B., and Kroemer-Elbert, K. E. (1994) *Ergonomics: How to Design for Ease and Efficiency*. Prentice Hall, Englewood Cliffs, NJ.

Kroemer, K. H. E., Kroemer, H. J., and Kroemer-Elbert, K. E. (1990) *Engineering Physiology—Bases of Human Factors/Ergonomics*, 2nd edn. Van Nostrand Reinhold, New York.

Kuhberger, A. (1995) The framing of decisions: A new look at old problems, *Organ Behav Human Dec*, 62, 230–240.

Kumar, S. (1994) A computer desk for bifocal lens wearers, with special emphasis on selected telecommunication tasks, *Ergonomics*, 37, 1669–1678.

Kundel, H. S. and LaFollette, P. S. (1972) Visual search patterns and experience with radiological images, *Radiology*, 103, 523–528.

Kuo, R. J. and Xue, K. C. (1998) A decision support system for sales forecasting through fuzzy neural networks with asymmetric fuzzy weights, *Dec Supp Syst*, 24(2), 105–126.

Kusiak, A. (Ed.) (1993) *Concurrent Engineering*. John Wiley & Sons, New York.

Kusiak, A., Larson, T. N., and Wang, J. R. (1994) Reengineering of design and manufacturing processes, *Comp Ind Eng*, 26(3), 521–536.

Kusiak, A. and Zakarian, A. (1996) Risk assessment of process models, *Comp Ind Eng*, 30(4), 599–610.

Kvalseth, T. O. (1981) An experimental paradigm for analyzing human information processing during motor control tasks. *Proceedings of the Human Factors Society 25th Annual Meeting*, Santa Monica, CA, pp. 581–585.

Lam, S.-H., Cao, J.-M., and Fan, H. (2003) Development of an intelligent agent for airport gate assignment, *J Air Transp*, 8(1), 103–114.

Langkilde, G., Alexandersen, K., Wyon, D. P., and Fanger, P. O. (1973) Mental performance during slight cool or warm discomfort, *Arch Des Sci Physiol*, 27, A511–A518.

Langolf, G. D., Chaffin, D. B., and Foulke, J. A. (1976) An investigation of Fitts' law using a wide range of movement amplitudes, *J Motor Behav*, 8, 113–128.

LaPlante, A. (1990) Bring in the expert: Expert systems can't solve all problems, but they're learning, *InfoWorld*, 12(40), 55–64.

Lari, A. (2003) A decision support system for solving quality problems using case-based reasoning, *Tot Qual Manage Bus*, 14(6), 733–745.

Larish, J. F. and Flach, J. M. (1990) Sources of optical information useful for perception of speed of rectilinear self-motion, *J Exp Psychol Hum Percep Perform*, 16, 295–302.

Lathrop, R. G. (1967) Perceived variability, *J Exp Psychol*, 73, 498–502.

Laughery, R., Drews, C., and Archer, R. (1986) A micro SAINT simulation analyzing operator workload in a future attack helicopter. *IEEE Proceedings*, NAECON, New York.

Laughlin, K. K. (1995) Increasing competitiveness with a simplified cellular process, *Ind Eng*, 27(4), 30–33, April.

Law, A. M. and Kelton, W. D. (2000) *Simulation Modeling and Analysis*. McGraw-Hill, Boston, MA.

Lee, D. (2008) Effects of special ability and richness of motion cues on learning mechanically complex domains. PhD dissertation, Purdue University, West Lafayette, IN.

Lehmann, G. (1958) Physiological measurement as a basis of work organization in industry, *Ergonomics*, 1, 328.

Lehto, M. R. (1985) A structured methodology for expert system development. Unpublished PhD dissertation, University of Michigan, Ann Arbor, MI.

Lehto, M. R. (1991) A proposed conceptual model of human behavior and its implications for the design of product warnings, *Percep Motor Skills*, 73, 595–611.

Lehto, M. R. (1992) Designing warning signs and warning labels: Part II—Scientific basis for initial guidelines, *Int J Ind Ergon*, 10, 115–138.

Lehto, M. R. (1996) Designing warning signs and labels: A theoretical/scientific framework, *J Consum Saf*, 3(4), 205–216.
Lehto, M. R. (2006) Human factors models. In Wogalter, M. S. (Ed.) *Handbook of Warnings*. Lawrence Erlbaum Associates, Inc., Mahwah, NJ, pp. 63–87, Chapter 6.
Lehto, M. R. and Buck, J. R. (1988) Status reporting of eyes on and peripheral displays during tracking, *Percep Mot Skills*, 67, 719–733.
Lehto, M. R. and Clark, D. R. (1990) Warning signs and labels in the workplace. In Karwowski, W. and Mital, A. (Eds.) *Workspace, Equipment and Tool Design*. Elsevier, Amsterdam, the Netherlands, pp. 303–344.
Lehto, M. R. and House, T. (1997) Evaluation of the comprehension of hazard communication phrases by chemical workers. *International Ergonomics Association 13th Triennial Congress*, Tampere, Finland, June 29–July 4.
Lehto, M. R., Lesch, M. F., and Horrey, W. S. (2000) Warnings and automation. In Nof, S. (Ed.) *Handbook of Automation*. Springer, New York, pp. 671–695.
Lehto, M. R. and Miller, J. M. (1986) *Warnings Volume I: Fundamentals, Design, and Evaluation Methodologies*. Fuller Technical Publications, Ann Arbor, MI.
Lehto, M. R. and Miller, J. M. (1988) The effectiveness of warning labels, *J Prod Liab*, 11, 225–270.
Lehto, M. R. and Miller, J. M. (1997) Principles of prevention: Safety information. In Stellman, J. M. (Ed.) *ILO Encyclopedia of Occupational Health and Safety*, 4th edn. International Labour Office, Geneva, Switzerland, pp. 56.33–56.38.
Lehto, M. R. and Papastavrou, J. (1993) Models of the warning process: Important implications towards effectiveness, *Saf Sci*, 16, 569–595.
Lehto, M. R., Boose, J., Sharit, J., and Salvendy, G. (1992) Knowledge acquisition. In Salvendy, G. (Ed.) *Handbook of Industrial Engineering*, 2nd edn. John Wiley & Sons, New York, pp. 1495–1545.
Lehto, M. R., House, T. E., and Papastavrou, J. D. (2000) Interpretation of fuzzy qualifiers by chemical workers, *Int J Cogn Ergon*, 4(1), 73–88.
Lehto, M. R., James, D. S., and Foley, J. P. (1994) Exploratory factor analysis of adolescent attitudes toward alcohol and risk, *J Safe Res*, 25, 197–213.
Lehto, M. R. and Lee, D. (2008) Learning in mechanically complex domains. Working Paper, Purdue University, West Lafayette, IN.
Lehto, M. R. and Miller, J. M. (1986) *Warnings Volume I. Fundamentals, Design, and Evaluation Methodologies*. Fuller Technical Publications, Ann Arbor, MI, p. 287.
Lehto, M. R. and Salvendy, G. (1991) Models of accident causation and their application: Review and reappraisal, *J Eng Technol Manage*, 8, 173–205.
Lehto, M. R., Sharit, J., and Salvendy, G. (1991) The application of cognitive simulation techniques to work measurement and methods analysis of production control tasks, *Int J Prod Res*, 29(8), 1565–1586.
Leonard, S. D., Otani, H., and Wogalter, M. S. (1999) Comprehension and memory. In Wogalter, M., DeJoy, D., and Laughery, K. (Eds.) *Warnings and Risk Communication*. Taylor & Francis, London, U.K., pp. 149–186.
Lerch, J. F. and Buck, J. R. (1975) An exponential learning curve experiment. *Proceedings of the Human Factors Society, 19th Annual Meeting*, Dallas, TX, pp. 116–172.
Levin, L. P. (1975) Information integration in numerical judgments and decision processes, *J Exp Psychol*, 104, 39–53.
Levy, F. K. (1965) Adaptation in the production process. *Manage Sci*, 11, 135–154.
Liberty Mutual. (2009) *Liberty Mutual Workplace Safety Index*. National Safety Council (2010), Hopkinton, MA., *Injury Facts*. National Safety Council, Chicago, IL.
Lichtenstein, S., Fischhoff, B., and Phillips, L. D. (1982) Calibration of probabilities: The state of the art to 1980. In Kahneman, D., Slovic, P., and Tversky, A. (Eds.) *Judgment under Uncertainty: Heuristics and Biases*. Cambridge University Press, Cambridge, NY, pp. 306–334.
Lichtenstein, S., Slovic, P., Fischhoff, B., Layman, M., and Coombs, B. (1978) Judged frequency of lethal events, *J Exp Psychol Hum Learn Mem*, 4, 551–578.
Lie, I. and Fostervold, K. I. (1994) VDT-work with different gaze inclinations. *Proceedings of the WWDU: 4th International Scientific Conference on Work with Display Units*, Milan, Italy, pp. 40–42, October 2–5.

Likert, R. and Likert, J. G. (1976) *New Ways of Managing Conflict*. McGraw-Hill, New York.

Lim, L.-H. and Benbasat, I. (1993) A theoretical perspective of negotiation support systems, *J Manage Inform Syst*, 9(3), 27–44.

Lintern, G. and Roscoe, S. N. (1980) Visual cue augmentation in contact flight simulation. In Roscoe, S. N. (Ed.) *Aviation Psychology*, Iowa State University Press, Ames, IA.

Lirtzman, S. I. (1984) Labels, perception, and psychometrics. In O'Connor, C. J. and Lirtzman, S. I. (Eds.) *Handbook of Chemical Industry Labeling*. Noyes, Park Ridge, IL, pp. 5–39.

Little, J. D. C. (1970) Models and managers: The concept of a decision calculus, *Manage Sci*, 16(8), 466–485.

Logan, L. R. (1993) Team members identify key ingredients for team-building success, *Nat Prod Rev*, 12(2), 209–223.

Lovetinsky, E. J. (1998) Use of ergonomic methods in an engineering environment. Unpublished master's thesis, Department of Industrial Engineering, The University of Iowa, Iowa City, IA.

Luczak, H. (1997) Task analysis. In Salvendy, G. (Ed.) *Handbook of Human Factors and Ergonomics*, 2nd edn. John Wiley & Sons, New York, Chapter 12.

MacKenzie, I. S. (1995) Movement time prediction in human-computer interfaces. In Baecker, R. M., Buxton, W. A. S., Grudin, J., and Greenberg, S. (Eds.) *Readings in Human-Computer Interaction*, 2nd edn. Kaufmann, Los Altos, CA, pp. 483–493. [reprint of MacKenzie, 1992.]

Mackworth, N. H. (1946) Effects of heat on wireless telegraphy operators hearing and recording Morse, *Brit J Ind Med*, 3, 143–158.

MacLeod, D. (1995) *The Ergonomics Edge: Improving Safety, Quality, and Productivity*. Van Nostrand Reinhold, New York.

Mahone, D. (1993) Work sampling as a method for assessing CTDS. *Proceedings of the Human Factors and Ergonomics Society, 37th Annual Meeting*, Seattle, WA, pp. 679–682.

Majoros, A. E. and Taylor, S. A. (1997, January) Work envelopes in equipment design, *Ergonomics Des*, 5(1), 18–24.

Mandal, A. C. (1984) What is the correct height of furniture? In Grandjean, E. (Ed.) *Ergonomics and Health in Modern Offices*. Taylor & Francis, Philadelphia, PA, pp. 465–470.

Mantei, M. M. and Teory, T. J. (1988) Cost/benefit analysis for incorporating human factors in the software design cycle, *Commun ACM*, 31, 428–439.

Manz, C. C. and Sims, H. P., Jr. (1987) Leading workers to lead themselves: The external leadership of self-managing work teams, *Admin Sci Q*, 32, 106–128.

Marks, M. L., Mirvis, P. H., Hackett, E. J., and Grady, J. F., Jr. (1986) Employee participation in quality circle program: Impact on quality of work life, productivity, and absenteeism, *J Appl Psychol*, 71, 61–69.

Martignon, L. and Krauss, S. (2003) Can L'Homme Eclaire be fast and frugal? Reconciling Bayesianism and bounded rationality. In Schnieder, S. and Shanteau, J. (Eds.) *Emerging Perspectives on Judgment and Decision Research*. Cambridge University Press, Cambridge, NY, pp. 108–122.

Martocchio, J. J., Webster, J., and Baker, C. R. (1993) Decision-making in management information systems research: The utility of policy capturing methodology, *Behav Info Technol*, 12, 238–248.

Matwin, S., Szpakowicz, S., Koperczak, Z., Kersten, G. E., and Michalowski, W. (1989) Negoplan: An expert system shell for negotiation support, *IEEE Exp*, 4(4), 50–62.

Maule, A. J. and Hockey, G. R. J. (1993) State, stress, and time pressure. In Svenson, O. and Maule, A. J. (Eds.) *Time Pressure and Stress in Human Judgment and Decision Making*. Plenum, New York, pp. 27–40.

Maynard, H. B., Stegemerten, G. J., and Schwab, J. L. (1948) *Methods-Time Measurement*. McGraw-Hill, New York.

Mayo, G. D. (1955) Effects of temperature upon technical training, *J Appl Psychol*, 39, 244–249.

McClelland, I. (1990) Product assessment and usability trials. In Wilson, J. R. and Corlett, E. N. (Eds.) *Evaluation of Human Work*. Taylor & Francis, New York, pp. 218–247, *Human Factors in Engineering and Design*. McGraw-Hill, New York.

McCormick, E. J. (1979) *Job Analysis: Methods and Applications*. AMACOM—American Management Associations, New York.

McFarling, L. H. and Heimstra, N. W. (1975) Pacing, product complexity, and task perceptions in simulated inspection, *Hum Fact*, 17, 361–367.

McGinty, R. L. and Hanke, J. (1989) Compensation management in practice = merit pay plans: Are they truly tied to performance? *Comp Benefit Rev*, 12(5), 12–16.

McGrath, J. E. (1964) Toward a "theory of method" for research in organizations. In Cooper, W. W., Leavitt, H. J., and Shelly, M. W. (Eds.) *New Perspectives in Organization Research*. John Wiley & Sons, New York, pp. 533–556.

McGrath, J. E. (1984) *Groups: Interaction and Performance*. Prentice Hall, Englewood Cliffs, NJ.

McGuire, W. J. (1966) Attitudes and opinions, *Annu Rev Psychol*, 17, 475–514.

McGuire, W. J. (1980) The communication-persuasion model and health-risk labeling. In Morris, L. A., Mazis, M. B., and Barofsky, I. (Eds.) *Product Labeling and Health Risks*, Banbury Report 6. Cold Spring Harbor Laboratory, pp. 99–122.

McKarns, J. S. and Brief, R. S. (1966) Nomographs give refined estimate of heat stress index, *Heat Pipe Air Cond*, 38, 113–116.

McKnight, A. J., Langston, E. A., McKnight, A. S., and Lange, J. E. (1995) The bases of decisions leading to alcohol impaired driving. In Kloeden, C. N. and McLean, A. J. (Eds.) *Proceedings of the 13th International Conference on Alcohol, Drugs, and Traffic Safety*, Adelaide, South Australia, August 13–18, pp. 143–147.

McLeod, P. L. (1992) An assessment of the experimental literature on electronic support of group work: Results of a meta-analysis, *Hum Comp Interact*, 7, 257–280.

McRuer, D. T. and Jex, H. R. (1967, September) A review of quasi-linear pilot models, *IEEE Trans Hum Fact Elect*, 8(3), 231–249.

Megaw, W. D. (Ed.) (1988) *Contemporary Ergonomics*. Taylor & Francis, Ltd., London, U.K.

Meister, D. (1971) *Human Factors: Theory and Practice*. John Wiley & Sons, New York.

Meister, D. (1991) *Psychology of System Design*. Elsevier, Amsterdam, the Netherlands.

Meister, D. and Rabideau, G. F. (1965) *Human Factors Evaluation in System Development*. John Wiley & Sons, New York.

Messick, D. M. (1991) Equality as a decision heuristic. In Mellers, B. (Ed.) *Psychological Issues in Distributive Justice*. Cambridge University Press, Cambridge, NY.

Meyer, D., Abrams, R., Kornblum, S., Wright, C., and Smith, J. E. (1988) Optimality in human motor performance: Ideal control of rapid aimed movements, *Psychol Rev*, 95(3), 340–370.

Meyer, J. and Seagull, F. J. (1996) When human factors meets marketing, *Ergonom Des*. 4, 22–55.

Milas, G. H. (1992) IE's role in implementing "The Americans With Disabilities Act." *Ind Eng*, January, 36–39.

MIL-HDBK-759C. (1995, July 31) *Department of Defense Handbook for Human Engineering Design Guidelines*.

Miller, J. M. and Lehto, M. R. (2001) *Warnings and Safety Instructions: The Annotated Bibliography*, 4th edn. Fuller Technical Publications, Ann Arbor, MI.

Miller, J. M., Rhoades, T. M., and Lehto, M. R. (1987) *Slip Resistance Predictions for Various Metal Step Materials, Shoe Soles and Contaminant Conditions*. Society of Automotive Engineers, Warrendale, PA.

Miller, R. B. (1962) Task description and analysis. In Gagné, R. M. (Ed.) *Psychological Principles in System Design*. Rinehart & Winston, New York, pp. 187–228.

Miller, R. B. (1973) Development of a taxonomy of human performance, *JSAS Cat Sel Doc Psychol*, 3, 29–30.

Miller, R. B. (1974) A method for determining task strategies. AFHRL-TR-74–26. American Institutes for Research, Washington, DC.

Mills, R. G. and Hatfield, S. A. (1974) Sequential task performance task module relationships, reliabilities, and times, *Hum Fact*, 16(2), 117–128.

Mills, R. G., Bachert, R. F., and Hatfield, S. A. (1975) Quantification and prediction of human performance: Sequential task performance reliability and time (AMRL Technical Report 74-48). Aerospace Medical Research Laboratory, Wright-Patterson AFB, OH (DTIC No. A017 333).

Minsky, M. (1975) A framework for representing knowledge. In Winston, P. (Ed.) *The Psychology of Computer Vision*. McGraw-Hill, New York, pp. 211–277.

Mital, A. and Wright, U. (1996, July) Weight limits for foldaway wheelchairs, *Ergon Des*, 4(3), 26–30.

Mizrahi, I., Isaacs, B., Barnea, T., Bentur, N., and Simkin, A. (1995) Adapting the chair, adapting the user, *Ergon Des*, 4, 27–32.

Monk, T. H. (1986) Advantages and disadvantages of rapidly rotating shift schedules-A circadian viewpoint, *Hum Fact*, 28(5), 553–557.
Montgomery, D. C. (2004) *Design and Analysis of Experiments*, 6th edn. John Wiley & Sons, New York.
Montgomery, H. (1989) From cognition to action: The search for dominance in decision making. In Montgomery, H. and Svenson, O. (Eds.) *Process and Structure in Human Decision Making*. John Wiley & Sons, Chichester, U.K.
Montgomery, H. and Willen, H. (Eds.) (1999) Decision making and action: The search for a good structure. In *Judgment and Decision Making: Neo-Brunswikian and Process-Tracing Approaches*. Lawrence Erlbaum Associates, Hillsdale, NJ, pp. 147–173.
Mood, A. M. (1950) *Introduction to the Theory of Statistics*. McGraw-Hill, New York.
Morata, T. C., Fiorini, A. C., Fischer, F. M., Krieg, E. F., Gozzoli, L., and Colacioppo, S. (2001) Factors affecting the use of hearing protectors in a population of printing workers, *Noise Health*, 4(13), 25–32.
Moray, N. (1986) Monitoring behavior and supervisory control. In Boff, K., Beatty, J., and Kaufmann, L. (Eds.) *Human Performance*. Wiley, New York, Chapter 40.
Morgan, C. T., Cook, J. S., III, Chapanis, A., and Lund, M. W. (1963) *Human Engineering Guide to Equipment Design*. McGraw-Hill Book Company, New York.
Morgan, B. B., Jr., Glickman, A. S., Woodard, E. A., Blaiwes, A. S., and Salas, E. (1986) Measurement of team behaviors in a Navy environment. Technical Report No. NTSC TR-86-014, Naval Training Systems Center, Orlando, FL.
Moscovici, S. (1976) *Social Influence and Social Change*. Academic Press, London, U.K.
Mottram, R. D. (1982) Team skills management, *J Manage Dev*, 1(1), 22–33.
Muckler, F. A., Nygaard, J. E., O'Kelly, L. L., and Williams, A. C. (1959) Psychological variables in the design of flight simulators for training. Technical Report 56-369, Wright Air Development Center, Wright Patterson Air Force Base, OH.
Mundell, M. (1978) *Motion and Time Study*. Prentice-Hall, Englewood Cliffs, NJ.
Mundell, M. E. (1985) *Motion and Time Study: Improving Productivity*, 6th edn. Prentice Hall, Englewood Cliffs, NJ.
Munro, H. P., Martin, F. W., and Roberts, M. C. (1968) How to use simulation techniques to determine the optimum manning levels of continuous process plants. In Edwards, E. and Lees, F. P. (Eds.) *The Human Operator in Process Control*. Taylor & Francis, London, U.K., pp. 320–326.
Murphy, A. H. and Winkler, R. L. (1974) Probability forecasts: A survey of National Weather Service Forecasters, *Bull Am Meteorol Soc*, 55, 1449–1453.
Murrell, K. F. H. (1965) *Ergonomics: Man in His Working Environment*. Chapman & Hall, London, U.K.
Murrell, K. F. H. (1969) *Ergonomics: Man in His Working Environment*. Chapman & Hall, London, U.K., pp. 372–378.
Murrell, K. F. H., Griew, S., and Tucker, W. A. (1957) Age structure in the engineering industry: A preliminary study, *Occup Psychol*, 31, 150.
Muth, J. F. (1986) Search theory and the manufacturing process function, *Manage Sci*, 32(8), 948–962.
Nadler, G. (1963) The measurement of three-dimensional human motion. In Bennett, E., Degan, J., and Spiegel, J. (Eds.) *Human Factors in Technology*. McGraw-Hill, New York, pp. 596–612.
Nadler, G. (1970) *Work Design: A Systems Concept*, Rev. edn. Richard D. Irwin, Inc., Homewood, IL.
Nadler, G. and Goldman, J. (1958, January–February) The UNOPAR, *J Ind Eng*, 9(1), 59–65.
Nadler, G. and Smith, W. (1963) Manufacturing progress functions for types of processes, *Int J Prod Res*, 2, 115–135.
Nah, F. and Benbasat, I. (2004) Knowledge-based support in a group decision making context: An expert-novice comparison, *J Assoc Inform Syst*, 5(3), 125–150.
Nah, F. H., Mao, J., and Benbasat, I. (1999) The effectiveness of expert support technology for decision making: Individuals versus small groups, *J Inform Technol*, 14(2), 137–147.
Nakajima, S., Yamashina, H., Kumagai, C., and Toyota, T. (1992) Maintenance management and control. In Salvendy, G. (Ed.) *Handbook of Industrial Engineering*, 2nd edn. John Wiley & Sons, New York, Chapter 73.
Nanda, R. and Adler, G. (Eds.) (1977) *Learning Curves Theory and Applications*. Monograph 6, American Institute of Industrial Engineers, Norcross, GA.

Naruo, N., Lehto, M. R., and Salvendy, G. (1990) Development of a knowledge-based decision support system for diagnosing malfunctions of advanced production equipment. *Int J Prod Res*, 28, 2259–2276.

National Highway Transportation Services Administration. (1995) A leadership guide to quality improvement for emergency medical services systems. Available online: http://www.nhtsa.gov/people/Injury/ems/leaderguide/ (accessed on July 11, 2012).

National Safety Council. (2010) *Injury Facts*. National Safety Council, Chicago, IL.

Natrella, M. (1963) *Experimental Statistics*, Handbook 91. Superintendent of Public Documents, Washington, DC.

Navon, D., and Gopher, D. (1979) On the economy of the human information-processing system, *Psychol Rev*, 86, 214–255.

Nayak, U. S. L. (1995) Elders-led design. *Ergon Des*, 1, 8–13.

Neelam, S. (1994) Using torque arms to reduce CTDs, *Ergon Des*, 2, 25–28.

Neibel, B. W. (1988) *Motion and Time Study*, 7th edn. Richard D. Irwin, Inc., Homewood, IL.

Neibel, B. W. and Freivalds, A. (2003) *Methods, Standards, and Work Design*. McGraw-Hill, New York.

Neilsen, J. (1993) *Usability Engineering*. Academic Press Professional, Boston, MA.

Nelson, J. B. and Barany, J. W. (1969) A dynamic visual recognition test, *AIIE Trans*, 1, 327–332.

New York Times. (2003) A leg with a mind of its own. (January 3).

von Neumann, J. (1951) Various techniques used in connection with random digits. (Summarized by George E. Forsythe), *J Res Nat Bur Stand Appl Math Ser*, 3, 36–38.

von Neumann, J. and Morgenstern, O. (1947) *Theory of Games and Economic Behavior*. Princeton University Press, Princeton, NJ.

Newell, A. F. and Cairns, A. Y. (1993) Designing for extraordinary users, *Ergon Des*, 10, 10–16.

Newell, A. and Simon, H. A. (1972) *Human Problem Solving*. Prentice-Hall, Englewood Cliffs, NJ.

Newell, A., Shaw, J. C., and Simon, H. A. (1958) Elements of a theory of human problem solving, *Psychol Rev*, 65, 151–166.

Newman, D. G. (1996) *Engineering Economic Analyses*, 6th edn. Engineering Press, San Jose, CA.

Newman, D. G., Eschenbach, T. G., and Lavelle, J. P. (2004) *Engineering Economic Analysis*, 9th edn. Engineering Press, Engineering Press, San Jose, CA.

Nielsen, J. (1994) Heuristic evaluation. In Nielsen, J. and Mack, R. L. (Eds.) *Usability Inspection Methods*. John Wiley & Sons, New York.

Nisbett, R. E. and Wilson, T. D. (1977) Telling more than we can know: Verbal reports on mental processes, *Psychol Rev*, 84, 231–259.

Nisbett, R. and Ross, L. (1980) *Human Inference: Strategies and Shortcomings of Social Judgment*. Prentice-Hall, Englewood Cliffs, NJ.

Nolan, J. A. (1960) Influence of classroom temperature on academic learning, *Auto Teach Bull*, 1, 12–20.

Norman, D. A. (1988) *The Psychology of Everyday Things*. Basic Books, Harper Collins, New York.

Norman, D. A. (1992) *Turn Signals Are the Facial Expressions of Automobiles*. Addison-Wesley, Reading, MA.

Nunamaker, J. F., Dennis, A. R., Valacich, J. S., Vogel, D. R., and George, J. F. (1991) Electronic meeting systems to support group work: Theory and practice at Arizona, *Commun ACM*, 34, 40–61.

O'Keefe, R., Kernaghan, J., and Rubenstein, A. (1975) Group cohesiveness: A factor in the adoption of innovations among scientific work groups, *Small Group Behav*, 5, 282–292.

O'Reilly, C. A. and Roberts, K. H. (1977) Task group structure, communications, and effectiveness in three organizations, *J Appl Psychol*, 62, 674–681.

Occupational Safety and Health Administration. (1999) *OSHA Technical Manual* (Directive # TED 01-00-015). OSHA, Washington, DC.

Olson, P. L. and Bernstein, A. (1979) The nighttime legibility of highway signs as a function of their luminance characteristics, *Hum Fact*, 21(2), 145–160.

Olson, D. and Mossman, C. (2003) Neural network forecasts of Canadian stock returns using accounting ratios, *Int J Forecast*, 19(3), 453–465.

Orasanu, J. (1997) Stress and naturalistic decision making: Strengthening the weak links. In Flin, R., Salas, E., Strub, M., and Martin, L. (Eds.) *Decision Making under Stress: Emerging Themes and Applications*. Ashgate, Aldershot, U.K.

Orasanu, J. and Salas, E. (1993) Team decision making in complex environments. In Klein, G. A., Orasanu, J., Calderwood, R., and Zsambok, E. (Eds.) *Decision Making in Action: Models and Methods*. Ablex, Norwood, NJ.

Orlansky, J. (1949) Psychological aspects of stick and rudder controls in aircraft, *Aero Eng Rev*, January, 22–31.

Osborn, F. (1937) *Applied Imagination*. Charles Scribner & Sons, New York.

Østerberg, G. (1935) Topography of the layer of rods and cones in the human retina, *Acta Ophthalmol*, Supplement, 6, 1–103.

Pachella, R. (1974) The use of reaction time measures in information processing research. In Kantowitz, B. H. (Ed.) *Human Information Processing*. Erlbaum Associates, Hillsdale, IL.

Paese, P. W., Bieser, M., and Tubbs, M. E. (1993) Framing effects and choice shifts in group decision making, *Organ Behav Hum Dec Proc*, 56, 149–165.

Papastavrou, J. and Lehto, M. R. (1995) A distributed signal detection theory model: Implications for the design of warnings, *Int J Occup Safe Ergon*, 1(3), 215–234.

Parasuraman, R. (1986) Vigilance, monitoring, and search. In Boff, K. R., Kaufman, L., and Thomas, J. P. (Eds.) *Handbook of Perception and Human Performance*. John Wiley & Sons, New York, Chapter 43.

Park, C. S. (2001) *Contemporary Engineering Economics*, 3rd edn. Addison-Wesley, Reading, MA, 2001.

Park, K. and Chaffin, D. (1974) Biomechanical evaluation of two methods of manual load lifting. *AIIE Trans*, 6(2), 105–113.

Parks, D. L. (1979) Current workload methods. In Moray, N. (Ed.) *Mental Workload: Its Theory and Measurement*. Plenum Press, New York.

Patton, J., Littlefield, C., and Self, S. (1964) *Job Evaluation Text and Cases*. Irwin, Homewood, IL.

Payne, J. W. (1973) Alternative approaches to decision making under risk: Moments versus risk dimensions, *Psychol Bull*, 80, 439–453.

Payne, J. W. (1980) Information processing theory: Some concepts and methods applied to decision research. In Wallsten, T. S. (Ed.) *Cognitive Processes in Choice and Decision Research*. Erlbaum, Hillsdale, NJ.

Payne, J. W., Bettman, J. R., and Johnson, E. J. (1993) *The Adaptive Decision Maker*. Cambridge University Press, Cambridge, NY.

Pegels, C. C. (1969) On startup of learning curves: An expanded view, *AIIE Trans*, 1(3), 216–222.

Penniman, W. (1975) A stochastic process analysis of online user behavior. *Proceedings of the 38th Annual Meeting of the American Society for Information Science*, Boston, MA, pp. 147–148.

Pennington, N. and Hastie, R. (1988) Explanation-based decision making: Effects of memory structure on judgment, *J Exp Psychol Learn Mem Cogn*, 14, 521–533.

Pepler, R. D. (1958) Warmth and performance: An investigation in the tropics. *Ergonomics*, 2, 63.

Pepler, R. D. (1964) Physiological effects of heat. In Leithhead, C. S. and Lind, A. R. (Eds.) *Heat Stress and Heat Disorders*. F. A. Davis Company, Philadelphia, PA.

Peterson, C. R. and Beach, L. R. (1967) Man as an intuitive statistician, *Psychol Bull*, 68, 29–46.

Petroski, H. (1993) *The Evolution of Useful Things*. Alfred A. Knopf, New York.

Pew, R. W. (1969) The speed-accuracy operating characteristic, *Acta Psychol*, 30, 16–26.

Pheasant, S. (1999) *Body Space: Anthropometry, Ergonomics and the Design of Work*, 2nd edn. Taylor & Francis, London, U.K.

Pheasant, S. (1990) Anthropometry and the design of workspaces. In Wilson, J. R. and Corlett, E. N. (Eds.) *Evaluation of Human Work*. Taylor & Francis, London, U.K., pp. 455–471.

Piso, E. (1981) Task analysis for process control tasks: The method of Annett et al. applied, *Occup Psychol*, 54, 247–254.

Pitz, G. F. (1980) The very guide of life: The use of probabilistic information for making decisions. In Wallsten, T. S. (Ed.) *Cognitive Processes in Choice and Decision Behavior*. Lawrence Erlbaum Associates, Mahwah, NJ.

Pliske, R. M., McCloskey, M. J., and Klein, G. (2001) Decision skills training: Facilitating learning from experience. In Salas, E. and Klien, G. (Eds.) *Linking Expertise and Naturalistic Decision Making*. Lawrence Erlbaum Associates Inc., Hillsdale, NJ, pp. 37–53.

Pliske, R. and Klein, G. (2003) The naturalistic decision-making perspective. In Schnieder, S. and Shanteau, J. (Eds.) *Emerging Perspectives on Judgment and Decision Research*. Cambridge University Press, Cambridge, NY, pp. 108–122.

References

Polk, E. J. (1984) *Methods Analysis and Work Measurement*. McGraw-Hill, New York.
Pope, M. H., Andersson, B. J., Frymoyer, J. W., and Chaffin, D. B. (1991) *Occupational Low Back Pain: Assessment, Treatment and Prevention*. Mosby Yearbook, St. Louis, MI.
Poulsen, E. and Jorgensen, K. (1971) Back muscle strength, lifting and stooped working postures, *Appl Ergon*, 2(3), 133–137.
Poulton, E. (1970) *Environment and Human Efficiency*. C. T. Thomas, Springfield, IL.
Povenmire, H. K. and Roscoe, S. N. (1973) Incremental transfer effectiveness of a ground-based general aviation trainer, *Hum Fact*, 13, 534–542.
Poza, E. J. and Markus, M. L. (1980) Success story: The team approach to work restructuring, *Organ Dyn*, Winter, 8, 3–25.
Presgrave, R. (1945) *The Dynamics of Time Study*. McGraw-Hill Book Company, Inc., New York, London, U.K.
Pritchard, R. D., Jones, S., Roth, P., Stuebing, K., and Ekeberg, S. (1988) Effects of group feedback, goal setting, and incentives on organizational productivity, *J Appl Psychol*, 73(2), 337–358.
Pritchett, P. and Muirhead, B. (1998) *The MARS PATHFINDER Approach to "Faster-Better-Cheaper."* Pritchett and Associates, Inc., Dallas, TX.
Pritsker, A. A. B. and Happ, W. W. (1966) GERT: Graphical evaluation and review technique, Part 1. Fundamental, *J Ind Eng*, 17(5), 267–274.
Proctor, R. and Vu, K. (2006) *Stimulus-Response Compatibility Principles: Stimulus-Response Compatibility Principles*. Taylor & Francis, CRC Press, New York, p. 360.
Proctor, S. L., Khalid, T. M., and Steve, C. C. (1971) A quantitative approach to performance valuation of man-machine systems having a stochastic environment, *Int J Man-Mach Stud*, 33, 127–140.
Putz-Anderson, V. (1988) *Cumulative Trauma Disorders—A Manual for Musculo-Skeletal Diseases of the Upper Limb*. Taylor & Francis, Inc., Bristol, PA.
Raiffa, H. (1968) *Decision Analysis*. Addison-Wesley, Reading, MA.
Raiffa, H. (1982) *The Art and Science of Negotiation*. Harvard University Press, Cambridge, MA.
Rajendra, N. (1987) The effects of task streams and mixtures on the performance time and accuracy of independent cognitive tasks. Unpublished master's thesis, University of Iowa, Iowa City, IA.
Ramsden, J. (1976) *The Safe Airline*. McDonnalds and James, London, U.K.
Ramsey, C. G. and Sleeper, H. R. (1932) *Architectural Graphic Standards*. John Wiley & Sons, New York.
Rasmussen, J. (1983) Skills, rules, knowledge: Signals, signs, and symbols and other distinctions in human performance models, *IEEE Trans Syst Man Cyber*, SMC-13(3), 257–267.
Rasmussen, J. (1985) *Information Processing and Human-Machine Interactions: As Approach to Cognitive Engineering*. North-Holland Publishing, New York.
Rasmussen, J. (1986) *Information Processing and Human-Machine Interaction*. North-Holland, Amsterdam, the Netherlands.
Reason, J. (1990) *Human Error*. Cambridge University Press, Cambridge, U.K.
Rempel, D., Horie, S., and Tal, R. (1994) Carpal tunnel pressure changes during keying. *Proceedings of the Marconi Keyboard Research Conference*, Ergonomics Laboratory, UC San Francisco, Berkeley, CA, pp. 1–3.
Rethans, A. J. (1980) Consumer perceptions of hazards. In *PLP-80 Proceedings*, pp. 25–29.
Reyna, V. F. and Brainerd, C. F. (1995) Fuzzy-trace theory: An interim synthesis, *Learn Individ Diff*, 7, 1–75.
Ridd, J. E. (1985) Spatial restraints and intra-abdominal pressure, *Ergonomics*, 28, 149–166.
Rigby, L. (1973) Why do people drop things? *Qual Prog*, 6, 16–19.
Rigby, L. V., Cooper, J. I., and Spickard, W. A. (1961) Guide to integrated system design for maintainability. USAF, ASD, TR 61–424.
Riley, M. W. and Cochran, D. J. (1984) Dexterity performance and reduced ambient temperature, *Hum Fact*, 26(2), 207–214.
Rizzi, A. M., Buck, J. R., and Anderson, V. L. (1979) Performance effects of variables in dynamic visual inspection, *AIIE Trans*, 11(4), 278–285.
Robert, H. M. (1990) *Robert's Rules of Order Newly Revised*, 9th edn. Scott, Foresman, Glenview, IL.
Roberts, E. B. (1976) *Symposium on: The Management of Innovation*. Reprinted by the MIT Alumni Center, Cambridge, MA.
Roberts, P. C. (1983) A theory of the learning process, *J Oper Res Soc Am*, 34, 71–79.

Robinson, D. W. and Dadson, R. S. (1957) Threshold of hearing and equal-loudness relations for pure tones and the loudness function, *J Accoust Soc Am*, 29, 1284–1288.

Robinson, G. H. (1964) Continuous estimation of a time-varying probability, *Ergonomics*, 7, 7–21.

Rodgers, S. H. (1986) *Ergonomic Design for People at Work*, Vol. 2. Van Nostrand Reinhold Company, New York, Chapter 7.

Roebuck, J. A., Jr. (1995) *Anthropometric Methods: Designing to Fit the Human Body. Monographs in Human Factors and Ergonomics*. Human Factors and Ergonomics Society, Santa Monica, CA.

Roebuck, J. A., Jr., Kroemer, K. H. E., and Thomson, W. G. (1975) *Engineering Anthropometry Methods*. John Wiley & Sons, New York.

Rohles, F. H., Jr., Hayter, R. B., and Milliken, B. (1975) Effective temperature (ET*) as a predictor of thermal comfort, *ASHRAE Trans*, 80(part 2), 148–156.

Roland, H. E. and Moriarty, B. (1990) *Systems Safety Engineering and Management*. Wiley, New York, p. 367.

Roscoe, S. N. (1980) *Aviation Psychology*. Iowa State University Press, Ames, IA.

Rose, L., Ericson, M., Glimskar, B., Nordgren, B., and Ortengren, R. (1992) Ergo-index. Development of a model to determine pause needs after fatigue and pain reactions during work. In Mattila, M. and Karwowski, W. (Eds.) *Computer Applications in Ergonomics, Occupational Safety and Health*. Elsevier Science Publishers B. V, Amsterdam, the Netherlands, pp. 461–468.

Rouse, W. B. (1980) *Systems Engineering Models of Human-Machine Interaction*. North Holland, New York.

Rouse, W. B. (1991) *Design for Success*. John Wiley & Sons, New York.

Rubenstein, A., Barth, R., and Douds, C. (1971) Ways to improve communications between R&D groups, *Res Manage*, 14, 49–59.

Rumelhart, D. E. and McClelland J. L. (1986) *Parallel Distributed Processing: Explorations in the Microstructure of Cognition*, Vol. 1. MIT Press, Cambridge, MA.

Russell, R. W. (1957) Effects of variations in ambient temperature on certain measures of tracking skill and sensory sensitivity. US Army Medical Research Laboratory, Fort Knox, KY. AMRL Report 300.

Russo, J., Johnson, E. J., and Stephens, D. L. (1989) The validity of verbal protocols, *Mem Cogn*, 17, 759–769.

Saaty, T. L. (1988) *Multicriteria Decision Making: The Analytic Hierarchy Process*. T. Saaty, Pittsburgh, PA.

Sage, A. (1981) Behavioral and organizational considerations in the design of information systems and processes for planning and decision support, *IEEE Trans Syst Man Cyber*, SMC-11, 61–70.

Sahal, D. (1979) A theory of process functions, *AIIE Trans*, 11, 23–29.

Sale, D. R. (1978, September) The effect of alternate periods of learning and forgetting on economic manufacturing quantity, *AIIE Trans*, 10(3), 338–343.

Salvendy, G. (2001) *Handbook of Industrial Engineering, Technology, and Operations Management*, 3rd edn. John Wiley, New York, pp. 2172–2223.

Salvendy, G. (2006) *Handbook of Human Factors and Ergonomics* 3rd edn. John Wiley, New York.

Salvendy, G. (2012) *Handbook of Human Factors and Ergonomics*, 4th edn. John Wiley, New York, p. 1732.

Sanders, M. S. and McCormick, E. J. (1987) *Human Factors in Engineering and Design*. McGraw-Hill, New York.

Sanderson, P. M., James, J. M., and Seidler, K. S. (1989) SHAPA: An interactive software environment for protocol analysis, *J Exp Psychol Learn Mem Cogn*, 15, 729–747.

Sapena, O., Botti, V., and Argente, E. (2003) Application of neural networks to stock prediction in 'pool' companies, *Appl Art Intell*, 17(7), 661–673.

Satish, U. and Streufert, S. (2002) Value of a cognitive simulation in medicine: Towards optimizing decision making performance of healthcare personnel, *Qual Saf Health Care*, 11(2), 163–167.

Satzinger, J. and Olfmann, L. (1995) Computer support for group work: Perceptions of the usefulness of support scenarios and specific tools, *J Manage Syst*, 11(4), 115–148.

Savage, L. J. (1954) *The Foundations of Statistics*. John Wiley & Sons, New York.

Schank, R. C. and Abelson, R. P. (1977) *Scripts, Plans, Goals, and Understanding*. Lawrence Erlbaum Associates, Hillsdale, NJ.

Schelling, T. (1960) *The Strategy of Conflict*. Harvard University Press, Cambridge, MA.

Schelling, T. (1978) *Micromotives and Macrobehavior*. W. W. Norton, New York.

Schmidt, D. K. (1978) A queuing analysis of the air traffic controller's work load, *IEEE Trans Syst Man Cyber*, 8(6), 492–498, June.
Schmidt, J. W. and Taylor, R. E. (1970) *Simulation and Analysis of Industrial Systems*. Richard D. Irwin, Homewood, IL.
Schmidtke, H. and Stier, F. (1961) An experimental evaluation of the validity of predetermined elemental time systems, *J Ind Eng*, 12(3), 182–204.
Schmitt, S. A. (1969) *Measuring Uncertainty: An Elementary Introduction to Bayesian Statistics*. Addison-Wesley Publishing Co., Reading, MA.
Schroeder, W., Lorensen, W., and Linthicum, S. (1994) Implicit modeling of swept surfaces and volumes. *Proceedings of Visualization '94*. IEEE Press, New York, pp. 40–45.
Schultz, G. M. (1961) Beware of diagonal lines in bar graphs, *Prof Geographer*, 13(4), 28–29.
Schweikart, R. and Fisher, D. (1987) Stochastic network models. In Salvendy, G. (Ed.) *Handbook of Human Factors*. John Wiley & Sons, New York, pp. 1177–1211.
Scott-Morton, M. S. (1977) *Management Decision Systems: Computer-Based Support for Decision Making*. Harvard University, Cambridge, MA.
Sellie, C. (1961) Comments on An experimental evaluation of the validity of predetermined elemental time systems, *J Ind Eng*, 13(5), 330–333.
Selye, H. (1936) A syndrome produced by noxious agents, *Nature*, 138, 32.
Selye, H. (1979) The stress concept and some of its implications. In Hamilton, V. and Warburton, D. M. (Eds.) *Human Stress and Cognition*. Wiley, New York.
Seminara, J. L. (1993) Taking control of controls, *Ergon Des*, July, 21–25.
Seymour, W. D. (1966) *Industrial Skills*. Sir Isaac Pitman & Sons Ltd., London, U.K.
Shannon, C. E. (1948) A mathematical theory of communication. *Bell Syst Technol J*, 27, 623–656.
Sharda, R., Barr, S. H., and McDonnell, J. C. (1988) Decision support system effectiveness: A review and an empirical test, *Manage Sci*, 34(2), 139–159.
Sheehan, J. J. and Drury, C. G. (1971) The analysis of industrial inspection, *Appl Ergon*, 2(2), 74–78.
Shehab, R. L. and Schlegel, R. E. (1993) A test of Fitts' law in a dual-task paradigm. *Proceedings of the Human Factors and Ergonomics Society, 37th Annual Meeting*, Santa Monica, CA, pp. 559–563.
Shepherd, A. (1989) Analysis and training in information tasks. In Diaper, D. (Ed.) *Task Analysis for Human-Computer Interaction*. Ellis Horwood, Chichester, U.K.
Sheridan, T. B. (1983) Supervisory control systems. In Committee on Human Factors (Eds.) *Research Needs for Human Factors*. National Academy Press, Washington, DC.
Sheridan, T. B. (2006) Supervisory control. In Salvendy, G. (Ed.) *Handbook of Human Factors and Ergonomics*, 3rd edn. John Wiley, New York, pp. 1025–1052.
Sheridan, T. B. (Ed.) (1984) *Supervisory Control*. National Academy Press, Washington, DC.
Sidowski, J. B. (1966) *Experimental Methods and Instrumentation in Psychology*. McGraw-Hill, New York.
Siebold, D. R. (1992) Making meetings more successful: Plans, formats, and procedures for group problem-solving. In Cathcart, R. and Samovar, L. (Eds.) *Small Group Communication*, 6th edn. Brown, Dubuque, IA, pp. 178–191.
Siegel, S. (1956) *Nonparametric Statistics for the Behavioral Sciences*. McGraw-Hill, New York.
Siegel, A. I. and Wolf, J. J. (1969) *Man-Machine Simulations Models: Psychosocial and Performance Interactions*. John Wiley & Sons, New York.
Simon, H. A. (1955) A behavioral model of rational choice, *Q J Econ*, 69, 99–118.
Simon, H. (1977) *The New Science of Management Decisions*. Prentice Hall, Englewood Cliffs, NJ.
Simon, H. A. (1983) *Alternative Visions of Rationality. Reason in Human Affairs*. Stanford University Press, Stanford, CA.
Simpson, G. C. (1993) Applying ergonomics in industry: Some lessons from the mining industry. In Lovesey, E. J. (Ed.) *Proceedings of the Ergonomics Society's 1993 Annual Conference on Contemporary Ergonomics*, Edinburgh, Scotland, April 13–16, 1993, pp. 490–503.
Sinclair, M. A. (1975) Questionnaire design, *Appl Ergon*, 6, 73–80.
Sinclair, M. A. (1990) Subjective assessment. In Wilson, J. R. and Corlett, E. N. (Eds.) *Evaluation of Human Work: A Practical Ergonomics Methodology*. Taylor & Francis, Ltd., London, U.K., pp. 58–88.
Singley, M. K. and Andersen, J. R. (1989) *The Transfer of Cognitive Skill*. Harvard University Press, Cambridge, MA.

Sipple, P. A. and Passel, C. F. (1945) Movement of dry atmospheric cooling in subfreezing temperatures, *Proc Am Philos Soc*, 89, 177–199.

Slovic, P. (1978) The psychology of protective behavior, *J Saf Res*, 10, 58–68.

Slovic, P. (1987) Perception of risk, *Science*, 236, 280–285.

Slovic, P., Fischhoff, B., and Lichtenstein, S. (1977) Behavioral decision theory. *Annu Rev Psychol*, 28, 1–39.

Smillie, R. J. (1985) Design strategies for job performance aids. In Duffy, T. M. and Waller, R. (Eds.) *Designing Usable Text*. Academic Press, Orlando, FL, pp. 213–242, 341–375, Chapter 10.

Smith, S. L. (1984) Letter size and legibility. In Easterby, R. and Zwaga, H. (Eds.) *Information Design*. John Wiley & Sons Ltd., New York, pp. 171–186.

Smith, L. A. and Barany, J. W. (1971) An elementary model of human performance on paced visual inspection, *AIIE Trans*, 4, 298–308.

Smode, A. F., Hall, E. R., and Meyer, D. E. (1966) An assessment of research relevant to pilot training. Report AMRL-TR-66-196. Aeromedical Research Laboratory, Wright-Patterson AFN, OH.

Sniezek, J. A. (1992) Groups under uncertainty: An examination of confidence in group decision making, *Organ Behav Hum Dec*, 52, 124–155.

Sniezek, J. A. and Henry, R. A. (1989) Accuracy and confidence in group judgment. *Organ Behav Hum Dec*, 43, 1–28.

Sniezek, J. A., Wilkins, D. C., Wadlington, P. L., and Baumann, M. R. (2002) Training for crisis decision-making: Psychological issues and computer-based solutions, *J Manage Inform Syst*, 18(4), 147–168.

Snoddy, G. S. (1926) Learning and stability, *J Appl Psychol*, 10, 1–36.

Snow, M. P., Kies, J. K., Neale, D. C., and Williges, R. C. (1996) A case study in participatory design, *Ergon Des*, 4(2), 18–24.

Speitel, K. F. (1992) Measurement assurance. In Salvendy, G. (Ed.) *Handbook of Industrial Engineering*, 2nd edn. John Wiley & Sons, New York, Chapter 86.

Squires, P. (1956) The shape of the normal work area. Report 275—Navy Department, Bureau of Medicine and Surgery, Medical Research Laboratories, New London, CT.

Stammers, R. B., Carey, M. E., and Astley, J. A. (1990) Task analysis. In Wilson, J. R. and Corlett, E. N. (Eds.) *Evaluation of Human Work: A Practical Ergonomics Methodology* (Reprinted in 1999). Taylor & Francis, London, U.K., pp. 134–160.

Standridge, C. R., Vaughan, D. K., LaVal, D. K., and Simpson, T. D. (1986) A tutorial on TESS: The extended simulation system. *Proceedings of the 18th Conference on Winter Simulation WSC'86*, Washington, DC, pp. 212–217.

Stanney, K. M., Smith, M. J., Carayon, P., and Salvendy, G. (2001) Human-computer interaction. In Salvendy, G. (Ed.) *Handbook of Industrial Engineering*, 3rd edn. John Wiley & Sons, New York, pp. 1192–1236.

Stanoulov, N. (1994) Expert knowledge and computer-aided group decision making: Some pragmatic reflections, *Ann Operat Res*, 51, 141–162.

Stasser, G. and Titus, W. (1985) Pooling of unshared information in group decision making: Biased information sampling during discussion, *J Person Soc Psychol*, 48, 1467–1478.

Steinburg, C. (1993) The downfall of teams, *Train Dev*, 46(12), 24–81.

Steiner, H. M. (2004) *Engineering Economic Principles*. McGraw-Hill, New York, 1992.

Sternberg, R. J. (1979) The nature of mental abilities, *Am Psychol*, 34(3), 214–230.

Stevenson, M. K., Busemeyer, J. R., and Naylor, J. C. (1993) Judgment and decision-making theory. In Dunnette, M. D. and Hough, L. M. (Eds.) *Handbook of Industrial and Organizational Psychology*, 2nd edn., Vol. 1. Consulting Psychologists Press, Palo Alto, CA.

Sticka, P. J. (1987) Models of procedural control for human performance simulation, *Hum Fact*, 29, 421–432.

Stoll, H. W. (1988) Design for manufacture, *Manuf Eng*, 100(1), 67–73.

Strata, R. (1989) Organization learning—The key to management innovation, *Sloan Manage Rev*, Spring, 30(3), 63ff.

Suokas, J. and Pyy, P. (1988) Evaluation of the validity of four hazard identification methods with event descriptions. Technical Research Centre of Finland, Research Report 516.

Suokas, J. and Rouhiainen, V. (1984) Work safety analysis: Method description and user's guide. (Research Report No. 314), Technical Research Centre of Finland, Tampere, Finland.

Susman, G. I. (1970) The impact of automation on work group autonomy land task specialization, *Hum Relat*, 23, 567–577.

Svenson, O. (1990) Some propositions for the classification of decision situations. In Borcherding, K., Larichev, O., and Messick, D. (Eds.) *Contemporary Issues in Decision Making*. North Holland, Amsterdam, the Netherlands, pp. 17–31.

Sviokla, J. J. (1986) PlanPower, Xcon and mudman: An in-depth analysis into three commercial expert systems in use. Unpublished PhD dissertation, Harvard University, Cambridge, MA.

Sviokla, J. J. (1989) Expert systems and their impact on the firm: The effects of PlanPower use on the information processing capacity of the financial collaborative, *J Manag Inform Syst*, 6(3), 65–84.

Sviokla, J. J. (1990) An examination of the impact of expert systems on the firm: The case of XCON, *MIS Q*, 14(2), 127–140.

Swain, A. D. and Guttman, H. (1983) *Handbook for Human Reliability Analysis with Emphasis on Nuclear Power Plant Applications*. Nuclear Regulatory Commission Washington, DC, NUREG/CR-1278.

Swain, A. D. and Guttmann, H. E. (1983) *Handbook of Human Reliability with Emphasis on Nuclear Power Plant Applications*. National Technical Information Service (NUREG/CR-1278), Springfield, VA

Swezey, R. W. and Pearlstein, R. B. (2001) Selection, training, and development of personnel. In Salvendy, G. (Ed.) *Handbook of Industrial Engineering*, 3rd edn. Wiley-Interscience, New York, pp. 920–947.

Taguchi, G. (1978) Off-line and on-line quality control. *Proceedings of International Conference on Quality Control*, Tokyo, Japan.

Tang, L. T., Tollison, P. S., and Whiteside, H. D. (1989) Quality circle productivity as related to upper-management attendance, circle initiations, and collar color, *J Manage*, 15, 101–113.

TarVip. (2004) *User Manual for the Target Visibility Predictor (Tarvip) Computer Model*. Operator Performance Laboratory (OPL), Center for Computer Aided Design (CCAD), The University of Iowa, Iowa City, IA.

Taylor, F. W. (1911) *The Principles of Scientific Management*. Harper & Rowe, New York.

Teichner, W. H. and Krebs, M. J. (1972) Laws of the simple visual reaction time, *Psychol Rev*, 79, 344–358.

Teichner, W. H. and Wehrkamp, R. F. (1954) Visual-motor performance as a function of short-duration ambient temperature, *J Exp Psychol*, 47, 447–450.

Thorndike, R. L. (1951) Reliability. In Lindquist, E. F. (Ed.) *Educational Measurement*. American Council on Education, Washington, DC, pp. 560–620.

Thuesen, G. J. and Fabrycky, W. J. (2001) *Engineering Economy*, 9th edn. Prentice Hall, Englewood Cliffs, NJ.

Tichauer, E. R. (1966) Some aspects of stress on forearm and hand in industry, *J Occup Med*, 8(2), 63–71.

Tippett, L. H. C. (1935) Statistical methods in textile research uses of the binomial and Poisson distributions, A snap-reading method of making time studies of machines and operatives in factory surveys, *J Textile Inst Trans*, 26, 51–55.

Todd, P. and Benbasat, I. (1991) An experimental investigation of the impact of computer based decision aids on decision making strategies, *Inform Syst Res*, 2(2), 87–115.

Todd, P. and Benbasat, I. (1992) The use of information in decision making: An experimental investigation of the impact of computer based DSS on processing effort, *MIS Q*, 16(3), 373–393.

Todd, P. and Benbasat, I. (1999) Evaluating the impact of DSS, Cognitive effort, and incentives on strategy selection, *Inform Syst Res*, 10(4), 356–374.

Towill, D. R. and Kaloo, U. (1978) Productivity drift in extended learning curves, *Omega*, 6(4), 295–304.

Trist, E. L., Higgins, G., Murry, H., and Pollack, A. (1963) *Organizational Choice*. Tavistock, London, U.K.

Trist, E. L., Higgins, G. R., and Brown, G. R. (1977) An experiment in autonomous working in an American underground coal mine, *Hum Relat*, 30, 201–236.

Tsang, P. S. and Vidulich, M. A. (2006) Mental workload and situation awareness. In Salvendy, G. (Ed.) *Handbook of Human Factors and Ergonomics, Engineering*, 3rd edn. John Wiley & Sons, New York, pp. 243–268.

Tuckman, B. W. (1965) Development sequence in small groups, *Psychol Bull*, 63, 289–399.
Turban, E., and Aronson, J. E. (2001) *Decision Support Systems and Intelligent Systems*, 6th edn. Prentice Hall, Upper Saddle River, NJ.
Turino, J. (1992) *Managing Concurrent Engineering: Buying Time to Market—A Definitive Guide to Improved Competitiveness in Electronics Design and Manufacturing*. Van Nostrand Reinhold, New York.
Tushman, M. L. (1977) Impacts of perceived environmental variability on patterns of work related communications, *Admin Sci Q*, 22, 587–605.
Tversky, A. (1969) Intransitivity of preferences, *Psychol Rev*, 76, 31–48.
Tversky, A. (1972) Elimination by aspects: A theory of choice, *Psychol Rev*, 79, 281–289.
Tversky, A. and Kahneman, D. (1973) Availability: A heuristic for judging frequency and probability, *Cogn Psychol*, 5, 207–232.
Tversky, A. and Kahneman, D. (1974) Judgment under uncertainty: Heuristics and biases, *Science*, 185, 1124–1131.
Tversky, A. and Kahneman, D. (1981) The framing of decisions and the psychology of choice, *Science*, 211, 453–458.
U.S. Department of Commerce (1991). *Standard Occupational Classification Manual (SOC)*, Appendix B, Occupational descriptions. National Technical Information Service, Washington, DC.
Valenzi, E. and Andrews, I. R. (1973) Individual differences in the decision processes of employment interviews, *J Appl Psychol*, 58, 49–53.
Van Cott, H. P. and Kincade, R. G. (1972) *Human Engineering Guide to Equipment Design*. U.S. Government Printing Office, Washington, DC.
Venema, A. (1989) Product information for the prevention of accidents in the home and during leisure activities. (SWOKA, Research Report No. 69), Institute for Consumer Research, Leiden, the Netherlands:.
Vernon, H. M. (1924) On the extent and effect of repetitive work. Pt. A. The degree of variety in repetitive industrial work. I.F.R.B. Rept. No. 26. H.M.S.O., London, U.K.
Vernon, H. M., and Bedford, T. (1931) Two studies of absenteeism in coal mines, I. Re absenteeism of miners in relation to short time and other conditions. (I.H.R.B. Report No. 62), H. M. Stationery Office, London, U.K.
Versace, J. (1959) Subjective measurements in engineering. Society of Automotive Engineers. *Annual Meeting Proceedings*, Detroit, MI.
Viscusi, W. K. (1991) *Reforming Products liability*. Harvard University Press, Cambridge, MA.
Viteles, M. S. (1953) *Motivation and Morale in Industry*. W. W. Norton, New York.
Waganaar, W. A. and Stakenburg, H. (1975) Paced and self-paced continuous reaction time, *Q J Exp Psychol*, 27, 559–563.
Wagenaar, W. A. (1992) Risk taking and accident causation. In Yates, J. F. (Ed.) *Risk-Taking Behavior*. John Wiley & Sons, New York, pp. 257–281.
Wagenaar, W. A. and Groeneweg, J. (1988) Accidents at sea: Multiple causes and impossible consequences. In Hollnagel, E., Mancini, G., and Woods, D. D. (Eds.) *Cognitive Engineering in Complex Dynamic Worlds*. Academic Press, London, U.K., pp. 133–144.
Waller, P. F. and Green, P. A. (1997) Human factors in transportation. In Salvendy, G. (Ed.) *Handbook of Industrial Engineering*, 3rd edn. John Wiley & Sons, New York, pp. 1972–2009.
Wallsten, T. S. (1972) Conjoint-measurement framework for the study of probabilistic information processing, *Psychol Rev*, 79, 245–260.
Wallsten, T. S. (1976) Using conjoint-measurement models to investigate a theory about probabilistic information processing, *J Math Psychol*, 14, 144–185.
Wallsten, T. S., Zwick, R., Kemp, S., and Budescu, D. V. (1993) Preferences and reasons for communicating probabilistic information in verbal and numerical terms, *Bull Psychol Soc*, 31, 135–138.
Walton, R. E. (1977) Work innovations at Topeka: After six years, *J Appl Behav Sci*, 13, 422–433.
Wandell, B. A. (1995) *Foundations of Vision*. Sinauer Associates, Sunderland, MA.
Wang, M. I. (1986) Cognitive tasks, micromethods and simulations. Unpublished master's thesis, The University of Iowa, Iowa City, IA.
Wang, Y., Das, B., and Sengupta, A. K. (1999) Normal horizontal working area: The concept of inner boundary, *Ergonomics*, 42, 628–646.

Ware, C. and Mikaelian, H. H. (1987) An evaluation of an eye tracker as a device for computer input. *Proceedings of the CHI + GI'87 Conference on Human Factors in Computing Systems and Graphics Interface.* ACM, New York, pp. 183–188.

Warrick, M. J. (1947) Direction of movement in the use of control knobs to position visual indicators. USAF AMc Report No. 694-4C.

Waters, T. R., Putz-Anderson, V. P., and Garg, A. (1994) Applications manual for the revised NIOSH lifting equation. U.S. Department of Health and Human Services (NIOSH) Publication No. 94–110.

Weber, E. (1994) From subjective probabilities to decision weights: The effect of asymmetric loss functions on the evaluation of uncertain outcomes and events, *Psychol Bull,* 115, 228–242.

Wegner, D. (1987) Transactive memory: A contemporary analysis of group mind. In Mullen, B. and Goethals, G. R. (Eds.) *Theories of Group Behavior.* Springer-Verlag, New York, pp. 185–208.

Weinstein, N. D. (1979) Seeking reassuring or threatening information about environmental cancer, *J Behav Med,* 2, 125–139.

Weinstein, N. D. and Klein, W. M. (1995) Resistance of personal risk perceptions to debiasing interventions, *Health Psychol,* 14, 132–140.

Welch, D. D. (1994) *Conflicting Agendas: Personal Morality in Institutional Settings.* Pilgrim Press, Cleveland, OH.

Welford, A. T. (1958) *Aging and Human Skill.* Oxford University Press, London, U.K.

Welford, A. T. (1976) *Skilled Performance: Perceptual and Motor Skills.* Scott, Foresman and Company, Palo Alto, CA.

Wentworth, R. N. and Buck, J. R. (1982) Presentation effects and eye-motion behaviors in dynamic visual inspection, *Hum Fact,* 24(6), 643–658.

Wever, R. A. (1985) Men in temporal isolation: Basic principles of the circadian system. In Folkard, S. and Monk, T. H. (Eds.) *Hours of Work.* John Wiley & Sons, Chichester, U.K.

Wherry, R. J. (1971) *The Human Operator Simulator.* Naval Air Development Center, Warminster, PA.

Wherry, R. J. (1976) The human operator simulator. In Sheridan, T. B. and Johannsen, G. Drury, C. G. and Fox, J. G. (Eds.) *Monitoring Behavior and Supervisory Control.* Plenum Press, New York.

Wickelgreen, W. (1977) Speed accuracy tradeoff and information processing dynamics, *Acta Psychol,* 41, 67–85.

Wickens, C. D. (1984) *Engineering Psychology and Human Performance.* Merrill Publishing Co., Columbus, OH.

Wickens, C. D. (1992) *Engineering Psychology and Human Performance,* 2nd edn. Harper Collins, New York.

Wickens, C. D. (1998, October) Commonsense statistics, *Ergon Des,* 6(4), 18–22.

Wickens, C. D. and Carswell, C. M. (1995) The proximity compatibility principle: Its psychological foundation and its relevance to display design, *Hum Fact,* 37(3), 473–494.

Wickens, C. D., Lee, J., Liu, Y., and Becker, S. (2004) *An Introduction to Human Factors Engineering.* Prentice Hall, Upper Saddle River, NJ.

Wiener, E. L. (1975) Individual and group differences in inspection. In (Eds.) *Human Reliability in Quality Control.* Taylor & Francis, Ltd., London, U.K, pp. 101–122.

Wilcox, S. B. (1994) Why forklifts need signaling devices, *Ergon Des,* 2, 17–20.

Williams, H. (1973) Developing a table of relaxation allowances, *Ind Eng,* 5(12), 18–22.

Williams, T. (1992) Fuzzy logic is anything but fuzzy, *Comp Des,* 31(4), 113–127.

Williges, R. H., Roscoe, S. N., and Williges, R. C. (1973) Synthetic flight training revisited, *Hum Fact,* 15, 543–560.

Wing, J. R. (1965) Upper thermal tolerance limits for unimportant mental performance, *Aero Med,* 36(10), 960–964.

Winkler, R. L. and Murphy, A. H. (1973) Experiments in the laboratory and the real world, *Organ Behav Hum Perform,* 10, 252–270.

Winterfeldt, D. V. and Edwards, W. (1986) *Decision Analysis and Behavioral Research.* Cambridge University Press, Cambridge, U.K.

Wogalter, M. S. and Leonard, S. D. (1999) Attention capture and maintenance. In Wogalter, M., DeJoy, D., and Laughery, K. (Eds.) *Warnings and Risk Communication.* Taylor & Francis, London, U.K., pp. 123–148.

Woodson, W. E. (1954) *Human Engineering Guide for Equipment Designers*. University of California Press, Berkeley, CA

Woodson, W. E. (1981) *Human Factors Design Handbook*. McGraw-Hill, New York.

Woodworth, R. (1899) The accuracy of voluntary movements, *Psychol Rev*, 3(13), 1–114.

Woodworth, R. S. and Schlosberg, H. (1954) *Exp psycho*, 2nd edn. Holt, Rinehart & Winston, New York.

Wortman, D. B., Duket, S. D., and Seifert, D. J. (1977) Modeling and analysis using SAINT: A combined discrete/continuous network simulation language. *Proceedings of the Winter Simulation Conference*, Gaithersburg, MD, pp. 528–534.

Wright, T. P. (1936) Factors affecting the cost of airplanes, *J Aero Sci*, 3, 122–128.

Wright, G. and Rea, M. (1984) Age, a human factor in lighting. In Attwood, W. and McCann, C. (Eds.) *Proceedings of the 1984 International Conference on Occupations Ergonomics*. Human Factors Association of Canada, Rexdale, ON, Canada, pp. 508–512.

Wulff, J. J. and Berry, P. C. (1962) Aids to job performance. In Gagné, R. M. (Ed.) *Psychological Principles in System Development*. Holt, Rinehart and Winston, New York, Chapter 8.

Wyatt, S. (1934) Incentives in repetitive work: A practical experiment in a factory. Industrial Health Research Report Number 69, London, U.K.

Wyatt, S. and Langdon, J. N. (1932) Inspection processes in industry. Industrial Health Research Board Report 63. Her Majesty's Stationery Office, London, U.K.

Wyon, D. P. (1977) Assessing productivity decrements in heat and cold: On site simulation. In Mangum, B. W. and Hill, J. E. (Eds.) *Thermal Analysis—Human Comfort—Indoor Environments, Proceedings of a Conference*, Washington, DC, NBS Special Publication 491.

Yang, Y. and Lehto, M. R. (2001) Increasing access of visually disabled users to the World Wide Web. Paper presented at the *HCI International '01, 9th International Conference on Human–Computer Interaction*, New Orleans, LA.

Yang, J., Marler, T., Kim, H., Farrell, K., Mathai, A., Beck, S. et al. (2005, April) Santos: A new generation of virtual humans. Paper presented at the *SAE 2005 World Congress*, Cobo Center, Detroit, MI.

Yarbus, A. L. (1967) *Eye Movement and Vision*. Plenum Press, New York.

Yates, J. F., Veinott, E. S., and Patalano, A. L. (2003) Hard decisions, bad decisions: On decision quality and decision aiding. In Schnieder, S. and Shanteau, J. (Eds.) *Emerging Perspectives on Judgment and Decision Research*. Cambridge University Press Cambridge, U.K., pp. 13–63.

Yerkes, R. M. and Dodson, J. D. (1908) The relation of strength of stimulus to rapidity of habit formation, *J Comp Neurol Psychol*, 18, 459–482.

Zadeh, L. A. (1965) Fuzzy sets, *Inform Cont*, 8, 338–353.

Zadeh, L. A. (1972) A fuzzy-set-theoretic interpretation of linguistic hedges, *J Cyber*, 2, 4–34.

Zadeh, L. A. (1975) The concept of a linguistic variable and its application to approximate reasoning: Part 1, 2, 3. *Inform Sci*, 8, 199–249, 301–357; 9, 43–80.

Zander, A. (1994) *Making Groups Effective*, 2nd edn. Jossey-Bass, San Francisco, CA.

Zandin, K. B. (1980) *MOST Work Measurement Systems*. Marcel Dekker, Inc., New York.

Zellers, S. M. and Buck, J. R. (1997) Device modeling and HCI: Modifications for rapid-aimed movement. *Proceedings of the 13th Triennial Congress of the International Ergonomics Association*, Tampere, Finland, Vol. 5, pp. 71–13.

Zhou, M. and Xu, S. (1999) Dynamic recurrent neural networks for a hybrid intelligent decision support system for the metallurgical industry, *Exp Syst*, 16(4), 240–247.

Zigurs, I. and Buckland, B. K. (1998) A theory of task/technology fit and group support systems effectiveness, *MIS Q*, 22(3), 313–334.

Zimmer, A. (1983) Verbal versus numerical processing of subjective probabilities. In Scholtz, R. W. (Ed.) *Decision Making under Uncertainty*. North Holland, Amsterdam, the Netherlands.

Zimolong, B. (1997) Occupational risk management. Salvendy, G. (Ed.) *Handbook of Human Factors and Ergonomics*, 2nd edn. Wiley, New York, pp. 989–1020, Chapter 31.

Index

A

Accommodation efforts characterization, 131
Actin and myosin, 47
Activity sampling
    advantages and disadvantages, 293, 294
    binomial sampling model, 295–297
    continuous/periodic observations, 293
    description, 292
    manual forms, observations, 295
    manual observation, 293
    nomographs, 297–299
    observation schedule development
        length, study, 299
            mapping, random numbers, 299, 300
            uniform sampling theory, 300
    performance standard estimation, 302–303
    pocket-carried reminder/recording devices, 294
    problem solving, 304–305
    random data collection times, 293
    recording and analysis, data
        Chi-square tests, 301
        simple form, activity analysis, 300, 301
        spreadsheet program, 301
    size calculation
    stationarity checking, 301–302
    steps, 293
ADEA, *see* Age Discrimination in Employment Act (ADEA)
Age Discrimination in Employment Act (ADEA), 685
AHS, *see* Automated highway system (AHS)
Allowances, performance leveling
    machine interference, 246
    ranges, 245, 246
    time standard, 245
    total allowance time, 246
    types, 245
    variable allowances, 246
    variable posture and motion allowance percentages, 246, 248
Americans with Disabilities Act, 1992, 27
Analysis of variance (ANOVA)
    effects tested, 374, 375
    forms, 374
    model, 374
    replications, 374
    strong and weak interactions, 374, 375
    treatments and methods, 374
ANOVA, *see* Analysis of variance (ANOVA)
Anthropometry; *see also* Applied anthropometry
    correlations, body measurements, 54, 57
    human body segment mass prediction, 53–54
    people stature prediction, 51–53
    planes, 54, 57
Applied anthropometry
    adjustability criteria, 88
    body type and clothing design, 86, 88
    clearance requirements and accessibility, 86
    databases, 88–89
    drafting templates, 89–90
    inaccessibility requirements, 88
    measures and design criteria, 86, 87
    visibility, postural, mechanical advantage criterias, 88
Aqueous humor, 60
Assembly, therbligs, *see* Therbligs
ATM, *see* Automated teller machine (ATM)
Attention models, human information processing
    attentional capacity, 396
    conspicuity, presented information, 397
    incoming sensory data, 395
    perceptual discriminability, 397
    stimulus salience, 397
    task-related workload, 396
Automated highway system (AHS), 371
Automated teller machine (ATM)
    Chapanis, 7
    customers, 359
    procedure, transaction initiation, 7

B

Basic oxygen furnace (BOF)
    computer replications, 365
    heat-to-heat time statistics, 366
    steel ingots, 365
Bayes' rule, 601–602
Beta (type 1) Bayesian (high density) 90%, 95%, and 99% confidence intervals, 716–718

BFOQ, *see* Bonafide occupational qualification (BFOQ)
Binomial sampling model
    close-approximation, 296
    industrial applications, 296
    mean and variance, 295, 297
    normal distribution, 297
    normal random deviate, 296
    sample size, 297
    selecting marbles, 295
Body movements
    ergonomic design, 54–59
    mobility, adult females and males, 57, 58
    muscular-skeletal system, levers, 57–59
    planes, 54, 57
BOF, *see* Basic oxygen furnace (BOF)
Bonafide occupational qualification (BFOQ), 645
Boolean algebra, 600
Brunswik lens model, human judgment
    cognitive continuum theory, 413
    cue utilization, 412
    Dawes' rule, 413
    probabilistic mental model, 413
    SJT, 412
Burn hazards, 185–186

C

CAD, *see* Computer-aided design (CAD)
Carpal tunnel syndrome, *see* Cumulative trauma disorders (CTDs)
Carrots and sticks, macroergonomics
    Americans with Disabilities Act, 1992, 27
    antidiscrimination laws, 26
    legislation, 26
Cartilage, 40
CAs, *see* Customer attributes (CAs)
Cathode-ray tube (CRT), 153–154
Cause-and-effect (C&E) diagrams
    bad emergency medical service run reports, 341, 342
    brainstorming, 341
CDC, *see* Centers for Disease Control (CDC)
Cellular manufacturing processes, 31
Centers for Disease Control (CDC), 618
Channel capacity, 383
Chapanis activities identification
    action/information requirements analysis, 7
    analysis/sampling, 4–6
    critical incident study/analysis, 7
    functional flow study and decision/action analysis, 7
    function allocation, 7–8
    link analysis and controlled experimentation, 8
    operational sequence analysis and workload assessment, 9
    similar systems analysis, 4
    task, failure mode and effects analysis, 8
Chi-square analysis, 536–537

Circulatory subsystems
    ECG/EKG, 78
    heart operation, 76, 77
    semi-realistic description, 76, 77
CIT, *see* Critical-incident technique (CIT)
Classical decision theory
    consequentialism, 405
    elementary probability theory, 405
    elements, 404–405
    non-compensatory decision rules, 405–406
    subjective expected utility model and prospect theory, 406–407
Cognitive tasks, model and evaluation
    communication theory, *see* Communication theory
    decision making, *see* Decision making
    GOMS, *see* GOMS model
    human information processing, *see* Human information processing
    SDT, *see* Signal detection theory (SDT)
Communication theory
    components, 380, 381
    contributions, 380
    information transmission, *see* Information transmission
    model, system, 381
    people's interactions, 380
    transmitter, 380
Computer-aided design (CAD), 277, 508
Computer-integrated production systems, 578–579
Computer languages
    AutoMod, 360, 361
    FlexSim simulation model, 360
    generation, random numbers, 362
    GUIs, 359
    health-care facility, 360, 361
    methods, output analysis, 362
    MicroSAINT approach, 360
    network, resources, 361
    system dynamics, 362
Computer modeling methods
    BOEMAN, 277
    CAD, 277, 278
    digital human modeling technology, 279
    SAMMIE, 277
    SANTOS™, 278
Computer simulation (CS)
    ANOVA, 374
    ATM, 359
    computer languages, 359–362
    confounding, 376
    development, human factors, 363
    economic constraints, 373
    ergonomics, 363–366
    factorial designs, 377
    fidelity, 357
    methods, creation, 357
    OLS, 366–373

PS, 357
randomization, 376
system dynamics, 362
time-sequenced simulation, 359
types, 358
validation, 362
work cell, FlexSim, 358
Concordance calculation, 546
Conditional standard errors of measurement (CSEM), 541, 542
Cone receptors, 61
Confidence intervals, sequential Bayesian work sampling
asymmetric beta distribution, 311, 312
definition, 311
extensive Bayesian, 315–316
tabled Bayesian, 312
Control devices
and displays (lights), 489
types, 483
Control error statistics
issues, 440
partial means, 441–444
root-mean-square (RMS) error, 440
time plots, 441
Control tasks and systems
conditioned signal, 425
control model, 425
elements, 424, 425
fuzzy control, *see* Fuzzy control
hierarchy, ship tending, 424
manual control, *see* Manual control
outer loop, control hierarchy, 424
supervisory control, *see* Supervisory control
Corneal bulge reflection method, 60, 62
Critical-incident technique (CIT), 629
CRT, *see* Cathode-ray tube (CRT)
CS, *see* Computer simulation (CS)
CSEM, *see* Conditional standard errors of measurement (CSEM)
CTDs, *see* Cumulative trauma disorders (CTDs)
Cumulative chi-square table, 719
Cumulative student t-distribution table, 719
Cumulative trauma disorders (CTDs)
defined, 43, 44
occupational tasks and risk factors, 183
prevention, 50
Customer attributes (CAs), 497–500
Customer requirements
assembly/setup stages, 503
automobiles, 504
marketing efforts and design features, 501
products, 500
purchasing stage, 503
seat cushion quality, 502–503

D

Dark adaptation, 154

Data analysis
chi-square analysis, 536–537
spearman rank correlation coefficient, 537–540
Decision making
Brunswik lens model, *see* Brunswik lens model, human judgment
classical decision theory, *see* classical decision theory
heuristics and biases, 411–412
human inference, 409–410
recognition-primed, 413–414
time pressure and stress models, 414–415
Declarative *vs.* procedural knowledge representations
ACT production system model, 401
schema/script, 401
DeQuervain's disease, 44
Designed experiments
dependent variables, 523–524
ergonomic experiments, 525
human subjects, 523
independent variables, 521–523
statistical significance, 524
Designs
controls
bandwidth, 483
devices, 483, 484
discrete/continuous movements, 483
and display relationships, 488 491
principles, 484–488
selection, 483
issues, 491
knobs and dials era, 458
linguistic modeling, 462–463
people's interactions, 457
principles, display design, *see* Display design, principles
text comprehension testing, 464–466
types, 466–468
user-centered interface, 458–461
visual characteristics, control, 457–458
Dials, 479
Display design, principles
information content and coding, 477–483
legibility, display elements, 472–477
location and layout, 470–472
sensory modality, 469–470
Drafting templates, 89–90
DVA tests, *see* Dynamic visual acuity tests
Dynamic visual acuity (DVA) tests, 66, 67

E

ECG, *see* Electrocardiogram (ECG)
Ecological validity, cue, 413
ECs, *see* Engineering characteristics (ECs)
EEOC, *see* Equal Employment Opportunity Commission (EEOC)
Effective temperature (ET), 171–172

EKG, *see* Electrocardiogram (ECG)
Electrical hazards, 184–185
Electrocardiogram (ECG), 78
Electromyography (EMG), 48–49
Elementary model, manual control
    description, 428
        periodic inputs, response delays, and phase angles, 429
EMG, *see* Electromyography (EMG)
Engineering characteristics (ECs), 497–500
Equal Employment Opportunity Commission (EEOC), 641
Ergonomic design, work and work spaces
    activities identification, Chapanis, 4–9
    beer cans, 4–6
    criteria, 15–16
    human-centered design, 9–13
    human performance
        factors, 2
        models, 16–24
    industrial engineering, 3
    macroergonomics, 24–28
    methods, *see* Ergonomic research methods
    military equipment design, 13–14
    organizations and information, 14
    people, 1
    scientific management, 2
    trends, industry, 28–32
Ergonomic research methods
    computer simulation, 35
    designers, 32
    differences, 35
    experimental simulations, 34
    field studies, 33–34
    laboratory experiments, 34–35
Ergonomics
    accuracy, 15
    amount, learning, 15
    complex products, 525
    definition, quality, 493
    designed experiments, *see* Designed experiments
    employee turnover and absenteeism, 16
    HOS, 363
    human factors, 363
    marketing people, 494
    model, aircraft pilot, 363, 364
    nuclear systems, 363
    product sales, 493
    PS, 364–366
    quality management, *see* Quality management, customer-driven design
    safety and health, 15
    SAINT, 364
    software design, 495, 496
    speed-accuracy trade-off, 15
    speed/performance time, 15
    time variability, 15
    usability analysis and testing, *see* Usability analysis and testing
    user, 493, 494
Eye-hand coordination
    components, switch actuation, 283
    left-and right-hand switch actuations, simulated flying task, 283, 284
    negative correlation, 284
    performance times, 284

F

Failure modes and effects analysis (FMEA)
    description, 339–340, 579
    variants, 340
Failure modes effects and criticality analysis (FMECA), 579
Fault tree analysis (FTA)
    actual alerting system, 339
    AND-gate and OR-gate, 338
    description, 338
    integrated definition (IDEF), 339
    numerical example, 339
    relationships, conditions and nodes, 338
    safety instrumented system (SIS), 340
F distribution table, 719
Fire-extinguishing methods, 188
Fire hazards
    autoignition temperature, 187
    carbon monoxide, 188
    compounds, 187, 188
    elements, 186
    flammable liquids, 186–187
    LFL and UFL, 186
    occurrence prevention and early detection, 188
    temperature, concentration, and flammability, 187
Fish bone diagram, *see* Cause-and-effect (C&E) diagrams
Fitting learning curves
    discrete exponential model, 230
    powerform model, 229–230
Fitts and Seeger experiment, 490
Fitts' law
    experimental setup, 282
    first-order lag, 438–439
    foot pedal design, 282
    hand path and accelerations, ballistic start, 280
    index of difficulty, 281, 385
    Kvalseth's model, 282–283
    layouts, tapping experiment, 281
    movement times (MT), 281, 282
Flow diagrams
    description, 332
    interferences, 333
    sheet metal stove pipe, manufacturing process, 332
Flow process charts
    additional and storage activities, 330

Index

description, 330
parts, movement, 332
timing gear cover plate assembling, 331
FMEA, *see* Failure modes and effects analysis (FMEA)
FMECA, *see* Failure modes effects and criticality analysis (FMECA)
Fovea, 60, 61
Function allocation, 352–353
Fuzzy control
    description, 446
    examples, 448–451
    logic, 447–448
    measurements, 446–447
"Fuzzy set" theory, 21

G

Gaussian probability density function, 20
GEMS, *see* Generic error modeling system (GEMS)
Generic error modeling system (GEMS), 401
Global positioning systems (GPS), 468
GOMS model
    goals and operators, 417, 418
    natural language, *see* Natural GOMS language (NGOMSL)
    skilled cognitive activity, 418
    task performance
        components, 416
        functional operators, 416, 417
        selection rules, 416
    UNIT-TASK model, 417
GPS, *see* Global positioning systems (GPS)
Graphic user interfaces (GUIs), 359, 362
Grip design properties
    adequate mechanical advantage, 114, 115
    bend, tool's handle, 114–117
    creation, tool orientation, 114
    screwdrivers, 113
    tool vibrations, 113–114
    vibration insulation, 114
Groupthink, 668
GUIs, *see* Graphic user interfaces (GUIs)
Guyon canal, 44

H

Hands and handedness, tools and equipment design
    dimensions, human, 109
    dominant and subdominant grip strength, 109, 110
    greater hand strength, 109, 111
    grip strength, male and female, 110, 111
    grip strengths, handle openings, 109–110
    measurements, male and female, 108, 109
    power grip, 111, 112
    precision grips, 112
Hazard communication
    information mapped, accident sequence, 634
    product safety labels, 636
    selected warning systems, 635, 637–638
    sources, information, 634
    warning signs, labels, and tags, 635–636
Hazard control, hierarchy
    eliminate human errors, 631
    nonbehavioral solutions, 632
    radar speed, 633
Hazards and control measures
    burn, 185–186
    crushing/pinching, 181–182
    CTDs, 183
    cutting/shearing, 180–181
    electrical, 184–185
    fall and impact, 179–180
    fire, *see* Fire hazards
    pressure, 183–184
    toxic materials, *see* Toxic materials hazards
    ultraviolet (UV) radiation, 186
    vibration, 182
Heat stress indexes (HSIs)
    calculation, 174
    physiological implications, 174, 175
    thermal sensation and ET, 172–174
Heat transfer, temperature and humidity
    conduction and convection, 165–166
    evaporation, 166–167
    metabolism, 163–165
    radiative effects, 166
Helson's hypotheses, human performance models, 17–18
HFES, *see* Human Factors and Ergonomics Society (HFES)
Hierarchical task analysis (HTA), 348–349
Hinge joints, 41
HOS, *see* Human operator simulator (HOS)
Hot and cold environment control strategy
    administrative, 169
    evaporative cooling, 167, 169
    heat-producing equipment and processes, 169
    protective, 169–170
    psychrometric chart, 167, 168
Housekeeping and maintenance, 136–137
HSIs, *see* Heat stress indexes (HSIs)
Human-centered design
    activities, 10–11
    attributes, requirements, 9
    avoidance, changing activities, 12
    changes, system functions, 11
    level and activities increment, 11–12
    operators, dissatisfaction, 11
    overcome, human limitations, 10
    people enjoyment, activities, 11
    people selection, fit machines and jobs, 9
    personnel involvement, 12, 13
    problem identification, 10
    system operations, 13
    training sessions, 12

Human controller
    error signal, 431
    first-order exponential lag, 432
    out-of-phase input and output functions, 430
    phase shift, response delay, 429
    tracking periodic inputs, 432–433
    visibility, 426
Human decision making, *see* Decision making
Human Factors and Ergonomics Society (HFES), 618
Human information processing
    attention models, *see* Attention models, human information processing
    declarative *vs.* procedural knowledge, 401
    judgment-based error, 404
    knowledge-based error, 403–404
    mode
        educational forms, information, 400
        judgment-based performance, 399
        operation visibility and consistency, 399–400
        procedural and declarative knowledge, 398
        signals, signs, symbols and values, 398–399
    rule-based error, 403
    sequential model, 395, 396
    skill-based error, 401–403
    stages, 395
    working memory, 397–398
Human inspection
    analysis and improvement, 566–567
    description, 552
    illumination levels
        supplemental lighting system, 563–565
        visual acuity, 563
    individual differences, 562–563
    SDT, *see* Signal detection theory (SDT)
    steps, 553
    strategies, *see* Inspection strategies
    task pacing and time pressure
        belt speeds, 561
        performance, 560–561
        rest breaks and job rotations, 561
    viewing time and movement, 559–560
    visual freedom, 565–566
Human learning
    models, 219
    performance criteria and experience units, 219–221
    reasons, usage, 219
    speed and accuracy, 218
Human operator simulator (HOS), 363
Human perception, sound
    anatomy, hearing, 70, 72
    calculation, sound intensity and loudness, 71
    conditions, sources, 68, 69
    intensity, 69–70
    loudness contours, pure tones, 70, 73
    loudness, pressure function, 70
    measurement values, situations, 71, 73
    physics, sound waves, 67, 68
    position and motion sensing, *see* Position and motion sensing, sound
    wavelength and frequency, 67
Human performance models
    and behavior, 17
    capacities, 18–19
    communications system, 19
    control system, 22
    digital computer and simulation languages, 22–23
    "fuzzy set" theory, 21
    Gaussian distribution, 20
    Helson's hypotheses, 17–18
    learning curves, 20
    limited resource, 20
    operator/machine interface, 19
    queuing system, 20–21
    SDT, 20
    software design, 23–24
    "young" person, 21
Human reliability analysis., 579
Human system
    anthropometry, 50–54
    body movement, 54–59
    components, 39, 40
    muscles, 47–50
    sensory subsystems, *see* Sensory subsystems
    skeletal subsystem, 40–47
    support subsystems, *see* Support subsystems, task performance

I

ICI, *see* International Commission on Illumination (ICI)
Indirect calorimetry, 81–82
Indirect glare, *see* Veiling glare
Industrial engineering, 3
Industrial ergonomics
    activity, *see* Activity sampling
    sequential Bayesian work sampling, *see* Sequential Bayesian work sampling
    stratified, *see* Stratified sampling
Industrial work teams
    categories, 673
    personnel involvement, 671
    quality circle, 671
    SAGs, 672
Industry trends, ergonomic design
    assembly line, 28–29
    cellular manufacturing processes, 31
    concurrent engineering design, 31
    design practices, objectives, 32
    JIT, 30
    make-buy decision, 30
    quality certification, 31
    smaller production runs, 30
    standardization, 29

Inference
    Bayesian inference, 409, 410
    description, 409
Information display, perceptual elements
    confusion matrices, 474
    eye movements, 474
    glance legibility, 473
    legibility distance, 474
    reaction time, 473
    tachistoscopic procedures and glance legibility, 473
Information theory, 386–387
Information transmission
    average uncertainty, 382
    conditional uncertainty, 382
    equivocation, 383
    noise, 383
    partitioning and hypothesis testing
        analysis of variance (ANOVA), 389
        average information, 388
        decision making, 388
        explained and unexplained proportion, 389
    rate, 383–388
    Shannon's Mathematical Theory of Communication, 381–382
    speedometer, 381
Inspection economics
    categories, 569–570
    costs, 567, 569
    learning and quality improvement, 572–573
    location, 570–572
    procedure, 569–570
Inspection strategies
    rejection attributes / features, 558–559
    tasks division, 558
    train people, visual search patterns, 559
Integrated definition (IDEF), 339
International Commission on Illumination (ICI), 64
Iowa model
    human reliability models, 271
    mental tasks, 270
    performance time
        and errors, 270
        information-seeking tasks, 271
Ischial tuberosities, 98

J

JIT, *see* "Just in time" (JIT)
Job classification methods
    FWS, 688, 689
    grade description, 688
Job evaluation and compensation
    ADEA, 685
    analysis, 686
    characteristics, 686
    classification methods, 688–689
    compensation levels and perceived equity, 683, 684
    factor comparison method, 693–696
    PAQ, 685
    point systems, 689–693
    ranking methods, 686–688
    systematic process, 683
    wage administration, 697–705
    wage and salary surveys, 705
    written job description, 686, 687
Job factor comparison method
    compensable factors contribution, wages, 696
    contribution, wages, 695
    monetary value, 693
    ordering, compensable factors, 695
    ranking, benchmark jobs, 694, 695
    wage curve, 693
Job performance aids (JPAs)
    inspection procedure, 595–596
    vs. training, 597
Job ranking methods
    pairwise ranking systems, 688
    ranking process, 687
    types, 686
Joint-related disorders, skeletal subsystem
    CTD, 43, 44
    sprain/dislocation, 42
    thoracic outlet syndrome, 44
    trigger finger, 44
    ulnar deviation, 44
JPAs, *see* Job performance aids (JPAs)
Judgment-based errors, 404
"Just in time" (JIT), 30, 31

K

Knowledge-based errors, 403–404

L

Lean revolution, manufacturing, 137
Learning
    description, 217
    experience curve, 218
    Fitting learning curves, *see* Fitting learning curves
    learning curve models
        discrete exponential, 225–228
        powerform, 221–225
    learning curves, 218
    mathematical models, 217
    modeling human learning, *see* Human learning
    powerform *vs.* discrete exponential, 231–232
LEDs, *see* Light-emitting diodes (LEDs)
LFL, *see* Lower flammability limit (LFL)
Lifting
    computer tools, task analysis, 216–217
    coupling multiplier (CM), 213, 215
    frequency multiplier, 213, 214
    hand-to-contour coupling classification, 213, 215
    NIOSH guidelines, 212–213

principles, 214–216
recommended weight limit (RWL), 213
styles, 211, 212
Light-emitting diodes (LEDs), 153–154
Lighting
  advantages and disadvantages, sources, 145–146
  avoidance/minimization, extreme transitions, 155–156
  categorization, 143–145
  color and rendering properties matching, 156
  CRTs and LEDs, 153–154
  elimination/minimization, direct sources, 154–155
  general illumination supply, 149–151
  illumination/contrast, 150–152
  increment, contrast and background/size, 153
  less illumination, 149
  and measurement, 140–143
  strategies, 146
  supply, 147–149
  surface illumination levels, conditions, 147
Linguistic modeling, 462–463
Link diagrams
  home kitchen layout, 335, 336
  pilot eye movement, 335–336
  relationships, 335
Long-term memory (LTM), 654
Lower flammability limit (LFL), 186, 188
LTM, see Long-term memory (LTM)

M

Macro-ergonomics
  carrots and sticks, 26–28
  combining activities, 327
  compatible and natural activities, 327
  computer languages, industrial process control, 24, 25
  design principles, 325
  flow of work, 328
  function allocation
  layouts and methods improvement, 325–326
  operational sequences, 327–328
  "resistance to change", 325
  safety and health, industry, 24–25
  shift design and setting levels, compensation, 26, 27
  simplification, 326–327
  stages, problem solving, 324
  task requirements determination
  tasks and jobs analysis, see Tasks and jobs analysis
  unnecessary elements elimination, 326
  visual graphs, operations, see Visual graphs, operations
Management oversight and risk tree (MORT), 624
Manual assembly
  description, 208–209
  principles, 209–211
  robots, 209

Manual control
  control error statistics, 439–444
  elementary model, 428
  feedforward/predictive information, 426
  human controller, 431–434
  naive theory, adaptive control, 434–437
  numerical simulation, 444–446
  positioning and movement time, 437–439
  tracking tasks, 428–431
Material safety data sheet (MSDS), 633
Mean time between failures (MTBF), 578
Mento-Factor system
  activities, 269
  elementary example, 270
  task analysis, 269
Message recall, 464
Metabolism, heat transfer
  activities measurements rate, 163, 164
  DuBois approximation formula, 163
  rates calculation, mets, 164–165
Metabolism, support subsystems
  chemistry, 79
  components, food, 78
  energy requirements, tasks, 80
  heat, 78
  oxygen consumption, 79
  runner's weight, 80
Methods
  description, 199
  initial standard, 199
  standard, 199
Methods-time measurement (MTM), 655
Micromotion study
  description, 202–203
  experts and novices, 203
  sports medicine, 203
Military equipment design, 13–14
Mockups and computer tools, 469
MORT, see Management oversight and risk tree (MORT)
Motion economy
  description, 200
  and micromotion study, 202–203
  principles, 201–202
Motion-time-measurement (MTM) system
  application data card, 258–260
  description, 258
  MTM-1, see MTM-1
  MTM-2, 263–265
  positioning, 258
  principal elements, 258
  structural basis, 258
Moving legends, 479
MSDS, see Material safety data sheet (MSDS)
MTBF, see Mean time between failures (MTBF)
MTM, see Methods-time measurement (MTM)
MTM-1
  analytical procedure, 261–262
  assumptions, 262–263

time predictions, 261
time standard setting, 263
MTM-2
  description, 263
  distance codes, 264
  elemental actions, 264, 265
  error, predictions, 264
  EYE action, 265
  FOOT action, 265
  GETs and PUTs, 265
  time requirements, 264
Multiple activity charts
  assembly line balancing, 334
  description, 333
  five-station bicycle assembly line, 333
Murell's principle application, 475
Murell's approximation, 205
Muscles
  defined, 47
  EMG, 48–49
  exercise, CTD prevention, 50
  fibers, 47–48
  gender, age and training effects, 50
  injuries and disorders, 49–50
  oxygen role, 49
  physiology, 48

N

Naive theory, adaptive control
  Crossman's waterbath experiment, 436
  crossover model, 434
  human controller, 437
  position controls, 434
  reaction times, control tasks, 433
  second order lag, 437
  sluggish systems, 436
  velocity and acceleration controls, 435
NASA, *see* National Aeronautics and Space Administration (NASA)
National Aeronautics and Space Administration (NASA), 32, 368
National Electric Manufacturers Association (NEMA) point system
  assignment, job grades, 692, 694
  compensable factors and subfactors, 692–694
  job demands and requirements, 691–692
  variations, 690
National Institute for Occupational Safety and Health (NIOSH), 616–617
National Safety Council (NSC), 613
Natural GOMS language (NGOMSL)
  action–object pairs, 418
  execution times, 419
  high-level operators, 419
  interface designs, 419
  predefined primitive mental operators, 419
  selection rules, 419
NGOMSL, *see* Natural GOMS language (NGOMSL)

NGT, *see* Nominal group technique (NGT)
NIOSH, *see* National Institute for Occupational Safety and Health (NIOSH)
Noise
  annoyance and effects, 159–160
  control strategies, 160
  health effects, 156–157
  hearing protection, 161–162
  OSHA exposure limits, 157–159
Nominal group technique (NGT), 545
Nomographs
  description, 297–298
  error tolerance, 298
  sample size calculation, 298
  turning line, 298–299
Normal (Gaussian) distribution tables
  correlated bivariate, 713–715
  cumulative, 711–712
  ordinates, 712–713
NSC, *see* National Safety Council (NSC)

O

Occupational Safety and Health Administration (OSHA); *see also* Occupational safety and health management
  calculation, dose (D), 157
  permissible, 157, 158, 172
  TWAN sound, 159
Occupational safety and health management
  accident reporting, 626–627
  application, 617
  applications, systems model, 623
  BLS, classification system, 618, 620
  classifications, occupational injuries and illnesses, 618–619
  control strategies, 630–639
  disabling injuries, 2007, 619, 621
  employment opportunity laws, 617
  Environmental Protection Act, 1969, 615
  hazard and task analysis, 629–630
  hazard types, 617
  human error and unsafe behavior, 622–623
  human factors engineering, 639
  industrial revolution, 614
  industry-wide statistics, 629
  injuries/illness types, 627
  labor movement, 614
  lost workdays rate, 628
  models and theories, 619, 621
  multifactor theories, 621–622
  prevention efforts, 613
  public acceptance, 614
  root causes, 614
  standards and codes, 624, 626
  timeline, laws and acts, 615
  worker injuries and illnesses, 614
OCT, *see* Optimum control theory (OCT)

Office design
    business week, 103
    chair design, 103
    computer workstations, 104–108
    hotelling, 104
OLS, see Operator-in-the-loop simulation (OLS)
Operation hazards
    description, 122
    guard/barrier installation, 122–123
    machine deactivation, 123–124
    machine feeding mechanisms, 123
    operators, tool/machine activation, 123
    warning, operator, 124–125
Operator-in-the-loop simulation (OLS)
    active time percentages, BOF, 366, 367
    AHS, 371
    computer-driven displays, 373
    description, 34
    entering maneuver, experiment, 372
    face validity, 368
    forms, training, 368
    ground vehicle simulators, 369–370
    industrial engineering, 373
    laboratory setup, 367
    models, human behavior, 368
    NASA, 368
    nuclear reactor, 366
    simulation fidelity, 370–371
    training simulators, 368–369
Optimum control theory (OCT), 454
OSHA, see Occupational Safety and Health Administration (OSHA)
Oygen debt phase, 49

P

PAQ, see Position aptitude questionnaire (PAQ)
People stature prediction
    cumulative normal distribution, 51, 52
    effect, age, 52, 53
    NASA and other organizations, 52
Performance leveling
    allowances, see Allowances, performance leveling
    description, 198, 241
    mean time, 240
    ratings
        alternatives, 244–245
        extracted basis, 244
        multifactor, 242–243
        subjective, 241
        synthetic, 243–244
Performance-shaping factors (PSFs), 583
Personnel selection, placement and training
    adaptive, 658
    aircraft electricians and test scores, 650, 651
    analogy/metaphor, 655
    attitude development, 658
    correlated statistics, 649
    crew proficiency training, 658
    designing, program, 653
    distribution, test scores, 650, 651
    economics, 659
    face validity, 647
    function, test validity, 652, 653
    Helson U-hypothesis, 652
    methods, 653
    personnel selection program, 652
    physical dexterity, 651
    power, 647
    reduction, concurrent task loads, 654
    reliability, 647
    repeatability, 647
    segmented and part-task training, 658
    statistical relationships, test scores, 647, 648
    stress, program, 655
    systems, 642–646
    television tapes, 659
    TER, 656
    training and orientation process, 660
    transfer effectiveness ratio function, 657
    types, 646
    types, training strategies, 657
    Z-scores, selection ratio and test validity, 647–649
Personnel systems
    equal employment opportunity law, 643–644
    flows, company, 642
    job applications, 642
    production department, 646
    relief and remedies, unlawful discrimination, 644–646
PHA, see Preliminary hazard analysis (PHA)
Phon, 71
Photopic vision, 61
Physical environment assessment and design
    cleanliness, clutter and disorder, 136–140
    defined, 135–136
    hazards and control measures, see Hazards and control measures
    lighting, see Lighting
    noise, 156–162
    temperature and humidity, see Temperature and humidity
Physical human performance modeling
    eye-hand coordination, 283–284
    speed-accuracy trade-offs, see Speed-accuracy trade-offs
    synthetic data systems, see Synthetic data systems
Pilot testing and data collection, 535
Pointers, 479
Point systems
    job ranking and classification systems, 689
    mapping, degrees, 690
    NEMA, 690–693
    task description, 690, 691

Index

Poisson activity sampling, 319
Position and motion sensing, sound
    accelerations, 73
    ear anatomy, 71–72, 74
    sensory processes, 74, 75
Position aptitude questionnaire (PAQ), 685
Positioning and movement time
    Fitts' law, first-order lag, 438–439
    overshooting, 439
    position controls, 438
    velocity controls, 439
Power tools
    advantage, 119–120
    disadvantage, 120
    monitoring, tool design, 120
    principles, 120–121
Power transmission hazards, 121–122
Precedence diagrams and matrices, 334–335
Predictive variables and normal performance, synthetic data systems
    coefficient of determination, 275
    computer-generated plots, 275
    correlated data, 274
    uncorrelated data, 274
    variables plotting, 273
Preliminary hazard analysis (PHA), 579
Pressure hazards, 183–184
Probability distributions, 708–709
Probability theory, 583
Process simulation (PS)
    BOF, 365–366
    industrial process, 365
    production process, 365
    system performance, 364–365
Product liability, *see* Carrots and sticks, macroergonomics
Protective equipment, operator
    eye and spectacles, 130
    gloves, 129–130
    hearing, 130
    helmets, 127–129
    safety shoes, 125–127
PS, *see* Process simulation (PS)
PSFs, *see* Performance-shaping factors (PSFs)
Psychometric scales, 464
"Purkinje shift", 61
P-wave, 78

Q

QFD, *see* Quality function deployment (QFD)
Quality control
    description, 550
    economics, *see* Inspection economics
    human, *see* Human inspection
    information-finding function, 573
    maintenance and safety inspections, 551–552
    people and materials, screening, 552
    quality assurance inspections, 551
    SDT, 550–551
    tampering detection, 552
Quality function deployment (QFD), 497–500
Quality management, customer-driven design
    CAs, 498
    customer-driven design, 495–497
    customer requirements, 500–505
    design requirements, 505–507
    door, 499
    ECs, 498–500
    house, 497, 498
    prototyping and testing, 507–508
    QFD, 497
    ratings and preference testing, customer, 508–510
Questionnaires and interviews
    assign numbers, 532
    data analysis, *see* Data analysis
    demographic information, 534–535
    description, 527–528
    "Does Not Apply" category, 532
    epilogue, 530
    interrater consistency, 545–546
    methods, 543–545
    numerous response categories, 534
    open-ended questions, 531
    personal nature and flexibility, 542
    pilot testing and data collection, 535
    planning, 528–529
    prologue, 530
    respondent demographics, 530
    right or wrong answers, 533
    sampling procedure, 529–530
    Wilcoxon sign test, 540–542

R

Raynaud's syndrome/dead Ÿngers, 182
R&D, *see* Research and development (R&D)
RDD, *see* Research, development, and design (RDD)
Readability indexes, 464, 465–466
Recognition/matching technique, 464
Red fibers, 47
Regression, 63
Research and development (R&D), 664
Research, development, and design (RDD)
    communications, 679, 680
    features, 679
    group cohesiveness, 679–680
Respiratory operations and mechanics, 75–76
Robotics, 578–579
Rotary switches, control positions, 487
Rule-based errors, 403

S

Saccades, 62
Saddle joint, 42
Safety and health management control strategy
    hazard communication, 633–639

hierarchy, hazard control, 631–633
job and process design, 630–631
SAGs, see Semi-autonomous groups (SAGs)
SAMMIE, see System for aiding man-machine interaction evaluation (SAMMIE)
Sampling methods, industrial ergonomics
   activity, see Activity sampling
   sequential Bayesian work sampling, see Sequential Bayesian work sampling
   stratified, see Stratified sampling
Scientific management, 2
"Scientific method", industry, 29
Scotopic vision, 61
SDT, see Signal detection theory (SDT)
Seating design
   chair, 100–102
   flat footrests, 100
   height, 98–99
   seat depth, 99–100
   sitter, posture changes, 98
   space, 100
   stool, 98, 99
   weight distribution, 98
Semi-autonomous groups (SAGs), 672
Sensory-motor tasks and cognitive tasks, 263
Sensory subsystems
   human perception, sound, 67–74
   visual, 59–67
Sequential Bayesian work sampling
   acceptable error, 314
   beta distribution and parameters, 310–311
   confidence intervals, see Confidence intervals, sequential Bayesian work sampling
   description, 308, 310
   number, observations, 315
   numerical example, activity analysis, 316–319
   plotting, sequential study vs. time, 314
   Poisson activity sampling, see Poisson activity sampling
   prior distribution determination, 313–314
   theorem and concept, 312–313
Signal detection theory (SDT)
   bias, 393
   d' and β calculations, 394–395, 553–555
   description, 20, 389
   discriminability, 392
   d' measures, 553, 553
   human inspection, 553
   measured/perceived strength, signal, 392
   nonparametric extensions, 556–558
   probability ordinate, 393
   random disturbances, 392
   receiver operating characteristic (ROC) curve, 393, 395
   relative frequencies, 390, 391
   type-I and type-II errors, 553
   visual inspection, 553, 554
Sirens, emergency vehicles, 482

Skeletal subsystem
   defined, 40
   extremities
      bones, wrist and hand, 42
      movement range, factors, 41
      pivot joints, 41
      saddle joint, 42
      synovial and cartilaginous joints, 40–41
   joint-related disorders, 42–44
   spine, 45–47
Skill-based errors
   capture errors, 402
   fixation errors, 402
   interruptions, 403
   transition, rule-based/higher level, 402
Skill-based task, cognitive work measurement, see GOMS model
Sluggish systems, 436
Software design model, human performance, 23–24
Sone, 71
Sound, see Human perception, sound
Spearman rank correlation coefficient, 537–540
Speed-accuracy operating characteristic (SAOC), 250, 279
Speed-accuracy trade-offs
   Fitts' law, 280–283
   SAOC, 279–280
Spine, skeletal subsystem
   back disorders, 46
   degenerative disorders, 47
   lumbar region, 45, 46
   skull downward, 45
5S programs, physical environment assessment and design
   arrangement, items, 139
   benefits, 138
   defined, 137
   shine, 139
   sorting, 138–139
   standardization, 139–140
   sustainment, 140
Standard data systems
   advantage, 272
   linear regression analysis, 272
   product families, 271
Statistical computer packages, regression
   confidence interval check, 276
   correlations, pairs of variable, 276
   F test values, 277
   residual plot, 277
   selected variables, step-wise regression, 276
   SPSS printout, variables, 276
   tracking task, 275
   types, 275
Stratified sampling
   application, 306
   binomial, 308, 310

Index

data analysis
  binomial data, 308
  steps, 308
designing, 306–308
numerical example, 308, 309
representative observation, 305–306
Subjective likelihood index methodology
    (SLIM-MAUD)
  probability estimation, human error, 581–583
  PSFs, 583
Supervisory control
  automated systems, 453
  benefits, 452
  description, 451
  function allocation, 452
  HIS and TIS, 451–452, 453
  optimum control theory (OCT), 454
  real signal, 454
  single feedback loop, 453
Support subsystems, task performance
  circulatory, 76–78
  defined, 74
  indirect calorimetry, 81–82
  metabolism, 78–81
  respiratory operations and mechanics, 75–76
Synthetic data systems
  computer modeling methods, *see* Computer modeling methods
  description, 257
  direct measurement and synthetic prediction, 257
  formula development
    principles, 272–273
    variables, 272
  Iowa model, 270–271
  mento-factor system, 269–270
  MTM, *see* Motion-time-measurement (MTM) system
  prediction systems, 257
  predictive variables and normal performance, 273–275
  standard data systems, 271–272
  statistical computer packages, regression, 275–277
  WF system, *see* Work-factor (WF) system
System for aiding man-machine interaction evaluation (SAMMIE), 277
System reliability and maintenance
  aircraft maintenance errors and safety, 591
  analytical methods, 600
  CAD-based approach, 608, 609
  definition, 578
  detection algorithms, 584
  error correction algorithms., 577
  expert system, 601
  fasteners and tools, 607–608
  fault and malfunction detection, 593–596
  fault tolerant computing, 585
  Fischertechnik® assembly kit, 603, 604
  group components, modules, 604–605

infant mortality, 584
logical connections, components, 599, 600
lost productivity, 587
malfunction symptoms, 602
management, 577
network logic, faulty nodes, 598, 599
nodes and branches, 603, 605
production machines, 577
quality assurance, 577–578
record keeping and measurement, 591–592
and repair, training, 590–591
sampling maintenance activities, 586–587
slotting, 587
standard data systems, 587
storing and delivering technical information, 589–590
symmetrical part mating, 606
timed replacement, 584
tools and diagnostic equipment, 589–590
tracking maintenance activity, time, 592
work envelope, 608–609

T

Take-the-best heuristic, 413
Task analytic methods, 511
Task requirements determination
  cognitive abilities, 350, 351
  cognitive activities, 349
  knowledge requirements, 350
  questionnaire-based methods, 352
  resource allocation models, 350
  sensory, cognitive processing and motor activities, 350
Tasks and jobs analysis
  coffee making, 346–347
  flow diagram, 347
  HTA, *see* Hierarchical task analysis (HTA)
  methods, 344
  position, 344
  subtasks and sub-subtasks, 343
Task time-line analysis charts
  pilot workload, 337
  time-line graph (TLG), 337
  uses, 337–338
TEAM, *see* Team evolution and maturation model (TEAM)
Team evolution and maturation model (TEAM), 667
Technique for human error and reliability prediction (THERP)
  probability estimation, human error, 580
  PSFs, 583
Temperature and humidity
  control strategies, hot and cold environments, 167–170
  ET, *see* Effective temperature (ET)
  factors, human body and environment, 170–171
  heat transfer, 163–167
  HSIs, *see* Heat stress indexes (HSIs)

measurement, 171
thermal conditions, task performance, 177–179
thermal regulation, 162–163
WBGT, see wet bulb global temperature index
windchill index, 172
TER, see Transfer effectiveness ratio (TER)
Therbligs
　description, 195
　Gilbreth's therbligs, 196
　productive and nonproductive, 196
THERP, see Technique for human error and reliability prediction (THERP)
Thoracic outlet syndrome, 44
Three-dimensional static strength prediction program (3DSSPP), 217
Time study
　elements, task, 234–235
　industrial purposes, 233
　procedure selection, 233–234
　recording forms, 235, 236
　sample size and time statistics
　　conservative calculation, 240
　　correlation, 237
　　electric's empirical estimates, 239, 240
　　mean and variance, performance time, 236
　　probability density function, 237
　　procedure, calculation, 238
　　range method, 239
　　sample mean and true mean, 237
　　standard deviations, 239, 240
　　time variance, successive elements, 237
　　variance, cycle time, 237
　selecting and training, observed operator, 235–236
　"should take" and "did take" time, 233
　steps, 232–233
Time-weighted average noise (TWAN), 158, 159
Tools and equipment design
　features, 117–118
　hands and handedness, 108–112
　properties, grip design, 113–118
　techniques, hand tool adequacy, 118–119
Toxic materials hazards
　carcinogenic substances, 190
　classification, 189
　depressants/narcotics, 189
　irritants, 189
　measures, exposure and limits, 190–192
　protection, airborne contaminants, 192
　simple and chemical asphyxiates, 189–190
　systemic poisons, 189
Traffic areas
　clearance requirements, passageways, 90, 91
　recommended dimensions, stairs, 90, 92
　wheelchair clearance dimensions, 92, 93
Transfer effectiveness ratio (TER), 656
"Trigger finger", 118
TWAN, see Time-weighted average noise (TWAN)

U

UFL, see Upper flammability limit (UFL)
Ultraviolet (UV) radiation hazards, 186
Upper flammability limit (UFL), 186, 188
Usability analysis and testing
　advantages and disadvantages, 511
　computer interfaces, 516
　control room, 518
　debriefing interview, 520
　designed interfaces, 510
　heuristic evaluation, 514
　instructions, 516
　laboratory tests, 519
　methods, 517
　model builders, 518
　multiple evaluators, 515
　principles, preference testing, 509
　professional raters, 514
　self-reporting, 519, 520
　task analytic methods, 511
　web-based and telephone interfaces, 510
User-centered interface design
　description, 458–459
　natural mappings, 459–461
　population stereotypes and stimulus, 461–462

V

Veiling glare, 153
Vibration hazards, 182
Vigilance decrement, 387
Visibility, control applications, 426
Visibility reference function, 149
Visual graphs, operations
　C&E diagrams, see Cause-and-effect (C&E) diagrams
　decision flow diagrams, 341, 343, 344
　fault trees, see Fault trees
　flow diagrams, 332–333
　flow process charts, 330–332
　FMEA, see Failure modes and effects analysis (FMEA)
　link diagrams, see Link diagrams
　multiple activity charts, 333–334
　operations process charts
　　activity nodes, 329
　　assembly, home furnace, 329, 330
　　description, 329
　precedence diagrams and matrices, 334–335
　task analysis, 328–329
　task time-line analysis charts, see Task time-line analysis charts
Visual sensory subsystems
　acuity, 65–66
　contrast sensitivity, 66
　depth perception, 66
　DVA, 66
　eye movements, 62–63

Index

people's dark adaptation rate, 67
perception, color and brightness, 64–65
reading tasks, 63
structure, eye, 60–62

W

Wage administration
  calibration curves, 699–700
  curves, 700–701
  forms, compensation, 697–698
  group level, 704–705
  individual level, 703–704
  organization level, 705
  rate ranges, 701–702
  and salary surveys, 699
WBGT index, see Wet bulb global temperature index
Weber-Fechner Law, 63, 65
Wet bulb global temperature (WBGT) index
  assumptions, 177
  calculation, 175, 176
  heat exposure threshold limit values, 175, 176
Wilcoxon sign test, 540–542
Windchill index, 172
Work areas and stations design
  seating, 98–102
  traffic areas, 90–92
  workplace dimensions and layout principles, 92–98
Work-factor (WF) system
  description, 265–266
  fully skilled and highly skilled term, 268
  structure
    elements, 266, 267
    grasp element types, 268
    manual control, 266
    pre-position element, 268
    time units (TUs), 266
    transport time values, 267
  walking, 268
Work groups and teams
  good teamwork, 663
  group effectiveness
    affective conflict, 665, 666
    elements, 664–665
    forming, storming, norming, and performing, 665
    influencing factors, 665
  group performance and biases
    Groupthink, 668
    information-processing limitations, 669
  industrial work teams, 671–673
  life cycle, teams and crews
    group cohesion and authority structure, 667–668
    limited life span, 665, 667
    TEAM, 667
  measuring team–crew work performance
    reliability, observer ratings, 670
    skill dimensions in team operations, 669–670
  mental and/or physical demands, job/task, 663
  observations from NASA, 680–681
  principles, team effectiveness
    feedback principle, 673–674
    social loafing, 674
  RDD, 679–680
  research, development, and design teams
    over the wall design, 675
    pie-in-the-sky concepts, 674
    problems, traditional serial design process, 675–676
  sports analogies, 664
  team roles
    bridge person, 679
    chairperson, 676
    company workers, 677
    completers, 678
    gatekeeper, 678–679
    innovators, 677
    monitor-evaluator, 677
    resource investigators, 677–678
    shaper, 676
    team workers, 677
Work measurement and analysis
  accuracy, performance, 250
  alternatives, learnability, 230–231
  assembly, 195
  forgetting curves, 232
  indirect performance measurement, 248–249
  learning, see Learning
  lifting, see Lifting
  manual assembly, see Manual assembly
  method
    description, 199
    initial standard, 199
    jobs, without standard time, 200
    operating personnel, 199–200
    standard, 199
  motion economy, see Motion economy
  normal distribution and central limit theorem, 198
  performance leveling, 198–199, 240–248
  performance time, 197
  Presgrave's observations, 199
  probability density function, 198
  probability, task, 197
  speed, 249
  standard operator, 197
  standards maintenance, 248
  therbligs, see Therbligs
  time study, see Time study
  work-physiology-related principles, 203–208
Work-physiology-related principles
  ballistic movements, 205–206
  eye fixations, 208

joints, midpoint of range of movement, 203
muscular force, 204
rest pauses, 204–205
unbalanced force, 206–208
uniform and natural rhythm, 208
work above heart level, 204
Workplace dimensions and layout principles
areas, 5th and 95th percentile U.S. females, 96
computation, reach, 96
determination, reach, 95
gravity feed bins and containers, 97
high precision work, 93, 95
seated and standing tasks, 93, 94
tools, materials and controls, 97–98
Work sampling
performance standard estimation
direct continuous observation, 302
normal time, 302
rated work sampling, 303
problem solving
maintenance operations, 304
productivity charting, time, 305

PGSTL 07/24/2017